INTERMEDIATE ALGEBRA

MARK DUGOPOLSKI

Southeastern Louisiana University

SECOND EDITION

ADDISON-WESLEY PUBLISHING COMPANY

Reading, Massachusetts • Menlo Park, California • New York • Don Mills, Ontario • Wokingham, England
Amsterdam • Bonn • Sydney • Singapore • Tokyo • Madrid • San Juan • Milan • Paris

Sponsoring Editor *Jason Jordan*
Managing Editor *Karen Guardino*
Art and Design Director *Karen Rappaport*
Development Editor *Roberta Lewis*
Production Supervisor *Karen Wernholm*
Prepress Buying Manager *Sarah McCracken*
Technical Art Buyer *Joseph Vetere*
Marketing Manager *Michelle Babinec*
Manufacturing Manager *Roy Logan*
Text Designer *Margaret Ong Tsao*
Art Editor *Alena Konecny*
Technical Illustrator *Burmar Technical Corporation*
Situational Illustrator *Scientific Illustrators*
Photo Researcher *Susan Van Etten*
Compositor/Production Coordination *Burmar Technical Corporation*

Photo Credits
Pages 1, 116, © Paul Conklin; page 53, Courtesy, U.S. Army Corps of Engineers; pages 28, 73, 191, 259, 301, 341, 387, 442, 483, 490, 577, © Susan Van Etten; page 245, © Herb Snitzer/Stock Boston; page 451, Molly McAllister; page 541, Concorde photograph copyright © 1995 by British Airways Plc. All rights reserved; page 556, Friedrich von Huene, von Huene Workshop

Library of Congress Cataloging-in-Publication Data

Dugopolski, Mark
    Intermediate algebra/Mark Dugopolski.—2nd ed.
        p.   cm.
    Includes index.
    ISBN 0-201-59564-8
    1. Algebra.   I. Title.
QA154.2.D84   1996
512.9—dc20
                                                                95-14618
                                                                CIP

1 2 3 4 5 6 7 8 9 10-DOW-98979695

To my wife and daughters,
Cheryl, Sarah, and Alisha

Contents

APPENDIXES   **A-1**

ANSWERS TO SELECTED EXERCISES   **A-7**

INDEX   **I-1**

Preface

I*ntermediate Algebra* is designed to provide students with the algebra background needed for further college-level mathematics courses. The unifying theme of this text is the development of the skills necessary for solving equations and inequalities, followed by the application of those skills to solving applied problems. My primary goal in writing the second edition of *Intermediate Algebra* has been to retain the features that made the first edition so successful, while incorporating more of the new standards in mathematics education and the comments and suggestions of first-edition users. I have endeavored to write a text that students can read, understand, and enjoy while they gain confidence in their ability to use mathematics. Although a complete development of each topic is provided in *Intermediate Algebra,* the text *Elementary Algebra* in this series would be more appropriate for students with no prior experience in algebra.

Content Changes

The arrangement of topics in this edition is primarily the same as that of the first edition. The topics that have been rearranged or added are as follows:

- Reading and interpreting graphs have been added to Chapter 1 so that students can work with real-data applications from the beginning of the course.

- More geometric applications and examples have been added throughout the book. This material not only demonstrates the relevance of the geometry and encourages skill development and practice, but also reviews basic geometric facts and provides visual applications of algebra, helping students make connections through spatial representations. A reference summary of geometric facts is found in Appendix A: Geometry Review.

- Scientific notation is now presented in Chapter 3 (Exponents and Polynomials) instead of in Chapter 5 (Rational Exponents and Radicals) so that students can study scientific notation immediately after studying integer exponents.

■ Synthetic division is now presented in Chapter 3 in the section on division of polynomials instead of in Chapter 8 in the section on the Factor Theorem.

■ Complex numbers are now presented in Chapter 5 (Rational Exponents and Radicals) instead of in Chapter 6 (Quadratic Equations and Inequalities) so that students can be introduced to the concept of a complex number before they study quadratic equations.

In addition to these changes, the exercise sets have been carefully revised throughout. Particular care has been given to achieving an appropriate balance of problems that progressively increase in difficulty from routine, practice exercises in the beginning of the set to more challenging exercises at the end of the set. Also, more simple fractions and decimals are used in the exercises and throughout the text discussions to help reinforce the basic arithmetic skills that are necessary for success in algebra.

Features

■ Each chapter now begins with a **Chapter Opener** that discusses a real application of algebra. The discussion is accompanied by a photograph and, in most cases, by a real-data application graph that helps students visualize algebra and more fully understand the underlying math. In addition, each chapter contains a **Math at Work** feature, which profiles a real person and the mathematics he or she uses on the job. These two features have corresponding real data exercises.

■ An increased emphasis on real-data applications is a focus of the second edition. Applications have been added throughout the text to help demonstrate concepts, motivate students, and give students practice using new skills. The exercise sets include exercises that involve real data and real situations. Many of these new, applied exercises also require students to read and interpret graphs containing real data in order to solve the problem. An **Index of Applications** listing applications by subject matter is included at the front of the text.

■ Every section of the second edition begins with **In This Section,** a list of topics that shows the student what will be covered. Since the topics correspond to the headings within each section, students will find it easy to locate and study specific concepts.

■ Important ideas, such as definitions, rules, summaries, and strategies, are set apart in boxes for quick reference. Color is used to highlight these boxes as well as other important points in the text.

■ A new **Calculator Close-up** feature, which is integrated throughout the text, gives instructions on the use of scientific and graphing calculators. Exercises requiring a scientific calculator are designated by a logo (▦). There are also groups of exercises specifically requiring graphing calculators (◿).

■ Each section ends with **Warm-up** exercises, a set of ten simple statements that are to be answered true or false. These exercises are designed to provide a smooth transition between the ideas and the exercise sets, and they can also be used to stimulate discussion of the concepts presented in the section. Many of the false statements point out examples of common student errors, which can help students identify areas they may need to work on. The answers to all Warm-ups are given in the answer section at the end of the book.

■ The end-of-section **Exercises** follow the same order as the textual material and contain exercises that are keyed to the examples, as well as numerous exercises that are not keyed to examples. This organization allows an instructor to cover only part of a section and easily determine which exercises are appropriate to assign. The keyed exercises give the student a place to start practicing and building confidence, whereas the nonkeyed exercises are designed to bring all of the chapter ideas together. New **Getting More Involved** exercises have been added to the end-of-section exercise sets to encourage critical thinking, writing, group discussion, exploration, and cooperative learning.

■ At the end of each chapter there is a **Collaborative Activities** feature, which is designed to encourage interaction and learning in groups. Research has shown that students who study together do better than those who try to make it on their own. Instructions and suggestions for using these activities and answers to all problems can be found in the Instructor's Solutions Manual.

■ Each chapter ends with a four-part **Wrap-up,** which includes the following:

The **Chapter Summary** lists important concepts along with brief illustrative examples.

The **Review Exercises** contain problems that are keyed to the sections of the chapter as well as numerous miscellaneous exercises.

The **Chapter Test** is designed to help the student assess his or her readiness for a test. To aid the student in working independently of the chapter examples and exercises, the Chapter Test has no keyed exercises.

The **Tying It All Together** exercises are designed to help students "tie together," or synthesize, the new material with ideas from previous chapters and, in some cases, review material necessary for success in the upcoming chapter. Each Tying It All Together section includes at least one applied exercise that requires ideas from one or more of the previous chapters.

Coverage

For those who wish to cover more on graphing earlier in their course, it is possible to present Chapter 7 (graphing linear equations in two variables) immediately after covering Chapter 2 (linear equations in one variable). In this case, it is recommended that the discussion of the midpoint formula and the distance formula presented in Section 7.1 be postponed until Chapter 11 (conic sections).

Supplements for the Instructor

Instructor's Solutions Manual

Prepared by Mark Dugopolski, this supplement contains detailed worked solutions to all the even-numbered exercises in the text. The solution processes match those shown in the text. Instructions and suggestions for using the

Collaborative Activities feature in the text are also included in the Instructor's Solutions Manual.

Answer Book

The Answer Book contains answers to all exercises in the text. Instructors may make quick reference to all answers, or have quantities of these booklets made available for sale if they want students to have all the answers. Contact your local Addison-Wesley representative for details or ordering information.

Printed Test Bank

Prepared by Rebecca Muller, the Printed Test Bank is an extensive collection of alternate chapter test forms, including the following:

- Two alternate test forms for each chapter, with free-response questions modeled after the Chapter Tests in the text.
- One test form for each chapter, with multiple-choice questions modeled after the Chapter Tests in the text.
- Four alternate forms of the final examination—two in free-response form, and two in multiple-choice form.
- Answer keys for all alternate test forms and final examinations.

Computerized Testing: OmniTest^3

Addison-Wesley's algorithm-driven computerized testing system for Macintosh and DOS computers features a brand new graphical user interface for the DOS version and a substantial increase in the number of test items available for each chapter of the text.

The new graphical user interface for DOS is a Windows look-alike. It allows users to choose items by test item number or by reviewing all the test items available for a specific text topic. Users can choose the exact iteration of the test item they wish to have on their test or allow the computer to generate iterations for them. Users can also preview all the items for a test on screen and make changes to them during the preview process. They can control the format of the test, including the appearance of the test header, the spacing between items, and the layout of the test and the answer sheet. In addition, users can now save the exact form of the test they have created so that they can modify it for later use. Users can also create their own items using OmniTest^3's WYSIWYG editor, and have access to a library of preloaded graphics.

Course Management and Testing System

InterAct Math Plus for Windows or Macintosh (available from Addison-Wesley) combines a powerful and flexible system for course management and on-line testing with the features of the basic tutorial software (see "Supplements for the Student") to create an invaluable teaching resource. Consult your Addison-Wesley representative for details.

Supplements for the Student

Student's Solutions Manual

Prepared by Mark Dugopolski, the Student's Solutions Manual contains complete worked-out solutions to all the odd-numbered exercises in the text. It may be purchased by your students from Addison-Wesley.

Math at Work Videotapes

The *Math at Work* videos are new to this edition of *Intermediate Algebra.* They feature an easy-to-follow *MathCad* narration of worked examples, many of which are taken from the exposition of the text, along with motivational math applications that are tied to the text. Question-and-answer narratives between a student and an instructor are used to highlight important points and to help make the videotapes interactive and engaging. A complete set of *Math at Work* videotapes is free to qualifying adopters.

InterAct Math Tutorial Software

InterAct Math Tutorial Software, new to this edition of *Intermediate Algebra,* has been developed and designed by professional software engineers working closely with a team of experienced developmental math teachers.

InterAct Math Tutorial Software includes exercises that are linked with the textbook and require the same computational and problem-solving skills as their companion exercises in the text. Each exercise has an example and an interactive guided solution that are designed to involve students in the solution process and to help them identify precisely where they are having trouble. In addition, the software recognizes common student errors and provides students with appropriate customized feedback.

Both free-response and multiple-choice items are contained in the tutorial. It also tracks student activity and scores for each section, which can then be printed out.

Available for both DOS-based and Macintosh computers, the software is free to qualifying adopters.

Acknowledgments

First of all I thank all of the students and professors who used the first edition of this text, for without their support there would not be a second edition. I sincerely appreciate the efforts of the reviewers who made many helpful suggestions for improving the first edition:

Nancy Angle, *Cerritos College*

Chris Barker, *DeAnza College*

Richard Basich, *Lakeland Community College*

Richard A. Butterworth, *Massasoit Community College*

Nancy Carpenter, *Johnson County Community College*

Florence Chambers, *Joliet Junior College*

Irene Doo, *Austin Community College*

David Dudley, *Phoenix College*

James Fryxell, *College of Lake County*

Terry Fung, *Kean College*

Jane Hammontree, *Tulsa Junior College*

Robert A. Hawes, *Northern Essex Community College*

Dale Hughes, *Johnson County Community College*

Kathy Kepner, *Paducah Community College*

Judith Lenk, *Ocean County College*

Mitchel Levy, *Broward Community College*

Joy McMullen, *Lakeland Community College*

Gael Mericle, *Mankato State University*

Jane Morrison, *South Suburban Community College*

Linda Padilla, *Joliet Junior College*

Sue Parsons, *Cerritos College*

Joanne Peeples, *El Paso Community College*

Rose L. Pugh, *Bellevue Community College*

Scott Reed, *College of Lake County*

Fred Russell, *Charles County Community College*

Howard Sorken, *Broward Community College*

Patricia Stanley, *Ball State University*

Eric Stietzel, *Foothill College*

Diane Tesar, *South Suburban College*

Charles Waiveris, *Central Connecticut State University*

I also thank Rebecca Muller, Southeastern Louisiana University, for error checking and writing the printed test bank. I thank Robert Brown, Southeastern Louisiana University, for error checking the printed test bank. I thank Elaine W. Cohen, New Mexico State University and Visioneering Research Laboratory, and Vickie J. Aldrich, Dona Aña Branch Community College, for writing the Collaborative Activities. I thank Doreen Clark and Bob Martin for error checking the manuscript.

I thank the staff at Addison-Wesley for all of their help and encouragement throughout the revision process, especially Jason Jordan, Greg Tobin, Elka Block, Bobbie Lewis, Helen Curtis, Karen Wernholm, Michelle Babinec, Kate Derrick, Barbara Willette, and Alena Konecny.

I also want to express my sincere appreciation to my wife, Cheryl, and my daughters, Sarah and Alisha, for their invaluable patience and support.

Hammond, Louisiana   M.D.

CHAPTER 5

Rational Exponents and Radicals

Just how cold is it in Fargo, North Dakota, in winter? According to local meteorologists, the mercury hit a low of −33°F on January 18, 1994. But air temperature alone is not always a reliable indicator of how cold you feel. On the same date, the average wind velocity was 13.8 miles per hour. This dramatically affected how cold people felt when they stepped outside. High winds along with cold temperatures make exposed skin feel colder because the wind significantly speeds up the loss of body heat. Meteorologists use the terms "wind chill factor," "wind chill index," and "wind chill temperature" to take both air temperature and wind velocity into account.

Through experimentation in Antarctica, Paul A. Siple developed a formula in the 1940s that measures the wind chill from the velocity of the wind and the air temperature. His complex formula involving the square root of the velocity of the wind is still used today to calculate wind chill temperatures. Siple's formula is unlike most scientific formulas in that it is not based on theory. Siple experimented with various formulas involving wind velocity and temperature until he found a formula that seemed to predict how cold the air felt. His formula is stated and used in Exercises 101 and 102 of Section 5.2.

Chapter Openers

Chapter openers use real-life examples of mathematical concepts discussed in the chapters. Several exercises within the chapter relate to the chapter opening example.

For a closer look at this feature see page 245.

...onential Functions **451**

...bled in 6.5 years. Let's
...f time by graphing the
...f we enter the function

...≤ 150,000 we get the
...e calculator we can see
...52) and (50, 141,200),
...and after 50 years it is
...ases slowly at first and

Math at Work

A motivational **Math at Work** feature in each chapter illustrates how people in various careers use the mathematics discussed in the chapter. Exercises relating to each Math at Work feature appear in exercise sets.

For a closer look at this feature see page 451.

MATH AT WORK

Geophysicist

Neal Driscoll, geophysicist at the Lamont-Doherty Earth Observatory of Columbia University, explores both the ocean and the continents to understand the processes that shape the earth. What he finds fascinating is the interaction between the ocean and the land—not just at the shoreline, but underneath the sea as well.

To get a preliminary picture of the ocean floor, Dr. Driscoll spent part of the past summer working with the U.S.G.S., studying the effects of storms on beaches and underwater landscapes. The results of these studies can be used as a baseline to provide help to coastal planners building waterfront homes. Other information obtained can be used to direct transporters of dredged material to places where the material is least likely to affect plant, fish, and human life. The most recent study found many different types of ocean floor, ranging from sand and mud to large tracts of algae.

Imaging the sea floor is a difficult problem. It can be a costly venture, and there are numerous logistical problems. Recently developed technology, such as towable undersea cameras and satellite position systems, has made the task easier. Dr. Driscoll and his team use this new technology and sound reflection to gather data about how the sediment on the ocean floor changes in response to storm events. This research is funded by the Office of Naval Research (ONR).

In Exercise 28 of the Tying It All Together exercises you will see how a geophysicist uses sound to measure the depth of the ocean.

Using a slope of $\frac{9}{5}$ and the point $(100, 212)$ in the point-slope formula, we get

$$F - 212 = \frac{9}{5}(C - 100).$$

We can solve this equation for F to get the familiar formula relating Celsius and Fahrenheit temperature:

$$F = \frac{9}{5}C + 32$$

Calculator Close-up

Xmin = –10   Ymin = –10
Xmax = 10   Ymax = 10

We can perform computations with either a scientific or a graphing calculator, but only a graphing calculator can graph an equation in its display. To graph $y = 2x - 3$, enter the equation into the calculator using the [Y=] key and then press the [GRAPH] key. (Because there are many differences in how different brands of graphing calculators graph equations, you should consult your manual at this point.) Your calculator might show a graph of this straight line in its viewing window as in Fig. 7.26. This viewing window has x-values ranging from a minimum of -10 to a maximum of 10 and y-values ranging from a minimum of -10 to a maximum of 10, the **standard viewing window.** If the line $y = 2x - 3$ does not appear, it is probably because your viewing window is set so that it does not contain any part of the line. Consult your manual to see how to adjust your viewing window. Exercises for graphing calculators are included in this section and in many other sections for the remainder of the text.

Warm-ups

True or false? Explain your answer.
1. There is exactly one line through a given point with a given slope.
2. The line $y - a = m(x - b)$ goes through (a, b) and has slope m.
3. The equation of the line through (a, b) with slope m is $y = mx + b$.
4. The x-coordinate of the y-intercept of a nonvertical line is 0.
5. The y-coordinate of the x-intercept of a nonhorizontal line is 0.
6. Every line in the xy-plane has an equation in slope-intercept form.
7. The line $2y + 3x = 7$ has slope $-\frac{3}{2}$.
8. The line $y = 3x - 1$ is perpendicular to the line $y = \frac{1}{3}x - 1$.
9. The line $2y - 3x + 5$ has a y-intercept of $(0, 5)$.
10. Every line in the xy-plane has an equation in standard form.

$$75 + 76 + 77 = 228.$$

The three consecutive integers that have a sum of 228 are 75, 76, and 77.

General Strategy for Problem Solving

The steps to follow in providing a complete solution to a verbal problem can be stated as follows.

Strategy for Solving Word Problems
1. Read the problem until you understand the problem.
2. If possible, draw a diagram to illustrate the problem.
3. Choose a variable and write down what it represents.
4. Represent any other unknowns in terms of that variable.
5. Write an equation that models the situation.
6. Solve the equation.
7. Be sure that your solution answers the question posed in the original problem.
8. Check your answer by using it to solve the original problem (not the equation).

We will now see how this strategy can be applied to various types of problems.

Calculator Close-ups

Calculator Close-ups—optional boxes occurring throughout the book—give students guidance on using either a scientific or graphing calculator.

Warm-ups

Warm-ups are sets of ten true/false statements that precede exercise sets in each section and are designed to check student understanding of concepts.

For a closer look at these features see page 366.

Strategy Boxes

Strategy boxes teach students how to approach new problems and techniques with a step-by-step method; some boxes are particularly helpful in learning how to solve word problems.

For a closer look at this feature see page 75.

Geometry

Geometry is integrated in the presentation of algebra throughout the book. A **Geometry Review** appendix is in the back of the book.

Real Data

Many of the exercises in the book reference **real data** to help motivate and interest students. An **Index of Applications** is also included.

Application Graphs

Four-color **application graphs** based on real data appear throughout the text. These help students visualize and understand the mathematics.

For a closer look at these features see page 71.

Getting More Involved

Getting More Involved exercise sets present exploratory, writing, discussion, and collaborative learning exercises to promote critical thinking and conceptualization.

Calculator Exercises

Numerous **calculator exercises** in the exercise sets are identified with an icon.

For a closer look at these features see page 63.

---

81. Radius of the earth. If the circumference of the earth is 25,000 miles, then what is the radius?

82. Altitude of a satellite. If a satellite travels 26,000 miles in each circular orbit of the earth, then how high above the earth is the satellite orbiting? See Exercise 81.

Figure for Exercise 82

83. Height of a can. If the volume of a can is 30 cubic inches and the diameter of the top is 2 inches, then what is the height of the can?

84. Height of a cylinder. If the volume of a cylinder is 6.3 cubic meters and the diameter of the lid is 1.2 meters, then what is the height of the cylinder?

85. Great Chicago flood. The great Chicago flood of April 1992 occurred when an old freight tunnel connecting buildings in the Loop ruptured. As shown in the figure, engineers plugged the tunnel with concrete on each side of the hole. They used the formula $F = WDA$ to find the force F of the water on the plug. In this formula the weight of water W is 62 lb/ft^3, the average depth D of the tunnel below the surface of the river is 32 ft, and the cross-sectional area A of the tunnel is 48 ft^2. Find the force on the plug.

Figure for Exercise 85

86. Will it hold? To plug the tunnel described in the previous exercise, engineers drilled a 5-ft-diameter shaft down to the tunnel. The concrete plug was made so that it extended up into the shaft. For the plug to remain in place, the shear strength of the concrete in the shaft would have to be greater than the force of

the water. The amount of force F that it would take for the water to shear the concrete in the shaft is given by $F = s\pi r^2$, where s is the shear strength of concrete and r is the radius of the shaft in inches. If the shear strength of concrete is 38 lb/in.^{2}, then what force of water would shear the concrete in the shaft? Use the result from Exercise 85 to determine whether the concrete would be strong enough to hold back the water.

87. Decreasing cholesterol. According to the National Center for Health Statistics, the annual number of deaths d in the United States due to coronary heart disease depends on the average adult cholesterol level (*Boston Globe*, June 16, 1993). This relationship can be modeled by the formula $d = 5000c - 525,000$, where c is the average cholesterol level measured in milligrams per deciliter of blood.

a) If d was 500,000 in 1990, then what was the average cholesterol level in 1990?

b) If the national goal of 200 by the year 2000 is reached, then how many lives will be saved in 2000 compared to 1990?

c) As the average cholesterol level increases, does the number of deaths due to coronary heart disease increase or decrease?

88. Cigarette usage. According the National Institute on Drug Abuse, the percentage of Americans aged 18 to 25 who use cigarettes has been decreasing at an approximately constant rate since 1974. The formula $P = 0.479 - 0.00949n$ can be used to estimate the percentage of smokers in this age group, n years after 1974. Use the formula to find the percentage of smokers in this age group in 1991. Use the accompanying graph to estimate the year in which smoking will be eliminated from this age group. Use the formula to find the year in which smoking will be eliminated from this age group.

Figure for Exercise 88

---

ate the first year in which the United ed over 100 million tons of munici- e. Find the year in which 13% of the id waste will be recovered by solving

$+ 3.78 = 0.13(3.14n + 87.1).$

Figure for Exercise 105

99. $5 - 2x = -3$, $bx - 1 = 3b$

100. $3 - x = -2x$, $b - x = 5b$

101. $2b - x = 4b$, $x - 6 = 5x$

102. $3b - 2 = 5x$, $2(x - 1) = 4 - 3x$

Solve each problem.

103. Public school enrollment. Enrollment in U.S. public schools has been increasing since 1985 (National Center for Education Statistics). The expression $0.45x + 39.05$ can be used to approximate in millions the total enrollment in public elementary and secondary schools in the year $1985 + x$. What was the public school enrollment in 1992? In which year will enrollment reach 45 million students?

104. Teacher's average salary. Average salary of public school teachers has been increasing steadily since 1985 (National Center for Education Statistics). The expression $553.37x + 27,966$ can be used to approximate the average annual salary in dollars of public school teachers in the year $1985 + x$. What was the average teacher's salary in 1993? In which year will the average salary reach $36,000?

105. Solid waste recovery. In 1960 the United States generated 87.1 million tons of municipal solid waste and recovered (or recycled) only 4.3% of it (U.S. EPA, 1990). The amount of municipal solid waste generated in the United States is given by $w = 3.14n + 87.1$, while the amount recovered is given by $w = 0.576n + 3.78$, where w is in millions of tons and n is the number of years since 1960. Use the

106. Recycling progress. Find the year in which 14% of the municipal solid waste will be recovered by solving

$$0.576n + 3.78 = 0.14(3.14n + 87.1).$$

See the previous exercise.

Getting More Involved

107. Exploration. If you solved the equations in the two previous exercises, then you found the years in which recovery of municipal solid waste will reach 13% and 14%.

a) Write equations corresponding to recovery rates of 15%, 16%, 17%, 18%, and 19%.

b) Solve your equations to find the years in which those recovery rates will be achieved.

c) Use your results to judge whether we are making real progress in recovery of municipal solid waste.

108. Writing. Explain how to eliminate the decimals in an equation that involves numbers with decimal points. Would you use the same technique when using a calculator?

109. Discussion. Explain why the multiplication property of equality does not allow us to multiply each side of an equation by zero.

─────── COLLABORATIVE ACTIVITIES ───────

Building a Room Addition

Grouping: 2 to 4 students per group
Topic: Applications of the quadratic formula

Leslie decides to build a rectangular studio on the south side of his house. The studio is attached so that the north side of the studio will be a portion of the current south side of the house. Because Leslie lives in the Southwest, he decides to make the room from rammed earth, thus making the walls of the studio 2 feet thick. Because of the cost of building materials and labor, Leslie decides to limit the total area of the studio. He also decides to make the studio's inside south wall twice as long as its inside west wall.

Leslie would also like to build a semicircular patio around the studio (sliding glass doors will open from the studio out onto the patio). The studio will be circumscribed by the semicircle. A large tree, which he would like to keep for the summer shade is 20 feet from the south side of the house. Below is a sketch Leslie made of the studio and patio.

1. Working in your groups, finish the drawing of the house and studio, adding the information above. Define your variable(s) and find the internal dimensions of the rectangular studio if Leslie wants the *external* area of the studio to be 400 square feet. Round your answers to the nearest foot.

2. Determine how far the semicircular patio extends past the house at the farthest point. Round your answer to the nearest tenth of a foot.

3. Will Leslie be able to build the patio without removing the tree? Give a reason for your answer.

Wrap-up CHAPTER 5

SUMMARY

Roots and Rational Exponents		**Examples**
Definition of nth roots	The number a is an nth root of b if $a^n = b$.	$3^4 = 81$, $(-3)^4 = 81$ Both 3 and -3 are fourth roots of 81.
Definition of $1/n$ as an exponent	If n is a positive *even* integer and a is positive, then $a^{1/n}$ denotes the positive real nth root of a. If n is a positive *odd* integer and a is any real number, then $a^{1/n}$ is the real nth root of a.	$81^{1/4} = 3$ $(-8)^{1/3} = -2$ $8^{1/3} = 2$
Definition of nth root of zero	If n is a positive integer, then $0^{1/n} = 0$.	$0^{1/6} = 0$
Definition of rational exponents	If m and n are positive integers, then $a^{m/n} = (a^{1/n})^m$, provided that $a^{1/n}$ is real. We can find the root first or the power first.	$8^{2/3} = (8^{1/3})^2$ $\quad\quad = 2^2 = 4$ $(-16)^{3/4}$ is not real.
Definition of negative rational exponents	If m and n are positive integers, then $$a^{-m/n} = \frac{1}{a^{m/n}},$$ provided that $a^{1/n}$ is defined and nonzero.	$8^{-2/3} = \dfrac{1}{4}$

Rules for Rational Exponents		**Examples**
If a and b are nonzero real numbers and r and s are rational numbers, then the following rules hold.		
Product rule	$a^r \cdot a^s = a^{r+s}$	$3^{1/4} \cdot 3^{1/2} = 3^{3/4}$
Quotient rule	$\dfrac{a^r}{a^s} = a^{r-s}$	$\dfrac{x^{3/4}}{x^{1/4}} = x^{1/2}$
Power of a power rule	$(a^r)^s = a^{rs}$	$(2^{1/2})^{-1/2} = 2^{-1/4}$ $(x^{3/4})^4 = x^3$
Power of a product rule	$(ab)^r = a^r b^r$	$(a^2 b^6)^{1/2} = ab^3$
Power of a quotient rule	$\left(\dfrac{a}{b}\right)^r = \dfrac{a^r}{b^r}$	$\left(\dfrac{8}{x^6}\right)^{2/3} = \dfrac{4}{x^4}$

Radicals		**Examples**
Definition of radicals	If n is a positive integer and a is a number for which $a^{1/n}$ is real, then $\sqrt[n]{a} = a^{1/n}$. If $n = 2$, omit the 2 and write $\sqrt[2]{a}$ as $\sqrt{a}$.	$8^{1/3} = \sqrt[3]{8}$ $4^{1/2} = \sqrt{4}$

REVIEW EXERCISES

5.1 *Simplify the expressions involving rational exponents. Assume that all variables represent positive numbers. Write your answers with positive exponents.*

1. $(-27)^{-2/3}$ 2. $-25^{3/2}$ 3. $(2^6)^{1/3}$
4. $(5^2)^{1/2}$ 5. $100^{-3/2}$ 6. $1000^{-2/3}$
7. $\dfrac{3x^{-1/2}}{3^{-2}x^{-1}}$ 8. $\dfrac{(x^2y^{-3}z)^{1/2}}{x^{1/2}yz^{-1/2}}$
9. $(a^{1/2}b)^3(ab^{1/4})^2$ 10. $(t^{-1/2})^{-2}(t^{-2}v^2)$
11. $(x^{1/2}y^{1/4})(x^{1/4}y)$ 12. $(a^{1/3}b^{1/6})^2(a^{1/3}b^{2/3})$

5.2 *Simplify each radical expression. Assume that all variables represent positive real numbers.*

13. $\sqrt{72x^5}$ 14. $\sqrt{90y^9z^4}$ 15. $\sqrt[3]{72x^5}$
16. $\sqrt[3]{81a^8b^9}$ 17. $\sqrt{2^6}$ 18. $\sqrt{2^7}$
19. $\sqrt{\dfrac{2}{5}}$ 20. $\sqrt{\dfrac{1}{6}}$ 21. $\sqrt[3]{\dfrac{2}{3}}$
22. $\sqrt[3]{\dfrac{1}{9}}$ 23. $\dfrac{2}{\sqrt{3x}}$ 24. $\dfrac{3}{\sqrt{2y}}$
25. $\dfrac{\sqrt{10y^3}}{\sqrt{6}}$ 26. $\dfrac{\sqrt{5x^5}}{\sqrt{8}}$ 27. $\dfrac{3}{\sqrt[3]{2a}}$
28. $\dfrac{a}{\sqrt[3]{a^2}}$ 29. $\dfrac{5}{\sqrt{3x^2}}$ 30. $\dfrac{b}{\sqrt[4]{a^2b^3}}$
31. $\sqrt[4]{48x^5y^{12}}$ 32. $\sqrt[5]{32x^{10}y^{12}}$

5.3 *Perform the operation... variables represent...*

33. $\sqrt{13}\cdot\sqrt{13}$
35. $\sqrt{27}+\sqrt{45}-\sqrt{\ldots}$
39. $3\sqrt{2}(5\ldots$
40. $-2\sqrt{a}(\sqrt{a}-\sqrt{\ldots})$
41. $(2-\sqrt{3})(3+\sqrt{2}\ldots)$
42. $(2\sqrt{x}-\sqrt{y})(\sqrt{x}\ldots$
43. $\sqrt[3]{40}-\sqrt[3]{5}$
44. $\sqrt[3]{54x^4}+x\sqrt[3]{16x}$

5.4 *Perform the operat...*

45. $5\div\sqrt{2}$
47. $(\sqrt{3})^4$
49. $\dfrac{2-\sqrt{8}}{2}$

51. $\dfrac{\sqrt{6}}{1-\sqrt{3}}$ 52. $\dfrac{\sqrt{15}}{2+\sqrt{5}}$
53. $\dfrac{2\sqrt{3}}{3\sqrt{6}-\sqrt{12}}$ 54. $\dfrac{-\sqrt{xy}}{3\sqrt{x}+\sqrt{xy}}$
55. $\left(2w^3\sqrt[3]{2w^2}\right)^6$
56. $\left(m\sqrt[4]{m^3}\right)^8$

5.5 *Find all real solutions to each equation.*

57. $x^2=16$ 58. $w^2=100$
59. $(a-5)^2=4$ 60. $(m-7)^2=25$
61. $(a+1)^2=5$ 62. $(x+5)^2=3$
63. $(m+1)^2=-8$ 64. $(w+4)^2=16$
65. $\sqrt{m-1}=3$ 66. $3\sqrt{x+5}=12$
67. $\sqrt[3]{2x+9}=3$ 68. $\sqrt[4]{2x-1}=2$
69. $w^{2/3}=4$ 70. $m^{-4/3}=16$
71. $(m+1)^{1/3}=5$ 72. $(w-3)^{-2/3}=4$
73. $\sqrt{x-3}=\sqrt{x+2}-1$
74. $\sqrt{x^2+3x+6}=4$
75. $\sqrt{5x-x^2}=\sqrt{6}$
76. $\sqrt{x+4}-2\sqrt{x-1}=-1$
77. $\sqrt{x+7}-2\sqrt{x}=-2$
78. $\sqrt{x}-\sqrt{x-1}=1$

Miscellaneous

Determine whether each equation is true or false and explain your answer. An equation involving variables should be marked true only if it is an identity. Do not use a calculator.

97. $2^3\cdot3^2=6^5$ 98. $16^{1/4}=4^{1/2}$
99. $(\sqrt{2})^3=2\sqrt{2}$ 100. $\sqrt[3]{9}=3$
101. $8^{200}\cdot8^{200}=64^{200}$ 102. $\sqrt{295}\cdot\sqrt{295}=295$
103. $4^{1/2}=\sqrt{2}$ 104. $\sqrt{a^2}=|a|$
105. $5^2\cdot5^2=25^4$ 106. $\sqrt{6}\div\sqrt{2}=\sqrt{3}$
107. $\sqrt{w^{10}}=w^5$ 108. $\sqrt{a^{16}}=a^4$
109. $\sqrt{x^6}=x^3$ 110. $\sqrt[8]{16}=\sqrt[4]{4}$
111. $\sqrt{x^8}=x^4$ 112. $\sqrt[9]{2^6}=2^{2/3}$
113. $\sqrt[6]{16}=2$ 114. $2^{1/2}\cdot2^{1/4}=2^{3/4}$
115. $2^{600}=4^{300}$ 116. $\sqrt{2}\cdot\sqrt{2}=\sqrt[6]{2}$
117. $\dfrac{2+\sqrt{6}}{2}=1+\sqrt{6}$ 118. $\dfrac{4+2\sqrt{3}}{2}=2+\sqrt{3}$
119. $\sqrt{\dfrac{4}{6}}=\dfrac{2}{3}$ 120. $8^{200}\cdot8^{200}=8^{400}$
121. $81^{2/4}=81^{1/2}$ 122. $(-64)^{2/6}=(-64)^{1/3}$
123. $(a^4b^2)^{1/2}=|a^2b|$ 124. $\left(\dfrac{a^2}{b^6}\right)^{1/2}=\dfrac{|a|}{b^3}$

In Exercises 125–136, solve each problem.

125. **Falling objects.** If we neglect air resistance, the number of feet s that an object falls from rest during t seconds is given by the equation $s=16t^2$. How long would it take the landing gear of an airplane to reach the earth if it fell off the airplane at 12,000 feet?

126. **Timber.** Anne is pulling on a 60-foot rope attached to the top of a 48-foot tree while Walter is cutting the tree at its base. How far from the base of the tree is Anne standing?

Figure for Exercise 126

127. **Guy wire.** If a guy wire of length 40 feet is attached to an antenna at a height of 30 feet, then how far from the base of the antenna is the wire attached to the ground?

Figure for Exercise 127

128. **Touchdown.** Suppose that at the kickoff of a football game, the receiver catches the football at the left side of the goal line and runs for a touchdown diagonally across the field. How many yards would he run? (A football field is 100 yards long and 160 feet wide.)

129. **Long guy wires.** The manufacturer of an antenna recommends that guy wires from the top of the antenna to the ground be attached to the ground at a distance from the base equal to the height of the antenna. How long would the guy wires be for a 200-foot antenna?

130. **Height of a post.** Betty observed that the lamp post in front of her house casts a shadow of length 8 feet when the angle of inclination of the sun is 60 degrees. How tall is the lamp post? (In a 30-60-90 right triangle, the side opposite 30 is one-half the length of the hypotenuse.)

Figure for Exercise 130

CHAPTER 5 TEST

Simplify each expression. Assume that all variables represent positive numbers.

1. $8^{2/3}$
2. $4^{-3/2}$
3. $\sqrt{21} \div \sqrt{7}$
4. $2\sqrt{5} \cdot 3\sqrt{5}$
5. $\sqrt{20} \div \sqrt{5}$
6. $\sqrt{5} + \dfrac{1}{\sqrt{6}}$
7. $2^{1/2} \cdot 2^{1/2}$
8. $\sqrt{72}$
9. $\sqrt{\dfrac{5}{12}}$
10. $\dfrac{6 + \sqrt{18}}{6}$
11. $(2\sqrt{3} + 1)(\sqrt{3} - 2)$
12. $\sqrt[4]{32a^5y^8}$
13. $\dfrac{1}{\sqrt[3]{2x^2}}$
14. $\sqrt{\dfrac{8a^9}{b^3}}$
15. $\sqrt[3]{-27x^9}$
16. $\sqrt{20m^3}$
17. $x^{1/2} \cdot x^{1/4}$
18. $(9y^4x^{1/2})^{1/2}$
19. $\sqrt[3]{40x^7}$
20. $(4 + \sqrt{3})^2$

Rationalize the denominator and simplify.

21. $\dfrac{2}{5 - \sqrt{3}}$
22. $\dfrac{\sqrt{6}}{4\sqrt{3} + \sqrt{2}}$

Write each expression in the form $a + bi$.

23. $(3 - 2i)(4 + 5i)$
24. $i^4 - i^5$
25. $\dfrac{3 - i}{1 + 2i}$
26. $\dfrac{-6 + \sqrt{-12}}{8}$

Find all real or imaginary solutions to each equation.

27. $(x - 2)^2 = 49$
28. $2\sqrt{x + 4} = 3$
29. $w^{2/3} = 4$
30. $9y^2 + 16 = 0$
31. $\sqrt{2x^2 + x - 12} = x$
32. $\sqrt{x - 1} + \sqrt{x + 4} = 5$

Show a complete solution to each problem.

33. Find the exact length of the side of a square whose diagonal is 3 feet.

34. Two positive numbers differ by 11, and their square roots differ by 1. Find the numbers.

35. If the perimeter of a rectangle is 20 feet and the diagonal is $2\sqrt{13}$ feet, then what are the length and width?

36. The average radius R of the orbit of a planet around the sun is determined by $R = T^{2/3}$, where T is the number of years for one orbit and R is measured in astronomical units (AU). If it takes Pluto 248.530 years to make one orbit of the sun, then what is the average radius of the orbit of Pluto? If the average radius of the orbit of Neptune is 30.08 AU, then how many years does it take Neptune to complete one orbit of the sun?

Tying It All Together   CHAPTERS 1–5

Find all real solutions to each equation or inequality. For the inequalities, also sketch the graph of the solution set.

1. $3(x - 2) + 5 = 7 - 4(x + 3)$
2. $\sqrt{6x + 7} = 4$
3. $|2x + 5| > 1$
4. $8x^3 - 27 = 0$
5. $2x - 3 > 3x - 4$
6. $\sqrt{2x - 3} - \sqrt{3x + 4} = 0$
7. $\dfrac{w}{3} + \dfrac{w - 4}{2} = \dfrac{11}{2}$
8. $2(x + 7) - 4 = x - (10 - x)$
9. $(x + 7)^2 = 25$
10. $a^{-1/2} = 4$
11. $x - 3 > 2$ or $x < 2x + 6$
12. $a^{-2/3} = 16$
13. $3x^2 - 1 = 0$
14. $5 - 2(x - 2) = 3x - 5(x - 2) - 1$
15. $|3x - 4| < 5$
16. $3x - 1 = 0$
17. $\sqrt{y - 1} = 9$
18. $|5(x - 2) + 1| = 3$
19. $0.06x - 0.04(x - 20) = 2.8$
20. $|3x - 1| > -2$
21. $\dfrac{3\sqrt{2}}{x} = \dfrac{\sqrt{3}}{4\sqrt{5}}$
22. $\dfrac{\sqrt{x} - 4}{x} = \dfrac{1}{\sqrt{x} + 5}$
23. $\dfrac{3\sqrt{2} + 4}{\sqrt{2}} = \dfrac{x\sqrt{18}}{3\sqrt{2} + 2}$
24. $\dfrac{x}{2\sqrt{5} - \sqrt{2}} = \dfrac{2\sqrt{5} + \sqrt{2}}{x}$
25. $\dfrac{\sqrt{2x} - 5}{x} = \dfrac{-3}{\sqrt{2x} + 5}$
26. $\dfrac{\sqrt{6} + 2}{x} = \dfrac{2}{\sqrt{6} + 4}$
27. $\dfrac{x - 1}{\sqrt{6}} = \dfrac{\sqrt{6}}{x}$
28. $\dfrac{x + 3}{\sqrt{10}} = \dfrac{\sqrt{10}}{x}$
29. $\dfrac{1}{x} - \dfrac{1}{x - 1} = -\dfrac{1}{6}$
30. $\dfrac{1}{x^2 - 2x} + \dfrac{1}{x} = \dfrac{2}{3}$

The expression $\dfrac{-b + \sqrt{b^2 - 4ac}}{2a}$ will be used in Chapter 6 to solve quadratic equations. Evaluate it for the given values of a, b, and c.

31. $a = 1, b = 2, c = -15$
32. $a = 1, b = 8, c = 12$
33. $a = 2, b = 5, c = -3$
34. $a = 6, b = 7, c = -3$

Solve each problem.

35. **Popping corn.** The results of an experiment by D.D. Metzger to determine the relationship between the moisture content of popcorn and the volume of popped corn are shown in the figure (*Cereal Chemistry*, 1989). The formula $v = -94.8 + 21.4x - 0.761x^2$ can be used to model the data shown in the graph. In the formula, v is the number of cubic centimeters of popped corn that result from popping 1 gram of corn with moisture content x% in a hot-air popper. Use the formula to find the volume that results when 1 gram of corn with a moisture content of 11% is popped.

Figure for Exercise 35 and 36

36. **Maximizing the volume of popped corn.** Use the graph to estimate the moisture content that will produce the maximum amount of popped corn. What is the maximum possible volume for popping 1 gram of corn in a hot-air popper?

Chapter Tests

The **Chapter Tests** are carefully developed to help students check their readiness for tests.

For a closer look at this feature see page 299.

Tying It All Together

The **Tying It All Together** feature at the end of each chapter serves as a cumulative review for all of the previous chapters of the text. At least one problem is a word problem to give students important and continual practice with problem solving.

For a closer look at this feature see page 300.

CHAPTER 1

Basics

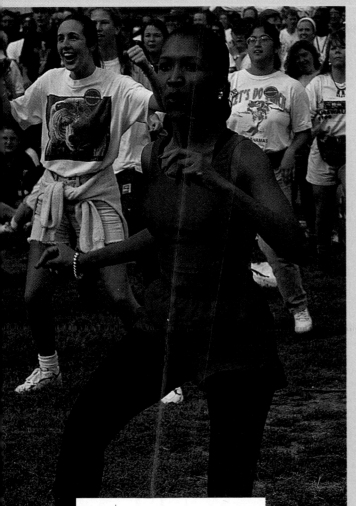

E verywhere you look, people are running, riding, dancing, and exercising their way to fitness. In the past year, more than $25 billion has been spent on sports equipment alone, and this amount is growing steadily.

Proponents of exercise claim that it can increase longevity, improve body image, decrease appetite, and generally enhance a person's health. While many sports activities can help you stay fit, experts have found that aerobic, or dynamic, workouts provide the most fitness benefit. Some of the best aerobic exercises include cycling, running, and even jumping rope. Whatever athletic activity you choose, trainers recommend that you set realistic goals and work your way toward them consistently and slowly. To achieve maximum health benefits, experts suggest that you exercise three to five times a week for 15 to 60 minutes at a time.

There are many different ways to measure exercise. One is to measure the energy used, or the rate of oxygen consumption. Since heart rate rises as a function of increased oxygen, another, easier measure of intensity of exercise is your heart rate during exercise. The desired heart rate, or target heart rate, for beneficial exercise varies for each individual depending on conditioning, age, and gender. In Exercises 103 and 104 of Section 1.4 you will see how an algebraic expression can determine your target heart rate for beneficial exercise.

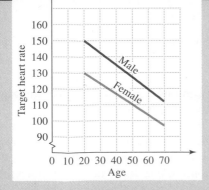

1.1 Sets

Every subject has its own terminology, and algebra is no different. In this section we will learn the basic terms and facts about sets.

Set Notation

A **set** is a collection of objects. At home you may have a set of dishes and a set of steak knives. In algebra we generally discuss sets of numbers. For example, we refer to the numbers 1, 2, 3, 4, 5, and so on as the set of **counting numbers** or **natural numbers.** Of course, these are the numbers that we use for counting.

The objects or numbers in a set are called the **elements** or **members** of the set. To describe sets with a convenient notation, we use braces, { }, and name the sets with capital letters. For example,

$$A = \{1, 2, 3\}$$

means that set A is the set whose members are the natural numbers 1, 2, and 3. The letter N is used to represent the entire set of natural numbers.

A set that has a fixed number of elements such as {1, 2, 3} is a **finite** set, while a set without a fixed number of elements such as the natural numbers is an **infinite** set. When listing the elements of a set, we use a series of three dots to indicate a continuing pattern. For example, the set of natural numbers is written as

$$N = \{1, 2, 3, \ldots\}.$$

The set of natural numbers *between* 4 and 40 can be written

$$\{5, 6, 7, 8, \ldots, 39\}.$$

Note that since the members of this set are *between* 4 and 40, it does not include 4 or 40.

Set-builder notation is another method of describing sets. In this notation we use a variable to represent the numbers in the set. A **variable** is a letter that is used to stand for some numbers. The set is then built from the variable and a description of the numbers that the variable represents. For example, the set

$$B = \{1, 2, 3, \ldots, 49\}$$

is written in set-builder notation as

$$B = \{x \mid x \text{ is a natural number less than 50}\}.$$

The set of numbers   such that   condition for membership

This notation is read as "B is the set of numbers x such that x is a natural number less than 50." Notice that the number 50 is not a member of set B.

The symbol $\in$ is used to indicate that a specific number is a member of a set, and $\notin$ indicates that a specific number is not a member of a set. For example, the statement $1 \in B$ is read as "1 is a member of B," "1 belongs to B," "1 is in B," or "1 is an element of B." The statement $0 \notin B$ is read as "0 is not a member of B," "0 does not belong to B," "0 is not in B," or "0 is not an element of B."

Two sets are **equal** if they contain exactly the same members. Otherwise, they are said to be not equal. To indicate equal sets, we use the symbol $=$. For sets that are not equal we use the symbol $\neq$. The elements in two equal sets do not need to be written in the same order. For example, $\{3, 4, 7\} = \{3, 4, 7\}$ and $\{2, 4, 1\} = \{1, 2, 4\}$, but $\{3, 5, 6\} \neq \{3, 5, 7\}$.

EXAMPLE 1   **Set notation**

Let $A = \{1, 2, 3, 5\}$ and $B = \{x \mid x$ is an even natural number less than 10$\}$. Determine whether each statement is true or false.

a) $3 \in A$ **b)** $5 \in B$ **c)** $4 \notin A$ **d)** $A = N$

e) $A = \{x \mid x$ is a natural number less than 6$\}$ **f)** $B = \{2, 4, 6, 8\}$

Solution

a) True, because 3 is a member of set A.

b) False, because 5 is not an even natural number.

c) True, because 4 is not a member of set A.

d) False, because A does not contain all of the natural numbers.

e) False, because 4 is a natural number less than 6, and $4 \notin A$.

f) True, because the even counting numbers less than 10 are 2, 4, 6, and 8. ∎

Union of Sets

Any two sets A and B can be combined to form a new set called their union that consists of all elements of A together with all elements of B.

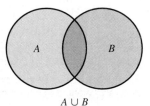

$A \cup B$

Figure 1.1

> **Union of Sets**
>
> If A and B are sets, the **union** of A and B, denoted $A \cup B$, is the set of all elements that are either in A, in B, or in both. In symbols,
>
> $$A \cup B = \{x \mid x \in A \text{ or } x \in B\}.$$

In mathematics the word "or" is always used in an inclusive manner (allowing the possibility of both alternatives). The diagram shown in Fig. 1.1, called a **Venn diagram,** illustrates the union of two sets A and B. The region inside each circle in the figure represents the members of each set.

EXAMPLE 2   **Union of sets**

Let $A = \{0, 2, 3\}$, $B = \{2, 3, 7\}$, and $C = \{7, 8\}$. List the elements in each of the following sets.

a) $A \cup B$ **b)** $A \cup C$

Solution

a) $A \cup B$ is the set of numbers that are in A, in B, or in both A and B.

$$A \cup B = \{0, 2, 3, 7\}$$

b) $A \cup C = \{0, 2, 3, 7, 8\}$ ∎

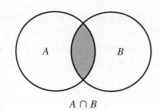

$A \cap B$

Figure 1.2

Intersection of Sets

Another way to form a new set from two known sets is by considering only those elements that the two sets have in common. The diagram shown in Fig. 1.2 illustrates the intersection of two sets A and B.

> **Intersection of Sets**
>
> If A and B are sets, the **intersection** of A and B, denoted $A \cap B$, is the set of all elements that are in both A and B. In symbols,
>
> $$A \cap B = \{x \,|\, x \in A \text{ and } x \in B\}.$$

It is possible for two sets to have no elements in common. A set with no members is called the **empty set** and is denoted by the symbol $\emptyset$. Note that $A \cup \emptyset = A$ and $A \cap \emptyset = \emptyset$ for any set A.

CAUTION The set $\{0\}$ is not the empty set. The set $\{0\}$ has one member, the number 0. ⊘

EXAMPLE 3 **Intersection of sets**

Let $A = \{0, 2, 3\}$, $B = \{2, 3, 7\}$, and $C = \{7, 8\}$. List the elements in each of the following sets.

a) $A \cap B$ **b)** $B \cap C$ **c)** $A \cap C$

Solution

a) $A \cap B$ is the set of all numbers that are in both A and B. So $A \cap B = \{2, 3\}$.

b) $B \cap C = \{7\}$

c) $A \cap C = \emptyset$ ∎

EXAMPLE 4 **Membership and equality**

Let $A = \{1, 2, 3, 5\}$, $B = \{2, 3, 7, 8\}$, and $C = \{6, 7, 8, 9\}$. Place one of the symbols $=, \neq, \in$, or $\notin$ in the blank to make each statement correct.

a) 5 ____ $A \cup B$ **b)** 5 ____ $A \cap B$

c) $A \cup B$ ____ $\{1, 2, 3, 5, 7, 8\}$ **d)** $A \cap B$ ____ $\{2\}$

Solution

a) $5 \in A \cup B$, because 5 is a member of A.

b) $5 \notin A \cap B$, because 5 must belong to *both* A and B to be a member of $A \cap B$.

c) $A \cup B = \{1, 2, 3, 5, 7, 8\}$, because the elements of A together with those of B are listed. Note that 2 and 3 are members of both sets but are listed only once.

d) $A \cap B \neq \{2\}$, because $A \cap B = \{2, 3\}$. ∎

$A \subseteq B$

Figure 1.3

Subsets

If every member of set A is also a member of set B, then we write $A \subseteq B$ and say that A is a **subset** of B. See Fig. 1.3. For example,

$$\{2, 3\} \subseteq \{2, 3, 4\}.$$

If set A is not a subset of B, we write $A \not\subseteq B$.

CAUTION   To claim that $A \not\subseteq B$, there *must* be an element of A that does *not* belong to B. For example,

$$\{1, 2\} \not\subseteq \{2, 3, 4\},$$

because 1 is a member of the first set but not the second.   ⊘

Is the empty set ($\varnothing$) a subset of $\{2, 3, 4\}$? If we say that $\varnothing$ is *not* a subset of $\{2, 3, 4\}$, then there must be an element of $\varnothing$ that does not belong to $\{2, 3, 4\}$. But that cannot happen, because $\varnothing$ is empty. So $\varnothing$ is a subset of $\{2, 3, 4\}$. In fact, by the same reasoning, *the empty set is a subset of every set.*

EXAMPLE 5   **Subsets**

Determine whether each statement is true or false.

a) $\{1, 2, 3\}$ is a subset of the set of natural numbers.

b)   The set of natural numbers is not a subset of $\{1, 2, 3\}$.

c) $\{1, 2, 3\} \not\subseteq \{2, 4, 6, 8\}$

d) $\{2, 6\} \subseteq \{1, 2, 3, 4, 5\}$

e) $\varnothing \subseteq \{2, 4, 6\}$

Solution

a)   True, because 1, 2, and 3 are natural numbers.

b)   True, because 5, for example, is a natural number and $5 \notin \{1, 2, 3\}$.

c)   True, because 1 is in the first set but not the second.

d)   False, because 6 is in the first set but not the second.

e)   True, because we cannot find anything in $\varnothing$ that fails to be in $\{2, 4, 6\}$.   ■

Combining Three or More Sets

We know how to find the union and intersection of two sets. For three or more sets we use parentheses to indicate which pair of sets to combine first. In the following example, notice that different results are obtained from different placements of the parentheses.

EXAMPLE 6   **Operations with three sets**

Let $A = \{1, 2, 3, 4\}$, $B = \{2, 5, 6, 8\}$, and $C = \{4, 5, 7\}$. List the elements of each of the following sets.

a) $(A \cup B) \cap C$

b) $A \cup (B \cap C)$

Solution

a) The parentheses indicate that the union of A and B is to be found first and then the result, $A \cup B$, is to be intersected with C.

$$A \cup B = \{1, 2, 3, 4, 5, 6, 8\}$$

Now examine $A \cup B$ and C to find the elements that belong to both sets:

$$A \cup B = \{1, 2, 3, 4, 5, 6, 8\}$$
$$C = \{4, 5, 7\}$$

The only numbers that are members of $A \cup B$ and C are 4 and 5. Thus

$$(A \cup B) \cap C = \{4, 5\}.$$

b) In $A \cup (B \cap C)$, first find $B \cap C$:

$$B \cap C = \{5\}$$

Now $A \cup (B \cap C)$ consist of all members of A together with 5 from $B \cap C$:

$$A \cup (B \cap C) = \{1, 2, 3, 4, 5\}$$ ■

Warm-ups

True or false? Explain your answer.
Let $A = \{1, 2, 3, 4\}$, $B = \{3, 4, 5\}$, and $C = \{3, 4\}$.

1. $A = \{x \mid x$ is a counting number$\}$
2. The set B has an infinite number of elements.
3. The set of counting numbers less than 50 million is an infinite set.
4. $1 \in A \cap B$ 5. $3 \in A \cup B$ 6. $A \cap B = C$
7. $C \subseteq B$ 8. $A \subseteq B$ 9. $\emptyset \subseteq C$
10. $A \nsubseteq C$

1.1 EXERCISES

For Exercises 1–46 let N represent the natural numbers,
$A = \{x \mid x$ is an odd counting number smaller than 10$\}$,
$B = \{2, 4, 6, 8\}$, and
$C = \{1, 2, 3, 4, 5\}$.

Determine whether each statement is true or false. Explain your answer. See Example 1.

1. $6 \in A$ 2. $8 \in A$
3. $A \neq B$ 4. $A = \{1, 3, 5, 7, \dots\}$
5. $3 \in C$ 6. $4 \notin B$
7. $A = \{1, 3, 7, 9\}$ 8. $B \neq C$
9. $0 \in N$ 10. $2.5 \in N$
11. $C = N$ 12. $N = A$

List the elements in each set. If the set is empty, write $\emptyset$. See Examples 2 and 3.

13. $A \cap B$ 14. $A \cup B$ 15. $A \cap C$
16. $A \cup C$ 17. $B \cup C$ 18. $B \cap C$
19. $A \cup \emptyset$ 20. $B \cup \emptyset$ 21. $A \cap \emptyset$
22. $B \cap \emptyset$ 23. $A \cap N$ 24. $A \cup N$

Use one of the symbols $\in, \notin, =, \neq, \cup,$ or $\cap$ in the blank of each statement to make it correct. See Example 4.

25. $A \cap B$ _____ $\emptyset$ 26. $A \cap C$ _____ $\emptyset$
27. A _____ $B = \{1, 2, 3, 4, 5, 6, 7, 8, 9\}$
28. A _____ $B = \emptyset$ 29. B _____ $C = \{2, 4\}$
30. B _____ $C = \{1, 2, 3, 4, 5, 6, 8\}$

31. 3 _____ $A \cap B$ **32.** 3 _____ $A \cap C$

33. 4 _____ $B \cap C$ **34.** 8 _____ $B \cup C$

Determine whether each statement is true or false. Explain your answer. See Example 5.

35. $A \subseteq N$ **36.** $B \subseteq N$ **37.** $\{2, 3\} \subseteq C$

38. $C \subseteq A$ **39.** $B \nsubseteq C$ **40.** $C \nsubseteq A$

41. $\varnothing \subseteq B$ **42.** $\varnothing \subseteq C$ **43.** $A \subseteq \varnothing$

44. $B \subseteq \varnothing$ **45.** $A \cap B \subseteq C$

46. $B \cap C \subseteq \{2, 4, 6, 8\}$

For Exercises 47–72, let

$$D = \{3, 5, 7\},$$

$$E = \{2, 4, 6, 8\}, and$$

$$F = \{1, 2, 3, 4, 5\}.$$

List the elements in each set. If the set is empty, write $\varnothing$. See Example 6.

47. $D \cup E$ **48.** $D \cap E$ **49.** $D \cap F$

50. $D \cup F$ **51.** $E \cup F$ **52.** $E \cap F$

53. $(D \cup E) \cap F$ **54.** $(D \cup F) \cap E$

55. $D \cup (E \cap F)$ **56.** $D \cup (F \cap E)$

57. $(D \cap F) \cup (E \cap F)$ **58.** $(D \cap E) \cup (F \cap E)$

59. $(D \cup E) \cap (D \cup F)$ **60.** $(D \cup F) \cap (D \cup E)$

Use one of the symbols $\in, \subseteq, =, \cup, or \cap$ in the blank of each statement to make it correct.

61. D _____ $\{x \mid x$ is an odd natural number$\}$

62. E _____ $\{x \mid x$ is an even natural number smaller than 9$\}$

63. 3 _____ D **64.** $\{3\}$ _____ D

65. D _____ $E = \varnothing$ **66.** $D \cap E$ _____ D

67. $D \cap F$ _____ F **68.** $3 \notin E$ _____ F

69. $E \nsubseteq E$ _____ F **70.** $E \subseteq E$ _____ F

71. D _____ $F = F \cup D$ **72.** E _____ $F = F \cap E$

List the elements in each set.

73. $\{x \mid x$ is an even natural number less than 20$\}$

74. $\{x \mid x$ is a natural number greater than 6$\}$

75. $\{x \mid x$ is an odd natural number greater than 11$\}$

76. $\{x \mid x$ is an odd natural number less than 14$\}$

77. $\{x \mid x$ is an even natural number between 4 and 79$\}$

78. $\{x \mid x$ is an odd natural number between 12 and 57$\}$

Write each set using set-builder notation.

79. $\{3, 4, 5, 6\}$ **80.** $\{1, 3, 5, 7\}$

81. $\{5, 7, 9, 11, \ldots\}$ **82.** $\{4, 5, 6, 7, \ldots\}$

83. $\{6, 8, 10, 12, \ldots, 82\}$ **84.** $\{9, 11, 13, 15, \ldots, 51\}$

Getting More Involved

85. *Writing.* Explain the difference between the intersection and the union of two sets.

86. *Discussion.* If A and B are finite sets, could $A \cup B$ be infinite? Explain.

87. *Cooperative learning.* Work with a small group to answer the following questions. If $A \subseteq B$ and $B \subseteq A$, then what can you conclude about A and B? If $(A \cup B) \subseteq (A \cap B)$, then what can you conclude about A and B?

88. *Discussion.* What is wrong with each statement? Explain.

a) $3 \subseteq \{1, 2, 3\}$ b) $\{3\} \in \{1, 2, 3\}$ c) $\varnothing = \{\varnothing\}$

89. *Exploration.* There are only two possible subsets of $\{1\}$, namely, $\varnothing$ and $\{1\}$.

a) List all possible subsets of $\{1, 2\}$. How many are there?

b) List all possible subsets of $\{1, 2, 3\}$. How many are there?

c) Guess how many subsets there are of $\{1, 2, 3, 4\}$. Verify your guess by listing all the possible subsets.

d) How many subsets are there for $\{1, 2, 3, \ldots, n\}$?

1.2 The Real Numbers

In this section:

▶ The Rational Numbers

▶ Graphing on the Number Line

▶ The Irrational Numbers

The set of real numbers is the basic set of numbers used in algebra. There are many different types of real numbers. To better understand the set of real numbers, we will study some of the subsets of numbers that make up this set.

The Rational Numbers

We have already introduced the most fundamental set of numbers, the **natural numbers.** The set of natural numbers, N, is written in symbols as follows:

The Natural Numbers

$$N = \{1, 2, 3, \ldots\}$$

The set of natural numbers together with the number 0 is called the set of **whole numbers.** The set of whole numbers, W, is written as follows:

The Whole Numbers

$$W = \{0, 1, 2, 3, \ldots\}$$

A negative number can be used to represent a loss or a debt. The number -10 could represent a debt of \$10, a temperature of 10° below zero, or an altitude of 10 feet below sea level. The whole numbers together with the negatives of the counting numbers form the set of **integers, J.**

The Integers

$$J = \{\ldots, -3, -2, -1, 0, 1, 2, 3, \ldots\}$$

To represent ratios or quotients, we use **rational numbers.** For example, if four out of five dentists recommend sugarless gum, then we say that $\frac{4}{5}$ of the dentists recommend sugarless gum. The set of rational numbers, Q, is written in set-builder notation as follows:

The Rational Numbers

$$Q = \left\{ \frac{a}{b} \,\middle|\, a \text{ and } b \text{ are integers, with } b \neq 0 \right\}$$

Examples of rational numbers are

$$7, \quad \frac{9}{4}, \quad -\frac{17}{10}, \quad 0, \quad \frac{0}{4}, \quad \frac{3}{1}, \quad -\frac{47}{3}, \quad \text{and} \quad \frac{-2}{-6}.$$

Note that the rational numbers are the numbers that can be expressed as a ratio (or quotient) of integers. The integer 7 is rational, because we can write it as $\frac{7}{1}$.

Another way to describe rational numbers is by using their decimal form. To obtain the decimal form, we divide the denominator into the numerator. For some rational numbers the division terminates, and for others it continues indefinitely. The following examples show some rational numbers and their equivalent decimal forms:

$$\frac{26}{100} = 0.26 \qquad \text{Terminating decimal}$$

$$\frac{4}{1} = 4.0 \qquad \text{Terminating decimal}$$

$$\frac{1}{4} = 0.25 \qquad \text{Terminating decimal}$$

$$\frac{2}{3} = 0.6666\ldots \qquad \text{The single digit 6 repeats.}$$

$$\frac{25}{99} = 0.252525\ldots \qquad \text{The pair of digits 25 repeats.}$$

$$\frac{4177}{990} = 4.2191919\ldots \qquad \text{The pair of digits 19 repeats.}$$

Rational numbers are defined as ratios of integers, but they can be described also by their decimal form. *The rational numbers are those decimal numbers whose digits either repeat or terminate.*

EXAMPLE 1   **Subsets of the rational numbers**

Determine whether each statement is true or false.

a) $0 \in W$ **b)** $N \subseteq J$ **c)** $0.75 \in J$ **d)** $J \subseteq Q$

Solution

a) True, because 0 is a whole number.

b) True, because every natural number is also a member of the set of integers.

c) False, because the rational number 0.75 is not an integer.

d) True, because the rational numbers include the integers.   ■

Graphing on the Number Line

To construct a number line, we draw a straight line and label any convenient point with the number 0. Now we choose any convenient length and use it to locate points to the right of 0 as points corresponding to the positive integers, and points to the left of 0 as points corresponding to the negative integers. See Figure 1.4. The numbers corresponding to the points on the line are called the **coordinates** of the points. The distance between two consecutive integers is called a **unit,** and it is the same for any two consecutive integers. The point with coordinate 0 is called the **origin.** The numbers on the number line increase in size from left to right. When we compare the size of any two numbers, the larger number lies to the right of the smaller on the number line.

Figure 1.4

It is often convenient to illustrate sets of numbers on a number line. The set of integers, *J*, is illustrated or **graphed** as in Fig. 1.5. The three dots to the right and left on the number line indicate that the integers go on indefinitely in both directions.

Figure 1.5

EXAMPLE 2 **Graphing on the number line**

List the elements of each set and graph each set on a number line.

a) {*x*|*x* is a whole number less than 4}

b) {*a*|*a* is an integer between 3 and 9}

c) {*y*|*y* is an integer greater than −3}

Solution

a) The whole numbers less than 4 are 0, 1, 2, and 3. Figure 1.6 shows the graph of this set.

Figure 1.6

b) The integers between 3 and 9 are 4, 5, 6, 7, and 8. The graph is shown in Fig. 1.7.

Figure 1.7

c) The integers greater than −3 are −2, −1, 0, 1, and so on. To indicate the continuing pattern, we use a series of dots on the graph in Fig. 1.8.

Figure 1.8   ■

When we draw a number line, we might label only the integers. But every point on the number line corresponds to a number. The set of all of these numbers is called the set of **real numbers,** *R*. It is easy to locate rational numbers on the number line. For example, the number $\frac{1}{2}$ corresponds to a point halfway between 0 and 1, and $\frac{3}{2}$ corresponds to a point halfway between 1 and 2, as shown in Fig. 1.9. However, on the number line, there are points such as $-\sqrt{2}$ that do not correspond to rational numbers. These points correspond to the set of **irrational numbers,** *I*. *The rational numbers Q and the* *irrational numbers I have no numbers in common, and together they form the set of real numbers R.*

Figure 1.9

The Irrational Numbers

Irrational numbers are those real numbers that cannot be expressed as a ratio of integers. However, it is difficult to decide whether a number is irrational by using this description. For an example of an irrational number, consider the positive square root of 2, written $\sqrt{2}$. If you multiply $\sqrt{2}$ by itself, the result is 2:

$$\sqrt{2} \cdot \sqrt{2} = 2 \quad \text{(The raised dot} \cdot \text{is used as a multiplication symbol.)}$$

Is $\sqrt{2}$ a rational number? A calculator or the square root table of *Appendix* B tells us that $\sqrt{2}$ is 1.414. But if we multiply 1.414 by itself, we get

$$(1.414)(1.414) = 1.999396.$$

So 1.414 is not really the positive square root of 2; it is the positive square root of 1.999396. The number 1.414 is an approximation for $\sqrt{2}$. There is no terminating or repeating decimal that will give us exactly 2 when we multiply it by itself. So we conclude that $\sqrt{2}$ is an irrational number. It can be shown that other square roots such as $\sqrt{3}$, $\sqrt{5}$, and $\sqrt{7}$ are also irrational numbers. But many square roots are rational numbers. For example, $\sqrt{4}$ is a rational number because $\sqrt{4} = 2$.

It is easy to write examples of irrational numbers in decimal form, because *as decimal numbers they have digits that neither repeat nor terminate.* Each of the following numbers has a continuing pattern that guarantees that its digits will neither repeat nor terminate:

$$0.606000600000600000006 \ldots$$
$$0.15115111511115 \ldots$$
$$3.12345678910111213 \ldots$$

Figure 1.10

So each of these numbers is an irrational number.

Since we generally work with rational numbers, the irrational numbers may seem to be unnecessary. However, irrational numbers occur in some very real situations. Over 2000 years ago, people in the Orient and Egypt observed that the ratio of the circumference and diameter is the same for any circle. This constant value was proven to be an irrational number by Johann Heinrich Lambert in 1767. Like other irrational numbers, it does not have any convenient representation as a decimal number. This number has been given the name π (Greek letter pi). See Fig. 1.10. The value of π rounded to nine decimal places is 3.141592654. When using π in computations, we frequently use the rational number 3.14 as an approximate value for π.

Figure 1.11 illustrates the subsets of the real numbers that we have been discussing.

Figure 1.11

EXAMPLE 3   **Classifying real numbers**

Determine which elements of the set

$$\left\{-\sqrt{7}, -\frac{1}{4}, 0, \sqrt{5}, \pi, 4.16, 12\right\}$$

are members of each of the following sets.

a) Real numbers   b) Rational numbers   c) Integers

Solution

a) All of the numbers are real numbers.

b) The numbers $-\frac{1}{4}$, 0, 4.16, and 12 are rational numbers.

c) The only integers in this set are 0 and 12.   ■

EXAMPLE 4   **Subsets of the real numbers**

Determine whether each of the following statements is true or false.

a) $\sqrt{7} \in Q$ b) $J \subseteq W$
c) $I \cap Q = \varnothing$ d) $-3 \in N$
e) $J \cap I = \varnothing$ f) $Q \subseteq R$
g) $R \subseteq N$ h) $\pi \in R$

Solution

a) False   b) False
c) True   d) False
e) True   f) True
g) False   h) True   ■

Warm-ups

True or false? Explain your answer.

1. The number π is a rational number.
2. The set of rational numbers is a subset of the set of real numbers.
3. Zero is the only number that is a member of both Q and I.
4. The set of real numbers is a subset of the set of irrational numbers.
5. The decimal number 0.44444 . . . is a rational number.
6. The decimal number 4.212112111211112 . . . is a rational number.
7. Every irrational number corresponds to a point on the number line.
8. If a real number is not irrational, then it is rational.
9. The set of counting numbers from 1 through 5 trillion is an infinite set.
10. There are infinitely many rational numbers.

1.2 EXERCISES

Determine whether each statement is true or false. Explain your answer. See Example 1.

1. $-6 \in Q$
2. $\frac{2}{7} \in Q$
3. $0 \notin Q$
4. $0 \notin N$
5. $0.6666 \ldots \in Q$
6. $0.00976 \notin Q$
7. $N \subseteq Q$
8. $Q \subseteq J$

List the elements in each set and graph each set on a number line. See Example 2.

9. $\{x \mid x$ is a whole number smaller than 6$\}$
10. $\{x \mid x$ is a natural number less than 7$\}$
11. $\{a \mid a$ is an integer greater than $-5\}$
12. $\{z \mid z$ is an integer between 2 and 12$\}$
13. $\{w \mid w$ is a natural number between 0 and 5$\}$
14. $\{y \mid y$ is a whole number greater than 0$\}$
15. $\{x \mid x$ is an integer between -3 and 5$\}$
16. $\{y \mid y$ is an integer between -4 and 7$\}$

Determine which elements of the set

$$A = \left\{ -\sqrt{10}, -3, -\frac{5}{2}, -0.025, 0, \sqrt{2}, 3\frac{1}{2}, \frac{8}{2} \right\}$$

are members of the following sets. See Example 3.

17. Real numbers
18. Natural numbers
19. Whole numbers
20. Integers
21. Rational numbers
22. Irrational numbers

Determine whether each statement is true or false. Explain. See Example 4.

23. $Q \subseteq R$
24. $I \subseteq Q$
25. $I \cap Q = \{0\}$
26. $J \subseteq Q$
27. $I \cup Q = R$
28. $J \cap Q = \varnothing$
29. $0.2121121112 \ldots \in Q$
30. $0.3333 \ldots \in Q$
31. $3.252525 \ldots \in I$
32. $3.1010010001 \ldots \in I$
33. $0.999 \ldots \in I$
34. $0.666 \ldots \in Q$
35. $\pi \in I$
36. $\pi \in Q$

Place one of the symbols $\subseteq$, $\nsubseteq$, $\in$, or $\notin$ in each blank so that each statement is true.

37. N _____ W
38. J _____ Q
39. J _____ N
40. Q _____ W
41. Q _____ R
42. I _____ R
43. $\varnothing$ _____ I
44. $\varnothing$ _____ Q
45. N _____ R
46. W _____ R
47. 5 _____ J
48. -6 _____ J
49. 7 _____ Q
50. 8 _____ Q
51. $\sqrt{2}$ _____ R
52. $\sqrt{2}$ _____ I
53. 0 _____ I
54. 0 _____ Q
55. $\{2, 3\}$ _____ Q
56. $\{0, 1\}$ _____ N
57. $\{3, \sqrt{2}\}$ _____ R
58. $\{3, \sqrt{2}\}$ _____ Q

Getting More Involved

59. *Writing.* What is the difference between a rational and an irrational number? Why is $\sqrt{9}$ rational and $\sqrt{3}$ irrational?

60. *Cooperative learning.* Work in a small group to make a list of the real numbers of the form $\sqrt{n}$, where n is a natural number between 1 and 100 inclusive. Decide on a method for determining which of these numbers are rational and find them. Compare your group's method and results with other groups' work.

61. *Exploration.* Find the decimal representations of

$$\frac{2}{9}, \quad \frac{2}{99}, \quad \frac{23}{99}, \quad \frac{23}{999}, \quad \frac{234}{999}, \quad \frac{23}{9999}, \quad \text{and} \quad \frac{1234}{9999}.$$

a) What do these decimals have in common?

b) What is the relationship between each fraction and its decimal representation?

1.3 Operations on the Set of Real Numbers

Computations in algebra are performed with positive and negative numbers. In this section we will extend the basic operations of arithmetic to the negative numbers.

In this section:

▶ Absolute Value
▶ Addition
▶ Subtraction
▶ Multiplication
▶ Division
▶ Division by Zero

Absolute Value

The real numbers are the coordinates of the points on the number line. However, we often refer to the points as numbers. For example, the numbers 5 and -5 are both five units away from 0 on the number line shown in Fig. 1.12. A number's *distance* from 0 on the number line is called the **absolute value** of the number. We write $|a|$ for "the absolute value of a." Therefore $|5| = 5$ and $|-5| = 5$.

Figure 1.12

EXAMPLE 1 **Absolute value**

Find the value of $|4|$, $|-4|$, and $|0|$.

Solution

Because both 4 and -4 are four units from 0 on the number line, we have $|4| = 4$ and $|-4| = 4$. Because the distance from 0 to 0 on the number line is 0, we have $|0| = 0$. ∎

Note that $|a|$ represents distance, and distance is never negative. So $|a|$ is greater than or equal to zero for any number a.

Two numbers that are located on opposite sides of zero and have the same absolute value are called **opposites** of each other. The opposite of zero is zero. Every number has a unique opposite. The numbers 9 and -9 are opposites of one another. The minus sign, $-$, is used to signify "opposite" in addition to "negative." When the minus sign is used in front of a number, it is read as "negative." When it is used in front of parentheses or a variable it is read as "opposite." For example,

$$-(9) = -9$$

is read as "the opposite of 9 is negative 9," and

$$-(-9) = 9$$

is read as "the opposite of negative 9 is 9." In general, $-a$ is read "the opposite of a." If a is positive, $-a$ is negative. If a is negative, $-a$ is positive. Opposites have the following property.

Opposite of an Opposite

For any number a,

$$-(-a) = a.$$

EXAMPLE 2   **Opposite of an opposite**
Evaluate.
a) $-(-12)$ b) $-(-(-8))$

Solution
a) The opposite of negative 12 is 12. So $-(-12) = 12$.
b) The opposite of the opposite of -8 is -8. So $-(-(-8)) = -8$. ∎

Remember that we have defined $|a|$ to be the distance between 0 and a on the number line. Using opposites, we can give a symbolic definition of absolute value.

Absolute Value

$$|a| = \begin{cases} a & \text{if } a \text{ is positive or zero} \\ -a & \text{if } a \text{ is negative.} \end{cases}$$

Using this definition, we write

$$|7| = 7,$$

because 7 is positive. To find the absolute value of -7, we use the second line of the definition and write

$$|-7| = -(-7) = 7.$$

Addition

A good way to understand positive and negative numbers is to think of the *positive numbers as assets* and the *negative numbers as debts*. For this illustration we can think of assets simply as cash. Think of debts as unpaid bills such as the electric bill, the phone bill, etc. If you have assets of $4 and $11 and no debts, then your net worth is $15. **Net worth** is the total of your debts and assets. If you have debts of $6 and $7 and no assets, then your net worth is −$13. In symbols,

$$(-6) \quad + \quad (-7) \quad = \quad -13.$$

$6 debt   Added to $7 debt $13 debt

We can think of this addition as adding the absolute values of −6 and −7 (that is, $6 + 7 = 13$) and then putting a negative sign on that result to get −13. These examples illustrate the following rule.

Sum of Two Numbers with Like Signs

To find the sum of two numbers with the same sign, add their absolute values. The sum has the same sign as the original numbers.

If you have a debt of $5 and have only $5 in cash, then your debts equal your assets (in absolute value), and your net worth is $0. In symbols,

$$-5 \quad + \quad 5 \quad = \quad 0.$$

Debt of $5   Asset of $5   Net worth

The number a and its opposite $-a$ have a sum of zero for any a. For this reason a and $-a$ are called **additive inverses** of each other. Note that the words "negative," "opposite," and "additive inverse" are often used interchangeably.

Additive Inverse Property

For any number a,

$$a + (-a) = 0 \quad \text{and} \quad (-a) + a = 0.$$

To understand the sum of a positive and a negative number, consider the following situation. If you have a debt of $7 and $10 in cash, you may have $10 in hand, but your net worth is only $3. Your assets exceed your debts (in absolute value), and you have a positive net worth. In symbols,

$$-7 + 10 = 3.$$

Note that to get 3, we actually subtract 7 from 10. If you have a debt of $8 but have only $5 in cash, then your debts exceed your assets (in absolute value). You have a net worth of −$3. In symbols,

$$-8 + 5 = -3.$$

Note that to get the 3 in the answer, we subtract 5 from 8.

As you can see from these examples, the sum of a positive number and a negative number (with different absolute values) may be either positive or negative. These examples illustrate the rule for adding numbers with unlike signs and different absolute values.

> **Sum of Two Numbers with Unlike Signs (and Different Absolute Values)**
>
> To find the sum of two numbers with unlike signs, subtract their absolute values.
>
> > The sum is positive if the number with the larger absolute value is positive.
> > The sum is negative if the number with the larger absolute value is negative.

EXAMPLE 3   **Adding signed numbers**

Find each sum.

a) $-6 + 13$ **b)** $-9 + (-7)$ **c)** $2 + (-2)$

d) $-35.4 + 2.51$ **e)** $-7 + 0.05$ **f)** $\dfrac{1}{5} + \left(-\dfrac{3}{4}\right)$

Solution

a) The absolute values of -6 and 13 are 6 and 13. Subtract 6 from 13 to get 7. Because the number with the larger absolute value is 13 and it is positive, the result is 7.

b) $-9 + (-7) = -16$

c) $2 + (-2) = 0$

d) Line up the decimal points and subtract 2.51 from 35.40 to get 32.89. Because 35.4 is larger than 2.51 and 35.4 has a negative sign, the answer is negative.

$$-35.4 + 2.51 = -32.89$$

e) Line up the decimal points and subtract 0.05 from 7.00 to get 6.95. Because 7.00 is larger than 0.05, and 7.00 has a negative sign, the answer is negative.

$$-7 + 0.05 = -6.95$$

f) $\dfrac{1}{5} + \left(-\dfrac{3}{4}\right) = \dfrac{4}{20} + \left(-\dfrac{15}{20}\right) = -\dfrac{11}{20}$ ■

Subtraction

Think of subtraction as removing debts or assets, and think of addition as receiving debts or assets. For example, if you have \$10 in cash and \$4 is taken from you, your resulting net worth is the same as if you have \$10 and a water bill for \$4 arrives in the mail. In symbols,

$$10 \quad - \quad 4 \quad = \quad 10 \quad + \quad (-4).$$

$\qquad\quad$ ↑ $\qquad$ ↑ $\qquad\qquad\qquad$ ↑ $\qquad\quad$ ↑

$\qquad$ Remove   Cash $\qquad\qquad\qquad$ Receive   Debt

Removing cash is equivalent to receiving a debt.

Suppose that you have $17 in cash but owe $7 in library fines. Your net worth is $10. If the debt of $7 is canceled or forgiven, your net worth will increase to $17, the same as if you received $7 in cash. In symbols,

$$10 \quad - \quad (-7) \quad = \quad 10 \quad + \quad 7.$$

	↑	↑		↑	↑
	Remove	Debt		Receive	Cash

Removing a debt is equivalent to receiving cash.

Notice that each subtraction problem above is equivalent to an addition problem in which we add the opposite of what we were going to subtract. These examples illustrate the definition of subtraction.

Subtraction of Real Numbers

For any real numbers a and b,

$$a - b = a + (-b).$$

EXAMPLE 4   **Subtracting signed numbers**

Find each difference.

a) $-7 - 3$ b) $7 - (-3)$ c) $48 - 99$

d) $-3.6 - (-7)$ e) $0.02 - 7$ f) $\dfrac{1}{3} - \left(-\dfrac{1}{6}\right)$

Solution

a) To subtract 3 from -7, add the opposite of 3 and -7:

$$-7 - 3 = -7 + (-3) = -10$$

b) To subtract -3 from 7, add the opposite of -3 and 7. The opposite of -3 is 3:

$$7 - (-3) = 7 + (3) = 10$$

c) To subtract 99 from 48, add -99 and 48:

$$48 - 99 = 48 + (-99) = -51$$

d) $-3.6 - (-7) = -3.6 + 7 = 3.4$

e) $0.02 - 7 = 0.02 + (-7) = -6.98$

f) $\dfrac{1}{3} - \left(-\dfrac{1}{6}\right) = \dfrac{1}{3} + \dfrac{1}{6} = \dfrac{2}{6} + \dfrac{1}{6} = \dfrac{3}{6} = \dfrac{1}{2}$ ■

Multiplication

The result of multiplying two numbers is called the **product** of the numbers. The numbers multiplied are **factors.** In algebra we use a raised dot to indicate multiplication, or we place symbols next to one another. For example, the product of a and b is written as $a \cdot b$ or ab. The product of 4 and x is $4x$. We also use parentheses to indicate multiplication. For example, the product of 4 and 3 is written as $4 \cdot 3$, $4(3)$, $(4)3$, or $(4)(3)$.

Multiplication is just a short way to do repeated additions. Adding five 2's gives

$$2 + 2 + 2 + 2 + 2 = 10.$$

So we have the multiplication fact $5 \cdot 2 = 10$. Adding together five negative 2's gives

$$(-2) + (-2) + (-2) + (-2) + (-2) = -10.$$

So we must have $5(-2) = -10$. We can think of $5(-2) = -10$ as saying that taking on five debts of \$2 each is equivalent to a debt of \$10. Losing five debts of \$2 each is equivalent to gaining \$10, so we must have $-5(-2) = 10$.

The rules for multiplying signed numbers are easy to state and remember.

Product of Signed Numbers

To find the product of two nonzero real numbers, multiply their absolute values.

The product is *positive* if the numbers have the *same* sign.
The product is *negative* if the numbers have *unlike* signs.

For example, to multiply -4 and -5, we multiply their absolute values $(4 \cdot 5 = 20)$. Since -4 and -5 have the same sign, $(-4)(-5) = 20$. To multiply -6 and 3, we multiply their absolute values $(6 \cdot 3 = 18)$. Since -6 and 3 have unlike signs, $-6 \cdot 3 = -18$.

EXAMPLE 5   **Multiplying signed numbers**

Find each product.

a) $(-3)(-6)$ **b)** $-4(10)$ **c)** $(-0.01)(0.02)$ **d)** $\dfrac{4}{9} \cdot \left(-\dfrac{1}{5}\right)$

Solution

a) First multiply the absolute values $(3 \cdot 6 = 18)$. Because -3 and -6 have the same sign, we get $(-3)(-6) = 18$.

b) $-4(10) = -40$ Opposite signs, negative result

c) When multiplying decimals, we total the number of decimal places used in the numbers multiplied to get the number of decimal places in the answer. Thus $(-0.01)(0.02) = -0.0002$.

d) $\dfrac{4}{9} \cdot \left(-\dfrac{1}{5}\right) = -\dfrac{4}{45}$ ∎

Division

Just as every real number has an additive inverse or opposite, every nonzero real number a has a **multiplicative inverse** or **reciprocal** $\dfrac{1}{a}$. The reciprocal of 3 is $\dfrac{1}{3}$, and

$$3 \cdot \dfrac{1}{3} = 1.$$

Multiplicative Inverse Property

For any nonzero real number a there is a number $\frac{1}{a}$ such that

$$a \cdot \frac{1}{a} = 1 \quad \text{and} \quad \frac{1}{a} \cdot a = 1.$$

EXAMPLE 6   **Finding multiplicative inverses**
Find the multiplicative inverse (reciprocal) of each number.

a) -2 　　　　　　**b)** $\frac{3}{8}$ 　　　　　　**c)** -0.2

Solution

a) The multiplicative inverse (reciprocal) of -2 is $-\frac{1}{2}$, because

$$-2\left(-\frac{1}{2}\right) = 1.$$

b) The reciprocal of $\frac{3}{8}$ is $\frac{8}{3}$, because

$$\frac{3}{8} \cdot \frac{8}{3} = 1.$$

c) First convert the decimal number -0.2 to a fraction:

$$-0.2 = -\frac{2}{10}$$

$$= -\frac{1}{5}$$

So the reciprocal of -0.2 is -5 and $-0.2(-5) = 1.$ ■

Note that the reciprocal of any negative number is negative.

Earlier, we defined subtraction for real numbers as addition of the additive inverse. We now define division for real numbers as multiplication by the multiplicative inverse (reciprocal).

Division of Real Numbers

For any real numbers a and b with $b \neq 0$,

$$a \div b = a \cdot \frac{1}{b}.$$

If $a \div b = c$, then a is called the **dividend,** b the **divisor,** and c the **quotient.** We also refer to $a \div b$ and $\frac{a}{b}$ as the quotient of a and b.

EXAMPLE 7   **Dividing signed numbers**
Find each quotient.

a) $-60 \div (-2)$ **b)** $-24 \div \dfrac{3}{8}$ **c)** $6 \div (-0.2)$

Solution

a) $-60 \div (-2) = -60 \cdot \left(-\dfrac{1}{2}\right) = 30$ Multiply by $-\frac{1}{2}$, the reciprocal of -2.

b) $-24 \div \dfrac{3}{8} = -24 \cdot \dfrac{8}{3} = -64$

c) $6 \div (-0.2) = 6(-5) = -30$ ∎

You can see from Examples 6 and 7 that a product or quotient is positive when the signs are the same and negative when the signs are opposite:

$$\text{same signs} \quad \leftrightarrow \quad \text{positive result,}$$
$$\text{opposite signs} \quad \leftrightarrow \quad \text{negative result.}$$

Even though all division can be done as multiplication by a reciprocal, we generally use reciprocals only when dividing fractions. Instead, we find quotients using our knowledge of multiplication and the fact that

$$a \div b = c \qquad \text{if and only if} \qquad c \cdot b = a.$$

For example, $-72 \div 9 = -8$ because $-8 \cdot 9 = -72$. Using long division or a calculator, you can get

$$-43.74 \div 1.8 = -24.3$$

and check that you have it correct by finding $-24.3 \cdot 1.8 = -43.74$.
We use the same rules for division when division is indicated by a fraction bar. For example,

$$\frac{-6}{3} = -2, \qquad \frac{6}{-3} = -2, \qquad \frac{-1}{3} = \frac{1}{-3} = -\frac{1}{3}, \qquad \text{and} \qquad \frac{-6}{-3} = 2.$$

Note that if one negative sign appears in a fraction, the fraction has the same value whether the negative sign is in the numerator, in the denominator, or in front of the fraction. If the numerator and denominator of a fraction are both negative, then the fraction has a positive value.

Division by Zero

Why do we omit division by zero from the definition of division? If we write $10 \div 0 = c$, we need to find c such that $c \cdot 0 = 10$. But there is no such number. If we write $0 \div 0 = c$, we need to find c such that $c \cdot 0 = 0$. But $c \cdot 0 = 0$ is true for any number c. Having $0 \div 0$ equal to any number would be confusing. Thus $a \div b$ is defined only for $b \neq 0$. Quotients such as

$$5 \div 0, \qquad 0 \div 0, \qquad \frac{7}{0}, \qquad \text{and} \qquad \frac{0}{0}$$

are said to be *undefined*.

Calculator Close-up

There are many different types of calculators available today. Calculators that are appropriate for use in an algebra course are usually called **scientific calculators.** The most modern scientific calculators, called **graphing calculators,** can draw graphs and perform many other functions that are not available on scientific calculators. Other kinds of special-purpose calculators may not be appropriate in an algebra class. If you have doubts about the kind of calculator to use, ask your instructor for advice.

No matter what kind of calculator you use, you will need to be able to enter signed numbers. For a scientific calculator, the number entered is positive unless you change the sign to negative. The sign-change key $\boxed{+/-}$ changes the sign of a displayed number. To find $3.2 - (-1.5)$ with a scientific calculator, try the following keystrokes:

$$3.2 \; \boxed{-} \; 1.5 \; \boxed{+/-} \; \boxed{=}$$

The display will read as follows:

$$\boxed{4.7}$$

With a graphing calculator, you enter the negative sign before the number using the negative-sign key $\boxed{(-)}$. To find $3.2 - (-1.5)$ with a graphing calculator you can use parentheses to make the display look just like the original problem, then press $\boxed{\text{ENTER}}$ or $\boxed{\text{EXE}}$:

$$3.2-(^{-}1.5) \; \boxed{\text{ENTER}}$$

$$\boxed{4.7}$$

On most graphing calculators you can also display $3.2 - {}^{-}1.5$ and press $\boxed{\text{ENTER}}$ to get 4.7. If you cannot perform this computation, see your calculator manual or ask your instructor for help.

Exercises that should be done with the use of a calculator are included in the exercises of this text. Always use the same calculator when doing these exercises, and you will soon become proficient with it.

Warm-ups

True or false? Explain your answer.

1. The additive inverse of -6 is 6.

2. The opposite of negative 5 is positive 5.

3. The absolute value of 6 is -6.

4. The result of a subtracted from b is the same as $b + (-a)$.

5. If a is positive and b is negative, then ab is negative.

6. If a is positive and b is negative, then $a + b$ is negative.

7. $(-3) - (-6) = -9$

8. $6 \div \left(-\dfrac{1}{2}\right) = -3$

9. $-3 \div 0 = 0$

10. $0 \div (-7) = 0$

1.3 EXERCISES

Evaluate. See Examples 1 and 2.

1. $|-34|$ **2.** $|17|$

3. $|0|$ **4.** $|-15|$

5. $|-6| - |-6|$ **6.** $|8| - |-8|$

7. $-|-9|$ **8.** $-|-3|$

9. $-(-9)$ **10.** $-(-(8))$

11. $-(-(-3))$ **12.** $-(-(-2))$

Find each sum. See Example 3.

13. $(-5) + 9$ **14.** $(-3) + 10$

15. $(-4) + (-3)$ **16.** $(-15) + (-11)$

17. $-6 + 4$ **18.** $5 + (-15)$

19. $7 + (-17)$ **20.** $-8 + 13$

21. $(-11) + (-15)$ **22.** $-18 + 18$

23. $18 + (-20)$ **24.** $7 + (-19)$

25. $-14 + 9$ **26.** $-6 + (-7)$

27. $-4 + 4$ **28.** $-7 + 9$

29. $-\dfrac{1}{10} + \dfrac{1}{5}$ **30.** $-\dfrac{1}{8} + \left(-\dfrac{1}{8}\right)$

31. $\dfrac{1}{2} + \left(-\dfrac{2}{3}\right)$ **32.** $\dfrac{3}{4} + \dfrac{1}{2}$

33. $-15 + 0.02$ **34.** $0.45 + (-1.3)$

35. $-2.7 + (-0.01)$ **36.** $0.8 + (-1)$

37. $47.39 + (-44.587)$ **38.** $0.65357 + (-2.375)$

39. $0.2351 + (-0.5)$ **40.** $-1.234 + (-4.756)$

Find each difference. See Example 4.

41. $7 - 10$ **42.** $8 - 19$ **43.** $-4 - 7$

44. $-5 - 12$ **45.** $7 - (-6)$ **46.** $3 - (-9)$

47. $-1 - 5$ **48.** $-4 - 6$ **49.** $-12 - (-3)$

50. $-15 - (-6)$ **51.** $20 - (-3)$ **52.** $50 - (-70)$

53. $\dfrac{9}{10} - \left(-\dfrac{1}{10}\right)$ **54.** $\dfrac{1}{8} - \dfrac{1}{4}$ **55.** $1 - \dfrac{3}{2}$

56. $-\dfrac{1}{2} - \left(-\dfrac{1}{3}\right)$ **57.** $2 - 0.03$ **58.** $-0.02 - 3$

59. $5.3 - (-2)$ **60.** $-4.1 - 0.13$

61. $-2.44 - 48.29$ **62.** $-8.8 - 9.164$

63. $-3.89 - (-5.16)$ **64.** $0 - (-3.5)$

Find each product. See Example 5.

65. $(25)(-3)$ **66.** $(5)(-7)$

67. $\left(-\dfrac{1}{3}\right)\left(-\dfrac{1}{2}\right)$ **68.** $\left(-\dfrac{1}{2}\right)\left(-\dfrac{6}{7}\right)$

69. $(0.3)(-0.3)$ **70.** $(-0.1)(-0.5)$

71. $(-0.02)(-10)$ **72.** $(0.05)(-2.5)$

Find the multiplicative inverse of each number. See Example 6.

73. 20 **74.** -5 **75.** $-\dfrac{6}{5}$

76. $-\dfrac{1}{8}$ **77.** -0.3 **78.** 0.125

Evaluate. See Example 7.

79. $-6 \div 3$ **80.** $84 \div (-2)$ **81.** $30 \div (-0.8)$

82. $(-9)(-6)$ **83.** $(-0.8)(0.1)$ **84.** $7 \div (-0.5)$

85. $(-0.1) \div (-0.4)$ **86.** $(-18) \div (-0.9)$

87. $9 \div \left(-\dfrac{3}{4}\right)$ **88.** $-\dfrac{1}{3} \div \left(-\dfrac{5}{8}\right)$

89. $-\dfrac{2}{3}\left(-\dfrac{9}{10}\right)$ **90.** $\dfrac{1}{2}\left(-\dfrac{2}{5}\right)$

91. $(0.25)(-365)$ **92.** $7.5 \div (-0.15)$

93. $(-51) \div (-0.003)$ **94.** $(-2.8)(5.9)$

Perform the following computations.

95. $-62 + 13$ **96.** $-88 + 39$

97. $-32 - (-25)$ **98.** $-71 - (-19)$

99. $|-15|$ **100.** $-|-75|$

101. $\dfrac{1}{2}(-684)$ **102.** $\dfrac{1}{3}(-123)$ **103.** $\dfrac{1}{2} - \left(-\dfrac{1}{4}\right)$

104. $\dfrac{1}{8} - \left(-\dfrac{1}{4}\right)$ **105.** $-\dfrac{1}{3}(-96)$ **106.** $-\dfrac{1}{3}(-36)$

107. $-0.2 + 19$ **108.** $-0.6 + 27$

109. $(-0.3) + (-0.03)$ **110.** $-0.02 + (-0.5)$

111. $-57 \div 19$ **112.** $0 \div (-36)$

113. $|-17| + |-3|$ **114.** $64 - |-12|$

115. $0 \div (-0.15)$ **116.** $-20 \div \left(-\dfrac{8}{3}\right)$

117. $27 \div (-0.15)$ **118.** $33 \div (-0.2)$

119. $-\dfrac{1}{3} + \dfrac{1}{6}$ **120.** $-\dfrac{2}{3} + \dfrac{1}{6}$ **121.** $-63 + |8|$

122. $|-34| - 27$ **123.** $-\dfrac{1}{2} + \left(-\dfrac{1}{2}\right)$

124. $-\dfrac{2}{3} + \left(-\dfrac{2}{3}\right)$ **125.** $-\dfrac{1}{2} - 19$ **126.** $-\dfrac{1}{3} - 22$

127. $28 - 0.01$

128. $55 - 0.1$

129. $-29 - 0.3$

130. $-0.241 - 0.3$

131. $(-2)(0.35)$

132. $(-3)(0.19)$

133. $(-10)(-0.2)$

134. $\left(-\dfrac{1}{2}\right)(-50)$

Use an operation with signed numbers to solve each problem.

135. *Net worth of a family.* The average American family has a \$45,000 mortgage, \$2,300 in credit card debt, \$1500 in other debts, \$1200 in savings, two cars worth \$3,500 each, and \$20,000 in home equity. What is the net worth of the average American family?

136. *Net worth of a bank.* Just before the recession, First Federal Homestead had \$15.6 million in mortgage loans, had \$23.3 million on deposit, and owned \$8.5 million worth of real estate. After the recession started, the value of the real estate decreased to \$4.8 million. What was the net worth of First Federal before the recession and after the recession started? (To a financial institution a loan is an asset and a deposit is a liability.)

137. *Warming up.* On January 11 the temperature at noon was 14°F in St. Louis and −6°F in Duluth. How much warmer was it in St. Louis?

Figure for Exercise 137

138. *Bitter cold.* On January 16, the temperature at midnight was −5°F in Chicago and −23°F in Minneapolis. How much warmer was it in Chicago?

139. *Below sea level.* The altitude of the floor of Death Valley is −282 feet (282 feet below sea level); the altitude of the shore of the Dead Sea is −1296 feet (*Rand McNally World Atlas*). How many feet above the shore of the Dead Sea is the floor of Death Valley?

140. *Highs and lows.* The altitude of the peak of Mt. Everest, the highest point on earth, is 29,028 feet. The world's greatest known ocean depth of −36,201 feet was recorded in the Marianas Trench (*Rand McNally World Atlas*). How many feet above the bottom of the Marianas Trench is a climber who has reached the top of Mt. Everest?

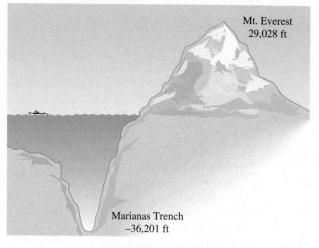

Figure for Exercise 140

Getting More Involved

141. *Discussion.* Why is it necessary to learn addition of signed numbers before learning subtraction of signed numbers and to learn multiplication of signed numbers before division of signed numbers?

142. *Writing.* Explain why 0 is the only real number that does not have a multiplicative inverse.

1.4 Evaluating Expressions

In algebra you will learn to work with variables. However, there is often nothing more important than finding a numerical answer to a question. This section is concerned with computation.

Arithmetic Expressions

The result of writing numbers in a meaningful combination with the ordinary operations of arithmetic is called an **arithmetic expression** or simply an **expression.** An expression that involves more than one operation is called a **sum, difference, product,** or **quotient** if the last operation to be performed is addition, subtraction, multiplication, or division, respectively. Parentheses are used as **grouping symbols** to indicate which operations are performed first. The expression

$$5 + (2 \cdot 3)$$

is a sum because the parentheses indicate that the product of 2 and 3 is to be found before the addition is performed. So we evaluate this expression as follows:

$$5 + (2 \cdot 3) = 5 + 6 = 11$$

If we write $(5 + 2)3$, the expression is a product and it has a different value.

$$(5 + 2)3 = 7 \cdot 3 = 21$$

Brackets [ ] are also used to indicate grouping. If an expression occurs within absolute value bars | |, it is evaluated before the absolute value is found. So absolute value bars also act as grouping symbols. Operations within the innermost grouping symbols are performed first.

EXAMPLE 1   **Grouping symbols**
Evaluate each expression.

a) $5[(2 \cdot 3) - 8]$ b) $2[(4 \cdot 5) - |3 - 6|]$

Solution

a) $5[(2 \cdot 3) - 8] = 5[6 - 8]$ Innermost grouping first.
$= 5[-2]$
$= -10$

b) $2[(4 \cdot 5) - |3 - 6|] = 2[20 - |-3|]$
$= 2[20 - 3]$
$= 2[17]$
$= 34$ ■

Exponential Expressions

We use the notation of exponents to simplify the writing of repeated multiplication. The product $5 \cdot 5 \cdot 5 \cdot 5$ is written as 5^4. The number 4 in 5^4 is called the exponent, and it indicates the number of times that the factor 5 occurs in the product.

> **Exponential Expression**
>
> For any natural number n and real number a,
>
> $$a^n = \underbrace{a \cdot a \cdot a \cdot \ldots \cdot a.}_{n \text{ factors of } a}$$
>
> We call a the **base,** n the **exponent,** and a^n an **exponential expression.**

We read a^n as "the nth power of a" or "a to the nth power." The exponential expressions 3^5 and 10^6 are read as "3 to the fifth power" and "10 to the sixth power." We can also use the words "squared" and "cubed" for the second and third powers, respectively. For example, 5^2 and 2^3 are read as "5 squared" and "2 cubed," respectively.

EXAMPLE 2   **Exponential expressions**

Evaluate.

a) 2^3 **b)** $(-3)^4$ **c)** $(0.5)^2$ **d)** $\left(-\frac{1}{2}\right)^5$

Solution

a) $2^3 = 2 \cdot 2 \cdot 2 = 8$

b) $(-3)^4 = (-3)(-3)(-3)(-3) = 81$

c) $(0.5)^2 = (0.5)(0.5) = 0.25$

d) $\left(-\frac{1}{2}\right)^5 = \left(-\frac{1}{2}\right)\left(-\frac{1}{2}\right)\left(-\frac{1}{2}\right)\left(-\frac{1}{2}\right)\left(-\frac{1}{2}\right) = -\frac{1}{32}$ ∎

Order of Operations

To simplify the writing of expressions, some grouping symbols are often omitted. If we saw the expression

$$5 + 2 \cdot 3$$

written without parentheses, we would not know how to evaluate it unless we had a rule for which operation to perform first. Expressions without grouping symbols are evaluated consistently by using a rule called the **order of operations.**

> **Order of Operations**
> 1. Evaluate each exponential expression (in order from left to right).
> 2. Perform multiplication and division (in order from left to right).
> 3. Perform addition and subtraction (in order from left to right).

When grouping symbols appear in an expression, we first evaluate within each set of grouping symbols, using the above order.

EXAMPLE 3   **Order of operations**

Evaluate each expression.

a) $5 + 2 \cdot 3$ **b)** $9 \cdot 2^3$ **c)** $(6 - 4^2)^2$ **d)** $40 \div 8 \cdot 2 \div 5 \cdot 3$

Solution

a) $5 + 2 \cdot 3 = 5 + 6$ Multiply first.

$= 11$ Then add.

b) $9 \cdot 2^3 = 9 \cdot 8$ Evaluate the exponential expression first.

$= 72$ Then multiply.

c) $(6 - 4^2)^2 = (6 - 16)^2$ Evaluate 4^2 within the parentheses first.

$= (-10)^2$ Then subtract.

$= 100$ $(-10)(-10) = 100$

d)   Multiplication and division are done from left to right.

$$40 \div 8 \cdot 2 \div 5 \cdot 3 = 5 \cdot 2 \div 5 \cdot 3$$
$$= 10 \div 5 \cdot 3$$
$$= 2 \cdot 3$$
$$= 6 \qquad \blacksquare$$

Consider the expression $-1 \cdot 3^2$. By the order of operations the exponential expression is evaluated first. Thus

$$-1 \cdot 3^2 = -1 \cdot 9 = -9.$$

In the expression -3^2 the negative sign has the same role as the -1 in the expression $-1 \cdot 3^2$. To evaluate -3^2, we find the square of 3 and then take the opposite of the result. So

$$-3^2 = -9.$$

The expression -3^2 should be read as "the opposite of 3^2." The square of -3 is written $(-3)^2$, and we have $(-3)^2 = 9$.

EXAMPLE 4   **The order of negative signs**

Evaluate each expression.

a) -2^4 **b)** $(-5)^2$ **c)** $-(3 - 5)^2$ **d)** $-(5^2 - 4 \cdot 7)^2$

Solution

a)   To evaluate -2^4, find 2^4 first and then take the opposite. So $-2^4 = -16$.

b) $(-5)^2 = (-5)(-5) = 25$

c)   Evaluate within the parentheses first, square that result, then take the opposite.

$$-(3 - 5)^2 = -(-2)^2$$ Evaluate within parentheses first.

$$= -4$$ Square -2 to get 4, then take the opposite of 4 to get -4.

d) $-(5^2 - 4 \cdot 7)^2 = -(25 - 28)^2$ Evaluate 5^2 within the parentheses first.

$$= -(-3)^2$$ Then subtract.

$$= -9$$ Square -3 to get 9, then take the opposite of 9 to get -9. $\blacksquare$

When an expression involves a fraction bar, the numerator and denominator are each treated as if they are in parentheses. The next example illustrates how the fraction bar groups the numerator and denominator.

EXAMPLE 5   **Order of operations in fractions**

Evaluate each quotient.

a) $\dfrac{10 - 8}{6 - 8}$ **b)** $\dfrac{-6^2 + 2 \cdot 7}{4 - 3 \cdot 2}$

Solution

a) $\dfrac{10 - 8}{6 - 8} = \dfrac{2}{-2}$ Evaluate the numerator and denominator separately.

$\phantom{\dfrac{10-8}{6-8}} = -1$ Then divide.

b) $\dfrac{-6^2 + 2 \cdot 7}{4 - 3 \cdot 2} = \dfrac{-36 + 14}{4 - 6}$

$\phantom{\dfrac{-6^2 + 2 \cdot 7}{4 - 3 \cdot 2}} = \dfrac{-22}{-2}$ Evaluate the numerator and denominator separately.

$\phantom{\dfrac{-6^2 + 2 \cdot 7}{4 - 3 \cdot 2}} = 11$ Then divide.   ■

Algebraic Expressions

The result of combining numbers and variables with the ordinary operations of arithmetic (in some meaningful way) is called an **algebraic expression.** For example,

$$2x - 5y, \quad 5x^2, \quad (x - 3)(x + 2), \quad b^2 - 4ac, \quad 5, \quad \text{and} \quad \dfrac{x}{2}$$

are algebraic expressions, or simply **expressions.** An expression such as $2x - 5y$ has no definite value unless we assign values to x and y. For example,

**MATH
AT
WORK**

**Financial Aid
Director**

Nancy Gittins, Assistant Director of Financial Aid at Babson College, helps graduate and undergraduate students achieve their goal of financing their educations. Because recent tuition and fees can be as high as $20,000 a year, many students need financial aid to help defray these expenses. Federal loans and state loans, as well as grants from the federal and state level, can be given to students who qualify. Ms. Gittins administers many of these loans and grants and helps students understand the different options that are available. The interest rate for these loans is now a variable rate that is tied to treasury bills. The rate can be as high as 9.0% and as low as 7.3%. In Exercise 109 of this section you will work with interest rates compounded annually.

if $x = 3$ and $y = 4$, then the value of $2x - 5y$ is found by replacing x with 3 and y with 4 and evaluating:

$$2x - 5y = 2(3) - 5(4) = 6 - 20 = -14$$

Note the importance of the order of operations in evaluating an algebraic expression.

To find the value of the difference $2x - 5y$ when $x = -2$ and $y = -3$, replace x and y by -2 and -3, respectively, and then evaluate.

$$2x - 5y = 2(-2) - 5(-3) = -4 - (-15) = -4 + 15 = 11$$

EXAMPLE 6 **Value of an algebraic expression**

Evaluate each expression for $a = 2$, $b = -3$, and $c = 4$.

a) $a - c^2$ **b)** $a - b^2$ **c)** $b^2 - 4ac$ **d)** $\dfrac{a - b}{c - b}$

Solution

a) Replace a by 2 and c by 4 in the expression $a - c^2$.

$$a - c^2 = 2 - 4^2 = 2 - 16 = -14$$

b) $a - b^2 = 2 - (-3)^2 = 2 - 9 = -7$

c) $b^2 - 4ac = (-3)^2 - 4(2)(4) = 9 - 32 = -23$

d) $\dfrac{a - b}{c - b} = \dfrac{2 - (-3)}{4 - (-3)} = \dfrac{5}{7}$ ∎

CAUTION When you replace a variable by a negative number, be sure to use parentheses around the negative number. If the parentheses were omitted in Example 6(c), we would get $-3^2 - 4(2)(4) = -41$ instead of -23. ⊘

A symbol such as y_1 is treated like any other variable. We read y_1 as "y one" or "y sub one." The 1 is called a **subscript.** We can think of y_1 as the "first y" and y_2 as the "second y." The subscript notation is used in the following example.

EXAMPLE 7 **An algebraic expression with subscripts**

Let $y_1 = -12$, $y_2 = -5$, $x_1 = -3$, and $x_2 = 4$. Find the value of $\dfrac{y_2 - y_1}{x_2 - x_1}$.

Solution

Substitute the appropriate values into the expression.

$$\frac{y_2 - y_1}{x_2 - x_1} = \frac{-5 - (-12)}{4 - (-3)} = \frac{7}{7} = 1$$ ∎

When we evaluate an algebraic expression involving only one variable for many values of that variable, we get a collection of data. A graph (picture) of this data can give us useful information.

EXAMPLE 8   **Reading a graph**

The expression $0.85(220 - A)$ gives the target heart rate for beneficial exercise for an athlete who is A years old. Use the graph in Fig. 1.13 to estimate the target heart rate for a 40-year-old athlete. Use the graph to estimate the age of an athlete with a target heart rate of 170.

Figure 1.13

Solution

To find the target heart rate for a 40-year-old athlete, first draw a vertical line from age 40 up to the graph as shown in Fig. 1.13. From the point of intersection, draw a horizontal line to the heart-rate scale. So the target heart rate for a 40-year-old athlete is about 150. To find the age corresponding to a heart rate of 170, first draw a horizontal line from heart rate 170 to the graph as shown in the figure. From the point of intersection, draw a vertical line down to the age scale. The heart rate of 170 corresponds to an age of about 20.   ■

Calculator Close-up

You can use your calculator to evaluate exponential expressions. For example, to evaluate 2^8 using a scientific calculator, try the following keystrokes:

$$2 \; \boxed{x^y} \; 8 \; \boxed{=}$$

The display will read as follows:

$$\boxed{\qquad\qquad 256\;}$$

If x^y is printed above another key, then it is the second function for that key, and you must press $\boxed{\text{2nd}}$ before pressing the key. On a graphing calculator the symbol $\wedge$ is used to indicate raising to a power. For example, 2^8 is found on a graphing calculator as follows:

$$2 \wedge 8 \; \boxed{\text{ENTER}}$$

$$\boxed{\qquad\qquad 256\;}$$

Even though the x^y key can be used to find any power of a number, many calculators also have an x^2 key, $\boxed{x^2}$, which can be used for squaring a number.

Most scientific and graphing calculators have parentheses that can be used for grouping, and they follow the order of operations when no parentheses are used. Use your calculator to verify the following equations:

$$5.6 + (3.4)(6.7) = 28.38$$
$$(5.6 + 3.4)6.7 = 60.3$$
$$3 \cdot 8.2^3 - 5.6^2 = 1622.744$$

Warm-ups

True or false? Explain your answer.

1. $2^3 = 6$ **2.** $-1 \cdot 2^2 = -4$

3. $-2^2 = -4$ **4.** $6 + 3 \cdot 2 = 18$

5. $(6 + 3) \cdot 2 = 81$ **6.** $(6 + 3)^2 = 18$

7. $6 + 3^2 = 15$ **8.** $(-3)^3 = -3^3$

9. $|-3 - (-2)| = 5$ **10.** $|7 - 8| = |7| - |8|$

1.4 EXERCISES

Evaluate each expression. See Example 1.

1. $(-3 \cdot 4) - (2 \cdot 5)$

2. $|-3 - 2| - |2 - 6|$

3. $4[5 - |3 - (2 \cdot 5)|]$

4. $-2|(-3 \cdot 4) - 6|$

5. $(6 - 8)(|2 - 3| + 6)$

6. $-5(6 + [(5 - 7) - 4])$

Evaluate each exponential expression. See Example 2.

7. 2^5 **8.** 3^4 **9.** $(-1)^4$

10. $(-1)^5$ **11.** $(-4)^2$ **12.** $\left(-\frac{1}{2}\right)^6$

13. $\left(-\frac{3}{4}\right)^3$ **14.** $\left(-\frac{1}{2}\right)^7$ **15.** $(0.1)^3$

16. $(0.1)^4$

Evaluate each expression. See Examples 3 and 4.

17. $4 - 6 \cdot 2$ **18.** $8 - 3 \cdot 9$

19. $5 - 6(3 - 5)$ **20.** $8 - 3(4 - 6)$

21. $\left(\frac{1}{3} - \frac{1}{2}\right)\left(\frac{1}{4} - \frac{1}{2}\right)$ **22.** $\left(\frac{1}{2} - \frac{1}{4}\right)\left(\frac{1}{2} - \frac{3}{4}\right)$

23. $7^2 + 9$ **24.** $-6^2 + (-3)^3$

25. $-(2 + 3)^2$ **26.** $-(1 - 3 \cdot 2)^3$

27. $-5^2 \cdot 2^3$ **28.** $8 - 2^3$

29. $5 \cdot 2^3$ **30.** $(8 - 2)^3$

31. $-(3^2 - 4)^2$ **32.** $-(6 - 2^3)^4$

33. $-60 \div 10 \cdot 3 \div 2 \cdot 5 \div 6$

34. $75 \div (-5)(-3) \div \frac{1}{2} \cdot 6$

35. $30 \cdot 2 - (-6) \div 3 \cdot 12$

36. $-48 \div 12 - 6 \div 3 - 4$

37. $5.5 - 2.3^4$

38. $5.3^2 - 4 \cdot 6.1$

39. $(1.3 - 0.31)(2.9 - 4.88)$

40. $(6.7 - 9.88)^3$

41. $-388.8 \div (13.5)(9.6)$

42. $(-4.3)(5.5) \div (3.2)(-1.2)$

Evaluate each expression. See Example 5.

43. $\frac{2 - 6}{9 - 7}$ **44.** $\frac{9 - 12}{4 - 5}$

45. $\frac{-3 - 5}{6 - (-2)}$ **46.** $\frac{-14 - (-2)}{-3 - 3}$

47. $\frac{4 + 2 \cdot 7}{3 \cdot 2 - 9}$ **48.** $\frac{-6 - 2(-3)}{8 - 3(-3)}$

49. $\frac{-3^2 - (-9)}{2 - 3^2}$ **50.** $\frac{-2^4 - 5}{3^2 - 2^4}$

Evaluate each expression for $a = -1$, $b = 3$, and $c = -4$. See Example 6.

51. $b^2 - 4ac$ **52.** $a^2 - 4bc$

53. $\frac{a - b}{a - c}$ **54.** $\frac{b - c}{b - a}$

55. $(a - b)(a + b)$ **56.** $(a - c)(a + c)$

57. $c^2 - 2c + 1$ **58.** $b^2 - 2b - 3$

59. $\frac{2}{a} + \frac{b}{c} - \frac{1}{c}$ **60.** $\frac{c}{a} + \frac{c}{b} - \frac{a}{b}$

61. $|a - b|$ **62.** $|b + c|$

Find the value of $\frac{y_2 - y_1}{x_2 - x_1}$ for each choice of y_1, y_2, x_1, and x_2. See Example 7.

63. $y_1 = 4$, $y_2 = -6$, $x_1 = 2$, $x_2 = -7$

64. $y_1 = -3$, $y_2 = -3$, $x_1 = 4$, $x_2 = -5$

65. $y_1 = -1$, $y_2 = 2$, $x_1 = -3$, $x_2 = 1$

66. $y_1 = -2$, $y_2 = 5$, $x_1 = 2$, $x_2 = 6$

67. $y_1 = 2.4$, $y_2 = 5.6$, $x_1 = 5.9$, $x_2 = 4.7$

68. $y_1 = -5.7$, $y_2 = 6.9$, $x_1 = 3.5$, $x_2 = 4.2$

Find the value of $b^2 - 4ac$ for each choice of a, b, and c.

69. $a = -1$, $b = -2$, $c = 3$ **70.** $a = 1$, $b = 2$, $c = 3$

71. $a = 2$, $b = 3$, $c = -5$ **72.** $a = 2$, $b = -3$, $c = -4$

73. $a = 0.5$, $b = 8$, $c = 2$ **74.** $a = 0.2$, $b = 3$, $c = 5$

75. $a = -1.2$, $b = 3.2$, $c = 5.6$

76. $a = 2.4$, $b = -8.5$, $c = -5.8$

Evaluate each expression.

77. $-2^2 + 5(3)^2$ **78.** $-3^2 + 3(6)^2$

79. $(-2 + 5)3^2$ **80.** $(-3 + 3)6^2$

81. $5^2 - 4(1)(6)$ **82.** $6^2 - 4(2)(3)$

83. $[13 + 2(-5)]^2$ **84.** $[6 + 2(-4)]^2$

85. $\dfrac{4 - (-1)}{-3 - 2}$ **86.** $\dfrac{2 - (-3)}{3 - 5}$

87. $3(-2)^2 - 5(-2) + 4$ **88.** $3(-1)^2 + 5(-1) - 6$

89. $-4\left(\dfrac{1}{2}\right)^2 + 3\left(\dfrac{1}{2}\right) - 2$ **90.** $8\left(\dfrac{1}{2}\right)^2 - 6\left(\dfrac{1}{2}\right) + 1$

91. $-\dfrac{1}{2}|6 - 2|$ **92.** $-\dfrac{1}{3}|9 - 6|$

93. $\dfrac{1}{2} - \dfrac{1}{3}\left|\dfrac{1}{4} - \dfrac{1}{2}\right|$ **94.** $\dfrac{1}{3} - \dfrac{1}{2}\left|\dfrac{1}{3} - \dfrac{1}{2}\right|$

95. $|6 - 3 \cdot 7| + |7 - 5|$

96. $|12 - 4| - |3 - 4 \cdot 5|$

97. $3 - 7[4 - (2 - 5)]$ **98.** $9 - 2[3 - (4 + 6)]$

99. $3 - 4(2 - |4 - 6|)$ **100.** $3 - (|-4| - |-5|)$

101. $4[2 - (5 - |-3|)^2]$

102. $[5 - (-3)]^2 + [4 - (-2)]^2$

Solve each problem. See Example 8.

103. *Female target heart rate.* The algebraic expression $0.65(220 - A)$ gives the target heart rate for beneficial exercise for women, where A is the age of the woman. How much larger is the target heart rate of a 25-year-old woman than that of a 65-year-old woman? Use the accompanying graph to estimate the age at which a woman's target heart rate is 115.

104. *Male target heart rate.* The algebraic expression $0.75(220 - A)$ gives the target heart rate for beneficial exercise for men, where A is the age of the man. Use the algebraic expression to find the target heart rate for a 20-year-old and a 50-year-old man. Use the accompanying graph to estimate the age at which a man's target heart rate is 115.

Figure for Exercises 103 and 104

Solve each problem.

105. *Perimeter of a pool.* The algebraic expression $2L + 2W$ gives the perimeter of a rectangle with length L and width W. Find the perimeter of a rectangular swimming pool that has length 34 feet and width 18 feet.

106. *Area of a lot.* The algebraic expression for the area of a trapezoid $0.5h(b_1 + b_2)$ gives the area of the property shown in the figure. Find the area if $h = 150$ feet, $b_1 = 260$ feet, and $b_2 = 220$ feet.

Figure for Exercise 106

107. *Saving for retirement.* Long-term corporate bonds have had an average yield of 6.2% annually over the last 40 years (Fidelity Investments, Boston). If you invest $10,000 in these bonds and average 6.2% annually for the next 30 years, then what will your investment amount to? The expression $P(1 + r)^n$ gives the amount of an investment of P dollars invested for n years at interest rate r compounded annually.

108. *Saving for college.* The average cost of a B.A. for students beginning college in the year 2005 will be $252,000 at an Ivy League school (*Fortune 1992 Investors Guide*). When Bill's daughter was born in 1987, he made an investment that was large enough to enable his daughter to attend an Ivy League school in 2005. If he gets 7% compounded annually on his investment for 18 years, then how much did he invest in 1987? The principal that you must invest today at interest rate r compounded annually to have A dollars in n years is given by the algebraic expression

$$\frac{A}{(1 + r)^n}.$$

109. *Student loan.* A college student borrowed $4,000 at 8% compounded annually in his freshman year and did not have to make payments until four years later. Use the accompanying graph to estimate the amount he owes at the time the payments start. Use the expression $P(1 + r)^n$ to find the actual amount of the debt at the time the payments start.

Figure for Exercise 109

110. *High cost of nursing care.* Statistics show that 33% of men and 52% of women will enter a nursing home at some time in their lives (Fortune 1992 Investors Guide). The average cost for a one-year stay in a nursing home in 1990 was $29,930, and conservative projections indicate that in n years from 1990 the average cost will be $29{,}930(1.05)^n$ dollars. Using the algebraic expression, find the projected cost for a one-year stay in 2005.

111. *Could be higher.* Some economists project that the average cost of a one-year stay in a nursing home n years from 1990 could be as high as $29{,}930(1.08)^n$ dollars. How much more would you pay for a one-year stay in 2005 using this formula rather than the formula in Exercise 110?

Getting More Involved

112. *Discussion.* Evaluate $5(5(5 \cdot 3 + 6) + 4) + 7$ and $3 \cdot 5^3 + 6 \cdot 5^2 + 4 \cdot 5 + 7$. Explain why these two expressions must have the same value.

113. *Cooperative learning.* Find some examples of algebraic expressions not mentioned in this text and explain to your class what they are used for.

1.5 Properties

In this section:
▶ Commutative Properties
▶ Associative Properties
▶ Distributive Properties
▶ Identity Properties
▶ Inverse Properties
▶ Multiplication Property of Zero

You know that the price of a hamburger plus the price of a Coke is the same as the price of a Coke plus the price of a hamburger. But do you know which property of the real numbers is at work in this situation? In arithmetic we may be unaware when we use properties of the real numbers, but in algebra we need a better understanding of those properties. In this section we will study the properties of the basic operations on the set of real numbers.

Commutative Properties

We get the same result whether we evaluate $3 + 7$ or $7 + 3$. This example illustrates the **commutative property of addition.** The fact that $4 \cdot 5$ and $5 \cdot 4$ are both 20 illustrates the **commutative property of multiplication.**

> **Commutative Properties**
> For any real numbers a and b,
> $$a + b = b + a \qquad \text{and} \qquad a \cdot b = b \cdot a.$$

In writing the product of a number and a variable, it is customary to write the number first. We write $3x$ rather than $x3$. In writing the product of two variables, it is customary to write them in alphabetical order. We write cd rather than dc.

Addition and multiplication are commutative operations, but what about subtraction and division? Because $7 - 3 = 4$ and $3 - 7 = -4$, subtraction is not commutative. To see that division is not commutative, consider the amount each person gets when a \$1 million lottery prize is divided between two people and when a \$2 prize is divided among 1 million people.

Associative Properties

Consider the expression $2 + 3 + 7$. Using the order of operations, we add from left to right to get 12. If we first add 3 and 7 to get 10 and then add 2 and 10, we also get 12. So

$$(2 + 3) + 7 = 2 + (3 + 7).$$

Now consider the expression $2 \cdot 3 \cdot 5$. Using the order of operations, we multiply from left to right to get 30. However, we can first multiply 3 and 5 to get 15 and then multiply by 2 to get 30. So

$$(2 \cdot 3) \cdot 5 = 2 \cdot (3 \cdot 5).$$

These examples illustrate the **associative property of addition** and the **associative property of multiplication.**

Associative Properties

For any real numbers a, b, and c,

$$(a + b) + c = a + (b + c) \qquad \text{and} \qquad (ab)c = a(bc).$$

Consider the expression

$$4 - 9 + 8 - 5 - 8 + 6 - 13.$$

According to the accepted order of operations, we could evaluate this expression by computing from left to right. However, if we use the definition of subtraction, we can rewrite this expression as

$$4 + (-9) + 8 + (-5) + (-8) + 6 + (-13).$$

The commutative and associative properties of addition allow us to add these numbers in any order we choose. A good way to add them is to add the positive numbers, add the negative numbers, and then combine the two totals:

$$4 + 8 + 6 + (-9) + (-5) + (-8) + (-13) = 18 + (-35) = -17$$

For speed we usually do not rewrite the expression. We just sum the positive numbers and the negative numbers, and then combine their totals.

EXAMPLE 1   **Using commutative and associative properties**
Evaluate.

a) $4 - 7 + 10 - 5$ **b)** $6 - 5 - 9 + 7 - 2 + 5 - 8$

Solution

a) $4 - 7 + 10 - 5 = 14 + (-12) = 2$

                     ↑   ↑

     Sum of the positive   Sum of the negative
        numbers   numbers

b) $6 - 5 - 9 + 7 - 2 + 5 - 8 = 18 + (-24) = -6$ ■

Not all operations are associative. Using subtraction, for example, we have

$$(8 - 4) - 1 \neq 8 - (4 - 1),$$

because $(8 - 4) - 1 = 3$ and $8 - (4 - 1) = 5$. For division we have

$$(8 \div 4) \div 2 \neq 8 \div (4 \div 2),$$

because $(8 \div 4) \div 2 = 1$ and $8 \div (4 \div 2) = 4$. So subtraction and division are not associative.

Distributive Properties

Using the order of operations, the product $3(6 + 4)$ is evaluated by first adding 6 and 4 and then multiplying by 3:

$$3(6 + 4) = 3 \cdot 10 = 30$$

Note that we also have

$$3 \cdot 6 + 3 \cdot 4 = 18 + 12 = 30.$$

Therefore

$$3(6 + 4) = 3 \cdot 6 + 3 \cdot 4.$$

Note that multiplication by 3 from outside the parentheses is *distributed* over each term inside the parentheses. This example illustrates the **distributive property.**

    You can check that multiplication is distributed over subtraction in the same manner as it is for addition. As an example, check that

$$3(6 - 4) = 3 \cdot 6 - 3 \cdot 4$$

by evaluating the expressions on each side of the equal sign.

Distributive Properties

For any real numbers a, b, and c,

$$a(b + c) = ab + ac \qquad \text{and} \qquad a(b - c) = ab - ac.$$

We used arithmetic expressions to illustrate the distributive properties, but we will use them most often with algebraic expressions. For example, if we start with $5(x + 4)$ and write

$$5(x + 4) = 5x + 20,$$

we are using the distributive property to multiply or remove parentheses. If we start with $6x - 18$ and write

$$6x - 18 = 6(x - 3),$$

we are using it to write the difference $6x - 18$ as a product $6(x - 3)$.

EXAMPLE 2   **Using the distributive property**

Use the distributive property to rewrite each sum or difference as a product, and each product as a sum or difference.

a) $9x - 9$ **b)** $b(2 - a)$ **c)** $3a + ac$ **d)** $-2(x - 3)$

Solution

a) $9x - 9 = 9(x - 1)$

b) $b(2 - a) = 2b - ab$ Note that $b \cdot 2 = 2b$ by the commutative property.

c) $3a + ac = a(3 + c)$

d) $-2(x - 3) = -2x - (-2)(3)$

$\qquad\qquad\quad = -2x - (-6)$

$\qquad\qquad\quad = -2x + 6$ ∎

Identity Properties

The numbers 0 and 1 have special properties. Addition of 0 to a number does not change the number, and multiplication of a number by 1 does not change the number. For this reason, 0 is called the **additive identity** and 1 is called the **multiplicative identity.**

> **Identity Properties**
> For any real number a,
> $$a + 0 = 0 + a = a \qquad \text{and} \qquad a \cdot 1 = 1 \cdot a = a.$$

Inverse Properties

The ideas of *additive inverses* and *multiplicative inverses* were introduced in Section 1.3. Every real number a has a unique additive inverse or opposite, $-a$, such that $a + (-a) = 0$. Every nonzero real number a also has a unique multiplicative inverse (reciprocal), written $\frac{1}{a}$, such that $a\left(\frac{1}{a}\right) = 1$. For rational numbers the multiplicative inverse is easy to find. For example, the multiplicative inverse of $\frac{2}{5}$ is $\frac{5}{2}$, because

$$\frac{2}{5} \cdot \frac{5}{2} = \frac{10}{10} = 1.$$

Inverse Properties

For any real number a there is a unique number $-a$, such that

$$a + (-a) = 0.$$

For any nonzero real number a there is a unique number $\frac{1}{a}$ such that

$$a \cdot \frac{1}{a} = 1.$$

Calculator Close-up

Most scientific calculators have a reciprocal key $\boxed{1/x}$ that gives the reciprocal of the number on the display. For example, you can evaluate.

$$\frac{1}{6.25} + \frac{1}{12.5}$$

on your scientific calculator as follows:

6.25 $\boxed{1/x}$ $\boxed{+}$ 12.5 $\boxed{1/x}$ $\boxed{=}$

The display will read

0.24

On a graphing calculator there is no reciprocal key, but you can enter the reciprocal of a number on the display as follows:

1/6.25+1/12.5 $\boxed{\text{ENTER}}$

0.24

Reciprocals are used in problems involving rates. For example, if you peel one potato in 0.01 hour, then you are peeling at the rate of $\frac{1}{0.01}$ or 100 potatoes per hour. If your helper peels one potato in 0.02 hour, then your helper is peeling at the rate of $\frac{1}{0.02}$ or 50 potatoes per hour. Together, you are peeling at the rate of 150 potatoes per hour. In general, if one task is completed in x hours, then the rate is $\frac{1}{x}$ tasks per hour.

▦ EXAMPLE 3   **Work rates**

An old computer system can process one water bill in 0.002 hour. A new computer system can process one water bill in 0.00125 hour. If the old system is used simultaneously with the new, then at what rate will the processing of the water bills be accomplished?

Solution

Since the old system does one bill in 0.002 hour, its rate is $\frac{1}{0.002}$ bills per hour.

Since the new system does one bill in 0.00125 hour, its rate is $\frac{1}{0.00125}$ bills per hour. Their rate when working together is the sum of their individual rates:

$$\frac{1}{0.002} + \frac{1}{0.00125} = 1300$$

They are working together at the rate of 1300 bills per hour.   ■

Multiplication Property of Zero

Zero has a property that no other number has. Multiplication involving zero always results in zero. It is the **multiplication property of zero** that prevents 0 from having a reciprocal.

> **Multiplication Property of Zero**
> For any real number a,
> $$0 \cdot a = 0 \qquad \text{and} \qquad a \cdot 0 = 0.$$

EXAMPLE 4   **Recognizing properties**

Identify the property that is illustrated in each case.

a) $5 \cdot 9 = 9 \cdot 5$ b) $3 \cdot \frac{1}{3} = 1$

c) $1 \cdot 865 = 865$ d) $3 + (5 + a) = (3 + 5) + a$

e) $4x + 6x = (4 + 6)x$ f) $7 + (x + 3) = 7 + (3 + x)$

g) $4567 \cdot 0 = 0$ h) $239 + 0 = 239$

i) $-8 + 8 = 0$ j) $-4(x - 5) = -4x + 20$

Solution

a) Commutative   b) Inverse

c) Identity   d) Associative

e) Distributive   f) Commutative

g) Multiplication property of zero   h) Identity

i) Inverse   j) Distributive   ■

Warm-ups

True or false? Explain your answer.

1. Addition is a commutative operation.
2. $8 \div (4 \div 2) = (8 \div 4) \div 2$
3. $10 \div 2 = 2 \div 10$
4. $5 - 3 = 3 - 5$
5. $10 - (7 - 3) = (10 - 7) - 3$
6. $4(6 \div 2) = (4 \cdot 6) \div (4 \cdot 2)$
7. The multiplicative inverse of 0.02 is 50.
8. Division is not an associative operation.
9. $3 + 2x = 5x$ for any value of x.
10. A machine that washes one car in 0.04 hour is washing at the rate of 25 cars per hour.

1.5 EXERCISES

Evaluate. See Example 1.

1. $9 - 4 + 6 - 10$
2. $-3 + 4 - 12 + 9$
3. $6 - 10 + 5 - 0 - 7$
4. $5 - 11 + 6 - 9 + 12 - 2$
5. $-4 - 11 + 6 - 8 + 13 - 20$
6. $-8 + 12 - 9 - 15 + 6 - 22 + 3$
7. $-3.2 + 1.4 - 2.8 + 4.5 - 1.6$
8. $4.4 - 5.1 + 3.6 - 2.3 + 8.1$
9. $3.27 - 11.41 + 5.7 - 12.36 - 5$
10. $4.89 - 2.1 + 7.58 - 9.06 - 5.34$

Use the distributive property to rewrite each sum or difference as a product and each product as a sum or difference. See Example 2.

11. $4(x - 6)$
12. $5(a - 1)$
13. $2m + 10$
14. $3y + 9$
15. $a(3 + t)$
16. $b(y + w)$
17. $-2(w - 5)$
18. $-4(m - 7)$
19. $-2(3 - y)$
20. $-5(4 - p)$
21. $5x - 5$
22. $3y + 3$
23. $-1(-2x - y)$
24. $-1(-4y - w)$

25. $-3(-2w + {}^{-}3y)$
26. $-4(-x - 6)$
27. $3y - 15$
28. $5x + 10$
29. $3a + 9$
30. $7b - 49$
31. $\frac{1}{2}(4x + 8)$
32. $\frac{1}{3}(3x + 6)$
33. $-\frac{1}{2}(2x - 4)$
34. $-\frac{1}{3}(9x - 3)$

Find the multiplicative inverse (reciprocal) of each number.

35. $\frac{1}{2}$
36. $\frac{1}{3}$
37. 1
38. -1
39. 6
40. 8
41. 0.25
42. 0.75
43. -0.7
44. -0.9
45. -1.8
46. -2.6

📱 *Use a calculator to evaluate each expression. Round answers to four decimal places.*

47. $\dfrac{1}{2.3} + \dfrac{1}{5.4}$
48. $\dfrac{1}{13.5} - \dfrac{1}{4.6}$
49. $\dfrac{\dfrac{1}{4.3}}{\dfrac{1}{5.6} + \dfrac{1}{7.2}}$
50. $\dfrac{\dfrac{1}{4.5} - \dfrac{1}{5.6}}{\dfrac{1}{3.2} + \dfrac{1}{2.7}}$

▦ *Solve each problem. See Example 3.*

51. *Fastest airliner.* The world's fastest airliner, the Concorde, travels one mile in 0.0006897 hour and carries 128 passengers (*The Doubleday Almanac*). Find its rate in miles per hour.

52. *Fastest jet plane.* The U.S. Lockheed SR-71 is the world's fastest jet plane. The SR-71 can travel one mile in 0.000456 hours (*The Doubleday Almanac*). Find its rate in miles per hour.

53. *Who's got the button.* A small clothing factory has three workers who attach buttons. Rita, Mary, and Sam can attach a single button in 0.01 hour, 0.02 hour, and 0.015 hour, respectively. At what hourly rate are they attaching buttons when working simultaneously?

54. *Modern art.* Emilio can paint the exterior of a certain house in 36.5 hours. Alex can paint the same house in 30 hours. If they work together without interfering with each other, then at what hourly rate will the house be painted?

Name the property that is illustrated in each case. See Example 4.

55. $3 + x = x + 3$ **56.** $x \cdot 5 = 5x$

57. $5(x - 7) = 5x - 35$ **58.** $a(3b) = (a \cdot 3)b$

59. $3(xy) = (3x)y$ **60.** $3(x - 1) = 3x - 3$

61. $4(0.25) = 1$ **62.** $0.3 + 9 = 9 + 0.3$

63. $y^3x = xy^3$ **64.** $0 \cdot 52 = 0$

65. $1 \cdot x = x$ **66.** $(0.1)(10) = 1$

67. $2x + 3x = (2 + 3)x$ **68.** $8 + 0 = 8$

69. $7 + (-7) = 0$ **70.** $1 \cdot y = y$

71. $(36 + 79)0 = 0$ **72.** $5x + 5 = 5(x + 1)$

73. $xy + x = x(y + 1)$ **74.** $ab + 3ac = a(b + 3c)$

Complete each statement using the property named.

75. $5 + w =$ _____ , commutative

76. $2x + 2 =$ _____ , distributive

77. $5(xy) =$ _____ , associative

78. $x + \dfrac{1}{2} =$ _____ , commutative

79. $\dfrac{1}{2}x - \dfrac{1}{2} =$ _____ , distributive

80. $3(x - 7) =$ _____ , distributive

81. $6x + 9 =$ _____ , distributive

82. $(x + 7) + 3 =$ _____ , associative

83. $8(0.125) =$ _____ , inverse property

84. $-1(a - 3) =$ _____ , distributive

85. $0 = 5($ _____ $)$, multiplication property of zero

86. $8 \cdot ($ _____ $) = 8$, identity property

87. $0.25($ _____ $) = 1$, inverse property

88. $45(1) =$ _____ , identity property

Getting More Involved

89. *Discussion.* Does the order in which your groceries are placed on the checkout counter make any difference in your total bill? Which properties are at work here?

90. *Discussion.* Suppose that you just bought 10 grocery items and paid a total bill that included 6% sales tax. Would there be any difference in your total bill if you purchased the items one at a time? Which property is at work here?

1.6 Using the Properties

In this section:

▶ Using the Properties in Computation
▶ Like Terms
▶ Combining Like Terms
▶ Products and Quotients
▶ Removing Parentheses

The properties of the real numbers can be helpful when we are doing computations. In this section we will see how the properties can be applied in arithmetic and algebra.

Using the Properties in Computation

Consider the product of 36 and 200. Using the associative property, we can write

$$(36)(200) = (36)(2 \cdot 100) = (36 \cdot 2)(100).$$

To find this product mentally, first multiply 36 by 2 to get 72, then multiply 72 by 100 to get 7200.

EXAMPLE 1   **Using properties in computation**

Evaluate each expression mentally by using an appropriate property.

a) $536 + 25 + 75$ b) $5 \cdot 426 \cdot \dfrac{1}{5}$ c) $7 \cdot 45 + 3 \cdot 45$

Solution

a) To perform this addition mentally, the associative property can be applied as follows:

$$536 + (25 + 75) = 536 + 100 = 636$$

b) Use the commutative and associative properties to mentally rearrange this product.

$$5 \cdot 426 \cdot \frac{1}{5} = 426 \cdot 5 \cdot \frac{1}{5} \qquad \text{Commutative property}$$

$$= 426\left(5 \cdot \frac{1}{5}\right) \qquad \text{Associative property}$$

$$= 426 \cdot 1 \qquad \text{Inverse property}$$

$$= 426$$

c) Use the distributive property to rewrite the expression, then evaluate it.

$$7 \cdot 45 + 3 \cdot 45 = (7 + 3)45 = 10 \cdot 45 = 450 \qquad \blacksquare$$

Like Terms

The properties of the real numbers are used also with algebraic expressions. Simple algebraic expressions such as

$$-2, \quad 4x, \quad -5x^2y, \quad b, \quad \text{and} \quad -abc$$

are called terms. A **term** is a single number or the product of a number and one or more variables raised to powers. The number preceding the variables in a term is called the **coefficient.** In the term $4x$ the coefficient of x is 4. In the term $-5x^2y$ the coefficient of x^2y is -5. In the term b the coefficient of b is 1, and in the term $-abc$ the coefficient of abc is -1. If two terms contain the same variables with the same powers, they are called **like terms.** For example, $3x^2$ and $-5x^2$ are like terms, while $3x^2$ and $-2x^3$ are not like terms.

Combining Like Terms

We can combine any two like terms involved in a sum by using the distributive property. For example,

$$2x + 5x = (2 + 5)x \qquad \text{Distributive property}$$

$$= 7x \qquad \text{Add 2 and 5.}$$

Because the distributive property is valid for any real numbers, we have $2x + 5x = 7x$ for any real number x.

We can also use the distributive property to combine any two like terms involved in a difference. For example,

$$-3xy - (-2xy) = [-3 - (-2)]xy \quad \text{Distributive property}$$

$$= -1xy \quad \text{Subtract.}$$

$$= -xy \quad \text{Multiplying by } -1 \text{ is the same as taking the opposite.}$$

Of course, we do not want to write out these steps every time we combine like terms. We can combine like terms as easily as we can add or subtract their coefficients.

EXAMPLE 2 **Combining like terms**
Perform the indicated operation.

a) $b + 3b$ **b)** $5x^2 - 7x^2$ **c)** $5xy - (-13xy)$ **d)** $-2a + (-9a)$

Solution

a) $b + 3b = 1b + 3b = 4b$ **b)** $5x^2 - 7x^2 = -2x^2$

c) $5xy - (-13xy) = 18xy$ **d)** $-2a + (-9a) = -11a$ ∎

CAUTION The distributive property allows us to combine only *like* terms. Expressions such as

$$3xw + 5, \quad 7xy + 9t, \quad 5b + 6a, \quad \text{and} \quad 6x^2 + 7x$$

do not contain like terms, so their terms cannot be combined. ⊘

Products and Quotients

We can use the associative property of multiplication to simplify the product of two terms. For example,

$$4(7x) = (4 \cdot 7)x \quad \text{Associative property}$$

$$= (28)x$$

$$= 28x \quad \text{Remove unnecessary parentheses.}$$

CAUTION Multiplication does not distribute over multiplication. For example, $2(3 \cdot 4) \neq 6 \cdot 8$ because $2(3 \cdot 4) = 2(12) = 24$. ⊘

In the next example we use the fact that dividing by 3 is equivalent to multiplying by $\frac{1}{3}$, the reciprocal of 3:

$$3\left(\frac{x}{3}\right) = 3\left(x \cdot \frac{1}{3}\right) \quad \text{Definition of division.}$$

$$= 3\left(\frac{1}{3} \cdot x\right) \quad \text{Commutative property}$$

$$= \left(3 \cdot \frac{1}{3}\right)x \quad \text{Associative property}$$

$$= 1 \cdot x \quad 3 \cdot \frac{1}{3} = 1 \quad \text{(Inverse property)}$$

$$= x \quad \text{Identity property}$$

To find the product $(3x)(5x)$, we use both the commutative and associative properties:

$$
\begin{aligned}
(3x)(5x) &= (3x \cdot 5)x & \text{Associative property}\\
&= (3 \cdot 5x)x & \text{Commutative property}\\
&= (3 \cdot 5)(x \cdot x) & \text{Associative property}\\
&= (15)(x^2) & \text{Simplify.}\\
&= 15x^2 & \text{Remove unnecessary parentheses.}
\end{aligned}
$$

All of the steps in finding the product $(3x)(5x)$ are shown here to illustrate that every step is justified by a property. However, you should write $(3x)(5x) = 15x^2$ without doing any intermediate steps.

EXAMPLE 3   **Multiplying terms**
Find each product.

a) $(-5)(6x)$ **b)** $(-3a)(-8a)$ **c)** $(-4y)(-6)$ **d)** $(-5a)\left(\dfrac{b}{5}\right)$

Solution

a) $-30x$ **b)** $24a^2$ **c)** $24y$ **d)** $-ab$ ■

In the next example we use the properties to find quotients. Try to identify the property that is used at each step.

EXAMPLE 4   **Dividing terms**
Find each quotient.

a) $\dfrac{5x}{5}$ **b)** $\dfrac{4x + 8}{2}$

Solution

a)   First use the definition of division to change the division by 5 to multiplication by $\frac{1}{5}$.

$$
\frac{5x}{5} = 5x \cdot \frac{1}{5} = \left(\frac{1}{5} \cdot 5\right)x = 1 \cdot x = x
$$

b)   First use the definition of division to change division by 2 to multiplication by $\frac{1}{2}$.

$$
\frac{4x + 8}{2} = (4x + 8) \cdot \frac{1}{2} = \frac{1}{2} \cdot (4x + 8) = 2x + 4
$$

Since both $4x$ and 8 are divided by 2, we could have written

$$
\frac{4x + 8}{2} = \frac{4x}{2} + \frac{8}{2} = 2x + 4
$$ ■

CAUTION   Do not divide a number into just one term of a sum. For example,

$$
\frac{2 + 7}{2} \neq 1 + 7.
$$ ⊘

Removing Parentheses

Multiplying a number by -1 merely changes the sign of the number. For example,

$$(-1)(6) = -6 \quad \text{and} \quad (-1)(-15) = 15.$$

Thus -1 times a number is the same as the *opposite* of the number. Using variables, we have

$$(-1)x = -x \quad \text{or} \quad -1(a + 2) = -(a + 2).$$

When a minus sign appears in front of a sum, we can think of it as multiplication by -1 and use the distributive property. For example,

$$
\begin{aligned}
-(a + 2) &= -1(a + 2) \\
&= (-1)a + (-1)2 \quad \text{Distributive property} \\
&= -a + (-2) \\
&= -a - 2.
\end{aligned}
$$

If a minus sign occurs in front of a difference, we can rewrite the expression as a sum. For example,

$$-(x - 5) = -1(x - 5) = (-1)x - (-1)5 = -x + 5.$$

Note that a minus sign in front of a set of parentheses affects each term in the parentheses, changing the sign of each term.

EXAMPLE 5   **Removing parentheses**
Simplify each expression.

a) $6 - (x + 8)$ **b)** $4x - 6 - (7x - 4)$ **c)** $3x - (-x + 7)$

Solution

a)
$$
\begin{aligned}
6 - (x + 8) &= 6 - x - 8 \quad &&\text{Change sign of each term in parentheses.} \\
&= 6 - 8 - x \quad &&\text{Rearrange the terms.} \\
&= -2 - x \quad &&\text{Combine like terms.}
\end{aligned}
$$

b)
$$
\begin{aligned}
4x - 6 - (7x - 4) &= 4x - 6 - 7x + 4 \quad &&\text{Remove parentheses.} \\
&= 4x - 7x - 6 + 4 \quad &&\text{Rearrange the terms.} \\
&= -3x - 2 \quad &&\text{Combine like terms.}
\end{aligned}
$$

c)
$$
\begin{aligned}
3x - (-x + 7) &= 3x + x - 7 \quad &&\text{Remove parentheses.} \\
&= 4x - 7 \quad &&\text{Combine like terms.}
\end{aligned}
$$
∎

The commutative and associative properties of addition allow us to rearrange the terms so that we may combine the like terms. However, it is not necessary to actually write down the rearrangement. We can identify the like terms and combine them without rearranging.

EXAMPLE 6   **More parentheses and like terms**
Simplify each expression.

a) $(-5x + 7) + (2x - 9)$ **b)** $-4x + 7x + 3(2 - 5x)$

c) $-3x(4x - 9) - (x - 5)$ **d)** $x - 0.03(x + 300)$

Solution

a) $(-5x + 7) + (2x - 9) = -3x - 2$ Combine like terms.

b) $-4x + 7x + 3(2 - 5x) = -4x + 7x + 6 - 15x$ Distributive property

$$= -12x + 6$$ Combine like terms.

$$-3x(4x - 9) - (x - 5) = -12x^2 + 27x - x + 5$$ Remove parentheses.

$$= -12x^2 + 26x + 5$$ Combine like terms.

d) $x - 0.03(x + 300) = 1x - 0.03x - 9$ Distributive property; $(-0.03)(300) = -9$

$$= 0.97x - 9$$ Combine like terms: $1.00 - 0.03 = 0.97$ ∎

Warm-ups

True or false? Explain your answer.
A statement involving variables should be marked true only if it is true for all values of the variable.

1. $5(x + 7) = 5x + 35$
2. $-4x + 8 = -4(x + 8)$
3. $-1(a - 3) = -(a - 3)$
4. $5y + 4y = 9y$
5. $(2x)(5x) = 10x$
6. $-2t(5t - 3) = -10t^2 + 6t$
7. $a + a = a^2$
8. $b \cdot b = 2b$
9. $1 + 7x = 8x$
10. $(3x - 4) - (8x - 1) = -5x - 3$

1.6 EXERCISES

Perform each computation. Make use of appropriate rules to simplify each problem. See Example 1.

1. $45(200)$
2. $25(300)$
3. $\frac{4}{3}(0.75)$
4. $5(0.2)$
5. $(427 + 68) + 32$
6. $(194 + 78) + 22$
7. $47 \cdot 4 + 47 \cdot 6$
8. $53 \cdot 3 + 53 \cdot 7$
9. $19 \cdot 5 \cdot 2 \cdot \frac{1}{5}$
10. $17 \cdot 4 \cdot 2 \cdot \frac{1}{4}$
11. $(120)(400)$
12. $150 \cdot 300$
13. $13 \cdot 377(-5 + 5)$
14. $(456 \cdot 8)\frac{1}{8}$
15. $(348 + 5) + 45$
16. $(135 + 38) + 12$
17. $\frac{2}{3}(1.5)$
18. $(1.25)(0.8)$
19. $17 \cdot 101 - 17 \cdot 1$
20. $33 \cdot 2 - 12 \cdot 33$
21. $354 + 7 + 8 + 3 + 2$
22. $564 + 35 + 65 + 72 + 28$
23. $(567 + 874)(-2 \cdot 4 + 8)$
24. $(567^2 + 48)[3(-5) + 15]$

Combine like terms where possible. See Example 2.

25. $-4n + 6n$
26. $-3a + 15a$
27. $3w - (-4w)$
28. $3b - (-7b)$
29. $4mw^2 - 15mw^2$
30. $2b^2x - 16b^2x$
31. $-5x - (-2x)$
32. $-11 - 7t$
33. $-4 - 7z$
34. $-19m - (-3m)$
35. $4t^2 + 5t^2$
36. $5a + 4a^2$
37. $-4ab + 3a^2b$
38. $-7x^2y + 5x^2y$
39. $9mn - mn$
40. $3cm - cm$
41. $x^3y - 3x^3y$
42. $s^4t - 5s^4t$
43. $-kz^6 - kz^6$
44. $m^7w - m^7w$

Find each product or quotient. See Examples 3 and 4.

45. $4(7t)$
46. $-3(4r)$
47. $(-2x)(-5x)$
48. $(-3h)(-7h)$
49. $(-h)(-h)$
50. $x(-x)$
51. $7w(-4)$
52. $-5t(-1)$
53. $-x(1 - x)$
54. $-p(p - 1)$
55. $(5k)(5k)$
56. $(-4y)(-4y)$

57. $3\left(\dfrac{y}{3}\right)$

58. $5z\left(\dfrac{z}{5}\right)$

59. $9\left(\dfrac{2y}{9}\right)$

60. $8\left(\dfrac{y}{8}\right)$

61. $\dfrac{6x^3}{2}$

62. $\dfrac{-8x^2}{4}$

63. $\dfrac{3x^2y + 15x}{3}$

64. $\dfrac{6xy^2 - 8w}{2}$

65. $\dfrac{2x - 4}{-2}$

66. $\dfrac{-6x - 9}{-3}$

67. $\dfrac{-xt + 10}{-2}$

68. $\dfrac{-2xt^2 + 8}{-4}$

Simplify each expression. See Example 5.

69. $a - (4a - 1)$

70. $5x - (2x - 7)$

71. $6 - (x - 4)$

72. $9 - (w - 5)$

73. $4m + 6 - (m + 5)$

74. $5 - 6t - (3t + 4)$

75. $-5b - (-at + 7b)$

76. $-4x^2 - (-7x^2 + 2y)$

77. $t^2 - 5w - (-2w - t^2)$

78. $n^2 - 6m - (-n^2 - 2m)$

79. $x^2 - (x^2 - y^2 - z)$

80. $5w - (6w - 3xy - zy)$

Simplify each expression. See Example 6.

81. $(2x + 7x) + (3 + 5)$

82. $(3x + 4x) + (5 + 12)$

83. $(-3x + 4) + (5x - 6)$

84. $(-4x + 11) + (6x - 8)$

85. $4a^2 - 5c - (6a^2 - 7c)$

86. $3x^2 - 4 - (x^2 - 5)$

87. $5(t^2 - 3w) - 2(-3w - t^2)$

88. $6(xy^2 + 2) - 5(-xy^2 - 1)$

89. $-7m + 3(m - 4) + 5m$

90. $-6m + 4(m - 3) + 7m$

91. $8 - 7(k^3 + 3) - 4$

92. $6 + 5(k^3 - 2) - k^3 + 5$

93. $x - 0.04(x + 50)$

94. $x - 0.03(x + 500)$

95. $0.1(x + 5) - 0.04(x + 50)$

96. $0.06x + 0.14(x + 200)$

97. $3k + 5 - 2(3k - 4) - k + 3$

98. $5w - 2 + 4(w - 3) - 6(w - 1)$

99. $5.7 - 4.5(x - 3.9) - 5.42$

100. $0.04(5.6x - 4.9) + 0.07(7.3x - 34)$

Simplify.

101. $3(1 - xy) - 2(xy - 5) - (35 - xy)$

102. $2(x^2 - 3) - (6x^2 - 2) + 2(-7x^2 - 4)$

103. $w \cdot 3w + 5w(-6w) - w(2w)$

104. $3w^3 + 5w^3 - 4w^3 + 12w^3 - 2w^2$

105. $3a^2w^2 - 5w^2 \cdot a^2 - 2aw \cdot 2aw$

106. $-3(aw^2 + 5a^2w) - 2(-a^2w - a^2w)$

107. $\dfrac{1}{6} - \dfrac{1}{3}\left(-6x^2y - \dfrac{1}{2}\right)$

108. $-\dfrac{1}{2}bc - \dfrac{1}{2}bc(3 - a)$

109. $-\dfrac{1}{2}m\left(-\dfrac{1}{2}m\right) - \dfrac{1}{2}m - \dfrac{1}{2}m$

110. $\dfrac{4wyt}{4} + \dfrac{-8wyt}{2} - \dfrac{-2wy}{2}$

111. $\dfrac{-8t^3 - 6t^2 + 2}{-2}$

112. $\dfrac{7x^3 - 5x^3 - 4}{-2}$

113. $\dfrac{-6xyz - 3xy + 9z}{-3}$

114. $\dfrac{20a^2b^4 - 10a^2b^4 + 5}{-5}$

Write an algebraic expression for each problem.

115. Triangle. The lengths of the sides of a triangular flower bed are s feet, $s + 2$ feet, and $s + 4$ feet. What is its perimeter?

116. Parallelogram. The lengths of the sides of a lot in the shape of a parallelogram are w feet and $w + 50$ feet. What is its perimeter? Is it possible to find the area from this information?

$w + 50$ ft

w ft w ft

$w + 50$ ft

Figure for Exercise 116

117. Parthenon. To obtain a pleasing rectangular shape, the ancient Greeks constructed buildings with a length that was about $\frac{1}{6}$ longer than the width. If the width of the Parthenon is x meters and its length is $x + \frac{1}{6}x$ meters, then what is the perimeter? What is the area?

118. Square. If the length of each side of a square sign is x inches, then what are the perimeter and area of the square?

——————————— C O L L A B O R A T I V E A C T I V I T I E S ———————————

OOOP! Order of Operations Game

In this game you will be reviewing the established order of operations for real numbers. You may evaluate any compound expression while playing this game. Use a new piece of paper for each game.

1. Before Play Begins

You will need three to five players on a team. Assign each player a role or operation: **E**—Exponents, **M**—Multiply/Divide (may have two players working as a team, one who multiplies and one who divides); **A**—Add/Subtract (may have two players working as a team, one who adds and one who subtracts).

2. Determining a Player's Turn at Play

All players working together analyze the expression and decide which part to complete. If there are parentheses, then the players decide whether what is inside the parentheses needs simplification. **E** *performs his or her operation before **M**, and **M** performs his or her operation before* **A.** Each player's turn ends when he or she encounters an operation that precedes his or hers or when he or she reaches the end of the expression.

Grouping: 3 to 5 students per group
Topic: Order of operations, learning to work in groups

3. Each Player's Task

E—Exponents: Working left to right, **E** evaluates exponential expressions in order.

M—Multiply/Divide: Working left to right, **M** performs multiplications and divisions in order. Multiplication and division may be done by two players working as a team. If these tasks are split, then the two players perform their assigned operations in order, taking turns as needed.

A—Add/Subtract: Working left to right, **A** performs additions and subtractions in order. Addition and subtraction may be done by two players working as a team. If these tasks are split, then the two players perform their assigned operations in order, taking turns as needed.

4. Recording Results

Results are recorded on one sheet of paper. As each player finishes his or her operation, she or he passes the paper to the player with the next task. Each player rewrites the new form of the expression on the next line of the page and initials his or her work with **E, M,** or **A.**

Wrap-up   CHAPTER 1

SUMMARY

Sets		Examples
Set-builder notation	Notation for describing a set using variables.	$C = \{x \mid x$ is a natural number smaller than 4$\}$ $D = \{3, 4\}$
Membership	The symbol $\in$ means "is an element of."	$1 \in C, 4 \notin C$
Union	$A \cup B = \{x \mid x \in A$ or $x \in B\}$	$C \cup D = \{1, 2, 3, 4\}$
Intersection	$A \cap B = \{x \mid x \in A$ and $x \in B\}$	$C \cap D = \{3\}$
Subset	A is a subset of B if every element of A is also an element of B.	$\{1, 2\} \subseteq C$
	$\emptyset \subseteq A$ for any set A.	$\emptyset \subseteq C, \emptyset \subseteq D$

Real Numbers		Examples
Rational numbers	$Q = \left\{ \dfrac{a}{b} \,\middle|\, a \text{ and } b \text{ are integers with } b \neq 0 \right\}$	$\dfrac{3}{2}, 5, -6, 0, 0.25252525\ldots$
Irrational numbers	$I = \{x \mid x \text{ is a real number that is not rational}\}$	$\sqrt{2}, \sqrt{3}, \pi, 0.1515515551\ldots$
Real numbers	$R = \{x \mid x \text{ is the coordinate of a point on the number line}\}$	$R = Q \cup I$

Operations with Real Numbers		Examples
Absolute value	$|a| = \begin{cases} a & \text{if } a \text{ is positive or zero} \\ -a & \text{if } a \text{ is negative} \end{cases}$	$|6| = 6, \ |0| = 0$ $|-6| = 6$
Addition and subtraction	To find the sum of two numbers with the same sign, add their absolute values. The sum has the same sign as the original numbers.	$-2 + (-7) = -9$
	To find the sum of two numbers with unlike signs, subtract their absolute values. The sum is positive if the number with the larger absolute value is positive. The sum is negative if the number with the larger absolute value is negative.	$-6 + 9 = 3$ $-9 + 6 = -3$
	Subtraction: $a - b = a + (-b)$ (Change the sign and add.)	$4 - 7 = 4 + (-7) = -3$ $5 - (-3) = 5 + 3 = 8$
Multiplication and division	To find the product or quotient of two numbers, multiply or divide their absolute values: Same signs $\leftrightarrow$ Positive result Opposite signs $\leftrightarrow$ Negative result	$(-4)(-2) = 8, \ (-4)(2) = -8$ $-8 \div (-2) = 4, -8 \div 2 = -4$
Exponential expressions	In the expression a^n, a is the base and n is the exponent.	$2^3 = 2 \cdot 2 \cdot 2 = 8$

Order of operations	In an expression without parentheses or absolute value:
	    1. Evaluate exponential expressions
	    2. Perform multiplication and division
	    3. Perform addition and subtraction
	 With parentheses or absolute value:
	    First evaluate within each set of parentheses
	    or absolute value, using the above order.
		$7 + 2^3 = 15$
		 $3 + 4 \cdot 6 = 27$
		 $5 + 4 \cdot 3^2 = 41$
		
		 $(2 + 4)(5 - 9) = -24$
		 $3 + 4|2 - 3| = 7$

Properties of Real Numbers		Examples
Commutative properties	For any real numbers a, b, and c: $a + b = b + a$ $a \cdot b = b \cdot a$	$3 + 7 = 7 + 3$ $4 \cdot 3 = 3 \cdot 4$
Associative properties	$a + (b + c) = (a + b) + c$ $a \cdot (b \cdot c) = (a \cdot b) \cdot c$	$1 + (3 + 5) = (1 + 3) + 5$ $3(5 \cdot 7) = (3 \cdot 5)7$
Distributive properties	$a(b + c) = ab + ac$ $a(b - c) = ab - ac$	$3(4 + x) = 12 + 3x$ $5x - 10 = 5(x - 2)$

Identity properties	$a + 0 = a$ and $0 + a = a$. $1 \cdot a = a$ and $a \cdot 1 = a$.	$6 + 0 = 0 + 6 = 6$ $6 \cdot 1 = 1 \cdot 6 = 6$
Inverse properties	$a + (-a) = 0$ and $-a + a = 0$. $a \cdot \dfrac{1}{a} = 1$ and $\dfrac{1}{a} \cdot a = 1$ for $a \neq 0$.	$8 + (-8) = 0, -8 + 8 = 0$ $8 \cdot \dfrac{1}{8} = 1, -2\left(-\dfrac{1}{2}\right) = 1$
Multiplication property of 0	$a \cdot 0 = 0$ and $0 \cdot a = 0$.	$9 \cdot 0 = 0$ $(0)(-4) = 0$

Algebraic Concepts		**Examples**
Algebraic expressions	Any meaningful combination of numbers, variables, and operations	$x^2 + y^2, -5abc$
Term	An expression containing a number or the product of a number and one or more variables	$3x^2, -7x^2y, 8$
Like terms	Terms with identical variable parts	$4bc - 8bc = -4bc$

REVIEW EXERCISES

1.1 *Let $A = \{1, 2, 3\}$, $B = \{3, 4, 5\}$, $C = \{1, 2, 3, 4, 5\}$, $D = \{3\}$, and $E = \{4, 5\}$. Determine whether each statement is true or false.*

1. $A \cap B = D$

2. $A \cap B = E$

3. $A \cup B = E$

4. $A \cup B = C$

5. $B \cup C = C$

6. $A \cap C = B$

7. $A \cap \varnothing = A$

8. $A \cup \varnothing = \varnothing$

9. $(A \cap B) \cup E = B$

10. $(C \cap B) \cap A = D$

11. $B \subseteq C$

12. $A \subseteq E$

13. $A = B$

14. $B = C$

15. $3 \in D$

16. $5 \notin A$

17. $0 \in E$

18. $D \subseteq \varnothing$

19. $\varnothing \subseteq E$

20. $1 \in A$

1.2 *Which elements of the set*

$$\left\{ -\sqrt{2}, -1, 0, 1, 1.732, \sqrt{3}, \pi, \frac{22}{7}, 31 \right\}$$

are members of the following sets?

21. Whole numbers

22. Natural numbers

23. Integers

24. Rational numbers

25. Irrational numbers

26. Real numbers

True or false? Explain.

27. The set of whole numbers is a subset of the set of natural numbers.

28. Zero is not a real number.

29. The set of natural numbers larger than -3 is $\{-2, -1, 0, 1, 2, 3, \ldots\}$.

30. The set of rational numbers is finite.

31. The ratio of the circumference to the diameter of any circle is exactly 3.14.

32. Every terminating decimal number is an integer.

33. Every repeating decimal number is a rational number.

34. The number $\sqrt{9}$ is an irrational number.

35. Zero is both rational and irrational.

36. The irrational numbers have no decimal representation.

1.3 *Evaluate.*

37. $-4 + 9$

38. $-3 + (-5)$

39. $25 - 37$

40. $-6 - 10$

41. $(-4)(6)$

42. $(-7)(-6)$

43. $(-8) \div (-4)$

44. $40 \div (-8)$

45. $-\dfrac{1}{4} + \dfrac{1}{12}$

46. $\dfrac{1}{3} - \left(-\dfrac{1}{12}\right)$

47. $\dfrac{-20}{-2}$

48. $\dfrac{30}{-6}$

49. $-0.04 + 10$

50. $-0.05 + (-3)$

51. $-6 - (-2)$

52. $-0.2 - (-0.04)$

53. $-0.5 + 0.5$

54. $-0.04 \div 0.2$

55. $3.2 \div (-0.8)$

56. $(0.2)(-0.9)$

57. $0 \div (-0.3545)$

58. $(-6)(-0.5)$

59. $\dfrac{1}{4}(-12)$

60. $-7 - (-9)$

1.4 *Evaluate each expression.*

61. $4 + 7(5)$

62. $(4 + 7)5$

63. $(4 + 7)^2$

64. $4 + 7^2$

65. $5 + 3 \cdot |6 - 4 \cdot 3|$

66. $6 - (7 - 8)$

67. $(6 - 8) - (5 - 9)$

68. $5 - 6 - 8 - 10$

69. $3 - 5(6 - 2 \cdot 5)$

70. $4^2 - 9 + 3^2$

71. $5^2 - (6 + 5)^2$

72. $|3 - 4 \cdot 2| - |5 - 8|$

73. $\dfrac{6}{2} + 2$

74. $\dfrac{6 + 3}{2}$

75. $\dfrac{-4 - 5}{7 - (-2)}$

76. $\dfrac{5 - 9}{2 - 4}$

77. $1 - (0.8)(0.3)$

78. $5 - (0.2)(0.1)$

79. $(-3)^2 - (4)(-1)(-2)$

80. $3^2 - 4(1)(-3)$

81. $3 - 2|4 - 7|$

82. $3|4 - 6| + 1$

83. $\dfrac{-5 - 3}{-4 + 2}$

84. $\dfrac{-6 + 3}{3 - 9}$

Let $a = -2$, $b = 3$, and $c = -1$. Find the value of each algebraic expression.

85. $b^2 - 4ac$

86. $a^2 - 4b$

87. $(c - b)(c + b)$

88. $(a + b)(a - b)$

89. $a^2 + 2ab + b^2$

90. $a^2 - 2ab + b^2$

91. $a^3 - b^3$

92. $a^3 + b^3$

93. $\dfrac{b + c}{a + b}$

94. $\dfrac{b - c}{2b - a}$

95. $|a - b|$

96. $|b - a|$

97. $(a + b)c$

98. $ac + bc$

1.5   *Name the property that justifies each equation.*

99. $a + x = x + a$

100. $0 \cdot 5 = 0$

101. $3(x - 1) = 3x - 3$

102. $10 + (-10) = 0$

103. $5(2x) = (5 \cdot 2)x$

104. $w + y = y + w$

105. $1 \cdot y = y$

106. $4 \cdot \dfrac{1}{4} = 1$

107. $5(0.2) = 1$

108. $3 \cdot 1 = 3$

109. $12 \cdot 0 = 0$

110. $x + 1 = 1 + x$

111. $18 + 0 = 18$

112. $2w + 2m = 2(w + m)$

113. $-5 + 5 = 0$

114. $2 + (3 + 4) = (2 + 3) + 4$

Use the distributive property to rewrite each sum or difference as a product, and each product as a sum or difference, whichever is appropriate.

115. $3x - 3a$

116. $5x - 5y$

117. $3(w + 1)$

118. $2(m + 14)$

119. $7x + 7$

120. $3w + 3$

121. $5(x - 5)$

122. $13(b - 3)$

123. $-3(2x - 5)$

124. $-2(5 - 4x)$

125. $p - pt$

126. $ab + b$

1.6 *Simplify each expression.*

127. $3a + 7 + 4a - 5$

128. $2m + 6 + m - 2$

129. $5(t - 4) - 3(2t - 6)$

130. $2(x - 3) + 2(3 - x)$

131. $-(a - 2) + 2 - a$

132. $-(w - y) + 3(y - w)$

133. $5 - 3(x - 2) + 7(x + 4) - 6$

134. $7 - 2(x - 7) + 7 - x$

135. $0.2(x + 0.1) - (x + 0.5)$

136. $0.1(x - 0.2) - (x + 0.1)$

137. $0.05(x + 3) - 0.1(x + 20)$

138. $0.02(x + 100) - 0.2(x + 50)$

139. $\dfrac{1}{2}(x + 4) - \dfrac{1}{4}(x - 8)$

140. $\dfrac{1}{2}(2x - 1) + \dfrac{1}{4}(x + 1)$

141. $\dfrac{-9x^2 - 6x + 3}{3}$

142. $\dfrac{4x - 2}{-2} + \dfrac{4x + 2}{2}$

Miscellaneous

Evaluate the following expressions for $w = 24$, $x = -6$, $y = 6$, and $z = 4$. Name the property or properties used.

143. $32z(x + y)$

144. $(wz)\dfrac{1}{w}$

145. $768z + 768y$

146. $28z + 28y$

147. $(12z + x) + y$

148. $(42 + x) + y$

149. $752x + 752y$

150. $37y + 37x$

151. $(47y)\dfrac{z}{w}$

152. $3w + 3y$

153. $(xw)\dfrac{1}{y}$

154. $(xz)\dfrac{1}{x}$

155. $5(x + y)(z + w)$

156. $(4x + 7y)(w + xz)$

Solve each problem.

157. *Teamwork.* Istvan can attach one shingle using a nail gun in 0.2 minute while Robin takes 0.5 minute to attach one shingle using a hammer. How many shingles per minute will they be attaching if they work together?

158. *Carpeting costs.* Write an algebraic expression for the cost of carpeting a rectangular room that is x yards by $x + 2$ yards if carpeting costs \$20 per square yard?

159. *Inflationary spiral.* If car prices increase 5% annually, then in n years a Mustang LX that currently costs P dollars will cost $P(1.05)^n$ dollars. What will the Mustang LX cost in the year 2000 if the 1994 model costs \$18,264 and the prices increase 5% annually? Use the accompanying graph to determine the first year in which the price of this car will be over \$21,000.

Figure for Exercise 159

160. *Lots of water.* The volume of water in a round swimming pool with radius r feet and depth h feet is $7.5\pi r^2 h$ gallons. Find the volume of water in a pool that has diameter 24 feet and depth 3 feet.

Figure for Exercise 160

CHAPTER 1 TEST

Let $A = \{2, 4, 6, 8, 10\}$, $B = \{3, 4, 5, 6, 7\}$, and $C = \{6, 7, 8, 9, 10\}$. List the elements in each of the following sets.

1. $A \cup B$ **2.** $B \cap C$ **3.** $A \cap (B \cup C)$

Which elements of $\{-4, -\sqrt{3}, -\frac{1}{2}, 0, 1.65, \sqrt{5}, \pi, 8\}$ are members of the following sets

4. Whole numbers   **5.** Integers

6. Rational numbers   **7.** Irrational numbers

Evaluate each expression.

8. $6 + 3(-5)$ **9.** $(-2)^2 - 4(2)(-5)$

10. $-5 + 6 - 12$ **11.** $0.02 - 2$

12. $\dfrac{-3 - (-7)}{3 - 5}$ **13.** $\dfrac{-6 - 2}{4 - 2}$

14. $\left(\dfrac{2}{3} - 1\right)\left(\dfrac{1}{3} - \dfrac{1}{2}\right)$ **15.** $-\dfrac{4}{7} - \dfrac{1}{2}\left(24 - \dfrac{8}{7}\right)$

16. $|3 - 5(2)|$

17. $5 + {}^-2|6 - 10|$

18. $(452 + 695)[2(-4) + 8]$

19. $478(8) + 478(2)$

20. $-8 \cdot 3 - 4(6 - 9 \cdot 2^3)$

21. $-3 + 5 - 6 \cdot 3 - 4 + 8 - 9 \cdot 2$

Evaluate each expressson for $a = -3$, $b = -4$, and $c = 2$.

22. $b^2 - 4ac$ **23.** $\dfrac{a^2 - b^2}{b - a}$ **24.** $\dfrac{ab - 6c}{b^2 - c^2}$

Identify the property that justifies each equation.

25. $2(5 + 7) = 10 + 14$ **26.** $57 \cdot 4 = 4 \cdot 57$

27. $2 + (6 + x) = (2 + 6) + x$

28. $-6 + 6 = 0$ **29.** $1 \cdot (-6) = (-6) \cdot 1$

30. $0 \cdot 28 = 0$

Simplify each expression.

31. $3(m - 5) - 4(-2m - 3)$ **32.** $x + 3 - 0.05(x + 2)$

33. $\dfrac{1}{2}(x - 4) + \dfrac{1}{4}(x + 3)$

34. $-3(x^2 - 2y) - 2(3y - 4x^2)$

35. $\dfrac{-6x^2 - 4x + 2}{-2}$ **36.** $\dfrac{-8xy}{2} - \dfrac{6xy}{2}$

Use the distributive property to rewrite each expression as a product.

37. $5x - 40$ **38.** $7t - 7$

Solve each problem.

39. If Celeste and her crew of loggers can cut and load one tree in 0.0625 hour, then how many trees per hour can they cut and load?

40. The rectangular table for table tennis is x feet long and $x - 4$ feet wide. Write algebraic expressions for the perimeter and the area of the table. Find the actual perimeter and area using $x = 9$.

41. If the population of the earth grows at 3% annually, then in n years the present population P will grow to $P(1.03)^n$. Assuming an annual growth rate of 3% and a present population of 6 billion people, what will the population be in 25 years?

CHAPTER 2

Linear Equations and Inequalities

WEST SIDE

CAGO

6th French

EAST SIDE

KINZIE STREET

RIVER

SHAFT No. 1

SHAFT No. 2

PLAN

20'

11.55'

0.2

5' DIA.

LEAK

BEAM

SANDBAGS / GRAVEL

FREIGHT TUNNEL

TREMIE CONCRETE

LAYER SILT/DEBRIS

SECTION THRU TUNNEL ₵

On April 13, 1992, the headline of *The Chicago Tri-bune* read, "Flood Cripples Loop Businesses." Workers driving pilings around a bridge had ruptured an abandoned freight tunnel under the Chicago River. Water was gushing into the 40 miles of open tunnels below the 12 square blocks of Chicago's downtown area called the Loop. The rapidly rising water entered basements, saturated foundations, and quickly forced the shutdown of most utilities. Some subway lines were closed, and eventually thousands of workers were evacuated. While divers were used to survey the problem, the Army Corps of Engineers was called in. Their solution was to seal off the portion of the tunnel that was ruptured, using a steel-reinforced concrete plug. Once the plug was in place, the engineers worked on reversing the flow of the water. For over a month, millions of gallons of water were drained off to a water reclamation plant, and the Loop slowly returned to normal.

In this chapter we will study algebraic equations and formulas. In Exercises 85 and 86 of Section 2.2, you will see how the engineers used very simple algebraic formulas to calculate the amount of force the water would have on the plug and how much force the concrete plug would withstand. The story of the Chicago flood is just one example of how algebraic formulas are used daily in practical situations.

2.1 Linear Equations in One Variable

In this section:

- ▶ Basic Ideas and Definitions
- ▶ Solving Equations
- ▶ Types of Equations
- ▶ Strategy for Solving Linear Equations
- ▶ Techniques

The applications of algebra often lead to equations. The skills that you learned in Chapter 1, such as combining like terms and performing operations with algebraic expressions, will now be used to solve equations.

Basic Ideas and Definitions

An **equation** is a sentence that expresses the equality of two algebraic expressions. Consider the equation

$$2x + 1 = 7.$$

Because $2(3) + 1 = 7$ is true, we say that 3 **satisfies** the equation. No other number in place of x will make the statement $2x + 1 = 7$ true. However, an equation might be satisfied by more than one number. For example, both 3 and -3 satisfy $x^2 = 9$. Any number that satisfies an equation is called a **solution** or **root** to the equation.

Solution Set

The set of all solutions to an equation is called the **solution set** to the equation.

The solution set to $2x + 1 = 7$ is $\{3\}$. To determine whether a number is in the solution set to an equation, we simply replace the variable by the number and see whether the equation is correct.

EXAMPLE 1

Satisfying an equation

Determine whether each equation is satisfied by the number following the equation.

a) $3x + 7 = -8, \quad -5$ **b)** $2(x - 1) = 2x + 3, \quad 4$

Solution

a) Replace x by -5 and evaluate each side of the equation.

$$3x + 7 = -8$$
$$3(-5) + 7 = -8$$
$$-15 + 7 = -8$$
$$-8 = -8 \quad \text{Correct}$$

Because -5 satisfies the equation, -5 is in the solution set to the equation.

b) Replace x by 4 and evaluate each side of the equation.

$$2(x - 1) = 2x + 3$$
$$2(4 - 1) = 2(4) + 3 \quad \text{Replace } x \text{ by 4.}$$
$$2(3) = 8 + 3$$
$$6 = 11 \quad \text{Incorrect}$$

The two sides of the equation have different values when $x = 4$. So 4 is *not* in the solution set to the equation. ∎

Solving Equations

To **solve** an equation means to find its solution set. It is easy to determine whether a given number is in the solution set of an equation as in Example 1, but that example does not provide a method for *solving* equations. The most basic method for solving equations involves the **properties of equality.**

> **Properties of Equality**
>
> **Addition-Subtraction Property of Equality**
> The same real number may be added to or subtracted from each side of an equation without changing the solution set.
>
> **Multiplication-Division Property of Equality**
> Each side of an equation may be multiplied by or divided by the same nonzero real number without changing the solution set.

Equations that have the same solution set are called **equivalent equations.** In the next example we use the properties of equality to solve an equation by writing an equivalent equation with x isolated on one side of the equation.

EXAMPLE 2   **Using the properties of equality**

Solve the equation $6 - 3x = 8 - 2x$.

Solution

We want to obtain an equivalent equation with only a single x on the left side and a number on the other side.

$$6 - 3x = 8 - 2x$$
$$6 - 3x - 6 = 8 - 2x - 6 \qquad \text{Subtraction property of equality}$$
$$-3x = 2 - 2x \qquad \text{Simplify.}$$
$$-3x + 2x = 2 - 2x + 2x \qquad \text{Add } 2x \text{ to each side.}$$
$$-x = 2 \qquad \text{Combine like terms.}$$
$$-1 \cdot (-x) = -1 \cdot 2 \qquad \text{Multiply each side by } -1.$$
$$x = -2$$

Replacing x by -2 in the original equation gives us

$$6 - 3(-2) = 8 - 2(-2),$$

which is correct. So the solution set to the original equation is $\{-2\}$.   ■

The addition property of equality allows us to add $2x$ to each side of the equation in Example 2, because $2x$ represents a real number.

CAUTION   If you add an expression to each side that does not always represent a real number, then the equations might not be equivalent. For example,

$$x = 0 \qquad \text{and} \qquad x + \frac{1}{x} = 0 + \frac{1}{x}$$

are *not* equivalent because 0 satisfies the first equation but not the second. (The expression $\frac{1}{x}$ is not a real number if x is 0.)   ⊘

To solve some equations, we must simplify the equation before using the properties of equality.

EXAMPLE 3   **Simplifying the equation first**

Solve the equation $2(x - 4) + 5x = 34$.

Solution

Before using the properties of equality, we simplify the expression on the left-hand side of the equation:

$$2(x - 4) + 5x = 34$$

$2x - 8 + 5x = 34$	Distributive property
$7x - 8 = 34$	Combine like terms.
$7x - 8 + 8 = 34 + 8$	Add 8 to each side.
$7x = 42$	Simplify.
$\dfrac{7x}{7} = \dfrac{42}{7}$	Divide each side by 7 to get a single x on the left side.
$x = 6$	

To check, we replace x by 6 in the original equation and simplify:

$$2(6 - 4) + 5 \cdot 6 = 34$$
$$2(2) + 30 = 34$$
$$34 = 34$$

The solution set to the equation is $\{6\}$. ∎

When an equation involves fractions, we can simplify it by multiplying each side by a number that is evenly divisible by all of the denominators. The smallest such number is called the **least common denominator (LCD).** Multiplying each side of the equation by the LCD will eliminate all of the fractions.

EXAMPLE 4   **An equation with fractions**

Find the solution set for the equation

$$\frac{x}{2} - \frac{1}{3} = \frac{x}{3} + \frac{5}{6}.$$

Solution

To solve this equation, we multiply each side by 6, the LCD for 2, 3, and 6:

$6\left(\dfrac{x}{2} - \dfrac{1}{3}\right) = 6\left(\dfrac{x}{3} + \dfrac{5}{6}\right)$	Multiply each side by 6.
$\dfrac{6x}{2} - \dfrac{6}{3} = \dfrac{6x}{3} + \dfrac{30}{6}$	Distributive property
$3x - 2 = 2x + 5$	Divide to eliminate the fractions.
$3x - 2 - 2x = 2x + 5 - 2x$	Subtract $2x$ from each side.
$x - 2 = 5$	Combine like terms.
$x - 2 + 2 = 5 + 2$	Add 2 to each side.
$x = 7$	Combine like terms.

Check 7 in the original equation. The solution set is $\{7\}$. ∎

Equations that involve decimal numbers can be solved like equations involving fractions. If we multiply a decimal number by 10, 100, or 1000, the decimal point is moved one, two, or three places to the right, respectively. If the decimal points are all moved far enough to the right, the decimal numbers will be replaced by whole numbers. The next example shows how to use the multiplication property of equality to eliminate decimal numbers in an equation.

EXAMPLE 5 **An equation with decimals**

Solve the equation $x - 0.1x = 0.75x + 4.5$.

Solution

Because the number with the most decimal places in this equation is 0.75 (75 hundredths), multiplying each side by 100 will eliminate all decimals.

$$100(x - 0.1x) = 100(0.75x + 4.5) \quad \text{Multiply each side by 100.}$$
$$100x - 10x = 75x + 450 \quad \text{Distributive property}$$
$$90x = 75x + 450 \quad \text{Combine like terms.}$$
$$90x - 75x = 75x + 450 - 75x \quad \text{Subtract } 75x \text{ from each side.}$$
$$15x = 450 \quad \text{Combine like terms.}$$
$$\frac{15x}{15} = \frac{450}{15} \quad \text{Divide each side by 15.}$$
$$x = 30$$

Check that 30 satisfies the original equation. The solution set is {30}. ∎

Types of Equations

We often think of an equation such as $3x + 4x = 7x$ as an "addition fact," because the equation is satisfied by all real numbers. However, some equations that we think of as facts are not satisfied by all real numbers. For example, $\frac{x}{x} = 1$ is satisfied by every real number except 0, because $\frac{0}{0}$ is undefined. The equation $x + 1 = x + 1$ is satisfied by all real numbers because both sides are identical. All of these equations are called *identities*.

The equation $2x + 1 = 7$ is true only on condition that we choose $x = 3$. For this reason it is called a *conditional equation*. The equations in Examples 2 through 5 are conditional equations.

Some equations are false no matter what value is used to replace the variable. For example, no number satisfies $x = x + 1$. The solution set to this *inconsistent* equation is the empty set, $\varnothing$.

Identity, Conditional Equation, Inconsistent Equation

An **identity** is an equation that is satisfied by every number for which both sides are defined.

A **conditional equation** is an equation that is satisfied by at least one number but is not an identity.

An **inconsistent equation** is an equation whose solution set is the empty set.

It is easy to classify $2x = 2x$ as an identity and $x = x + 2$ as an inconsistent equation, but some equations must be simplified before they can be classified.

EXAMPLE 6 **An inconsistent equation and an identity**

Solve each equation.

a) $8 - 3(x - 5) + 7 = 3 - (x - 5) - 2(x - 11)$

b) $5 - 3(x - 6) = 4(x - 9) - 7x$

Solution

a) First simplify each side of the equation:

$$8 - 3(x - 5) + 7 = 3 - (x - 5) - 2(x - 11)$$

$8 - 3x + 15 + 7 = 3 - x + 5 - 2x + 22$ Distributive property

$30 - 3x = 30 - 3x$ Combine like terms.

This last equation is satisfied by any value of x because the two sides are identical. Because the last equation is equivalent to the original equation, the original equation is satisfied by any value of x and is an identity. The solution set is R, the set of all real numbers.

b) First simplify each side of the equation.

$$5 - 3(x - 6) = 4(x - 9) - 7x$$

$5 - 3x + 18 = 4x - 36 - 7x$ Distributive property

$23 - 3x = -36 - 3x$ Combine like terms.

$23 - 3x + 3x = -36 - 3x + 3x$ Add $3x$ to each side.

$23 = -36$ Combine like terms.

The equation $23 = -36$ is false for any choice of x. Because these equations are all equivalent, the original equation is also false for any choice of x. The solution set to this inconsistent equation is the empty set $\varnothing$. ∎

Strategy for Solving Linear Equations

The most basic equations of algebra are linear equations. In Chapter 7 we will see a connection between linear equations in one variable and straight lines.

> **Linear Equation in One Variable**
>
> A **linear equation in one variable** x is an equation of the form $ax + b = 0$, where a and b are real numbers, with $a \neq 0$.

The equations in Examples 2 through 5 are called linear equations in one variable, or simply linear equations, because they could all be rewritten in the form $ax + b = 0$. At first glance the equations in Example 6 appear to be linear equations. However, they cannot be written in the form $ax + b = 0$, with $a \neq 0$, so they are not linear equations. A linear equation has exactly one solution. The strategy that we use for solving linear equations is summarized in the following box.

> ### Strategy for Solving a Linear Equation
> 1. If fractions are present, multiply each side by the LCD to eliminate them.
> 2. Use the distributive property to remove parentheses.
> 3. Combine any like terms.
> 4. Use the addition and subtraction properties to get all variables on one side and numbers on the other.
> 5. Use the multiplication or division property to get a single variable on one side.
> 6. Check by replacing the variable in the original equation with your solution.

Note that not all equations require all of the steps.

EXAMPLE 7   **Using the equation-solving strategy**

Solve the equation $\frac{y}{2} - \frac{y-4}{5} = \frac{23}{10}$.

Solution

We first multiply each side of the equation by 10, the LCD for 2, 5, and 10. However, we do not have to write down that step. We can simply use the distributive property to multiply each term of the equation by 10.

$$\frac{y}{2} - \frac{y-4}{5} = \frac{23}{10}$$

$$\overset{5}{\cancel{10}}\left(\frac{y}{\cancel{2}}\right) - \overset{2}{\cancel{10}}\left(\frac{y-4}{\cancel{5}}\right) = \cancel{10}\left(\frac{23}{\cancel{10}}\right) \qquad \text{Multiply each side by 10.}$$

$$5y - 2(y-4) = 23 \qquad \begin{array}{l}\text{Divide each denominator into}\\ \text{10 to eliminate fractions.}\end{array}$$

$$5y - 2y + 8 = 23 \qquad \begin{array}{l}\text{Be careful to change all signs:}\\ -2(y-4) = -2y+8.\end{array}$$

$$3y + 8 = 23 \qquad \text{Combine like terms.}$$

$$3y + 8 - 8 = 23 - 8 \qquad \text{Subtract 8 from each side.}$$

$$3y = 15 \qquad \text{Simplify.}$$

$$\frac{3y}{3} = \frac{15}{3} \qquad \text{Divide each side by 3.}$$

$$y = 5$$

Check that 5 satisfies the original equation. The solution set is {5}.   ■

Techniques

Writing down every step when solving an equation is not always necessary. Solving an equation is often part of a larger problem, and anything that we can do to make the process more efficient will make solving the entire problem faster and easier. For example, we can combine some steps.

<table>
<tr><td align="center">***Combining Steps***</td><td align="center">***Writing Every Step***</td></tr>
</table>

Combining Steps		***Writing Every Step***
$4x - 5 = 23$		$4x - 5 = 23$
$4x = 28$ Add 5 to each side.		$4x - 5 + 5 = 23 + 5$
$x = 7$ Divide each side by 4.		$4x = 28$
		$\dfrac{4x}{4} = \dfrac{28}{4}$
		$x = 7$

The same steps are used in each of the solutions. However, when 5 is added to each side in the solution on the left, only the result is written. When each side is divided by 4, only the result is written.

The equation $-x = -5$ says that the additive inverse of x is -5. Since the additive inverse of 5 is -5, we conclude that x is 5. So instead of multiplying each side of $-x = -5$ by -1, we solve the equation as follows:

$$-x = -5$$
$$x = 5 \quad \text{Additive inverse property}$$

It is often simpler to isolate x on the right-hand side of the equation:

$$3x + 1 = 4x - 5$$
$$6 = x \quad \text{Subtract } 3x \text{ from each side and add 5 to each side.}$$

Now read the equation $6 = x$ from right to left as "x equals 6."

For some equations with fractions it is more efficient to multiply by a multiplicative inverse instead of multiplying by the LCD:

$$-\frac{2}{3}x = \frac{1}{2}$$
$$-\frac{3}{2}\left(-\frac{2}{3}x\right) = -\frac{3}{2}\left(\frac{1}{2}\right) \quad \text{Multiply each side by } -\frac{3}{2}, \text{ the reciprocal of } -\frac{2}{3}.$$
$$x = -\frac{3}{4}$$

The techniques shown here should not be attempted until you have become proficient at solving equations by writing out every step. The more efficient techniques shown here are not a requirement of algebra, but they can be a labor-saving tool that will be useful when we solve more complicated problems.

EXAMPLE 8 **Efficient solutions**

Solve each equation.

a) $3x + 4 = 0$

b) $2 - (x + 5) = -2(3x - 1) + 6x$

Solution

a) Combine steps to solve the equation efficiently.

$$3x + 4 = 0$$

$$3x = -4 \qquad \text{Subtract 4 from each side.}$$

$$x = -\frac{4}{3} \qquad \text{Divide each side by 3.}$$

Check $-\frac{4}{3}$ in the original equation:

$$3\left(-\frac{4}{3}\right) + 4 = -4 + 4 = 0$$

The solution set is $\left\{-\frac{4}{3}\right\}$.

b) $2 - (x + 5) = -2(3x - 1) + 6x$

$$-x - 3 = 2 \qquad \text{Simplify each side.}$$

$$-x = 5 \qquad \text{Add 3 to each side.}$$

$$x = -5 \qquad \text{Additive inverse property}$$

Check that -5 satisfies the original equation. The solution set is $\{-5\}$. ■

Warm-ups

True or false? Explain your answer.

1. The equation $-2x + 3 = 8$ is equivalent to $-2x = 11$.
2. The equation $x - (x - 3) = 5x$ is equivalent to $3 = 5x$.
3. To solve $\frac{3}{4}x = 12$, we should multiply each side by $\frac{3}{4}$.
4. The equation $-x = -6$ is equivalent to $x = 6$.
5. To eliminate fractions, we multiply each side of an equation by the LCD.
6. The solution set to $3x + 5 = 7$ is $\left\{\frac{2}{3}\right\}$.
7. The equation $2(3x + 4) = 6x + 12$ is an inconsistent equation.
8. The equation $4(x + 3) = x + 3$ is a conditional equation.
9. The equation $x - 0.2x = 0.8x$ is an identity.
10. The equation $3x - 5 = 7$ is a linear equation.

2.1 EXERCISES

Determine whether each equation is satisfied by the given number. See Example 1.

1. $3x + 7 = -5, \quad -4$
2. $-3x - 5 = 13, \quad -6$
3. $\frac{1}{2}x - 4 = \frac{1}{3}x - 2, \quad 12$
4. $\dfrac{y - 7}{2} - \dfrac{1}{3} = \dfrac{y - 7}{3}, \quad 9$
5. $0.2(x - 50) = 20 - 0.05x, \quad 200$
6. $0.12x - (4 - x) = 1.02x + 1, \quad 50$
7. $0.1x - 30 = 16 - 0.06x, \quad 80$
8. $0.08x + 3.2 = 0.1x + 2.8, \quad 20$

Solve each linear equation. Show your work and check your answer. See Examples 2 and 3.

9. $-72 - x = 15$ **10.** $51 - x = -9$

11. $-3x - 19 = 5 - 2x$ **12.** $-5x + 4 = -9 - 4x$

13. $2x - 3 = 0$ **14.** $5x + 7 = 0$

15. $-2x + 5 = 7$ **16.** $-3x - 4 = 11$

17. $-12x - 15 = 21$ **18.** $-13x + 7 = -19$

19. $26 = 4x + 16$ **20.** $14 = -5x - 21$

21. $-3(x - 16) = 12 - x$ **22.** $-2(x + 17) = 13 - x$

23. $2(x + 9) - x = 36$ **24.** $3(x - 13) - x = 9$

25. $2 + 3(x - 1) = x - 1$ **26.** $x + 9 = 1 - 4(x - 2)$

Solve each equation. See Example 4.

27. $-\dfrac{3}{7}x = 4$ **28.** $\dfrac{5}{6}x = -2$

29. $-\dfrac{5}{7}x - 1 = 3$ **30.** $4 - \dfrac{3}{5}x = -6$

31. $\dfrac{x}{3} + \dfrac{1}{2} = \dfrac{7}{6}$ **32.** $\dfrac{1}{4} + \dfrac{1}{5} = \dfrac{x}{2}$

33. $\dfrac{2}{3}x + 5 = -\dfrac{1}{3}x + 17$ **34.** $\dfrac{1}{4}x - 6 = -\dfrac{3}{4}x + 14$

35. $\dfrac{1}{2}x + \dfrac{1}{4} = \dfrac{1}{4}(x - 6)$ **36.** $\dfrac{1}{3}(x - 2) = \dfrac{2}{3}x - \dfrac{13}{3}$

37. $8 - \dfrac{x - 2}{2} = \dfrac{x}{4}$ **38.** $\dfrac{x}{3} - \dfrac{x - 5}{5} = 3$

39. $\dfrac{y - 3}{3} - \dfrac{y - 2}{2} = -1$ **40.** $\dfrac{x - 2}{2} - \dfrac{x - 3}{4} = \dfrac{7}{4}$

Solve each equation. See Example 5.

41. $x - 0.2x = 72$ **42.** $x - 0.1x = 63$

43. $0.03(x + 200) + 0.05x = 86$

44. $0.02(x - 100) + 0.06x = 62$

45. $0.1x + 0.05(x - 300) = 105$

46. $0.2x - 0.05(x - 100) = 35$

Solve each equation. Identify each as a conditional equation, an inconsistent equation, or an identity. See Example 6.

47. $2(x + 1) = 2(x + 3)$ **48.** $2x + 3x = 6x$

49. $x + x = 2x$ **50.** $4x - 3x = x$

51. $x + x = 2$ **52.** $4x - 3x = 5$

53. $\dfrac{4x}{4} = x$ **54.** $5x \div 5 = x$

55. $x \cdot x = x^2$ **56.** $\dfrac{2x}{2x} = 1$

57. $2(x + 3) - 7 = 5(5 - x) + 7(x + 1)$

58. $2(x + 4) - 8 = 2x + 1$

59. $2\left(\dfrac{1}{2}x + \dfrac{3}{2}\right) - \dfrac{7}{2} = \dfrac{3}{2}(x + 1) - \left(\dfrac{1}{2}x + 2\right)$

60. $2\left(\dfrac{1}{4}x + 1\right) - 2 = \dfrac{1}{2}x$

61. $2(0.5x + 1.5) - 3.5 = 3(0.5x + 0.5)$

62. $2(0.25x + 1) - 2 = 0.75x - 1.75$

Solve each equation. See Example 7.

63. $4 - 6(2x - 3) + 1 = 3 + 2(5 - x)$

64. $3x - 5(6 - 2x) = 4(x - 8) + 3$

65. $\dfrac{1}{2}\left(y - \dfrac{1}{6}\right) + \dfrac{2}{3} = \dfrac{5}{6} + \dfrac{1}{3}\left(\dfrac{1}{2} - 3y\right)$

66. $\dfrac{3}{4} - \dfrac{1}{3}\left(\dfrac{1}{2}y - 2\right) = 3\left(y - \dfrac{1}{4}\right)$

67. $8 - \dfrac{2}{3}(60x - 900) = \dfrac{1}{2}(400x + 6)$

68. $\dfrac{40x - 5}{2} + \dfrac{5}{2} = \dfrac{33 - 2x}{3} - 11$

69. $\dfrac{a - 3}{4} - \dfrac{2a - 5}{2} = \dfrac{a + 1}{3} - \dfrac{1}{6}$

70. $\dfrac{1}{2}\left(\dfrac{b}{3} - \dfrac{4b}{5}\right) + \dfrac{1}{6} = \dfrac{1}{3} - \dfrac{b - 1}{2}$

71. $1.3 - 0.2(6 - 3x) = 0.1(0.2x + 3)$

72. $0.01(500 - 30x) = 5.4x + 200$

Solve each equation. Practice combining some steps. Look for more efficient ways to solve each equation. See Example 8.

73. $3x - 9 = 0$ **74.** $5x + 1 = 0$

75. $7 - z = -9$ **76.** $-3 - z = 3$

77. $\dfrac{2}{3}x = \dfrac{1}{2}$ **78.** $\dfrac{3}{2}x = -\dfrac{9}{5}$

79. $-\dfrac{3}{5}y = 9$ **80.** $-\dfrac{2}{7}w = 4$

81. $3y + 5 = 4y - 1$ **82.** $2y - 7 = 3y + 1$

83. $5x + 10(x + 2) = 110$

84. $1 - 3(x - 2) = 4(x - 1) - 3$

Solve each equation.

85. $\dfrac{P + 7}{3} - \dfrac{P - 2}{5} = \dfrac{7}{3} - \dfrac{P}{15}$

86. $\dfrac{w - 3}{8} - \dfrac{5 - w}{4} = \dfrac{4w - 1}{8} - 1$

87. $5 - [4 - (t - 6)] = 17 + 2[t + 3(t - 6)]$

88. $[-3 - 5(2 - a)] + 6a = -6 - 2[-1 + 4(a - 6)]$

89. $0.3 - [4 - (t - 0.2)] = [t - (2t + 6)] + 3 - t$

90. $x - (0.3 + x) = -10.9 - 2[x - 6(1 - x)]$

91. $x - 0.06x = 50,000$

92. $x - 0.05x = 800$

93. $2.365x + 3.694 = 14.8095$

94. $-3.48x + 6.981 = 4.329x - 6.851$

95. $5.39 - [4.21 - (x - 3.52)] = 17.6 + 2(x + 13.581)$

96. $5.024 - 3.67(x - 1.09) = 12.5x - 3(1.4x + 10.72)$

Find the value of b that makes each pair of equations equivalent.

97. $2x + 1 = 7, \ x - 1 = b$

98. $3 - 2x = 5, \ 2b - x = 11$

99. $5 - 2x = -3, \ bx - 1 = 3b$

100. $3 - x = -2x, \ b - x = 5b$

101. $2b - x = 4b, \ x - 6 = 5x$

102. $3b - 2 = 5x, \ 2(x - 1) = 4 - 3x$

Solve each problem.

103. *Public school enrollment.* Enrollment in U.S. public schools has been increasing since 1985 (National Center for Education Statistics). The expression $0.45x + 39.05$ can be used to approximate in millions the total enrollment in public elementary and secondary schools in the year $1985 + x$. What was the public school enrollment in 1992? In which year will enrollment reach 45 million students?

104. *Teacher's average salary.* Average salary of public school teachers has been increasing steadily since 1985 (National Center for Education Statistics). The expression $553.37x + 27,966$ can be used to approximate the average annual salary in dollars of public school teachers in the year $1985 + x$. What was the average teacher's salary in 1993? In which year will the average salary reach $36,000?

105. *Solid waste recovery.* In 1960 the United States generated 87.1 million tons of municipal solid waste and recovered (or recycled) only 4.3% of it (U.S. EPA, 1990). The amount of municipal solid waste generated in the United States is given by $w = 3.14n + 87.1$, while the amount recovered is given by $w = 0.576n + 3.78$, where w is in millions of tons and n is the number of years since 1960. Use the graph to estimate the first year in which the United States generated over 100 million tons of municipal solid waste. Find the year in which 13% of the municipal solid waste will be recovered by solving

$$0.576n + 3.78 = 0.13(3.14n + 87.1).$$

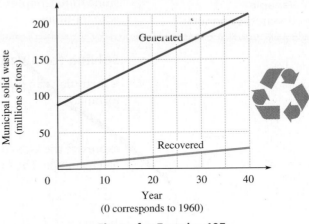

Figure for Exercise 105

106. *Recycling progress.* Find the year in which 14% of the municipal solid waste will be recovered by solving

$$0.576n + 3.78 = 0.14(3.14n + 87.1).$$

See the previous exercise.

Getting More Involved

107. *Exploration.* If you solved the equations in the two previous exercises, then you found the years in which recovery of municipal solid waste will reach 13% and 14%.

   a) Write equations corresponding to recovery rates of 15%, 16%, 17%, 18%, and 19%.

   b) Solve your equations to find the years in which those recovery rates will be achieved.

   c) Use your results to judge whether we are making real progress in recovery of municipal solid waste.

108. *Writing.* Explain how to eliminate the decimals in an equation that involves numbers with decimal points. Would you use the same technique when using a calculator?

109. *Discussion.* Explain why the multiplication property of equality does not allow us to multiply each side of an equation by zero.

2.2 Formulas

A real-life problem may involve many variable quantities that are related to each other. The relationship between these variables may be expressed by a formula. In this section we will combine our knowledge of evaluating expressions from Chapter 1 with the equation-solving skills of Section 2.1 to work with formulas.

Solving for a Variable

A **formula** is an equation involving two or more variables. For example, the formula

$$A = LW$$

expresses the known relationship between the length L, width W, and area A of a rectangle. The formula

$$C = \frac{5}{9}(F - 32)$$

expresses the relationship between the Fahrenheit and Celsius measurements of temperature. For example, if the Fahrenheit temperature F is 95°, we can use the formula to find the Celsius temperature C as follows:

$$C = \frac{5}{9}(95 - 32) = \frac{5}{9}(63) = 35$$

A temperature of 95°F is equivalent to 35°C.

It is sometimes necessary to solve a formula for one variable in terms of the others without substituting numbers for the variables. We will use the same steps in solving for a particular variable as we did in solving linear equations.

EXAMPLE 1   **Solving for a variable**

Solve the formula $C = \frac{5}{9}(F - 32)$ for F.

Solution

To solve the formula for F, we isolate F on one side of the equation. We can eliminate both the 9 and the 5 from the right side of the equation by multiplying by $\frac{9}{5}$, the reciprocal of $\frac{5}{9}$:

$$C = \frac{5}{9}(F - 32)$$

$$\frac{9}{5}C = \frac{9}{5} \cdot \frac{5}{9}(F - 32) \qquad \text{Multiply each side by } \frac{9}{5}.$$

$$\frac{9}{5}C = F - 32 \qquad\qquad \frac{9}{5} \cdot \frac{5}{9} = 1$$

$$\frac{9}{5}C + 32 = F \qquad\qquad \text{Add 32 to each side.}$$

This formula is usually written as $F = \frac{9}{5}C + 32$. ∎

Note that both $F = \frac{9}{5}C + 32$ and $C = \frac{5}{9}(F - 32)$ express the relationship between C and F. If we substitute 35 for C in $F = \frac{9}{5}C + 32$, we get

$$F = \frac{9}{5}(35) + 32 = 63 + 32 = 95.$$

In Chapter 7 we will study linear equations involving x and y. We often need to solve such equations for one of the variables.

EXAMPLE 2   **Equations involving x and y**

Solve $3x - 2y = 6$ for y.

Solution

We can isolate y on the left side:

$$3x - 2y = 6$$
$$-2y = -3x + 6 \qquad \text{Subtract } 3x \text{ from each side.}$$
$$\frac{-2y}{-2} = \frac{-3x + 6}{-2} \qquad \text{Divide each side by } -2.$$
$$y = \frac{3}{2}x - 3 \qquad\qquad\qquad ■$$

The formula $A = P + Prt$ is used to find the amount of an investment. In the next example we solve this formula for P. Note that the variable P occurs twice in the formula.

EXAMPLE 3   **Specified variable occurring twice**

Solve $A = P + Prt$ for P.

Solution

We can use the distributive property to write the sum $P + Prt$ as a product of P and $1 + rt$:

$$A = P + Prt$$
$$A = P(1 + rt) \qquad \text{Distributive property}$$
$$\frac{A}{1 + rt} = \frac{P(1 + rt)}{1 + rt} \qquad \text{Divide each side by } 1 + rt.$$
$$\frac{A}{1 + rt} = P$$

Note that parentheses are not needed around the expression $1 + rt$ in the denominator because the fraction bar acts as a grouping symbol.   ■

CAUTION   If you write $A = P + Prt$ as $P = A - Prt$, then you have not solved the formula for P. When a formula is solved for a specified variable, that variable must be isolated on one side, and it must not occur on the other side.   ⊘

When the variable for which we are solving occurs on opposite sides of the equation, we must move all terms involving that variable to the same side and then use the distributive property to write the expression as a product.

EXAMPLE 4   **Specified variable occurring on both sides**

Solve $3a + 7 = -5ab + b$ for a.

Solution

Get all terms involving a onto one side and all other terms onto the other side:

$$3a + 7 = -5ab + b$$

$$3a + 5ab + 7 = b \qquad \text{Add } 5ab \text{ to each side.}$$

$$3a + 5ab = b - 7 \qquad \text{Subtract 7 from each side.}$$

$$a(3 + 5b) = b - 7 \qquad \begin{array}{l}\text{Use the distributive property to write} \\ \text{the left side as a product.}\end{array}$$

$$\frac{a(3 + 5b)}{3 + 5b} = \frac{b - 7}{3 + 5b} \qquad \text{Divide each side by } 3 + 5b.$$

$$a = \frac{b - 7}{3 + 5b}$$

∎

Calculator Close-up

Not only is it important to know how to use a calculator, but it is equally important to know when to use a calculator. When solving an equation that contains many decimal numbers, we use a calculator to carry out the arithmetic. However, if you use a calculator at every step and round off your results, the final answer can differ greatly from the correct answer. You will get more accurate results if you solve such equations for the variable, as we do when rewriting formulas, and use your calculator only on the final step.

▦ EXAMPLE 5   **Doing computations last**

Solve $3.24x - 6.78 = 6.31(x + 23.45)$.

Solution

Use the distributive property on the right side, but simply write $(6.31)(23.45)$ rather than the result obtained on a calculator.

$$3.24x - 6.78 = 6.31x + (6.31)(23.45) \qquad \text{Distributive property}$$

$$3.24x - 6.31x = (6.31)(23.45) + 6.78 \qquad \text{Get all } x\text{-terms on the left.}$$

$$(3.24 - 6.31)x = (6.31)(23.45) + 6.78 \qquad \text{Distributive property}$$

$$x = \frac{(6.31)(23.45) + 6.78}{3.24 - 6.31} \qquad \text{Divide each side by } (3.24 - 6.31).$$

$$\approx -50.407 \qquad \text{Round to three decimal places.}$$

Check -50.407 in the original equation. When you check an approximate answer, you will get approximately the same value for each side of the equation. The solution set is $\{-50.407\}$.   ∎

Finding the Value of a Variable

In many situations we know the values of all the variables in a formula except one. We can use the formula to determine the unknown value. A list of common formulas and their meanings is given at the back of the text. This list may be helpful for doing the exercises at the end of this section.

EXAMPLE 6   **Finding the value of a variable**

Use the formula $-2x + 3y = 9$ to find y given that $x = -3$.

Solution

To find y, we solve the formula for y:

$$-2x + 3y = 9 \qquad \text{Original equation}$$
$$3y = 2x + 9 \qquad \text{Add } 2x \text{ to each side.}$$
$$y = \frac{2}{3}x + 3 \qquad \text{Divide each side by 3.}$$

Now replace x by -3:

$$y = \frac{2}{3}(-3) + 3$$
$$y = 1 \qquad\qquad\qquad\blacksquare$$

EXAMPLE 7   **Finding the interest rate**

The simple interest on a loan is \$50, the principal is \$500, and the time is 2 years. What is the simple interest rate?

Solution

The formula $I = Prt$ gives the relationship between interest I, principal P, rate r, and time t. To find the rate, we first solve the formula for r, then insert the values of P, I, and t:

$$Prt = I$$
$$\frac{Prt}{Pt} = \frac{I}{Pt} \qquad \text{Divide each side by } Pt.$$
$$r = \frac{I}{Pt}$$
$$r = \frac{50}{500(2)} \qquad \text{Substitute values for } I, \ P, \text{ and } t.$$
$$r = 0.05 = 5\% \qquad \text{A rate is usually written as a percent.} \qquad\blacksquare$$

In Example 7 we solved the formula for r and then inserted the values of the other variables. If we had to find the interest rate for many different loans, this method would be the most direct. But, we could also have inserted the values of I, P, and t into the original formula and then solved for r. Examples 8 and 9 illustrate this second approach.

Geometric Formulas

There are many well-known geometric formulas in Appendix A that are used to solve problems involving geometric shapes.

EXAMPLE 8 **Area of a trapezoid**

The wildlife sanctuary shown in the Fig. 2.1 has a trapezoidal shape with an area of 30 square kilometers. If one base, b_1, of the trapezoid is 10 kilometers and its height is 5 kilometers, find the length of the other base, b_2.

Solution

In any geometric problem it is helpful to have a diagram, as in Fig. 2.1. Substitute the given values into the formula for the area of a trapezoid, found in Appendix A, and then solve for b_2:

$$A = \frac{1}{2}h(b_1 + b_2)$$

$30 = \frac{1}{2} \cdot 5(10 + b_2)$ Substitute given values into the formula for the area of a trapezoid.

$60 = 5(10 + b_2)$ Multiply each side by 2.

$12 = 10 + b_2$ Divide each side by 5.

$2 = b_2$ Subtract 10 from each side.

Figure 2.1

The length of the base b_2 is 2 kilometers. ∎

EXAMPLE 9 **Volume of a rectangular solid**

Millie has just completed pouring 14 cubic yards of concrete to construct a rectangular driveway. If the concrete is 4 inches thick and the driveway is 18 feet wide, then how long is her driveway?

Solution

First draw a diagram as in Figure 2.2. The driveway is a rectangular solid. The formula for volume of a rectangular solid is $V = LWH$ (from Appendix A). Before we insert the values of the variables into the formula, we must convert all of them to the same unit of measurement. We will convert feet and inches to yards:

Figure 2.2

$$4 \text{ inches} = 4 \text{ in.} \cdot \frac{1 \text{ yd}}{36 \text{ in.}} = \frac{1}{9} \text{ yard}$$

$$18 \text{ feet} = 18 \text{ ft} \cdot \frac{1 \text{ yd}}{3 \text{ ft}} = 6 \text{ yards}$$

Now replace W, H, and V by the appropriate values:

$$V = LWH$$

$$14 = L \cdot 6 \cdot \frac{1}{9}$$

$\frac{9}{6} \cdot 14 = L$ Multiply each side by $\frac{9}{6}$.

$$21 = L$$

The length of the driveway is 21 yards, or 63 feet. ∎

Warm-ups

True or false? Explain your answer.

1. The formula $A = P + Prt$ solved for P is $P = A - Prt$.
2. In solving $A = P + Prt$ for P, we do not need the distributive property.
3. Solving $I = Prt$ for t gives us $t = I - Pr$.
4. If $a = \dfrac{bh}{2}$, $b = 5$, and $h = 6$, then $a = 15$.
5. The perimeter of a rectangle is found by multiplying its length and width.
6. The volume of a rectangular box is the product of its length, width, and height.
7. The area of a trapezoid with parallel sides b_1 and b_2 is $\frac{1}{2}(b_1 + b_2)$.
8. Solving $x - y = 5$ for y gives us $y = x - 5$.
9. If $x = -3$ and $y = -2x - 4$, then $y = 2$.
10. The area of a rectangle is the total distance around the outside edge.

2.2 EXERCISES

Solve each formula for the specified variable. See Example 1.

1. $I = Prt$ for t
2. $d = rt$ for r
3. $F = \dfrac{9}{5}C + 32$ for C
4. $A = \dfrac{1}{2}bh$ for h
5. $A = LW$ for W
6. $C = 2\pi r$ for r
7. $A = \dfrac{1}{2}(b_1 + b_2)$ for b_1
8. $A = \dfrac{1}{2}(b_1 + b_2)$ for b_2

Solve for y. See Example 2.

9. $2x + 3y = 9$
10. $4y + 5x = 8$
11. $x - y = 4$
12. $y - x = 6$
13. $4x - 2y = 6$
14. $2y + 4x = -8$
15. $x + 3y = -6$
16. $3x - 2y = 12$
17. $x = 2y + 5$
18. $x = 4y - 8$
19. $x = \dfrac{1}{2}y + 3$
20. $x = \dfrac{1}{3}y + 2$
21. $\dfrac{1}{2}x - \dfrac{1}{3}y = 2$
22. $\dfrac{1}{3}x - \dfrac{1}{4}y = 1$
23. $y - 2 = \dfrac{1}{2}(x - 3)$
24. $y - 3 = \dfrac{1}{3}(x - 4)$
25. $y + 5 = 4(x - 6)$
26. $y + 2 = -3(x + 6)$
27. $0.1725x + 0.0575y = 56.58$
28. $0.0875x - 0.04375y = 87.5$

Solve for the specified variable. See Examples 3 and 4.

29. $A = P + Prt$ for t
30. $A = P + Prt$ for r
31. $ab + a = 1$ for a
32. $y - wy = m$ for y
33. $xy + 5 = y - 7$ for y
34. $xy + 5 = x + 7$ for x
35. $xy^2 + xz^2 = xw^2 - 6$ for x
36. $xz^2 + xw^2 = xy^2 + 5$ for x
37. $\dfrac{1}{R} = \dfrac{1}{R_1} + \dfrac{1}{R_2}$ for R_1
38. $\dfrac{1}{a} + \dfrac{1}{b} = \dfrac{1}{2}$ for a

Solve each equation. Use a calculator only on the last step. Round answers to three decimal places and use your calculator to check your answer. See Example 5.

39. $3.35x - 54.6 = 44.3 - 4.58x$
40. $-4.487x - 33.41 = 55.83 - 22.49x$
41. $4.59x - 66.7 = 3.2(x - 5.67)$
42. $457(36x - 99) = 34(28x - 239)$
43. $\dfrac{x}{19} - \dfrac{3}{23} = \dfrac{4}{31} - \dfrac{3x}{7}$
44. $\dfrac{1}{8} - \dfrac{5}{7}\left(x - \dfrac{5}{22}\right) = \dfrac{4x}{9} + \dfrac{1}{12}$

Find y given that x = 3. See Example 6.

45. $2x - 3y = 5$ **46.** $-3x - 4y = 4$

47. $-4x + 2y = 1$ **48.** $x - y = 7$

49. $y = -2x + 5$ **50.** $y = -3x - 6$

51. $-x + 2y = 5$ **52.** $-x - 3y = 6$

53. $y - 1.046 = 2.63(x - 5.09)$

54. $y - 2.895 = -1.07(x - 2.89)$

Find x in each formula given that y = 2, z = −3, and w = 4. See Example 6.

55. $wxy = 5$ **56.** $wxz = 4$

57. $x + xz = 7$ **58.** $xw - x = 3$

59. $w(x - z) = y(x - 4)$ **60.** $z(x - y) = y(x + 5)$

61. $w = \dfrac{1}{2}xz$ **62.** $y = \dfrac{1}{2}wx$

63. $\dfrac{1}{w} + \dfrac{1}{x} = \dfrac{1}{y}$ **64.** $\dfrac{1}{w} + \dfrac{1}{y} = \dfrac{1}{x}$

Solve each problem. Draw a diagram for each geometric problem. See Examples 7–9.

65. *Simple interest.* If the simple interest on $1,000 for two years is $300, then what is the rate?

66. *Finding time.* If the simple interest on $2,000 at 18% is $180, then what is the time?

67. *Rectangular room.* The area of a rectangular room is 23 square yards. The width is 4 yards. Find the length.

68. *Rectangular garden.* The area of a rectangular garden is 55 square meters. The length is 7 meters. Find the width.

69. *Ice sculpture.* The volume of a rectangular block of ice is 36 cubic feet. The bottom is 2 feet by 2.5 feet. Find the height of the block.

70. *Cardboard box.* A shipping box has a volume of 2.5 cubic meters. The box measures 1 meter high by 1.25 meters wide. How long is the box?

71. *Fish tank.* The volume of a rectangular aquarium is 900 gallons. The bottom is 4 feet by 6 feet. Find the height of the tank. (*Hint:* There are 7.5 gallons per cubic foot.) See figure for Exercise 71.

72. *Reflecting pool.* A rectangular reflecting pool with a horizontal bottom holds 60,000 gallons of water. If the pool is 40 feet by 100 feet, how deep is the water?

73. *Area of a triangle.* The area of a triangle is 30 square feet. If the base is 4 feet, then what is the height?

74. *Larger triangle.* The area of a triangle is 40 square meters. If the height is 10 meters, then what is the length of the base?

Figure for Exercise 71

75. *Second base.* The area of a trapezoid is 300 square inches. If the height is 20 inches and the lower base is 16 inches, then what is the length of the upper base?

Figure for Exercise 75

76. *Height of a trapezoid.* The area of a trapezoid is 200 square centimeters. The bases are 16 centimeters and 24 centimeters. Find the height.

77. *Fencing.* If it takes 600 feet of fence to enclose a rectangular lot that is 132 feet wide, then how deep is the lot?

78. *Football.* The perimeter of a football field in the NFL, excluding the end zones, is $306\frac{2}{3}$ yards. How wide is the field?

Figure for Exercise 78

79. *Radius of a circle.* If the circumference of a circle is 3π meters, then what is the radius?

80. *Diameter of a circle.* If the circumference of a circle is 12π inches, then what is the diameter?

81. *Radius of the earth.* If the circumference of the earth is 25,000 miles, then what is the radius?

82. *Altitude of a satellite.* If a satellite travels 26,000 miles in each circular orbit of the earth, then how high above the earth is the satellite orbiting? See Exercise 81.

Figure for Exercise 82

83. *Height of a can.* If the volume of a can is 30 cubic inches and the diameter of the top is 2 inches, then what is the height of the can?

84. *Height of a cylinder.* If the volume of a cylinder is 6.3 cubic meters and the diameter of the lid is 1.2 meters, then what is the height of the cylinder?

85. *Great Chicago flood.* The great Chicago flood of April 1992 occurred when an old freight tunnel connecting buildings in the Loop ruptured. As shown in the figure, engineers plugged the tunnel with concrete on each side of the hole. They used the formula $F = WDA$ to find the force F of the water on the plug. In this formula the weight of water W is 62 lb/ft^3, the average depth D of the tunnel below the surface of the river is 32 ft, and the cross-sectional area A of the tunnel is 48 ft^2. Find the force on the plug.

river bottom

32 ft

5 ft

hole

48 ft^2 concrete plug   tunnel

Figure for Exercise 85

86. *Will it hold?* To plug the tunnel described in the previous exercise, engineers drilled a 5-ft-diameter shaft down to the tunnel. The concrete plug was made so that it extended up into the shaft. For the plug to remain in place, the shear strength of the concrete in the shaft would have to be greater than the force of the water. The amount of force F that it would take for the water to shear the concrete in the shaft is given by $F = s\pi r^2$, where s is the shear strength of concrete and r is the radius of the shaft in inches. If the shear strength of concrete is 38 lb/in.^{2}, then what force of water would shear the concrete in the shaft? Use the result from Exercise 85 to determine whether the concrete would be strong enough to hold back the water.

87. *Decreasing cholesterol.* According to the National Center for Health Statistics, the annual number of deaths d in the United States due to coronary heart disease depends on the average adult cholesterol level (*Boston Globe*, June 16, 1993). This relationship can be modeled by the formula $d = 5000c - 525{,}000$, where c is the average cholesterol level measured in milligrams per deciliter of blood.

a) If d was 500,000 in 1990, then what was the average cholesterol level in 1990?

b) If the national goal of 200 by the year 2000 is reached, then how many lives will be saved in 2000 compared to 1990?

c) As the average cholesterol level increases, does the number of deaths due to coronary heart disease increase or decrease?

88. *Cigarette usage.* According the National Institute on Drug Abuse, the percentage of Americans aged 18 to 25 who use cigarettes has been decreasing at an approximately constant rate since 1974. The formula $P = 0.479 - 0.00949n$ can be used to estimate the percentage of smokers in this age group, n years after 1974. Use the formula to find the percentage of smokers in this age group in 1991. Use the accompanying graph to estimate the year in which smoking will be eliminated from this age group. Use the formula to find the year in which smoking will be eliminated from this age group.

Figure for Exercise 88

89. *Distance between streets.* Harold Johnson lives on a four-sided, 50,000-square-foot, lot that is bounded on two sides by parallel streets. The city has assessed him $1000 for curb repair, $2 for each foot of property bordering these two streets. How far apart are the streets?

Figure for Exercises 89–91

90. *Assessed for repairs.* Harold's sister Maude lives next door on a triangular lot of 25,000 square feet that also extends from street to street but has frontage only on one street. What will her assessment be? (See Exercise 89.)

91. *Juniper's lot.* Harold's other sister, Juniper, lives on the other side of him on a lot of 60,000 square feet in the shape of a parallelogram. What will her assessment be? (See Exercise 89.)

92. *Mother's driveway.* Harold's mother, who lives across the street, is pouring a concrete driveway, 12 feet wide and 4 inches thick, from the street straight to her house. This is too much work for Harold to do in one day, so his mother has agreed to buy 4 cubic yards of concrete each Saturday for three consecutive Saturdays. How far is it from the street to her house?

Getting More Involved

93. *Exploration.* Electric companies often point out the low cost of electricity in performing common household tasks.

a) Find the cost of a kilowatt-hour of electricity in your area.

b) Write a formula for finding the cost of electricity for a household appliance to perform a certain task and explain what each variable represents.

c) Use your formula to find the cost in your area for baking a $1\frac{1}{2}$-pound loaf of bread for 5 hours in a 750-watt Welbilt breadmaker.

2.3   Applications

We often use algebra to solve problems by translating them into algebraic equations. Sometimes we can use formulas such as those in Appendix A. More often, we have to set up a new equation that represents or **models** the problem. We begin with translating verbal expressions into algebraic expressions.

Writing Algebraic Expressions

Consider the three consecutive integers 5, 6, and 7. Note that each integer is 1 larger than the previous integer. To represent three *unknown* consecutive integers, we let

$$x = \text{the first integer,}$$
$$x + 1 = \text{the second integer,}$$
and $\qquad x + 2 = \text{third integer.}$

Consider the three consecutive odd integers 7, 9, and 11. Note that each odd integer is 2 larger than the previous odd integer. To represent three *unknown* consecutive odd integers, we let

$$x = \text{the first odd integer}$$
$$x + 2 = \text{the second odd integer}$$
and $\qquad x + 4 = \text{the third odd integer.}$

Note that consecutive even integers as well as consecutive odd integers differ by 2. So the same expressions are used in either case.

How would we represent two numbers that have a sum of 8? If one of the numbers is 2, the other is certainly 6, or $8 - 2$. So if x is one of the numbers, then $8 - x$ is the other. The expressions x and $8 - x$ have a sum of 8 for any value of x.

EXAMPLE 1   **Writing algebraic expressions**

Write algebraic expressions to represent each verbal expression.

a)   Two numbers that differ by 12

b)   Two consecutive even integers

c)   Two investments that total $5000

d)   The length of a rectangle if the width is x meters and the perimeter is 10 meters

Solution

a)   The expressions x and $x + 12$ differ by 12. Note that we could also use x and $x - 12$ for two numbers that differ by 12.

b)   The expressions x and $x + 2$ represent two consecutive even integers.

c)   If x represents the amount of one investment, then $5000 - x$ represents the amount of the other investment.

d)   Because the perimeter is 10 meters and $P = 2L + 2W = 2(L + W)$, the sum of the length and width is 5 meters. Because the width is x, the length is $5 - x$. ∎

Many verbal phrases occur repeatedly in applications. The following list of some frequently occurring verbal phrases and their translations into algebraic expressions will help you to translate words into algebra.

▶ **Translating Words into Algebra** ◀

	Verbal Phrase	Algebraic Expression
Addition:	The sum of a number and 8	$x + 8$
	Five is added to a number	$x + 5$
	Two more than a number	$x + 2$
	A number increased by 3	$x + 3$
Subtraction:	Four is subtracted from a number	$x - 4$
	Three less than a number	$x - 3$
	The difference between 7 and a number	$7 - x$
	Some number decreased by 2	$x - 2$
	A number less 5	$x - 5$
Multiplication:	The product of 5 and a number	$5x$
	Seven times a number	$7x$
	Twice a number	$2x$
	One-half of a number	$\frac{1}{2}x \left(\text{or } \frac{x}{2} \right)$
Division:	The ratio of a number to 6	$\frac{x}{6}$
	The quotient of 5 and a number	$\frac{5}{x}$
	Three divided by some number	$\frac{3}{x}$

More than one operation can be combined in a single expression. For example, 7 less than twice a number is written as $2x - 7$.

Solving Problems

We will now see how algebraic expressions can be used to form an equation. If the equation correctly models a problem, then we may be able to solve the equation to get the solution to the problem. Some problems in this section could be solved without using algebra. However, the purpose of this section is to gain experience in setting up equations and using algebra to solve problems. We will show a complete solution to each problem so that you can gain the experience needed to solve more complex problems. We begin with a simple number problem.

EXAMPLE 2   **A number problem**

The sum of three consecutive integers is 228. Find the integers.

Solution

We first represent the unknown quantities with variables. The unknown quantities are the three consecutive integers. Let

$$x = \text{the first integer,}$$
$$x + 1 = \text{the second integer,}$$

and $\qquad\qquad\quad x + 2 = \text{the third integer.}$

Since the sum of these three expressions for the consecutive integers is 228, we can write the following equation and solve it:

$$x + (x + 1) + (x + 2) = 228 \qquad \text{The sum of the integers is 228.}$$
$$3x + 3 = 228$$
$$3x = 225$$
$$x = 75$$
$$x + 1 = 76 \qquad \text{Identify the other unknown quantities.}$$
$$x + 2 = 77$$

To verify that these values are the correct integers, we compute

$$75 + 76 + 77 = 228.$$

The three consecutive integers that have a sum of 228 are 75, 76, and 77.   ■

General Strategy for Problem Solving

The steps to follow in providing a complete solution to a verbal problem can be stated as follows.

> ### Strategy for Solving Word Problems
>
> 1. Read the problem until you understand the problem.
> 2. If possible, draw a diagram to illustrate the problem.
> 3. Choose a variable and write down what it represents.
> 4. Represent any other unknowns in terms of that variable.
> 5. Write an equation that models the situation.
> 6. Solve the equation.
> 7. Be sure that your solution answers the question posed in the original problem.
> 8. Check your answer by using it to solve the original problem (not the equation).

We will now see how this strategy can be applied to various types of problems.

Geometric Problems

Any problem that involves a geometric figure may be referred to as a **geometric problem.** For geometric problems the equation is often a geometric formula.

EXAMPLE 3   **Finding the length and width of a rectangle**

The length of a rectangular piece of property is 1 foot more than twice the width. If the perimeter is 302 feet, find the length and width.

Solution

First draw a diagram as in Fig. 2.3. Because the length is 1 foot more than twice the width, we let

$$x = \text{the width}$$

and

$$2x + 1 = \text{the length.}$$

x

$2x + 1$

Figure 2.3

The perimeter of a rectangle is modeled by the equation $2L + 2W = P$:

$$2L + 2W = P$$
$$2(2x + 1) + 2(x) = 302 \qquad \text{Replace } L \text{ by } 2x + 1 \text{ and } W \text{ by } x.$$
$$4x + 2 + 2x = 302 \qquad \text{Remove the parentheses.}$$
$$6x = 300$$
$$x = 50$$
$$2x + 1 = 101 \qquad \text{Because } 2(50) + 1 = 101$$

Because $P = 2(101) + 2(50) = 302$ and 101 is 1 more than twice 50, we can be sure that the answer is correct. So the length is 101 feet, and the width is 50 feet. ■

Investment Problems

Investment problems involve sums of money invested at various interest rates. In this chapter we consider simple interest only.

EXAMPLE 4   **Investing at two rates**

Greg Smith invested some money in a certificate of deposit with an annual yield of 9%. He invested twice as much money in a mutual fund with an annual yield of 12%. His interest from the two investments at the end of the year was $396. How much money was invested at each rate?

Solution

Recall the formula $I = Prt$. In this problem the time t is 1 year, so $I = Pr$. If we let x represent the amount invested at the 9% rate, then $2x$ is the amount invested at 12%. The interest on these investments is the principal times the rate, or $0.09x$ and $0.12(2x)$. It is often helpful to make a table for the unknown quantities.

	Amount	Rate	Interest
Certificate of deposit	x	9%	$0.09x$
Mutual fund	$2x$	12%	$0.12(2x)$

The fact that the total interest from the investments was $396 is expressed in the following equation:

$$0.09x + 0.12(2x) = 396$$
$$0.09x + 0.24x = 396 \qquad \text{We could multiply each side by 100}$$
$$\text{to eliminate the decimals.}$$
$$0.33x = 396$$
$$x = \frac{396}{0.33}$$
$$x = 1200$$
$$2x = 2400$$

To check this answer, we find that 0.09($1200) = $108 and 0.12($2400) = $288. Now $108 + $288 = $396. So Greg invested $1200 at 9% and $2400 at 12%. ∎

Mixture Problems

Mixture problems involve solutions containing various percentages of a particular ingredient.

EXAMPLE 5   **Mixing milk**

How many gallons of milk containing 5% butterfat must be mixed with 90 gallons of 1% milk to obtain 2% milk?

Solution

If x represents the number of gallons of 5% milk, then $0.05x$ represents the amount of fat in that milk. If we mix x gallons of 5% milk with 90 gallons of 1% milk, we will have $x + 90$ gallons of 2% milk. See Fig. 2.4. We can make a table to classify all of the unknown quantities.

	Quantity	% fat	Amount of fat
5% milk	x	5%	$0.05x$
1% milk	90	1%	$0.01(90)$
2% milk	$x + 90$	2%	$0.02(x + 90)$

Figure 2.4

In mixture problems we always write an equation that accounts for one of the ingredients in the process. In this case we write an equation to express the fact that the total amount of fat from the first two types of milk is the same as the amount of fat in the mixture.

$$0.05x + 0.01(90) = 0.02(x + 90)$$
$$0.05x + 0.9 = 0.02x + 1.8 \qquad \text{Remove parentheses.}$$
$$\text{Note that we chose to work with the decimals}$$
$$0.03x = 0.9 \qquad \text{rather than eliminate them.}$$
$$x = 30$$

We should use 30 gallons of 5% milk. There are 1.5 gallons of fat in the 30 gallons of 5% milk. The 1% milk will contribute 0.9 gallon of fat, and there will be 2.4 gallons of fat in 120 gallons of 2% milk. Because 1.5 + 0.9 = 2.4, we have the correct solution. ∎

EXAMPLE 6 **Blending gasoline**

A dealer has 10,000 gallons of unleaded gasoline. He wants to add just enough ethanol to make the fuel a 10% ethanol mixture. How many gallons of ethanol should be added?

Solution

Let x represent the number of gallons of 100% pure ethanol that should be added. (The original gasoline has no ethanol in it.) We can classify all of this information in a table.

	Amount	% ethanol	Amount of ethanol
Gasoline	10,000	0%	0
Ethanol	x	100%	x
Mixture	$x + 10,000$	10%	$0.1(x + 10,000)$

We can write $0 + x = 0.1(x + 10,000)$ to express the fact that the amount of ethanol in the gasoline plus the amount of ethanol added is equal to the amount of ethanol in the final mixture. This model accounts for all of the ethanol in the process.

$$0 + x = 0.1(x + 10,000)$$

$$x = 0.1x + 1000 \quad \text{Remove parentheses.}$$

$$0.9x = 1000 \quad \text{Subtract } 0.1x \text{ from each side of the equation.}$$

$$x = \frac{1000}{0.9} \quad \text{Divide each side by 0.9.}$$

$$x = 1,111.1 \text{ gallons}$$

The amount of ethanol has been rounded to the nearest tenth of a gallon, so we cannot expect checking to be exact. If we combine 10,000 gallons of gasoline with 1,111.1 gallons of ethanol, we obtain 11,111.1 gallons of fuel. Notice that the amount of ethanol is 10% of the total mixture, 10% of $(10,000 + x)$. ∎

Uniform Motion Problems

Problems that involve motion at a constant rate are referred to as **uniform motion problems.**

EXAMPLE 7 **Driving Miss Jennifer**

Jennifer drove her car for 3 hours in a dust storm. When the skies cleared, she increased her speed by 30 miles per hour and drove for 4 more hours, completing her 295-mile trip. How fast did she travel during the dust storm?

Solution

If x was Jennifer's speed during the dust storm, then her speed under clear skies was $x + 30$. For problems involving motion we use the formula $D = RT$ (distance equals rate times time). It is again helpful to make a table to classify the given information.

	Rate	Time	Distance
Dust storm	x	3	$3x$
Clear skies	$x + 30$	4	$4(x + 30)$

The following equation indicates that the total distance traveled was 295 miles:

$$3x + 4(x + 30) = 295$$
$$3x + 4x + 120 = 295 \quad \text{Remove parentheses.}$$
$$7x = 175$$
$$x = 25 \text{ miles per hour}$$

Check this answer in the original problem. Jennifer traveled 25 mph during the storm. ∎

Commission Problems

When property is sold, the percentage of the selling price that the selling agent receives is the **commission.**

EXAMPLE 8 **Selling price of a house**

Sonia is selling her house through a real estate agent whose commission is 6% of the selling price. What should be the selling price so that Sonia can get $84,600?

Solution

Let x be the selling price. The commission is 6% of x (not 6% of $84,600). Sonia receives the selling price less the sales commission.

$$\text{Selling price} - \text{commission} = \text{Sonia's share}$$
$$x - 0.06x = 84{,}600$$
$$0.94x = 84{,}600$$
$$x = \frac{84{,}600}{0.94}$$
$$= \$90{,}000$$

The commission is 0.06($90,000), or $5,400. Sonia's share is $90,000 − $5,400, or $84,600. The house should sell for $90,000. ∎

Warm-ups

True or false? Explain your answer.

1. The recommended first step in solving a word problem is to write the equation.
2. When solving word problems, always write what the variable stands for.
3. Any solution to your equation must solve the word problem.
4. To represent two consecutive odd integers, we use x and $x + 1$.
5. We can represent two numbers that have a sum of 6 by x and $6 - x$.
6. Two numbers that differ by 7 can be represented by x and $x + 7$.
7. If $5x$ ft is 2 ft more than $3(x + 20)$ ft, then $5x + 2 = 3(x + 20)$.
8. If x is the selling price and the commission is 8% of the selling price, then the commission is $0.08x$.
9. If you need $80,000 for your house and the agent gets 10% of the selling price, then the agent gets $8,000, and the house sells for $88,000.
10. When we mix a 10% acid solution with a 14% acid solution, we can get as high as 24% acid.

2.3 EXERCISES

Find algebraic expressions for each of the following. See Example 1.

1. Two consecutive even integers
2. Two consecutive odd integers
3. Two numbers with a sum of 10
4. Two numbers with a difference of 3
5. Eighty-five percent of the selling price
6. The product of a number and 3
7. The distance traveled in 3 hours at x miles per hour
8. The time it takes to travel 100 miles at $x + 5$ miles per hour
9. The perimeter of a rectangle if the width is x feet and the length is 5 feet longer than the width
10. The width of a rectangle if the length is x meters and the perimeter is 20 meters

Show a complete solution for each number problem. See Example 2.

11. The sum of three consecutive integers is 84. Find the integers.
12. Find three consecutive integers whose sum is 171.
13. Find three consecutive even integers whose sum is 252.
14. Find three consecutive even integers whose sum is 84.
15. Two consecutive odd integers have a sum of 128. What are the integers?
16. Four consecutive odd integers have a sum of 56. What are the integers?

Show a complete solution to each geometric problem. See Example 3.

17. **Length and width.** If the perimeter of a rectangle is 278 meters and the length is 1 meter longer than twice the width, then what are the length and width?
18. **Dimensions of a frame.** A frame maker made a large picture frame using 10 feet of frame molding. If the length of the finished frame was 2 feet more than the width, then what were the dimensions of the frame?
19. **Perimeter of a lot.** Having finished fencing the perimeter of a triangular piece of land, Lance observed that the second side was just 10 feet short of being twice as long as the first side, and the third side was exactly 50 feet longer than the first side. If he used 684 feet of fencing, what are the lengths of the three sides?
20. **Isosceles triangle.** A flag in the shape of an isosceles triangle has a base that is 3.5 inches shorter than either of the equal sides. If the perimeter of the triangle is 49 inches, what is the length of the equal sides? See the diagram on the next page.

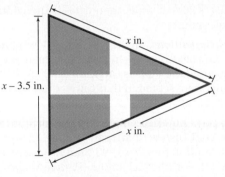

Figure for Exercise 20

21. *Hog heaven.* Farmer Hodges has 50 feet of fencing to make a rectangular hog pen beside a very large barn. He needs to fence only three sides, because the barn will form the fourth side. Studies have shown that under those conditions the side parallel to the barn should be 5 feet longer than twice the width. If Farmer Hodges uses all of the fencing, what should the dimensions be?

Figure for Exercise 21

22. *Doorway dimensions.* A carpenter made a doorway that is 1 foot taller than twice the width. If she used three pieces of door edge molding with a total length of 17 feet, then what are the approximate dimensions of the doorway?

Show a complete solution to each investment problem. See Example 4.

23. *Investing money.* Mr. and Mrs. Jackson invested some money at 6% simple interest and some money at 10% simple interest. In the second investment they put $1,000 more than they put in the first. If the income from both investments for one year was $340, then how much did they invest at each rate?

24. *Sibling rivalry.* Samantha lent her brother some money at 9% simple interest and her sister one-half as much money at 16% simple interest. If she received a total of 34 cents in interest, then how much did she lend to each one?

25. *Investing inheritance.* Norman invested one-half of his inheritance in a certificate of deposit that had a 10% annual yield. He lent one-quarter of his inheritance to his brother-in-law at 12% simple interest. His income from these two investments was $6,400 for one year. How much was the inheritance?

26. *Insurance settlement.* Gary invested one-third of his insurance settlement in a certificate of deposit that yielded 12%. He also invested one-third in Tara's computer business. Tara paid Gary 15% on this investment. If Gary's total income from these investments was $10,800 for one year, then what was the amount of his insurance settlement?

Show a complete solution to each mixture problem. See Examples 5 and 6.

27. *Acid solutions.* How many gallons of 5% acid solution should be mixed with 20 gallons of a 10% acid solution to obtain an 8% acid solution?

28. *Alcohol solutions.* How many liters of a 10% alcohol solution should be mixed with 12 liters of a 20% alcohol solution to obtain a 14% alcohol solution?

29. *Increasing acidity.* A gallon of Del Monte White Vinegar is labeled 5% acidity. How many fluid ounces of pure acid must be added to get 6% acidity?

30. *Chlorine bleach.* A gallon of Clorox bleach is labeled "5.25% sodium hypochlorite by weight." If a gallon of bleach weighs 8.3 pounds, then how many ounces of sodium hypochlorite must be added so that the bleach will be 6% sodium hypochlorite?

Show a complete solution to each uniform motion problem. See Example 7.

31. *Driving in a fog.* Carlo drove for 3 hours in a fog, then increased his speed by 30 miles per hour and drove 6 more hours. If his total trip was 540 miles, then what was his speed in the fog?

32. *Walk, don't run.* Louise walked for 2 hours then ran for $1\frac{1}{2}$ hours. If she runs twice as fast as she walks and the total trip was 20 miles, then how fast does she run?

33. *Commuting to work.* A commuter bus takes 2 hours to get downtown; an express bus, averaging 25 mph faster, takes 45 minutes to cover the same route. What is the average speed for the commuter bus?

Figure for Exercise 33

34. *Passengers versus freight.* A freight train takes $1\frac{1}{4}$ hours to get to the city; a passenger train averaging 40 mph faster takes only 45 minutes to cover the same distance. What is the average speed of the passenger train?

Show a complete solution to each problem. See Example 8.

35. *Listing a house.* Karl wants to get $80,000 for his house. The real estate agent charges 8% of the selling price for selling the house. What should the selling price be?

36. *Hot tamales.* Martha sells hot tamales at a sidewalk stand. Her total receipts including the 5% sales tax were $915.60. What amount of sales tax did she collect?

37. *Mustang Sally.* Sally bought a used Mustang. The selling price plus the 7% state sales tax was $9,041.50. What was the selling price?

38. *Choosing a selling price.* Roy is selling his car through a broker. Roy wants to get $3,000 for himself, but the broker gets a commission of 10% of the selling price. What should the selling price be?

Show a complete solution to each problem.

39. *Tennis.* The distance from the base line to the service line on a tennis court is 3 feet longer than the distance from the service line to the net. If the distance from the base line to the net is 39 feet, then what is the distance from the service line to the net?

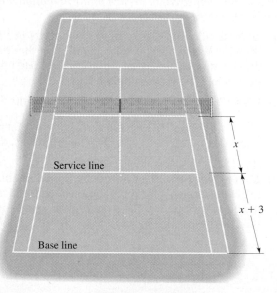

Service line

x

$x + 3$

Base line

Figure for Exercise 39

40. *Mixed doubles.* The doubles court in tennis is one-third wider than the singles court. If the doubles court is 36 feet wide, then what is the width of the singles court?

41. *First Super Bowl.* In the first Super Bowl game in the Los Angeles Coliseum in 1967, the Green Bay Packers outscored the Kansas City Chiefs by 25 points. If 45 points were scored in that game, then what was the final score?

42. *Top pay.* America's highest-paid corporate executive, Walt Disney Chairman Michael Eisner, earned $167 million more in 1993 than America's highest-paid athlete, Michael Jordan (*Parade Magazine*, June 26, 1994). The total earnings of these two men was $239 million. How much did each one earn in 1993?

43. *Blending coffee.* Mark blends $\frac{3}{4}$ of a pound of premium Brazilian coffee with $1\frac{1}{2}$ pounds of standard Colombian coffee. If the Brazilian coffee sells for $10 per pound and the Colombian coffee sells for $8 per pound, then what should the price per pound be for the blended coffee?

$10/lb
COFFEE
$\frac{3}{4}$ lb

+

$8/lb
COFFEE
$1\frac{1}{2}$ lb

=

?
COFFEE
$2\frac{1}{4}$ lb

Figure for Exercise 43

44. *'Tis the seasoning.* Cheryl's Famous Pumpkin Pie Seasoning consists of a blend of cinnamon, nutmeg, and cloves. When Cheryl mixes up a batch, she uses 200 ounces of cinnamon, 100 ounces of nutmeg, and 100 ounces of cloves. If cinnamon sells for $1.80 per ounce, nutmeg sells for $1.60 per ounce, and cloves sell for $1.40 per ounce, what should be the price per ounce of the mixture?

45. *Health food mix.* Dried bananas sell for $0.80 per quarter-pound, and dried apricots sell for $1.00 per quarter-pound. How many pounds of apricots should be mixed with 10 pounds of bananas to get a mixture that sells for $0.95 per quarter-pound?

46. *Mixed nuts.* Cashews sell for $1.20 per quarter-pound, and Brazil nuts sell for $1.50 per quarter-pound. How many pounds of cashews should be mixed with 20 pounds of Brazil nuts to get a mix that sells for $1.30 per quarter-pound?

47. *Antifreeze mixture.* A mechanic finds that a car with a 20-quart radiator has a mixture containing 30% antifreeze in it. How much of this mixture would he have to drain out and replace with pure antifreeze to get a 50% antifreeze mixture?

48. *Increasing the percentage.* A mechanic has found that a car with a 16-quart radiator has a 40% antifreeze mixture in the radiator. She has on hand a 70% antifreeze solution. How much of the 40% solution would she have to replace with the 70% solution to get the solution in the radiator up to 50%?

49. *Jobs in retail sales.* The number of U.S. jobs in retail sales is projected to be 4,180,000 in the year 2005 (U.S. Bureau of Labor Statistics, *Monthly Labor Review,* Nov. 1991). If this figure represents a 15% increase over the number in retail sales in 1990, then how many jobs were there in retail sales in 1990?

50. *Registered nurses.* The number of jobs for registered nurses in 2005 could be as high as 2,648,000, a 53% increase over the 1990 figure (U.S. Bureau of Labor Statistics, *Monthly Labor Review,* Nov. 1991). How many jobs for registered nurses were there in 1990?

51. *Dividing the estate.* Uncle Albert's estate is to be divided among his three nephews. The will specifies that Daniel receive one-half of the amount that Brian receives and that Raymond receive $1,000 less than one-third of the amount that Brian receives. If the estate amounts to $25,400, then how much does each inherit?

52. *Mary's assets.* Mary Hall's will specifies that her lawyer is to liquidate her assets and divide the proceeds among her three sisters. Lena's share is to be one-half of Lisa's, and Lisa's share is to be one-half of Lauren's. If the lawyer has agreed to a fee that is equal to 10% of the largest share and the proceeds amount to $164,428, then how much does each person get?

53. *Missing integers.* If the larger of two consecutive integers is subtracted from twice the smaller, the result is 21. Find the integers.

54. *Really odd integers.* If the smaller of two consecutive odd integers is subtracted from twice the larger, the result is 13. Find the integers.

55. *Highway miles.* Berenice and Jarrett drive a rig for Continental Freightways. In one day Berenice averaged 50 mph and Jarrett averaged 56 mph, but Berenice drove for two more hours than Jarrett. If together they covered 683 miles, then for how many hours did Berenice drive?

56. *Spring break.* Fernell and Dabney shared the driving to Florida for spring break. Fernell averaged 50 mph, and Dabney averaged 64 mph. If Fernell drove for 3 hours longer than Dabney but covered 18 miles less than Dabney, then for how many hours did Fernell drive?

57. *Stacy's square.* Stacy has 70 meters of fencing and plans to make a square pen. In one side she is going to leave an opening that is one-half the length of the side. If she uses all 70 meters of fencing, how large can the square be?

58. *Shawn's shed.* Shawn is building a tool shed with a square foundation and has enough siding to cover 32 linear feet of walls. If he leaves a 4-foot space for a door, then what size foundation would use up all of his siding?

Figure for Exercise 58

59. *Splitting investments.* Joan had $3000 to invest. She invested part of it in an investment paying 8% and the remainder in an investment paying 10%. If the total income on these investments was $290, then how much did she invest at each rate?

60. *Financial independence.* Dorothy had $8000 to invest. She invested part of it in an investment paying 6% and the rest in an investment paying 9%. If the total income from these investments was $690, then how much did she invest at each rate?

61. *Alcohol solutions.* Amy has two solutions available in the laboratory, one with 5% alcohol and the other with 10% alcohol. How much of each should she mix together to obtain 5 gallons of an 8% solution?

62. *Alcohol and water.* Joy has a solution containing 12% alcohol. How much of this solution and how much water must she use to get 6 liters of a solution containing 10% alcohol?

63. *Chance meeting.* In 6 years, Todd will be twice as old as Darla was when they met 6 years ago. If their ages total 78 years, then how old are they now?

64. *Centennial Plumbing Company.* The three Hoffman brothers advertise that together they have a century of plumbing experience. Bart has twice the experience of Al, and in 3 years, Carl will have twice the experience that Al had a year ago. How many years of experience does each of them have?

2.4 Inequalities

So far, we have been working with equations in this chapter. Equations express the equality of two algebraic expressions. But we are often concerned with two algebraic expressions that are not equal, one expression being greater than or less than the other. In this section we will begin our study of inequalities.

Basic Ideas

Statements that express the inequality of algebraic expressions are called **inequalities.** The symbols that we use to express inequality are given below with their meanings.

Inequality Symbols	
Symbol	**Meaning**
$<$	Is less than
$\leq$	Is less than or equal to
$>$	Is greater than
$\geq$	Is greater than or equal to

It is clear that 5 is less than 10, but how do we compare -5 and -10? If we think of negative numbers as debts, we would say that -10 is the larger debt. However, in algebra the size of a number is determined only by its position on the number line. For two numbers a and b we say that a *is less than* b if and only if a is to the *left* of b on the number line. To compare -5 and -10, we locate each point on the number line in Fig. 2.5. Because -10 is to the left of -5 on the number line, we say that -10 is less than -5, in symbols,

$$-10 < -5.$$

$$\begin{array}{ccccccccccc} \bullet & | & | & | & | & \bullet & | & | & | & | & \\ -10 & -9 & -8 & -7 & -6 & -5 & -4 & -3 & -2 & -1 & 0 \end{array}$$

Figure 2.5

We say that a is greater than b if and only if a is to the *right* of b on the number line. Thus, we can also write

$$-5 > -10.$$

The statement $a \leq b$ is true if a is less than b or if a is equal to b. The statement $a \geq b$ is true if a is greater than b or if a equals b. For example, the statement $3 \leq 5$ is true, and so is the statement $5 \leq 5$.

EXAMPLE 1   **Inequalities**
Determine whether each statement is true or false.

a) $-5 < 3$ **b)** $-9 > -6$ **c)** $-3 \leq 2$ **d)** $4 \geq 4$

Solution

a) The statement $-5 < 3$ is true, because -5 is to the left of 3 on the number line. In fact, any negative number is less than any positive number.

b) The statement $-9 > -6$ is false, because -9 lies to the left of -6.

c) The statement $-3 \le 2$ is true, because -3 is less than 2.

d) The statement $4 \ge 4$ is true, because $4 = 4$ is true. ■

Graphing Inequalities

The statement $x < 3$ means that x is a real number to the left of 3 on the number line. The numbers 1.5, 0, and -2 are all to the left of 3 on the number line, and so the statements

$$1.5 < 3, \qquad 0 < 3, \qquad \text{and} \qquad -2 < 3$$

are true statements. The set of all numbers that give a true statement when used as a replacement for x is called the **solution set** to the inequality.

The solution set to $x < 3$ is the set of all real numbers to the left of 3 on the number line. Figure 2.6(a) shows the graph. The parenthesis at 3 on the graph indicates that 3 is not in the solution set. We write this set in *set notation* as $\{x \mid x < 3\}$ and in **interval notation** as $(-\infty, 3)$. Note that $-\infty$ (negative infinity) is not used to represent a specific number, but only to indicate that there is no end to the numbers less than 3. If an endpoint of an interval is to be included in the interval, brackets are used in the interval notation and on the graph. For example, the graph of $\{x \mid x \ge 1\}$ is shown in Fig. 2.6(b). This set is written in interval notation as $[1, \infty)$.

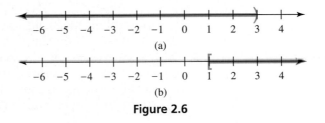

(a)

(b)

Figure 2.6

EXAMPLE 2   **Interval notation and graphs**

Write the solution set to each inequality in interval notation and graph it.

a) $x > -5$ **b)** $x \le 2$

Solution

a) The solution set to the inequality $x > -5$ is $\{x \mid x > -5\}$. The solution set is the interval of all numbers to the right of -5 on the number line. This set is written in interval notation as $(-5, \infty)$, and it is graphed in Fig. 2.7.

Figure 2.7

b) The solution set to $x \le 2$ is $\{x \mid x \le 2\}$. This set includes 2 and all real numbers to the left of 2. Because 2 is included, we use a bracket at 2. The interval notation for this set is $(-\infty, 2]$. The graph is shown in Fig. 2.8.

Figure 2.8 ■

Solving Linear Inequalities

In Section 2.1 we defined a linear equation as an equation of the form $ax + b = 0$. If we replace the equality symbol in a linear equation with an inequality symbol, we have a linear inequality.

Linear Inequality

A **linear inequality** in one variable x is any inequality of the form $ax + b < 0$, where a and b are real numbers, with $a \neq 0$. In place of $<$ we may also use $\leq$, $>$, or $\geq$.

Inequalities that can be *rewritten* in the form of a linear inequality are also called linear inequalities.

Before we solve linear inequalities, let's examine the results of performing various operations on each side of an inequality. If we start with the inequality $2 < 6$ and add 2 to each side, we get the true statement $4 < 8$. Examine the results in the following table.

Perform these operations on each side of $2 < 6$:

	Add 2	Subtract 2	Multiply by 2	Divide by 2
Resulting inequality	$4 < 8$	$0 < 4$	$4 < 12$	$1 < 3$

All of the resulting inequalities are correct. However, if we perform operations on each side of $2 < 6$ using -2, the situation is not as simple. For example, $-2 \cdot 2 = -4$ and $-2 \cdot 6 = -12$, but -4 is greater than -12. To get a correct inequality when each side is multiplied or divided by -2, we must reverse the inequality symbol, as shown in the following table.

Perform these operations on each side of $2 < 6$:

	Add -2	Subtract -2	Multiply by -2	Divide by -2
Resulting inequality	$0 < 4$	$4 < 8$	$-4 > -12$	$-1 > -3$

Inequality reverses

These examples illustrate the properties that we use for solving inequalities.

Properties of Inequality

Addition-Subtraction Property of Inequality
The same real number may be added to or subtracted from each side of an inequality without changing the solution set.

Multiplication-Division Property of Equality
Each side of an inequality may be multiplied by or divided by the same nonzero real number without changing the solution set. *If the inequality is multiplied by or divided by a negative number, the inequality symbol is reversed.*

Equivalent inequalities are inequalities with the same solution set. We find the solution to a linear inequality by using the properties to convert it into an equivalent inequality with an obvious solution set, just as we do when solving equations.

EXAMPLE 3   **Solving inequalities**

Solve the inequality $5 - 3x < 11$. State and graph the solution set.

Solution

We proceed exactly as we do when solving an equation.

$$5 - 3x < 11$$
$$-3x < 6 \qquad \text{Subtract 5 from each side.}$$
$$x > -2 \qquad \text{Divide each side by } -3 \text{ and reverse the inequality symbol.}$$

The solution set is written in set notation as $\{x \mid x > -2\}$ and in interval notation as $(-2, \infty)$. The graph is shown in Fig. 2.9.

Figure 2.9

EXAMPLE 4   **Solving inequalities**

Solve the inequality. State and graph the solution set.

$$\frac{8 + 3x}{-5} \geq -4$$

Solution

$$-5\left(\frac{8 + 3x}{-5}\right) \leq -5(-4) \qquad \text{Multiply each side by } -5 \text{ and reverse the inequality symbol.}$$
$$8 + 3x \leq 20 \qquad \text{Simplify.}$$
$$3x \leq 12 \qquad \text{Subtract 8 from each side.}$$
$$x \leq 4 \qquad \text{Divide each side by 3.}$$

The solution set is $(-\infty, 4]$, and its graph is shown in Fig. 2.10.

Figure 2.10

EXAMPLE 5   **An inequality with fractions**

Solve the inequality. State and graph the solution set.

$$\frac{1}{2}x - \frac{2}{3} \leq x + \frac{4}{3}$$

Solution

First multiply each side of the inequality by 6, the LCD:

$$6\left(\frac{1}{2}x - \frac{2}{3}\right) \leq 6\left(x + \frac{4}{3}\right)$$ Multiplying by positive 6 does not reverse the inequality.

$$3x - 4 \leq 6x + 8$$ Distributive property

$$3x \leq 6x + 12$$ Add 4 to each side.

$$-3x \leq 12$$ Subtract $6x$ from each side.

$$x \geq -4$$ Divide each side by -3 and reverse the inequality.

The solution set is the interval $[-4, \infty)$. Its graph is shown in Fig. 2.11.

Figure 2.11

Applications

Inequalities have applications just as equations do. To use inequalities, we must be able to translate a verbal problem into an algebraic inequality. Inequality can be expressed verbally in a variety of ways.

EXAMPLE 6 **Writing inequalities**

Identify the variable and write an inequality that describes the situation.

a) Chris paid more than $200 for a suit.

b) A candidate for president must be at least 35 years old.

c) The capacity of an elevator is at most 1500 pounds.

d) The company must hire no fewer than 10 programmers.

Solution

a) If c is the cost of the suit in dollars, then $c > 200$.

b) If a is the age of the candidate in years, then $a \geq 35$.

c) If x is the capacity of the elevator in pounds, then $x \leq 1500$.

d) If n represents the number of programmers and n is not less than 10, then $n \geq 10$.

In Example 6(d) we knew that n was not less than 10. So there were exactly two other possibilities: n was greater than 10 or equal to 10. The fact that there are only three possible ways to position two real numbers on a number line is called the **trichotomy property.**

Trichotomy Property

For any two real numbers a and b, exactly one of the following is true:

$$a < b, \qquad a = b, \qquad \text{or} \qquad a > b.$$

We follow the same steps to solve problems involving inequalities as we do to solve problems involving equations.

EXAMPLE 7   **Price range**

Lois plans to spend less than $500 on an electric dryer, including the 9% sales tax and a $64 setup charge. In what range is the selling price of the dryer that she can afford?

Solution

If we let x represent the selling price in dollars for the dryer, then the amount of sales tax is $0.09x$. Because her total cost must be less than $500, we can write the following inequality:

$$x + 0.09x + 64 < 500$$
$$1.09x < 436 \qquad \text{Subtract 64 from each side.}$$
$$x < \frac{436}{1.09} \qquad \text{Divide each side by 1.09.}$$
$$x < 400$$

The selling price of the dryer must be less than $400. ■

Note that if we had written the equation $x + 0.09x + 64 = 500$ for the last example, we would have gotten $x = 400$. We could then have concluded that the selling price must be less than $400. This would certainly solve the problem, but it would not illustrate the use of inequalities. The original problem describes an inequality, and we should solve it as an inequality.

EXAMPLE 8   **Paying off the mortgage**

Tessie owns a piece of land on which she owes $12,760 to a bank. She wants to sell the land for enough money to at least pay off the mortgage. The real estate agent gets 6% of the selling price, and her city has a $400 real estate transfer tax paid by the seller. What should the range of the selling price be for Tessie to get at least enough money to pay off her mortgage?

Solution

If x is the selling price in dollars, then the commission is $0.06x$. We can write an inequality expressing the fact that the selling price minus the real estate commission minus the $400 tax must be at least $12,760:

$$x - 0.06x - 400 \geq 12{,}760$$
$$0.94x - 400 \geq 12{,}760 \qquad 1 - 0.06 = 0.94$$
$$0.94x \geq 13{,}160 \qquad \text{Add 400 to each side.}$$
$$x \geq \frac{13{,}160}{0.94} \qquad \text{Divide each side by 0.94.}$$
$$x \geq 14{,}000$$

The selling price must be at least $14,000 for Tessie to pay off the mortgage. ■

Warm-ups

True or false? Explain your answer.

1. $0 < 0$
2. $-300 > -2$
3. $-60 \leq -60$
4. The inequality $6 < x$ is equivalent to $x < 6$.
5. The inequality $-2x < 10$ is equivalent to $x < -5$.
6. The solution set to $3x \geq -12$ is $(-\infty, -4]$.
7. The solution set to $-x > 4$ is $(-\infty, -4)$.
8. If x is no larger than 8, then $x \leq 8$.
9. If m is any real number, then exactly one of the following is true: $m < 0$, $m = 0$, or $m > 0$.
10. The number -2 is a member of the solution set to the inequality $3 - 4x \leq 11$.

2.4 EXERCISES

Determine whether each inequality is true or false. See Example 1.

1. $-3 < -9$
2. $-8 > -7$
3. $0 \leq 8$
4. $-6 \geq -8$
5. $(-3)20 > (-3)40$
6. $(-1)(-3) < (-1)(5)$
7. $9 - (-3) \leq 12$
8. $(-4)(-5) + 2 \geq 21$

Determine whether each inequality is satisfied by the given number.

9. $2x - 4 < 8, -3$
10. $5 - 3x > -1, 6$
11. $2x - 3 \leq 3x - 9, 5$
12. $6 - 3x \geq 10 - 2x, -4$
13. $5 - x < 4 - 2x, -1$
14. $3x - 7 \geq 3x - 10, 9$

Write the solution set in interval notation and graph it. See Example 2.

15. $x \leq -1$
16. $x \geq -7$
17. $x > 20$
18. $x < 30$
19. $3 \leq x$
20. $-2 > x$
21. $x < 2.3$
22. $x \leq 4.5$

Rewrite each set in interval notation.

23. $\{x \mid x > 1\}$
24. $\{x \mid x < 3\}$
25. $\{x \mid x \leq -3\}$
26. $\{x \mid x \geq -2\}$
27. $\{x \mid x < 5\}$
28. $\{x \mid x > -7\}$
29. $\{x \mid x \geq -4\}$
30. $\{x \mid x \leq -9\}$

Fill in the blank with an inequality symbol so that the two statements are equivalent.

31. $x + 5 > 12$
    $x \_\_ 7$
32. $2x - 3 \leq -4$
    $2x \_\_ -1$
33. $-x < 6$
    $x \_\_ -6$
34. $-5 \geq -x$
    $5 \_\_ x$
35. $-2x \geq 8$
    $x \_\_ -4$
36. $-5x > -10$
    $x \_\_ 2$
37. $4 < x$
    $x \_\_ 4$
38. $-9 \leq -x$
    $x \_\_ 9$

Solve each of the following inequalities. Express the solution set in interval notation and graph it. See Examples 3–5.

39. $7x > -14$
40. $4x \leq -8$
41. $-3x \leq 12$
42. $-2x > -6$
43. $2x - 3 > 7$
44. $3x - 2 < 6$
45. $3 - 5x \leq 18$
46. $5 - 4x \geq 19$
47. $\dfrac{x - 3}{-5} < -2$
48. $\dfrac{2x - 3}{4} > 6$
49. $\dfrac{5 - 3x}{4} \leq 2$
50. $\dfrac{7 - 5x}{-2} \geq -1$
51. $3x - 8 > 0$
52. $7x < 0$
53. $4 - 3x < 0$
54. $9 - 2x \leq 0$
55. $-\dfrac{5}{6}x \geq -10$
56. $-\dfrac{2}{3}x \geq -8$

57. $3 - \dfrac{1}{4}x \geq 2$

58. $5 - \dfrac{1}{3}x > 2$

59. $\dfrac{1}{4}x - \dfrac{1}{2} < \dfrac{1}{2}x - \dfrac{2}{3}$

60. $\dfrac{1}{3}x - \dfrac{1}{6} < \dfrac{1}{6}x - \dfrac{1}{2}$

61. $\dfrac{y-3}{2} > \dfrac{1}{2} - \dfrac{y-5}{4}$

62. $\dfrac{y-1}{3} - \dfrac{y+1}{5} > 1$

Solve each inequality and graph the solution set.

63. $2x + 3 > 2(x - 4)$

64. $-2(5x - 1) \leq -5(5 + 2x)$

65. $-4(2x - 5) \leq 2(6 - 4x)$

66. $-3(2x - 1) \leq 2(5 - 3x)$

67. $-\dfrac{1}{2}(x - 6) < \dfrac{1}{2}x + 2$

68. $-3\left(\dfrac{1}{2}x - \dfrac{1}{4}\right) > \dfrac{x}{2} - \dfrac{1}{4}$

69. $-\dfrac{1}{2}(2x - 3) + \dfrac{1}{3}(4 - 6x) \geq \dfrac{1}{4}(7 - 2x) - 3$

70. $\dfrac{3}{5}(x - 3) - \dfrac{1}{4}(7 - 5x) < \dfrac{2}{3}(3 - x) - 5$

71. $\dfrac{1}{2}\left(\dfrac{3}{5}x - \dfrac{1}{8}\right) + 1 > \dfrac{1}{8}\left(\dfrac{3}{4}x + \dfrac{1}{10}\right)$

72. $-\dfrac{2}{3}\left(5x - \dfrac{3}{4}\right) + \dfrac{1}{8}\left(\dfrac{1}{2}x - 2\right) > \dfrac{3}{8}$

73. $4.273 + 2.8x \leq 10.985$

74. $1.064 < 5.94 - 3.2x$

75. $3.25x - 27.39 > 4.06 + 5.1x$

76. $4.86(3.2x - 1.7) > 5.19 - x$

Identify the variable and write an inequality that describes each situation. See Example 6.

77. Tony is taller than 6 feet.

78. Glenda is under 60 years old.

79. Wilma makes less than $80,000 per year.

80. Bubba weighs over 80 pounds.

81. The maximum speed for the Concorde is 1450 mph.

82. The minimum speed on the freeway is 45 mph.

83. Julie can afford at most $400 per month.

84. Fred must have at least a 3.2 grade point average.

85. Burt is no taller than 5 feet.

86. Ernie cannot run faster than 10 mph.

87. Tina makes no more than $8.20 per hour.

88. Rita will not take less than $12,000 for the car.

Solve each problem by using an inequality. See Examples 7 and 8.

89. *Car shopping.* Jennifer is shopping for a new car. In addition to the price of the car, there is an 8% sales tax and a $172 title and license fee. If Jennifer decides that she will spend less than $10,000 total, then what is the price range for the car?

90. *Sewing machines.* Charles wants to buy a sewing machine in a city with a 10% sales tax. He has at most $700 to spend. In what price range should he look?

91. *Truck shopping.* Linda and Bob are shopping for a new truck in a city with a 9% sales tax. There is also an $80 title and license fee to pay. They want to get a good truck and plan to spend at least $10,000. What is the price range for the truck?

92. *Curly's contribution.* Larry, Curly, and Moe are going to buy their mother a color television set. Larry has a better job than Curly and agrees to contribute twice as much as Curly. Moe is unemployed and can spare only $50. If the kind of television Mama wants costs at least $600, then what is the price range for Curly's contribution?

93. *Bachelor's degrees.* The accompanying graph shows the number of bachelor's degrees awarded in the United States each year since 1985 (National Center for Education Statistics).

a) Has the number of bachelor's degrees been increasing or decreasing since 1985?

b) The formula $B = 16.45n + 980.20$ can be used to approximate the number of degrees awarded in thousands in the year $1985 + n$. What is the first year in which the number of bachelor's degrees will exceed 1.2 million?

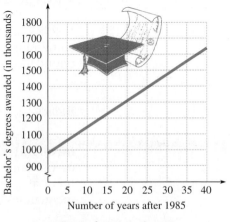

Figure for Exercise 93

94. *Master's degrees.* In 1985, 15.9% of all degrees awarded in U.S. higher education were master's degrees (National Center for Education Statistics). If the formulas $M = 7.79n + 287.87$ and $T = 30.95n + 1808.22$ give the number of master's degrees and the total number of higher education degrees awarded in thousands, respectively, in the year $1985 + n$, then what is the first year in which more than 18% of all degrees awarded will be master's degrees?

95. *Weighted average.* Professor Jorgenson gives only a midterm exam and a final exam. The semester average is computed by taking $\frac{1}{3}$ of the midterm exam score plus $\frac{2}{3}$ of the final exam score. The grade is determined from the semester average by using the grading scale given in the table. If Stanley scored only 56 on the midterm, then for what range of scores on the final exam would he get a C or better in the course?

Grading Scale

90–100	A
80–89	B
70–79	C
60–69	D

Table for Exercises 95 and 96

96. *C or better.* Professor Brown counts her midterm as $\frac{2}{3}$ of the grade and her final as $\frac{1}{3}$ of the grade. Wilbert scored only 56 on the midterm. If Professor Brown also uses the grading scale given in the table, then what range of scores on the final exam would give Wilbert a C or better in the course?

97. *Designer jeans.* A pair of ordinary jeans at A-Mart costs $50 less than a designer pair of jeans at Enrico's. In fact, you can buy four pairs of A-Mart jeans for less than one pair of Enrico's jeans. What is the price range for a pair of A-Mart jeans?

98. *United Express.* Al and Rita both drive parcel delivery trucks for United Express. Al averages 20 miles per hour less than Rita. In fact, Al is so slow that in 5 hours he covered fewer miles than Rita did in 3 hours. What are the possible values for Al's rate of speed?

Getting More Involved

99. *Discussion.* If 3 is added to every number in $(4, \infty)$, the resulting set is $(7, \infty)$. In each of the following cases, write the resulting set of numbers in interval notation. Explain your results.
   a) The number -6 is subtracted from every number in $[2, \infty)$.
   b) Every number in $(-\infty, -3)$ is multiplied by 2.
   c) Every number in $(8, \infty)$ is divided by 4.
   d) Every number in $(6, \infty)$ is multiplied by -2.
   e) Every number in $(-\infty, -10)$ is divided by -5.

100. *Writing.* Explain why saying that x is *at least* 9 is equivalent to saying that x is *greater than or equal to* 9. Explain why saying that x is *at most* 5 is equivalent to saying that x is *less than or equal to* 5.

2.5 Compound Inequalities

In this section:
▶ Basics
▶ Graphing the Solution Set
▶ Applications

In this section we will use the ideas of union and intersection from Chapter 1 along with our knowledge of inequalities from Section 2.4 to work with compound inequalities.

Basics

The inequalities that we studied in the last section are referred to as **simple inequalities.** If we join two simple inequalities with the connective "and" or the connective "or," we get a **compound inequality.** A compound inequality using the connective "and" is true if and only if *both* simple inequalities are true.

EXAMPLE 1   **Compound inequalities using the connective "and"**

Determine whether each compound inequality is true.

a) $3 > 2$ and $3 < 5$ **b)** $6 > 2$ and $6 < 5$

Solution

a) The compound inequality is true, because $3 > 2$ is true and $3 < 5$ is true.

b) The compound inequality is false, because $6 < 5$ is false. ■

A compound inequality using the connective "or" is true if one or the other or both of the simple inequalities are true. It is false only if both simple inequalities are false.

EXAMPLE 2   **Compound inequalities using the connective "or"**

Determine whether each compound inequality is true.

a) $2 < 3$ or $2 > 7$ **b)** $4 < 3$ or $4 \geq 7$

Solution

a) The compound inequality is true, because $2 < 3$ is true.

b) The compound inequality is false, because both $4 < 3$ and $4 \geq 7$ are false. ■

If a compound inequality involves a variable, then we are interested in the solution set to the inequality. The solution set to an "and" inequality consists of all numbers that satisfy both simple inequalities, while the solution set to an "or" inequality consists of all numbers that satisfy at least one of the simple inequalities.

EXAMPLE 3   **Solutions of compound inequalities**

Determine whether 5 satisfies each compound inequality.

a) $x < 6$ and $x < 9$ **b)** $2x - 9 \leq 5$ or $-4x \geq -12$

Solution

a) Because $5 < 6$ and $5 < 9$ are both true, 5 satisfies the compound inequality.

b) Because $2 \cdot 5 - 9 \leq 5$ is true, it does not matter that $-4 \cdot 5 \geq -12$ is false. So 5 satisfies the compound inequality. ■

Graphing the Solution Set

The solution set to a compound inequality such as

$$x > 2 \text{ and } x < 5$$

consists of all numbers that are in the solution sets to both simple inequalities. So the solution set to this compound inequality is the intersection of those two solution sets. In symbols,

$$\{x \mid x > 2 \text{ and } x < 5\} = \{x \mid x > 2\} \cap \{x \mid x < 5\}.$$

EXAMPLE 4   **Graphing compound inequalities**

Graph the solution set to the compound inequality $x > 2$ and $x < 5$.

Solution

We first sketch the graph of $x > 2$ and then the graph of $x < 5$, as shown in the top two number lines in Fig. 2.12. The intersection of these two solution sets is the portion of the number line that is shaded on both graphs, just the part between 2 and 5, not including the endpoints. The graph of $\{x \mid x > 2 \text{ and } x < 5\}$ is shown at the bottom of Fig. 2.12. We write this set in interval notation as $(2, 5)$.

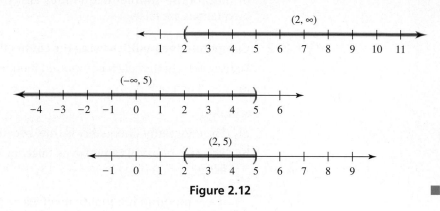

Figure 2.12

The solution set to a compound inequality such as

$$x > 4 \text{ or } x < -1$$

consists of all numbers that satisfy one or the other or both of the simple inequalities. So the solution set to the compound inequality is the union of the solution sets to the simple inequalities. In symbols,

$$\{x \mid x > 4 \text{ or } x < -1\} = \{x \mid x > 4\} \cup \{x \mid x < -1\}.$$

EXAMPLE 5   **Graphing compound inequalities**

Graph the solution set to the compound inequality $x > 4$ or $x < -1$.

Solution

To find the union of the solution sets to the simple inequalities, we sketch their graphs as shown at the top of Fig. 2.13. We graph the union of these two sets by putting both shaded regions together on the same line as shown in the bottom graph in Fig. 2.13. This set is written in interval notation as $(-\infty, -1) \cup (4, \infty)$.

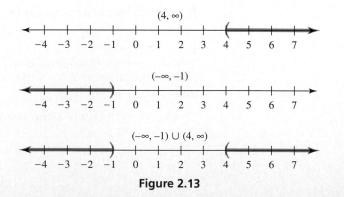

Figure 2.13

CAUTION When graphing the intersection of two simple inequalities, do not draw too much. For the intersection, graph only numbers that satisfy *both* inequalities. Omit numbers that satisfy one but not the other. Graphing a union is usually easier, because we can simply draw both solution sets on the same number line. ⊘

It is not always necessary to graph the solution set to each simple inequality before graphing the solution set to the compound inequality. We can save time and work if we learn to think of the two preliminary graphs but draw only the final one.

EXAMPLE 6 **Overlapping intervals**

Sketch the graph and write the solution set in interval notation to each compound inequality.

a) $x < 3$ and $x < 5$ **b)** $x > 4$ or $x > 0$

Solution

a) To graph $x < 3$ and $x < 5$, we shade only the numbers that are both less than 3 and less than 5. So numbers between 3 and 5 are not shaded in Fig. 2.14. The compound inequality $x < 3$ and $x < 5$ is equivalent to the simple inequality $x < 3$. The solution set can be written as $(-\infty, 3)$.

Figure 2.14

b) To graph $x > 4$ or $x > 0$, we shade both regions on the same number line as shown in Fig. 2.15. The compound inequality $x > 4$ or $x > 0$ is equivalent to the simple inequality $x > 0$. The solution set is $(0, \infty)$.

Figure 2.15 ▪

The next example shows a compound inequality that has no solution and one that is satisfied by every real number.

EXAMPLE 7 **All or nothing**

Sketch the graph and write the solution set in interval notation to each compound inequality.

a) $x < 2$ and $x > 6$ **b)** $x < 3$ or $x > 1$

Solution

a) A number satisfies $x < 2$ and $x > 6$ if it is both less than 2 *and* greater than 6. There are no such numbers. The solution set is the empty set, $\varnothing$.

b) To graph $x < 3$ or $x > 1$, we shade both regions on the same number line as shown in Fig. 2.16. Since the two regions cover the entire line, the solution set is the set of all real numbers $(-\infty, \infty)$.

Figure 2.16 ▪

If we start with a more complicated compound inequality, we first simplify each part of the compound inequality and then find the union or intersection.

EXAMPLE 8   **Intersection**

Solve $x + 2 > 3$ and $x - 6 < 7$. Graph the solution set.

Solution

First simplify each simple inequality:

$$x + 2 - 2 > 3 - 2 \quad \text{and} \quad x - 6 + 6 < 7 + 6$$
$$x > 1 \quad \text{and} \quad x < 13$$

The intersection of these two solution sets is the set of numbers between (but not including) 1 and 13. Its graph is shown in Fig. 2.17. The solution set is written in interval notation as $(1, 13)$.

Figure 2.17

EXAMPLE 9   **Union**

Graph the solution set to the inequality

$$5 - 7x \geq 12 \quad \text{or} \quad 3x - 2 < 7.$$

Solution

First solve each of the simple inequalities:

$$5 - 7x - 5 \geq 12 - 5 \quad \text{or} \quad 3x - 2 + 2 < 7 + 2$$
$$-7x \geq 7 \quad \text{or} \quad 3x < 9$$
$$x \leq -1 \quad \text{or} \quad x < 3$$

The union of the two solution intervals is $(-\infty, 3)$. The graph is shown in Fig. 2.18.

Figure 2.18

An inequality may be read from left to right or from right to left. Consider the inequality $1 < x$. If we read it in the usual way, we say, "1 is less than x." The meaning is clearer if we read the variable first. Reading from right to left, we say, "x is greater than 1."

Another notation is commonly used for the compound inequality

$$x > 1 \text{ and } x < 13.$$

This compound inequality can also be written as

$$1 < x < 13.$$

Reading from left to right, we read $1 < x < 13$ as "1 is less than x is less than 13." The meaning of this inequality is clearer if we read the variable first and read the first inequality symbol from right to left. Reading the variable first, $1 < x < 13$ is read as "x is greater than 1 and less than 13." So x is between 1 and 13, and reading x first makes it clear.

CAUTION   We write $a < x < b$ only if $a < b$, and we write $a > x > b$ only if $a > b$. Similar rules hold for $\le$ and $\ge$. So $4 < x < 9$ and $-6 \ge x \ge -8$ are correct uses of this notation, but $5 < x < 2$ is not. Also, the inequalities should *not* point in opposite directions as in $5 < x > 7$.   ⊘

EXAMPLE 10   **Another notation**

Solve the inequality and graph the solution set:

$$-2 \le 2x - 3 < 7$$

Solution

This inequality could be written as the compound inequality

$$2x - 3 \ge -2 \quad \text{and} \quad 2x - 3 < 7.$$

However, there is no need to rewrite it, because we can solve it in its original form.

$$-2 + 3 \le 2x - 3 + 3 < 7 + 3 \qquad \text{Add 3 to each part.}$$

$$1 \le 2x < 10$$

$$\frac{1}{2} \le \frac{2x}{2} < \frac{10}{2} \qquad \text{Divide each part by 2.}$$

$$\frac{1}{2} \le x < 5$$

The solution set is $\left[\frac{1}{2}, 5\right)$, and its graph is shown in Fig. 2.19.

Figure 2.19   ■

EXAMPLE 11   **Solving a compound inequality**

Solve the inequality $-1 < 3 - 2x < 9$ and graph the solution set.

Solution

$$-1 - 3 < 3 - 2x - 3 < 9 - 3 \qquad \text{Subtract 3 from each part of the inequality.}$$

$$-4 < -2x < 6$$

$$2 > x > -3 \qquad \text{Divide each part by } -2 \text{ and reverse both inequality symbols.}$$

$$-3 < x < 2 \qquad \text{Rewrite the inequality with the smallest number on the left.}$$

The solution set is $(-3, 2)$, and its graph is shown in Fig. 2.20.

Figure 2.20   ■

Applications

When final exams are approaching, students are often interested in finding the final exam score that would give them a certain grade for a course.

EXAMPLE 12   **Final exam scores**

Fiana made a score of 76 on her midterm exam. For her to get a B in the course, the average of her midterm exam and final exam must be between 80 and 89 inclusive. What possible scores on the final exam would give Fiana a B in the course?

Solution

Let x represent her final exam score. Between 80 and 89 inclusive means that an average between 80 and 89 as well as an average of exactly 80 or 89 will get a B. So the average of the two scores must be greater than or equal to 80 and less than or equal to 89.

$$80 \leq \frac{x + 76}{2} \leq 89$$

$$160 \leq x + 76 \leq 178 \qquad \text{Multiply by 2.}$$

$$160 - 76 \leq x \leq 178 - 76 \qquad \text{Subtract 76.}$$

$$84 \leq x \leq 102$$

If Fiana scores between 84 and 102 inclusive, she will get a B in the course. ■

Warm-ups

True or false? Explain your answer.

1. $3 < 5$ and $3 \leq 10$
2. $3 < 5$ or $3 < 10$
3. $3 > 5$ and $3 < 10$
4. $3 \geq 5$ or $3 \leq 10$
5. $4 < 8$ and $4 > 2$
6. $4 < 8$ or $4 > 2$
7. $-3 < 0 < -2$
8. $(3, \infty) \cap (8, \infty) = (8, \infty)$
9. $(3, \infty) \cup [8, \infty) = [8, \infty)$
10. $(-2, \infty) \cap (-\infty, 9) = (-2, 9)$

2.5 EXERCISES

Determine whether each compound inequality is true. See Examples 1 and 2.

1. $-6 < 5$ and $-6 > -3$
2. $3 < 5$ or $0 < -3$
3. $4 \leq 4$ and $-4 \leq 0$
4. $1 < 5$ and $1 > -3$
5. $6 < 5$ or $-4 > -3$
6. $4 \leq -4$ or $0 \leq 0$

Determine whether -4 satisfies each compound inequality. See Example 3.

7. $x < 5$ and $x > -3$
8. $x < 5$ or $x > -3$
9. $x - 3 \geq -7$ or $x + 1 > 1$
10. $2x \leq -8$ and $5x \leq 0$

11. $2x - 1 < -7$ or $-2x > 18$
12. $-3x > 0$ and $3x - 4 < 11$

Graph the solution set to each compound inequality. See Examples 4–7.

13. $x > -1$ and $x < 4$
14. $x \leq 3$ and $x \leq 0$
15. $x \geq 2$ or $x \geq 5$
16. $x < -1$ or $x < 3$
17. $x \leq 6$ or $x > -2$
18. $x > -2$ and $x \leq 4$
19. $x \leq 6$ and $x > 9$
20. $x < 7$ or $x > 0$
21. $x \leq 6$ or $x > 9$
22. $x \geq 4$ and $x \leq -4$
23. $x \geq 6$ and $x \leq 1$
24. $x > 3$ or $x < -3$

Solve each compound inequality. Write the solution set using interval notation and graph it. See Examples 8 and 9.

25. $x - 3 > 7$ or $3 - x > 2$

26. $x - 5 > 6$ or $2 - x > 4$

27. $3 < x$ and $1 + x > 10$

28. $-0.3x < 9$ and $0.2x > 2$

29. $\frac{1}{2}x > 5$ or $-\frac{1}{3}x < 2$

30. $5 < x$ or $3 - \frac{1}{2}x < 7$

31. $2x - 3 \le 5$ and $x - 1 > 0$

32. $\frac{3}{4}x < 9$ and $-\frac{1}{3}x \le -15$

33. $\frac{1}{2}x - \frac{1}{3} \ge -\frac{1}{6}$ or $\frac{2}{7}x \le \frac{1}{10}$

34. $\frac{1}{4}x - \frac{1}{3} > -\frac{1}{5}$ and $\frac{1}{2}x < 2$

35. $0.5x < 2$ and $-0.6x < -3$

36. $0.3x < 0.6$ or $0.05x > -4$

37. $5 + 4(x - 1) < 3$ and $1 + 3(x - 1) < 1$

38. $3 - 2(x - 4) > 1$ and $2(x - 2) > x - 2$

39. $0.2(x - 3) > 0.5$ or $0.05x < 2$

40. $0.1(x - 3) > 0.3$ or $0.01x < 2$

Solve each compound inequality. Write the solution set in interval notation and graph it. See Examples 10 and 11.

41. $5 < 2x - 3 < 11$

42. $-2 < 3x + 1 < 10$

43. $-1 < 5 - 3x \le 14$

44. $-1 \le 3 - 2x < 11$

45. $-3 < \dfrac{3m + 1}{2} \le 5$

46. $0 \le \dfrac{3 - 2x}{2} < 5$

47. $-2 < \dfrac{1 - 3x}{-2} < 7$

48. $-3 < \dfrac{2x - 1}{3} < 7$

49. $3 \le 3 - 5(x - 3) \le 8$

50. $2 \le 4 - \dfrac{1}{2}(x - 8) \le 10$

51. $-1.2 \le 1 - 0.02(x - 6.1) < 1.4$

52. $0.58 < 1.2 + 3.1(1 - 0.5x) \le 1.44$

Write each union or intersection of intervals as a single interval if possible.

53. $(2, \infty) \cup (4, \infty)$

54. $(-3, \infty) \cup (-6, \infty)$

55. $(-\infty, 5) \cap (-\infty, 9)$

56. $(-\infty, -2) \cap (-\infty, 1)$

57. $(-\infty, 4] \cap [2, \infty)$

58. $(-\infty, 8) \cap [3, \infty)$

59. $(-\infty, 5) \cup [-3, \infty)$

60. $(-\infty, -2] \cup (2, \infty)$

61. $(3, \infty) \cap (-\infty, 3]$

62. $[-4, \infty) \cap (-\infty, -6]$

63. $(3, 5) \cap [4, 8)$

64. $[-2, 4] \cap (0, 9]$

65. $[1, 4) \cup (2, 6]$

66. $[1, 3) \cup (0, 5)$

Write either a simple or a compound inequality that has the given graph as its solution set.

67.

68.

69.

70.

71.

72.

73.

74.

75.

76.

Solve each compound inequality and write the solution set using interval notation.

77. $2 < x < 7$ and $2x > 10$

78. $3 < 5 - x < 8$ or $-3x < 0$

79. $-1 < 3x + 2 \le 5$ or $\frac{3}{2}x - 6 > 9$

80. $0 < 5 - 2x \le 10$ and $-6 < 4 - x < 0$

81. $-3 < \dfrac{x - 1}{2} < 5$ and $-1 < \dfrac{1 - x}{2} < 2$

82. $-3 < \dfrac{3x - 1}{5} < \dfrac{1}{2}$ and $\dfrac{1}{3} < \dfrac{3 - 2x}{6} < \dfrac{9}{2}$

83. $-0.02 \le \dfrac{0.05x}{8} < 0.03$ or $-4.07 \le \dfrac{5.3 - 2.1x}{2} < 1.39$

84. $0.35 \le 1 - \dfrac{x}{2} < 1.6$ or $-0.175 \le \dfrac{x}{4} - 0.5 \le 0.9$

Solve each problem by using a compound inequality. See Example 12.

85. Aiming for a C. Professor Johnson gives only a midterm exam and a final exam. The semester average is computed by taking $\frac{1}{3}$ of the midterm exam score plus $\frac{2}{3}$ of the final exam score. To get a C, Beth must have a semester average between 70 and 79 inclusive. If Beth scored only 64 on the midterm, then for what range of scores on the final exam would Beth get a C?

86. Two tests only. Professor Davis counts his midterm as $\frac{2}{3}$ of the grade, and his final as $\frac{1}{3}$ of the grade. Jason scored only 64 on the midterm. What range of scores on the final exam would put Jason's average between 70 and 79 inclusive?

87. Keep on truckin'. Abdul is shopping for a new truck in a city with an 8% sales tax. There is also an $84 title and license fee to pay. He wants to get a good truck, and he plans to spend at least $12,000 but not more than $15,000. What is the price range for the truck?

88. Selling-price range. Renee wants to sell her car through a broker who charges a commission of 10% of the selling price. The book value of the car is $14,900, but Renee still owes $13,104 on it. Although the car is in only fair condition and will not sell for more than the book value, Renee must get enough to at least pay off the loan. What is the range of the selling price?

89. Hazardous to her health. Trying to break her smoking habit, Jane calculates that she smokes only three full cigarettes a day, one after each meal. The rest of the time she smokes on the run and smokes only half of the cigarette. She estimates that she smokes the equivalent of 5 to 12 cigarettes per day. How many times a day does she light up on the run?

90. Possible width. The length of a rectangle is 20 meters longer than the width. The perimeter must be between 80 and 100 meters. What are the possible values for the width of the rectangle?

91. Higher education. The number of degrees awarded in U.S. higher education is expected to increase steadily into the twenty-first century (National Center for Education Statistics). The formulas

$B = 16.45n + 980.20$ and $M = 7.79n + 287.87$ can be used to approximate the number of bachelor's and master's degrees in thousands, respectively, awarded in the year $1985 + n$. What is the first year in which both B is greater than 1.3 million and M is greater than 0.5 million? What is the first year in which either B is greater than 1.3 million or M is greater than 0.5 million?

92. Senior citizens. The number of senior citizens (65 years old and over) in the United States has been increasing steadily since 1970 (U.S. Bureau of the Census). The size of this group in millions in the year $1970 + n$ can be estimated by using the formula $S = 0.48n + 19.71$. The percentage of senior citizens living below the poverty level has been decreasing steadily from the 1970 figure of 24.2%. The formula $p = -0.72n + 24.2$ can be used to estimate that percentage in the year $1970 + n$. What is the first year in which we can expect both the number of senior citizens to be greater than 36 million and fewer than 5% living below the poverty level?

Getting More Involved

93. Discussion. If $-x$ is between a and b, then what can you say about x?

94. Discussion. For which of the inequalities is the notation used correctly?

a) $-2 \le x < 3$ b) $-4 \ge x < 7$
c) $-1 \le x > 0$ d) $6 < x \le -8$
e) $5 \ge x \ge -9$

95. Discussion. In each case, write the resulting set of numbers in interval notation. Explain your answers.

a) Every number in $(3, 8)$ is multiplied by 4.
b) Every number in $[-2, 4)$ is multiplied by -5.
c) Three is added to every number in $(-3, 6)$.
d) Every number in $[3, 9]$ is divided by -3.

96. Discussion. Write the solution set using interval notation for each of the following inequalities in terms of s and t. State any restrictions on s and t. For what values of s and t is the solution set empty?

a) $x > s$ and $x < t$ b) $x > s$ and $x > t$

2.6 Absolute Value Equations and Inequalities

In this section:

▶ Absolute Value Equations
▶ Absolute Value Inequalities
▶ Applications

In Chapter 1 we learned that absolute value measures the distance of a number from 0 on the number line. In this section we will learn to solve equations and inequalities involving absolute value.

Absolute Value Equations

Solving equations involving absolute value requires some techniques that are different from those studied in previous sections. For example, the solution set to the equation

$$|x| = 5$$

is $\{-5, 5\}$, because both 5 and -5 are five units from 0 on the number line, as shown in Fig. 2.21. So $|x| = 5$ is equivalent to the compound equation

$$x = 5 \text{ or } x = -5.$$

Figure 2.21

The equation $|x| = 0$ is equivalent to the equation $x = 0$, because 0 is the only number whose distance from 0 is zero. The solution set to $|x| = 0$ is $\{0\}$.

The equation $|x| = -7$ is inconsistent because absolute value measures distance, and distance is never negative. So the solution set is empty. These ideas are summarized as follows.

Basic Absolute Value Equations		
Absolute Value Equation	**Equivalent Equation**	**Solution Set**
$\|x\| = k \, (k > 0)$	$x = k \text{ or } x = -k$	$\{k, -k\}$
$\|x\| = 0$	$x = 0$	$\{0\}$
$\|x\| = k \, (k < 0)$		$\varnothing$

We can use these ideas to solve more complicated absolute value equations.

EXAMPLE 1   **Absolute value equations**

Solve each equation.

a) $|x - 7| = 2$ **b)** $|3x - 5| = 7$

Solution

a) First rewrite $|x - 7| = 2$ without absolute value:

$$x - 7 = 2 \quad \text{or} \quad x - 7 = -2 \quad \text{Equivalent equation}$$
$$x = 9 \quad \text{or} \quad x = 5$$

The solution set is $\{5, 9\}$. The distance from 5 to 7 or from 9 to 7 is 2 units.

b) First rewrite $|3x - 5| = 7$ without absolute value:

$$3x - 5 = 7 \quad \text{or} \quad 3x - 5 = -7 \quad \text{Equivalent equation}$$
$$3x = 12 \quad \text{or} \quad 3x = -2$$
$$x = 4 \quad \text{or} \quad x = -\frac{2}{3}$$

The solution set is $\left\{-\frac{2}{3}, 4\right\}$. ∎

EXAMPLE 2 **Absolute value equal to zero**

Solve $|2(x - 6) + 7| = 0$.

Solution

Since 0 is the only number whose absolute value is 0, the expression within the absolute value bars must be 0.

$$2(x - 6) + 7 = 0 \quad \text{Equivalent equation}$$
$$2x - 12 + 7 = 0$$
$$2x - 5 = 0$$
$$2x = 5$$
$$x = \frac{5}{2}$$

The solution set is $\left\{\frac{5}{2}\right\}$. ∎

EXAMPLE 3 **Isolating the absolute value**

Solve $-5|3x - 6(5 - x)| + 4 = 14$.

Solution

First subtract 4 from each side to isolate the absolute value expression:

$$-5|3x - 6(5 - x)| + 4 = 14$$
$$-5|3x - 6(5 - x)| = 10 \quad \text{Subtract 4 from each side.}$$
$$|3x - 6(5 - x)| = -2 \quad \text{Divide each side by } -5.$$

There is no solution, because no number has a negative absolute value. ∎

The equation in the next example has absolute value on both sides.

EXAMPLE 4 **Absolute value on both sides**

Solve $|2x - 1| = |x + 3|$.

Solution

Two quantities have the same absolute value only if they are equal or opposites. So we can write an equivalent compound equation:

$$2x - 1 = x + 3 \quad \text{or} \quad 2x - 1 = -(x + 3)$$
$$x - 1 = 3 \quad \text{or} \quad 2x - 1 = -x - 3$$
$$x = 4 \quad \text{or} \quad 3x = -2$$
$$x = 4 \quad \text{or} \quad x = -\frac{2}{3}$$

Check that both 4 and $-\frac{2}{3}$ satisfy the original equation. The solution set is $\left\{ -\frac{2}{3}, 4 \right\}$. ■

Absolute Value Inequalities

Since absolute value measures distance from 0 on the number line, $|x| > 5$ indicates that x is more than five units from 0. Any number on the number line to the right of 5 or to the left of -5 is more than five units from 0. So $|x| > 5$ is equivalent to

$$x > 5 \text{ or } x < -5.$$

The solution set to this inequality is the union of the solution sets to the two simple inequalities. The solution set is $(-\infty, -5) \cup (5, \infty)$. The graph of $|x| > 5$ is shown in Fig. 2.22.

Figure 2.22

The inequality $|x| \leq 3$ indicates that x is less than or equal to three units from 0. Any number between -3 and 3 inclusive satisfies that condition. So $|x| \leq 3$ is equivalent to

$$-3 \leq x \leq 3.$$

The graph of $|x| \leq 3$ is shown in Figure 2.23. These examples illustrate the basic types of absolute value inequalities.

Figure 2.23

Basic Absolute Value Inequalities (k > 0)			
Absolute Value Inequality	Equivalent Inequality	Solution Set	Graph of Solution Set
$\|x\| > k$	$x > k$ or $x < -k$	$(-\infty, -k) \cup (k, \infty)$	
$\|x\| \geq k$	$x \geq k$ or $x \leq -k$	$(-\infty, -k] \cup [k, \infty)$	
$\|x\| < k$	$-k < x < k$	$(-k, k)$	
$\|x\| \leq k$	$-k \leq x \leq k$	$[-k, k]$	

We can solve more complicated inequalities in the same manner as simple ones.

EXAMPLE 5 **Absolute value inequality**

Solve $|x - 9| < 2$ and graph the solution set.

Solution

Because $|x| < k$ is equivalent to $-k < x < k$, we can rewrite $|x - 9| < 2$ as follows:

$$-2 < x - 9 < 2$$
$$-2 + 9 < x - 9 + 9 < 2 + 9 \qquad \text{Add 9 to each part of the inequality.}$$
$$7 < x < 11$$

The graph of the solution set $(7, 11)$ is shown in Figure 2.24. Note that the graph consists of all real numbers that are within two units of 9.

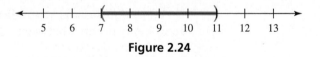

Figure 2.24

EXAMPLE 6 **Absolute value inequality**

Solve $|3x + 5| > 2$ and graph the solution set.

Solution

$$
\begin{array}{lcll}
3x + 5 > 2 & \text{or} & 3x + 5 < -2 & \text{Equivalent compound inequality} \\
3x > -3 & \text{or} & 3x < -7 & \\
x > -1 & \text{or} & x < -\dfrac{7}{3} &
\end{array}
$$

The solution set is $\left(-\infty, -\frac{7}{3}\right) \cup (-1, \infty)$, and its graph is shown in Fig. 2.25.

Figure 2.25

EXAMPLE 7   **Absolute value inequality**

Solve $|5 - 3x| \leq 6$ and graph the solution set.

Solution

$$-6 \leq 5 - 3x \leq 6 \qquad \text{Equivalent inequality.}$$

$$-11 \leq -3x \leq 1 \qquad \text{Subtract 5 from each part.}$$

$$\frac{11}{3} \geq x \geq -\frac{1}{3} \qquad \text{Divide by } -3 \text{ and reverse each inequality symbol.}$$

$$-\frac{1}{3} \leq x \leq \frac{11}{3} \qquad \text{Write } -\frac{1}{3} \text{ on the left, because it is smaller than } \frac{11}{3}.$$

The solution set is $\left[-\frac{1}{3}, \frac{11}{3}\right]$ and its graph is shown in Fig. 2.26.

Figure 2.26

There are a few absolute value inequalities that do not fit the above categories. They are easy to solve using the definition of absolute value.

EXAMPLE 8   **Special case**

Solve $3 + |7 - 2x| \geq 3$.

Solution

Subtract 3 from each side to isolate the absolute value expression.

$$|7 - 2x| \geq 0$$

Because the absolute value of any real number is greater than or equal to 0, the solution set is R, the set of all real numbers.

EXAMPLE 9   **An impossible case**

Solve $|5x - 12| < -2$.

Solution

We write an equivalent inequality only when the value of k is positive. With -2 on the right side, we do not write an equivalent inequality. Since the absolute value of any quantity is greater than or equal to 0, no value for x can make this absolute value less than -2. The solution set is $\varnothing$, the empty set.

Applications

A simple example will show how absolute value inequalities can be used in applications.

EXAMPLE 10 **Controlling water temperature**

The water temperature in a certain manufacturing process must be kept at 143°F. The computer is programmed to shut the process down if the water temperature is more than 7 degrees away from what it is supposed to be. For what temperature readings is the process shut down?

Solution

If we let x represent the water temperature, then $x - 143$ represents the difference between the actual temperature and the desired temperature. The quantity $x - 143$ could be positive or negative. The process is shut down if the absolute value of $x - 143$ is greater than 7.

$$|x - 143| > 7$$
$$x - 143 > 7 \quad \text{or} \quad x - 143 < -7$$
$$x > 150 \quad \text{or} \quad x < 136$$

The process is shut down for temperatures greater than 150°F or less than 136°F. ∎

Warm-ups

True or false? Explain your answer.

1. The equation $|x| = 2$ is equivalent to $x = 2$ or $x = -2$.
2. All absolute value equations have two solutions.
3. The equation $|2x - 3| = 7$ is equivalent to $2x - 3 = 7$ or $2x + 3 = 7$.
4. The inequality $|x| > 5$ is equivalent to $x > 5$ or $x < -5$.
5. The equation $|x| = -5$ is equivalent to $x = 5$ or $x = -5$.
6. There is only one solution to the equation $|3 - x| = 0$.
7. We should write the inequality $x > 3$ or $x < -3$ as $3 < x < -3$.
8. The inequality $|x| < 7$ is equivalent to $-7 \le x \le 7$.
9. The equation $|x| + 2 = 5$ is equivalent to $|x| = 3$.
10. The inequality $|x| < -2$ is equivalent to $x < 2$ and $x > -2$.

2.6 EXERCISES

Solve each absolute value equation. See Examples 1–3.

1. $|a| = 5$
2. $|x| = 2$
3. $|x - 3| = 1$
4. $|x - 5| = 2$
5. $|3 - x| = 6$
6. $|7 - x| = 6$
7. $|3x - 4| = 12$
8. $|5x + 2| = -3$
9. $\left|\frac{2}{3}x - 8\right| = 0$
10. $\left|3 - \frac{3}{4}x\right| = \frac{1}{4}$
11. $|6 - 0.2x| = 10$
12. $|5 - 0.1x| = 0$
13. $|7(x - 6)| = -3$
14. $|2(a + 3)| = 15$
15. $|2(x - 4) + 3| = 5$
16. $|3(x - 2) + 7| = 6$
17. $|7.3x - 5.26| = 4.215$
18. $|5.74 - 2.17x| = 10.28$

Solve each absolute value equation. See Examples 3 and 4.

19. $3 + |x| = 5$

20. $|x| - 10 = -3$

21. $2 - |x + 3| = -6$

22. $4 - 3|x - 2| = -8$

23. $5 - \dfrac{|3 - 2x|}{3} = 4$

24. $3 - \dfrac{1}{2}\left|\dfrac{1}{2}x - 4\right| = 2$

25. $|x - 5| = |2x + 1|$

26. $|w - 6| = |3 - 2w|$

27. $\left|\dfrac{5}{2} - x\right| = \left|2 - \dfrac{x}{2}\right|$

28. $\left|x - \dfrac{1}{4}\right| = \left|\dfrac{1}{2}x - \dfrac{3}{4}\right|$

29. $|x - 3| = |3 - x|$

30. $|a - 6| = |6 - a|$

31. $|0.5w + 3| = |0.5w - 3|$

32. $|0.5m + 1.25| = |0.5m - 1.25|$

Write an absolute value inequality whose solution set is shown by the graph. See Examples 5–7.

33.

34.

35.

36.

37.

38.

39.

40.

Determine whether each absolute value inequality is equivalent to the inequality following it. See Examples 5–7.

41. $|x| < 3, \ x < 3$

42. $|x| > 3, \ x > 3$

43. $|x - 3| > 1, \ x - 3 > 1$ or $x - 3 < -1$

44. $|x - 3| \le 1, \ -1 \le x - 3 \le 1$

45. $|x - 3| \ge 1, \ x - 3 \ge 1$ or $x - 3 \le 1$

46. $|x - 3| > 0, \ x - 3 > 0$

47. $|4 - x| < 1, \ 4 - x < 1$ and $-(4 - x) < 1$

48. $|4 - x| > 1, \ 4 - x > 1$ or $-(4 - x) > 1$

49. $|4 - x| \le 1, \ -1 < 4 - x < 1$

50. $|4 - x| \ge 1, \ 4 - x > 1$ or $4 - x \le -1$

Solve each absolute value inequality and graph the solution set. See Examples 5–7.

51. $|x| > 6$

52. $|w| > 3$

53. $|2a| < 6$

54. $|3x| < 21$

55. $|x - 2| \ge 3$

56. $|x - 5| \ge 1$

57. $\dfrac{1}{5}|2x - 4| < 1$

58. $\dfrac{1}{3}|2x - 1| < 1$

59. $-2|5 - x| \ge -14$

60. $-3|6 - x| \ge -3$

61. $2|3 - 2x| - 6 \ge 18$

62. $2|5 - 2x| - 15 \ge 5$

63. $\left|\dfrac{2w + 7}{3}\right| > 1$

64. $\left|\dfrac{2y + 3}{2}\right| > 1$

Solve each absolute value inequality and graph the solution set. See Examples 8 and 9.

65. $|x - 2| > 0$

66. $|6 - x| \ge 0$

67. $|x - 5| \ge 0$

68. $|3x - 7| \ge -3$

69. $-2|3x - 7| > 6$

70. $-3|7x - 42| > 18$

71. $|2x + 3| + 6 > 0$

72. $|5 - x| + 5 > 5$

Solve each inequality. Write the solution set using interval notation.

73. $1 < |x + 2|$

74. $5 \ge |x - 4|$

75. $5 > |x| + 1$

76. $4 \le |x| - 6$

77. $3 - 5|x| > -2$

78. $1 - 2|x| < -7$

79. $|5.67x - 3.124| < 1.68$

80. $|4.67 - 3.2x| \ge 1.43$

81. $|2x - 1| < 3$ and $2x - 3 > 2$

82. $|5 - 3x| \ge 3$ and $5 - 2x > 3$

83. $|x - 2| < 3$ and $|x - 7| < 3$

84. $|x - 5| < 4$ and $|x - 6| > 2$

Solve each equation or inequality.

85. $|t - 5| = t$

86. $|w - 6| = w + 2$

87. $|7 - w| = w + 5$

88. $|w + 2| = -w$

89. $x + |x| = 3$

90. $x - |x| = 0$

91. $|x| > x$

92. $|x| > x + 6$

93. $|x - 5| < 2x + 1$

94. $|x - 2| > 1 - 3x$

Solve each problem by using an absolute value equation or inequality. See Example 10.

95. Famous battles. In the Hundred Years' War, Henry V defeated a French army in the battle of Agincourt and Joan of Arc defeated an English army in the battle of Orleans *(The Doubleday Almanac).* Suppose that you know only that these two famous battles were 14 years apart and that the battle of Agincourt occurred in 1415. Use an absolute value equation to find the possibilities for the year in which the battle of Orleans occurred?

96. World records. In July 1985, Steve Cram of Great Britain set a world record of 3 min 29.67 sec for the 1500-meter race and a world record of 3 min 46.31 sec for the 1-mile race *(The Doubleday Almanac).* Suppose that you know only that these two events occurred 11 days apart and that the 1500 meter record was set on July 16. Use an absolute value equation to find the possible dates for the 1-mile record run?

97. Weight difference. Research at a major university has shown that identical twins generally differ by less than 6 pounds in body weight. If Kim weighs 127 pounds, then in what range is the weight of her identical twin sister Kathy?

98. Intelligence quotient. Jude's IQ score is more than 15 points away from Sherry's. If Sherry scored 110, then in what range is Jude's score?

99. Unidentified flying objects. Two balls are tossed into the air simultaneously, one from the ground at 50 feet per second and one from a height of 10 feet at 40 feet per second. For what values of t will their heights above the ground differ by less than 5 feet (while they are both in the air)? The formula $S = -16t^2 + v_0 t + s_0$ gives the height in feet above the earth at time t seconds for an object projected into the air with initial velocity v_0 feet per second from initial height s_0 feet.

100. Playing catch. A circus clown at the top of a 60-foot platform is playing catch with another clown on the ground. The clown on the platform drops a ball at the same time as the one on the ground tosses a ball upward at 80 ft per second. For what length of time is the distance between the balls less than or equal to 10 feet? *(Hint:* Use the formula given in the Exercise 99. The initial velocity of a ball that is dropped is 0 ft/sec.)

0 ft/sec

80 ft/sec

60 ft

Figure for Exercise 100

Getting More Involved

101. Discussion. For which real numbers m and n is each equation satisfied?

a) $|m - n| = |n - m|$

b) $|mn| = |m| \cdot |n|$

c) $\left|\dfrac{m}{n}\right| = \dfrac{|m|}{|n|}$

102. Exploration. a) Evaluate $|m + n|$ and $|m| + |n|$ for

i) $m = 3$ and $n = 5$

ii) $m = -3$ and $n = 5$

iii) $m = 3$ and $n = -5$

iv) $m = -3$ and $n = -5$

b) What can you conclude about the relationship between $|m + n|$ and $|m| + |n|$?

──────────────COLLABORATIVE ACTIVITIES──────────────

Everyday Algebra

Grouping: 2 to 4 students per group
Topic: Use of algebra in common occurrences

Every day, people use algebra without even knowing it. Any time you solve for an unknown quantity, you are using algebra. Here is an example of a simple problem that you could solve without even thinking of algebra.

• While shopping, Joe notices a store brand available at a lower price than the name brand. If the name brand product costs $3.79 and Joe has a coupon for $0.50, what would the store brand price need to be for Joe to save money with the coupon?

This is a problem you could solve mentally by subtracting the two quantities and finding that if the store brand costs $3.29 or more, Joe would save money with the coupon. If the store brand is less than $3.29, then Joe would save money by buying it instead. The beauty of algebra is not apparent in cases like this because you already know what operation to perform to find the unknown quantity. Algebra becomes useful when it is not clear what to do.

We will consider another situation in which we want to find the best price of an item. For this situation we will use the following formulas.

The formula for markup of an item is

$$P = C + rC,$$

where P is the price of the item, C is the wholesale cost of the item, and r is the percent of markup.

The formula for discounting an item is

$$S = P - dP,$$

where S is the discounted price and d is the percent discount.

• Lane belongs to a wholesale buying club. She can order items through the club with a markup of 8% above the wholesale cost. She also can buy the same items in a store where she can get a 10% discount off of the shelf price. Lane wants to know what the store markup on any particular item must be for it to be cheaper to order through the club.

Form groups of two to four people. Assign a role to each person: **Recorder, Moderator, Messenger,** or **Quality Manager** (roles may be combined if there are fewer than four people in a group). See the Preface for a description of these roles. In your groups:

1. Decide how to rewrite Lane's problem using algebra. Decide what the unknown quantities are and assign them variable names.

2. Write the equations or inequalities on your paper using the variables you defined above.

3. Solve the problem and state the group's decision on what Lane should do.

Extension: Pick a similar problem from your own lives to use algebra to solve.

Wrap-up   CHAPTER 2

SUMMARY

Equations		Examples
Solution set	The set of all numbers that satisfy an equation (or inequality)	$x + 2 = 6$ has solution set {4}.
Equivalent equations	Equations with the same solution set	$2x + 1 = 5$ $2x = 4$

Properties of equality	We may perform the same operation $(+, -, \cdot, \div)$ with the same real number on each side of an equation without changing the solution set (excluding multiplication and division by 0).	$x = 4$ $x + 1 = 5$ $x - 1 = 3$ $2x = 8$ $\dfrac{x}{2} = 2$
Identity	An equation that is satisfied by every number for which both sides are defined	$x + x = 2x$
Conditional equation	An equation whose solution set contains at least one real number but is not an identity	$5x - 10 = 0$
Inconsistent equation	An equation whose solution set is $\varnothing$	$x = x + 1$
Linear equation in one variable	An equation of the form $ax + b = 0$ with $a \neq 0$	$3x + 8 = 0$ $5x - 1 = 2x - 9$
Strategy for solving a linear equation	1. If fractions are present, multiply each side by the LCD to eliminate the fractions. 2. Use the distributive property to remove parentheses. 3. Combine any like terms. 4. Use the addition and subtraction properties to get all variables on one side and numbers on the other. 5. Use the multiplication or division property to get a single variable on one side. 6. Check your work by replacing the variable in the original equation with your solution.	
Strategy for solving word problems	1. Read the problem until you understand the problem. 2. If possible, draw a diagram to illustrate the problem. 3. Choose a variable and write down what it represents. 4. Represent any other unknowns in terms of that variable. 5. Write an equation that models the situation. 6. Solve the equation. 7. Be sure that your solution answers the question posed in the original problem. 8. Check your answer by using it to solve the original problem (not the equation).	

Inequalities		**Examples**
Linear inequality in one variable	Any inequality of the form $ax + b < 0$ with $a \neq 0$ In place of $<$ we can use $\leq$, $>$ or $\geq$.	$2x + 9 < 0$ $x - 2 \geq 7$ $-3x - 1 \geq 2x + 5$
Properties of inequality	We may perform the same operation $(+, -, \cdot, \div)$ on each side of an inequality just as we do in solving equations, with one exception: When multiplying or dividing by a negative number, the inequality symbol is reversed.	$-3x > 6$ $x < -2$

Trichotomy property	For any two real numbers a and b, exactly one of the following statements is true: $a < b$, $a = b$, or $a > b$.	If w is not greater than 7, then $w \le 7$.
Compound inequality	Two simple inequalities connected with the word "and" or "or" *And* corresponds to *intersection.* *Or* corresponds to *union.*	$x > 1$ and $x < 5$ $x > 3$ or $x < 1$

Absolute value

	Absolute Value Equation	Equivalent Equation	Solution Set	
Basic absolute value equations	$\|x\| = k \ (k > 0)$ $\|x\| = 0$ $\|x\| = k \ (k < 0)$	$x = k$ or $x = -k$ $x = 0$	$\{k, -k\}$ $\{0\}$ $\varnothing$	

	Absolute Value Inequality	Equivalent Inequality	Solution Set	Graph of Solution Set
Basic absolute value inequalities $(k > 0)$	$\|x\| > k$	$x > k$ or $x < -k$	$(-\infty, -k) \cup (k, \infty)$	
	$\|x\| \ge k$	$x \ge k$ or $x \le -k$	$(-\infty, -k] \cup [k, \infty)$	
	$\|x\| < k$	$-k < x < k$	$(-k, k)$	
	$\|x\| \le k$	$-k \le x \le k$	$[-k, k]$	

REVIEW EXERCISES

2.1 *Solve each equation.*

1. $2x - 7 = 9$

2. $5x - 7 = 38$

3. $11 = 5 - 4x$

4. $-8 = 7 - 3x$

5. $x - 6 - (x - 6) = 0$

6. $x - 6 - 2(x - 3) = 0$

7. $2(x - 3) - 5 = 5 - (3 - 2x)$

8. $2(x - 4) + 5 = -(3 - 2x)$

9. $\dfrac{3}{17}x = 0$

10. $-\dfrac{3}{8}x = \dfrac{1}{2}$

11. $\dfrac{1}{4}x - \dfrac{1}{5} = \dfrac{1}{5}x + \dfrac{4}{5}$

12. $\dfrac{1}{2}x - 1 = \dfrac{1}{3}x$

13. $\dfrac{t}{2} - \dfrac{t - 2}{3} = \dfrac{3}{2}$

14. $\dfrac{y + 1}{4} - \dfrac{y - 1}{6} = y + 5$

15. $1 - 0.4(x - 4) + 0.6(x - 7) = -0.6$

16. $0.04x - 0.06(x - 8) = 0.1x$

2.2 *Solve each equation for x.*

17. $ax + b = 0$

18. $mx + c = d$

19. $ax + 2 = cx$

20. $mx = 3 - x$

21. $mwx = P$

22. $xyz = 2$

23. $\dfrac{1}{x} + \dfrac{1}{2} = w$

24. $\dfrac{1}{x} + \dfrac{1}{a} = 2$

Solve each equation for y.

25. $3x - 2y = -6$

26. $4x - 3y + 9 = 0$

27. $y - 2 = -\dfrac{1}{3}(x - 6)$

28. $y + 6 = \dfrac{1}{2}(x - 4)$

29. $\dfrac{1}{2}x - \dfrac{1}{4}y = 5$

30. $-\dfrac{x}{3} + \dfrac{y}{2} = \dfrac{5}{8}$

2.3 *Solve each problem.*

31. Legal pad. If the perimeter of a legal-size note pad is 45 inches and the pad is 5.5 inches longer than it is wide, then what are its length and width?

32. Area of a trapezoid. The height of a trapezoid is 5 feet, and the upper base is 2 feet shorter than the lower base. If the area of the trapezoid is 45 square feet, then how long is the lower base?

x − 2 ft

5 ft

x ft

Figure for Exercise 32

33. Saving for retirement. Roy makes $8,000 more per year than his wife. Roy saves 10% of his income for retirement, and his wife saves 8%. Together they save $5,660 per year. How much does each make?

34. Charitable contributions. Duane makes $1,000 less per year than his wife. Duane gives 5% of his income to charity, and his wife gives 10% of her income to charity. Together they contribute $2,500 to charity. How much does each make?

35. Dealer discounts. Sarah is buying a car for $7,600. The dealer gave her a 20% discount off the list price. What was the list price?

36. Gold sale. At 25% off, a jeweler is selling a gold chain for $465. What was the list price?

37. Nickels and dimes. Rebecca has 15 coins consisting of dimes and nickels. The total value of the coins is $0.95. How many of each does she have?

38. Nickels, dimes, and quarters. Camille has 19 coins consisting of nickels, dimes, and quarters. The value of the coins is $1.60. If she has six times as many nickels as quarters, then how many of each does she have?

39. Tour de desert. On a recent bicycle trip across the desert, Barbara rode for 5 hours. Her bicycle then developed mechanical difficulties, and she walked the bicycle for 3 hours to the nearest town. Altogether, she covered 85 miles. If she rides 9 miles per hour faster than she walks, then how far did she walk?

40. Motor city. Delmas flew to Detroit in 90 minutes and drove his new car back home in 6 hours. If he drove 150 miles per hour slower than he flew, then how fast did he fly?

Speed to Detroit = *x* mph

Speed from Detroit = *x* − 150 mph

Figure for Exercise 40

2.4 *Solve each inequality. State the solution set using interval notation and graph it.*

41. $3 - 4x < 15$　　　　**42.** $5 - 6x > 35$

43. $2(x - 3) > -6$　　　**44.** $4(5 - x) < 20$

45. $-\frac{3}{4}x \geq 6$　　　　**46.** $-\frac{2}{3}x \leq 4$

47. $3(x + 2) > 5(x - 1)$　**48.** $4 - 2(x - 3) < 0$

49. $\frac{1}{2}x + 7 \leq \frac{3}{4}x - 5$　**50.** $\frac{5}{6}x - 3 \geq \frac{2}{3}x + 7$

2.5 *Solve each compound inequality. State the solution set using interval notation and graph it.*

51. $x + 2 > 3$ or $x - 6 < -10$

52. $x - 2 > 5$ or $x - 2 < -1$

53. $x > 0$ and $x - 6 < 3$　**54.** $x \leq 0$ and $x + 6 > 3$

55. $6 - x < 3$ or $-x < 0$　**56.** $-x > 0$ or $x + 2 < 7$

57. $2x < 8$ and $2(x - 3) < 6$

58. $\frac{1}{3}x > 2$ and $\frac{1}{4}x > 2$

59. $x - 6 > 2$ and $6 - x > 0$

60. $-\frac{1}{2}x < 6$ or $\frac{2}{3}x < 4$

61. $0.5x > 10$ or $0.1x < 3$

62. $0.02x > 4$ and $0.2x < 3$

63. $-2 \leq \dfrac{2x - 3}{10} \leq 1$　**64.** $-3 < \dfrac{4 - 3x}{5} < 2$

Write each union or intersection of intervals as a single interval.

65. $[1, 4) \cup (2, \infty)$　　**66.** $(2, 5) \cup (-1, \infty)$

67. $(3, 6) \cap [2, 8]$　　　**68.** $[-1, 3] \cap [0, 8]$

69. $(-\infty, 5) \cup [5, \infty)$　**70.** $(-\infty, 1) \cup (0, \infty)$

71. $(-3, -1] \cap [-2, 5]$　**72.** $[-2, 4] \cap (4, 7]$

2.6 *Solve each absolute value equation or inequality and graph the solution set.*

73. $|2x| \geq 8$

74. $|5x - 1| \leq 14$

75. $|x| = -5$

76. $|x| + 2 = 16$

77. $\left|1 - \dfrac{x}{5}\right| > \dfrac{9}{5}$

78. $\left|1 - \dfrac{1}{6}x\right| < \dfrac{1}{2}$

79. $|x - 3| < -3$

80. $|x - 7| \leq -4$

81. $\left|\dfrac{x}{2}\right| - 5 = -1$

82. $\left|\dfrac{x}{2} - 5\right| = -1$

83. $|x + 4| \geq -1$

84. $|6x - 1| \geq 0$

85. $1 - \dfrac{3}{2}|x - 2| < -\dfrac{1}{2}$

86. $1 > \dfrac{1}{2}|6 - x| - \dfrac{3}{4}$

Miscellaneous

Solve each problem by using equations or inequalities.

87. *Rockbuster video.* Stephen plans to open a video rental store. Industry statistics show that 45% of the rental price goes for overhead. If the maximum that anyone will pay to rent a tape is $5 and Stephen wants a profit of at least $1.65 per tape, then in what range should the rental price be?

88. *Working girl.* Regina makes $6.80 per hour working in the snack bar. To keep her grant, she may not earn more than $51 per week. What is the range of the number of hours per week that she may work?

89. *Skeletal remains.* Forensic scientists use the formula $h = 60.089 + 2.238F$ to predict the height h (in centimeters) for a male whose femur measures F centimeters. (See the accompanying figure.) In what range is the length of the femur for males between 150 cm and 180 cm in height? Round to the nearest tenth of a centimeter.

Figure for Exercises 89 and 90

90. *Female femurs.* Forensic scientists use the formula $h = 61.412 + 2.317F$ to predict the height h (in cen-

timeters) for a female whose femur measures F centimeters. In what range is the length of the femur for females who are over 170 centimeters tall?

91. *Car trouble.* Dane's car was found abandoned at mile marker 86 on the interstate. If Dane was picked up by the police on the interstate exactly 5 miles away, then at what mile marker was he picked up?

92. *Comparing scores.* Scott scored 72 points on the midterm, and Katie's score was more than 16 points away from Scott's. What was Katie's score?

93. *Year-end bonus.* A law firm has agreed to distribute 20% of its profits to its employees as a year-end bonus. To the firm's accountant, the bonus is an expense that must be used to determine the profit. That is, bonus = 20% × (profit before bonus − bonus). Given that the profit before the bonus is $300,000, find the amount of the bonus using the accountant's method. How does this answer compare to 20% of $300,000, which is what the employees want?

94. *Higher rate.* Suppose that the employees in Exercise 93 got the bonus that they wanted. To the accountant, what percent of the profits was given in bonuses?

95. *Dairy cattle.* Thirty percent of the dairy cattle in Washington County are Holsteins, while 60% of the dairy cattle in neighboring Cade County are Holsteins. In the combined two-county area, 50% of the 3,600 dairy cattle are Holsteins. How many dairy cattle are in each county?

96. *Profitable business.* United Home Improvement makes 20% profit on its good grade of vinyl siding, 30% profit on its better grade, and 60% profit on its best grade. So far this year, UHI has $40,000 in sales of good siding and $50,000 in sales of better siding. The company goal is to have at least an overall profit of 50% of total sales. What would the sales figures for the best grade of siding have to be to reach this goal?

For each graph in Exercises 97–114, write an equation or inequality that has the solution set shown by the graph. Use absolute value when possible.

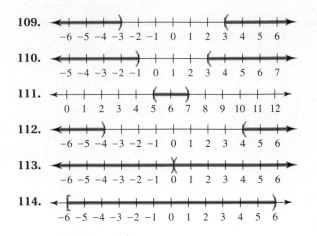

CHAPTER 2 TEST

Solve each equation.

1. $-10x - 5 + 4x = -4x + 3$

2. $\dfrac{y}{2} - \dfrac{y-3}{3} = \dfrac{y+6}{6}$

3. $|w| + 3 = 9$

4. $|3 - 2(5 - x)| = 3$

Solve for the indicated variable.

5. $2x - 5y = 20$ for y

6. $a = axt + S$ for a

Solve each inequality. State the solution set using interval notation and graph the solution set.

7. $|m - 6| \leq 2$

8. $2|x - 3| - 5 > 15$

9. $2 - 3(w - 1) < -2w$

10. $2 < \dfrac{5 - 2x}{3} < 7$

11. $3x - 2 < 7$ and $-3x \leq 15$

12. $\dfrac{2}{3}y < 4$ or $y - 3 < 12$

Solve each equation or inequality.

13. $|2x - 7| = 3$

14. $x - 4 > 1$ or $x < 12$

15. $3x < 0$ and $x - 5 > 2$

16. $|2x - 5| \leq 0$

17. $|x - 3| < 0$

18. $x + 3x = 4x$

19. $2(x + 7) = 2x + 9$

20. $|x - 6| > -6$

21. $x - 0.04(x - 10) = 96.4$

Write a complete solution to each problem.

22. The perimeter of a rectangle is 84 meters. If the width is 16 meters less than the length, then what is the width of the rectangle?

23. If the area of a triangle is 21 square inches and the base is 3 inches, then what is the height?

24. Joan bought a gold chain marked 30% off. If she paid $210, then what was the original price?

25. How many liters of an 11% alcohol solution should be mixed with 60 liters of a 5% alcohol solution to obtain a mixture that is 7% alcohol?

26. Al and Brenda do the same job, but their annual salaries differ by more than $3000. Assume that Al makes $28,000 per year and write an absolute value inequality to describe this situation. What are the possibilities for Brenda's salary?

Simplify each expression.

1. $5x + 6x$
2. $5x \cdot 6x$
3. $\dfrac{6x + 2}{2}$
4. $5 - 4(2 - x)$
5. $(30 - 1)(30 + 1)$
6. $(30 + 1)^2$
7. $(30 - 1)^2$
8. $(2 + 3)^2$
9. $2^2 + 3^2$
10. $(8 - 3)(3 - 8)$
11. $(-1)(3 - 8)$
12. -2^2
13. $3x + 8 - 5(x - 1)$
14. $(-6)^2 - 4(-3)2$
15. $3^2 \cdot 2^3$
16. $4(-6) - (-5)(3)$
17. $-3x \cdot x \cdot x$
18. $(-1)(-1)(-1)(-1)(-1)(-1)$

Solve each equation.

19. $5x + 6x = 8x$
20. $5x + 6x = 11x$
21. $5x + 6x = 0$
22. $5x + 6 = 11x$
23. $3x + 1 = 0$
24. $5 - 4(2 - x) = 1$
25. $3x + 6 = 3(x + 2)$
26. $x - 0.01x = 990$
27. $|5x + 6| - 11$

Solve each problem.

28. **Cost analysis.** A company can rent a copy machine for five years from American Business Supply for $75 per month plus 6 cents per copy. The same copy machine can be purchased for $8000, but then it costs only 2 cents per copy for supplies and maintenance. For what number of copies in the 5-year period would the total cost be the same? If this number is exceeded, then is renting or buying cheaper? (Assume that the purchased copier has no value after 5 years.)

Exponents and Polynomials

One statistic that can be used to measure the general health of a nation or a group within a nation is life expectancy. This data is considered more accurate than many other statistics because it is easy to determine the precise number of years in a person's lifetime.

The *World Health Organization Statistical Yearbook* published in May 1994 states that Japanese men have an average life expectancy of 76.3 years, followed by 75.1 years in Israel and 74.9 years in Australia. In the United States, men now have a life expectancy of 72.7 years, while women can hope to live at least until the age of 79.6. In the next few years, thanks in part to advances in health care and science, longevity is expected to increase significantly worldwide. There have been some authenticated cases of people who have reached the age of 120 years! Some scientists have arbitrarily placed the maximum age of a human being at 150 years.

In this chapter we study algebraic expressions involving exponents. In Exercises 95 and 96 of Section 3.2 you will see how formulas involving exponents can be used to find the life expectancy of men and women.

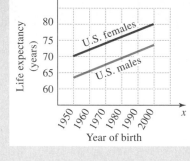

3.1 Integral Exponents

In Chapter 1 we defined positive integral exponents and learned to evaluate expressions involving exponents. In this section we will extend the definition of exponents to include all integers and learn some rules for working with integral exponents. In Chapter 5 we will see that any rational number can be used as an exponent.

Positive and Negative Exponents

Positive integral exponents provide a convenient way to write repeated multiplication or very large numbers. For example,

$$2 \cdot 2 \cdot 2 = 2^3, \qquad y \cdot y \cdot y \cdot y = y^4, \qquad \text{and} \qquad 1{,}000{,}000{,}000 = 10^9.$$

We refer to 2^3 as "2 cubed," "2 raised to the third power," or "a power of 2."

Positive Integral Exponents

If a is a nonzero real number and n is a positive integer, then

$$a^n = \underbrace{a \cdot a \cdot a \cdot \ldots \cdot a.}_{n \text{ factors of } a}$$

In the **exponential expression** a^n, the **base** is a, and the **exponent** is n.

We use 2^{-3} to represent the reciprocal of 2^3. Because $2^3 = 8$, we have $2^{-3} = \frac{1}{8}$. In general, a^{-n} is defined as the reciprocal of a^n.

Negative Integral Exponents

If a is a nonzero real number and n is a positive integer, then

$$a^{-n} = \frac{1}{a^n}. \qquad \text{(If } n \text{ is positive, } -n \text{ is negative.)}$$

To evaluate 2^{-3}, you can first cube 2 to get 8 and then find the reciprocal to get $\frac{1}{8}$, or you can first find the reciprocal of 2 $\left(\text{which is } \frac{1}{2}\right)$ and then cube $\frac{1}{2}$ to get $\frac{1}{8}$. So

$$2^{-3} = \left(\frac{1}{2}\right)^3.$$

The power and the reciprocal can be found in either order. If the exponent is -1, we simply find the reciprocal. For example,

$$2^{-1} = \frac{1}{2}, \qquad \left(\frac{2}{3}\right)^{-1} = \frac{3}{2}, \qquad \text{and} \qquad \left(-\frac{1}{4}\right)^{-1} = -4.$$

Because 2^3 and 2^{-3} are reciprocals of each other, we have

$$2^{-3} = \frac{1}{2^3} \quad \text{and} \quad \frac{1}{2^{-3}} = 2^3.$$

These examples illustrate the following rules.

Rules for Negative Exponents

If a is a nonzero real number and n is a positive integer, then

$$a^{-n} = \left(\frac{1}{a}\right)^n, \quad a^{-1} = \frac{1}{a}, \quad \text{and} \quad \frac{1}{a^{-n}} = a^n.$$

EXAMPLE 1   **Negative exponents**

Evaluate each expression.

a) 3^{-2} **b)** $(-3)^{-2}$ **c)** -3^{-2} **d)** $\left(\frac{3}{4}\right)^{-3}$ **e)** $\frac{1}{5^{-3}}$

Solution

a) $3^{-2} = \frac{1}{3^2} = \frac{1}{9}$ Definition of negative exponent

b) $(-3)^{-2} = \frac{1}{(-3)^2} = \frac{1}{9}$ Definition of negative exponent

c) $-3^{-2} = -\frac{1}{3^2} = -\frac{1}{9}$ Evaluate 3^{-2}, then take the opposite.

d) $\left(\frac{3}{4}\right)^{-3} = \left(\frac{4}{3}\right)^3$ The reciprocal of $\frac{3}{4}$ is $\frac{4}{3}$.

       $= \frac{64}{27}$ The cube of $\frac{4}{3}$ is $\frac{64}{27}$.

e) $\frac{1}{5^{-3}} = 5^3 = 125$ The reciprocal of 5^{-3} is 5^3.   ■

CAUTION In Chapter 1 we agreed to evaluate -3^2 by squaring 3 first and then taking the opposite. So $-3^2 = -9$, while $(-3)^2 = 9$. The same agreement also holds for negative exponents. That is why the answer to Example 1(c) is negative.   ⊘

Product Rule

We can simplify an expression such as $2^3 \cdot 2^5$ using the definition of exponents.

$$2^3 \cdot 2^5 = \underbrace{\overbrace{(2 \cdot 2 \cdot 2)}^{\text{three 2's}}\overbrace{(2 \cdot 2 \cdot 2 \cdot 2 \cdot 2)}^{\text{five 2's}}}_{\text{eight 2's}} = 2^8$$

Notice that the exponent 8 is the sum of the exponents 3 and 5. This example illustrates the **product rule for exponents.**

> **Product Rule for Exponents**
> If m and n are integers and $a \neq 0$, then
> $$a^m \cdot a^n = a^{m+n}.$$

EXAMPLE 2 **Using the product rule**

Simplify each expression. Write answers with positive exponents and assume that all variables represent nonzero real numbers.

a) $3^4 \cdot 3^6$ **b)** $4x^{-3} \cdot 5x$ **c)** $-2y^{-3}(-5y^{-4})$

Solution

a) $3^4 \cdot 3^6 = 3^{4+6} = 3^{10}$ Product rule

b) $4x^{-3} \cdot 5x = 4 \cdot 5 \cdot x^{-3} \cdot x^1$

$\qquad = 20x^{-2}$ Product rule: $x^{-3} \cdot x^1 = x^{-3+1} = x^{-2}$

$\qquad = \dfrac{20}{x^2}$ Definition of negative exponent

c) $-2y^{-3}(-5y^{-4}) = (-2)(-5)y^{-3}y^{-4}$

$\qquad = 10y^{-7}$ Product rule: $-3 + (-4) = -7$

$\qquad = \dfrac{10}{y^7}$ Definition of negative exponent ∎

CAUTION The product rule cannot be applied to $2^3 \cdot 3^2$, because the bases are not identical. Even when the bases are identical, we do not multiply the bases. For example, $2^5 \cdot 2^4 \neq 4^9$. Using the rule correctly, we get $2^5 \cdot 2^4 = 2^9$. ⊘

Zero Exponent

We have used positive and negative integral exponents, but we have not yet seen the integer 0 used as an exponent. Note that the product rule was stated to hold for *any* integers m and n. If we use the product rule on $2^3 \cdot 2^{-3}$, we get

$$2^3 \cdot 2^{-3} = 2^0.$$

Because $2^3 = 8$ and $2^{-3} = \frac{1}{8}$, we must have $2^3 \cdot 2^{-3} = 1$. So for consistency we define 2^0 and the zero power of any nonzero number to be 1.

> **Zero Exponent**
> If a is any nonzero real number, then $a^0 = 1$.

EXAMPLE 3   **Using zero as an exponent**

Simplify each expression. Write answers with positive exponents and assume that all variables represent nonzero real numbers.

a) -3^0 **b)** $\left(\frac{1}{4} - \frac{3}{2}\right)^0$ **c)** $-2a^5b^{-6} \cdot 3a^{-5}b^2$

Solution

a) To evaluate -3^0, we find 3^0 and then take the opposite. So $-3^0 = -1$.

b) $\left(\frac{1}{4} - \frac{3}{2}\right)^0 = 1$ Definition of zero exponent

c) $-2a^5b^{-6} \cdot 3a^{-5}b^2 = -6a^5 \cdot a^{-5} \cdot b^{-6} \cdot b^2$

$\qquad\qquad\qquad = -6a^0b^{-4}$ Product rule

$\qquad\qquad\qquad = -\dfrac{6}{b^4}$ Definitions of negative and zero exponent   ■

Changing the Sign of an Exponent

Because a^{-n} and a^n are reciprocals of each other, we know that

$$a^{-n} = \frac{1}{a^n} \quad \text{and} \quad \frac{1}{a^{-n}} = a^n.$$

So a negative exponent in the numerator or denominator can be changed to positive by relocating the exponential expression. In the next example we use these facts to remove negative exponents from exponential expressions.

EXAMPLE 4   **Simplifying expressions with negative exponents**

Write each expression without negative exponents and simplify. All variables represent nonzero real numbers.

a) $\dfrac{5a^{-3}}{a^2 \cdot 2^{-2}}$ **b)** $\dfrac{x^{-3}}{y^{-2}z^3}$

Solution

a) $\dfrac{5a^{-3}}{a^2 \cdot 2^{-2}} = 5 \cdot a^{-3} \cdot \dfrac{1}{a^2} \cdot \dfrac{1}{2^{-2}}$ Rewrite division as multiplication.

$\qquad\qquad = 5 \cdot \dfrac{1}{a^3} \cdot \dfrac{1}{a^2} \cdot 2^2$ Change the signs of the negative exponents.

$\qquad\qquad = \dfrac{20}{a^5}$ Product rule: $a^3 \cdot a^2 = a^5$

Note that in $5a^{-3}$ the negative exponent applies only to a.

b) $\dfrac{x^{-3}}{y^{-2}z^3} = x^{-3} \cdot \dfrac{1}{y^{-2}} \cdot \dfrac{1}{z^3}$ Rewrite as multiplication.

$\qquad\qquad = \dfrac{1}{x^3} \cdot y^2 \cdot \dfrac{1}{z^3}$ Change signs of the negative exponents.

$\qquad\qquad = \dfrac{y^2}{x^3z^3}$ Simplify.   ■

In Example 4 we showed more steps than are necessary. For instance, in part (b) we could simply write

$$\frac{x^{-3}}{y^{-2}z^3} = \frac{y^2}{x^3z^3}.$$

Exponential expressions (that are factors) can be moved from numerator to denominator (or vice versa) as long as we change the sign of the exponent.

CAUTION If an exponential expression is not a factor, you *cannot* move it from numerator to denominator (or vice versa). For example,

$$\frac{2^{-1} + 1^{-1}}{1^{-1}} \neq \frac{1}{2 + 1}.$$

Because $2^{-1} = \frac{1}{2}$ and $1^{-1} = 1$, we get

$$\frac{2^{-1} + 1^{-1}}{1^{-1}} = \frac{\frac{1}{2} + 1}{1} = \frac{\frac{3}{2}}{1} = \frac{3}{2} \quad \text{not} \quad \frac{1}{2 + 1} = \frac{1}{3}. \qquad \oslash$$

Quotient Rule

We can use arithmetic to simplify the quotient of two exponential expressions. For example,

$$\frac{2^5}{2^3} = \frac{\cancel{2} \cdot \cancel{2} \cdot \cancel{2} \cdot 2 \cdot 2}{\cancel{2} \cdot \cancel{2} \cdot \cancel{2}} = 2^2.$$

There are five 2's in the numerator and three 2's in the denominator. After dividing, two 2's remain. The exponent in 2^2 can be obtained by subtracting the exponents 3 and 5. This example illustrates the **quotient rule for exponents.**

> **Quotient Rule for Exponents**
> If m and n are any integers and $a \neq 0$, then
> $$\frac{a^m}{a^n} = a^{m-n}.$$

EXAMPLE 5 **Using the quotient rule**

Simplify each expression. Write answers with positive exponents only. All variables represent nonzero real numbers.

a) $\dfrac{2^9}{2^4}$　　　　b) $\dfrac{m^5}{m^{-3}}$　　　　c) $\dfrac{y^{-4}}{y^{-2}}$

Solution

a) $\dfrac{2^9}{2^4} = 2^{9-4} = 2^5$ 　　　　　b) $\dfrac{m^5}{m^{-3}} = m^{5-(-3)} = m^8$

c) $\dfrac{y^{-4}}{y^{-2}} = y^{-4-(-2)} = y^{-2} = \dfrac{1}{y^2}$ 　　　　　■

The next example further illustrates the rules of exponents. Remember that the bases must be identical for the quotient rule or the product rule.

EXAMPLE 6   **Using the product and quotient rules**

Use the rules of exponents to simplify each expression. Write answers with positive exponents only. All variables represent nonzero real numbers.

a) $\dfrac{2x^{-7}}{x^{-7}}$ **b)** $\dfrac{w(2w^{-4})}{3w^{-2}}$ **c)** $\dfrac{x^{-1}x^{-3}y^5}{x^{-2}y^2}$

Solution

a) $\dfrac{2x^{-7}}{x^{-7}} = 2x^0$ Quotient rule: $-7 - (-7) = 0$

   $\quad\quad\ = 2$ Definition of zero exponent

b) $\dfrac{w(2w^{-4})}{3w^{-2}} = \dfrac{2w^{-3}}{3w^{-2}}$ Product rule: $w^1 \cdot w^{-4} = w^{-3}$

   $\quad\quad\quad\ = \dfrac{2w^{-1}}{3}$ Quotient rule: $-3 - (-2) = -1$

   $\quad\quad\quad\ = \dfrac{2}{3w}$ Definition of negative exponent

c) $\dfrac{x^{-1}x^{-3}y^5}{x^{-2}y^2} = \dfrac{x^{-4}y^5}{x^{-2}y^2} = x^{-2}y^3 = \dfrac{y^3}{x^2}$ ∎

Scientific Notation

Many of the numbers that are encountered in science are either very large or very small. For example, the distance from the earth to the sun is 93,000,000 miles, and a hydrogen atom weighs 0.0000000000000000000000017 gram. Scientific notation provides a convenient way of writing very large and very small numbers. In scientific notation the distance from the earth to the sun is 9.3×10^7 miles and a hydrogen atom weighs 1.7×10^{-24} gram. In scientific notation the times symbol, $\times$, is used to indicate multiplication. Converting a number from scientific notation to standard notation is simply a matter of multiplication.

EXAMPLE 7   **Scientific notation to standard notation**

Write each number using standard notation.

a) 7.62×10^5 **b)** 6.35×10^{-4}

Solution

a) Multiplying a number by 10^5 moves the decimal point five places to the right:

$$7.62 \times 10^5 = 762000. = 762,000$$

b) Multiplying a number by 10^{-4} or 0.0001 moves the decimal point four places to the left:

$$6.35 \times 10^{-4} = 0.000635 = 0.000635$$ ∎

The procedure for converting a number from scientific notation to standard notation is summarized as follows.

Strategy for Converting from Scientific Notation to Standard Notation

1. Determine the number of places to move the decimal point by examining the exponent on the 10.
2. Move to the right for a positive exponent and to the left for a negative exponent.

A positive number in scientific notation is written as a product of a number between 1 and 10, and a power of 10. Numbers in scientific notation are written with only one digit to the left of the decimal point. A number larger than 10 is written with a positive power of 10, and a positive number smaller than 1 is written with a negative power of 10. Numbers between 1 and 10 are not written in scientific notation. To convert to scientific notation, we reverse the strategy for converting from scientific notation.

Strategy for Converting from Standard Notation to Scientific Notation

1. Count the number of places (n) that the decimal point must be moved so that it will follow the first nonzero digit of the number.
2. If the original number was larger than 10, use 10^n.
3. If the original number was smaller than 1, use 10^{-n}.

EXAMPLE 8   **Standard notation to scientific notation**

Convert each number to scientific notation.

a) 934,000,000

b) 0.0000025

Solution

a) In 934,000,000 the decimal point must be moved eight places to the left to get it to follow 9, the first nonzero digit.

$$934,000,000 = 9.34 \times 10^8 \quad \text{Use 8 because } 934,000,000 > 10.$$

b) The decimal point in 0.0000025 must be moved six places to the right to get the 2 to the left of the decimal point.

$$0.0000025 = 2.5 \times 10^{-6} \quad \text{Use } -6 \text{ because } 0.0000025 < 1. \qquad ■$$

We can perform computations with numbers in scientific notation by using the rules of exponents on the powers of 10.

EXAMPLE 9   **Using scientific notation in computations**

Evaluate $\dfrac{(10,000)(0.000025)}{0.000005}$ by first converting each number to scientific notation.

Solution

$$\frac{(10,000)(0.000025)}{0.000005} = \frac{(1 \times 10^4)(2.5 \times 10^{-5})}{5 \times 10^{-6}}$$

$$= \frac{2.5}{5} \cdot \frac{10^4 \cdot 10^{-5}}{10^{-6}} \qquad \text{Commutative and associative properties}$$

$$= 0.5 \times 10^5$$

$$= 5 \times 10^{-1} \times 10^5 \qquad \text{Write 0.5 in scientific notation.}$$

$$= 5 \times 10^4 \qquad\qquad\qquad\blacksquare$$

Calculator Close-up

A scientific calculator automatically converts into scientific notation any result that will not fit onto its display screen. Most calculators are limited to a power of 10 from -99 to 99. To enter a number in scientific notation into a calculator, we usually use a key labeled $\boxed{\text{EXP}}$ or $\boxed{\text{EE}}$ (enter exponent). For example, the product $(6.3 \times 10^{20})(9.5 \times 10^{-6})$ is found on a scientific calculator as follows:

$$6.3 \; \boxed{\text{EE}} \; 20 \; \boxed{\times} \; 9.5 \; \boxed{\text{EE}} \; 6 \; \boxed{+/-} \; \boxed{=}$$

The display will read

$$\boxed{\qquad 5.985 \; \text{E15} \qquad}$$

This result indicates that the answer is 5.985×10^{15}. The exponent 15 may appear in another manner on your calculator. On a graphing calculator your display should look like the following:

$$6.3\text{E}20*9.5\text{E}^-6 \; \boxed{\text{ENTER}}$$

$$\boxed{\qquad 5.985 \; \text{E15} \qquad}$$

So the answer is again 5.985×10^{15}.

EXAMPLE 10   **Counting hydrogen atoms**

Determine the number of hydrogen atoms in one kilogram of hydrogen.

Solution

One hydrogen atom weighs 1.7×10^{-24} gram. One kilogram is 1000, or 1×10^3, grams. So to obtain the number of hydrogen atoms in one kilogram of hydrogen, we divide 1×10^3 by 1.7×10^{-24}:

$$\frac{1 \times 10^3}{1.7 \times 10^{-24}} \approx 5.9 \times 10^{26}$$

So there are 5.9×10^{26} hydrogen atoms in one kilogram of hydrogen. $\blacksquare$

Warm-ups

True or false? Explain your answer.

1. $3^5 \cdot 3^4 = 3^9$ **2.** $3 \cdot 3 \cdot 3^{-1} = \dfrac{1}{27}$ **3.** $10^{-3} = 0.0001$

4. $1^{-1} = 1$ **5.** $\dfrac{2^5}{2^{-2}} = 2^7$ **6.** $2^3 \cdot 5^2 = 10^5$

7. $-2^{-2} = -\dfrac{1}{4}$ **8.** $46.7 \times 10^5 = 4.67 \times 10^6$

9. $0.512 \times 10^{-3} = 5.12 \times 10^{-4}$ **10.** $\dfrac{8 \times 10^{30}}{2 \times 10^{-5}} = 4 \times 10^{25}$

3.1 EXERCISES

For all exercises in this section, assume that the variables represent nonzero real numbers and use only positive exponents in your answers.

Evaluate each expression. See Example 1.

1. 4^2 **2.** 3^3 **3.** 4^{-2}

4. 3^{-3} **5.** $\left(\dfrac{1}{4}\right)^{-2}$ **6.** $\left(\dfrac{3}{5}\right)^{-1}$

7. $\left(-\dfrac{3}{4}\right)^{-2}$ **8.** $10^2 \cdot \left(-\dfrac{1}{10}\right)^{-2}$ **9.** $-2^{-1} \cdot (-6)^2$

10. $-2^{-3} \cdot (-4)^2$ **11.** $3 \cdot 10^{-3}$ **12.** $5 \cdot 10^4$

13. $\dfrac{1}{2^{-4}}$ **14.** $\dfrac{2}{2^{-3}}$

Simplify. See Examples 2 and 3.

15. $2^5 \cdot 2^{12}$ **16.** $2x^2 \cdot 3x^3$

17. $2y^3 \cdot y^2 \cdot y^{-7}$ **18.** $2 \cdot 10^6 \cdot 5 \cdot 10^{-7}$

19. $5x^{-3} \cdot 6x^{-4}$ **20.** $4a^{-13} \cdot a^{-4}$

21. $-8^0 - 8^0$ **22.** $(-8 - 8)^0$

23. $(x - y)^0$ **24.** $(x^2 - y^2)^0$

25. $2w^{-3}(w^7 \cdot w^{-4})$ **26.** $5y^2z(y^{-3}z^{-1})$

Write each expression without negative exponents and simplify. See Example 4.

27. $\dfrac{2}{4^{-2}}$ **28.** $\dfrac{5}{10^{-3}}$ **29.** $\dfrac{3^{-1}}{10^{-2}}$

30. $\dfrac{2y^{-2}}{3^{-1}}$ **31.** $\dfrac{2x^{-3}(4x)}{5y^{-2}}$ **32.** $\dfrac{5^{-2}xy^{-3}}{3x^{-2}}$

33. $\dfrac{4^{-2}x^3x^{-6}}{3x^{-3}x^2}$ **34.** $\dfrac{3y^{-4}y^{-6}}{2^{-3}y^2y^{-7}}$

Simplify each expression. See Examples 5 and 6.

35. $\dfrac{x^5}{x^3}$ **36.** $\dfrac{a^8}{a^3}$ **37.** $\dfrac{3^6}{3^{-2}}$

38. $\dfrac{6^2}{6^{-5}}$ **39.** $\dfrac{4a^{-5}}{12a^{-2}}$ **40.** $\dfrac{-3a^{-3}}{-21a^{-4}}$

41. $\dfrac{-6w^{-5}}{2w^3}$ **42.** $\dfrac{10x^{-6}}{-2x^2}$ **43.** $\dfrac{3^3w^{-2}w^5}{3^{-5}w^{-3}}$

44. $\dfrac{2^{-3}w^5}{2^5w^3w^{-7}}$ **45.** $\dfrac{3x^{-6} \cdot x^2y^{-1}}{6x^{-5}y^{-2}}$ **46.** $\dfrac{2r^{-3}t^{-1}}{10r^5t^2 \cdot t^{-3}}$

Use the rules of exponents to simplify each expression.

47. $3^{-1}\left(\dfrac{1}{3}\right)^{-3}$ **48.** $2^{-2}\left(\dfrac{1}{4}\right)^{-3}$

49. $-2^4 + \left(\dfrac{1}{2}\right)^{-1}$ **50.** $-3^4 - (-3)^4$

51. $-(-2)^{-3} \cdot 2^{-1}$ **52.** $-(-3)^{-1} \cdot 9^{-1}$

53. $(1 + 2^{-1})^{-2}$ **54.** $(2^{-1} + 2^{-1})^{-2}$

55. $-5a^2 \cdot 3a^2$ **56.** $2x^2 \cdot 5x^{-5}$

57. $5a \cdot a^2 + 3a^{-2} \cdot a^5$ **58.** $2x^2 \cdot 5y^{-5}$

59. $\dfrac{-3a^5(-2a^{-1})}{6a^3}$ **60.** $\dfrac{6a(-ab^{-2})}{-2a^2b^{-3}}$

61. $\dfrac{(-3x^3y^2)(-2xy^{-3})}{-9x^2y^{-5}}$ **62.** $\dfrac{(-2x^{-5}y)(-3xy^6)}{-6x^{-6}y^2}$

63. $\dfrac{2^{-1} + 3^{-1}}{2^{-1}}$ **64.** $\dfrac{3^{-1} + 4^{-1}}{12^{-1}}$

65. $\dfrac{(2 + 3)^{-1}}{2^{-1} + 3^{-1}}$ **66.** $\dfrac{(3 - 4)^{-1}}{3 \cdot 2^{-1}}$

For each equation, find the integer that can be used as the exponent to make the equation correct.

67. $8 = 2^?$ **68.** $27 = 3^?$ **69.** $\dfrac{1}{4} = 2^?$

70. $\dfrac{1}{125} = 5^?$ **71.** $16 = \left(\dfrac{1}{2}\right)^?$ **72.** $81 = \left(\dfrac{1}{3}\right)^?$

73. $10^? = 0.001$ **74.** $10^? = 10,000$

Write each number in standard notation. See Example 7.

75. 4.86×10^8 **76.** 3.80×10^2 **77.** 2.37×10^{-6}

78. 1.62×10^{-3} **79.** 4×10^6 **80.** 496×10^3

81. 5×10^{-6} **82.** 48×10^{-3}

Write each number in scientific notation. See Example 8.

83. $320,000$ **84.** $43,298,000$ **85.** 0.00000071

86. 0.00000894 **87.** 0.00007 **88.** $8,295,100$

89. 235×10^5 **90.** 0.43×10^{-9}

Evaluate each expression using scientific notation without a calculator. See Example 9.

91. $\dfrac{(5,000,000)(0.0003)}{2000}$ **92.** $\dfrac{(6000)(0.00004)}{(30,000)(0.002)}$

93. $\dfrac{6 \times 10^{40}}{2 \times 10^{18}}$ **94.** $\dfrac{4.6 \times 10^{12}}{2.3 \times 10^5}$

95. $\dfrac{(-4 \times 10^5)(6 \times 10^{-9})}{2 \times 10^{-16}}$

96. $\dfrac{(4.8 \times 10^{-3})(5 \times 10^{-8})}{(1.2 \times 10^{-6})(2 \times 10^{12})}$

Evaluate each expression using a calculator. Write answers in scientific notation. Round the decimal part to three decimal places. See Example 10.

97. $(4.3 \times 10^9)(3.67 \times 10^{-5})$

98. $(2.34 \times 10^6)(8.7 \times 10^5)$

99. $(4.37 \times 10^{-6}) + (8.75 \times 10^{-5})$

100. $(6.72 \times 10^5) + (8.98 \times 10^6)$

101. $(9.27 \times 10^{80})(6.43 \times 10^{76})$

102. $(1.35 \times 10^{66})(2.7 \times 10^{74})$

103. $\dfrac{(5.6 \times 10^{14})^2(3.2 \times 10^{-6})}{(6.4 \times 10^{-3})^3}$

104. $\dfrac{(3.51 \times 10^{-6})^3(4000)^5}{2\pi}$

Solve each problem.

105. *Distance to the sun.* The distance from the earth to the sun is 93 million miles. Express this distance in feet using scientific notation (1 mile = 5280 feet).

106. *Traveling time.* The speed of light is 9.83569×10^8 feet per second. How long does it take light to get from the sun to the earth? (See Exercise #105.)

107. *Space travel.* How long does it take a spacecraft traveling 1.2×10^5 kilometers per second to travel 4.6×10^{12} kilometers?

108. *Diameter of a dot.* If the circumference of a very small circle is 2.35×10^{-8} meter, then what is the diameter of the circle?

109. *Solid waste per person.* According to the Environmental Protection Agency (EPA), in 1960 the 1.80863×10^8 people in the United States generated 8.71×10^7 tons of municipal solid waste. How many pounds per person per day were generated in 1960?

Figure for Exercises 109 and 110

110. *An increasing problem*. According to the EPA, in 1988 the 2.46027×10^8 people in the United States generated 3.592×10^{11} pounds of solid municipal waste. How many pounds per person per day were generated in 1988? Use the graph to predict the number of pounds per person per day that will be generated in the year 2000.

Getting More Involved

111. *Exploration.* **a)** Using pairs of integers, find values for m and n for which $2^m \cdot 3^n = 6^{m+n}$. **b)** For which values of m and n is it true that $2^m \cdot 3^n \neq 6^{m+n}$?

112. *Cooperative learning.* Work in a group to find the units digit of 3^{99} and explain how you found it.

113. *Discussion.* What is the difference between $-a^n$ and $(-a)^n$, where n is an integer? For which values of a and n do they have the same value, and for which values of a and n do they have different values?

114. *Exploration.* If $a + b = a$, then what can you conclude about b? Use scientific notation on your calculator to find $5 \times 10^{20} + 3 \times 10^6$. Explain why your calculator displays the answer that it gets.

3.2   **The Power Rules**

In Section 3.1 you learned some of the basic rules for working with exponents. All of the rules of exponents are designed to make it easier to work with exponential expressions. In this section we will extend our list of rules to include three new ones.

Raising an Exponential Expression to a Power

An expression such as $(x^3)^2$ consists of the exponential expression x^3 raised to the power 2. We can use known rules to simplify this expression.

$$(x^3)^2 = x^3 \cdot x^3 \quad \text{Exponent 2 indicates two factors of } x^3.$$
$$= x^6 \quad \text{Product rule: } 3 + 3 = 6$$

Note that the exponent 6 is the *product* of the exponents 2 and 3. This example illustrates the **power of a power rule.**

Power of a Power Rule

If m and n are any integers and $a \neq 0$, then

$$(a^m)^n = a^{mn}.$$

EXAMPLE 1   **Using the power of a power rule**

Use the rules of exponents to simplify each expression. Write the answer with positive exponents only. Assume that all variables represent nonzero real numbers.

a) $(2^3)^5$

b) $(x^2)^{-6}$

c) $3(y^{-3})^{-2}y^{-5}$

d) $\dfrac{(x^2)^{-1}}{(x^{-3})^3}$

Solution

a) $(2^3)^5 = 2^{15}$ Power of a power rule

b) $(x^2)^{-6} = x^{-12}$ Power of a power rule

$\qquad\qquad = \dfrac{1}{x^{12}}$ Definition of negative exponent

c) $3(y^{-3})^{-2}y^{-5} = 3y^6y^{-5}$ Power of a power rule

$\qquad\qquad\qquad = 3y$ Product rule

d) $\dfrac{(x^2)^{-1}}{(x^{-3})^3} = \dfrac{x^{-2}}{x^{-9}}$ Power of a power rule

$\qquad\qquad = x^7$ Quotient rule

■

Raising a Product to a Power

Consider how we would simplify a product raised to a positive power and a product raised to a negative power using known rules.

$$(2x)^3 = \overbrace{2x \cdot 2x \cdot 2x}^{3 \text{ factors of } 2x} = 2^3 \cdot x^3 = 8x^3$$

$$(ay)^{-3} = \frac{1}{(ay)^3} = \frac{1}{(ay)(ay)(ay)} = \frac{1}{a^3y^3} = a^{-3}y^{-3}$$

In each of these cases the original exponent is applied to each factor of the product. These examples illustrate the **power of a product rule.**

> **Power of a Product Rule**
>
> If a and b are nonzero real numbers and n is any integer, then
> $$(ab)^n = a^n \cdot b^n.$$

EXAMPLE 2   **Using the power of a product rule**

Simplify. Assume that the variables represent nonzero real numbers. Write the answers with positive exponents only.

a) $(-3x)^4$ **b)** $(-2x^2)^3$ **c)** $(3x^{-2}y^3)^{-2}$

Solution

a) $(-3x)^4 = (-3)^4x^4$ Power of a product rule
$$= 81x^4$$

b) $(-2x^2)^3 = (-2)^3(x^2)^3$ Power of a product rule
$$= -8x^6 \qquad \text{Power of a power rule}$$

c) $(3x^{-2}y^3)^{-2} = (3)^{-2}(x^{-2})^{-2}(y^3)^{-2} = \frac{1}{9}x^4y^{-6} = \frac{x^4}{9y^6}$ ∎

Raising a Quotient to a Power

Now consider an example of applying known rules to a power of a quotient:

$$\left(\frac{x}{5}\right)^3 = \frac{x}{5} \cdot \frac{x}{5} \cdot \frac{x}{5} = \frac{x^3}{5^3}$$

We get a similar result with a negative power:

$$\left(\frac{x}{5}\right)^{-3} = \left(\frac{5}{x}\right)^3 = \frac{5}{x} \cdot \frac{5}{x} \cdot \frac{5}{x} = \frac{5^3}{x^3} = \frac{x^{-3}}{5^{-3}}$$

In each of these cases the original exponent applies to both the numerator and denominator. These examples illustrate the **power of a quotient rule.**

> **Power of a Quotient Rule**
>
> If a and b are nonzero real numbers and n is any integer, then
> $$\left(\frac{a}{b}\right)^n = \frac{a^n}{b^n}.$$

EXAMPLE 3   **Using the power of a quotient rule**

Use the rules of exponents to simplify each expression. Write your answers with positive exponents only. Assume that the variables are nonzero real numbers.

a) $\left(\dfrac{x}{2}\right)^3$ b) $\left(-\dfrac{2x^3}{3y^2}\right)^3$

c) $\left(\dfrac{x^{-2}}{2^3}\right)^{-1}$ d) $\left(-\dfrac{3}{4x^3}\right)^{-2}$

Solution

a) $\left(\dfrac{x}{2}\right)^3 = \dfrac{x^3}{2^3}$ Power of a quotient rule

$= \dfrac{x^3}{8}$

b) $\left(-\dfrac{2x^3}{3y^2}\right)^3 = \dfrac{(-2)^3 x^9}{3^3 y^6}$ Because $(x^3)^3 = x^9$ and $(y^2)^3 = y^6$

$= \dfrac{-8x^9}{27y^6} = -\dfrac{8x^9}{27y^6}$

c) $\left(\dfrac{x^{-2}}{2^3}\right)^{-1} = \dfrac{x^2}{2^{-3}} = 8x^2$

d) $\left(-\dfrac{3}{4x^3}\right)^{-2} = \dfrac{(-3)^{-2}}{4^{-2}x^{-6}} = \dfrac{4^2 x^6}{(-3)^2} = \dfrac{16x^6}{9}$ ■

A fraction to a negative power can be simplified by using the power of a quotient rule as in Example 3. Another method is to find the reciprocal of the fraction first, then use the power of a quotient rule as shown in the next example.

EXAMPLE 4   **Negative powers of fractions**

Simplify. Assume that the variables are nonzero real numbers and write the answers with positive exponents only.

a) $\left(\dfrac{3}{4}\right)^{-3}$ b) $\left(\dfrac{x^2}{5}\right)^{-2}$ c) $\left(-\dfrac{2y^3}{3}\right)^{-2}$

Solution

a) $\left(\dfrac{3}{4}\right)^{-3} = \left(\dfrac{4}{3}\right)^3$ The reciprocal of $\frac{3}{4}$ is $\frac{4}{3}$.

$= \dfrac{4^3}{3^3}$ Power of a quotient rule

$= \dfrac{64}{27}$

b) $\left(\dfrac{x^2}{5}\right)^{-2} = \left(\dfrac{5}{x^2}\right)^2 = \dfrac{5^2}{(x^2)^2} = \dfrac{25}{x^4}$

c) $\left(-\dfrac{2y^3}{3}\right)^{-2} = \left(-\dfrac{3}{2y^3}\right)^2 = \dfrac{9}{4y^6}$ ■

Variable Exponents

So far, we have used the rules of exponents only on expressions with integral exponents. However, we can use the rules to simplify expressions having variable exponents that represent integers.

EXAMPLE 5 **Expressions with variables as exponents**

Simplify. Assume that the variables represent integers.

a) $3^{4y} \cdot 3^{5y}$ **b)** $(5^{2x})^{3x}$ **c)** $\left(\dfrac{2^n}{3^m}\right)^{5n}$

Solution

a) $3^{4y} \cdot 3^{5y} = 3^{9y}$ Product rule: $4y + 5y = 9y$

b) $(5^{2x})^{3x} = 5^{6x^2}$ Power of a power rule: $2x \cdot 3x = 6x^2$

c) $\left(\dfrac{2^n}{3^m}\right)^{5n} = \dfrac{(2^n)^{5n}}{(3^m)^{5n}}$ Power of a quotient rule

$= \dfrac{2^{5n^2}}{3^{5mn}}$ Power of a power rule ∎

Summary of the Rules

The definitions and rules that were introduced in the last two sections are summarized in the following box.

Rules for Integral Exponents

For these rules, m and n are integers and a and b are nonzero real numbers.

1. $a^{-n} = \dfrac{1}{a^n}$ Definition of negative exponent

2. $a^{-n} = \left(\dfrac{1}{a}\right)^n$, $a^{-1} = \dfrac{1}{a}$, and $\dfrac{1}{a^{-n}} = a^n$ Negative exponent rules

3. $a^0 = 1$ Definition of zero exponent

4. $a^m a^n = a^{m+n}$ Product rule

5. $\dfrac{a^m}{a^n} = a^{m-n}$ Quotient rule

6. $(a^m)^n = a^{mn}$ Power of a power rule

7. $(ab)^n = a^n b^n$ Power of a product rule

8. $\left(\dfrac{a}{b}\right)^n = \dfrac{a^n}{b^n}$ Power of a quotient rule

Calculator Close-up

Expressions with negative exponents can be evaluated by entering a negative exponent or by using the reciprocal key [1/x]. For example, to evaluate 2^{-3} on a scientific calculator, try the following:

$$2 \; \boxed{x^y} \; 3 \; \boxed{\pm} \; \boxed{=} \qquad \text{or} \qquad 2 \; \boxed{x^y} \; 3 \; \boxed{=} \; \boxed{1/x}$$

In each case the display will read

$$\boxed{\qquad\qquad 0.125\ }$$

With a graphing calculator, 2^{-3} should appear on the display as follows:

$$2 \wedge {}^{-}3 \; \boxed{\text{ENTER}}$$

$$\boxed{\qquad\qquad 0.125\ }$$

Applications

Both positive and negative exponents occur in formulas used in investment situations. The amount of money invested is the **principal,** and the value of the principal after a certain time period is the **amount.** Interest rates are annual percentage rates.

> **Amount Formula**
>
> The amount A of an investment of P dollars with interest rate r compounded annually for n years is given by the formula
>
> $$A = P(1 + r)^n.$$

EXAMPLE 6   **Finding the amount**

According to Fidelity Investments of Boston, U.S. common stocks have returned an average of 10% annually since 1926. If your great-grandfather had invested $100 dollars in the stock market in 1926 and obtained the average increase each year, then how much would the investment be worth in 1996 after 70 years of growth?

Solution

Use $n = 70$, $P = \$100$, and $r = 0.10$ in the amount formula:

$$A = P(1 + r)^n$$
$$A = 100(1 + 0.10)^{70}$$
$$= 100(1.1)^{70}$$
$$= 78{,}974.70 \quad \text{Use a calculator with an exponent key.}$$

So $100 invested in 1926 would have amounted to $78,974.70 in 1996.   ∎

When we are interested in the principal that must be invested today to grow to a certain amount, the principal is called the **present value** of the investment. We can find a formula for present value by solving the amount formula for P:

$$A = P(1 + r)^n$$

$$P = \frac{A}{(1 + r)^n} \qquad \text{Divide each side by } (1 + r)^n.$$

$$P = A(1 + r)^{-n} \qquad \text{Definition of negative exponent}$$

Present-value Formula

The present value P that will amount to A dollars after n years with interest compounded annually at annual interest rate r is given by

$$P = A(1 + r)^{-n}.$$

EXAMPLE 7   **Finding the present value**

If your great-grandfather wanted you to have $1,000,000 in 1996, then how much could he have invested in the stock market in 1926 to achieve this goal. Assume that he could get the average annual return of 10% (from Example 6) for 70 years.

Solution

Use $r = 0.10$, $n = 70$, and $A = 1,000,000$ in the present value formula:

$$P = A(1 + r)^{-n}$$

$$P = 1,000,000(1 + 0.10)^{-70}$$

$$P = 1,000,000(1.1)^{-70} \qquad \text{Use a calculator with an exponent key.}$$

$$P = 1,266.23$$

A deposit of $1,266.23 in 1926 would have grown to $1,000,000 in 70 years at a rate of 10% compounded annually. ∎

Warm-ups

True or false? Explain your answer. Assume that all variables represent nonzero real numbers.

1. $(2^2)^3 = 2^5$

2. $(2^{-3})^{-1} = 8$

3. $(x^{-3})^3 = x^{-9}$

4. $(2^3)^3 = 2^{27}$

5. $(2x)^3 = 6x^3$

6. $(-3y^3)^2 = 9y^9$

7. $\left(\frac{2}{3}\right)^{-1} = \frac{3}{2}$

8. $\left(\frac{2}{3}\right)^3 = \frac{8}{27}$

9. $\left(\frac{x^2}{2}\right)^3 = \frac{x^6}{8}$

10. $\left(\frac{2}{x}\right)^{-2} = \frac{x^2}{4}$

3.2 EXERCISES

For all exercises in this section, assume that the variables represent nonzero real numbers and use positive exponents only in your answers.

Use the rules of exponents to simplify each expression. See Example 1.

1. $(2^2)^3$ **2.** $(3^2)^2$ **3.** $(y^2)^5$

4. $(x^6)^2$ **5.** $(x^2)^{-4}$ **6.** $(x^{-2})^7$

7. $(m^{-3})^{-6}$ **8.** $(a^{-3})^{-3}$

9. $(x^{-2})^3(x^{-3})^{-2}$ **10.** $(m^{-3})^{-1}(m^2)^{-4}$

11. $\dfrac{(x^3)^{-4}}{(x^2)^{-5}}$ **12.** $\dfrac{(a^2)^{-3}}{(a^{-2})^4}$

Simplify. See Example 2.

13. $(-9y)^2$ **14.** $(-2a)^3$ **15.** $(-5w^3)^2$

16. $(-2w^{-5})^3$ **17.** $(x^3y^{-2})^3$ **18.** $(a^2b^{-3})^2$

19. $(3ab^{-1})^{-2}$ **20.** $(2x^{-1}y^2)^{-3}$ **21.** $\dfrac{2xy^{-2}}{(3x^2y)^{-1}}$

22. $\dfrac{3ab^{-1}}{(5ab^2)^{-1}}$ **23.** $\dfrac{(2ab)^{-2}}{2ab^2}$ **24.** $\dfrac{(3xy)^{-3}}{3xy^3}$

Simplify. See Example 3.

25. $\left(\dfrac{w}{2}\right)^3$ **26.** $\left(\dfrac{m}{5}\right)^2$ **27.** $\left(-\dfrac{3a}{4}\right)^3$

28. $\left(-\dfrac{2}{3b}\right)^4$ **29.** $\left(\dfrac{2x^{-1}}{y}\right)^{-2}$ **30.** $\left(\dfrac{2a^2b}{3}\right)^{-3}$

31. $\left(\dfrac{-3x^3}{y}\right)^{-2}$ **32.** $\left(\dfrac{-2y^2}{x}\right)^{-3}$

Simplify. See Example 4

33. $\left(\dfrac{2}{5}\right)^{-2}$ **34.** $\left(\dfrac{3}{4}\right)^{-2}$ **35.** $\left(-\dfrac{1}{2}\right)^{-2}$

36. $\left(-\dfrac{2}{3}\right)^{-2}$ **37.** $\left(-\dfrac{2x}{3}\right)^{-3}$ **38.** $\left(-\dfrac{ab}{c}\right)^{-1}$

39. $\left(\dfrac{2x^2}{3y}\right)^{-3}$ **40.** $\left(\dfrac{ab^{-3}}{a^2b}\right)^{-2}$

Simplify each expression. Assume that the variables represent integers. See Example 5.

41. $5^{2t} \cdot 5^{4t}$ **42.** $3^{2n-3} \cdot 3^{4-2n}$

43. $(2^{-3w})^{-2w}$ **44.** $6^{8x} \cdot (6^{2x})^{-3}$

45. $\dfrac{7^{2m+6}}{7^{m+3}}$ **46.** $\dfrac{4^{-3p}}{4^{-4p}}$

47. $8^{2a-1} \cdot (8^{a+4})^3$ **48.** $(5^{4-3y})^3(5^{y-2})^2$

49. $\dfrac{2^{3m-1}(2^{5-m})^2}{(2^{1-m})^3(2^{m+1})^{-1}}$ **50.** $\left(\dfrac{3^{n-1}}{2^{n+5}}\right)^2\left(\dfrac{2^{n+4}}{3^{2n-1}}\right)^{-3}$

Use the rules of exponents to simplify each expression. If possible, write down only the answer.

51. $3x^4 \cdot 2x^5$ **52.** $(3x^4)^2$ **53.** $(-2x^2)^3$

54. $3x^2 \cdot 2x^{-4}$ **55.** $\dfrac{3x^{-2}y^{-1}}{z^{-1}}$ **56.** $\dfrac{2^{-1}x^2}{y^{-2}}$

57. $\left(\dfrac{-2}{3}\right)^{-1}$ **58.** $\left(\dfrac{-1}{5}\right)^{-1}$ **59.** $\left(\dfrac{2x^3}{3}\right)^2$

60. $\left(\dfrac{-2y^4}{x}\right)^3$ **61.** $(-2x^{-2})^{-1}$ **62.** $(-3x^{-2})^3$

Use the rules of exponents to simplify each expression.

63. $\left(\dfrac{2x^2y}{xy^2}\right)^{-3}$ **64.** $\left(\dfrac{2x^3y^2}{3xy^3}\right)^{-1}$

65. $\dfrac{(5a^{-1}b^2)^3}{(5ab^{-2})^4}$ **66.** $\dfrac{(2m^2n^{-3})^4}{mn^5}$

67. $\dfrac{(2x^{-2}y)^{-3}}{(2xy^{-1})^2}(2x^2y^{-7})$ **68.** $\dfrac{(3x^{-1}y^3)^{-2}}{(3xy^{-1})^3}(9x^{-9}y^5)$

69. $\left(\dfrac{6a^{-2}b^3}{2c^4}\right)^{-2}(3a^{-1}b^2)^3$ **70.** $(7xz^2)^{-4}\left(\dfrac{7xy^{-1}}{z}\right)^3$

71. $\left(\dfrac{3a^{-2}b^3}{a^{-3}b^5}\right)^4\left(\dfrac{9a^{-5}b^4a^2}{b^{-6}a^5}\right)^{-3}$

72. $\dfrac{(2x^2y^3z^{-3})^5(3x^{-4}y^3z^{-1})^{-1}}{(-3x^2y^4)^{-2}(-2xyz)^{-3}}$

73. $\dfrac{(-2x^{-3}yz)^{-3}(-xy^{-1}z^2)^2(3^{-1}xyz^2)^2}{(2^{-1}x^2y)^3(3^{-2}xy^{-1})^{-1}(-xz^2)^{-3}}$

74. $\left(\dfrac{2^{-1}ab^{-2}}{3a^2b^{-3}}\right)^{-2}\left(\dfrac{-5a^{-1}b^3}{4ab^{-3}}\right)^2\left(\dfrac{-2^{-1}ab}{3a^{-4}b}\right)^3$

Write each expression as 2 raised to a power. Assume that the variables represent integers.

75. $32 \cdot 64$ **76.** 8^{20} **77.** $81 \cdot 6^{-4}$

78. $10^{-6} \cdot 20^6$ **79.** 4^{3n} **80.** $6^{n-5} \cdot 3^{5-n}$

81. $\left(\dfrac{1}{16^{-m}}\right)^m$ **82.** $\dfrac{32^n}{128^{3n}}$

⊞ *Use a calculator to evaluate each expression. Round approximate answers to three decimal places.*

83. $\dfrac{1}{5^{-2}}$

84. $\dfrac{(2.5)^{-3}}{(2.5)^{-5}}$

85. $2^{-1} + 2^{-2}$

86. $\left(\dfrac{2}{3}\right)^{-1} + 2^{-1}$

87. $(0.036)^{-2} + (4.29)^3$

88. $3(4.71)^2 - 5(0.471)^{-3}$

89. $\dfrac{(5.73)^{-1} + (4.29)^{-1}}{(3.762)^{-1}}$

90. $[5.29 + (0.374)^{-1}]^3$

Solve each problem. See Examples 6 and 7.

91. *Deeper in debt.* Melissa borrowed $40,000 at 12% compounded annually and made no payments for 3 years. How much did she owe the bank at the end of the 3 years? (Use the compound interest formula.)

92. *Comparing stocks and bonds.* According to Fidelity Investments of Boston, throughout the 1980s and early 1990s, annual returns on common stocks averaged 18%, while annual returns on bonds averaged 12%. If you had invested $10,000 in bonds in 1980 and achieved the average return, then what would your investment be worth after 13 years in 1993? How much more would your investment be worth in 1993 if you had invested your $10,000 in stocks in 1980?

Figure for Exercise 92

93. *Saving for college.* Mr. Watkins wants to have $10,000 in a savings account when his little Wanda is ready for college. How much must he deposit today in an account paying 7% compounded annually to have $10,000 in 18 years?

94. *Saving for retirement.* In the early 1990s, returns on treasury bills fell to an average of 4.5% per year (Fidelity Investments). Wilma wants to have $2,000,000 when she retires in 45 years. If she assumes an average annual return of 4.5%, then how much must she invest now in treasury bills to achieve her goal?

⊞ **95.** *Life expectancy of white males.* Strange as it may seem, your life expectancy increases as you get older ("Monitoring Your Health", *Readers Digest*, 1991). The formula

$$L = 67.0166(1.00308)^a$$

can be used to model life expectancy L for U.S. white males with present age a. How long can a 20-year-old white male expect to live? How long can a 70-year-old white male expect to live? (See also Chapter Review Exercises 153 and 154.)

⊞ **96.** *Life expectancy of white females.* Life expectancy improved more for females than males during the 1940s and 1950s owing to medical advances that caused a dramatic decrease in maternal mortality rates ("Monitoring You Health," *Reader's Digest,* 1991). The formula

$$L = 74.9742(1.00201)^a$$

can be used to model life expectancy L for U.S. white females with present age a. How long can a 20-year-old white female expect to live? If Bob, 30, and Ashley, 26, are an average white couple, then how many years can Ashley expect to live as a widow? (See Exercise 95.)

Getting More Involved

97. *Discussion.* For which values of a and b is it true that $(ab)^{-1} = a^{-1}b^{-1}$? Find a pair of nonzero values for a and b for which $(a + b)^{-1} \neq a^{-1} + b^{-1}$.

98. *Writing.* Explain how to evaluate $\left(-\dfrac{2}{3}\right)^{-3}$ in three different ways.

99. *Discussion.* Which of the following expressions has a value different from the others? Explain.

a) -1^{-1} **b)** -3^0 **c)** $-2^{-1} - 2^{-1}$

d) $(-1)^{-2}$ **e)** $(-1)^{-3}$

3.3 Addition, Subtraction, and Multiplication of Polynomials

A polynomial is a particular type of algebraic expression that serves as a fundamental building block in algebra. We used polynomials in Chapters 1 and 2, but we did not identify them as polynomials. In this section you will learn to recognize polynomials and to add, subtract, and multiply them.

Polynomials

The expression $3x^3 - 15x^2 + 7x - 2$ is an example of a polynomial in one variable. Because this expression could be written as

$$3x^3 + (-15x^2) + 7x + (-2),$$

we say that this polynomial is a sum of four terms:

$$3x^3, \quad -15x^2, \quad 7x, \quad \text{and} \quad -2.$$

A **term** of a polynomial is a single number or the product of a number and one or more variables raised to whole-number powers. The number preceding the variable in each term is called the **coefficient** of that variable. In $3x^3 - 15x^2 + 7x - 2$ the coefficient of x^3 is 3, the coefficient of x^2 is -15, and the coefficient of x is 7. In algebra a number is frequently referred to as a **constant,** and so the last term -2 is called the **constant term.** A **polynomial** is defined as a single term or a sum of a finite number of terms.

EXAMPLE 1   **Identifying polynomials**

Determine whether each algebraic expression is a polynomial.

a) -3 b) $3x + 2^{-1}$ c) $3x^{-2} + 4y^2$

d) $\dfrac{1}{x} + \dfrac{1}{x^2}$ e) $x^{49} - 8x^2 + 11x - 2$

Solution

a)   The number -3 is a polynomial of one term, a constant term.

b)   Since $3x + 2^{-1}$ can be written as $3x + \frac{1}{2}$, it is a polynomial of two terms.

c)   The expression $3x^{-2} + 4y^2$ is not a polynomial because x has a negative exponent.

d)   If this expression is rewritten as $x^{-1} + x^{-2}$, then it fails to be a polynomial because of the negative exponents. So a polynomial does not have variables in denominators, and

$$\frac{1}{x} + \frac{1}{x^2}$$

is not a polynomial.

e)   The expression $x^{49} - 8x^2 + 11x - 2$ is a polynomial.   ■

For simplicity we usually write polynomials in one variable with the exponents in decreasing order from left to right. Thus we would write

$$3x^3 - 15x^2 + 7x - 2 \qquad \text{rather than} \qquad -15x^2 - 2 + 7x + 3x^3.$$

When a polynomial is written in decreasing order, the coefficient of the first term is called the **leading coefficient.**

Certain polynomials have special names depending on the number of terms. A **monomial** is a polynomial that has one term, a **binomial** is a polynomial that has two terms, and a **trinomial** is a polynomial that has three terms. The **degree** of a polynomial in one variable is the highest power of the variable in the polynomial. The number 0 is considered to be a monomial without degree, because $0 = 0x^n$, where n could be any number.

EXAMPLE 2   **Identifying coefficients and degree**

State the degree of each polynomial and the coefficient of x^2. Determine whether the polynomial is monomial, binomial, or trinomial.

a) $\dfrac{x^2}{3} - 5x^3 + 7$ **b)** $x^{48} - x^2$ **c)** 6

Solution

a) The degree of this trinomial is 3, and the coefficient of x^2 is $\frac{1}{3}$.

b) The degree of this binomial is 48, and the coefficient of x^2 is -1.

c) Because $6 = 6x^0$, the number 6 is a monomial with degree 0. Because x^2 does not appear in this polynomial, the coefficient of x^2 is 0. ∎

Although we are mainly concerned here with polynomials in one variable, we will also encounter polynomials in more than one variable, such as

$$4x^2 - 5xy + 6y^2, \qquad x^2 + y^2 + z^2, \qquad \text{and} \qquad ab^2 - c^3.$$

In a term containing more than one variable, the coefficient of a variable consists of all other numbers and variables in the term. For example, the coefficient of x in $-5xy$ is $-5y$, and the coefficient of y is $-5x$.

Evaluating Polynomials

The formula $D = -16t^2 + v_0 t + s_0$ is used to model the effect of gravity on an object tossed straight upward with initial velocity v_0 from an initial height of s_0. For example, if a ball is tossed into the air at 64 feet per second from a height of 4 feet, then $D = -16t^2 + 64t + 4$ gives the ball's distance above the ground in feet, t seconds after it is tossed. The values of t range from $t = 0$ when the ball is tossed to the time when it hits the ground. To emphasize that the value of D depends on t, we often write $D(t)$ in place of D:

$$D(t) = -16t^2 + 64t + 4$$

We read $D(t)$ as "D of t." The expression $D(t)$ is the value of the polynomial at time t. To find the value when $t = 2$, replace t by 2:

$$D(2) = -16 \cdot 2^2 + 64 \cdot 2 + 4$$
$$= -16 \cdot 4 + 128 + 4$$
$$= 68$$

The statement $D(2) = 68$ means that the ball is 68 feet above the ground 2 seconds after the ball was tossed upward. Note that $D(2)$ does not mean D times 2.

EXAMPLE 3 **Finding the value of a polynomial**

Suppose $Q(x) = 2x^3 - 3x^2 - 7x - 6$. Find $Q(3)$ and $Q(-1)$.

Solution

To find $Q(3)$, replace x by 3 in $Q(x) = 2x^3 - 3x^2 - 7x - 6$:

$$Q(3) = 2 \cdot 3^3 - 3 \cdot 3^2 - 7 \cdot 3 - 6$$
$$= 54 - 27 - 21 - 6$$
$$= 0$$

To find $Q(-1)$, replace x by -1 in $Q(x) = 2x^3 - 3x^2 - 7x - 6$:

$$Q(-1) = 2(-1)^3 - 3(-1)^2 - 7(-1) - 6$$
$$= -2 - 3 + 7 - 6$$
$$= -4$$

So $Q(3) = 0$ and $Q(-1) = -4$. ∎

Addition and Subtraction of Polynomials

When we evaluate a polynomial, we get a real number. So the operations that we perform with real numbers can be performed with polynomials. Actually, we have been adding and subtracting polynomials since Chapter 1. To add two polynomials, we simply add the like terms.

EXAMPLE 4 **Adding polynomials**

Find the sums.

a) $(x^2 - 5x - 7) + (7x^2 - 4x + 10)$

b) $(3x^3 - 5x^2 - 7) + (4x^2 - 2x + 3)$

Solution

a) $(x^2 - 5x - 7) + (7x^2 - 4x + 10) = 8x^2 - 9x + 3$ Combine like terms.

b) For illustration we will write this addition vertically:

$$
\begin{array}{l}
3x^3 - 5x^2 \qquad - 7 \\
\underline{\qquad\quad 4x^2 - 2x + 3} \quad \text{Line up like terms.} \\
3x^3 - \quad x^2 - 2x - 4 \quad \text{Add.}
\end{array}
$$
∎

When we subtract polynomials, we subtract like terms. Because $a - b = a + (-b)$, we often perform subtraction by changing signs and adding.

EXAMPLE 5 **Subtracting polynomials**

Find the differences.

a) $(x^2 - 7x - 2) - (5x^2 + 6x - 4)$

b) $(6y^3z - 5yz + 7) - (4y^2z - 3yz - 9)$

Solution

a) We find the first difference horizontally:

$$(x^2 - 7x - 2) - (5x^2 + 6x - 4) = x^2 - 7x - 2 - 5x^2 - 6x + 4 \qquad \text{Change signs.}$$
$$= -4x^2 - 13x + 2 \qquad\qquad \begin{array}{l}\text{Combine like}\\ \text{terms.}\end{array}$$

b) For illustration we write $(6y^3z - 5yz + 7) - (4y^2z - 3yz - 9)$ vertically:

$$
\begin{array}{ll}
6y^3z \qquad\quad - 5yz + \ \ 7 & \\
\underline{-4y^2z + 3yz + \ \ 9} & \text{Change signs.} \\
6y^3z - 4y^2z - 2yz + 16 & \text{Add.}
\end{array}
$$
■

It is certainly not necessary to write out all of the steps shown in Examples 4 and 5, but we must use the following rule.

> **Addition and Subtraction of Polynomials**
>
> To add two polynomials, add the like terms.
> To subtract two polynomials, subtract the like terms.

Multiplication of Polynomials

We learned how to multiply monomials when we learned the product rule in Section 3.1. For example,

$$-2x^3 \cdot 4x^2 = -8x^5.$$

To multiply a monomial and a polynomial of two or more terms, we apply the distributive property. For example,

$$3x(x^3 - 5) = 3x^4 - 15x.$$

EXAMPLE 6 **Multiplying a monomial and a polynomial**

Find the products.

a) $2ab^2 \cdot 3a^2b$ **b)** $(-1)(5 - x)$ **c)** $(x^3 - 5x + 2)(-3x)$

Solution

a) $2ab^2 \cdot 3a^2b = 6a^3b^3$

b) $(-1)(5 - x) = -5 + x = x - 5$

c) Each term of $x^3 - 5x + 2$ is multiplied by $-3x$:

$$(x^3 - 5x + 2)(-3x) = -3x^4 + 15x^2 - 6x$$
■

Note what happened to the binomial in Example 6(b) when we multiplied it by -1. If we multiply any difference by -1, we get the same type of result:

$$-1(a - b) = -a + b = b - a.$$

Because multiplying by -1 is the same as taking the opposite, we can write this equation as

$$-(a - b) = b - a.$$

This equation says that $a - b$ and $b - a$ are opposites or additive inverses of each other. Note that the opposite of $a + b$ is $-a - b$, not $a - b$.

To multiply a binomial and a trinomial, we can use the distributive property or we can set it up like multiplication of whole numbers.

EXAMPLE 7   **Multiplying a binomial and a trinomial**

Find the product $(x + 2)(x^2 + 3x - 5)$.

Solution

We can find this product by apply the distributive property twice. First we multiply the binomial and each term of the trinomial:

$$
\begin{aligned}
(x + 2)(x^2 + 3x - 5) &= (x + 2)x^2 + (x + 2)3x + (x + 2)(-5) && \text{Distributive property} \\
&= x^3 + 2x^2 + 3x^2 + 6x - 5x - 10 && \text{Distributive property} \\
&= x^3 + 5x^2 + x - 10 && \text{Combine like terms.}
\end{aligned}
$$

We could have found this product vertically:

$$
\begin{array}{r}
x^2 + 3x - 5 \\
x + 2 \\
\hline
2x^2 + 6x - 10 \\
x^3 + 3x^2 - 5x \\
\hline
x^3 + 5x^2 + x - 10
\end{array}
$$

$2(x^2 + 3x - 5) = 2x^2 + 6x - 10$

$x(x^2 + 3x - 5) = x^3 + 3x^2 - 5x$

Add.   ■

Multiplication of Polynomials

To multiply polynomials, multiply each term of the first polynomial by each term of the second polynomial and then combine like terms.

In the next example we multiply binomials.

EXAMPLE 8   **Multiplying binomials**

Find the products.

a) $(x + y)(z + 4)$ **b)** $(x - 3)(2x + 5)$

Solution

a)
$$
\begin{aligned}
(x + y)(z + 4) &= (x + y)z + (x + y)4 && \text{Distributive property} \\
&= xz + yz + 4x + 4y && \text{Distributive property}
\end{aligned}
$$

Notice that this product does not have any like terms to combine.

b) Multiply:

$$
\begin{array}{r}
x - 3 \\
2x + 5 \\
\hline
5x - 15 \\
2x^2 - 6x \\
\hline
2x^2 - x - 15
\end{array}
$$

■

Warm-ups

True or false? Explain your answers.

1. The expression $3x^{-2} - 5x + 2$ is a trinomial.
2. In the polynomial $3x^2 - 5x + 3$ the coefficient of x is 5.
3. The degree of the polynomial $x^2 + 3x - 5x^3 + 4$ is 2.
4. If $C(x) = x^2 - 3$, then $C(5) = 22$.
5. If $P(t) = 30t + 10$, then $P(0) = 40$.
6. $(2x^2 - 3x + 5) + (x^2 + 5x - 7) = 3x^2 + 2x - 2$ for any value of x.
7. $(x^2 - 5x) - (x^2 - 3x) = -8x$ for any value of x.
8. $-2x(3x - 4x^2) = 8x^3 - 6x^2$ for any value of x.
9. $-(x - 7) = 7 - x$ for any value of x.
10. The opposite of $y + 5$ is $y - 5$ for any value of y.

3.3 EXERCISES

Determine whether each algebraic expression is a polynomial. See Example 1.

1. $3x$
2. -9
3. $x^{-1} + 4$
4. $3x^{-3} + 4x - 1$
5. $x^2 - 3x + 1$
6. $\dfrac{x^3}{3} - \dfrac{3x^2}{2} + 4$
7. $\dfrac{1}{x} + y - 3$
8. $x^{50} - \dfrac{9}{y^2}$

State the degree of each polynomial and the coefficient of x^3. Determine whether each polynomial is a monomial, binomial, or trinomial. See Example 2.

9. $x^4 - 8x^3$
10. $15 - x^3$
11. -8
12. 17
13. $\dfrac{x^7}{15}$
14. $5x^4$
15. $x^3 + 3x^4 - 5x^6$
16. $\dfrac{x^3}{2} + \dfrac{5x}{2} - 7$

For each given polynomial, find the indicated value of the polynomial. See Example 3.

17. $P(x) = x^4 - 1$, $P(3)$
18. $P(x) = x^2 - x - 2$, $P(-1)$
19. $M(x) = -3x^2 + 4x - 9$, $M(-2)$
20. $C(w) = 3w^2 - w$, $C(0)$
21. $R(x) = x^5 - x^4 + x^3 - x^2 + x - 1$, $R(1)$
22. $T(a) = a^7 + a^6$, $T(-1)$

Perform the indicated operations. See Examples 4 and 5.

23. $(2a - 3) + (a + 5)$
24. $(2w - 6) + (w + 5)$
25. $(7xy + 30) - (2xy + 5)$
26. $(5ab + 7) - (3ab + 6)$
27. $(x^2 - 3x) + (-x^2 + 5x - 9)$
28. $(2y^2 - 3y - 8) + (y^2 + 4y - 1)$
29. $(2x^3 - 4x - 3) - (x^2 + 2x + 5)$
30. $(2x - 5) - (x^2 - 3x + 2)$

Perform the indicated operations vertically. See Examples 4 and 5.

31. Add
$x^3 + 3x^2 - 5x - 2$
$-x^3 + 8x^2 + 3x - 7$

32. Add
$x^2 - 3x + 7$
$-2x^2 - 5x + 2$

33. Subtract
$5x + 2$
$4x - 3$

34. Subtract
$4x + 3$
$2x - 6$

35. Subtract
$-x^2 + 3x - 5$
$5x^2 - 2x - 7$

36. Subtract
$-3x^2 + 5x - 2$
$x^2 - 5x - 6$

37. Add
$x - y$
$x + y$

38. Add
$-w + 4$
$2w - 3$

Find each product. See Examples 6–8.

39. $-3x^2 \cdot 5x^4$

40. $(-ab^5)(-2a^2b)$

41. $-1(3x - 2)$

42. $-1(-x^2 + 3x - 9)$

43. $5x^2y^3(3x^2y - 4x)$

44. $3y^4z(8y^2z^2 - 3yz + 2y)$

45. $(x - 2)(x + 2)$

46. $(x - 1)(x + 1)$

47. $(x^2 + x + 2)(2x - 3)$

48. $(x^2 - 3x + 2)(x - 4)$

Find each product vertically. See Examples 6–8.

49. Multiply
$$2x - 3$$
$$\underline{-5x}$$

50. Multiply
$$3a^3 - 5a^2 + 7$$
$$\underline{-2a}$$

51. Multiply
$$x + 5$$
$$\underline{x + 5}$$

52. Multiply
$$a + b$$
$$\underline{a - b}$$

53. Multiply
$$x + 6$$
$$\underline{2x - 3}$$

54. Multiply
$$3x^2 + 2$$
$$\underline{2x^2 - 5}$$

55. Multiply
$$x^2 + xy + y^2$$
$$\underline{x - y}$$

56. Multiply
$$a^2 - ab + b^2$$
$$\underline{a + b}$$

Perform the indicated operations.

57. $(x - 7) + (2x - 3) + (5 - x)$

58. $(5x - 3) + (x^3 + 3x - 2) + (-2x - 3)$

59. $(a^2 - 5a + 3) + (3a^2 - 6a - 7)$

60. $(w^2 - 3w + 2) + (2w - 3 + w^2)$

61. $(w^2 - 7w - 2) + (w + 3w^2 + 5)$

62. $(a^3 - 3a) - (1 - a - 2a^2)$

63. $(x - 2)(x^2 + 2x + 4)$

64. $(a - 3)(a^2 + 3a + 9)$

65. $(x - w)(z + 2w)$

66. $(w^2 - a)(t^2 + 3)$

67. $(2xy - 1)(3xy + 5)$

68. $(3ab - 4)(ab + 8)$

⊞ *Perform the following operations using a calculator.*

69. $(2.31x - 5.4)(6.25x + 1.8)$

70. $(x - 0.28)(x^2 - 34.6x + 21.2)$

71. $(3.759x^2 - 4.71x + 2.85)$
$\quad + (11.61x^2 + 6.59x - 3.716)$

72. $(43.19x^3 - 3.7x^2 - 5.42x + 3.1)$
$\quad - (62.7x^3 - 7.36x - 12.3)$

Perform the indicated operations.

73. $\left(\frac{1}{2}x + 2\right) + \left(\frac{1}{4}x - \frac{1}{2}\right)$

74. $\left(\frac{1}{3}x + 1\right) + \left(\frac{1}{3}x - \frac{3}{2}\right)$

75. $\left(\frac{1}{2}x^2 + \frac{1}{3}x - \frac{1}{5}\right) - \left(x^2 - \frac{2}{3}x - \frac{1}{5}\right)$

76. $\left(\frac{2}{3}x^2 - \frac{1}{3}x + \frac{1}{6}\right) - \left(-\frac{1}{3}x^2 + x + 1\right)$

77. $[x^2 - 3 - (x^2 + 5x - 4)] - [x - 3(x^2 - 5x)]$

78. $[x^3 - 4x(x^2 - 3x + 2) - 5x] + [x^2 - 5(4 - x^2) + 3]$

79. $[5x - 4(x - 3)][3x - 7(x + 2)]$

80. $[x^2 - (5x - 2)][x^2 + (5x - 2)]$

81. $[x^2 - (m + 2)][x^2 + (m + 2)]$

82. $[3x^2 - (x - 2)][3x^2 + (x + 2)]$

83. $2x(5x - 4) - 3x[5x^2 - 3x(x - 7)]$

84. $-3x(x - 2) - 5[2x - 4(x + 6)]$

Perform the indicated operations. A variable used in an exponent represents an integer; a variable used as a base represents a nonzero real number.

85. $(a^{2m} + 3a^m - 3) + (-5a^{2m} - 7a^m + 8)$

86. $(b^{3z} - 6) - (4b^{3z} - b^{2z} - 7)$

87. $(x^n - 1)(x^n + 3)$

88. $(2y^t - 3)(4y^t + 7)$

89. $z^{3w} - z^{2w}(z^{1-w} - 4z^w)$

90. $(w^p - 1)(w^{2p} + w^p + 1)$

91. $(x^{2r} + y)(x^{4r} - x^{2r}y + y^2)$

92. $(2x^a - z)(2x^a + z)$

Solve each problem. See Exercises 93–99.

93. *Cost of gravel.* The cost in dollars of x cubic yards of gravel is given by the formula
$$C(x) = 20x + 15.$$
Find $C(3)$, the cost of 3 cubic yards of gravel.

94. *Annual bonus.* Sheila's annual bonus in dollars for selling n life insurance policies is given by the formula
$$B(n) = 0.1n^2 + 3n + 50.$$
Find $B(20)$, her bonus for selling 20 policies.

95. *Marginal cost.* A company uses the formula $C(n) = 50n - 0.01n^4$ to find the daily cost in dollars of manufacturing n aluminum windows. The *marginal cost* of the nth window is the additional cost incurred for manufacturing that window. For example, the marginal cost of the third window is $C(3) - C(2)$. Find the marginal cost for manufacturing the third window. What is the marginal cost for manufacturing the tenth window?

96. *Marginal profit.* A company uses the formula $P(n) = 4n + 0.9n^3$ to estimate its daily profit in dollars for producing n automatic garage door openers. The *marginal profit* of the nth opener is the amount of additional profit made for that opener. For example, the marginal profit for the fourth opener is $P(4) - P(3)$. Find the marginal profit for the fourth opener. What is the marginal profit for the tenth opener? Use the bar graph to explain why the marginal profit increases as production goes up.

Figure for Exercise 96

97. *Male life expectancy at birth.* Since 1950 the life expectancy of a U.S. male born in year y is modeled by the formula

$$L(y) = 0.15863y - 244.25497$$

("Monitoring Your Health," *Reader's Digest*, 1991). How much greater is the life expectancy of a male born in 1990 than a male born in 1960?

98. *Female life expectancy at birth.* Since 1950 the life expectancy of a U.S. female born in year y is modeled by the formula

$$L(y) = 0.203424y - 325.66257$$

("Monitoring Your Health," *Reader's Digest*, 1991). How many more years is a female born in 1990 expected to live than a male born in 1990? (See Exercise 97.) Use the accompanying graph to estimate the year of birth for which a female's life expectancy will exceed a male's life expectancy by 10 years. Use the formulas to find the year of birth for which a female's life expectancy will exceed a male's life expectancy by 10 years.

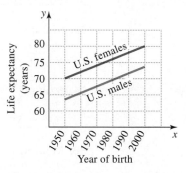

Figure for Exercises 97 and 98

Getting More Involved

99. *Discussion.* Is it possible for a binomial to have degree 4? If so, give an example.

100. *Discussion.* Give an example of two fourth-degree trinomials whose sum is a third-degree binomial?

101. *Cooperative learning.* Work in a group to find the product $(a + b)(c + d)$. How many terms does it have? Find the product $(a + b)(c + d)(e + f)$. How many terms does it have? How many terms are there in a product of four binomials in which there are no like terms to combine? How many terms are there in a product of n binomials in which there are no like terms?

3.4 Multiplying Binomials

In this section:
▶ The FOIL Method
▶ The Square of a Binomial
▶ Product of a Sum and a Difference

In Section 3.3 you learned to multiply polynomials. In this section you will learn rules to make multiplication of binomials simpler.

The FOIL Method

Consider how we find the product of two binomials $x + 3$ and $x + 5$ using the distributive property twice:

$$(x + 3)(x + 5) = (x + 3)x + (x + 3)5 \quad \text{Distributive property}$$
$$= x^2 + 3x + 5x + 15 \quad \text{Distributive property}$$
$$= x^2 + 8x + 15 \quad \text{Combine like terms.}$$

There are four terms in the product. The term x^2 is the product of the first term of each binomial. The term $5x$ is the product of the two outer terms, 5 and x. The term $3x$ is the product of the two inner terms, 3 and x. The term 15 is the product of the last two terms in each binomial, 3 and 5. It may be helpful to connect the terms multiplied by lines.

$$(x + 3)(x + 5) \qquad \begin{aligned} F &= \text{First terms} \\ O &= \text{Outer terms} \\ I &= \text{Inner terms} \\ L &= \text{Last terms} \end{aligned}$$

So instead of writing out all of the steps in using the distributive property, we can get the result by finding the products of the first, outer, inner, and last terms. This method is called the **FOIL method.**

For example, let's apply FOIL to the product $(x - 3)(x + 4)$:

$$(x - 3)(x + 4) = x^2 + 4x - 3x - 12 = x^2 + x - 12$$

If the outer and inner products are like terms, you can save a step by writing down only their sum.

EXAMPLE 1 **Multiplying binomials**

Use FOIL to find the products of the binomials.

a) $(2x - 3)(3x + 4)$ **b)** $(2x^3 + 5)(2x^3 - 5)$

c) $(m + w)(2m - w)$ **d)** $(a + b)(a - 3)$

Solution

$$\overset{\quad F \quad\quad O \quad\quad I \quad\quad L}{}$$

a) $(2x - 3)(3x + 4) = 6x^2 + 8x - 9x - 12 = 6x^2 - x - 12$

b) $(2x^3 + 5)(2x^3 - 5) = 4x^6 - 10x^3 + 10x^3 - 25 = 4x^6 - 25$

c) $(m + w)(2m - w) = 2m^2 - mw + 2mw - w^2 = 2m^2 + mw - w^2$

d) $(a + b)(a - 3) = a^2 - 3a + ab - 3b$ There are no like terms. ∎

The Square of a Binomial

To find $(a + b)^2$, the square of a sum, we can use FOIL on $(a + b)(a + b)$:

$$(a + b)(a + b) = a^2 + ab + ab + b^2$$
$$= a^2 + 2ab + b^2$$

You can use the result $a^2 + 2ab + b^2$ that we obtained from FOIL to quickly find the square of *any* sum. *To square a sum, we square the first term* (a^2), *add twice the product of the two terms* $(2ab)$, *then add the square of the last term* (b^2).

Rule for the Square of a Sum

$$(a + b)^2 = a^2 + 2ab + b^2$$

EXAMPLE 2   **Squaring a binomial**

Square each sum, using the new rule.

a) $(x + 5)^2$ **b)** $(2w + 3)^2$ **c)** $(2y^4 + 3)^2$

Solution

a) $(x + 5)^2 = x^2 + 2(x)(5) + 5^2 = x^2 + 10x + 25$

$\qquad\qquad\quad \uparrow \qquad\quad \uparrow \qquad\quad \uparrow$

$\qquad\qquad$ Square   Twice   Square
$\qquad\qquad$ of   the   of
$\qquad\qquad$ first   product   last

b) $(2w + 3)^2 = (2w)^2 + 2(2w)(3) + 3^2 = 4w^2 + 12w + 9$

c) $(2y^4 + 3)^2 = (2y^4)^2 + 2(2y^4)(3) + 3^2 = 4y^8 + 12y^4 + 9$ ■

CAUTION   Squaring $x + 5$ correctly, as in Example 2a, gives us the identity

$$(x + 5)^2 = x^2 + 10x + 25,$$

which is satisfied by any x. If you forget the middle term and write $(x + 5)^2 = x^2 + 25$, then you have an equation that is satisfied only if $x = 0$.   ⊘

To find $(a - b)^2$, the square of a difference, we can use FOIL:

$$(a - b)(a - b) = a^2 - ab - ab + b^2$$
$$= a^2 - 2ab + b^2$$

As in squaring a sum, it is better to simply remember the result of using FOIL. *To square a difference, square the first term, subtract twice the product of the two terms, and add the square of the last term.*

Rule for the Square of a Difference

$$(a - b)^2 = a^2 - 2ab + b^2$$

EXAMPLE 3   **Squaring a binomial**

Square each difference, using the new rule.

a) $(x - 6)^2$ **b)** $(3w - 5y)^2$ **c)** $(-4 - st)^2$ **d)** $(3 - 5a^3)^2$

Solution

a) $(x - 6)^2 = x^2 - 2(x)(6) + 6^2$ For the middle term, subtract twice
the product: $2(x)(6)$

$= x^2 - 12x + 36$

b) $(3w - 5y)^2 = (3w)^2 - 2(3w)(5y) + (5y)^2$

$= 9w^2 - 30wy + 25y^2$

c) $(-4 - st)^2 = (-4)^2 - 2(-4)(st) + (st)^2$

$= 16 + 8st + s^2t^2$

d) $(3 - 5a^3)^2 = 3^2 - 2(3)(5a^3) + (5a^3)^2$

$= 9 - 30a^3 + 25a^6$ ■

Product of a Sum and a Difference

If we multiply the sum $a + b$ and the difference $a - b$ by using FOIL, we get

$$(a + b)(a - b) = a^2 - ab + ab - b^2$$
$$= a^2 - b^2.$$

The inner and outer products add up to zero, canceling each other out. So *the product of a sum and a difference is the difference of two squares*, as shown in the following rule.

> **Rule for the Product of a Sum and a Difference**
>
> $$(a + b)(a - b) = a^2 - b^2$$

EXAMPLE 4   **Finding the product of a sum and a difference**

Find the products.

a) $(x + 3)(x - 3)$ **b)** $(a^3 + 8)(a^3 - 8)$ **c)** $(3x^2 - y^3)(3x^2 + y^3)$

Solution

a) $(x + 3)(x - 3) = x^2 - 9$

b) $(a^3 + 8)(a^3 - 8) = a^6 - 64$

c) $(3x^2 - y^3)(3x^2 + y^3) = 9x^4 - y^6$ ■

The square of a sum, the square of a difference, and the product of a sum and a difference are referred to as **special products.** Although the special products can be found by using the distributive property or FOIL, they occur so frequently in algebra that it is essential to learn the new rules. In the next example we use the special product rules to multiply two trinomials and to square a trinomial.

EXAMPLE 5   **Using special product rules to multiply trinomials**
Find the products.

a) $[(x + y) + 3][(x + y) - 3]$ **b)** $[(m - n) + 5]^2$

Solution

a) Use the rule $(a + b)(a - b) = a^2 - b^2$ with $a = x + y$ and $b = 3$:

$$[(x + y) + 3][(x + y) - 3] = (x + y)^2 - 3^2$$
$$= x^2 + 2xy + y^2 - 9$$

b) Use the rule $(a + b)^2 = a^2 + 2ab + b^2$ with $a = m - n$ and $b = 5$:

$$[(m - n) + 5]^2 = (m - n)^2 + 2(m - n)5 + 5^2$$
$$= m^2 - 2mn + n^2 + 10m - 10n + 25$$ ■

Warm-ups

True or false? Explain your answer.

1. $(x + 2)(x + 5) = x^2 + 7x + 10$ for any value of x.
2. $(2x - 3)(3x + 5) = 6x^2 + x - 15$ for any value of x.
3. $(2 + 3)^2 = 2^2 + 3^2$
4. $(x + 7)^2 = x^2 + 14x + 49$ for any value of x.
5. $(8 - 3)^2 = 64 - 9$
6. The product of a sum and a difference of the same two terms is equal to the difference of two squares.
7. $(60 - 1)(60 + 1) = 3600 - 1$
8. $(x - y)^2 = x^2 - 2xy + y^2$ for any values of x and y.
9. $(x - 3)^2 = x^2 - 3x + 9$ for any value of x.
10. The expression $3x \cdot 5x$ is a product of two binomials.

3.4 EXERCISES

Find each product. When possible, write down only the answer. See Example 1.

1. $(x - 2)(x + 4)$
2. $(x - 3)(x + 5)$
3. $(2x + 1)(x + 3)$
4. $(2y + 3)(y + 2)$
5. $(-2a - 3)(-a + 5)$
6. $(-3x - 5)(-x + 6)$
7. $(2x^2 - 7)(2x^2 + 7)$
8. $(3y^3 + 8)(3y^3 - 8)$
9. $(2x^3 - 1)(x^3 + 4)$
10. $(3t^2 - 4)(2t^2 + 3)$
11. $(6z + w)(w - z)$
12. $(4y + w)(w - 2y)$
13. $(3k - 2t)(4t + 3k)$
14. $(7a - 2x)(x + a)$
15. $(x - 3)(y + w)$
16. $(z - 1)(y + 2)$
17. $\left(\frac{1}{2} + t\right)\left(\frac{1}{3} + t\right)$
18. $\left(\frac{1}{3} + z\right)\left(\frac{2}{3} + z\right)$
19. $\left(\frac{1}{2}x^2 - 4\right)\left(\frac{1}{2}x^2 + 6\right)$
20. $\left(\frac{1}{3}a^2 - 3\right)\left(\frac{1}{3}a^2 + 6\right)$

Find the square of each sum or difference. When possible, write down only the answer. See Examples 2 and 3.

21. $(m + 3)^2$
22. $(a + 2)^2$
23. $(a - 4)^2$
24. $(b - 3)^2$
25. $(2w + 1)^2$
26. $(3m + 4)^2$
27. $(3t - 5u)^2$
28. $(3w - 2x)^2$
29. $(-x - 1)^2$
30. $(-d - 5)^2$
31. $(a - 3y^3)^2$
32. $(3m - 5n^3)^2$
33. $\left(\frac{1}{2}x - 1\right)^2$
34. $\left(\frac{2}{3}x - \frac{1}{2}\right)^2$
35. $\left(\frac{3}{4}a + \frac{1}{3}\right)^2$
36. $\left(\frac{3}{2}x + \frac{1}{2}\right)^2$
37. $(-2 + gt)^2$
38. $(-1 + 3rs)^2$
39. $(8y^2z^4 + 3)^2$
40. $(3x^2y^3 + 9)^2$

Find each product. See Example 4.

41. $(w - 9)(w + 9)$ **42.** $(m - 4)(m + 4)$

43. $(w^3 + y)(w^3 - y)$ **44.** $(a^3 - x)(a^3 + x)$

45. $(2x - 7)(2x + 7)$ **46.** $(5x + 3)(5x - 3)$

47. $(3x^2 - 2)(3x^2 + 2)$ **48.** $(4y^2 + 1)(4y^2 - 1)$

Use the special product rules to find each product. See Example 5.

49. $[(m + t) + 5][(m + t) - 5]$

50. $[(2x + 3) - y][(2x + 3) + y]$

51. $[y - (r + 5)][y + (r + 5)]$

52. $[x + (3 - k)][x - (3 - k)]$

53. $[(2y - t) + 3]^2$ **54.** $[(u - 3v) - 4]^2$

55. $[3h + (k - 1)]^2$ **56.** $[2p - (3q + 6)]^2$

Perform the operations and simplify.

57. $(x - 6)(x + 9)$ **58.** $(2x^2 - 3)(3x^2 + 4)$

59. $(5 - x)(5 + x)$ **60.** $(4 - ab)(4 + ab)$

61. $(3x - 4a)(2x + 5a)$ **62.** $(x^5 + 2)(x^5 - 2)$

63. $(2t - 3)(t + w)$ **64.** $(5x - 9)(ax + b)$

65. $(3x^2 + 2y^3)^2$ **66.** $(5a^4 - 2b)^2$

67. $(2y + 2)(3y - 5)$ **68.** $(3b - 3)(2b + 3)$

69. $(50 + 2)(50 - 2)$ **70.** $(100 - 1)(100 + 1)$

71. $(3 + 7x)^2$ **72.** $(1 - pq)^2$

73. $4y\left(3y + \dfrac{1}{2}\right)^2$ **74.** $25y\left(2y - \dfrac{1}{5}\right)^2$

75. $(a + h)^2 - a^2$ **76.** $\dfrac{(x + h)^2 - x^2}{h}$

77. $(x + 2)(x + 2)^2$ **78.** $(a + 1)^2(a + 1)^2$

79. $(y - 3)^3$ **80.** $(2b + 1)^4$

▦ *Use a calculator to help you to perform the following operations.*

81. $(3.2x - 4.5)(5.1x + 3.9)$

82. $(5.3x - 9.2)^2$ **83.** $(3.6y + 4.4)^2$

84. $(3.3a - 7.9b)(3.3a + 7.9b)$

Find the products. Assume that all variables are nonzero and that variables used in exponents represent integers.

85. $(x^m + 2)(x^{2m} + 3)$ **86.** $(a^n - b)(a^n + b)$

87. $a^{n+1}(a^{2n} + a^n - 3)$ **88.** $x^{3b}(x^{-3b} + 3x^{-b} + 5)$

89. $(a^m + a^n)^2$ **90.** $(x^w - x^t)^2$

91. $(5y^m + 8z^k)(3y^{2m} + 4z^{3-k})$

92. $(4x^{a-1} + 3y^{b+5})(x^{2a-3} - 2y^{4-b})$

Solve each problem.

93. *Area of a room.* Suppose that the length of a rectangular room is $x + 3$ meters and the width is $x + 1$ meters. Find a trinomial that can be used to represent the area of the room.

94. *House plans.* Barbie and Ken planned to build a square house with area x^2 square feet. Then they revised the plan so that one side was lengthened by 20 feet and the other side was shortened by 6 feet. Find a trinomial that can be used to represent the area of the revised house.

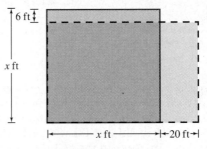

Figure for Exercise 94

95. *Available habitat.* A wild animal will generally stay more than x kilometers from the edge of a forest preserve. So the available habitat for the animal excludes an area of uniform width x on the edge of the rectangular forest preserve shown in the figure. The value of x depends on the animal. Find a trinomial in x that gives the area of the available habitat in square kilometers for the forest preserve shown. What is the available habitat in this forest preserve for a bobcat for which $x = 0.4$ km.

Figure for Exercise 95

96. *Cubic coating.* A cubic metal box x inches on each side is designed for transporting frozen specimens. The box is surrounded on all sides by a 2-inch-thick layer of styrofoam insulation. Find a polynomial that represents the total volume of the cube and styrofoam.

97. *Overflow pan.* An air conditioning contractor makes an overflow pan for a condenser by cutting squares with side of length x ft from the corners of a 4-ft by 6-ft piece of galvanized sheet metal as shown in the figure. The sides are then folded up, and the corners are sealed. Find a polynomial that gives the volume of the pan in cubic feet. What is the volume of the pan if $x = 4$ in.?

Figure for Exercise 97

98. *Energy efficient.* A manufacturer of mobile homes makes a custom model that is x feet long, 12 feet wide, and 8 feet high (all inside dimensions). The insulation is 3 in. thick in the walls, 6 in. thick in the floor, and 8 in. thick in the ceiling. Given that the insulation costs the manufacturer 25 cents per cubic foot and doors and windows take up 80 square feet of wall space, find a polynomial in x that gives the cost in dollars for insulation in this model. State any assumptions that you are making to solve this problem.

Getting More Involved

99. *Exploration.* a) Find $(a + b)^3$ by multiplying $(a + b)^2$ by $a + b$. **b)** Next find $(a + b)^4$ and $(a + b)^5$. **c)** How many terms are in each of these powers of $a + b$ after combining like terms? **d)** Make a general statement about the number of terms in $(a + b)^n$.

100. *Cooperative learning.* Make a four-column table with columns for a, b, $(a + b)^2$, and $a^2 + b^2$. Work with a group to fill in the table with five pairs of numbers for a and b for which $(a + b)^2 \neq a^2 + b^2$. For what values of a and b does $(a + b)^2 = a^2 + b^2$?

101. *Discussion.* The area of the large square shown in the figure is $(a + b)^2$. Find the area of each of the four smaller regions in the figure, and then find the sum of those areas. What conclusion can you draw from these areas about $(a + b)^2$?

Figure for Exercise 101

<hr>

3.5 Division of Polynomials

In this section:

▶ Dividing a Polynomial by a Monomial

▶ Dividing a Polynomial by a Binomial

▶ Synthetic Division

▶ Division and Factoring

We began our study of polynomials in Section 3.3 by learning how to add, subtract, and multiply polynomials. In this section we will study division of polynomials.

Dividing a Polynomial by a Monomial

You learned how to divide monomials in Section 3.1. For example,

$$6x^3 \div (3x) = \frac{6x^3}{3x} = 2x^2.$$

We check by multiplying. Because $2x^2 \cdot 3x = 6x^3$, this answer is correct. Recall that $a \div b = c$ if and only if $c \cdot b = a$. We call a the **dividend,** b the **divisor,** and c the **quotient.** We may also refer to $a \div b$ and $\frac{a}{b}$ as quotients.

We can use the distributive property to find that

$$3x(2x^2 + 5x - 4) = 6x^3 + 15x^2 - 12x.$$

So if we divide $6x^3 + 15x^2 - 12x$ by the monomial $3x$, we must get $2x^2 + 5x - 4$. We can perform this division by dividing $3x$ into *each term* of $6x^3 + 15x^2 - 12x$:

$$\frac{6x^3 + 15x^2 - 12x}{3x} = \frac{6x^3}{3x} + \frac{15x^2}{3x} - \frac{12x}{3x}$$

$$= 2x^2 + 5x - 4$$

In this case the divisor is $3x$, the dividend is $6x^3 + 15x^2 - 12x$, and the quotient is $2x^2 + 5x - 4$.

EXAMPLE 1   **Dividing polynomials**

Find the quotient.

a) $-12x^5 \div (2x^3)$ **b)** $(-20x^6 + 8x^4 - 4x^2) \div (4x^2)$

Solution

a)   When dividing x^5 by x^3, we subtract the exponents:

$$-12x^5 \div (2x^3) = \frac{-12x^5}{2x^3} = -6x^2$$

The quotient is $-6x^2$. Check:

$$-6x^2 \cdot 2x^3 = -12x^5$$

b)   Divide each term of $-20x^6 + 8x^4 - 4x^2$ by $4x^2$:

$$\frac{-20x^6 + 8x^4 - 4x^2}{4x^2} = \frac{-20x^6}{4x^2} + \frac{8x^4}{4x^2} - \frac{4x^2}{4x^2}$$

$$= -5x^4 + 2x^2 - 1$$

The quotient is $-5x^4 + 2x^2 - 1$. Check:

$$4x^2(-5x^4 + 2x^2 - 1) = -20x^6 + 8x^4 - 4x^2$$ ■

Dividing a Polynomial by a Binomial

We can multiply $x - 2$ and $x + 5$ to get

$$(x - 2)(x + 5) = x^2 + 3x - 10.$$

So if we divide $x^2 + 3x - 10$ by the factor $x - 2$, we should get the other factor $x + 5$. This division is not done like division by a monomial; it is done like long division of whole numbers. We get the first term of the quotient by dividing the first term of $x - 2$ into the first term of $x^2 + 3x - 10$. Divide x^2 by x to get x.

$$
\begin{array}{r}
x \\
x - 2 \overline{)x^2 + 3x - 10} \\
\underline{x^2 - 2x} \\
5x
\end{array}
$$

$x^2 \div x = x$

Multiply: $x(x - 2) = x^2 - 2x$

Subtract: $3x - (-2x) = 5x$

Now bring down -10. We get the second term of the quotient (below) by dividing the first term of $x - 2$ into the first term of $5x - 10$. Divide $5x$ by x to get 5.

$$
\begin{array}{r}
x + 5 \\
x - 2 \overline{)x^2 + 3x - 10} \\
\underline{x^2 - 2x} \\
5x - 10 \\
\underline{5x - 10} \\
0
\end{array}
$$

$5x \div x = 5$

Multiply: $5(x - 2) = 5x - 10$

Subtract: $-10 - (-10) = 0$

So the quotient is $x + 5$ and the **remainder** is 0. If the remainder is not 0, then

$$\textbf{dividend} = \textbf{(divisor)(quotient)} + \textbf{(remainder).}$$

If we divide each side of this equation by the divisor, we get

$$\frac{\textbf{dividend}}{\textbf{divisor}} = \textbf{quotient} + \frac{\textbf{remainder}}{\textbf{divisor}}.$$

When dividing polynomials, we must write the terms of the divisor and the dividend in descending order of the exponents. If any terms are missing, as in the next example, we insert terms with a coefficient of 0 as placeholders. When dividing polynomials, we stop the process when the degree of the remainder is smaller than the degree of the divisor.

EXAMPLE 2   **Dividing polynomials**

Find the quotient and remainder for $(3x^4 - 2 - 5x) \div (x^2 - 3x)$.

Solution

Rearrange $3x^4 - 2 - 5x$ as $3x^4 - 5x - 2$ and insert the terms $0x^3$ and $0x^2$:

$$
\begin{array}{r}
3x^2 + 9x + 27 \\
x^2 - 3x \overline{)3x^4 + 0x^3 + 0x^2 - 5x - 2} \\
\underline{3x^4 - 9x^3} \\
9x^3 + 0x^2 \\
\underline{9x^3 - 27x^2} \\
27x^2 - 5x \\
\underline{27x^2 - 81x} \\
76x - 2
\end{array}
$$

$0x^3 - (-9x^3) = 9x^3$

The quotient is $3x^2 + 9x + 27$, and the remainder is $76x - 2$. Note that the degree of the remainder is 1, and the degree of the divisor is 2. To check, verify that

$$(x^2 - 3x)(3x^2 + 9x + 27) + 76x - 2 = 3x^4 - 5x - 2. \qquad \blacksquare$$

EXAMPLE 3   **Rewriting a ratio of two polynomials**

Write $\dfrac{4x^3 - x - 9}{2x - 3}$ in the form

$$\text{quotient} + \frac{\text{remainder}}{\text{divisor}}.$$

Solution

Divide $4x^3 - x - 9$ by $2x - 3$. Insert $0 \cdot x^2$ for the missing term.

$$
\begin{array}{r}
2x^2 + 3x + 4 \\
2x - 3 \overline{\smash{\big)}\,4x^3 + 0x^2 - x - 9} \\
\underline{4x^3 - 6x^2} \\
6x^2 - x \\
\underline{6x^2 - 9x} \\
8x - 9 \\
\underline{8x - 12} \\
3
\end{array}
$$

$4x^3 \div (2x) = 2x^2$

$0x^2 - (-6x^2) = 6x^2$

$-x - (-9x) = 8x$

$-9 - (-12) = 3$

Since the quotient is $2x^2 + 3x + 4$ and the remainder is 3, we have

$$\frac{4x^3 - x - 9}{2x - 3} = 2x^2 + 3x + 4 + \frac{3}{2x - 3}.$$

To check the answer, we must verify that

$$(2x - 3)(2x^2 + 3x + 4) + 3 = 4x^3 - x - 9. \qquad \blacksquare$$

Synthetic Division

When dividing a polynomial by a binomial of the form $x - c$, we can use **synthetic division** to speed up the process. For synthetic division we write only the essential parts of ordinary division. For example, to divide $x^3 - 5x^2 + 4x - 3$ by $x - 2$, we write only the coefficients of the dividend $1, -5, 4$, and -3 in order of descending exponents. From the divisor $x - 2$ we use 2 and start with the following arrangement:

$$
2 \,\big|\; 1 \quad -5 \quad 4 \quad -3 \qquad (1 \cdot x^3 - 5x^2 + 4x - 3) \div (x - 2)
$$

Next we bring the first coefficient, 1, straight down:

$$
2 \,\big|\; 1 \quad -5 \quad 4 \quad -3
$$
$\downarrow$ Bring down
$$
1
$$

We then multiply the 1 by the 2 from the divisor, place the answer under the -5, and then add that column. Using 2 for $x - 2$ allows us to add the column rather than subtract as in ordinary division:

$$
2 \,\big|\; 1 \quad -5 \quad 4 \quad -3 \quad \text{Add}
$$
Multiply
$$
\qquad\quad 2
$$
$$
1 \quad -3
$$

We then repeat the multiply-and-add step for each of the remaining columns:

$$
2 \,\big|\; 1 \quad -5 \quad 4 \quad -3
$$
Multiply
$$
\qquad\quad 2 \quad -6 \quad -4
$$
$$
1 \quad -3 \quad -2 \quad -7 \quad \leftarrow \text{Remainder}
$$
$$
\underbrace{\qquad\qquad}_{\text{Quotient}}
$$

From the bottom row we can read the quotient and remainder. Since the degree of the quotient is one less than the degree of the dividend, the quotient is $1x^2 - 3x - 2$. The remainder is -7.

The strategy for getting the quotient $Q(x)$ and remainder R by synthetic division can be stated as follows.

Strategy for Using Synthetic Division

1. List the coefficients of the polynomial (the dividend).
2. Be sure to include zeros for any missing terms in the dividend.
3. For dividing by $x - c$, place c to the left.
4. Bring the first coefficient down.
5. Multiply by c and add for each column.
6. Read $Q(x)$ and R from the bottom row.

CAUTION Synthetic division is used only for dividing a polynomial by the binomial $x - c$, where c is a constant. If the binomial is $x - 7$, then $c = 7$. For the binomial $x + 7$ we have $x + 7 = x - (-7)$ and $c = -7$. ⊘

EXAMPLE 4   **Using synthetic division**

Find the quotient and remainder when $2x^4 - 5x^2 + 6x - 9$ is divided by $x + 2$.

Solution

Since $x + 2 = x - (-2)$, we use -2 for the divisor. Because x^3 is missing in the dividend, use a zero for the coefficient of x^3:

$$
\begin{array}{r|rrrrr}
-2 & 2 & 0 & -5 & 6 & -9 \\
   &   & -4 & 8 & -6 & 0 \\
\hline
   & 2 & -4 & 3 & 0 & -9
\end{array}
$$

Multiply / Add ← $2x^4 + 0 \cdot x^3 - 5x^2 + 6x - 9$ ← Quotient and remainder

Because the degree of the dividend is 4, the degree of the quotient is 3. The quotient is $2x^3 - 4x^2 + 3x$, and the remainder is -9. We can also express the results of this division in the form quotient $+ \dfrac{\text{remainder}}{\text{divisor}}$:

$$\frac{2x^4 - 5x^2 + 6x - 9}{x + 2} = 2x^3 - 4x^2 + 3x + \frac{-9}{x + 2}$$

Division and Factoring

To **factor** a polynomial means to write it as a product of two or more simpler polynomials. If we divide two polynomials and get 0 remainder, then we can write

$$\text{dividend} = (\text{divisor})(\text{quotient})$$

and we have factored the dividend. *The dividend factors as the divisor times the quotient if and only if the remainder is 0.* We can use division to help us discover factors of polynomials. To use this idea, however, we must know a factor or a possible factor to use as the divisor.

EXAMPLE 5 **Using synthetic division to determine factors**

Is $x - 1$ a factor of $6x^3 - 5x^2 - 4x + 3$?

Solution

We can use synthetic division to divide $6x^3 - 5x^2 - 4x + 3$ by $x - 1$:

$$
\begin{array}{r|rrrr}
1 & 6 & -5 & -4 & 3 \\
  & \downarrow & 6 & 1 & -3 \\
\hline
  & 6 & 1 & -3 & 0
\end{array}
$$

Because the remainder is 0, $x - 1$ is a factor, and

$$6x^3 - 5x^2 - 4x + 3 = (x - 1)(6x^2 + x - 3).$$

EXAMPLE 6 **Using division to determine factors**

Is $a - b$ a factor of $a^3 - b^3$?

Solution

Divide $a^3 - b^3$ by $a - b$. Insert zeros for the missing a^2b- and ab^2-terms.

$$
\begin{array}{r}
a^2 + ab + b^2 \\
a - b{\overline{\smash{\big)}\,a^3 +\ \ 0 +\ \ 0 -\ \ b^3}} \\
\underline{a^3 - a^2b} \\
a^2b +\ \ 0 \\
\underline{a^2b - ab^2} \\
ab^2 - b^3 \\
\underline{ab^2 - b^3} \\
0
\end{array}
$$

Because the remainder is 0, $a - b$ is a factor, and

$$a^3 - b^3 = (a - b)(a^2 + ab + b^2).$$

Warm-ups

True or false? Explain your answer.

1. If $a \div b = c$, then c is the dividend.
2. The quotient times the dividend plus the remainder equals the divisor.
3. $(x + 2)(x + 3) + 1 = x^2 + 5x + 7$ is true for any value of x.
4. The quotient of $(x^2 + 5x + 7) \div (x + 3)$ is $x + 2$.
5. If $x^2 + 5x + 7$ is divided by $x + 2$, the remainder is 1.
6. To divide $x^3 - 4x + 1$ by $x - 3$, we use -3 in synthetic division.
7. We can use synthetic division to divide $x^3 - 4x^2 - 6$ by $x^2 - 5$.
8. If $3x^5 - 4x^2 - 3$ is divided by $x + 2$, the quotient has degree 4.
9. If the remainder is zero, then the divisor is a factor of the dividend.
10. If the remainder is zero, then the quotient is a factor of the dividend.

3.5 EXERCISES

Find the quotient. See Example 1.

1. $36x^7 \div (3x^3)$
2. $-30x^3 \div (-5x)$
3. $16x^2 \div (-8x^2)$
4. $-22a^3 \div (11a^2)$
5. $(6b - 9) \div 3$
6. $(8x^2 - 6x) \div 2$
7. $(3x^2 + 6x) \div (3x)$
8. $(5x^3 - 10x^2 + 20x) \div (5x)$
9. $(10x^4 - 8x^3 + 6x^2) \div (-2x^2)$
10. $(-9x^3 + 6x^2 - 12x) \div (-3x)$
11. $(7x^3 - 4x^2) \div (2x)$
12. $(6x^3 - 5x^2) \div (4x^2)$

Find the quotient and remainder as in Example 2. Check by using the formula

$$\text{dividend} = (\text{divisor})(\text{quotient}) + \text{remainder}.$$

13. $(x^2 + 8x + 13) \div (x + 3)$
14. $(x^2 + 5x + 7) \div (x + 3)$
15. $(x^2 - 2x) \div (x + 2)$
16. $(3x) \div (x - 1)$
17. $(x^3 + 8) \div (x + 2)$
18. $(y^3 - 1) \div (y - 1)$
19. $(a^3 + 4a - 5) \div (a - 2)$
20. $(w^3 + w^2 - 3) \div (w - 2)$
21. $(x^3 - x^2 + x - 3) \div (x + 1)$
22. $(a^3 - a^2 + a - 4) \div (a + 2)$
23. $(x^4 - x + x^3 - 1) \div (x - 2)$
24. $(3x^4 + 6 - x^2 + 3x) \div (x + 2)$
25. $(5x^2 - 3x^4 + x - 2) \div (x^2 - 2)$
26. $(x^4 - 2 + x^3) \div (x^2 + 3)$
27. $(x^2 - 4x + 2) \div (2x - 3)$
28. $(x^2 - 5x + 1) \div (3x + 6)$
29. $(2x^2 - x + 6) \div (3x - 2)$
30. $(3x^2 + 4x - 1) \div (2x + 1)$

Write each expression in the form

$$\text{quotient} + \frac{\text{remainder}}{\text{divisor}}.$$

See Example 3.

31. $\dfrac{2x}{x - 5}$
32. $\dfrac{x}{x - 1}$
33. $\dfrac{x^2}{x + 1}$

34. $\dfrac{x^2 + 9}{x + 3}$
35. $\dfrac{x^3}{x + 2}$
36. $\dfrac{x^3 - 1}{x - 2}$
37. $\dfrac{x^3 + 2x}{x^2}$
38. $\dfrac{2x^2 + 3}{2x}$
39. $\dfrac{x^2 - 4x + 9}{x + 2}$
40. $\dfrac{x^2 - 5x - 10}{x - 3}$
41. $\dfrac{3x^3 - 4x^2 + 7}{x - 1}$
42. $\dfrac{-2x^3 + x^2 - 3}{x + 2}$

Use synthetic division to find the quotient and remainder when the first polynomial is divided by the second. See Example 4.

43. $x^3 - 5x^2 + 6x - 3, \quad x - 2$
44. $x^3 + 6x^2 - 3x - 5, \quad x - 3$
45. $2x^2 - 4x + 5, \quad x + 1$
46. $3x^2 - 7x + 4, \quad x + 2$
47. $3x^4 - 15x^2 + 7x - 9, \quad x - 3$
48. $-2x^4 + 3x^2 - 5, \quad x - 2$
49. $x^5 - 1, \quad x - 1$
50. $x^6 - 1, \quad x + 1$
51. $x^3 - 5x + 6, \quad x + 2$
52. $x^3 - 3x - 7, \quad x - 4$
53. $2.3x^2 - 0.14x + 0.6, \quad x - 0.32$
54. $1.6x^2 - 3.5x + 4.7, \quad x + 1.8$

For each pair of polynomials, determine whether the first polynomial is a factor of the second. Use synthetic division when possible. See Examples 5 and 6.

55. $x + 4, \quad x^3 + x^2 - 11x + 8$
56. $x + 4, \quad x^3 + x^2 + x + 48$
57. $x - 4, \quad x^3 - 13x - 12$
58. $x - 1, \quad x^3 + 3x^2 - 5x$
59. $2x - 3, \quad 2x^3 - 3x^2 - 4x + 6$
60. $3x - 5, \quad 6x^2 - 7x - 6$
61. $3w + 1, \quad 27w^3 + 1$
62. $2w + 3, \quad 8w^3 + 27$
63. $a - 5, \quad a^3 - 125$
64. $a - 2, \quad a^6 - 64$
65. $x^2 - 2, \quad x^4 + 3x^3 - 6x - 4$
66. $x^2 - 3, \quad x^4 + 2x^3 - 4x^2 - 6x + 3$

Factor each polynomial given that the binomial following each polynomial is a factor of the polynomial.

67. $x^2 - 6x + 8$, $x - 4$ **68.** $x^2 + 3x - 40$, $x + 8$

69. $w^3 - 27$, $w - 3$ **70.** $w^3 + 125$, $w + 5$

71. $x^3 - 4x^2 + 6x - 4$, $x - 2$

72. $2x^3 + 5x + 7$, $x + 1$

73. $z^2 + 6z + 9$, $z + 3$

74. $4a^2 - 20a + 25$, $2a - 5$

75. $6y^2 + 5y + 1$, $2y + 1$

76. $12y^2 - y - 6$, $4y - 3$

Solve each problem

77. *Average cost.* The weekly cost in dollars for manufacturing x professional racing bicycles is given by the polynomial $C(x) = 0.03x^2 + 300x$. The average cost per bicycle is given by $AC(x) = \dfrac{C(x)}{x}$. Find $AC(x)$. Find the average cost per bicycle when 12 bicycles are manufactured.

78. *Average profit.* The weekly profit in dollars for manufacturing x bicycles is given by the polynomial $P(x) = 100x + 2x^2$. The average profit per bicycle is given by $AP(x) = \dfrac{P(x)}{x}$. Find $AP(x)$. Find the average profit per bicycle when 12 bicycles are manufactured.

79. *Area of a poster.* The area of a rectangular poster advertising a Pearl Jam concert is $x^2 - 1$ square feet. If the length is $x + 1$ feet, then what is the width?

80. *Volume of a box.* The volume of shipping crate is $h^3 + 5h^2 + 6h$. If the height is h and the length is $h + 2$, then what is the width?

Figure for Exercise 80

81. *Volume of a pyramid.* Ancient Egyptian pyramid builders knew that the volume of the truncated pyramid shown in the figure is given by

$$V = \frac{H(a^3 - b^3)}{3(a - b)},$$

where a^2 is the area of the square base, b^2 is the area of the square top, and H is the distance from the base to the top. Find the volume of a truncated pyramid that has a base of 900 square meters, a top of 400 square meters, and a height H of 10 meters.

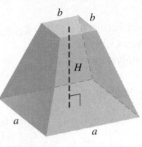

Figure for Exercise 81

82. *Egyptian pyramid formula.* Rewrite the formula of the previous exercise so that the denominator contains the number 3 only.

Getting More Involved

83. *Discussion.* On a test a student divided $3x^3 - 5x^2 - 3x + 7$ by $x - 3$ and got a quotient of $3x^2 + 4x$ and remainder $9x + 7$. Verify that the divisor times the quotient plus the remainder is equal to the dividend. Why was the student's answer incorrect?

84. *Exploration.* Use synthetic division to find the quotient when $x^5 - 1$ is divided by $x - 1$ and the quotient when $x^6 - 1$ is divided by $x - 1$. Observe the pattern in the first two quotients and then write the quotient for $x^9 - 1$ divided by $x - 1$ without dividing.

3.6 Factoring Polynomials

In Section 3.5 you learned that a polynomial could be factored by using division: If we know one factor of a polynomial, then we can use it as a divisor to obtain the other factor, the quotient. However, this technique is not very practical because the division process can be somewhat tedious, and it is not easy to obtain a factor to use as the divisor. In this section and the next two sections we will develop better techniques for factoring polynomials. These techniques will be used for solving equations and problems in the last section of this chapter.

Factoring Out the Greatest Common Factor (GCF)

A natural number larger than 1 that has no factors other than itself and 1 is called a **prime number.** The numbers

$$2, 3, 5, 7, 11, 13, 17, 19, 23$$

are the first nine prime numbers. There are infinitely many prime numbers.

To factor a natural number **completely** means to write it as a product of prime numbers. In factoring 12 we might write $12 = 4 \cdot 3$. However, 12 is not factored completely as $4 \cdot 3$ because 4 is not a prime. To factor 12 completely, we write $12 = 2 \cdot 2 \cdot 3$ (or $2^2 \cdot 3$).

We use the distributive property to multiply a monomial and a binomial:

$$6x(2x - 1) = 12x^2 - 6x$$

If we start with $12x^2 - 6x$, we can use the distributive property to get

$$12x^2 - 6x = 6x(2x - 1).$$

We have **factored out** $6x$, which is a common factor of $12x^2$ and $-6x$. We could have factored out just 3 to get

$$12x^2 - 6x = 3(4x^2 - 2x),$$

but this would not be factoring out the *greatest* common factor. The **greatest common factor** (GCF) is a monomial that includes every number or variable that is a factor of all terms of the polynomial.

We can use the following strategy for finding the greatest common factor of a group of terms.

> ◤ **Strategy for Finding the Greatest Common Factor (GCF)** ◥
>
> 1. Factor each term completely.
> 2. Write a product using each factor that is common to all of the terms.
> 3. On each of these factors, use an exponent equal to the smallest exponent that appears on that factor in any of the terms.

EXAMPLE 1   **The greatest common factor**

Find the greatest common factor for each group of terms.

a) $8x^2y$, $20xy^3$ **b)** $30a^2$, $45a^3b^2$, $75a^4b$

Solution

a) First factor each term completely:

$$8x^2y = 2^3x^2y$$
$$20xy^3 = 2^2 \cdot 5xy^3$$

The factors common to both terms are 2, x, and y. In the greatest common factor we use the smallest exponent that appears on each factor in either of the terms. So the greatest common factor is 2^2xy or $4xy$.

b) First factor each term completely:

$$30a^2 = 2 \cdot 3 \cdot 5a^2$$
$$45a^3b^2 = 3^2 \cdot 5a^3b^2$$
$$75a^4b = 3 \cdot 5^2a^4b$$

The greatest common factor is $3 \cdot 5a^2$ or $15a^2$. ∎

To factor out the greatest common factor from a polynomial, find the greatest common factor for the terms, then use the distributive property to factor it out.

EXAMPLE 2   **Factoring out the greatest common factor**

Factor each polynomial by factoring out the greatest common factor.

a) $5x^4 - 10x^3 + 15x^2$ **b)** $8xy^2 + 20x^2y$

c) $60x^5 + 24x^3 + 36x^2$

Solution

a) First factor each term completely:

$$5x^4 = 5x^4, \qquad 10x^3 = 2 \cdot 5x^3, \qquad 15x^2 = 3 \cdot 5x^2.$$

The greatest common factor of the three terms is $5x^2$. Now factor $5x^2$ out of each term:

$$5x^4 - 10x^3 + 15x^2 = 5x^2(x^2 - 2x + 3)$$

b) The greatest common factor for $8xy^2$ and $20x^2y$ is $4xy$:

$$8xy^2 + 20x^2y = 4xy(2y + 5x)$$

c) First factor each coefficient in $60x^5 + 24x^3 + 36x^2$:

$$60 = 2^2 \cdot 3 \cdot 5, \qquad 24 = 2^3 \cdot 3, \qquad 36 = 2^2 \cdot 3^2.$$

The greatest common factor of the three terms is $2^2 \cdot 3x^2$ or $12x^2$:

$$60x^5 + 24x^3 + 36x^2 = 12x^2(5x^3 + 2x + 3)$$ ∎

In the next example, the common factor in each term is a binomial.

EXAMPLE 3 **Factoring out a binomial**

Factor $(x + 3)w + (x + 3)a$.

Solution

Although $x + 3$ is not a monomial, it can be treated like one for factoring:

$$(x + 3)w + (x + 3)a = (x + 3)(w + a).$$ ∎

Factoring Out the Opposite of the GCF

The GCF, greatest common factor, for $-6x^2 - 4x$ is $2x$, but we can factor out either $2x$ or its opposite, $-2x$:

$$-6x^2 - 4x = 2x(-3x - 2)$$
$$= -2x(3x + 2)$$

In Example 8 of this section it will be necessary to factor out the opposite of the greatest common factor.

EXAMPLE 4 **Factoring out the opposite of the GCF**

Factor out the greatest common factor, then factor out the opposite of the greatest common factor.

a) $5x - 5y$ **b)** $-x^2 - 3$ **c)** $-x^3 + 3x^2 - 5x$

Solution

a) $5x - 5y = 5(x - y)$ Factor out 5.

$\quad\quad\quad\quad = -5(-x + y)$ Factor out -5.

b) $-x^2 - 3 = 1(-x^2 - 3)$ The greatest common factor is 1.

$\quad\quad\quad\quad = -1(x^2 + 3)$ Factor out -1.

c) $-x^3 + 3x^2 - 5x = x(-x^2 + 3x - 5)$ Factor out x.

$\quad\quad\quad\quad\quad\quad = -x(x^2 - 3x + 5)$ Factor out $-x$. ∎

Factoring the Difference of Two Squares

A first-degree polynomial in one variable, such as $3x - 5$, is called a linear polynomial. (The equation $3x - 5 = 0$ is a linear equation.)

> **Linear Polynomial**
>
> If a and b are real numbers with $a \neq 0$, then $ax + b$ is called a **linear polynomial.**

A second-degree polynomial such as $x^2 + 5x - 6$ is called a quadratic polynomial.

> **Quadratic Polynomial**
>
> If a, b, and c are real numbers with $a \neq 0$, then $ax^2 + bx + c$ is called a **quadratic polynomial.**

One of the main goals of this chapter is to write a quadratic polynomial (when possible) as a product of linear factors.

Consider the quadratic polynomial $x^2 - 25$. We recognize that $x^2 - 25$ is a difference of two squares, $x^2 - 5^2$. We recall that the product of a sum and a difference is a difference of two squares: $(a + b)(a - b) = a^2 - b^2$. If we reverse this special product rule, we get a rule for factoring the difference of two squares.

> **Factoring the Difference of Two Squares**
>
> $$a^2 - b^2 = (a + b)(a - b)$$

The difference of two squares factors as the product of a sum and a difference. To factor $x^2 - 25$, we replace a by x and b by 5 to get

$$x^2 - 25 = (x + 5)(x - 5).$$

This equation expresses a quadratic polynomial as a product of two linear factors.

EXAMPLE 5   **Factoring the difference of two squares**

Factor each polynomial.

a) $y^2 - 36$ $\qquad\qquad\qquad\qquad$ **b)** $9x^2 - 1$

c) $4x^2 - y^2$

Solution

Each of these binomials is a difference of two squares. Each binomial factors into a product of a sum and a difference.

a) $y^2 - 36 = (y + 6)(y - 6)$ We could also write $(y - 6)(y + 6)$, because the factors can be written in any order.

b) $9x^2 - 1 = (3x + 1)(3x - 1)$

c) $4x^2 - y^2 = (2x + y)(2x - y)$ ∎

Factoring Perfect Square Trinomials

The trinomial that results from squaring a binomial is called a **perfect square trinomial.** We can reverse the rules from Section 3.4 for the square of a sum or difference to get rules for factoring.

Factoring Perfect Square Trinomials

$$a^2 + 2ab + b^2 = (a + b)^2$$
$$a^2 - 2ab + b^2 = (a - b)^2$$

Consider the polynomial $x^2 + 6x + 9$. If we recognize that

$$x^2 + 6x + 9 = x^2 + 2 \cdot x \cdot 3 + 3^2,$$

then we can see that it is a perfect square trinomial. It fits the rule if $a = x$ and $b = 3$:

$$x^2 + 6x + 9 = (x + 3)^2$$

Perfect square trinomials can be identified by using the following strategy.

Strategy for Identifying Perfect Square Trinomials

A trinomial is a perfect square trinomial if
1. the first and last terms are of the form a^2 and b^2,
2. the middle term is 2 or -2 times the product of a and b.

We use this strategy in the next example.

EXAMPLE 6   **Factoring perfect square trinomials**
Factor each polynomial.

a) $x^2 - 8x + 16$ b) $a^2 + 14a + 49$ c) $4x^2 + 12x + 9$

Solution

a) Because the first term is x^2, the last is 4^2, and $-2(x)(4)$ is equal to the middle term $-8x$, the trinomial $x^2 - 8x + 16$ is a perfect square trinomial.

$$x^2 - 8x + 16 = (x - 4)^2$$

b) Because $49 = 7^2$ and $14a = 2(a)(7)$, we have a perfect square trinomial.

$$a^2 + 14a + 49 = (a + 7)^2$$

c) Because $4x^2 = (2x)^2$, $9 = 3^2$, and the middle term $12x$ is equal to $2(2x)(3)$, the trinomial $4x^2 + 12x + 9$ is a perfect square trinomial.

$$4x^2 + 12x + 9 = (2x + 3)^2$$ ∎

Factoring a Difference or Sum of Two Cubes

In Example 6 of Section 3.5 we divided $a^3 - b^3$ by $a - b$ to get the quotient $a^2 + ab + b^2$ and no remainder. So $a - b$ is a factor of $a^3 - b^3$, a difference of two cubes. If you divide $a^3 + b^3$ by $a + b$, you will get the quotient

$a^2 - ab + b^2$ and no remainder. Try it. So $a + b$ is a factor of $a^3 + b^3$, a sum of two cubes. These results give us two more factoring rules.

Factoring a Difference or Sum of Two Cubes

$$a^3 - b^3 = (a - b)(a^2 + ab + b^2)$$
$$a^3 + b^3 = (a + b)(a^2 - ab + b^2)$$

EXAMPLE 7   **Factoring a difference or sum of two cubes**

Factor each polynomial.

a) $x^3 - 8$ **b)** $y^3 + 1$ **c)** $8z^3 - 27$

Solution

a) Because $8 = 2^3$, we can use the formula for factoring the difference of two cubes. In the formula $a^3 - b^3 = (a - b)(a^2 + ab + b^2)$, let $a = x$ and $b = 2$:

$$x^3 - 8 = (x - 2)(x^2 + 2x + 4)$$

b) $y^3 + 1 = y^3 + 1^3$ Recognize a sum of two cubes.

$\qquad = (y + 1)(y^2 - y + 1)$ Let $a = y$ and $b = 1$ in the formula for the sum of two cubes.

c) $8z^3 - 27 = (2z)^3 - 3^3$ Recognize a difference of two cubes.

$\qquad = (2z - 3)(4z^2 + 6z + 9)$ Let $a = 2z$ and $b = 3$ in the formula for a difference of two cubes. ■

Factoring a Polynomial Completely

Polynomials that cannot be factored are called **prime polynomials.** Because binomials such as $x + 5$, $a - 6$, and $3x + 1$ cannot be factored, they are prime polynomials. Any monomial is also considered a prime polynomial. A polynomial is **factored completely** when it is written as a product of prime polynomials. To factor completely, always factor out the GCF (or its opposite) first. Then continue to factor until all of the factors are prime.

EXAMPLE 8   **Factoring completely**

Factor each polynomial completely.

a) $5x^2 - 20$ **b)** $3a^3 - 30a^2 + 75a$ **c)** $-2b^4 + 16b$

Solution

a) $5x^2 - 20 = 5(x^2 - 4)$ Greatest common factor

$\qquad = 5(x - 2)(x + 2)$ Difference of two squares

b) $3a^3 - 30a^2 + 75a = 3a(a^2 - 10a + 25)$ Greatest common factor

$\qquad = 3a(a - 5)^2$ Perfect square trinomial

c) $-2b^4 + 16b = -2b(b^3 - 8)$ Factor out $-2b$ to make the next step easier.

$\qquad = -2b(b - 2)(b^2 + 2b + 4)$ Difference of two cubes ■

Factoring by Substitution

So far, the polynomials that we have factored, without common factors, have all been of degree 2 or 3. Some polynomials of higher degree can be factored by substituting a single variable for a variable with a higher power. After factoring we replace the single variable by the higher-power variable. This method is called **substitution.**

EXAMPLE 9 **Factoring by substitution**

Factor each polynomial.

a) $x^4 - 9$

b) $y^8 - 14y^4 + 49$

Solution

a) We recognize $x^4 - 9$ as a difference of two squares in which $x^4 = (x^2)^2$ and $9 = 3^2$. If we let $w = x^2$, then $w^2 = x^4$. So we can replace x^4 by w^2 and factor:

$$x^4 - 9 = w^2 - 9 \qquad \text{Replace } x^4 \text{ by } w^2$$
$$= (w + 3)(w - 3) \qquad \text{Difference of two squares}$$
$$= (x^2 + 3)(x^2 - 3) \qquad \text{Replace } w \text{ by } x^2.$$

b) We recognize $y^8 - 14y^4 + 49$ as a perfect square trinomial in which $y^8 = (y^4)^2$ and $49 = 7^2$. We let $w = y^4$ and $w^2 = y^8$:

$$y^8 - 14y^4 + 49 = w^2 - 14w + 49 \qquad \text{Replace } y^4 \text{ by } w \text{ and } y^8 \text{ by } w^2.$$
$$= (w - 7)^2 \qquad \text{Perfect square trinomial}$$
$$= (y^4 - 7)^2 \qquad \text{Replace } w \text{ by } y^4$$

∎

CAUTION   The polynomials that we factor by substitution must contain just the right powers of the variable. We can factor $y^8 - 14y^4 + 49$ because $(y^4)^2 = y^8$, but we cannot factor $y^7 - 14y^4 + 49$ by substitution.   ⊘

In the next example we use substitution to factor polynomials that have variables as exponents.

EXAMPLE 10 **Polynomials with variable exponents**

Factor completely. The variables used in the exponents represent positive integers.

a) $x^{2m} - y^2$

b) $z^{2n+1} - 6z^{n+1} + 9z$

Solution

a) Notice that $x^{2m} = (x^m)^2$. So if we let $w = x^m$, then $w^2 = x^{2m}$.

$$x^{2m} - y^2 = w^2 - y^2 \qquad \text{Substitution}$$
$$= (w + y)(w - y) \qquad \text{Difference of two squares}$$
$$= (x^m + y)(x^m - y) \qquad \text{Replace } w \text{ by } x^m.$$

b) First factor out the common factor z:

$$z^{2n+1} - 6z^{n+1} + 9z = z(z^{2n} - 6z^n + 9)$$
$$= z(a^2 - 6a + 9) \qquad \text{Let } a = z^n.$$
$$= z(a - 3)^2 \qquad \text{Perfect square trinomial}$$
$$= z(z^n - 3)^2 \qquad \text{Replace } a \text{ by } z^n.$$

∎

Warm-ups

True or false? Explain your answer.

1. For the polynomial $3x^2y - 6xy^2$ we can factor out either $3xy$ or $-3xy$.
2. The greatest common factor for the polynomial $8a^3 - 15b^2$ is 1.
3. $2x - 4 = -2(2 - x)$ for any value of x.
4. $x^2 - 16 = (x - 4)(x + 4)$ for any value of x.
5. The polynomial $x^2 + 6x + 36$ is a perfect square trinomial.
6. The polynomial $y^2 + 16$ is a perfect square trinomial.
7. $9x^2 + 21x + 49 = (3x + 7)^2$ for any value of x.
8. The polynomial $x + 1$ is a factor of $x^3 + 1$.
9. $x^3 - 27 = (x - 3)(x^2 + 6x + 9)$ for any value of x.
10. $x^3 - 8 = (x - 2)^3$ for any value of x.

3.6 EXERCISES

Find the greatest common factor for each group of terms. See Example 1.

1. 48, 36x
2. $42a$, $28a^2$
3. $9wx$, $21wy$, $15xy$
4. $70x^2$, $84x$, $42x^3$
5. $24x^2y$, $42xy^2$, $66xy^3$
6. $60a^2b^5$, $140a^9b^2$, $40a^3b^6$

Factor out the greatest common factor in each expression. See Examples 2 and 3.

7. $x^3 - 5x$
8. $10x^2 - 20y^3$
9. $48wx + 36wy$
10. $42wz + 28wa$
11. $2x^3 - 4x^2 + 6x$
12. $6x^3 - 12x^2 + 18x$
13. $36a^3b^6 - 24a^4b^2 + 60a^5b^3$
14. $44x^8y^6z - 110x^6y^9z^2$
15. $(x - 6)a + (x - 6)b$
16. $(y - 4)3 + (y - 4)b$
17. $(y - 1)^2y + (y - 1)^2z$
18. $(w - 2)^2 \cdot w + (w - 2)^2 \cdot 3$

Factor out the greatest common factor, then factor out the opposite of the greatest common factor. See Example 4.

19. $2x - 2y$
20. $-3x + 6$
21. $6x^2 - 3x$
22. $10x^2 + 5x$
23. $-w^3 + 3w^2$
24. $-2w^4 + 6w^3$
25. $-a^3 + a^2 - 7a$
26. $-2a^4 - 4a^3 + 6a^2$

Factor each polynomial. See Example 5.

27. $x^2 - 100$
28. $81 - y^2$
29. $4y^2 - 49$
30. $16b^2 - 1$
31. $9x^2 - 25a^2$
32. $121a^2 - b^2$
33. $144w^2z^2 - 1$
34. $x^2y^2 - 9c^2$

Factor each polynomial. See Example 6.

35. $x^2 - 20x + 100$
36. $y^2 + 10y + 25$
37. $4m^2 - 4m + 1$
38. $9t^2 + 30t + 25$
39. $w^2 - 2wt + t^2$
40. $4r^2 + 20rt + 25t^2$

Factor. See Example 7.

41. $a^3 - 1$
42. $w^3 + 1$
43. $w^3 + 27$
44. $x^3 - 64$
45. $8x^3 - 1$
46. $27x^3 + 1$
47. $a^3 + 8$
48. $m^3 - 8$

Factor each polynomial completely. See Example 8.

49. $2x^2 - 8$
50. $3x^3 - 27x$
51. $x^3 + 10x^2 + 25x$
52. $5a^4m - 45a^2m$
53. $4x^2 + 4x + 1$
54. $ax^2 - 8ax + 16a$
55. $(x + 3)x + (x + 3)7$
56. $(x - 2)x - (x - 2)5$
57. $6y^2 + 3y$
58. $4y^2 - y$
59. $4x^2 - 20x + 25$
60. $a^3x^3 - 6a^2x^2 + 9ax$
61. $2m^4 - 2mn^3$
62. $5x^3y^2 - y^5$
63. $(2x - 3)x - (2x - 3)2$
64. $(2x + 1)x + (2x + 1)3$
65. $9a^3 - aw^2$
66. $2bn^2 - 4b^2n + 2b^3$
67. $-5a^2 + 30a - 45$
68. $-2x^2 + 50$
69. $16 - 54x^3$
70. $27x^2y - 64x^2y^4$
71. $-3y^3 - 18y^2 - 27y$
72. $-2m^2n - 8mn - 8n$
73. $-7a^2b^2 + 7$
74. $-17a^2 - 17a$

Factor each polynomial completely. See Example 9.

75. $x^{10} - 9$

76. $y^8 - 4$

77. $z^{12} - 6z^6 + 9$

78. $a^6 + 10a^3 + 25$

79. $2x^7 + 8x^4 + 8x$

80. $x^{13} - 6x^7 + 9x$

81. $4x^5 + 4x^3 + x$

82. $18x^6 + 24x^3 + 8$

83. $x^6 - 8$

84. $y^6 - 27$

85. $2x^9 + 16$

86. $x^{13} + x$

Factor each polynomial completely. The variables used as exponents represent positive integers. See Example 10.

87. $a^{2n} - 1$

88. $b^{4n} - 9$

89. $a^{2r} + 6a^r + 9$

90. $u^{6n} - 4u^{3n} + 4$

91. $x^{3m} - 8$

92. $y^{3n} + 1$

93. $a^{3m} - b^3$

94. $r^{3m} + 8t^3$

95. $k^{2w+1} - 10k^{w+1} + 25k$

96. $4a^{2t+1} + 4a^{t+1} + a$

97. $uv^{6k} - 2u^2v^{4k} + u^3v^{2k}$

98. $u^{3m}v^n + 2u^{2m}v^{2n} + u^m v^{3n}$

Replace k in each trinomial by a number that makes the trinomial a perfect square trinomial.

99. $x^2 + 6x + k$

100. $y^2 - 8y + k$

101. $4a^2 - ka + 25$

102. $9u^2 + kuv + 49v^2$

103. $km^2 - 24m + 9$

104. $kz^2 + 40z + 16$

105. $81y^2 - 180y + k$

106. $36a^2 + 60a + k$

Solve each problem.

107. *Volume of a bird cage.* The volume of a rectangular-shaped bird cage with a square bottom is $16b^3 + 24b^2 + 9b$ cubic centimeters. If the height is b centimeters, then what is the length of a side of the square bottom?

108. *Pyramid power.* A powerful crystal pyramid has a square base and a volume of $3y^3 + 12y^2 + 12y$ cubic centimeters. If its height is y centimeters, then what polynomial represents the length of a side of the square base? (The volume of a pyramid with a square base of area a^2 and height h is given by $V = \dfrac{ha^2}{3}$.)

Figure for Exercise 108

Getting More Involved

109. *Cooperative learning.* List the perfect square trinomials corresponding to $(x + 1)^2$, $(x + 2)^2$, $(x + 3)^2, \ldots, (x + 12)^2$. Use your list to quiz a classmate. Read a perfect square trinomial at random from your list and ask your classmate to write its factored form. Repeat until both of you have mastered these 12 perfect square trinomials.

3.7 Factoring $ax^2 + bx + c$

In this section:

▶ Factoring Trinomials with Leading Coefficient 1

▶ Factoring Trinomials with Leading Coefficient Not 1

▶ Trial and Error

▶ Higher Degrees and Variable Exponents

In Section 3.5 you learned to factor certain special polynomials. In this section you will learn to factor general quadratic polynomials. We first factor $ax^2 + bx + c$ with $a = 1$, and then we consider the case $a \neq 1$.

Factoring Trinomials with Leading Coefficient 1

Let's look closely at an example of finding the product of two binomials using the distributive property:

$$(x + 3)(x + 4) = (x + 3)x + (x + 3)4 \quad \text{Distributive property}$$
$$= x^2 + 3x + 4x + 12 \quad \text{Distributive property}$$
$$= x^2 + 7x + 12$$

To factor $x^2 + 7x + 12$, we need to reverse these steps. First observe that the coefficient 7 is the sum of two numbers that have a product of 12. The only numbers that have a product of 12 and a sum of 7 are 3 and 4. So write $7x$ as $3x + 4x$:

$$x^2 + 7x + 12 = x^2 + 3x + 4x + 12$$

Now factor the common factor x out of the first two terms and the common factor 4 out of the last two terms. This method is called **factoring by grouping.**

$$\overbrace{}^{\text{Factor out } x}\ \overbrace{}^{\text{Factor out } 4}$$

$$
\begin{aligned}
x^2 + 7x + 12 &= x^2 + 3x\ +\ 4x + 12 && \text{Rewrite } 7x \text{ as } 3x + 4x. \\
&= (x + 3)x + (x + 3)4 && \text{Factor out common factors.} \\
&= (x + 3)(x + 4) && \text{Factor out the common factor } x + 3.
\end{aligned}
$$

EXAMPLE 1   **Factoring a trinomial by grouping**

Factor $x^2 + 9x + 18$.

Solution

Find two integers that have a product of 18 and a sum of 9. For a product of 18 we could use 1 and 18, 2 and 9, or 3 and 6. Only 3 and 6 have a sum of 9.

$$
\begin{aligned}
x^2 + 9x + 18 &= x^2 + 3x + 6x + 18 && \text{Replace } 9x \text{ by } 3x + 6x. \\
&= (x + 3)x + (x + 3)6 && \text{Factor out common factors.} \\
&= (x + 3)(x + 6) && \text{Check by using FOIL.} \quad ■
\end{aligned}
$$

We can actually skip most of the steps shown in Example 1. Once we discover that 3 and 6 have a product of 18 and a sum of 9, we can simply write

$$x^2 + 9x + 18 = (x + 3)(x + 6).$$

In the next example we use this shortcut.

EXAMPLE 2   **Factoring $ax^2 + bx + c$ with $a = 1$**

Factor each quadratic polynomial.

a) $x^2 + 4x + 3$ **b)** $x^2 + 3x - 10$ **c)** $a^2 - 5a + 6$

Solution

a) Two integers with a product of 3 and a sum of 4 are 1 and 3:

$$x^2 + 4x + 3 = (x + 1)(x + 3).$$

Check by using FOIL.

b) Two integers with a product of -10 and a sum of 3 are 5 and -2:

$$x^2 + 3x - 10 = (x + 5)(x - 2)$$

Check by using FOIL.

c) Two integers with a product of 6 and a sum of -5 are -3 and -2:

$$a^2 - 5a + 6 = (a - 3)(a - 2)$$

Check by using FOIL.   ■

Factoring Trinomials with Leading Coefficient Not 1

If the leading coefficient of a quadratic trinomial is not 1, we can again use grouping to factor the trinomial. However, the procedure is slightly different.

Consider the trinomial $2x^2 + 11x + 12$, for which $a = 2$, $b = 11$, and $c = 12$. First find ac, the product of the leading coefficient and the constant term. In this case, $ac = 2 \cdot 12 = 24$. Now find two integers with a product of 24 and a sum of 11. The pairs of integers with a product of 24 are 1 and 24, 2 and 12, 3 and 8, and 4 and 6. Only 3 and 8 have a product of 24 and a sum of 11. Now replace $11x$ by $3x + 8x$ and factor by grouping:

$$2x^2 + 11x + 12 = 2x^2 + 3x + 8x + 12$$
$$= (2x + 3)x + (2x + 3)4$$
$$= (2x + 3)(x + 4)$$

This strategy for factoring a quadratic trinomial, known as the **ac method,** is summarized in the following box. The ac method works also when $a = 1$.

Strategy for Factoring $ax^2 + bx + c$ by the ac Method

To factor the trinomial $ax^2 + bx + c$

1. find two integers that have a product equal to ac and a sum equal to b,

2. replace bx by two terms using the two new integers as coefficients,

3. then factor the resulting four-term polynomial by grouping.

EXAMPLE 3   **Factoring $ax^2 + bx + c$ with $a \neq 1$**

Factor each trinomial.

a) $2x^2 + 9x + 4$

b) $2x^2 + 5x - 12$

Solution

a)   Because $2 \cdot 4 = 8$, we need two numbers with a product of 8 and a sum of 9. The numbers are 1 and 8. Replace $9x$ by $x + 8x$ and factor by grouping.

$$2x^2 + 9x + 4 = 2x^2 + x + 8x + 4$$
$$= (2x + 1)x + (2x + 1)4$$
$$= (2x + 1)(x + 4) \quad \text{Check by FOIL.}$$

Note that if you start with $2x^2 + 8x + x + 4$, you get the same factors if you factor by grouping.

b)   Because $2(-12) = -24$, we need two numbers with a product of -24 and a sum of 5. The pairs of numbers with a product of 24 are 1 and 24, 2 and 12, 3 and 8, and 4 and 6. To get a product of -24, one of the numbers must be negative and the other positive. To get a sum of positive 5, we need -3 and 8.

$$2x^2 + 5x - 12 = 2x^2 - 3x + 8x - 12$$
$$= (2x - 3)x + (2x - 3)4$$
$$= (2x - 3)(x + 4) \quad \text{Check by FOIL.} \quad \blacksquare$$

Trial and Error

After we have gained some experience at factoring by grouping, we can often find the factors without going through the steps of grouping. Consider the polynomial

$$2x^2 - 7x + 6.$$

The factors of $2x^2$ can only be $2x$ and x. The factors of 6 could be 2 and 3 or 1 and 6. We can list all of the possibilities that give the correct first and last terms, without putting in the signs:

$$(2x \quad 2)(x \quad 3) \qquad (2x \quad 6)(x \quad 1)$$
$$(2x \quad 3)(x \quad 2) \qquad (2x \quad 1)(x \quad 6)$$

Before actually trying these out we make an important observation. If $(2x \quad 2)$ or $(2x \quad 6)$ were one of the factors, then there would be a common factor 2 in the original trinomial, but there is not. *If the original trinomial has no common factor, there can be no common factor in either of its linear factors.* Since 6 is positive and the middle term is $-7x$, both of the missing signs must be negative. So the only possibilities are $(2x - 1)(x - 6)$ and $(2x - 3)(x - 2)$. The middle term of the first product is $-13x$, and the middle term of the second product is $-7x$. So we have found the factors:

$$2x^2 - 7x + 6 = (2x - 3)(x - 2)$$

Even though there may be many possibilities in some factoring problems, it is often possible to find the correct factors without writing down every possibility. We can use a bit of guesswork in factoring trinomials. *Try* whichever possibility you think might work. *Check* it by multiplying. If it is not right, then *try again.* That is why this method is called **trial and error.**

EXAMPLE 4 **Trial and error**

Factor each quadratic trinomial using trial and error.

a) $2x^2 + 5x - 3$ **b)** $3x^2 - 11x + 6$

Solution

a) Because $2x^2$ factors only as $2x \cdot x$ and 3 factors only as $1 \cdot 3$, there are only two possible ways to factor this trinomial to get the correct first and last terms:

$$(2x \quad 1)(x \quad 3) \qquad \text{and} \qquad (2x \quad 3)(x \quad 1)$$

Because the last term of the trinomial is negative, one of the missing signs must be $+$, and the other must be $-$. Now we try the various possibilities until we get the correct middle term:

$$(2x + 1)(x - 3) = 2x^2 - 5x - 3$$
$$(2x + 3)(x - 1) = 2x^2 + x - 3$$
$$(2x - 1)(x + 3) = 2x^2 + 5x - 3$$

Since the last product has the correct middle term, the trinomial is factored as

$$2x^2 + 5x - 3 = (2x - 1)(x + 3).$$

b) There are four possible ways to factor $3x^2 - 11x + 6$:

$$(3x \quad 1)(x \quad 6) \qquad (3x \quad 2)(x \quad 3)$$
$$(3x \quad 6)(x \quad 1) \qquad (3x \quad 3)(x \quad 2)$$

Because the last term is positive and the middle term is negative, both signs must be negative. Now try possible factors until we get the correct middle term:

$$(3x - 1)(x - 6) = 3x^2 - 19x + 6$$
$$(3x - 2)(x - 3) = 3x^2 - 11x + 6$$

The trinomial is factored correctly as

$$3x^2 - 11x + 6 = (3x - 2)(x - 3). \qquad \blacksquare$$

Higher Degrees and Variable Exponents

It is not necessary to always use substitution to factor polynomials with higher degrees or variable exponents as we did in Section 3.6. In the next example we use trial and error to factor two polynomials of higher degree and one with variable exponents. Remember that if there is a common factor to all terms, factor it out first.

EXAMPLE 5   **Higher-degree and variable-exponent trinomials**

Factor each polynomial completely. Variables used as exponents represent positive integers.

a) $x^8 - 2x^4 - 15$

b) $-18y^7 + 21y^4 + 15y$

c) $2u^{2m} - 5u^m - 3$

Solution

a) To factor by trial and error, notice that $x^8 = x^4 \cdot x^4$. Now 15 is $3 \cdot 5$ or $1 \cdot 15$. Using 1 and 15 will not give the required -2 for the coefficient of the middle term. So choose 3 and -5 to get the -2 in the middle term:

$$x^8 - 2x^4 - 15 = (x^4 - 5)(x^4 + 3)$$

b) $-18y^7 + 21y^4 + 15y = -3y(6y^6 - 7y^3 - 5)$ Factor out the common factor $-3y$ first.

$$= -3y(2y^3 + 1)(3y^3 - 5)$$ Factor the trinomial by trial and error.

c) Notice that $2u^{2m} = 2u^m \cdot u^m$ and $3 = 3 \cdot 1$. Using trial and error, we get

$$2u^{2m} - 5u^m - 3 = (2u^m + 1)(u^m - 3). \qquad \blacksquare$$

Warm-ups

True or false? Answer true if the polynomial is factored correctly and false otherwise.

1. $x^2 + 9x + 18 = (x + 3)(x + 6)$
2. $y^2 + 2y - 35 = (y + 5)(y - 7)$
3. $x^2 + 4 = (x + 2)(x + 2)$
4. $x^2 - 5x - 6 = (x - 3)(x - 2)$
5. $x^2 - 4x - 12 = (x - 6)(x + 2)$
6. $x^2 + 15x + 36 = (x + 4)(x + 9)$
7. $3x^2 + 4x - 15 = (3x + 5)(x - 3)$
8. $4x^2 + 4x - 3 = (4x - 1)(x + 3)$
9. $4x^2 - 4x - 3 = (2x + 1)(2x - 3)$
10. $4x^2 + 8x + 3 = (2x + 1)(2x + 3)$

3.7 EXERCISES

Factor each polynomial. See Examples 1 and 2.

1. $x^2 + 4x + 3$
2. $y^2 + 5y + 6$
3. $a^2 + 15a + 50$
4. $t^2 + 11t + 24$
5. $y^2 - 5y - 14$
6. $x^2 - 3x - 18$
7. $x^2 - 6x + 8$
8. $y^2 - 13y + 30$
9. $a^2 - 12a + 27$
10. $x^2 - x - 30$
11. $a^2 + 7a - 30$
12. $w^2 + 29w - 30$

Factor each polynomial using the ac method. See Example 3.

13. $6w^2 + 5w + 1$
14. $4x^2 + 11x + 6$
15. $2x^2 - 5x - 3$
16. $2a^2 + 3a - 2$
17. $4x^2 + 16x + 15$
18. $6y^2 + 17y + 12$
19. $6x^2 - 5x + 1$
20. $6m^2 - m - 12$
21. $12y^2 + y - 1$
22. $12x^2 + 5x - 2$
23. $6a^2 + a - 5$
24. $30b^2 - b - 3$

Factor each polynomial using trial and error. See Example 4.

25. $2x^2 + 15x - 8$
26. $3a^2 + 20a + 12$
27. $3b^2 - 16b - 35$
28. $2y^2 - 17y + 21$
29. $6w^2 - 35w + 36$
30. $15x^2 - x - 6$
31. $4x^2 - 5x + 1$
32. $4x^2 + 7x + 3$
33. $5m^2 + 13m - 6$
34. $5t^2 - 9t - 2$
35. $6y^2 - 7y - 20$
36. $7u^2 + 11u - 6$

Factor each polynomial completely. See Example 5. The variables used in exponents represent positive integers.

37. $x^6 - 2x^3 - 35$
38. $x^4 + 7x^2 - 30$
39. $a^{20} - 20a^{10} + 100$
40. $b^{16} + 22b^8 + 121$
41. $-12a^5 - 10a^3 - 2a$
42. $-4b^7 + 4b^4 + 3b$
43. $x^{2a} + 2x^a - 15$
44. $y^{2b} + y^b - 20$
45. $x^{2a} - y^{2b}$
46. $w^{4m} - a^2$
47. $x^8 - x^4 - 6$
48. $m^{10} - 5m^5 - 6$
49. $x^{a+2} - x^a$
50. $y^{2a+1} - y$
51. $x^{2a} + 6x^a + 9$
52. $x^{2a} - 2x^a y^b + y^{2b}$
53. $4y^3 z^6 + 5z^3 y^2 - 6y$
54. $2u^6 v^6 + 5u^4 v^3 - 12u^2$

Factor each polynomial completely.

55. $2x^2 + 20x + 50$
56. $3a^2 + 6a + 3$
57. $a^3 - 36a$
58. $x^3 - 5x^2 - 6x$
59. $5a^2 + 25a - 30$
60. $2a^2 - 2a - 84$
61. $2x^2 - 128y^2$
62. $a^3 - 6a^2 + 9a$
63. $-3x^2 + 3x + 36$
64. $xy^2 - 3xy - 70x$
65. $m^5 + 20m^4 + 100m^3$
66. $4a^2 - 16a + 16$
67. $6x^2 + 23x + 20$
68. $2y^2 - 13y + 6$
69. $y^2 - 12y + 36$
70. $m^2 - 2m + 1$
71. $9m^2 - 25n^2$
72. $m^2 n^2 - 2mn^3 + n^4$
73. $5a^2 + 20a - 60$
74. $-3y^2 + 9y + 30$
75. $-2w^2 + 18w + 20$
76. $x^2 z + 2xyz + y^2 z$

77. $w^2x^2 - 100x^2$ **78.** $9x^2 + 30x + 25$

79. $9x^2 - 1$ **80.** $6w^2 - 19w - 36$

81. $8x^2 - 2x - 15$ **82.** $4w^2 + 12w + 9$

83. $4x^2 - 20x + 25$ **84.** $9m^2 - 121$

85. $9a^2 + 60a + 100$ **86.** $10w^2 + 20w - 80$

87. $4a^2 + 24a + 32$ **88.** $20x^2 - 23x + 6$

89. $3m^4 - 24m$ **90.** $6w^3z + 6z$

Getting More Involved

91. *Discussion.* Which of the following is not a perfect square trinomial? Explain.

a) $4a^6 - 6a^3b^4 + 9b^8$ b) $1000x^2 + 200ax + a^2$

c) $900y^4 - 60y^2 + 1$ d) $36 - 36z^7 + 9z^{14}$

92. *Discussion.* Which of the following is not a difference of two squares? Explain.

a) $16a^8y^4 - 25c^{12}$ b) $a^9 - b^4$

c) $t^{90} - 1$ d) $x^2 - 196$

3.8 Factoring Strategy

In this section:

▸ Prime Polynomials

▸ Factoring Polynomials Completely

▸ Factoring Polynomials with Four Terms

▸ Summary

In previous sections we established the general idea of factoring and many special cases. In this section we will see that a polynomial can have as many factors as its degree, and we will factor higher-degree polynomials completely. We will also see a general strategy for factoring polynomials.

Prime Polynomials

Monomials are considered prime polynomials. Binomials with no common factors, such as $2x + 1$ and $a - 3$, are also prime polynomials. To determine whether a polynomial such as $x^2 + 1$ is a prime polynomial, we must try all possibilities for factoring it. If $x^2 + 1$ could be factored as a product of two binomials, the only possibilities that would give a first term of x^2 and a last term of 1 are $(x + 1)(x + 1)$ and $(x - 1)(x - 1)$. However,

$$(x + 1)(x + 1) = x^2 + 2x + 1 \quad \text{and} \quad (x - 1)(x - 1) = x^2 - 2x + 1.$$

Both products have an x-term. Of course $(x + 1)(x - 1)$ has no x-term, but

$$(x + 1)(x - 1) = x^2 - 1.$$

Because none of these possibilities results in $x^2 + 1$, the polynomial $x^2 + 1$ is a prime polynomial. Note that $x^2 + 1$ is a sum of two squares. A sum of two squares (that have no common factor) is always a prime polynomial.

EXAMPLE 1   **Prime polynomials**

Determine whether the polynomial $x^2 + 3x + 4$ is a prime polynomial.

Solution

To factor $x^2 + 3x + 4$, we must find two integers with a product of 4 and a sum of 3. The only pairs of positive integers with a product of 4 are 1 and 4, and 2 and 2. Because the product is positive 4, both numbers must be negative or both positive. Under these conditions it is impossible to get a sum of positive 3. The polynomial is prime. ∎

Factoring Polynomials Completely

So far, a typical polynomial has been a product of two factors, with possibly a common factor removed first. However, it is possible that the factors can still be factored again. A polynomial in a single variable may have as many factors as its degree. We have factored a polynomial completely when all of the factors are prime polynomials.

EXAMPLE 2 **Factoring higher-degree polynomials completely**

Factor $x^4 + x^2 - 2$ completely.

Solution

Two numbers with a product of -2 and a sum of 1 are 2 and -1:

$$x^4 + x^2 - 2 = (x^2 + 2)(x^2 - 1)$$
$$= (x^2 + 2)(x - 1)(x + 1) \quad \text{Difference of two squares}$$

Since $x^2 + 2$, $x - 1$, and $x + 1$ are prime, the polynomial is factored completely. ∎

In the next example we factor a sixth-degree polynomial.

EXAMPLE 3 **Factoring completely**

Factor $3x^6 - 3$ completely.

Solution

To factor $3x^6 - 3$, we must first factor out the common factor 3 and then recognize that x^6 is a perfect square: $x^6 = (x^3)^2$:

$$3x^6 - 3 = 3(x^6 - 1) \qquad\qquad \text{Factor out the common factor.}$$
$$= 3((x^3)^2 - 1) \qquad\quad \text{Write } x^6 \text{ as a perfect square.}$$
$$= 3(x^3 - 1)(x^3 + 1) \qquad \text{Difference of two squares}$$
$$= 3(x - 1)(x^2 + x + 1)(x + 1)(x^2 - x + 1) \quad \begin{array}{l}\text{Difference of two cubes and}\\ \text{sum of two cubes}\end{array}$$

Since $x^2 + x + 1$ and $x^2 - x + 1$ are prime, the polynomial is factored completely. ∎

In Example 3 we recognized $x^6 - 1$ as a difference of two squares. However, $x^6 - 1$ is also a difference of two cubes, and we can factor it using the rule for the difference of two cubes:

$$x^6 - 1 = (x^2)^3 - 1 = (x^2 - 1)(x^4 + x^2 + 1)$$

Now we can factor $x^2 - 1$, but it is difficult to see how to factor $x^4 + x^2 + 1$. (It is not prime.) Although x^6 can be thought of as a perfect square or a perfect cube, in this case it is better to think of it as a perfect square.

In the next example we use substitution to simplify the polynomial before factoring. This fourth-degree polynomial has four factors.

EXAMPLE 4 **Using substitution to simplify**

Factor $(w^2 - 1)^2 - 11(w^2 - 1) + 24$ completely.

Solution

Let $a = w^2 - 1$ to simplify the polynomial:

$$(w^2 - 1)^2 - 11(w^2 - 1) + 24 = a^2 - 11a + 24 \qquad \text{Replace } w^2 - 1 \text{ by } a.$$
$$= (a - 8)(a - 3)$$
$$= (w^2 - 1 - 8)(w^2 - 1 - 3) \qquad \text{Replace } a \text{ by } w^2 - 1.$$
$$= (w^2 - 9)(w^2 - 4)$$
$$= (w + 3)(w - 3)(w + 2)(w - 2) \qquad ■$$

Factoring Polynomials with Four Terms

In Section 3.6 we rewrote a trinomial as a polynomial with four terms and then used factoring by grouping. Factoring by grouping can also be used on other types of polynomials with four terms.

EXAMPLE 5 **Polynomials with four terms**

Use grouping to factor each polynomial completely.

a) $x^3 + x^2 + 4x + 4$ **b)** $3x^3 - x^2 - 27x + 9$ **c)** $ax - bw + bx - aw$

Solution

a) Note that the first two terms of $x^3 + x^2 + 4x + 4$ have a common factor of x^2, and the last two terms have a common factor of 4.

$$x^3 + x^2 + 4x + 4 = x^2(x + 1) + 4(x + 1) \qquad \text{Factor by grouping.}$$
$$= (x^2 + 4)(x + 1) \qquad \text{Factor out } x + 1.$$

Since $x^2 + 4$ is a sum of two squares, it is prime and the polynomial is factored completely.

b) We can factor x^2 out of the first two terms of $3x^3 - x^2 - 27x + 9$ and 9 or -9 from the last two terms. We choose -9 to get the factor $3x - 1$ in each case.

$$3x^3 - x^2 - 27x + 9 = x^2(3x - 1) - 9(3x - 1) \qquad \text{Factor by grouping.}$$
$$= (x^2 - 9)(3x - 1) \qquad \text{Factor out } 3x - 1.$$
$$= (x - 3)(x + 3)(3x - 1) \qquad \text{Difference of two squares}$$

This third-degree polynomial has three factors.

c) First rearrange the terms so that the first two and the last two have common factors:

$$ax - bw + bx - aw = ax + bx - aw - bw \qquad \text{Rearrange the terms.}$$
$$= x(a + b) - w(a + b) \qquad \text{Common factors}$$
$$= (x - w)(a + b) \qquad \text{Factor out } a + b. \qquad ■$$

Summary

A strategy for factoring polynomials is given in the following box.

> ### Strategy for Factoring Polynomials
>
> 1. If there are any common factors, factor them out first.
> 2. When factoring a binomial, look for the special cases: difference of two squares, difference of two cubes, and sum of two cubes. Remember that a sum of two squares is prime.
> 3. When factoring a trinomial, check to see whether it is a perfect square trinomial.
> 4. When factoring a trinomial that is not a perfect square, use grouping or trial and error.
> 5. When factoring a polynomial of high degree, use substitution to get a polynomial of degree 2 or 3, or use trial and error.
> 6. If the polynomial has four terms, try factoring by grouping

EXAMPLE 6   **Using the factoring strategy**

Factor each polynomial completely.

a) $3w^3 - 3w^2 - 18w$

b) $10x^2 + 160$

c) $16a^2b - 80ab + 100b$

d) $aw + mw + az + mz$

Solution

a) The greatest common factor for the three terms is $3w$:

$$3w^3 - 3w^2 - 18w = 3w(w^2 - w - 6) \quad \text{Factor out } 3w.$$
$$= 3w(w - 3)(w + 2) \quad \text{Factor completely.}$$

b) The greatest common factor in $10x^2 + 160$ is 10:

$$10x^2 + 160 = 10(x^2 + 16)$$

Because $x^2 + 16$ is prime, the polynomial is factored completely.

c) The greatest common factor in $16a^2b - 80ab + 100b$ is $4b$:

$$16a^2b - 80ab + 100b = 4b(4a^2 - 20a + 25)$$
$$= 4b(2a - 5)^2$$

d) The polynomial has four terms, and we can factor it by grouping:

$$aw + mw + az + mz = w(a + m) + z(a + m)$$
$$= (w + z)(a + m)$$

Warm-ups

True or false? Explain your answer.

1. $x^2 - 9 = (x - 3)^2$ for any value of x.
2. The polynomial $4x^2 + 12x + 9$ is a perfect square trinomial.
3. A sum of two squares (with no common factor) is prime.
4. The polynomial $x^4 - 16$ is factored completely as $(x^2 - 4)(x^2 + 4)$.
5. $y^3 - 27 = (y + 3)(y^2 + 3y - 9)$ for any value of y.
6. The polynomial $y^6 - 1$ is a difference of two squares.
7. The polynomial $2x^2 + 2x - 12$ is factored completely as $(2x - 4)(x + 3)$.
8. The polynomial $x^2 - 4x - 4$ is a prime polynomial.
9. The polynomial $a^6 - 1$ is the difference of two cubes.
10. The polynomial $x^2 + 3x - ax + 3a$ can be factored by grouping.

3.8 EXERCISES

Determine whether each polynomial is a prime polynomial. See Example 1.

1. $y^2 + 100$
2. $3x^2 + 27$
3. $-9w^2 - 9$
4. $25y^2 + 36$
5. $x^2 - 2x - 3$
6. $x^2 - 2x + 3$
7. $x^2 + 2x + 3$
8. $x^2 + 4x + 3$
9. $x^2 - 4x - 3$
10. $x^2 + 4x - 3$
11. $6x^2 + 3x - 4$
12. $4x^2 - 5x - 3$

Factor each polynomial completely. See Examples 2 – 4.

13. $a^4 - 10a^2 + 25$
14. $9y^4 + 12y^2 + 4$
15. $x^4 - 6x^2 + 8$
16. $x^6 + 2x^3 - 3$
17. $(3x - 5)^2 - 1$
18. $(2x + 1)^2 - 4$
19. $2y^6 - 128$
20. $6 - 6y^6$
21. $32a^4 - 18$
22. $2a^4 - 32$
23. $x^4 - (x - 6)^2$
24. $y^4 - (2y + 1)^2$
25. $(m + 2)^2 + 2(m + 2) - 3$
26. $(2w - 3)^2 - 2(2w - 3) - 15$
27. $3(y - 1)^2 + 11(y - 1) - 20$
28. $2(w + 2)^2 + 5(w + 2) - 3$
29. $(y^2 - 3)^2 - 4(y^2 - 3) - 12$
30. $(m^2 - 8)^2 - 4(m^2 - 8) - 32$

Use grouping to factor each polynomial completely. See Example 5.

31. $ax + ay + bx + by$
32. $7x + 7z + kx + kz$
33. $x^3 + x^2 - 9x - 9$
34. $x^3 + x^2 - 25x - 25$
35. $aw - bw - 3a + 3b$
36. $wx - wy + 2x - 2y$
37. $a^4 + 3a^3 + 27a + 81$
38. $ac - bc - 5a + 5b$
39. $y^4 - 5y^3 + 8y - 40$
40. $x^3 + ax - 3a - 3x^2$
41. $ady + d - w - awy$
42. $xy + by + ax + ab$
43. $x^2y - a + ax^2 - y$
44. $a^2d - b^2c + a^2c - b^2d$
45. $y^4 + b + by^3 + y$
46. $ab + mw^2 + bm + aw^2$

Use the factoring strategy to factor each polynomial completely. See Example 6.

47. $9x^2 - 24x + 16$
48. $-3x^2 + 18x + 48$
49. $12x^2 - 13x + 3$
50. $2x^2 - 3x - 6$
51. $3a^4 + 81a$
52. $-a^3 + 25a$
53. $32 + 2x^2$
54. $x^3 + 4x^2 + 4x$
55. $6x^2 - 5x + 12$
56. $x^4 + 2x^3 - x - 2$
57. $(x + y)^2 - 1$
58. $x^3 + 9x$
59. $a^3b - ab^3$
60. $2m^3 - 250$
61. $x^4 + 2x^3 - 8x - 16$
62. $(x + 5)^2 - 4$
63. $m^2n + 2mn^2 + n^3$
64. $a^2b - 6ab + 9b$
65. $2m + wn + 2n + wm$
66. $aw - 5b + bw - 5a$
67. $4w^2 + 4w - 3$
68. $4w^2 + 8w - 63$
69. $t^4 + 4t^2 - 21$
70. $m^4 + 5m^2 + 4$
71. $-a^3 - 7a^2 + 30a$
72. $2y^4 + 3y^3 - 20y^2$

73. $(y + 5)^2 - 2(y + 5) - 3$

74. $(2t - 1)^2 + 7(2t - 1) + 10$

75. $-2w^4 + 1250$ **76.** $5a^5 - 5a$

77. $8a^3 + 8a$ **78.** $awx + ax$

79. $(w + 5)^2 - 9$ **80.** $(a - 6)^2 - 1$

81. $4aw^2 - 12aw + 9a$ **82.** $9an^3 + 15an^2 - 14an$

83. $x^2 - 6x + 9$ **84.** $x^3 + 12x^2 + 36x$

85. $3x^4 - 75x^2$ **86.** $3x^2 + 9x + 12$

87. $m^3n - n$ **88.** $m^4 + 16m^2$

89. $12x^2 + 2x - 30$ **90.** $90x^2 + 3x - 60$

91. $2a^3 - 32$ **92.** $12x^2 - 28x + 15$

Factor completely. Assume that variables used as exponents represent positive integers.

93. $a^{3m} - 1$ **94.** $x^{6a} + 8$ **95.** $a^{3w} - b^{6n}$

96. $x^{2n} - 9$ **97.** $t^{8n} - 16$ **98.** $a^{3n+2} + a^2$

99. $a^{2n+1} - 2a^{n+1} - 15a$

100. $x^{3m} + x^{2m} - 6x^m$

101. $a^{2n} - 3a^n + a^n b - 3b$

102. $x^m z + 5z + x^{m+1} + 5x$

Getting More Involved

103. *Cooperative learning.* Write down 10 trinomials of the form $ax^2 + bx + c$ "at random" using integers for a, b, and c. What percent of your 10 trinomials are prime? Would you say that prime trinomials are the exception or the rule? Compare your results with those of your classmates.

104. *Writing.* The polynomial $x^5 + x^4 - 9x^3 - 13x^2 + 8x + 12$ is a product of five factors of the form $x \pm n$, where n is a natural number smaller than 4. Factor this polynomial completely and explain your procedure.

3.9

Solving Equations by Factoring

In this section:

▶ The Zero Factor Property

▶ Applications

The techniques of factoring can be used to solve equations involving polynomials that cannot be solved by the other methods that you have learned. After you learn to solve equations by factoring, we will use this technique to solve some new applied problems in this section and in Chapters 4 and 6.

The Zero Factor Property

The equation $ab = 0$ indicates that the product of two unknown numbers is 0. But the product of two real numbers is zero only when one or the other of the numbers is 0. So even though we do not know exactly the values of a and b from $ab = 0$, we do know that $a = 0$ or $b = 0$. This idea is called the **zero factor property.**

> **Zero Factor Property**
>
> The equation $ab = 0$ is equivalent to the compound equation
>
> $$a = 0 \quad \text{or} \quad b = 0.$$

The next example shows how to use the zero factor property to solve an equation in one variable.

EXAMPLE 1   **Using the zero factor property**

Solve $x^2 + x - 12 = 0$.

Solution

We factor the left side of the equation to get a product of two factors equal to 0. Then we write an equivalent equation using the zero factor property.

$$x^2 + x - 12 = 0$$

$$(x + 4)(x - 3) = 0 \quad \text{Factor the left side.}$$

$$x + 4 = 0 \quad \text{or} \quad x - 3 = 0 \quad \text{Zero factor property}$$

$$x = -4 \quad \text{or} \quad x = 3 \quad \text{Solve each part of the compound equation.}$$

Check that -4 and 3 both satisfy $x^2 + x - 12 = 0$. If $x = -4$, we get

$$(-4)^2 + (-4) - 12 = 16 - 4 - 12 = 0.$$

If $x = 3$, we get

$$(3)^2 + 3 - 12 = 9 + 3 - 12 = 0.$$

So the solution set is $\{-4, 3\}$. ∎

MATH AT WORK

Bodyboard Designer

Seamas Mercado, professional bodyboarder and 1988 National Champion, charges the waves off of Hawaii, Tahiti, Indonesia, Mexico, and California. In choosing a board for competition and for the maneuvers he wants to perform, Mercado factors in his height and weight as well as the size, power, and temperature of the waves he will be riding. In colder water a softer, more flexible board is used; in warmer water a stiffer board is chosen. When waves crash on shore, the ride usually lasts 3 to 5 seconds, and a shorter board with a narrow tail is chosen for greater control. When waves break along a sand bar or reef, the ride can sometimes last as long as 2 minutes, and a straighter board with more surface area is chosen so that the board will move faster and allow the rider to pull off more maneuvers. Basic maneuvers include bottom turns, aerials, forward and reverse 360's, and el rollos.

As one of the top ten bodyboarders in the world, Mercado helps to design the boards he uses. "Performance levels are greatly increased with fine-tuned equipment and techniques," he says. In Exercise 57 of Section 3.9 you will find the dimensions of a given bodyboard.

The zero factor property is used only in solving polynomial equations that have zero on one side and a polynomial that can be factored on the other side. The polynomials that we factored most often were the quadratic polynomials. The equations that we will solve most often using the zero factor property will be quadratic equations.

> **Quadratic Equation**
>
> If a, b, and c are real numbers, with $a \neq 0$, then the equation
>
> $$ax^2 + bx + c = 0$$
>
> is called a **quadratic equation.**

In Chapter 6 we will study quadratic equations further and solve quadratic equations that cannot be solved by factoring. Keep the following strategy in mind when solving equations by factoring.

> ► **Strategy for Solving Equations by Factoring** ◄
>
> 1. Write the equation with 0 on the right side.
> 2. Factor the left side.
> 3. Use the zero factor property to get simpler equations. (Set each factor equal to 0.)
> 4. Solve the simpler equations.
> 5. Check the answers in the original equation.

EXAMPLE 2   **Solving a quadratic equation by factoring**

Solve $10x^2 = 5x$.

Solution

Use the steps in the strategy for solving equations by factoring:

$$10x^2 = 5x$$

$$10x^2 - 5x = 0 \qquad \text{Rewrite the equation with zero on the right side.}$$

$$5x(2x - 1) = 0 \qquad \text{Factor the left side.}$$

$$5x = 0 \quad \text{or} \quad 2x - 1 = 0 \qquad \text{Zero factor property}$$

$$x = 0 \quad \text{or} \quad x = \frac{1}{2} \qquad \text{Solve the simpler equations.}$$

The solution set is $\left\{0, \frac{1}{2}\right\}$. Check each solution in the original equation. ■

CAUTION   If we divide each side of $10x^2 = 5x$ by $5x$, we get $2x = 1$, or $x = \frac{1}{2}$. We do not get $x = 0$. By dividing by $5x$ we have lost one of the factors and one of the solutions. $\oslash$

In the next example there are more than two factors, but we can still write an equivalent equation by setting each factor equal to zero.

EXAMPLE 3   **Solving a cubic equation by factoring**

Solve $2x^3 - 3x^2 - 8x + 12 = 0$.

Solution

First notice that the first two terms have the common factor x^2 and the last two terms have the common factor -4.

$$x^2(2x - 3) - 4(2x - 3) = 0 \quad \text{Factor by grouping.}$$
$$(x^2 - 4)(2x - 3) = 0 \quad \text{Factor out } 2x - 3.$$
$$(x - 2)(x + 2)(2x - 3) = 0 \quad \text{Factor completely.}$$

$x - 2 = 0 \quad$ or $\quad x + 2 = 0 \quad$ or $\quad 2x - 3 = 0 \quad$ Set each factor equal to 0.

$$x = 2 \quad \text{or} \quad x = -2 \quad \text{or} \quad x = \frac{3}{2}$$

The solution set is $\left\{-2, \frac{3}{2}, 2\right\}$. Check each solution in the original equation. ■

The equation in the next example involves absolute value.

EXAMPLE 4   **Solving an absolute value equation by factoring**

Solve $|x^2 - 2x - 16| = 8$.

Solution

First write an equivalent compound equation without absolute value:

$$x^2 - 2x - 16 = 8 \quad \text{or} \quad x^2 - 2x - 16 = -8$$
$$x^2 - 2x - 24 = 0 \quad \text{or} \quad x^2 - 2x - 8 = 0$$
$$(x - 6)(x + 4) = 0 \quad \text{or} \quad (x - 4)(x + 2) = 0$$

$x - 6 = 0 \quad$ or $\quad x + 4 = 0 \quad$ or $\quad x - 4 = 0 \quad$ or $\quad x + 2 = 0$

$x = 6 \quad$ or $\quad x = -4 \quad$ or $\quad x = 4 \quad$ or $\quad x = -2$

The solution set is $\{-2, -4, 4, 6\}$. Check each solution. ■

Applications

Many applied problems can be solved by using equations like those we have been solving.

EXAMPLE 5   **Area of a room**

Ronald's living room is 2 feet longer than it is wide, and its area is 168 square feet. What are the dimensions of the room?

Solution

Let x be the width and $x + 2$ be the length. See Figure 3.1. Because the area of a rectangle is the length times the width, we can write the equation

$$x(x + 2) = 168.$$

Figure 3.1

We solve the equation by factoring:

$$x^2 + 2x - 168 = 0$$
$$(x - 12)(x + 14) = 0$$
$$x - 12 = 0 \quad \text{or} \quad x + 14 = 0$$
$$x = 12 \quad \text{or} \quad x = -14$$

Because the width of a room is a positive number, we disregard the solution $x = -14$. We use $x = 12$ and get a width of 12 feet and a length of 14 feet. Check this answer by multiplying 12 and 14 to get 168. ■

Applications involving quadratic equations often require a theorem called the **Pythagorean theorem.** This theorem states that *in any right triangle the sum of the squares of the lengths of the legs is equal to the length of the hypotenuse squared.*

The Pythagorean Theorem

The triangle shown is a right triangle if and only if

$$a^2 + b^2 = c^2.$$

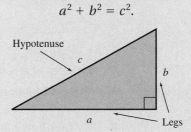

We use the Pythagorean theorem in the next example.

EXAMPLE 6   **Using the Pythagorean theorem**

Shirley used 14 meters of fencing to enclose a rectangular region. To be sure that the region was a rectangle, she measured the diagonals and found that they were 5 meters each. (If the opposite sides of a quadrilateral are equal and the diagonals are equal, then the quadrilateral is a rectangle.) What are the length and width of the rectangle?

Solution

Figure 3.2

The perimeter of a rectangle is twice the length plus twice the width, $P = 2L + 2W$. Because the perimeter is 14 meters, the sum of one length and one width is 7 meters. If we let x represent the width, then $7 - x$ is the length. We use the Pythagorean theorem to get a relationship between the length, width, and diagonal. See Figure 3.2.

$$x^2 + (7 - x)^2 = 5^2 \qquad \text{Pythagorean theorem}$$
$$x^2 + 49 - 14x + x^2 = 25 \qquad \text{Simplify.}$$
$$2x^2 - 14x + 24 = 0 \qquad \text{Simplify.}$$
$$x^2 - 7x + 12 = 0 \qquad \text{Divide each side by 2.}$$
$$(x - 3)(x - 4) = 0 \qquad \text{Factor the left side.}$$

$$x - 3 = 0 \quad \text{or} \quad x - 4 = 0 \qquad \text{Zero factor property}$$
$$x = 3 \quad \text{or} \quad x = 4$$
$$7 - x = 4 \quad \text{or} \quad 7 - x = 3$$

Solving the equation gives two possible rectangles: a 3 by 4 rectangle or a 4 by 3 rectangle. However, those are identical rectangles. The rectangle is 3 meters by 4 meters. ∎

Warm-ups

True or false? Explain your answer.

1. The equation $(x - 1)(x + 3) = 12$ is equivalent to $x - 1 = 3$ or $x + 3 = 4$.

2. Equations solved by factoring may have two solutions.

3. The equation $c \cdot d = 0$ is equivalent to $c = 0$ or $d = 0$.

4. The equation $|x^2 + 4| = 5$ is equivalent to the compound equation $x^2 + 4 = 5$ or $x^2 - 4 = 5$.

5. The solution set to the equation $(2x - 1)(3x + 4) = 0$ is $\left\{\frac{1}{2}, -\frac{4}{3}\right\}$.

6. The Pythagorean theorem states that the sum of the squares of any two sides of any triangle is equal to the square of the third side.

7. If the perimeter of a rectangular room is 38 feet, then the sum of the length and width is 19 feet.

8. Two numbers that have a sum of 8 can be represented by x and $8 - x$.

9. The solution set to the equation $x(x - 1)(x - 2) = 0$ is $\{1, 2\}$.

10. The solution set to the equation $3(x + 2)(x - 5) = 0$ is $\{3, -2, 5\}$.

3.9 EXERCISES

Solve each equation. See Examples 1 – 3.

1. $(x - 5)(x + 4) = 0$
2. $(a - 6)(a + 5) = 0$
3. $(2x - 5)(3x + 4) = 0$
4. $(3k + 8)(4k - 3) = 0$
5. $w^2 + 5w - 14 = 0$
6. $t^2 - 6t - 27 = 0$
7. $m^2 - 7m = 0$
8. $h^2 - 5h = 0$
9. $a^2 - a = 20$
10. $p^2 - p = 42$
11. $3x^2 - 3x - 36 = 0$
12. $-2x^2 - 16x - 24 = 0$
13. $z^2 + \dfrac{3}{2}z = 10$
14. $m^2 + \dfrac{11}{3}m = -2$
15. $x^3 - 4x = 0$
16. $16x - x^3 = 0$
17. $w^3 + 4w^2 - 25w - 100 = 0$
18. $a^3 + 2a^2 - 16a - 32 = 0$
19. $n^3 - 2n^2 - n + 2 = 0$
20. $w^3 - w^2 - 25w + 25 = 0$

Solve each equation. See Example 4.

21. $|x^2 - 5| = 4$
22. $|x^2 - 17| = 8$
23. $|x^2 + 2x - 36| = 12$
24. $|x^2 + 2x - 19| = 16$
25. $|x^2 + 4x + 2| = 2$
26. $|x^2 + 8x + 8| = 8$
27. $|x^2 + 6x + 1| = 8$
28. $|x^2 - x - 21| = 9$

Solve each equation.

29. $2x^2 - x = 6$
30. $3x^2 + 14x = 5$
31. $|x^2 + 5x| = 6$
32. $|x^2 + 6x - 4| = 12$
33. $x^2 + 5x = 6$
34. $x + 5x = 6$
35. $(x + 2)(x + 1) = 12$
36. $(x + 2)(x + 3) = 20$
37. $y^3 + 9y^2 + 20y = 0$
38. $m^3 - 2m^2 - 3m = 0$
39. $5a^3 = 45a$
40. $5x^3 = 125x$
41. $(2x - 1)(x^2 - 9) = 0$
42. $(x - 1)(x + 3)(x - 9) = 0$
43. $4x^2 - 12x + 9 = 0$
44. $16x^2 + 8x + 1 = 0$

Solve each equation for y. Assume that a and b are positive numbers.

45. $y^2 + by = 0$
46. $y^2 + ay + by + ab = 0$
47. $a^2y^2 - b^2 = 0$
48. $9y^2 + 6ay + a^2 = 0$
49. $4y^2 + 4by + b^2 = 0$
50. $y^2 - b^2 = 0$
51. $ay^2 + 3y - ay = 3$
52. $a^2y^2 + 2aby + b^2 = 0$

Solve each problem. See Examples 5 and 6.

53. **Color print.** The length of a new "super size" color print is 2 inches more than the width. If the area is 24 square inches, what are the length and width?

54. **Tennis court dimensions.** In singles competition, each player plays on a rectangular area of 117 square yards. Given that the length of that area is 4 yards greater than its width, find the length and width.

55. **Missing numbers.** The sum of two numbers is 13, and their product is 36. Find the numbers.

56. **More missing numbers.** The sum of two numbers is 6.5, and their product is 9. Find the numbers.

57. **Bodyboarding.** The Seamas Channel pro bodyboard shown in the figure has a length that is 21 inches greater than its width. Any rider weighing up to 200 pounds can use it because its surface area is 946 square inches. Find the length and width.

$x + 21$ in.

x in.

Figure for Exercise 57

58. **New dimensions in gardening.** Mary Gold has a rectangular flower bed that measures 4 feet by 6 feet. If she wants to increase the length and width by the same amount to have a flower bed of 48 square feet, then what will the new dimensions be?

x ft

4 ft

6 ft

x ft

Figure for Exercise 58

59. *Time for a toss.* A baseball player tosses a ball into the air at 64 feet per second. The distance D (in feet) that the ball is above the earth at any time t (in seconds) is given by the formula $D = -16t^2 + 64t$. Determine the length of time that the ball is in the air.

60. *Time until impact.* If an object is dropped from a height of s_0 feet, then its altitude after t seconds is given by the formula $S = -16t^2 + s_0$. If a pack of emergency supplies is dropped from an airplane at a height of 1600 feet, then how long does it take for it to reach the ground?

61. *Yolanda's closet.* The length of Yolanda's closet is 2 feet longer than twice its width. If the diagonal measures 13 feet, then what are the length and width?

13 ft

2x + 2 ft

x ft

Figure for Exercise 61

62. *Ski jump.* The base of a ski ramp forms a right triangle. One leg of the triangle is 2 meters longer than the other. If the hypotenuse is 10 meters, then what are the lengths of the legs?

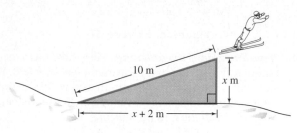

10 m

x m

x + 2 m

Figure for Exercise 62

63. *Trimming a gate.* A total of 34 feet of 1×4 lumber is used around the perimeter of the gate shown in the figure. If the diagonal brace is 13 feet long, then what are the length and width of the gate?

13 ft

Figure for Exercise 63

64. *Perimeter of a rectangle.* The perimeter of a rectangle is 28 inches, and the diagonal measures 10 inches. What are the length and width of the rectangle?

65. *Consecutive integers.* The sum of the squares of two consecutive integers is 25. Find the integers.

66. *Pete's garden.* Each row in Pete's garden is 3 feet wide. If the rows run north and south, he can have two more rows than if they run east and west. If the area of Pete's garden is 135 square feet, then what are the length and width?

67. *House plans.* In the plans for their dream house the Baileys have a master bedroom that is 240 square feet in area. If they increase the width by 3 feet, they must decrease the length by 4 feet to keep the original area. What are the original dimensions of the bedroom?

68. *Arranging the rows.* Mr. Converse has 112 students in his algebra class with an equal number in each row. If he arranges the desks so that he has one fewer rows, he will have two more students in each row. How many rows did he have originally?

Getting More Involved

69. *Writing.* If you divide each side of $x^2 = x$ by x, you get $x = 1$. If you subtract x from each side of $x^2 = x$, you get, $x^2 - x = 0$, which has two solutions. Which method is correct? Explain.

70. *Cooperative learning.* Work with a group to examine the following solution to $x^2 - 2x = -1$:

$$x(x - 2) = -1$$

$$x = -1 \quad \text{or} \quad x - 2 = -1$$

$$x = -1 \quad \text{or} \quad x = 1$$

Is this method correct? Explain.

71. *Cooperative learning.* Work with a group to examine the following steps in the solution to $5x^2 - 5 = 0$

$$5(x^2 - 1) = 0$$

$$5(x - 1)(x + 1) = 0$$

$$x - 1 = 0 \quad \text{or} \quad x + 1 = 0$$

$$x = 1 \quad \text{or} \quad x = -1$$

What happened to the 5? Explain.

──────────── C O L L A B O R A T I V E   A C T I V I T I E S ────────────

Magic Tricks

Jim and Sadar are talking one day after class.

Sadar: Jim, I have a trick for you. Think of a number between 1 and 10. I will ask you to do some things to this number. Then at the end tell me your result, and I will tell you your number.

Jim: Oh, yeah you probably rig it so the result is my number.

Sadar: Come on Jim, give it a try and see.

Jim: Okay, okay, I thought of a number.

Sadar: Good, now write it down, and don't let me see your paper. Now add x. Got that? Now multiply everything by 2.

Jim: Hey, I didn't know you were going to make me think! This is algebra!

Sadar: I know, now just do it. Okay, now square the polynomial. Got that? Now subtract $4x^2$.

Jim: How did you know I had a $4x^2$? I told you this was rigged!

Sadar: Of course it's rigged, or it wouldn't work. Do you want to finish or not?

Jim: Yeah, I guess so. Go ahead, what do I do next?

Sadar: Divide by 4. Okay, now subtract the x-term.

Grouping: 2 students per group
Topic: Practice with exponent rules, multiplying polynomials

Jim: Just any old x-term? Got any particular coefficient in mind?

Sadar: Now stop teasing me. I know you only have one x-term left, so subtract it.

Jim: Ha, ha, I *could* give you a hint about the coefficient, but that wouldn't be fair, would it?

Sadar: Well you could, and then I could tell you your number, or your could just tell me the number you have left after subtracting.

Jim: Okay, the number I had left at the end was 25. Let's see if you can tell me what the coefficient of the x-term I subtracted is.

Sadar: Aha, then the number you chose at the beginning was 5, and the coefficient was 10!

Jim: Hey, you're right! How did you do that?

In your group, follow Sadar's instructions, and determine why she knew Jim's number. Make up another set of instructions to use as a magic trick. Be sure to use variables and some of the exponent rules or rules for multiplying polynomials you learned in this chapter. Exchange instructions with another group and see whether you can figure out how their trick works.

Wrap-up   CHAPTER 3

SUMMARY

Definitions		Examples
Definition of negative integral exponents	If a is a nonzero real number and n is a positive integer, then $$a^{-n} = \frac{1}{a^n}.$$	$2^{-3} = \frac{1}{2^3} = 8$
Definition of zero exponent	If a is any nonzero real number, then $a^0 = 1$. The expression 0^0 is undefined.	$3^0 = 1$

Rules of Exponents		**Examples**
If a and b are nonzero real numbers and m and n are integers, then the following rules hold.		
Negative exponent rules	$a^{-n} = \left(\dfrac{1}{a}\right)^n, a^{-1} = \dfrac{1}{a}$, and $\dfrac{1}{a^{-n}} = a^n$	$5^{-1} = \dfrac{1}{5}, \dfrac{1}{5^{-3}} = 5^3$
	Find the power and reciprocal in either order.	$\left(\dfrac{2}{3}\right)^{-2} = \left(\dfrac{3}{2}\right)^{2}$
Product rule	$a^m \cdot a^n = a^{m+n}$	$3^5 \cdot 3^7 = 3^{12}, 2^{-3} \cdot 2^{10} = 2^7$
Quotient rule	$\dfrac{a^m}{a^n} = a^{m-n}$	$\dfrac{x^8}{x^5} = x^3, \dfrac{5^4}{5^{-7}} = 5^{11}$
Power of a power rule	$(a^m)^n = a^{mn}$	$(5^2)^3 = 5^6$
Power of a product rule	$(ab)^n = a^n b^n$	$(2x)^3 = 8x^3$ $\quad(2x^3)^4 = 16x^{12}$
Power of a quotient rule	$\left(\dfrac{a}{b}\right)^n = \dfrac{a^n}{b^n}$	$\left(\dfrac{x}{3}\right)^2 = \dfrac{x^2}{9}$

Scientific Notation		**Examples**
Converting from scientific notation	1. Determine the number of places to move the decimal point by examining the exponent on the 10.	$4 \times 10^3 = 4000$
	2. Move to the right for a positive exponent and to the left for a negative exponent.	$3 \times 10^{-4} = 0.0003$
Converting to scientific notation	1. Count the number of places (n) that the decimal point must be moved so that it will follow the first nonzero digit of the number.	
	2. If the original number was larger than 10, use 10^n.	$67{,}000 = 6.7 \times 10^4$
	3. If the original number was smaller than 1, use 10^{-n}.	$0.009 = 9 \times 10^{-3}$

Polynomials		**Examples**
Term of a polynomial	The product of a number (coefficient) and one or more variables raised to whole number powers	$3x^4, -2xy^2, 5$
Polynomial	A single term or a finite sum of terms	$x^5 - 3x^2 + 7$
Adding or subtracting polynomials	Add or subtract the like terms.	$(x + 3) + (x - 7) = 2x - 4$ $(x^2 - 2x) - (3x^2 - x) = -2x^2 - x$
Multiplying two polynomials	Multiply each term of the first polynomial by each term of the second polynomial, then combine like terms.	$(x^2 + 2x + 3)(x + 1)$ $= (x^2 + 2x + 3)x + (x^2 + 2x + 3)1$ $= x^3 + 2x^2 + 3x + x^2 + 2x + 3$ $= x^3 + 3x^2 + 5x + 3$

Dividing polynomials	Ordinary division or long division	

$$\text{dividend} = (\text{quotient})(\text{divisor}) + \text{remainder}$$

$$\frac{\text{dividend}}{\text{divisor}} = \text{quotient} + \frac{\text{remainder}}{\text{divisor}}$$

$$\begin{array}{r} x-7 \\ x+2\overline{)x^2-5x-14} \\ x^2+2x \\ \hline -7x-14 \\ -7x-14 \\ \hline 0 \end{array}$$

Synthetic division A condensed version of long division, used only for dividing by a polynomial of the form $x - c$

$$\begin{array}{r} -2\ \big|\ \ 1\ \ -5\ \ -14 \\ \ \ \ \ \ \ \ \ 2\ \ \ \ 14 \\ \hline \ \ 1\ \ -7\ \ \ \ 0 \end{array}$$

If the remainder is 0, then the dividend factors as

$$x^2 - 5x - 14 = (x-7)(x+2)$$

$$\text{dividend} = (\text{quotient})(\text{divisor}).$$

Shortcuts for Multiplying Two Binomials | Examples

FOIL	The product of two binomials can be found quickly by multiplying their **F**irst, **O**uter, **I**nner, and **L**ast terms.	$(x+2)(x+3) = x^2 + 5x + 6$
Square of a sum	$(a+b)^2 = a^2 + 2ab + b^2$	$(x+5)^2 = x^2 + 10x + 25$
Square of a difference	$(a-b)^2 = a^2 - 2ab + b^2$	$(m-3)^2 = m^2 - 6m + 9$
Product of a sum and a difference	$(a+b)(a-b) = a^2 - b^2$	$(x+3)(x-3) = x^2 - 9$

Factoring | Examples

Factoring a polynomial	Write a polynomial as a product of two or more polynomials. A polynomial is factored completely if it is a product of prime polynomials.	$3x^2 - 3 = 3(x^2 - 1)$ $= 3(x+1)(x-1)$
Common factors	Factor out the greatest common factor (GCF).	$2x^3 - 6x = 2x(x^2 - 3)$
Difference of two squares	$a^2 - b^2 = (a+b)(a-b)$ (A sum of two squares is prime.)	$m^2 - 25 = (m+5)(m-5)$ $m^2 + 25$ is prime.
Perfect square trinomials	$a^2 + 2ab + b^2 = (a+b)^2$ $a^2 - 2ab + b^2 = (a-b)^2$	$x^2 + 10x + 25 = (x+5)^2$ $x^2 - 6x + 9 = (x-3)^2$
Difference of two cubes	$a^3 - b^3 = (a-b)(a^2 + ab + b^2)$	$x^3 - 8 = (x-2)(x^2 + 2x + 4)$
Sum of two cubes	$a^3 + b^3 = (a+b)(a^2 - ab + b^2)$	$x^3 + 27 = (x+3)(x^2 - 3x + 9)$
Grouping	Factor out common factors from groups of terms.	$3x + 3w + bx + bw$ $= 3(x+w) + b(x+w)$ $= (3+b)(x+w)$

Factoring $ax^2 + bx + c$	By the *ac* method: 1. Find two numbers that have a product equal to *ac* and a sum equal to *b*. 2. Replace *bx* by two terms using the two new numbers as coefficients. 3. Factor the resulting four-term polynomial by grouping.	$2x^2 + 7x + 3$ $ac = 6, b = 7, 1 \cdot 6 = 6, 1 + 6 = 7$ $2x^2 + 7x + 3$ $= 2x^2 + x + 6x + 3$ $= (2x + 1)x + (2x + 1)3$ $= (2x + 1)(x + 3)$
	By trial and error: Try possibilities by considering factors of the first term and factors of the last term. Check them by FOIL.	$12x^2 + 19x - 18$ $= (3x - 2)(4x + 9)$
Substitution	Use substitution on higher-degree polynomials to reduce the degree to 2 or 3.	$x^4 - 3x^2 - 18$ Let $a = x^2$. $a^2 - 3a - 18$

Solving Equations by Factoring

Examples

Strategy	1. Write the equation with 0 on the right side. 2. Factor the left side. 3. Set each factor equal to zero. 4. Solve the simpler equations. 5. Check the answers in the original equation.	$x^2 - 3x - 18 = 0$ $(x - 6)(x + 3) = 0$ $x - 6 = 0$ or $x + 3 = 0$ $x = 6$ or $x = -3$ $6^2 - 3(6) - 18 = 0$ $(-3)^2 - 3(-3) - 18 = 0$

REVIEW EXERCISES

3.1 *Simplify each expression. Assume that all variables represent nonzero real numbers. Write your answers with positive exponents.*

1. $2 \cdot 2 \cdot 2^{-1}$ **2.** $5^{-1} \cdot 5$ **3.** $2^2 \cdot 3^2$

4. $3^2 \cdot 5^2$ **5.** $(-3)^{-3}$ **6.** $(-2)^{-2}$

7. $-(-1)^{-3}$ **8.** $3^4 \cdot 3^7$ **9.** $2x^3 \cdot 4x^{-6}$

10. $-3a^{-3} \cdot 4a^{-4}$ **11.** $\dfrac{y^{-5}}{y^{-3}}$ **12.** $\dfrac{w^3}{w^{-3}}$

13. $\dfrac{a^5 \cdot a^{-2}}{a^{-4}}$ **14.** $\dfrac{2m^3 \cdot m^6}{2m^{-2}}$ **15.** $\dfrac{6x^{-2}}{3x^2}$

16. $\dfrac{-5y^2x^{-3}}{5y^{-2}x^7}$

Write each number in standard notation.

17. 8.36×10^6 **18.** 3.4×10^7

19. 5.7×10^{-4} **20.** 4×10^{-3}

Write each number in scientific notation.

21. 8,070,000   **22.** 90,000

23. 0.000709   **24.** 0.0000005

Perform each computation without a calculator. Write the answer in scientific notation.

25. $\dfrac{(4,000,000,000)(0.0000006)}{(0.000012)(2,000,000)}$

26. $\dfrac{(1.2 \times 10^{32})(2 \times 10^{-5})}{4 \times 10^{-7}}$

3.2 *Simplify each expression. Assume that all variables represent nonzero real numbers. Write your answers with positive exponents.*

27. $(a^{-3})^{-2} \cdot a^{-7}$ **28.** $(-3x^{-2}y)^{-4}$

29. $(m^2n^3)^{-2}(m^{-3}n^2)^4$ **30.** $(w^{-3}xy)^{-1}(wx^{-3}y)^2$

31. $\left(\dfrac{2}{3}\right)^{-4}$ **32.** $\left(\dfrac{a^4}{3}\right)^{-2}$ **33.** $\left(\dfrac{1}{2} + \dfrac{1}{3}\right)^2$

34. $\left(\dfrac{1}{2} - \dfrac{1}{3}\right)^{-2}$ **35.** $\left(-\dfrac{3a}{4b^{-1}}\right)^{-1}$ **36.** $\left(-\dfrac{4x^5}{5y^{-3}}\right)^{-1}$

37. $\dfrac{(a^{-3}b)^4}{(ab^2)^{-5}}$ **38.** $\dfrac{(2x^3)^3}{(3x^2)^{-2}}$

Simplify each expression. Assume that the variables represent integers.

39. $5^{2w} \cdot 5^{4w} \cdot 5^{-1}$ **40.** $3^y(3^{2y})^3$

41. $\left(\dfrac{7^{3a}}{7^8}\right)^5$ **42.** $\left(\dfrac{2^{6-k}}{2^{2-3k}}\right)^3$

3.3 *Perform the indicated operations.*

43. $(2w - 3) + (6w + 5)$

44. $(3a - 2xy) + (5xy - 7a)$

45. $(x^2 - 3x - 4) - (x^2 + 3x - 7)$

46. $(7 - 2x - x^2) - (x^2 - 5x + 6)$

47. $(x^2 - 2x + 4)(x - 2)$ **48.** $(x + 5)(x^2 - 2x + 10)$

49. $xy + 7z - 5(xy - 3z)$ **50.** $7 - 4(x - 3)$

51. $m^2(5m^3 - m + 2)$ **52.** $(a + 2)^3$

3.4 *Perform the following computations mentally. Write down only the answers.*

53. $(x - 3)(x + 7)$ **54.** $(k - 5)(k + 4)$

55. $(z - 5y)(z + 5y)$ **56.** $(m - 3)(m + 3)$

57. $(m + 8)^2$ **58.** $(b + 2a)^2$

59. $(w - 6x)(w - 4x)$ **60.** $(2w - 3)(w + 6)$

61. $(k - 3)^2$ **62.** $(n - 5)^2$

63. $(m^2 - 5)(m^2 + 5)$ **64.** $(3k^2 - 5t)(2k^2 + 6t)$

3.5 *Find the quotient and remainder.*

65. $(x^3 + x^2 - 11x + 10) \div (x - 2)$

66. $(2x^3 + 5x^2 + 9) \div (x + 3)$

67. $(m^4 - 1) \div (m + 1)$ **68.** $(x^4 - 1) \div (x - 1)$

69. $(a^9 - 8) \div (a^3 - 2)$ **70.** $(a^2 - b^2) \div (a - b)$

71. $(3m^3 + 6m^2 - 18m) \div (3m)$

72. $(w - 3) \div (3 - w)$

Rewrite each expression in the form

$$quotient + \frac{remainder}{divisor}.$$

Use synthetic division.

73. $\dfrac{x^2 - 5}{x - 1}$ **74.** $\dfrac{x^2 + 3x + 2}{x + 3}$

75. $\dfrac{3x}{x - 2}$ **76.** $\dfrac{4x}{x - 5}$

Determine whether the first polynomial is a factor of the second. Use synthetic division when possible.

77. $x + 2, \quad x^3 - 2x^2 + 3x + 22$

78. $x - 2, \quad x^3 + x - 10$

79. $x - 5, \quad x^3 - x - 120$

80. $x + 3, \quad x^3 + 2x + 15$

81. $x - 1, \quad x^3 + x^2 - 3$

82. $x - 1, \quad x^3 + 1$

83. $x^2 + 2, \quad x^4 + x^3 + 5x^2 + 2x + 6$

84. $x^2 + 1, \quad x^4 - 1$

3.6 *Complete the factoring by filling in the parentheses.*

85. $3x - 6 = 3(\quad)$ **86.** $7x^2 - x = x(\quad)$

87. $4a - 20 = -4(\quad)$ **88.** $w^2 - w = -w(\quad)$

89. $3w - w^2 = -w(\quad)$ **90.** $3x - 6 = (\quad)(2 - x)$

Factor each polynomial.

91. $y^2 - 81$ **92.** $r^2t^2 - 9v^2$

93. $4x^2 + 28x + 49$ **94.** $y^2 - 20y + 100$

95. $t^2 - 18t + 81$ **96.** $4w^2 + 4ws + s^2$

97. $t^3 - 125$ **98.** $8y^3 + 1$

3.7 *Factor each polynomial.*

99. $x^2 - 7x - 30$ **100.** $y^2 + 4y - 32$

101. $w^2 - 3w - 28$ **102.** $6t^2 - 5t + 1$

103. $2m^2 + 5m - 7$ **104.** $12x^2 - 17x + 6$

105. $m^7 - 3m^4 - 10m$ **106.** $6w^5 - 7w^3 - 5w$

3.8 *Factor each polynomial completely.*

107. $5x^3 + 40$ **108.** $w^3 - 6w^2 + 9w$

109. $9x^2 + 9x + 2$ **110.** $ax^3 + a$

111. $x^3 + x^2 - x - 1$ **112.** $16x^2 - 4x - 2$

113. $-x^2y + 16y$ **114.** $-5m^2 + 5$

115. $-a^3b^2 + 2a^2b^2 - ab^2$ **116.** $-2w^2 - 16w - 32$

117. $x^3 - x^2 + 9x - 9$ **118.** $w^4 + 2w^2 - 3$

119. $x^4 - x^2 - 12$ **120.** $8x^3 - 1$

121. $a^6 - a^3$ **122.** $a^2 - ab + 2a - 2b$

123. $-8m^2 - 24m - 18$

124. $-3x^2 - 9x + 30$

125. $(2x - 3)^2 - 16$

126. $(m - 6)^2 - (m - 6) - 12$

127. $x^6 + 7x^3 - 8$

128. $32a^5 - 2a$

129. $(a^2 - 9)^2 - 5(a^2 - 9) + 6$

130. $x^3 - 9x + x^2 - 9$

Factor each polynomial completely. Variables used as exponents represent positive integers.

131. $x^{2k} - 49$ **132.** $x^{6k} - 1$

133. $m^{2a} - 2m^a - 3$ **134.** $2y^{2n} - 7y^n + 6$

135. $9z^{2k} - 12z^k + 4$

136. $25z^{6m} + 20z^{3m} + 4$

137. $y^{2a} - by^a + cy^a - bc$

138. $x^3y^b - xy^b + 2x^3 - 2x$

3.9 *Solve each equation.*

139. $x^3 - 5x^2 = 0$

140. $2m^2 + 10m + 12 = 0$

141. $(a - 2)(a - 3) = 6$

142. $(w - 2)(w + 3) = 50$

143. $2m^2 - 9m - 5 = 0$

144. $m^3 + 4m^2 - 9m - 36 = 0$

145. $w^3 + 5w^2 - w - 5 = 0$

146. $12x^2 + 5x - 3 = 0$

147. $|x^2 - 5| = 4$

148. $|x^2 - 3x - 7| = 3$

Miscellaneous

Solve each problem.

149. *Roadrunner and the coyote.* The roadrunner has just taken a position atop a giant saguaro cactus. While positioning a 10-foot Acme ladder against the cactus, Wile E. Coyote notices a warning label on the ladder. For safety, Acme recommends that the distance from the ground to the top of the ladder, measured vertically along the cactus, must be 2 feet longer than the distance between the bottom of the ladder and the cactus. How far from the cactus should he place the bottom of this ladder?

150. *Three consecutive integers.* Find three consecutive integers such that the sum of their squares is 50.

151. *Perimeter of a square.* If the area of a square field is $9a^2 + 6a + 1$ square kilometers, then what is its perimeter?

152. *Landscape design.* Rico planted red tulips in a square flower bed with an area of x^2 square feet. He plans to surround the tulips with daffodils in a uniform border with a width of 3 feet. Write a polynomial that represents the area planted in daffodils?

153. *Life expectancy of black males.* The age at which people die is precisely measured and provides an indication of the health of the population as a whole ("Monitoring Your Health," *Readers Digest*, 1991). The formula

$$L = 59.4022(1.00459)^a$$

can be used to model life expectancy L for U.S. black males with present age a. How long can a 20-year-old black male expect to live? How many more years is a 20-year-old white male expected to live than a 20-year-old black male? (See Excercise 95 in Section 3.2.)

154. *Life expectancy of black females.* The formula

$$L = 68.7732(1.00303)^a$$

can be used to model life expectancy for U.S. black females with present age a. How long can a 20-year-old black female expect to live?

155. *Golden years.* A single man earning $50,000 per year at age 65 should have $172,000 saved at age 65 to continue his standard of living into retirement (*Fortune 1992 Investor's Guide*). What annual savings for 20 years would amount to $172,000, assuming an annual return of 7%? The formula

$$S = R \cdot \frac{(1 + i)^n - 1}{i}$$

gives the total value S at the end of n years of n annual investments of R dollars earning annual interest rate i.

156. *Ivy League.* The total cost of a B.A. at an Ivy League school for freshmen beginning in 2005 will be about $252,000 (*Fortune 1992 Investor's Guide*). Use the formula in the previous exercise to find the annual savings for 18 years that would amount to $252,000, assuming an annual return of 8%?

CHAPTER 3 TEST

Simplify each expression. Assume that all variables represent nonzero real numbers. Exponents in your answers should be positive exponents.

1. 3^{-2}

2. $\frac{1}{6^{-2}}$

3. $\left(\frac{1}{2}\right)^{-3}$

4. $3x^4 \cdot 4x^3$

5. $\frac{8y^9}{2y^{-3}}$

6. $(4a^2b)^3$

7. $\left(\frac{x^2}{3}\right)^{-3}$

8. $\frac{(2^{-1}a^2b)^{-3}}{4a^{-9}}$

Convert to standard notation.

9. 3.24×10^9

10. 8.673×10^{-4}

Perform each computation by converting each number to scientific notation. Give the answer in scientific notation.

11. $\frac{(80,000)(0.0006)}{2,000,000}$

12. $\frac{(0.00006)^2(500)}{(30,000)^2(0.01)}$

Perform the indicated operations.

13. $(3x^3 - x^2 + 6) + (4x^2 - 2x - 3)$

14. $(x^2 - 6x - 7) - (3x^2 + 2x - 4)$

15. $(x^2 - 3x + 7)(x - 2)$

16. $(x^3 + 7x^2 + 7x - 15) \div (x + 3)$

17. $(x - 2)^3$

18. $(x - 3) \div (3 - x)$

Find the products.

19. $(x - 7)(2x + 3)$

20. $(x - 6)^2$

21. $(2x + 5)^2$

22. $(3y^2 - 5)(3y^2 + 5)$

Rewrite each expression in the form

$$quotient + \frac{remainder}{divisor}.$$

Use synthetic division.

23. $\dfrac{5x}{x + 3}$

24. $\dfrac{x^2 + 3x - 6}{x - 2}$

Factor completely.

25. $a^2 - 2a - 24$

26. $4x^2 + 28x + 49$

27. $3m^3 - 24$

28. $2x^2y - 32y$

29. $2xa + 3a - 10x - 15$

30. $x^4 + 3x^2 - 4$

Solve each equation.

31. $2m^2 + 7m - 15 = 0$

32. $x^3 - 4x = 0$

33. $|x^2 + x - 9| = 3$

Write a complete solution for each problem.

34. A portable television is advertised as having a 10-inch diagonal measure screen. If the width of the screen is 2 inches more than the height, then what are the dimensions of the screen?

35. The maternal mortality rate for the United States is the number of deaths of the mother per 100,000 live births due to complications of pregnancy or childbirth. The maternal mortality rate has fallen dramatically since 1900 ("Monitoring Your Health," *Readers Digest*, 1991). The formula

$$d = (1.2527 \times 10^{69})(1.08243)^{-y}$$

gives the maternal mortality rate d for the year y in this century. What was the maternal mortality rate in 1900? 1950? 1990?

Tying It All Together

Simplify each expression.

1. 4^2

2. $4(-2)$

3. 4^{-2}

4. $2^3 \cdot 4^{-1}$

5. $2^{-1} + 2^{-1}$

6. $2^{-1} \cdot 3^{-1}$

7. $3^{-1} - 2^{-2}$

8. $3^2 - 4(5)(-2)$

9. $2^7 - 2^6$

10. $0.08(32) + 0.08(68)$

11. $3 - 2 \, |5 - 7 \cdot 3|$

12. $5^{-1} + 6^{-1}$

Solve each equation.

13. $0.05a - 0.04(a - 50) = 4$

14. $15b - 27 = 0$

15. $2c^2 + 15c - 27 = 0$

16. $2t^2 + 15t = 0$

17. $|15u - 27| = 3$

18. $|15v - 27| = 0$

19. $|15x - 27| = -78$

20. $|x^2 + x - 4| = 2$

21. $(2x - 1)(x + 5) = 0$

22. $|3x - 1| + 6 = 9$

23. $(1.5 \times 10^{-4})w - 5 \times 10^5 = 7 \times 10^6$

24. $(3 \times 10^7)(y - 5 \times 10^3) = 6 \times 10^{12}$

Solve each problem.

25. **Social Security.** According to the Social Security Administration a person retiring in the year 2005 who earned a lifetime average annual salary of $25,000 will receive a benefit based on age. For ages 62 through 64 the benefit in dollars will be $7000 + 500(a - 62)$, for ages 65 through 67 the benefit will be $10,000 + 667(a - 67)$, and for ages 68 through 70 the benefit will be $10,000 + 800(a - 67)$, based on current legislation. What would be the annual Social Security benefit for a person who retires in 2005 at age 64? If a person retires in 2005 and gets an $11,600 benefit, then what is the age of that person in 2005?

Rational Expressions

Information is everywhere—in the newspapers and magazines we read, the televisions we watch, and the computers we use. And now people are talking about the Information Superhighway, which will deliver vast amounts of information directly to consumers' homes. In the future, the combination of telephone, television, and computer will give us on-the-spot health care recommendations, video conferences, home shopping, and perhaps even electronic voting and driver's license renewal, to name just a few. There is even talk of 500 television channels!

Some experts are concerned that the consumer will give up privacy for this technology. Others worry about regulation, access, and content of the enormous international computer network.

Whatever the future of this technology, few people understand how all their electronic devices work. However, this vast array of electronics rests on physical principles, which are described by mathematical formulas. In Exercises 49 and 50 of Section 4.6 we will see that the formula governing resistance for receivers connected in parallel involves rational expressions, which are the subject of this chapter.

A ratio of two integers is called a rational number; a ratio of two polynomials is called a rational expression. Rational expressions are as fundamental to algebra as rational numbers are to arithmetic. In this section we look carefully at some of the properties of rational numbers and see how they extend to rational expressions.

Definition of Rational Expressions

A **rational expression** is the ratio of two polynomials with the denominator not equal to 0. For example,

$$\frac{2}{3}, \quad 3a + 5, \quad \frac{x-3}{2x^2 - 2}, \quad \frac{y+2}{5y}, \quad \text{and} \quad \frac{x-2}{x+1}$$

are rational expressions. Because an integer is a monomial, a rational number is also a rational expression. If the denominator of a rational expression is 1, it is usually omitted, as in the expression $3a + 5$.

Domain

The **domain** of a rational expression is the set of all real numbers that can be used in place of the variable. Because the denominator of a rational expression cannot be 0, the domain of a rational expression consists of the set of real numbers except those that cause the denominator to be zero. The domain of

$$\frac{x}{x+5}$$

is the set of all real numbers excluding -5. In set-builder notation this set is written as

$$\{x \mid x \neq -5\},$$

and in interval notation it is written as

$$(-\infty, -5) \cup (-5, \infty).$$

EXAMPLE 1   **Domain**

Find the domain of each rational expression.

a) $\dfrac{x-2}{x+9}$

b) $\dfrac{y+2}{5y}$

c) $\dfrac{x-3}{2x^2 - 2}$

Solution

a) The denominator is zero if $x + 9 = 0$, or $x = -9$. The domain is $\{x \mid x \neq -9\}$, or

$$(-\infty, -9) \cup (-9, \infty).$$

b) The denominator is zero if $5y = 0$, or $y = 0$. The domain is $\{y \mid y \neq 0\}$, or

$$(-\infty, 0) \cup (0, \infty).$$

c) The denominator is zero if $2x^2 - 2 = 0$. Solve this equation.

$$2x^2 - 2 = 0$$
$$2(x^2 - 1) = 0 \quad \text{Factor out 2.}$$
$$2(x + 1)(x - 1) = 0 \quad \text{Factor completely.}$$
$$x + 1 = 0 \quad \text{or} \quad x - 1 = 0 \quad \text{Zero factor property}$$
$$x = -1 \quad \text{or} \quad x = 1$$

The domain is the set of all real numbers except -1 and 1. This set is written as $\{x \mid x \neq -1 \text{ and } x \neq 1\}$, or in interval notation as

$$(-\infty, -1) \cup (-1, 1) \cup (1, \infty). \qquad \blacksquare$$

CAUTION The numbers that you find when you set the denominator equal to zero and solve for x are *not* in the domain of the rational expression. The solutions to that equation are excluded from the domain. $\oslash$

Reducing to Lowest Terms

Each rational number can be written in infinitely many equivalent forms. For example,

$$\frac{2}{3} = \frac{4}{6} = \frac{6}{9} = \frac{8}{12} = \frac{10}{15} = \cdots.$$

Each equivalent form of $\frac{2}{3}$ is obtained from $\frac{2}{3}$ by multiplying both numerator and denominator by the same nonzero number. For example,

$$\frac{2}{3} = \frac{2}{3} \cdot 1 = \frac{2}{3} \cdot \frac{2}{2} = \frac{4}{6} \quad \text{and} \quad \frac{2}{3} = \frac{2}{3} \cdot \frac{3}{3} = \frac{6}{9}.$$

Note that we are actually multiplying $\frac{2}{3}$ by equivalent forms of 1, the multiplicative identity.

If we start with $\frac{4}{6}$ and convert it into $\frac{2}{3}$, we are simplifying by *reducing* $\frac{4}{6}$ to *lowest terms*. We can reduce as follows:

$$\frac{4}{6} = \frac{\cancel{2} \cdot 2}{\cancel{2} \cdot 3} = \frac{2}{3}$$

A rational number is expressed in lowest terms when the numerator and denominator have no common factors other than 1. In reducing $\frac{4}{6}$, we divide the numerator and denominator by the common factor 2, or "divide out" the common factor 2. We can multiply or divide both numerator and denominator of a rational number by the same nonzero number without changing the value of the rational number. This fact is called the **basic principle of rational numbers.**

Basic Principle of Rational Numbers

If $\dfrac{a}{b}$ is a rational number and c is a nonzero real number, then

$$\frac{a}{b} = \frac{ac}{bc}.$$

CAUTION   Although it is true that

$$\frac{5}{6} = \frac{2+3}{2+4}$$

we cannot divide out the 2's in this expression because the 2's are not factors. We can divide out only common *factors* when reducing fractions.   ⊘

Just as a rational number has infinitely many equivalent forms, a rational expression also has infinitely many equivalent forms. To reduce rational expressions to lowest terms, we follow exactly the same procedure as we do for rational numbers: *Factor the numerator and denominator completely, then divide out all common factors.*

EXAMPLE 2   **Reducing**

Reduce each rational expression to lowest terms.

a) $\dfrac{18}{42}$

b) $\dfrac{a^7 b}{a^2 b^3}$

Solution

a) Factor 18 as $2 \cdot 3^2$ and 42 as $2 \cdot 3 \cdot 7$:

$$\frac{18}{42} = \frac{2 \cdot 3^2}{2 \cdot 3 \cdot 7} \qquad \text{Factor.}$$

$$= \frac{3}{7} \qquad \text{Divide out the common factors.}$$

b) Because this expression is already factored, we use the quotient rule for exponents to reduce:

$$\frac{a^7 b}{a^2 b^3} = \frac{a^5}{b^2}$$ ■

In the next example we use the techniques for factoring polynomials that we learned in Chapter 3.

EXAMPLE 3   **Reducing**

Reduce each rational expression to lowest terms.

a) $\dfrac{2x^2 - 18}{x^2 + x - 6}$

b) $\dfrac{2a^3 - 16}{16 - 4a^2}$

Solution

a) $\dfrac{2x^2 - 18}{x^2 + x - 6} = \dfrac{2(x^2 - 9)}{(x - 2)(x + 3)}$ Factor.

$\phantom{\dfrac{2x^2 - 18}{x^2 + x - 6}} = \dfrac{2(x - 3)\cancel{(x + 3)}}{(x - 2)\cancel{(x + 3)}}$ Factor completely.

$\phantom{\dfrac{2x^2 - 18}{x^2 + x - 6}} = \dfrac{2x - 6}{x - 2}$ Divide out the common factors.

b) $\dfrac{2a^3 - 16}{16 - 4a^2} = \dfrac{2(a^3 - 8)}{-4(a^2 - 4)}$ Factoring out -4 will give the common factor $a - 2$.

$\phantom{\dfrac{2a^3 - 16}{16 - 4a^2}} = \dfrac{2\cancel{(a - 2)}(a^2 + 2a + 4)}{-2 \cdot 2\cancel{(a - 2)}(a + 2)}$ Difference of two cubes, difference of two squares

$\phantom{\dfrac{2a^3 - 16}{16 - 4a^2}} = -\dfrac{a^2 + 2a + 4}{2a + 4}$ Divide out common factors. ∎

The rational expressions in Example 3(a) are equivalent because they have the same value for any replacement of the variables, provided that the replacement is in the domain of both expressions. In other words, the equation

$$\frac{2x^2 - 18}{x^2 + x - 6} = \frac{2x - 6}{x - 2}$$

is an identity. It is true for any value of x except 2 and -3.

The main points to remember for reducing rational expressions are summarized as follows.

▶ **Strategy for Reducing Rational Expressions** ◀

1. All reducing is done by dividing out common factors.
2. Factor the numerator and denominator completely to see the common factors.
3. Use the quotient rule to reduce a ratio of two monomials involving exponents.
4. We may have to factor out a common factor with a negative sign to get identical factors in the numerator and denominator.

Building Up the Denominator

In Section 4.3 we will see that only rational expressions with identical denominators can be added or subtracted. Fractions without identical denominators can be converted to equivalent fractions with a common denominator by reversing the procedure for reducing fractions to lowest terms. This procedure is called **building up the denominator.**

Consider converting the fraction $\frac{1}{3}$ into an equivalent fraction with a denominator of 51. Any fraction that is equivalent to $\frac{1}{3}$ can be obtained by multiplying the numerator and denominator of $\frac{1}{3}$ by the same nonzero number. Because $51 = 3 \cdot 17$, we multiply the numerator and denominator of $\frac{1}{3}$ by 17 to get an equivalent fraction with a denominator of 51:

$$\frac{1}{3} = \frac{1}{3} \cdot 1 = \frac{1}{3} \cdot \frac{17}{17} = \frac{17}{51}$$

EXAMPLE 4   **Building up the denominator**

Convert each rational expression into an equivalent rational expression that has the indicated denominator.

a) $\dfrac{2}{7}, \dfrac{?}{42}$ b) $\dfrac{5}{3a^2b}, \dfrac{?}{9a^3b^4}$

Solution

a) Factor 42 as $42 = 2 \cdot 3 \cdot 7$, then multiply the numerator and denominator of $\frac{2}{7}$ by the missing factors, 2 and 3:

$$\frac{2}{7} = \frac{2 \cdot 2 \cdot 3}{7 \cdot 2 \cdot 3} = \frac{12}{42}$$

b) Because $9a^3b^4 = 3ab^3 \cdot 3a^2b$, we multiply numerator and denominator by $3ab^3$:

$$\frac{5}{3a^2b} = \frac{5 \cdot 3ab^3}{3a^2b \cdot 3ab^3}$$
$$= \frac{15ab^3}{9a^3b^4}$$

■

When building up a denominator to match a more complicated denominator, we factor both denominators completely to see which factors are missing from the simpler denominator. Then we multiply the numerator and denominator of the simpler expression by the missing factors.

EXAMPLE 5   **Building up the denominator**

Convert each rational expression into an equivalent rational expression that has the indicated denominator.

a) $\dfrac{5}{2a - 2b}, \dfrac{?}{6b - 6a}$

b) $\dfrac{x + 2}{x + 3}, \dfrac{?}{x^2 + 7x + 12}$

Solution

a) Factor both $2a - 2b$ and $6b - 6a$ to see which factor is missing in $2a - 2b$. Note that we factor out -6 from $6b - 6a$ to get the factor $a - b$:

$$2a - 2b = 2(a - b)$$
$$6b - 6a = -6(a - b) = -3 \cdot 2(a - b)$$

Now multiply the numerator and denominator by the missing factor, -3:

$$\frac{5}{2a - 2b} = \frac{5(-3)}{(2a - 2b)(-3)} = \frac{-15}{6b - 6a}$$

b) Because $x^2 + 7x + 12 = (x + 3)(x + 4)$, multiply numerator and denominator by $x + 4$:

$$\frac{x + 2}{x + 3} = \frac{(x + 2)(x + 4)}{(x + 3)(x + 4)} = \frac{x^2 + 6x + 8}{x^2 + 7x + 12}$$ ∎

Evaluating a Rational Expression

A rational expression is evaluated in the same way that a polynomial is evaluated. We use the notation $R(x)$, read "R of x," to name a rational expression and to indicate its value for any number in the domain of the rational expression.

EXAMPLE 6 **Evaluating a rational expression**

Find $R(3)$, $R(-1)$, and $R(2)$ for the rational expression

$$R(x) = \frac{3x - 1}{x^2 - 4}.$$

Solution

To find $R(3)$, replace x by 3 in the rational expression:

$$R(3) = \frac{3 \cdot 3 - 1}{3^2 - 4} = \frac{8}{5}$$

To find $R(-1)$, replace x by -1 in the rational expression:

$$R(-1) = \frac{3(-1) - 1}{(-1)^2 - 4}$$

$$= \frac{-4}{-3} = \frac{4}{3}$$

We cannot find $R(2)$ because 2 is not in the domain of the rational expression. ∎

Applications

A rational expression can occur in finding an average cost. The average cost of making a product is the total cost divided by the number of products made.

EXAMPLE 7   **Average cost**

An automobile manufacturer plans to spend $30,000,000 to develop a new model and $15,000 to manufacture each car. Find a rational expression that gives the average cost per vehicle for developing and producing x vehicles. Compare the average cost per vehicle for production levels of 10,000 cars and 100,000 cars.

Solution

The polynomial $15,000x + 30,000,000$ gives the cost in dollars of development and production for x vehicles. The average cost per vehicle when x vehicles are produced is given by the rational expression

$$AC(x) = \frac{15,000x + 30,000,000}{x}.$$

If the production is 10,000 vehicles, then

$$AC(10,000) = \frac{15,000(10,000) + 30,000,000}{10,000} = 18,000.$$

If the production is 100,000 vehicles, then

$$AC(100,000) = \frac{15,000(100,000) + 30,000,000}{100,000} = 15,300.$$

The average cost per vehicle when production is 10,000 cars is $18,000, while at a production level of 100,000 cars it is $15,300. ∎

Warm-ups

True or false? Explain.

1. A rational number is a rational expression.

2. The expression $\frac{2 + x}{x - 1}$ is a rational expression.

3. The domain of the rational expression $\frac{3}{x - 2}$ is $\{2\}$.

4. The domain of $\frac{2x + 5}{(x - 9)(2x + 1)}$ is $\left\{x \mid x \neq 9 \text{ and } x \neq -\frac{1}{2}\right\}$.

5. The domain of $\frac{x - 1}{x + 2}$ is $(-\infty, -2) \cup (-2, 1) \cup (1, \infty)$.

6. The rational expression $\frac{5x + 2}{15}$ reduces to $\frac{x + 2}{3}$.

7. Multiplying numerator and denominator of $\frac{x}{x - 1}$ by x yields $\frac{x^2}{x^2 - 1}$.

8. The expression $\frac{2}{3 - x}$ is equivalent to $\frac{-2}{x - 3}$.

9. The equation $\frac{4x^3}{6x} = \frac{2x^2}{3}$ is an identity.

10. The expression $\frac{x^2 - y^2}{x - y}$ reduced to lowest terms is $x - y$.

4.1 EXERCISES

Find the domain of each rational expression. See Example 1.

1. $\dfrac{3x}{x-1}$ **2.** $\dfrac{x}{x+5}$ **3.** $\dfrac{2z-5}{7z}$

4. $\dfrac{z-12}{4z}$ **5.** $\dfrac{5y-1}{y^2-4}$ **6.** $\dfrac{2y-1}{y^2-9}$

7. $\dfrac{2a-3}{a^2+5a+6}$ **8.** $\dfrac{3b+1}{2b^2-7b-4}$ **9.** $\dfrac{x-1}{x^2+4x}$

10. $\dfrac{2x}{3x^2+9x}$ **11.** $\dfrac{x+1}{x^3+x^2-6x}$ **12.** $\dfrac{x^2-3x-4}{2x^5-2x}$

Reduce each rational expression to lowest terms. See Examples 2 and 3.

13. $\dfrac{6}{57}$ **14.** $\dfrac{14}{91}$ **15.** $\dfrac{42}{210}$

16. $\dfrac{242}{154}$ **17.** $\dfrac{2x+2}{4}$ **18.** $\dfrac{3a+3}{3}$

19. $\dfrac{3x-6y}{10y-5x}$ **20.** $\dfrac{5b-10a}{2a-b}$ **21.** $\dfrac{ab^2}{a^3b}$

22. $\dfrac{36y^3z^8}{54y^2z^9}$ **23.** $\dfrac{-2w^2x^3y}{6wx^5y^2}$ **24.** $\dfrac{6a^3b^{12}c^5}{-8ab^4c^9}$

25. $\dfrac{a^3b^2}{a^3+a^4}$ **26.** $\dfrac{b^8-ab^5}{ab^5}$ **27.** $\dfrac{a-b}{2b-2a}$

28. $\dfrac{2m-2n}{4n-4m}$ **29.** $\dfrac{3x+6}{3x}$ **30.** $\dfrac{7x-14}{7x}$

31. $\dfrac{a^3-b^3}{a-b}$ **32.** $\dfrac{27x^3+y^3}{6x+2y}$ **33.** $\dfrac{4x^2-4}{4x^2+4}$

34. $\dfrac{2a^2-2b^2}{2a^2+2b^2}$ **35.** $\dfrac{2x^2+2x-12}{4x^2-36}$

36. $\dfrac{2x^2+10x+12}{2x^2-8}$ **37.** $\dfrac{x^3+7x^2-4x}{x^3-16x}$

38. $\dfrac{2x^4-32}{4x-8}$ **39.** $\dfrac{ab+3a-by-3y}{a^2-y^2}$

40. $\dfrac{2x^2-5x-3}{2x^2+11x+5}$

Convert each rational expression into an equivalent rational expression that has the indicated denominator. See Examples 4 and 5.

41. $\dfrac{1}{5},\dfrac{?}{50}$ **42.** $\dfrac{2}{3},\dfrac{?}{9}$ **43.** $\dfrac{1}{x},\dfrac{?}{3x^2}$

44. $\dfrac{3}{ab^2},\dfrac{?}{a^3b^5}$ **45.** $\dfrac{5}{x-1},\dfrac{?}{x^2-2x+1}$

46. $\dfrac{x}{x-3},\dfrac{?}{x^2-9}$ **47.** $\dfrac{1}{2x+2},\dfrac{?}{-6x-6}$

48. $\dfrac{-2}{-3x+4},\dfrac{?}{15x-20}$ **49.** $5,\dfrac{?}{a}$

50. $3,\dfrac{?}{a+1}$ **51.** $\dfrac{x+2}{x+3},\dfrac{?}{x^2+2x-3}$

52. $\dfrac{x}{x-5},\dfrac{?}{x^2-x-20}$ **53.** $\dfrac{7}{x-1},\dfrac{?}{1-x}$

54. $\dfrac{1}{a-b},\dfrac{?}{2b-2a}$ **55.** $\dfrac{3}{x+2},\dfrac{?}{x^3+8}$

56. $\dfrac{x}{x-2},\dfrac{?}{x^3-8}$ **57.** $\dfrac{x+2}{3x-1},\dfrac{?}{6x^2+13x-5}$

58. $\dfrac{a}{2a+1},\dfrac{?}{4a^2-16a-9}$

Find the indicated value for each given rational expression. See Example 6.

59. $R(x)=\dfrac{3x-5}{x+4}$, $R(3)$ **60.** $T(x)=\dfrac{5-x}{x-5}$, $T(-9)$

61. $H(y)=\dfrac{y^2-5}{3y-4}$, $H(-2)$ **62.** $G(a)=\dfrac{3-5a}{2a+7}$, $G(5)$

63. $W(b)=\dfrac{4b^3-1}{b^2-b-6}$, $W(-2)$

64. $N(x)=\dfrac{x+3}{x^3-2x^2-2x-3}$, $N(3)$

In place of each question mark in Exercises 65–84, put an expression that will make the rational expressions equivalent.

65. $\dfrac{1}{3}=\dfrac{?}{21}$ **66.** $4=\dfrac{?}{3}$ **67.** $5=\dfrac{10}{?}$

68. $\dfrac{3}{4}=\dfrac{12}{?}$ **69.** $\dfrac{3}{a}=\dfrac{?}{a^2}$ **70.** $\dfrac{5}{y}=\dfrac{10}{?}$

71. $\dfrac{2}{a-b}=\dfrac{?}{b-a}$ **72.** $\dfrac{3}{x-4}=\dfrac{?}{4-x}$

73. $\dfrac{2}{x-1}=\dfrac{?}{x^2-1}$ **74.** $\dfrac{5}{x+3}=\dfrac{?}{x^2-9}$

75. $\dfrac{2}{w-3}=\dfrac{-2}{?}$ **76.** $\dfrac{-2}{5-x}=\dfrac{2}{?}$

77. $\dfrac{2x+4}{6}=\dfrac{?}{3}$ **78.** $\dfrac{2x-3}{4x-6}=\dfrac{1}{?}$

79. $\dfrac{x+4}{x^2-16}=\dfrac{1}{?}$ **80.** $\dfrac{2x+2}{2x}=\dfrac{x+1}{?}$

81. $\dfrac{3a + 3}{3a} = \dfrac{?}{a}$

82. $\dfrac{x - 3}{x^2 - 9} = \dfrac{1}{?}$

83. $\dfrac{1}{x - 1} = \dfrac{?}{x^3 - 1}$

84. $\dfrac{x^2 + 2x + 4}{x + 2} = \dfrac{?}{x^2 - 4}$

Reduce each rational expression to lowest terms. Variables used in exponents represent integers.

85. $\dfrac{x^{2a} - 4}{x^a + 2}$

86. $\dfrac{x^{2b} + 3x^b - 18}{x^{2b} - 36}$

87. $\dfrac{x^a + m + wx^a + wm}{x^{2a} - m^2}$

88. $\dfrac{x^{3a} - 8}{x^{2a} + 2x^a + 4}$

89. $\dfrac{x^{3b+1} - x}{x^{2b+1} - x}$

90. $\dfrac{2x^{2a+1} + 3x^{a+1} + x}{4x^{2a+1} - x}$

Solve each problem. See Example 7.

91. Driving speed. If Jeremy drives 500 miles in $2x$ hours, then what rational expression represents his speed in miles per hour?

92. Filing suit. If Marsha files 48 suits in $2x + 2$ work days, then what rational expression represents the rate (in suits per day) at which she is filing suits?

93. Wedding bells. Wheeler Printing Co. charges $45 plus $0.50 per invitation to print wedding invitations. Find a rational expression that gives the average cost in dollars per invitation for printing n invitations. How much more does it cost per invitation to print 200 invitations rather than 300 invitations?

94. Rose Bowl bound. A travel agent offers a Rose Bowl package including hotel, tickets, and transportation. It costs the travel agent $50,000 plus $300 per person to charter the airplane. Find a rational expression that gives the average cost in dollars per person for the charter flight. How much lower is the average cost per person when 200 people go compared to 100 people?

95. Solid waste recovery. The amount of municipal solid waste generated in the United States in the year 1960 + n is given by the polynomial $3.14n + 87.1$, while the amount recovered is given by the polynomial $0.576n + 3.78$, where the amounts are in millions of tons (U.S. EPA, 1990). Write a rational expression $p(n)$ that gives the fraction of solid waste that is recovered in the year 1960 + n. Find $p(0)$, $p(30)$, and $p(60)$.

96. Higher education. The total number of degrees awarded in U.S. higher education in the year 1985 + n is given in thousands by the polynomial $30.96n + 1808.22$, while the number of bachelor's degrees awarded is given in thousands by the polynomial $16.45n + 980.20$ (U.S. National Center for Education Statistics). Write a rational expression $p(n)$ that gives the percentage of bachelor's degrees among the total number of degrees conferred for the year 1985 + n. What fraction of the degrees awarded were bachelor's degrees in 1985? What fraction of the degrees awarded will be bachelor's degrees in 2010?

Getting More Involved

97. Exploration. Use a calculator to find $R(2)$, $R(30)$, $R(500)$, $R(9000)$, and $R(80,000)$ for the rational expression

$$R(x) = \dfrac{x - 3}{2x + 1},$$

Round answers to four decimal places. What can you conclude about the value of $R(x)$ as x gets larger and larger without bound?

98. Exploration. Use a calculator to find $H(1,000)$, $H(100,000)$, $H(1,000,000)$, and $H(10,000,000)$ for the rational expression

$$H(x) = \dfrac{7x - 50}{3x + 91},$$

Round answers to four decimal places. What can you conclude about the value of $R(x)$ as x gets larger and larger without bound?

4.2 Multiplication and Division

In Chapter 3 you learned to add, subtract, multiply, and divide polynomials. In this chapter you will learn to perform the same operations with rational expressions. We begin in this section with multiplication and division.

Multiplying Rational Expressions

We multiply two rational numbers by multiplying their numerators and multiplying their denominators. For example,

$$\frac{6}{7} \cdot \frac{14}{15} = \frac{84}{105} = \frac{21 \cdot 4}{21 \cdot 5} = \frac{4}{5}.$$

Instead of reducing the rational number after multiplying, it is often easier to reduce before multiplying. We first factor all terms, then divide out the common factors, then multiply:

$$\frac{6}{7} \cdot \frac{14}{15} = \frac{2 \cdot 3}{7} \cdot \frac{2 \cdot 7}{3 \cdot 5} = \frac{4}{5}.$$

When we multiply rational numbers, we use the following definition.

Multiplication of Rational Numbers

If $\dfrac{a}{b}$ and $\dfrac{c}{d}$ are rational numbers, then $\dfrac{a}{b} \cdot \dfrac{c}{d} = \dfrac{ac}{bd}$.

We multiply rational expressions in the same way that we multiply rational numbers: Factor all polynomials, divide out the common factors, then multiply the remaining factors.

EXAMPLE 1 **Multiplying rational expressions**

Find each product of rational expressions.

a) $\dfrac{3a^8b^3}{6b} \cdot \dfrac{10a}{a^2b^6}$

b) $\dfrac{x^2 + 7x + 12}{x^2 + 3x} \cdot \dfrac{x^2}{x^2 - 16}$

Solution

a) First factor the coefficients in each numerator and denominator:

$$\frac{3a^8b^3}{6b} \cdot \frac{10a}{a^2b^6} = \frac{3a^8b^3}{2 \cdot 3b} \cdot \frac{2 \cdot 5a}{a^2b^6} \qquad \text{Factor.}$$

$$= \frac{5a^9b^3}{a^2b^7} \qquad \text{Divide out the common factors.}$$

$$= \frac{5a^7}{b^4} \qquad \text{Quotient rule}$$

b) $\dfrac{x^2 + 7x + 12}{x^2 + 3x} \cdot \dfrac{x^2}{x^2 - 16} = \dfrac{(x+3)(x+4)}{x(x+3)} \cdot \dfrac{x^2}{(x-4)(x+4)} = \dfrac{x}{x-4}$ ∎

CAUTION Do not attempt to divide out the x in $\dfrac{x}{x-4}$. This expression cannot be reduced because x is not a factor of *both* terms in the denominator. Compare this expression to the following:

$$\frac{3x}{x-xy} = \frac{x \cdot 3}{x(1-y)} = \frac{3}{1-y} \qquad \oslash$$

In Example 2(a) we will multiply a rational expression and a polynomial. For Example 2(b) we will use the rule for factoring the difference of two cubes.

EXAMPLE 2 **Multiplying rational expressions**

Find each product.

a) $(a^2 - 1) \cdot \dfrac{6}{2a^2 + 4a + 2}$
b) $\dfrac{a^3 - b^3}{b-a} \cdot \dfrac{6}{2a^2 + 2ab + 2b^2}$

Solution

a) First factor the polynomials completely:

$$(a^2 - 1) \cdot \frac{6}{2a^2 + 4a + 2} = \frac{(a+1)(a-1)}{1} \cdot \frac{2 \cdot 3}{2(a+1)(a+1)}$$

$$= \frac{3(a-1)}{a+1} \qquad \text{Divide out the common factors.}$$

$$= \frac{3a-3}{a+1} \qquad \text{Multiply.}$$

b) Note that $a - b$ is a factor of $a^3 - b^3$ and $b - a$ occurs in the denominator. We can factor $b - a$ as $-1(a - b)$ to get a common factor:

$$\frac{a^3 - b^3}{b-a} \cdot \frac{6}{2a^2 + 2ab + 2b^2} = \frac{(a-b)(a^2+ab+b^2)}{-1(a-b)} \cdot \frac{2 \cdot 3}{2(a^2+ab+b^2)}$$

$$= \frac{3}{-1}$$

$$= -3 \qquad \blacksquare$$

Dividing $a - b$ by $b - a$

In Example 2(b) we factored $b - a$ as $-1(a - b)$ to get the common factor $a - b$. Instead of factoring out -1, we could have used the fact that $(a - b) \div (b - a) = -1$, as shown in the next example.

EXAMPLE 3 **Dividing $a - b$ by $b - a$**

Find the product:

$$\frac{m-4}{3} \cdot \frac{6}{4-m}$$

Solution

Instead of factoring -1 out of $m - 4$, we divide $m - 4$ by $4 - m$ to get -1:

$$\frac{m-4}{3} \cdot \frac{6}{4-m} = \frac{\overset{-1}{\cancel{m-4}}}{\cancel{3}} \cdot \frac{\overset{2}{\cancel{6}}}{\cancel{4-m}} \qquad \text{Note that } (m-4) \div (4-m) = -1.$$
$$= -2 \qquad\qquad\qquad\qquad\qquad\qquad\qquad\qquad \blacksquare$$

Dividing Rational Expressions

We divide rational numbers by multiplying by the reciprocal or multiplicative inverse of the divisor. For example,

$$\frac{3}{4} \div \frac{15}{2} = \frac{3}{4} \cdot \frac{2}{15} = \frac{\cancel{3}}{2 \cdot \cancel{2}} \cdot \frac{\cancel{2} \cdot 1}{\cancel{3} \cdot 5} = \frac{1}{10}.$$

When we divide rational numbers, we use the following definition.

> **Division of Rational Numbers**
>
> If $\dfrac{a}{b}$ and $\dfrac{c}{d}$ are rational numbers with $\dfrac{c}{d} \neq 0$, then
>
> $$\frac{a}{b} \div \frac{c}{d} = \frac{a}{b} \cdot \frac{d}{c}.$$

We use the same method to divide rational expressions: We invert the divisor and multiply.

EXAMPLE 4   **Dividing rational expressions**

Find each quotient.

a) $\dfrac{10}{3x} \div \dfrac{6}{5x}$ **b)** $\dfrac{5a^2b^8}{c^3} \div (4ab^3c)$

Solution

a) The reciprocal of the divisor $\dfrac{6}{5x}$ is $\dfrac{5x}{6}$.

$$\frac{10}{3x} \div \frac{6}{5x} = \frac{10}{3x} \cdot \frac{5x}{6} \qquad \text{Invert and multiply.}$$
$$= \frac{\cancel{2} \cdot 5}{3\cancel{x}} \cdot \frac{5\cancel{x}}{\cancel{2} \cdot 3} = \frac{25}{9}$$

b) The reciprocal of $4ab^3c$ is $\dfrac{1}{4ab^3c}$.

$$\frac{5a^2b^8}{c^3} \div (4ab^3c) = \frac{5a^2b^8}{c^3} \cdot \frac{1}{4ab^3c} = \frac{5ab^5}{4c^4} \qquad \text{Quotient rule} \qquad \blacksquare$$

In the next example we factor the polynomials in the rational expressions.

EXAMPLE 5   **Dividing rational expressions**

Find the quotient:

$$\frac{25 - x^2}{x^2 + x} \div \frac{x - 5}{x^2 - 1}$$

Solution

$$\frac{25 - x^2}{x^2 + x} \div \frac{x - 5}{x^2 - 1} = \frac{25 - x^2}{x^2 + x} \cdot \frac{x^2 - 1}{x - 5}$$ Invert and multiply.

$$= \frac{\overset{-1}{\cancel{(5 - x)}}(5 + x)}{x\cancel{(x + 1)}} \cdot \frac{\cancel{(x + 1)}(x - 1)}{\cancel{x - 5}}$$ $(5 - x) \div (x - 5) = -1.$

$$= \frac{-1(5 + x)(x - 1)}{x}$$ Divide out the common factors.

$$= \frac{-x^2 - 4x + 5}{x}$$ Multiply the factors in the numerator. ■

CAUTION   When dividing rational expressions, you can factor the polynomials at any time, but do not reduce until after you have inverted the divisor.   ⊘

In the next example, division is indicated by a fraction bar.

EXAMPLE 6   **Dividing rational expressions**

Perform the operations indicated.

a) $\dfrac{\dfrac{a + b}{3}}{\dfrac{1}{2}}$ b) $\dfrac{\dfrac{x^2 - 4}{2}}{\dfrac{x - 2}{3}}$ c) $\dfrac{\dfrac{m^2 + 1}{5}}{3}$

Solution

a) $\dfrac{\dfrac{a + b}{3}}{\dfrac{1}{2}} = \dfrac{a + b}{3} \div \dfrac{1}{2}$

$$= \frac{a + b}{3} \cdot \frac{2}{1}$$ Invert the divisor.

$$= \frac{2a + 2b}{3}$$ Multiply.

b) $\dfrac{\dfrac{x^2-4}{2}}{\dfrac{x-2}{3}} = \dfrac{x^2-4}{2} \cdot \dfrac{3}{x-2}$ Invert and multiply.

$$= \dfrac{\cancel{(x-2)}(x+2)}{2} \cdot \dfrac{3}{\cancel{x-2}} \qquad \text{Factor.}$$

$$= \dfrac{3x+6}{2} \qquad\qquad\quad \text{Reduce.}$$

c) $\dfrac{\dfrac{m^2+1}{5}}{3} = \dfrac{m^2+1}{5} \cdot \dfrac{1}{3} = \dfrac{m^2+1}{15}$ Multiply by $\frac{1}{3}$, the reciprocal of 3.   ■

Warm-ups

True or false? Explain.

1. We can multiply only fractions that have identical denominators.

2. $\dfrac{2}{7} \cdot \dfrac{3}{7} = \dfrac{6}{7}$

3. To divide rational expressions, invert the divisor and multiply.

4. $a \div b = \dfrac{1}{a} \cdot b$ for any nonzero a and b.

5. $\dfrac{1}{2x} \cdot 8x^2 = 4x$ for any nonzero real number x.

6. One-half of one-third is one-sixth.

7. One-third divided by one-half is two-thirds.

8. The quotient of $w - z$ divided by $z - w$ is -1, provided that $z - w \neq 0$.

9. $\dfrac{x}{3} \div 2 = \dfrac{x}{6}$ for any real number x.

10. $\dfrac{a}{b} \div \dfrac{b}{a} = 1$ for any nonzero real numbers a and b.

4.2 EXERCISES

In Exercises 1–16, perform the indicated operations. See Examples 1–3.

1. $\dfrac{12}{42} \cdot \dfrac{35}{22}$

2. $\dfrac{3}{8} \cdot \dfrac{20}{21}$

5. $\dfrac{3x-3}{6} \cdot \dfrac{x}{x^2-x}$

6. $\dfrac{-2x-4}{2} \cdot \dfrac{6}{3x+6}$

3. $\dfrac{3a}{10b} \cdot \dfrac{5b^2}{6}$

4. $\dfrac{3x}{7y} \cdot \dfrac{14y^2}{9x}$

7. $\dfrac{10x+5}{5x^2+5} \cdot \dfrac{2x^2+x-1}{4x^2-1}$

8. $\dfrac{x^3+x}{5} \cdot \dfrac{5x-5}{x^3-x}$

9. $\dfrac{ax + aw + bx + bw}{x^2 - w^2} \cdot \dfrac{x - w}{a^2 - b^2}$

10. $\dfrac{3a - 3y}{3a - 3y - ab + by} \cdot \dfrac{b^2 - 9}{6b + 18}$

11. $\dfrac{a^2 - 2a + 4}{a^3 + 8} \cdot \dfrac{(a + 2)^3}{2a + 4}$

12. $\dfrac{w^3 - 1}{(w - 1)^2} \cdot \dfrac{w^2 - 1}{w^2 + w + 1}$ **13.** $\dfrac{x - 9}{12y} \cdot \dfrac{8y}{9 - x}$

14. $\dfrac{19x^2}{12y - 1} \cdot \dfrac{1 - 12y}{3x}$ **15.** $(a^2 - 4) \cdot \dfrac{7}{2 - a}$

16. $\dfrac{10x - 4x^2}{4x^2 - 20x + 25} \cdot (2x^3 - 5x^2)$

Perform the indicated operations. See Examples 4 and 5.

17. $\dfrac{15}{17} \div \dfrac{10}{17}$ **18.** $\dfrac{3}{4} \div \dfrac{1}{8}$ **19.** $\dfrac{36x}{5y} \div \dfrac{20x}{35y}$

20. $\dfrac{18a^3b^4}{c^9} \div \dfrac{12ab^6}{7c^2}$ **21.** $\dfrac{24a^5b^2}{5c^3} \div (4a^5bc^5)$

22. $\dfrac{60x^9y^2}{z} \div (48x^4y^3)$ **23.** $(w + 1) \div \dfrac{w^2 - 1}{w}$

24. $(a - 3) \div \dfrac{9 - a^2}{4}$

25. $\dfrac{x - y}{5} \div \dfrac{x^2 - 2xy + y^2}{10}$

26. $\dfrac{x^2 + 6x + 9}{18} \div \dfrac{(x + 3)^2}{36}$

27. $\dfrac{4x - 2}{x^2 - 5x} \div \dfrac{2x^2 + 9x - 5}{x^2 - 25}$

28. $\dfrac{2x^2 - 5x - 12}{6 + 4x} \div \dfrac{x^2 - 16}{2}$

Perform the indicated operations. See Example 6.

29. $\dfrac{\frac{x - y}{3}}{\frac{1}{6}}$ **30.** $\dfrac{\frac{2a - b}{10}}{\frac{1}{5}}$ **31.** $\dfrac{\frac{x^2 - 25}{3}}{\frac{x - 5}{6}}$

32. $\dfrac{\frac{3x^2 + 3}{5}}{\frac{3x + 3}{5}}$ **33.** $\dfrac{\frac{a - b}{2}}{3}$ **34.** $\dfrac{\frac{10}{a + b}}{5}$

35. $\dfrac{\frac{a^2 - b^2}{a + b}}{3}$ **36.** $\dfrac{\frac{x^2 + 5x + 6}{x + 2}}{x + 3}$

Perform the indicated operations. When possible write down only the answer.

37. $\dfrac{5x}{2} \div 3$ **38.** $\dfrac{x}{a} \div 2$ **39.** $\dfrac{3}{4} \div \dfrac{1}{4}$

40. $\dfrac{1}{4} \div \dfrac{1}{2}$ **41.** One-half of $\dfrac{1}{6}$

42. One-half of $\dfrac{b}{a}$ **43.** One-half of $\dfrac{4x}{3}$

44. One-third of $\dfrac{6x}{y}$ **45.** $(a - b) \div (b - a)$

46. $(a - b) \div (-1)$ **47.** $\dfrac{x - y}{3} \cdot \dfrac{6}{y - x}$

48. $\dfrac{5x - 5y}{x} \cdot \dfrac{1}{x - y}$ **49.** $\dfrac{2a + 2b}{a} \cdot \dfrac{1}{2}$

50. $\dfrac{x - y}{y - x} \cdot \dfrac{1}{2}$ **51.** $\dfrac{a + b}{\frac{1}{2}}$ **52.** $\dfrac{x + 3}{\frac{1}{3}}$

53. $\dfrac{\frac{3x}{5}}{y}$ **54.** $\dfrac{\frac{b^2 - 4a}{2}}{a}$ **55.** $\dfrac{\frac{3a}{5b}}{2}$ **56.** $\dfrac{\frac{6x}{a}}{x}$

Perform the indicated operations.

57. $\dfrac{3x^2 + 13x - 10}{x} \cdot \dfrac{x^3}{9x^2 - 4} \cdot \dfrac{7x - 35}{x^2 - 25}$

58. $\dfrac{x^2 + 5x + 6}{x} \cdot \dfrac{x^2}{3x + 6} \cdot \dfrac{9}{x^2 - 4}$

59. $\dfrac{(a^2b^3c)^2}{(-2ab^2c)^3} \cdot \dfrac{(a^3b^2c)^3}{(abc)^4}$

60. $\dfrac{(-wy^2)^3}{3w^2y} \cdot \dfrac{(2wy)^2}{4wy^3}$

61. $\dfrac{(2mn)^3}{6mn^2} \div \dfrac{2m^2n^3}{(m^2n)^4}$ **62.** $\dfrac{(rt)^3}{rt^4} \div \dfrac{(rt^2)^3}{r^2t^3}$

63. $\dfrac{2x^2 + 7x - 15}{4x^2 - 100} \cdot \dfrac{2x^2 - 9x - 5}{4x^2 - 1}$

64. $\dfrac{x^3 + 1}{x^2 - 1} \cdot \dfrac{3x - 3}{x^3 - x^2 + x}$

65. $\dfrac{k^2 + 2km + m^2}{k^2 - 2km + m^2} \cdot \dfrac{m^2 + 3m - mk - 3k}{m^2 + mk + 3m + 3k}$

66. $\dfrac{a^2 + 2ab + b^2}{ac + bc - ad - bd} \div \dfrac{ac + ad - bc - bd}{c^2 - d^2}$

67. $\dfrac{15w^2 - wt - 2t^2}{27w^3 + t^3} \cdot \dfrac{27w^2 - 9wt + 3t^2}{25w^2 - 4t^2}$

$$\div \dfrac{10w + 4t}{25w^2 + 20wt + 4t^2}$$

68. $\dfrac{6k^2 - k - 2}{2k^2 + 3kp - 2p^2} \cdot \dfrac{2k^2 + kp - p^2}{3k^2 - 10k + 3}$

$$\div \dfrac{2k^2 + k + p + 2kp}{k^2 + 2kp - 3k - 6p}$$

Perform the indicated operations. Variables in exponents represent integers.

69. $\dfrac{x^a}{y^2} \cdot \dfrac{y^{b+2}}{x^{2a}}$

70. $\dfrac{x^{3a+1}}{y^{2b-3}} \cdot \dfrac{y^{3b+4}}{x^{2a-1}}$

71. $\dfrac{x^{2a} + x^a - 6}{x^{2a} + 6x^a + 9} \div \dfrac{x^{2a} - 4}{x^{2a} + 2x^a - 3}$

72. $\dfrac{w^{2b} + 2w^b - 8}{w^{2b} + 3w^b - 4} \div \dfrac{w^{2b} - w^b - 2}{w^{2b} - 1}$

73. $\dfrac{m^k v^k + 3v^k - 2m^k - 6}{m^{2k} - 9} \cdot \dfrac{m^{2k} - 2m^k - 3}{v^k m^k - 2m^k + 2v^k - 4}$

74. $\dfrac{m^{3k} - 1}{m^{3k} + 1} \cdot \dfrac{m^{2k+1} - m^{k+1} + m}{m^{3k} + m^{2k} + m^k}$

Solve each problem.

75. *Secondary school enrollment.* In 1990, $\frac{1}{40}$ of the children enrolled in U.S. schools were enrolled in private secondary schools (National Center for Education Statistics). Use the accompanying figure to determine the percentage of secondary school children who were in private schools.

1990 distribution of students in U.S. schools

Secondary schools $\frac{9}{25}$

Elementary schools $\frac{16}{25}$

Figure for Exercise 75

76. *The golden state.* In 1990, $\frac{3}{25}$ of the U.S. population lived in California (U.S. Census Bureau). Use the accompanying figure to determine the percentage of

the population of the western region of the United States who lived in California in 1990.

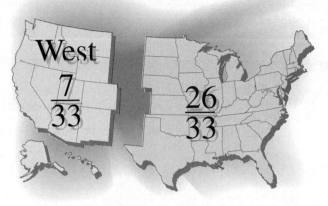

West $\dfrac{7}{33}$ $\dfrac{26}{33}$

Figure for Exercise 76

77. *Distance traveled.* Bonita drove 100 miles in x hours. Assuming that she continued to drive at the same speed, write a rational expression for the distance that she traveled in the next $\frac{3}{4}$ of an hour.

78. *Increasing speed.* Before lunch, Avonda drove 200 miles at x miles per hour. After lunch, she drove 250 miles in the same amount of time. Write a rational expression for her speed after lunch.

Getting More Involved

79. *Discussion.* Which of the following expressions is not equivalent to $x \div \frac{y}{z}$? Explain.

a) $x \cdot \dfrac{z}{y}$ **b)** $\dfrac{x}{y} \cdot z$ **c)** $zx \cdot \dfrac{1}{y}$

d) $\dfrac{\frac{x}{y}}{z}$ **e)** $\dfrac{xy}{z}$

80. *Discussion.* Which of the following equations is not an identity? Explain.

a) $\dfrac{x^2 - 1}{2} \cdot \dfrac{2}{x - 1} = x + 1$

b) $\dfrac{x - 1}{x^2 - 1} = x + 1$

c) $x^2 - 1 = (x - 1)(x + 1)$

d) $\dfrac{1}{x^2 - 1} \div \dfrac{1}{x + 1} = \dfrac{1}{x - 1}$

4.3 Addition and Subtraction

We can multiply or divide any rational expressions, but we add or subtract only rational expressions with identical denominators. So when the denominators are not the same, we must find equivalent forms of the expressions that have identical denominators. In this section, we review the idea of least common denominator and learn to use it for addition and subtraction of rational expressions.

Adding and Subtracting with Identical Denominators

It is easy to add or subtract fractions with identical denominators. For example,

$$\frac{1}{7} + \frac{3}{7} = \frac{4}{7} \quad \text{and} \quad \frac{3}{5} - \frac{2}{5} = \frac{1}{5}.$$

In general, we have the following definition.

> **Addition and Subtraction of Rational Numbers**
> If $b \neq 0$, then
> $$\frac{a}{b} + \frac{c}{b} = \frac{a+c}{b} \quad \text{and} \quad \frac{a}{b} - \frac{c}{b} = \frac{a-c}{b}.$$

Rational expressions with identical denominators are added or subtracted in the same manner as fractions.

EXAMPLE 1 **Identical denominators**

Perform the indicated operations.

a) $\dfrac{3}{2x} + \dfrac{5}{2x}$ **b)** $\dfrac{5x-3}{x-1} + \dfrac{5-7x}{x-1}$ **c)** $\dfrac{x^2+4x+7}{x^2-1} - \dfrac{x^2-2x+1}{x^2-1}$

Solution

a) $\dfrac{3}{2x} + \dfrac{5}{2x} = \dfrac{8}{2x}$ Add the numerators.

$\qquad\qquad = \dfrac{4}{x}$ Reduce.

b) $\dfrac{5x-3}{x-1} + \dfrac{5-7x}{x-1} = \dfrac{5x-3+5-7x}{x-1}$ Add the numerators.

$\qquad\qquad = \dfrac{-2x+2}{x-1}$ Combine like terms.

$\qquad\qquad = \dfrac{-2(x-1)}{x-1}$ Factor.

$\qquad\qquad = -2$ Reduce to lowest terms.

c) The polynomials in the numerators are treated as if they were in parentheses:

$$\frac{x^2 + 4x + 7}{x^2 - 1} - \frac{x^2 - 2x + 1}{x^2 - 1} = \frac{x^2 + 4x + 7 - (x^2 - 2x + 1)}{x^2 - 1}$$

$$= \frac{x^2 + 4x + 7 - x^2 + 2x - 1}{x^2 - 1}$$

$$= \frac{6x + 6}{x^2 - 1}$$

$$= \frac{6(x + 1)}{(x + 1)(x - 1)} = \frac{6}{x - 1} \qquad \blacksquare$$

Least Common Denominator

To add fractions with denominators that are not identical, we use the basic principle of rational numbers to build up the denominators to the **least common denominator (LCD).** For example,

$$\frac{1}{4} + \frac{1}{6} = \frac{1 \cdot 3}{4 \cdot 3} + \frac{1 \cdot 2}{6 \cdot 2} = \frac{3}{12} + \frac{2}{12} = \frac{5}{12}.$$

The LCD 12 is the **least common multiple (LCM)** of the numbers 4 and 6.

Finding the LCM for a pair of large numbers such as 24 and 126 will help you to understand the procedure for finding the LCM for any polynomials. First factor the numbers completely:

$$24 = 2^3 \cdot 3$$
$$126 = 2 \cdot 3^2 \cdot 7$$

Any number that is a multiple of both 24 and 126 must have all of the factors of 24 and all of the factors of 126 in its factored form. So in the LCM we use the factors 2, 3, and 7, and for each factor we use the highest power that appears on that factor. The highest power of 2 is 3, the highest power of 3 is 2, and the highest power of 7 is 1. So the LCM is $2^3 \cdot 3^2 \cdot 7$. If we write this product without exponents, we can see clearly that it is a multiple of both 24 and 126:

$$\underbrace{2 \cdot 2 \cdot \overbrace{2 \cdot 3}^{126} \cdot 3 \cdot 7}_{24} = 504 \qquad \begin{array}{l} 504 = 126 \cdot 4 \\ 504 = 24 \cdot 21 \end{array}$$

The strategy for finding the LCM for a group of polynomials can be stated as follows.

Strategy for Finding the LCM for Polynomials

1. Factor each polynomial completely. Use exponents to express repeated factors.
2. Write the product of all of the different factors that appear in the polynomials.
3. For each factor, use the highest power of that factor in any of the polynomials.

EXAMPLE 2   **Finding the LCM**

Find the least common multiple for each group of polynomials.

a) $4x^2y, 6y$ **b)** a^2bc, ab^3c^2, a^3bc

c) $x^2 + 5x + 6, x^2 + 6x + 9$

Solution

a) Factor $4x^2y$ and $6y$ as follows:

$$4x^2y = 2^2 \cdot x^2y, \qquad 6y = 2 \cdot 3y$$

To get the LCM, we use 2, 3, x, and y the maximum number of times that each appears in either of the expressions. The LCM is $2^2 \cdot 3 \cdot x^2y$, or $12x^2y$.

b) The expressions a^2bc, ab^3c^2, and a^3bc are already factored. To get the LCM, we use a, b, and c the maximum number of times that each appears in any of the expressions. The LCM is $a^3b^3c^2$.

c) Factor $x^2 + 5x + 6$ and $x^2 + 6x + 9$ completely:

$$x^2 + 5x + 6 = (x + 2)(x + 3), \qquad x^2 + 6x + 9 = (x + 3)^2$$

The LCM is $(x + 2)(x + 3)^2$.   ■

Adding and Subtracting with Different Denominators

To add or subtract rational expressions with different denominators, we must build up each rational expression to equivalent forms with identical denominators, as we did in Section 4.1. Of course, it is most efficient to use the LCD as in the following examples.

EXAMPLE 3   **Different denominators**

Perform the indicated operations.

a) $\dfrac{3}{a^2b} + \dfrac{5}{ab^3}$

b) $\dfrac{x + 1}{6} - \dfrac{2x - 3}{4}$

Solution

a) The LCD for a^2b and ab^3 is a^2b^3. To build up each denominator to a^2b^3, multiply the numerator and denominator of the first expression by b^2, and multiply the numerator and denominator of the second expression by a:

$$\frac{3}{a^2b} + \frac{5}{ab^3} = \frac{3(b^2)}{a^2b(b^2)} + \frac{5(a)}{ab^3(a)} \qquad \text{Build up each denominator to the LCD.}$$

$$= \frac{3b^2}{a^2b^3} + \frac{5a}{a^2b^3}$$

$$= \frac{3b^2 + 5a}{a^2b^3} \qquad \text{Add the numerators.}$$

b) $\dfrac{x+1}{6} - \dfrac{2x-3}{4} = \dfrac{(x+1)(2)}{6(2)} - \dfrac{(2x-3)(3)}{4(3)}$ Build up each denominator to the LCD 12.

$$= \dfrac{2x+2}{12} - \dfrac{6x-9}{12}$$ Distributive property

$$= \dfrac{2x+2-(6x-9)}{12}$$ Subtract the numerators. Note that $6x-9$ is put in parentheses.

$$= \dfrac{2x+2-6x+9}{12}$$ Remove the parentheses.

$$= \dfrac{-4x+11}{12}$$ Combine like terms.   ■

CAUTION   Before you add or subtract rational expressions, they must be written with identical denominators. For multiplication and division it is not necessary to have identical denominators.   ⊘

In the next example we must first factor polynomials to find the LCD.

EXAMPLE 4   **Different denominators**

Perform the indicated operations.

a) $\dfrac{1}{x^2-1} + \dfrac{2}{x^2+x}$

b) $\dfrac{5}{a-2} - \dfrac{3}{2-a}$

Solution

a)   Because $x^2 - 1 = (x+1)(x-1)$ and $x^2 + x = x(x+1)$, the LCD is $x(x-1)(x+1)$. The first denominator is missing the factor x, and the second denominator is missing the factor $x - 1$.

$$\dfrac{1}{x^2-1} + \dfrac{2}{x^2+x} = \dfrac{1}{\underbrace{(x-1)(x+1)}_{\text{Missing } x}} + \dfrac{2}{\underbrace{x(x+1)}_{\text{Missing } x-1}}$$ The LCD is $x(x-1)(x+1)$.

$$= \dfrac{1(x)}{(x-1)(x+1)(x)} + \dfrac{2(x-1)}{x(x+1)(x-1)}$$ Build up the denominators to the LCD.

$$= \dfrac{x}{x(x-1)(x+1)} + \dfrac{2x-2}{x(x-1)(x+1)}$$

$$= \dfrac{3x-2}{x(x-1)(x+1)}$$ Add the numerators.

For this type of answer, we usually leave the denominator in factored form. That way, if we need to work with the expression further, we do not have to factor the denominator again.

b) Because $-1(2 - a) = a - 2$, we can convert the denominator $2 - a$ to $a - 2$.

$$\frac{5}{a - 2} - \frac{3}{2 - a} = \frac{5}{a - 2} - \frac{3(-1)}{(2 - a)(-1)}$$

$$= \frac{5}{a - 2} - \frac{-3}{a - 2} \qquad \text{The LCD is } a - 2.$$

$$= \frac{5 - (-3)}{a - 2} \qquad \text{Subtract the numerators.}$$

$$= \frac{8}{a - 2} \qquad \text{Simplify.}$$

Note that if we had changed the denominator of the first expression to $2 - a$, we would have gotten the answer

$$\frac{-8}{2 - a},$$

but this rational expression is equivalent to the first answer. ∎

Shortcuts

Consider the following addition:

$$\frac{a}{b} + \frac{c}{d} = \frac{a(d)}{b(d)} + \frac{c(b)}{d(b)} = \frac{ad + bc}{bd} \qquad \text{The LCD is } bd.$$

We can use this result as a rule for adding simple fractions in which the LCD is the product of the denominators. A similar rule works for subtraction.

> **Adding or Subtracting Simple Fractions**
> If $b \neq 0$ and $d \neq 0$, then
> $$\frac{a}{b} + \frac{c}{d} = \frac{ad + bc}{bd} \qquad \text{and} \qquad \frac{a}{b} - \frac{c}{d} = \frac{ad - bc}{bd}.$$

EXAMPLE 5 **Adding and subtracting simple fractions**

Use the rules for adding and subtracting simple fractions to find the sums and differences.

a) $\frac{1}{2} + \frac{1}{3}$ **b)** $\frac{1}{a} - \frac{1}{x}$

c) $\frac{a}{5} + \frac{a}{3}$ **d)** $x - \frac{2}{3}$

Solution

a) For the numerator, compute $ad + bc = 1 \cdot 3 + 2 \cdot 1 = 5$. Use $2 \cdot 3$ or 6 for the denominator:

$$\frac{1}{2} + \frac{1}{3} = \frac{5}{6}$$

b) $\dfrac{1}{a} - \dfrac{1}{x} = \dfrac{1 \cdot x - 1 \cdot a}{ax} = \dfrac{x - a}{ax}$

c) $\dfrac{a}{5} + \dfrac{a}{3} = \dfrac{3a + 5a}{15} = \dfrac{8a}{15}$

d) $x - \dfrac{2}{3} = \dfrac{x}{1} - \dfrac{2}{3} = \dfrac{3x - 2}{3}$ $\qquad\qquad$ ■

CAUTION   The rules for adding or subtracting simple fractions can be applied to any rational expressions, but they work best when the LCD is the product of the two denominators. Always make sure that the answer is in lowest terms. If the product of the two denominators is too large, these rules are not helpful because then reducing can be difficult.   ⊘

Applications

Rational expressions occur often in expressing rates. For example, if you can process one application in 2 hours, then you are working at the rate of $\frac{1}{2}$ of an application per hour. If you can complete one task in x hours, then you are working at the rate of $\frac{1}{x}$ task per hour.

EXAMPLE 6   **Work rates**

Susan takes an average of x hours to process a mortgage application, while Betty's average is 1 hour longer. Write a rational expression for the number of applications that they can process in 40 hours.

Solution

The number of applications processed by Susan is the product of her rate and her time:

$$\frac{1}{x} \frac{\text{application}}{\text{hr}} \cdot 40 \text{ hr} = \frac{40}{x} \text{ applications}$$

The number of applications processed by Betty is the product of her rate and her time:

$$\frac{1}{x + 1} \frac{\text{application}}{\text{hr}} \cdot 40 \text{ hr} = \frac{40}{x + 1} \text{ applications}$$

Find the sum of the rational expressions:

$$\frac{40}{x} + \frac{40}{x + 1} = \frac{40x + 40 + 40x}{x(x + 1)} = \frac{80x + 40}{x(x + 1)}$$

So together in 40 hours they process $\dfrac{80x + 40}{x(x + 1)}$ applications.   ■

Warm-ups

True or false? Explain.

1. The LCM of 6 and 10 is 60.

2. The LCM of $6a^2b$ and $8ab^3$ is $24ab$.

3. The LCM of $x^2 - 1$ and $x - 1$ is $x^2 - 1$.

4. The LCD for the rational expressions $\dfrac{5}{x}$ and $\dfrac{x-3}{x+1}$ is $x + 1$.

5. $\dfrac{1}{2} + \dfrac{2}{3} = \dfrac{3}{5}$

6. $5 + \dfrac{1}{x} = \dfrac{6}{x}$ for any nonzero real number x.

7. $\dfrac{7}{a} + 3 = \dfrac{7 + 3a}{a}$ for any $a \neq 0$.

8. $\dfrac{c}{3} - \dfrac{d}{5} = \dfrac{5c - 3d}{15}$ for any real numbers c and d.

9. $\dfrac{2}{3} + \dfrac{3}{4} = \dfrac{17}{12}$

10. If Jamal uses x reams of paper in one day, then he uses $\dfrac{1}{x}$ ream per day.

4.3 EXERCISES

Perform the indicated operations. Reduce answers to lowest terms. See Example 1.

1. $\dfrac{3x}{2} + \dfrac{5x}{2}$

2. $\dfrac{5x^2}{3} + \dfrac{4x^2}{3}$

3. $\dfrac{x-3}{2x} - \dfrac{3x-5}{2x}$

4. $\dfrac{9-4y}{3y} - \dfrac{6-y}{3y}$

5. $\dfrac{3x-4}{2x-4} + \dfrac{2x-6}{2x-4}$

6. $\dfrac{a^3}{a+b} + \dfrac{b^3}{a+b}$

7. $\dfrac{x^2+4x-6}{x^2-9} - \dfrac{x^2+2x-12}{x^2-9}$

8. $\dfrac{x^2+3x-3}{x-4} - \dfrac{x^2+4x-7}{x-4}$

Find the least common multiple for each group of polynomials. See Example 2.

9. $24, 20$
10. $12, 18, 22$
11. $10x^3y, 15x$
12. $12a^3b^2, 18ab^5$
13. a^3b, ab^4c, ab^5c^2
14. x^2yz, xy^2z^3, xy^6
15. $x, x+2, x-2$
16. $y, y-5, y+2$
17. $4a+8, 6a+12$
18. $4a-6, 2a^2-3a$
19. x^2-1, x^2+2x+1
20. $y^2-2y-15, y^2+6y+9$
21. x^2-4x, x^2-16, x^2+6x+8
22. $z^2-25, 5z-25, 5z+25$

Perform the indicated operations. Reduce answers to lowest terms. See Examples 3 and 4.

23. $\dfrac{1}{28} + \dfrac{3}{35}$

24. $\dfrac{7}{48} - \dfrac{5}{36}$

25. $\dfrac{7}{24} - \dfrac{4}{15}$

26. $\dfrac{7}{52} + \dfrac{3}{40}$

27. $\dfrac{3}{wz^2} + \dfrac{5}{w^2z}$

28. $\dfrac{2}{a^2b} - \dfrac{3}{ab^2}$

29. $\dfrac{2x-3}{8} - \dfrac{x-2}{6}$

30. $\dfrac{a-5}{10} + \dfrac{3-2a}{15}$

31. $\dfrac{x}{2a} + \dfrac{3x}{5a}$

32. $\dfrac{x}{6y} - \dfrac{3x}{8y}$

33. $\dfrac{9}{4y} - x$

34. $\dfrac{b^2}{4a} - c$

35. $\dfrac{5}{a+2} - \dfrac{7}{a}$

36. $\dfrac{2}{x+1} - \dfrac{3}{x}$

37. $\dfrac{1}{a-b} + \dfrac{2}{a+b}$

38. $\dfrac{5}{x+2} + \dfrac{3}{x-2}$

39. $\dfrac{x}{x^2-9} + \dfrac{3}{x-3}$

40. $\dfrac{x}{x^2-25} + \dfrac{5}{x-5}$

41. $\dfrac{1}{a-b} + \dfrac{1}{b-a}$

42. $\dfrac{3}{x-5} + \dfrac{7}{5-x}$

43. $\dfrac{5}{2x-4} - \dfrac{3}{2-x}$

44. $\dfrac{4}{3x-9} - \dfrac{7}{3-x}$

45. $\dfrac{5}{x^2+x-2} - \dfrac{6}{x^2+2x-3}$

46. $\dfrac{2}{x^2-4} - \dfrac{5}{x^2-3x-10}$

47. $\dfrac{x}{x^2-9} + \dfrac{6}{x^2+4x+3}$

48. $\dfrac{2x-1}{x^2-x-12} + \dfrac{x+5}{x^2+5x+6}$

49. $\dfrac{1}{x} + \dfrac{2}{x-1} - \dfrac{3}{x+2}$

50. $\dfrac{2}{a} - \dfrac{3}{a+1} + \dfrac{5}{a-1}$

Perform the following operations. Write down only the answer. See Example 5.

51. $\dfrac{1}{3} + \dfrac{1}{4}$

52. $\dfrac{3}{5} + \dfrac{1}{4}$

53. $\dfrac{1}{8} - \dfrac{3}{5}$

54. $\dfrac{a}{2} + \dfrac{5}{3}$

55. $\dfrac{x}{3} + \dfrac{x}{2}$

56. $\dfrac{y}{4} - \dfrac{y}{3}$

57. $\dfrac{a}{b} - \dfrac{2}{3}$

58. $\dfrac{3}{x} + \dfrac{1}{9}$

59. $a + \dfrac{2}{3}$

60. $\dfrac{m}{3} + y$

61. $\dfrac{3}{a} + 1$

62. $\dfrac{1}{x} + 1$

63. $\dfrac{3+x}{x} - 1$

64. $\dfrac{a+2}{a} + 3$

65. $\dfrac{2}{3} + \dfrac{1}{4x}$

66. $\dfrac{1}{5} + \dfrac{1}{5x}$

Perform the indicated operations.

67. $\dfrac{w^2-3w+6}{w-5} + \dfrac{9-w^2}{w-5}$

68. $\dfrac{2z^2-3z+6}{z^2-1} - \dfrac{z^2-5z+9}{z^2-1}$

69. $\dfrac{1}{x+2} - \dfrac{2}{x+3}$

70. $\dfrac{2}{a+3} - \dfrac{3}{a+1}$

71. $\dfrac{1}{a^3-1} - \dfrac{1}{a^3+1}$

72. $\dfrac{1}{x^3-8} + \dfrac{1}{x^3+8}$

73. $\dfrac{(a^2b^3)^4}{(ab^4)^3} \cdot \dfrac{(ab)^3}{(a^4b)^2}$

74. $\dfrac{(ab)^2}{(a+b)^2} \cdot \dfrac{(a+b)^3}{(ab)^3}$

75. $\dfrac{x^2-3x}{x^3-1} + \dfrac{4}{x-1}$

76. $\dfrac{4a}{a^2+2a+1} - \dfrac{3}{a+1}$

77. $\dfrac{x^2+25}{x^2-25} \cdot \dfrac{x^2+10x+25}{x^2+5x}$

78. $\dfrac{a^2+a-2}{a^3-1} \cdot \dfrac{a^2+a+1}{a^2-4}$

79. $\dfrac{w^2-3}{3w^3+81} - \dfrac{2}{6w+18} - \dfrac{w-4}{w^2-3w+9}$

80. $\dfrac{a-3}{a^3+8} - \dfrac{2}{a+2} - \dfrac{a-3}{a^2-2a+4}$

81. $\dfrac{a^2-6a+9}{a^3-8} \div \dfrac{a^2-a-6}{a^2-4}$

82. $\dfrac{1}{z^2+4} \div \dfrac{z^3-8}{z^4-16}$

83. $\dfrac{w^2+3}{w^3-8} - \dfrac{2w}{w^2-4}$

84. $\dfrac{x+5}{x^3+27} - \dfrac{x-1}{x^2-9}$

85. $\dfrac{1}{x^3-1} - \dfrac{1}{x^2-1} + \dfrac{1}{x-1}$

86. $\dfrac{x-4}{x^3-1} + \dfrac{x-2}{x^2-1}$

In Exercises 87–92, solve each problem. See Example 6.

87. **Processing.** Joe takes x hours on the average to process a claim, while Ellen averages $x+1$ hours to process a claim. Write a rational expression for the number of claims that they will process while working together for an 8-hour shift.

88. **Roofing.** Bill attaches one bundle of shingles in an average of x minutes using a hammer, while Julio can attach one bundle in an average of $x-6$ minutes using a pneumatic stapler. Write a rational expression for the number of bundles that they can attach while working together for 10 hours.

89. *Selling.* George sells one magazine subscription every 20 minutes, while Theresa sells one every x minutes. Write a rational expression for the number of magazine subscriptions that they will sell when working together for 1 hour.

90. *Painting.* Harry can paint his house by himself in 6 days. His wife Judy can paint the house by herself in x days. Write a rational expression for the portion of the house that they paint when working together for 2 days.

91. *Driving.* Joan drove for 100 miles at one speed and then increased her speed by 5 miles per hour and drove 200 additional miles. Write a rational expression for her total travel time.

92. *Running.* Willard jogged for 3 miles at one speed and then doubled his speed for an additional mile. Write a rational expression for his total running time.

Getting More Involved

93. *Discussion.* Explain why fractions must have common denominators for addition but not for multiplication.

94. *Discussion.* Find each "infinite sum" and explain your answer.

a) $\dfrac{3}{10} + \dfrac{3}{10^2} + \dfrac{3}{10^3} + \dfrac{3}{10^4} + \cdots$

b) $\dfrac{9}{10} + \dfrac{9}{10^2} + \dfrac{9}{10^3} + \dfrac{9}{10^4} + \cdots$

4.4 Complex Fractions

In this section:
▶ Simplifying Complex Fractions
▶ Simplifying Expressions with Negative Exponents
▶ Applications

In this section we will use the techniques of Section 4.3 to simplify complex fractions. As their name suggests, complex fractions are rather messy-looking expressions.

Simplifying Complex Fractions

A **complex fraction** is a fraction that has rational expressions in the numerator, the denominator, or both. For example,

$$\frac{\dfrac{1}{2} + \dfrac{1}{3}}{\dfrac{1}{4} + \dfrac{1}{5}}, \qquad \frac{3 - \dfrac{2}{x}}{\dfrac{1}{x^2} - \dfrac{1}{4}}, \qquad \text{and} \qquad \frac{\dfrac{x+2}{x^2-9}}{\dfrac{x}{x^2-6x+9} + \dfrac{4}{x-3}}$$

are complex fractions. In the next example we show two methods for simplifying a complex fraction.

EXAMPLE 1 **A complex fraction without variables**

Simplify $\dfrac{\dfrac{1}{2} + \dfrac{1}{3}}{\dfrac{1}{4} + \dfrac{1}{5}}$.

Solution

Method A For this method we perform the computations of the numerator and denominator separately and then divide:

$$\frac{\dfrac{1}{2}+\dfrac{1}{3}}{\dfrac{1}{4}+\dfrac{1}{5}} = \frac{\dfrac{5}{6}}{\dfrac{9}{20}} = \frac{5}{6} \div \frac{9}{20} = \frac{5}{6} \cdot \frac{20}{9} = \frac{5 \cdot \cancel{2} \cdot 10}{\cancel{2} \cdot 3 \cdot 9} = \frac{50}{27}$$

Method B For this method we find the LCD for all of the fractions in the complex fraction. Then we multiply the numerator and denominator of the complex fraction by the LCD. The LCD for the denominators 2, 3, 4, and 5 is 60. So we multiply the numerator and denominator of the complex fraction by 60:

$$\frac{\dfrac{1}{2}+\dfrac{1}{3}}{\dfrac{1}{4}+\dfrac{1}{5}} = \frac{\left(\dfrac{1}{2}+\dfrac{1}{3}\right)60}{\left(\dfrac{1}{4}+\dfrac{1}{5}\right)60} = \frac{30+20}{15+12} = \frac{50}{27} \qquad \begin{array}{l}\frac{1}{2}\cdot 60 = 30, \frac{1}{3}\cdot 60 = 20 \\ \frac{1}{4}\cdot 60 = 15, \frac{1}{5}\cdot 60 = 12\end{array} \quad \blacksquare$$

In most cases, Method B of Example 1 is the faster method for simplifying complex fractions, and we will continue to use it.

EXAMPLE 2   **A complex fraction with variables**

Simplify $\dfrac{3 - \dfrac{2}{x}}{\dfrac{1}{x^2} - \dfrac{1}{4}}$.

Solution

The LCD of x, x^2, and 4 is $4x^2$. Multiply the numerator and denominator by $4x^2$:

$$\frac{3 - \dfrac{2}{x}}{\dfrac{1}{x^2} - \dfrac{1}{4}} = \frac{\left(3 - \dfrac{2}{x}\right)(4x^2)}{\left(\dfrac{1}{x^2} - \dfrac{1}{4}\right)(4x^2)}$$

$$= \frac{3(4x^2) - \dfrac{2}{x}(4x^2)}{\dfrac{1}{x^2}(4x^2) - \dfrac{1}{4}(4x^2)} \qquad \text{Distributive property}$$

$$= \frac{12x^2 - 8x}{4 - x^2} \qquad\qquad\qquad \blacksquare$$

EXAMPLE 3   **More complicated denominators**

Simplify $\dfrac{\dfrac{x+2}{x^2-9}}{\dfrac{x}{x^2-6x+9}+\dfrac{4}{x-3}}$.

Solution

Because $x^2 - 9 = (x-3)(x+3)$ and $x^2 - 6x + 9 = (x-3)^2$, the LCD is $(x-3)^2(x+3)$. Multiply the numerator and denominator by the LCD:

$$\frac{\dfrac{x+2}{x^2-9}}{\dfrac{x}{x^2-6x+9}+\dfrac{4}{x-3}}=\frac{\dfrac{x+2}{(x-3)(x+3)}(x-3)^2(x+3)}{\dfrac{x}{(x-3)^2}(x-3)^2(x+3)+\dfrac{4}{x-3}(x-3)^2(x+3)}$$

$$=\frac{(x+2)(x-3)}{x(x+3)+4(x-3)(x+3)} \qquad \text{Simplify.}$$

$$=\frac{(x+2)(x-3)}{(x+3)[x+4(x-3)]} \qquad \text{Factor out } x+3.$$

$$=\frac{(x+2)(x-3)}{(x+3)(5x-12)} \qquad\blacksquare$$

Simplifying Expressions with Negative Exponents

Consider the expression

$$\frac{3a^{-1}-2^{-1}}{1-b^{-1}}.$$

Using the definition of negative exponents, we can rewrite this expression as a complex fraction:

$$\frac{3a^{-1}-2^{-1}}{1-b^{-1}}=\frac{\dfrac{3}{a}-\dfrac{1}{2}}{1-\dfrac{1}{b}}$$

The LCD for the complex fraction is $2ab$. Note that $2ab$ could also be obtained from the bases of the expressions with the negative exponents. To simplify the complex fraction, we could use Method B as we have been doing. However, it is not necessary to rewrite the original expression as a complex fraction. The next example shows how to use Method B with the original expression.

EXAMPLE 4   **A complex fraction with negative exponents**

Simplify the complex fraction $\dfrac{3a^{-1}-2^{-1}}{1-b^{-1}}$.

Solution

Multiply the numerator and denominator by $2ab$, the LCD of the fractions. Remember that $a^{-1} \cdot a = a^0 = 1$.

$$\frac{3a^{-1} - 2^{-1}}{1 - b^{-1}} = \frac{(3a^{-1} - 2^{-1})2ab}{(1 - b^{-1})2ab}$$

$$= \frac{3a^{-1}(2ab) - 2^{-1}(2ab)}{1(2ab) - b^{-1}(2ab)} \qquad \text{Distributive property}$$

$$= \frac{6b - ab}{2ab - 2a} \qquad \blacksquare$$

EXAMPLE 5   **A complex fraction with negative exponents**

Simplify the complex fraction $\dfrac{a^{-1} + b^{-2}}{ab^{-2} + ba^{-3}}$.

Solution

If we rewrote a^{-1}, b^{-2}, b^{-2}, and a^{-3}, then the denominators would be a, b^2, b^2, and a^3. So the LCD is a^3b^2. If we multiply the numerator and denominator by a^3b^2, the negative exponents will be eliminated:

$$\frac{a^{-1} + b^{-2}}{ab^{-2} + ba^{-3}} = \frac{(a^{-1} + b^{-2})a^3b^2}{(ab^{-2} + ba^{-3})a^3b^2}$$

$$= \frac{a^{-1} \cdot a^3b^2 + b^{-2} \cdot a^3b^2}{ab^{-2} \cdot a^3b^2 + ba^{-3} \cdot a^3b^2} \qquad \text{Distributive property}$$

$$= \frac{a^2b^2 + a^3}{a^4 + b^3} \qquad \begin{array}{l} b^{-2}b^2 = b^0 = 1, \\ a^{-3}a^3 = a^0 = 1 \end{array}$$

Note that the positive exponents of a^3b^2 are just large enough to eliminate all of the negative exponents when we multiply. $\qquad \blacksquare$

The next example is not exactly a complex fraction, but we can use the same technique as in the previous example.

EXAMPLE 6   **More negative exponents**

Eliminate negative exponents and simplify $p + p^{-1}q^{-2}$.

Solution

If we multiply the numerator and denominator by pq^2, we will eliminate the negative exponents:

$$p + p^{-1}q^{-2} = \frac{(p + p^{-1}q^{-2})}{1} \cdot \frac{pq^2}{pq^2}$$

$$= \frac{p^2q^2 + 1}{pq^2} \qquad \begin{array}{l} p \cdot pq^2 = p^2q^2, \\ p^{-1}q^{-2} \cdot pq^2 = 1 \end{array} \qquad \blacksquare$$

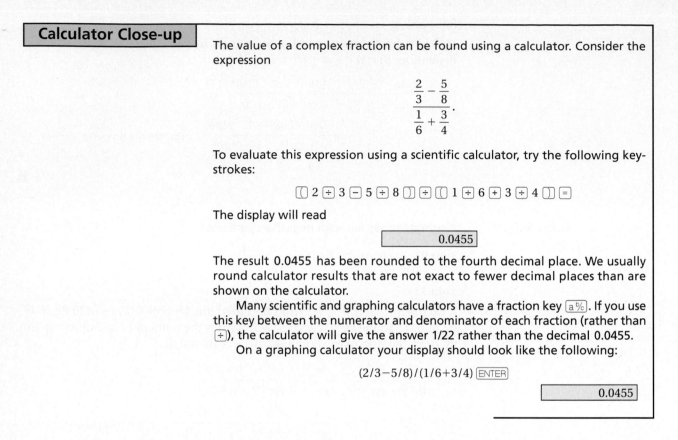

Calculator Close-up

The value of a complex fraction can be found using a calculator. Consider the expression

$$\frac{\frac{2}{3} - \frac{5}{8}}{\frac{1}{6} + \frac{3}{4}}.$$

To evaluate this expression using a scientific calculator, try the following keystrokes:

⟮ 2 ÷ 3 − 5 ÷ 8 ⟯ ÷ ⟮ 1 ÷ 6 + 3 ÷ 4 ⟯ =

The display will read

0.0455

The result 0.0455 has been rounded to the fourth decimal place. We usually round calculator results that are not exact to fewer decimal places than are shown on the calculator.

Many scientific and graphing calculators have a fraction key a%. If you use this key between the numerator and denominator of each fraction (rather than ÷), the calculator will give the answer 1/22 rather than the decimal 0.0455.

On a graphing calculator your display should look like the following:

(2/3−5/8)/(1/6+3/4) ENTER

0.0455

Applications

The next example illustrates how complex fractions can occur in a problem.

EXAMPLE 7   **An application of complex fractions**

Eastside Elementary has the same number of students as Westside Elementary. One-half of the students at Eastside ride buses to school, and two-thirds of the students at Westside ride buses to school. One-sixth of the students at Eastside are female, and one-third of the students at Westside are female. If all of the female students ride the buses, then what percentage of the students who ride the buses are female?

Solution

To find the required percentage, we must divide the number of females who ride the buses by the total number of students who ride the buses. Let

$$x = \text{the number of students at Eastside.}$$

Because the number of students at Westside is also x, we have

$$\frac{1}{2}x + \frac{2}{3}x = \text{the total number of students who ride the buses}$$

and

$$\frac{1}{6}x + \frac{1}{3}x = \text{the total number of female students.}$$

Because all of the female students ride the buses, we can express the percentage of riders who are female by the following rational expression:

$$\dfrac{\dfrac{1}{6}x + \dfrac{1}{3}x}{\dfrac{1}{2}x + \dfrac{2}{3}x}$$

Multiply the numerator and denominator by 6, the LCD for 2, 3, and 6:

$$\dfrac{\left(\dfrac{1}{6}x + \dfrac{1}{3}x\right)6}{\left(\dfrac{1}{2}x + \dfrac{2}{3}x\right)6} = \dfrac{x + 2x}{3x + 4x} = \dfrac{3x}{7x} = \dfrac{3}{7} \approx 0.43 = 43\%$$

So 43% of the students who ride the buses are female. ■

Warm-ups

True or false? Explain.

1. The LCM for 2, x, 6, and x^2 is $6x^3$.

2. The LCM for $a - b$, $2b - 2a$, and 6 is $6a - 6b$.

3. The LCD is the LCM of the denominators.

4. $\dfrac{\dfrac{1}{2} + \dfrac{1}{3}}{1 + \dfrac{1}{2}} = \dfrac{5}{6} \div \dfrac{3}{2}$

5. $2^{-1} + 3^{-1} = (2 + 3)^{-1}$

6. $(2^{-1} + 3^{-1})^{-1} = 2 + 3$

7. $2 + 3^{-1} = 5^{-1}$

8. $x + 2^{-1} = \dfrac{x}{2}$ for any real number x.

9. To simplify $\dfrac{a^{-1} - b^{-1}}{a - b}$, multiply numerator and denominator by ab.

10. To simplify $\dfrac{ab^{-2} + a^{-5}b^2}{a^{-3}b - a^5b^{-1}}$, multiply numerator and denominator by a^5b^2.

4.4 EXERCISES

Simplify each complex fraction. Use either method. See Example 1.

1. $\dfrac{\dfrac{1}{2} - \dfrac{1}{3}}{\dfrac{1}{4} - \dfrac{1}{5}}$

2. $\dfrac{\dfrac{1}{3} + \dfrac{1}{4}}{\dfrac{1}{5} + \dfrac{1}{6}}$

3. $\dfrac{\dfrac{2}{3} + \dfrac{5}{6} - \dfrac{1}{2}}{\dfrac{1}{8} - \dfrac{1}{3} + \dfrac{1}{12}}$

4. $\dfrac{\dfrac{2}{5} - \dfrac{x}{9} - \dfrac{1}{3}}{\dfrac{1}{3} + \dfrac{x}{5} + \dfrac{2}{15}}$

Simplify the complex fractions. Use Method B. See Example 2.

5. $\dfrac{\dfrac{a+b}{b}}{\dfrac{a-b}{ab}}$

6. $\dfrac{\dfrac{m-n}{m^2}}{\dfrac{m-3}{mn^3}}$

7. $\dfrac{a+\dfrac{3}{b}}{\dfrac{b}{a}+\dfrac{1}{b}}$

8. $\dfrac{m-\dfrac{2}{n}}{\dfrac{1}{m}-\dfrac{3}{n}}$

9. $\dfrac{\dfrac{x-3y}{xy}}{\dfrac{1}{x}+\dfrac{1}{y}}$

10. $\dfrac{\dfrac{2}{w}+\dfrac{3}{t}}{\dfrac{w-t}{4wt}}$

11. $\dfrac{3-\dfrac{m-2}{6}}{\dfrac{4}{9}+\dfrac{2}{m}}$

12. $\dfrac{6-\dfrac{2-z}{z}}{\dfrac{1}{3z}-\dfrac{1}{6}}$

13. $\dfrac{\dfrac{a^2-b^2}{a^2b^3}}{\dfrac{a+b}{a^3b}}$

14. $\dfrac{\dfrac{4x^2-1}{x^2y}}{\dfrac{4x-2}{xy^2}}$

15. $\dfrac{\dfrac{1}{x^2y^2}+\dfrac{1}{xy^3}}{\dfrac{1}{x^3y}-\dfrac{1}{xy}}$

16. $\dfrac{\dfrac{1}{2a^3b}-\dfrac{1}{ab^4}}{\dfrac{1}{6a^2b^2}+\dfrac{1}{3a^4b}}$

Simplify each complex fraction. See Examples 1–3.

17. $\dfrac{x+\dfrac{4}{x+4}}{x-\dfrac{4x+4}{x+4}}$

18. $\dfrac{x-\dfrac{x+6}{x+2}}{x-\dfrac{4x+15}{x+2}}$

19. $\dfrac{1-\dfrac{1}{y-1}}{3+\dfrac{1}{y+1}}$

20. $\dfrac{2-\dfrac{3}{a-2}}{4-\dfrac{1}{a+2}}$

21. $\dfrac{\dfrac{2}{3-x}-4}{\dfrac{1}{x-3}-1}$

22. $\dfrac{\dfrac{x}{x-5}-2}{\dfrac{2x}{5-x}-1}$

23. $\dfrac{\dfrac{w+2}{w-1}-\dfrac{w-3}{w}}{\dfrac{w+4}{w}+\dfrac{w-2}{w-1}}$

24. $\dfrac{\dfrac{x-1}{x+2}-\dfrac{x-2}{x+3}}{\dfrac{x-3}{x+3}+\dfrac{x+1}{x+2}}$

25. $\dfrac{\dfrac{1}{a-b}-\dfrac{3}{a+b}}{\dfrac{2}{b-a}+\dfrac{4}{b+a}}$

26. $\dfrac{\dfrac{3}{2+x}-\dfrac{4}{2-x}}{\dfrac{1}{x+2}-\dfrac{3}{x-2}}$

27. $\dfrac{3-\dfrac{4}{a-1}}{5-\dfrac{3}{1-a}}$

28. $\dfrac{\dfrac{x}{3}-\dfrac{x-1}{9-x}}{\dfrac{x}{6}-\dfrac{2-x}{x-9}}$

29. $\dfrac{\dfrac{2}{m-3}+\dfrac{4}{m}}{\dfrac{3}{m-2}+\dfrac{1}{m}}$

30. $\dfrac{\dfrac{1}{y+2}-\dfrac{4}{3y}}{\dfrac{3}{y}-\dfrac{2}{y+3}}$

31. $\dfrac{\dfrac{3}{x^2-1}-\dfrac{x-2}{x^3-1}}{\dfrac{3}{x^2+x+1}+\dfrac{x-3}{x^3-1}}$

32. $\dfrac{\dfrac{2}{a^3+8}-\dfrac{3}{a^2-2a+4}}{\dfrac{4}{a^2-4}+\dfrac{a-3}{a^3+8}}$

Simplify. See Examples 4–6.

33. $\dfrac{w^{-1}+y^{-1}}{z^{-1}+y^{-1}}$

34. $\dfrac{a^{-1}-b^{-1}}{a^{-1}+b^{-1}}$

35. $\dfrac{1-x^{-1}}{1-x^{-2}}$

36. $\dfrac{4-a^{-2}}{2-a^{-1}}$

37. $\dfrac{a^{-2}+b^{-2}}{a^{-1}b}$

38. $\dfrac{m^{-3}+n^{-3}}{mn^{-2}}$

39. $1-a^{-1}$

40. $m^{-1}-a^{-1}$

41. $\dfrac{x^{-1}+x^{-2}}{x+x^{-2}}$

42. $\dfrac{x-x^{-2}}{1-x^{-2}}$

43. $\dfrac{2m^{-1}-3m^{-2}}{m^{-2}}$

44. $\dfrac{4x^{-3}-6x^{-5}}{2x^{-5}}$

45. $\dfrac{a^{-1}-b^{-1}}{a-b}$

46. $\dfrac{a^2-b^2}{a^{-2}-b^{-2}}$

47. $\dfrac{x^3-y^3}{x^{-3}-y^{-3}}$

48. $\dfrac{(a-b)^2}{a^{-2}-b^{-2}}$

49. $\dfrac{1-8x^{-3}}{x^{-1}+2x^{-2}+4x^{-3}}$

50. $\dfrac{a+27a^{-2}}{1-3a^{-1}+9a^{-2}}$

51. $(x^{-1}+y^{-1})^{-1}$

52. $(a^{-1}-b^{-1})^{-2}$

Use a calculator to evaluate each complex fraction. Round answers to four decimal places. If your calculator does fractions, then also find the fractional answer.

53. $\dfrac{\dfrac{5}{3} - \dfrac{4}{5}}{\dfrac{1}{3} - \dfrac{5}{6}}$

54. $\dfrac{\dfrac{1}{12} + \dfrac{1}{2} - \dfrac{3}{4}}{\dfrac{3}{5} + \dfrac{5}{6}}$

55. $\dfrac{4^{-1} - 9^{-1}}{2^{-1} + 3^{-1}}$

56. $\dfrac{2^{-1} + 3^{-1} - 6^{-1}}{3^{-1} - 5^{-1} + 4^{-1}}$

Solve each problem. See Example 7.

57. *Racial balance.* Clarksville has three elementary schools. Northside has one-half as many students as Central, and Southside has two-thirds as many students as Central. One-third of the students at Northside are African-American, three-fourths of the students at Central are African-American, and one-sixth of the students at Southside are African-American. What percent of the city's elementary students are African-American?

58. *Explosive situation.* All of the employees at Acme Explosives are in either development, manufacturing, or sales. One-fifth of the employees in development are women, one-third of the employees in manufacturing are women, and one-half of the employees in sales are women. Use the accompanying figure to determine the percentage of workers at Acme who are women. What percent of the women at Acme are in sales?

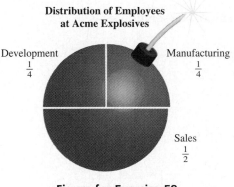

Distribution of Employees at Acme Explosives

Development $\frac{1}{4}$

Manufacturing $\frac{1}{4}$

Sales $\frac{1}{2}$

Figure for Exercise 58

59. *Average speed.* Mary drove from Clarksville to Leesville at 45 miles per hour. At Leesville she discovered that she had forgotten her purse. She immediately returned to Clarksville at 55 miles per hour. What was her average speed for the entire trip? (The answer is *not* 50 miles per hour.)

60. *Average price.* On her way to New York, Jenny spent the same amount for gasoline each of the three times that she filled up. She paid 99.9 cents per gallon the first time, 109.9 cents per gallon the second time, and 119.9 cents per gallon the third time. What was the average price per gallon to the nearest tenth of a cent for the gasoline that she bought?

Getting More Involved

61. *Cooperative learning.* Write a step-by-step strategy for simplifying complex fractions with negative exponents. Have a classmate use your strategy to simplify some complex fractions from Exercises 33–52.

62. *Discussion.* **a)** Find the exact value of each expression.

i) $\dfrac{1}{1 + \dfrac{1}{1 + \dfrac{1}{1 + \dfrac{1}{2}}}}$

ii) $\dfrac{1}{1 + \dfrac{1}{1 + \dfrac{1}{1 + \dfrac{1}{3}}}}$

b) Explain why in each case the exact value must be less than 1.

63. *Cooperative Learning.* Work with a group to simplify the complex fraction. For what values of x is the complex fraction undefined?

$$\dfrac{1}{1 + \dfrac{1}{1 + \dfrac{1}{1 + \dfrac{1}{x}}}}$$

4.5 Solving Equations Involving Rational Expressions

In this section:
▶ Multiplying by the LCD
▶ Extraneous Roots
▶ Proportions

Many problems in algebra are modeled by equations involving rational expressions. In this section you will learn how to solve equations that have rational expressions, and in Section 4.6 we will solve problems using these equations.

Multiplying by the LCD

To solve equations having rational expressions, we multiply each side of the equation by the LCD of the rational expressions.

EXAMPLE 1   **An equation with rational expressions**

Solve $\dfrac{1}{x} + \dfrac{1}{4} = \dfrac{1}{6}$.

Solution

The LCD for the denominators 4, 6, and x is $12x$:

$$12x\left(\frac{1}{x} + \frac{1}{4}\right) = 12x\left(\frac{1}{6}\right) \qquad \text{Multiply each side by } 12x.$$

$$12\cancel{x} \cdot \frac{1}{\cancel{x}} + \overset{3}{\cancel{12}}x \cdot \frac{1}{\cancel{4}} = \overset{2}{\cancel{12}}x \cdot \frac{1}{\cancel{6}} \qquad \text{Distributive property}$$

$$12 + 3x = 2x \qquad\qquad \text{Divide out the common factors.}$$

$$12 + x = 0$$

$$x = -12$$

Check -12 in the original equation. The solution set is $\{-12\}$. ∎

EXAMPLE 2   **An equation with rational expressions**

Solve $\dfrac{1}{x} + \dfrac{2}{3x} = \dfrac{1}{5}$.

Solution

Multiply each side by $15x$, the LCD for x, $3x$, and 5:

$$15x\left(\frac{1}{x} + \frac{2}{3x}\right) = 15x\left(\frac{1}{5}\right)$$

$$15x \cdot \frac{1}{x} + 15x \cdot \frac{2}{3x} = 3x$$

$$15 + 10 = 3x$$

$$25 = 3x$$

$$\frac{25}{3} = x$$

Check $\dfrac{25}{3}$ in the original equation. The solution set is $\left\{\dfrac{25}{3}\right\}$. ∎

CAUTION   To solve an equation with rational expressions, we do *not* convert the rational expressions to ones with a common denominator. Instead, we multiply each side by the LCD to *eliminate* the denominators.   ⊘

EXAMPLE 3   **An equation with two solutions**

Solve $\dfrac{200}{x} + \dfrac{300}{x + 20} = 10$.

Solution

$$x(x + 20)\left(\dfrac{200}{x} + \dfrac{300}{x + 20}\right) = x(x + 20)\,10$$ Multiply each side by $x(x + 20)$.

$$x(x + 20)\dfrac{200}{x} + x(x + 20)\dfrac{300}{x + 20} = x(x + 20)10$$ Distributive property

$$(x + 20)200 + x(300) = (x^2 + 20x)10$$ Simplify.

$$200x + 4000 + 300x = 10x^2 + 200x$$

$$4000 + 300x = 10x^2$$ Combine like terms.

$$400 + 30x = x^2$$ Divide each side by 10.

$$0 = x^2 - 30x - 400$$

$$0 = (x - 40)(x + 10)$$ Factor.

$$x - 40 = 0 \quad \text{or} \quad x + 10 = 0$$ Set each factor equal to 0.

$$x = 40 \quad \text{or} \quad x = -10$$

Check these values in the original equation. The solution set is $\{-10, 40\}$.   ∎

Extraneous Roots

Because equations involving rational expressions have variables in denominators, a root to the equation might cause a zero to appear in a denominator. In this case the root does not satisfy the original equation, and so it is called an **extraneous root.**

EXAMPLE 4   **An equation with an extraneous root**

Solve $\dfrac{3}{x} + \dfrac{6}{x - 2} = \dfrac{12}{x^2 - 2x}$.

Solution

Because $x^2 - 2x = x(x - 2)$, the LCD for x, $x - 2$, and $x^2 - 2x$ is $x(x - 2)$.

$$\cancel{x}(x - 2)\dfrac{3}{\cancel{x}} + x\cancel{(x - 2)}\dfrac{6}{\cancel{x - 2}} = \cancel{x(x - 2)}\dfrac{12}{\cancel{x(x - 2)}}$$ Multiply each side by $x(x - 2)$.

$$3(x - 2) + 6x = 12$$

$$3x - 6 + 6x = 12$$

$$9x - 6 = 12$$

$$9x = 18$$

$$x = 2$$

If $x = 2$, then the value of the denominator $x - 2$ is 0. So 2 is an extraneous root. The solution set is the empty set, $\varnothing$.   ∎

EXAMPLE 5   **An equation with an extraneous root**

Solve $x + 2 + \dfrac{x}{x-2} = \dfrac{2}{x-2}$.

Solution

Because the LCD is $x - 2$, we multiply each side by $x - 2$:

$$(x-2)(x+2) + (x-2)\frac{x}{x-2} = (x-2)\frac{2}{x-2}$$
$$x^2 - 4 + x = 2$$
$$x^2 + x - 6 = 0$$
$$(x+3)(x-2) = 0$$
$$x + 3 = 0 \quad \text{or} \quad x - 2 = 0$$
$$x = -3 \quad \text{or} \quad x = 2$$

Check that -3 satisfies the original equation and 2 is an extraneous root. The solution set is $\{-3\}$. ∎

Proportions

An equation that expresses the equality of two rational expressions is called a **proportion.** The equation

$$\frac{a}{b} = \frac{c}{d}$$

is a proportion. The terms in the position of b and c are called the **means.** The terms in the position of a and d are called the **extremes.** If we multiply this proportion by the LCD, bd, we get

$$bd \cdot \frac{a}{b} = bd \cdot \frac{c}{d}$$

or

$$ad = bc.$$

The equation $ad = bc$ says that *the product of the extremes is equal to the product of the means.* When solving a proportion, we can omit multiplication by the LCD and just remember the result, $ad = bc$, as the **extremes-means property.**

> **Extremes-Means Property**
>
> If $\dfrac{a}{b} = \dfrac{c}{d}$, then $ad = bc$.

The extremes-means property makes it easier to solve proportions.

EXAMPLE 6   **A proportion with one solution**

Solve $\dfrac{20}{x} = \dfrac{30}{x+20}$.

Solution

Rather than multiplying by the LCD, we use the extremes-means property to eliminate the denominators:

$$\frac{20}{x} = \frac{30}{x + 20}$$

$$20(x + 20) = 30x \qquad \text{Extremes-means property}$$

$$20x + 400 = 30x$$

$$400 = 10x$$

$$40 = x$$

Check 40 in the original equation. The solution set is {40}. ■

EXAMPLE 7 **A proportion with two solutions**

Solve $\dfrac{2}{x} = \dfrac{x + 3}{5}$.

Solution

Use the extremes-means property to write an equivalent equation:

$$x(x + 3) = 2 \cdot 5 \qquad \text{Extremes-means property}$$

$$x^2 + 3x = 10$$

$$x^2 + 3x - 10 = 0$$

$$(x + 5)(x - 2) = 0 \qquad \text{Factor.}$$

$$x + 5 = 0 \quad \text{or} \quad x - 2 = 0 \qquad \text{Zero factor property}$$

$$x = -5 \quad \text{or} \quad x = 2$$

Both -5 and 2 satisfy the original equation. The solution set is $\{-5, 2\}$. ■

MATH AT WORK

Marine Insurance Agent

Cargo has been lost, or the hull of a ship has been damaged. What is the amount of money that should be paid to the insured party? Lisa M. Paccione, Ocean Marine Claim Representative for the St. Paul Insurance Company, investigates, evaluates, resolves, and pays these types of claims. Ms. Paccione does this by gathering data, occasionally doing a visual inspection, interviewing witnesses, and negotiating with attorneys.

Decisions about losses are based on the insured party's individual policy as well as traditional marine practices and maritime law. When consignees suffer a cargo loss, they not only are compensated for the actual amount of the damaged goods, but also receive an additional "advance" in the settlement. Customarily, the advance is 10% over the value of the goods. The amount that St. Paul pays the insured party for a valid claim is computed by using a proportion. In Exercises 51 and 52 of this section you will solve problems involving this proportion.

CAUTION   Use the extremes-means property only when solving a *proportion*. It cannot be used on an equation such as

$$\frac{3}{x} = \frac{2}{x+1} + 5. \qquad \oslash$$

EXAMPLE 8   **Ratios and proportions**

The ratio of men to women at a football game was 4 to 3. If there were 12,000 more men than women in attendance, then how many men and how many women were in attendance?

Solution

Let x represent the number of men in attendance and $x - 12,000$ represent the number of women in attendance. Because the ratio of men to women was 4 to 3, we can write the following proportion:

$$\frac{4}{3} = \frac{x}{x - 12,000}$$
$$4x - 48,000 = 3x$$
$$x = 48,000$$

So there were 48,000 men and 36,000 women at the game.   ∎

Warm-ups

True or false? Explain.

1. In solving an equation involving rational expressions, multiply each side by the LCD for all of the denominators.

2. To solve $\frac{1}{x} + \frac{1}{2x} = \frac{1}{3}$, first change each rational expression to an equivalent rational expression with a denominator of $6x$.

3. Extraneous roots are not real numbers.

4. To solve $\frac{1}{x-2} + 3 = \frac{1}{x+2}$, multiply each side by $x^2 - 4$.

5. The solution set to $\frac{x}{3x+4} - \frac{6}{2x+1} = \frac{7}{5}$ is $\left\{-\frac{4}{3}, -\frac{1}{2}\right\}$.

6. The solution set to $\frac{3}{x} = \frac{2}{5}$ is $\left\{\frac{15}{2}\right\}$.

7. We should use the extremes-means property to solve $\frac{x-2}{x+3} + 1 = \frac{1}{x}$.

8. The equation $x^2 = x$ is equivalent to the equation $x = 0$.

9. The solution set to $(2x - 3)(3x + 4) = 0$ is $\left\{\frac{3}{2}, \frac{4}{3}\right\}$.

10. The equation $\frac{2}{x+1} = \frac{x-1}{4}$ is equivalent to $x^2 - 1 = 8$.

4.5 EXERCISES

Find the solution set to each equation. See Examples 1–5.

1. $\dfrac{1}{x} + \dfrac{1}{6} = \dfrac{1}{8}$

2. $\dfrac{3}{x} + \dfrac{1}{5} = \dfrac{1}{2}$

3. $\dfrac{2}{3x} + \dfrac{1}{15x} = \dfrac{1}{2}$

4. $\dfrac{5}{6x} - \dfrac{1}{8x} = \dfrac{17}{24}$

5. $\dfrac{3}{x-2} + \dfrac{5}{x} = \dfrac{10}{x}$

6. $\dfrac{5}{x-1} + \dfrac{1}{2x} = \dfrac{1}{x}$

7. $\dfrac{x}{x-2} + \dfrac{3}{x} = 2$

8. $\dfrac{x}{x-5} + \dfrac{5}{x} = \dfrac{11}{6}$

9. $\dfrac{100}{x} = \dfrac{150}{x+5} - 1$

10. $\dfrac{30}{x} = \dfrac{50}{x+10} + \dfrac{1}{2}$

11. $\dfrac{3x-5}{x-1} = 2 - \dfrac{2x}{x-1}$

12. $\dfrac{x-3}{x+2} = 3 - \dfrac{1-2x}{x+2}$

13. $x + 1 + \dfrac{2x-5}{x-5} = \dfrac{x}{x-5}$

14. $\dfrac{x-3}{2} - \dfrac{1}{x-3} = \dfrac{8-3x}{x-3}$

Find the solution set to each equation. See Examples 6 and 7.

15. $\dfrac{2}{x} = \dfrac{3}{4}$

16. $\dfrac{5}{x} = \dfrac{7}{9}$

17. $\dfrac{a}{3} = \dfrac{-1}{4}$

18. $\dfrac{b}{5} = \dfrac{-3}{7}$

19. $-\dfrac{5}{7} = \dfrac{2}{x}$

20. $-\dfrac{3}{8} = \dfrac{5}{x}$

21. $\dfrac{10}{x} = \dfrac{20}{x+20}$

22. $\dfrac{x}{5} = \dfrac{x+2}{3}$

23. $\dfrac{2}{x+1} = \dfrac{x-1}{4}$

24. $\dfrac{3}{x-2} = \dfrac{x+2}{7}$

25. $\dfrac{x}{6} = \dfrac{5}{x-1}$

26. $\dfrac{x+5}{2} = \dfrac{3}{x}$

27. $\dfrac{x}{x-3} = \dfrac{x+2}{x}$

28. $\dfrac{x+1}{x-5} = \dfrac{x+2}{x-4}$

29. $\dfrac{x-2}{x-3} = \dfrac{x+5}{x+2}$

30. $\dfrac{x}{x+5} = \dfrac{x}{x-2}$

Solve each equation.

31. $\dfrac{a}{9} = \dfrac{4}{a}$

32. $\dfrac{y}{3} = \dfrac{27}{y}$

33. $\dfrac{1}{2x-4} + \dfrac{1}{x-2} = \dfrac{1}{4}$

34. $\dfrac{7}{3x-9} - \dfrac{1}{x-3} = \dfrac{4}{9}$

35. $\dfrac{x-2}{4} = \dfrac{x-2}{x}$

36. $\dfrac{y+5}{2} = \dfrac{y+5}{y}$

37. $\dfrac{5}{2x+4} - \dfrac{1}{x-1} = \dfrac{3}{x+2}$

38. $\dfrac{5}{2w+6} - \dfrac{1}{w-1} = \dfrac{1}{w+3}$

39. $\dfrac{5}{x-3} = \dfrac{x}{x-3}$

40. $\dfrac{6}{a+2} = \dfrac{a}{a+2}$

41. $\dfrac{w}{6} = \dfrac{3}{2w}$

42. $\dfrac{2m}{5} = \dfrac{10}{m}$

43. $\dfrac{5}{4x-2} - \dfrac{1}{1-2x} = \dfrac{7}{3x+6}$

44. $\dfrac{5}{x+1} - \dfrac{1}{1-x} = \dfrac{1}{x^2-1}$

45. $\dfrac{5}{x} = \dfrac{2}{5}$

46. $\dfrac{-3}{2x} = \dfrac{1}{-5}$

47. $\dfrac{5}{x^2-9} + \dfrac{2}{x+3} = \dfrac{1}{x-3}$

48. $\dfrac{1}{x-2} - \dfrac{2}{x+3} = \dfrac{11}{x^2+x-6}$

49. $\dfrac{9}{x^3-1} - \dfrac{1}{x-1} = \dfrac{2}{x^2+x+1}$

50. $\dfrac{x+4}{x^3+8} + \dfrac{x+2}{x^2-2x+4} = \dfrac{11}{2x+4}$

In Exercises 51–60, solve each problem. See Example 8.

51. ***Maritime losses.*** The amount paid to an insured party by the American Insurance Company is computed by using the proportion

$$\frac{\text{value shipped}}{\text{amount of loss}} = \frac{\text{amount of declared premium}}{\text{amount insured party gets paid}}.$$

If the value shipped was \$300,000, the amount of loss was \$250,000, and the amount of declared premium was \$200,000, then what amount is paid to the insured party?

52. ***Maritime losses.*** Suppose that the value shipped was \$400,000, the amount of loss was \$300,000, and the amount that the insured party got paid was \$150,000. Use the proportion of Exercise 51 to find the amount of declared premium.

53. *Capture-recapture method.* To estimate the size of the grizzly bear population in a national park, rangers tagged and released 12 bears. Later, it was observed that in 23 sightings of grizzly bears, only two had been tagged. Assuming that the proportion of tagged bears in the later sightings is the same as the proportion of tagged bears in the population, estimate the number of bears in the population.

54. *Please rewind.* In a sample of 24 returned video-tapes, it was found that only three were rewound as requested. If 872 videos are returned in a day, then how many of them would you expect to find not rewound?

55. *Pleasing painting.* The ancient Greeks often used the ratio of length to width for a rectangle as 7 to 6 to give the rectangle a pleasing shape. If the length of a pleasantly shaped Greek painting is 22 cm longer then its width, then what are its length and width?

56. *Pickups and cars.* The ratio of pickups to cars sold at a dealership is 2 to 3. If the dealership sold 142 more cars than pickups in 1994, then how many of each did it sell?

57. *Cleaning up the river.* Pollution in the Tickfaw River has been blamed primarily on pesticide runoff from area farms. The formula

$$C = \frac{4,000,000p}{100 - p}$$

has been used to model the cost in dollars for removing p% of the pollution in the river. If the state gets a $1 million federal grant for cleaning up the river, then what percentage of the pollution can be removed? Use the bar graph to estimate the percentage that can be cleaned up with a $100 million grant.

Figure for Exercise 57

58. *Campaigning for governor.* A campaign manager for a gubernatorial candidate estimates that the cost in dollars for an advertising campaign that will get his candidate p% of the votes is given by

$$C = \frac{1,000,000 + 2,000,000p}{100 - p}.$$

If the candidate can spend only $2 million for advertising, then what percentage of the votes can she expect to receive? Use the bar graph to estimate the percentage of votes expected if $4 million is spent.

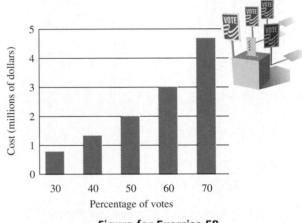

Figure for Exercise 58

59. *Wealth-building portfolio.* Misty decided to invest her annual bonus in a wealth-building portfolio as shown in the figure (*Fidelity Focus*, Winter 1994). If the amount that she invested in stocks was $20,000 greater than her investment in bonds, then how much did she invest in bonds? What was the amount of her annual bonus?

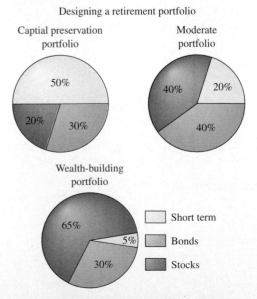

Figure for Exercises 59 and 60

60. *Capital preservation portfolio.* Aaron invested his inheritance in a capital preservation portfolio as shown in the figure (*Fidelity Focus*, Winter 1994). If the amount that he invested in short-term instruments was $13,000 greater than the amount that he invested in bonds, then how much did he invest in bonds? What was the amount of his inheritance?

Getting More Involved

61. *Writing.* In this chapter the LCD is used to add rational expressions and to solve equations. Explain the difference between using the LCD to solve the equation

$$\frac{3}{x-2} + \frac{7}{x+2} = 2$$

and using the LCD to find the sum

$$\frac{3}{x-2} + \frac{7}{x+2}.$$

62. *Discussion.* For each equation, find the values for x that *cannot* be solutions to the equation. Do not solve the equations.

a) $\dfrac{1}{x} + \dfrac{1}{x-1} = \dfrac{1}{2}$

b) $\dfrac{x}{x-1} = \dfrac{1}{2}$

c) $\dfrac{1}{x^2+1} = \dfrac{1}{x+1}$

4.6 Applications

In this section we use the techniques of Section 4.5 to rewrite formulas involving rational expressions and to solve some problems.

Formulas

Rewriting formulas having rational expressions is similar to solving equations having rational expressions. Generally, the first step is to multiply each side by the LCD for the rational expressions.

EXAMPLE 1 **Solving a formula**

In Chapter 7 we will see that an equation of the form

$$\frac{y - y_1}{x - x_1} = m$$

is an equation of a straight line. Solve the formula for y.

Solution

$$(x - x_1)\frac{y - y_1}{x - x_1} = (x - x_1)m \quad \text{Multiply each side by the denominator } x - x_1.$$
$$y - y_1 = (x - x_1)m \quad \text{Reduce.}$$
$$y = (x - x_1)m + y_1$$

∎

In the next example we solve for a variable that occurs twice in the original formula. Remember that when a formula is solved for a certain variable, that variable must appear only once in the final formula.

EXAMPLE 2 **Solving for a variable**

The formula

$$\frac{P}{P_W} = \frac{2L}{2L + d}$$

is used in physics to find the relative density of a substance. Solve it for L.

Solution

$$\frac{P}{P_W} = \frac{2L}{2L + d}$$

$P(2L + d) = P_W(2L)$ The extremes-means property

$2PL + Pd = 2LP_W$ Simplify.

$Pd = 2LP_W - 2PL$ Get all terms involving L onto the same side.

$Pd = (2P_W - 2P)L$ Factor out L.

$$\frac{Pd}{2P_W - 2P} = L$$ ∎

In the next example we find the value of one variable when given the values of the remaining variables.

EXAMPLE 3 **Evaluating a formula**

Find x if $x_1 = 2$, $y_1 = -3$, $y = -1$, $m = \frac{1}{2}$, and

$$\frac{y - y_1}{x - x_1} = m.$$

Solution

Substitute all of the values into the formula and solve for x:

$$\frac{-1 - (-3)}{x - 2} = \frac{1}{2}$$ Substitute.

$$\frac{2}{x - 2} = \frac{1}{2}$$

$x - 2 = 4$ Extremes-means property

$x = 6$ Check in the original formula. ∎

Uniform Motion Problems

The uniform motion problems here are similar to those of Chapter 2, but in this chapter the equations have rational expressions.

EXAMPLE 4   **Uniform motion**

Michele drove her empty rig 300 miles to Salina to pick up a load of cattle. When her rig was fully loaded, her average speed was 10 miles per hour less than when the rig was empty. If the return trip took her 1 hour longer, then what was her average speed with the rig empty? (See Fig. 4.1.)

Solution

Let x be Michele's average speed empty and let $x - 10$ be her average speed full. Because the time can be determined from the distance and the rate, $T = \dfrac{D}{R}$, we can make the following table.

300 miles
Speed = x mph

300 miles
Speed = $x - 10$ mph

Figure 4.1

	Rate	Time	Distance
Empty	$x\,\dfrac{\text{mi}}{\text{hr}}$	$\dfrac{300}{x}$ hr	300 mi
Full	$x - 10\,\dfrac{\text{mi}}{\text{hr}}$	$\dfrac{300}{x-10}$ hr	300 mi

We now write an equation expressing the fact that her time empty was 1 hour less than her time full:

$$\frac{300}{x} = \frac{300}{x - 10} - 1$$

$$x(x - 10)\frac{300}{x} = x(x - 10)\frac{300}{x - 10} - x(x - 10)1 \qquad \text{Multiply each side by } x(x - 10).$$

$$300x - 3000 = 300x - x^2 + 10x \qquad \text{Reduce.}$$

$$-3000 = -x^2 + 10x$$

$$x^2 - 10x - 3000 = 0 \qquad \text{Get 0 on one side.}$$

$$(x + 50)(x - 60) = 0 \qquad \text{Factor.}$$

$$x + 50 = 0 \qquad \text{or} \qquad x - 60 = 0 \qquad \text{The zero factor property.}$$

$$x = -50 \qquad \text{or} \qquad x = 60$$

The equation is satisfied if $x = -50$, but because -50 is negative, it cannot be the speed of the truck. Michele's average speed empty was 60 miles per hour. Checking this answer, we find that if she traveled 300 miles at 60 miles per hour, it would take her 5 hours. If she traveled 300 miles at 50 miles per hour with the loaded rig, it would take her 6 hours. Because Michele's time with the empty rig was 1 hour less than her time with the loaded rig, 60 miles per hour is the correct answer. ■

Work Problems

Problems involving different rates for completing a task are referred to as **work problems.** We did not solve work problems earlier because they usually require equations with rational expressions. Work problems are similar to uniform motion problems in which $RT = D$. The product of a person's time and rate is the amount of work completed. For example, if your puppy gains 1 pound every 3 days, then he is growing at the rate of $\frac{1}{3}$ pound per day. If he grows at the rate of $\frac{1}{3}$ pound per day for a period of 30 days, then he gains 10 pounds.

EXAMPLE 5 **Working together**

Linda can mow a certain lawn with her riding lawn mower in 4 hours. When Linda uses the riding mower and Rebecca operates the push mower, it takes them 3 hours to mow the lawn. How long would it take Rebecca to mow the lawn by herself using the push mower?

Solution

If x is the number of hours it takes for Rebecca to complete the lawn alone, then her rate is $\frac{1}{x}$ of the lawn per hour. Because Linda can mow the entire lawn in 4 hours, her rate is $\frac{1}{4}$ of the lawn per hour. In the 3 hours that they work together, Rebecca completes $\frac{3}{x}$ of the lawn while Linda completes $\frac{3}{4}$ of the lawn. We can classify all of the necessary information in a table that looks a lot like the table we used in Example 4.

	Rate	Time	Amount of Work
Linda	$\frac{1}{4}\frac{\text{lawn}}{\text{hr}}$	3 hr	$\frac{3}{4}$ lawn
Rebecca	$\frac{1}{x}\frac{\text{lawn}}{\text{hr}}$	3 hr	$\frac{3}{x}$ lawn

Because the lawn is finished in 3 hours, the two portions of the lawn (in the work column) mowed by each girl have a sum of 1:

$$\frac{3}{4} + \frac{3}{x} = 1$$

$$4x \cdot \frac{3}{4} + 4x \cdot \frac{3}{x} = 4x \cdot 1 \qquad \text{Multiply each side by } 4x.$$

$$3x + 12 = 4x$$

$$12 = x$$

If $x = 12$, then in the 3 hours that they work together, Rebecca does $\frac{3}{12}$ or $\frac{1}{4}$ of the job while Linda does $\frac{3}{4}$ of the job. So it would take Rebecca 12 hours to mow the lawn by herself using the push mower. ∎

Miscellaneous Problems

EXAMPLE 6 **Hamburger and steak**

Patrick bought 50 pounds of meat consisting of hamburger and steak. Steak costs twice as much per pound as hamburger. If he bought $30 worth of hamburger and $90 worth of steak, then how many pounds of each did he buy?

Solution

Let x be the number of pounds of hamburger and $50 - x$ be the number of pounds of steak. Because Patrick got x pounds of hamburger for $30, he paid $\frac{30}{x}$ dollars per pound for the hamburger. We can classify all of the given information in a table.

	Price per pound	Amount	Total price
Hamburger	$\dfrac{30}{x}$ $\dfrac{\text{dollars}}{\text{lb}}$	x lb	30 dollars
Steak	$\dfrac{90}{50-x}$ $\dfrac{\text{dollars}}{\text{lb}}$	$50-x$ lb	90 dollars

Because the price per pound of steak is twice that of hamburger, we can write the following equation:

$$2\left(\frac{30}{x}\right) = \frac{90}{50-x}$$

$$\frac{60}{x} = \frac{90}{50-x}$$

$$90x = 3000 - 60x \qquad \text{The extremes-means property}$$

$$150x = 3000$$

$$x = 20$$

$$50 - x = 30$$

Patrick purchased 20 pounds of hamburger and 30 pounds of steak. Check this answer. ∎

Warm-ups

True or false? Explain.

1. The formula $w = \dfrac{1-t}{t}$, solved for t, is $t = \dfrac{1-t}{w}$.

2. To solve $\dfrac{1}{p} + \dfrac{1}{q} = \dfrac{1}{s}$ for s, multiply each side by pqs.

3. If 50 pounds of steak cost x dollars, then the price is $\dfrac{50}{x}$ dollars per pound.

4. If Claudia drives x miles in 3 hours, then her rate is $\dfrac{x}{3}$ miles per hour.

5. If Takenori mows his entire lawn in $x + 2$ hours, then he mows $\dfrac{1}{x+2}$ of the lawn per hour.

6. If Kareem drives 200 nails in 12 hours, then he is driving $\dfrac{200}{12}$ nails per hour.

7. If x hours is 1 hour less than y hours, then $x - 1 = y$.

8. If $A = \dfrac{mv^2}{B}$ and m and B are nonzero, then $v^2 = \dfrac{AB}{m}$.

9. If a and y are nonzero and $a = \dfrac{x}{y}$, then $y = ax$.

10. If x hours is 3 hours more than y hours, then $x + 3 = y$.

4.6 EXERCISES

Solve each equation for y. See Example 1.

1. $\dfrac{y-3}{x-2} = 5$

2. $\dfrac{y-4}{x-7} = -6$

3. $\dfrac{y+1}{x-6} = -\dfrac{1}{3}$

4. $\dfrac{y+7}{x-2} = \dfrac{-2}{3}$

5. $\dfrac{y-a}{x-b} = m$

6. $\dfrac{y-h}{x-k} = a$

7. $\dfrac{y-2}{x+5} = -\dfrac{7}{3}$

8. $\dfrac{y-3}{x+1} = -\dfrac{9}{4}$

Solve each formula for the indicated variable. See Example 2.

9. $M = \dfrac{F}{f}$ for f

10. $P = \dfrac{A}{1+rt}$ for A

11. $A = \dfrac{\pi}{4} \cdot D^2$ for D^2

12. $V = \pi r^2 h$ for r^2

13. $F = k\dfrac{m_1 m_2}{r^2}$ for m_1

14. $F = \dfrac{mv^2}{r}$ for v^2

15. $\dfrac{1}{p} + \dfrac{1}{q} = \dfrac{1}{f}$ for q

16. $\dfrac{1}{R} = \dfrac{1}{R_1} + \dfrac{1}{R_2}$ for R_1

17. $e^2 = 1 - \dfrac{b^2}{a^2}$ for a^2

18. $e^2 = 1 - \dfrac{b^2}{a^2}$ for b^2

19. $\dfrac{P_1 V_1}{T_1} = \dfrac{P_2 V_2}{T_2}$ for T_1

20. $\dfrac{P_1 V_1}{T_1} = \dfrac{P_2 V_2}{T_2}$ for P_2

21. $V = \dfrac{4}{3}\pi r^2 h$ for h

22. $h = \dfrac{S - 2\pi r^2}{2\pi r}$ for S

Use the formula from the indicated exercise to find the value of the indicated variable. See Example 3. For calculator problems, round answers to three decimal places.

23. If $M = 10$ and $F = 5$ in Exercise 9, find f.

24. If $A = 550$, $P = 500$, and $t = 2$ in Exercise 10, find r.

25. If $A = 6\pi$ in Exercise 11, find D^2.

26. If $V = 12\pi$ and $r = 3$ in Exercise 12, find h.

27. If $F = 32$, $r = 4$, $m_1 = 6$, and $m_2 = 8$ in Exercise 13, find k.

28. If $F = 10$, $m = 8$, and $v = 6$ in Exercise 14, find r.

29. If $f = 2.3$ and $q = 1.7$ in Exercise 15, find p.

30. If $R = 1.29$ and $R_1 = 0.045$ in Exercise 16, find R_2.

31. If $e = 0.62$ and $b = 3.5$ in Exercise 17, find a^2.

32. If $a = 3.61$ and $e = 2.4$ in Exercise 18, find b^2.

33. If $V = 25.6$ and $h = 3.2$ in Exercise 21, find r^2.

34. If $h = 3.6$ and $r = 2.45$ in Exercise 22, find S.

Solve each problem. See Examples 4–6.

35. ***Walking and riding.*** Karen can ride her bike from home to school in the same amount of time as she can walk from home to the post office. She rides 10 mph faster than she walks. The distance from her home to school is 7 miles, and the distance from her home to the post office is 2 miles. How fast does Karen walk?

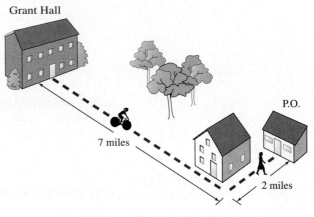

Grant Hall

P.O.

7 miles

2 miles

Figure for Exercise 35

36. ***Fast driving.*** Beverly can drive 600 miles in the same time as it takes Susan to drive 500 miles. If Beverly drives 10 mph faster than Susan, then how fast does Beverly drive?

37. ***Faster driving.*** Patrick drives 40 miles to work, and Guy drives 60 miles to work. Guy claims that he drives at the same speed as Patrick, but it takes him only 12 minutes longer to get to work. If this is true, then how long does it take each of them to get to work? What are their speeds?

38. ***Route drivers.*** David and Keith are route drivers for a fast-photo company. David's route is 80 miles, and Keith's is 100 miles. Keith averages 10 mph more than David and finishes his route 10 minutes before David. What is David's speed?

39. ***Physically fit.*** Every morning, Yong Yi runs 5 miles, then walks one mile. He runs 6 mph faster than he walks. If his total time yesterday was 45 minutes, then how fast did he run?

40. ***Row, row, row your boat.*** Norma can row her boat 12 miles in the same time as it takes Marietta to cover 36 miles in her motorboat. If Marietta's boat travels 15 mph faster than Norma's boat, then how fast is Norma rowing her boat?

41. Pumping out the pool. A large pump can drain an 80,000-gallon pool in 3 hours. With a smaller pump also operating, the job takes only 2 hours. How long would it take the smaller pump to drain the pool by itself?

42. Trimming hedges. Lourdes can trim the hedges around her property in 8 hours by using an electric hedge trimmer. Rafael can do the same job in 15 hours by using a manual trimmer. How long would it take them to trim the hedges working together?

43. Filling the tub. It takes 10 minutes to fill Alisha's bathtub and 12 minutes to drain the water out. How long would it take to fill it with the drain accidentally left open?

Figure for Exercise 43

44. Eating machine. Charles can empty the cookie jar in $1\frac{1}{2}$ hours. It takes his mother 2 hours to bake enough cookies to fill it. If the cookie jar is full when Charles comes home from school, and his mother continues baking and restocking the cookie jar, then how long will it take him to empty the cookie jar?

45. Filing the invoices. It takes Gina 90 minutes to file the monthly invoices. If Hilda files twice as fast as Gina does, how long will it take them working together?

46. Painting alone. Julie can paint a fence by herself in 12 hours. With Betsy's help, it takes only 5 hours. How long would it take Betsy by herself?

47. Buying fruit. Molly bought $5.28 worth of oranges and $8.80 worth of apples. She bought 2 more pounds of oranges than apples. If apples cost twice as much per pound as oranges, then how many pounds of each did she buy?

48. Raising rabbits. Luke raises rabbits and raccoons to sell for meat. The price of raccoon meat is three times the price of rabbit meat. One day, Luke sold 160 pounds of meat, $72 worth of each type. What is the price per pound of each type of meat?

49. Total resistance. If two receivers with resistances R_1 and R_2 are connected in parallel, then the formula

$$\frac{1}{R} = \frac{1}{R_1} + \frac{1}{R_2}$$

relates the total resistance for the circuit R with R_1 and R_2. Given that R_1 is 3 ohms and R is 2 ohms, find R_2.

Figure for Exercise 49

50. More resistance. Use the formula from Exercise 49 to find R_1 and R_2 given that the total resistance is 1.2 ohms and R_1 is 1 ohm larger than R_2.

51. Las Vegas vacation. Brenda of Horizon Travel Agency has arranged a charter flight to Las Vegas. The flight will cost $24,000. If Brenda can get 40 more people to go on the trip, the airfare per person will decrease by $100. How many people are in the original group?

52. White-water rafting. Adventures, Inc. has a $1,500 group rate for an overnight rafting trip on the Colorado River. For the last trip, five people failed to show, causing the price per person to increase by $25. How many were originally scheduled for the trip?

53. Doggie bag. Muffy can eat a 25-pound bag of dog food in 28 days, while Missy eats a 25-pound bag in 23 days. How many days would it take them together to finish a 50-pound bag of dog food.

54. Rodent food. A pest-control specialist has found that 6 rats can eat an entire box of sugar-coated breakfast cereal in 13.6 minutes, and it takes a dozen mice 34.7 minutes to devour the same size box of cereal. How long would it take all 18 rodents, in a cooperative manner, to finish off a box of cereal?

─────── C O L L A B O R A T I V E   A C T I V I T I E S ───────

Beorg's Business

In manufacturing or other businesses in which time is money and tasks are easily shared, problems involving work appear. An owner or manager who wants to know how to bid a job often develops a table of times needed to complete the job as determined by how much work is needed and who could be assigned to the job.

Beorg owns a kaleidoscope-manufacturing company with two employees, Scott and Salina. It takes Scott one hour to make one kaleidoscope, and it takes Salina $\frac{1}{2}$ hour to make one kaleidoscope. Beorg wants to know how long it would take to complete a certain number of kaleidoscopes. Using the information given and answering the questions below, fill in a table like the following one for Beorg.

Name of Employee	Time for one kaleidoscope	Time for 20 kaleidoscopes
Scott	1 hr	
Salina	$\frac{1}{2}$ hr	
Scott & Salina		
Sammy		
Scott & Sammy	$\frac{3}{4}$ hr	
Salina & Sammy		
Scott, Salina & Sammy		

Grouping: 4 students per group
Topic: Applications of work problems

1. How long will it take Scott and Salina working together to make one kaleidoscope?

2. Beorg hires a third person, Sammy, and has him and Scott make one kaleidoscope. Working together, it takes them $\frac{3}{4}$ hour to make one kaleidoscope. How long would it take Sammy by himself to make one kaleidoscope?

3. How long would it take Salina and Sammy working together to make one kaleidoscope? How long for all three working together?

Now Beorg wants to finish his time table. He would like to have 20 kaleidoscopes completed each day.

4. Finish the table above, and find the best combination or combinations of employees to use to have 20 kaleidoscopes at the end of an eight-hour day.

Extension: Is Sammy in the combination or combinations you found in the last question? Is it worth having Sammy work? Remember that when someone is starting a new job, he or she may work more slowly until he or she learns how to do the job more efficiently. Find out how fast Sammy would need to be for production to double (40 kaleidoscopes in an eight-hour day).

Wrap-up   CHAPTER 4

SUMMARY

Rational Expressions		Examples
Rational expression	The ratio of two polynomials with the denominator not equal to 0	$\dfrac{x^2-1}{2x-3}$
Domain of a rational expression	The set of all possible numbers that can be used as replacements for the variable	$D=\left\{x \mid x \neq \dfrac{3}{2}\right\}$

Operations with Rational Expressions

		Examples
Basic principle of rational numbers	If $\frac{a}{b}$ is a rational number and c is a nonzero real number, then $$\frac{a}{b} = \frac{ac}{bc}.$$	Used for reducing: $$\frac{14}{16} = \frac{2 \cdot 7}{2 \cdot 8} = \frac{7}{8}$$ Used for building: $$\frac{2}{x} = \frac{2 \cdot 3}{x \cdot 3} = \frac{6}{3x}$$
Multiplication of rational numbers	If $\frac{a}{b}$ and $\frac{c}{d}$ are rational numbers, then $$\frac{a}{b} \cdot \frac{c}{d} = \frac{ac}{bd}.$$	$$\frac{3}{x} \cdot \frac{6}{x^2} = \frac{18}{x^3}$$
Division of rational numbers	If $\frac{a}{b}$ and $\frac{c}{d}$ are rational numbers with $\frac{c}{d} \neq 0$, then $$\frac{a}{b} \div \frac{c}{d} = \frac{a}{b} \cdot \frac{d}{c}. \text{ (Invert and multiply.)}$$	$$\frac{a}{x} \div \frac{5}{4x} = \frac{a}{x} \cdot \frac{4x}{5} = \frac{4a}{5}$$
Least common multiple	The LCM is the product of all of the different factors that appear in the polynomials. The exponent on each factor is the highest power that occurs on that factor in any of the polynomials.	$4a^3b, 6ab^2$ LCM $= 12a^3b^2$
Least common denominator	The LCD for a group of denominators is the LCM of the denominators.	$$\frac{1}{4a^3b} + \frac{1}{6ab^2}$$ LCD $= 12a^3b^2$
Addition and subtraction	If $b \neq 0$, then $$\frac{a}{b} + \frac{c}{b} = \frac{a+c}{b} \text{ and } \frac{a}{b} - \frac{c}{b} = \frac{a-c}{b}$$ If the denominators are not identical, we must build up each fraction to an equivalent fraction with the LCD as denominator.	$$\frac{2x}{x-3} + \frac{7x}{x-3} = \frac{9x}{x-3}$$ $$\frac{1}{2} + \frac{1}{x} = \frac{x}{2x} + \frac{2}{2x} = \frac{x+2}{2x}$$
Rules for adding and subtracting simple fractions	If $b \neq 0$ and $d \neq 0$, then $$\frac{a}{b} + \frac{c}{d} = \frac{ad+bc}{bd}$$ and $$\frac{a}{b} - \frac{c}{d} = \frac{ad-bc}{bd}.$$	$$\frac{1}{2} + \frac{1}{3} = \frac{5}{6}$$ $$\frac{2}{5} - \frac{3}{7} = \frac{-1}{35}$$
Simplifying complex fractions	Multiply the numerator and denominator by the LCD.	$$\frac{\left(\frac{1}{2}+\frac{1}{x}\right)6x}{\left(\frac{1}{x}-\frac{1}{3}\right)6x} = \frac{3x+6}{6-2x}$$

Equations with Rational Expressions		Examples

Solving equations with rational expressions	Multiply each side by the LCD to eliminate all denominators.	$\dfrac{1}{x} - \dfrac{1}{3} = \dfrac{1}{2x} - \dfrac{1}{6}$
		$6x\left(\dfrac{1}{x} - \dfrac{1}{3}\right) = 6x\left(\dfrac{1}{2x} - \dfrac{1}{6}\right)$
		$6 - 2x = 3 - x$
Solving proportions by the extremes-means property	If $\dfrac{a}{b} = \dfrac{c}{d}$, then $ad = bc$.	$\dfrac{2}{x-3} = \dfrac{5}{6}$
		$12 = 5x - 15$

REVIEW EXERCISES

4.1 *State the domain of each rational expression.*

1. $\dfrac{5-x}{3x-3}$

2. $\dfrac{x-4}{x^2-25}$

3. $\dfrac{x}{x^2-x-2}$

4. $\dfrac{1}{x^3-x^2}$

Reduce each rational expression to lowest terms.

5. $\dfrac{a^3bc^3}{a^5b^2c}$

6. $\dfrac{x^4-1}{3x^2-3}$

7. $\dfrac{68x^3}{51xy}$

8. $\dfrac{5x^2-15x+10}{5x-10}$

4.2 *Perform the indicated operations.*

9. $\dfrac{a^3b^2}{b^3a} \cdot \dfrac{ab-b^2}{ab-a^2}$

10. $\dfrac{x^3-1}{3x} \cdot \dfrac{6x^2}{x-1}$

11. $\dfrac{w-4}{3w} \div \dfrac{2w-8}{9w}$

12. $\dfrac{x^3-xy^2}{y} \div \dfrac{x^3+2x^2y+xy^2}{3y}$

4.3 *Find the least common multiple for each group of polynomials.*

13. $6x,\ 3x-6,\ x^2-2x$

14. $x^3-8,\ x^2-4,\ 2x+8$

15. $6ab^3,\ 4a^5b^2$

16. $4x^2-9,\ 4x^2+12x+9$

Perform the indicated operations.

17. $\dfrac{3}{2x-6} + \dfrac{1}{x^2-9}$

18. $\dfrac{3}{x-3} - \dfrac{5}{x+4}$

19. $\dfrac{w}{ab^2} - \dfrac{5}{a^2b}$

20. $\dfrac{x}{x-1} + \dfrac{3x}{x^2-1}$

4.4 *Simplify the complex fractions.*

21. $\dfrac{\dfrac{3}{2x} - \dfrac{4}{5x}}{\dfrac{1}{3} - \dfrac{2}{x}}$

22. $\dfrac{\dfrac{5}{x-2} - \dfrac{4}{4-x^2}}{\dfrac{3}{x+2} - \dfrac{1}{2-x}}$

23. $\dfrac{\dfrac{1}{y-2} - 3}{\dfrac{5}{y-2} + 4}$

24. $\dfrac{\dfrac{a}{b^2} - \dfrac{b}{a^3}}{\dfrac{a}{b} + \dfrac{b}{a^2}}$

25. $\dfrac{a^{-2} - b^{-3}}{a^{-1}b^{-2}}$

26. $p^{-1} + pq^{-2}$

4.5 *Solve each equation.*

27. $\dfrac{-3}{8} = \dfrac{2}{x}$

28. $\dfrac{2}{x} + \dfrac{5}{2x} = 1$

29. $\dfrac{15}{a^2-25} + \dfrac{1}{a-5} = \dfrac{6}{a+5}$

30. $2 + \dfrac{3}{x-5} = \dfrac{x-1}{x-5}$

4.6 *Solve each formula for the indicated variable.*

31. $\dfrac{y-b}{m} = x$ for y

32. $\dfrac{2A}{h} = b_1 + b_2$ for A

33. $F = \dfrac{mv^2}{r}$ for m

34. $P = \dfrac{A}{1+rt}$ for r

35. $A = \dfrac{2}{3}\pi rh$ for r

36. $\dfrac{a}{w^2} = \dfrac{2}{b}$ for b

37. $\dfrac{y+3}{x-7} = 2$ for y

38. $\dfrac{y-5}{x+4} = \dfrac{-1}{2}$ for y

Miscellaneous

Either perform the indicated operation or solve the equation, whichever is appropriate.

39. $\dfrac{1}{x} + \dfrac{1}{3x}$

40. $\dfrac{1}{y} + \dfrac{3}{2y} = 5$

41. $\dfrac{5}{3xy} + \dfrac{7}{6x}$

42. $\dfrac{2}{x-2} - \dfrac{3}{x}$

43. $\dfrac{5}{a-5} - \dfrac{3}{-a-5}$

44. $\dfrac{2}{x-2} - \dfrac{3}{x} = \dfrac{-1}{5x}$

45. $\dfrac{1}{x-2} - \dfrac{1}{x+2} = \dfrac{1}{15}$

46. $\dfrac{2}{x-3} \cdot \dfrac{6x-18}{30}$

47. $\dfrac{-3}{x+2} \cdot \dfrac{5x+10}{10}$

48. $\dfrac{x}{10} = \dfrac{10}{x}$

49. $\dfrac{x}{-3} = \dfrac{-27}{x}$

50. $\dfrac{x^2-4}{x} \div \dfrac{x^3-8}{x}$

51. $\dfrac{wx+wm+3x+3m}{w^2-9} \div \dfrac{x^2-m^2}{w-3}$

52. $\dfrac{-5}{7} = \dfrac{3}{x}$

53. $\dfrac{5}{a^2-25} + \dfrac{3}{a^2-4a-5}$

54. $\dfrac{3}{w^2-1} + \dfrac{2}{2w+2}$

55. $\dfrac{-7}{2a^2-18} - \dfrac{4}{a^2+5a+6}$

56. $\dfrac{-5}{3a^2-12} - \dfrac{1}{a^2-3a+2}$

57. $\dfrac{7}{a^2-1} + \dfrac{2}{1-a} = \dfrac{1}{a+1}$

58. $2 + \dfrac{4}{x-1} = \dfrac{3x+1}{x-1}$

59. $\dfrac{2x}{x-3} + \dfrac{3}{x-2} = \dfrac{6}{(x-2)(x-3)}$

60. $\dfrac{a-3}{a+3} \div \dfrac{9-a^2}{3}$

61. $\dfrac{x-2}{6} \div \dfrac{2-x}{2}$

62. $\dfrac{x}{x+4} - \dfrac{2}{x+1} = \dfrac{-2}{(x+1)(x+4)}$

63. $\dfrac{x-3}{x^2+3x+2} \cdot \dfrac{x^2-4}{3x-9}$

64. $\dfrac{x^2-1}{x^2+2x+1} \cdot \dfrac{x^3+1}{2x-2}$

65. $\dfrac{a+4}{a^3-8} - \dfrac{3}{2-a}$

66. $\dfrac{x+2}{5} = \dfrac{3}{x}$

67. $\dfrac{x^3-9x}{1-x^2} \div \dfrac{x^3+6x^2+9x}{x-1}$

68. $\dfrac{x+3}{2x+3} = \dfrac{x-3}{x-1}$

69. $\dfrac{a^2+3a+3w+aw}{a^2+6a+8} \cdot \dfrac{a^2-aw-2w+2a}{a^2+3a-3w-aw}$

70. $\dfrac{3}{4-2y} + \dfrac{6}{y^2-4} + \dfrac{3}{2+y}$

71. $\dfrac{5}{x} - \dfrac{4}{x+2} = \dfrac{1}{5} + \dfrac{1}{5x}$

72. $\dfrac{1}{x} + \dfrac{1}{x-5} = \dfrac{2x+1}{x^2-25} + \dfrac{9}{x^2+5x}$

Replace each question mark by an expression that makes the equation an identity.

73. $\dfrac{6}{x} = \dfrac{?}{3x}$

74. $\dfrac{?}{a} = \dfrac{8}{4a}$

75. $\dfrac{3}{a-b} = \dfrac{?}{b-a}$

76. $\dfrac{-2}{a-x} = \dfrac{2}{?}$

77. $4 = \dfrac{?}{x}$

78. $5a = \dfrac{?}{b}$

79. $5x \div \dfrac{1}{2} = ?$

80. $3a \div \dfrac{1}{a} = ?$

81. $4a \div ? = 12a$

82. $14x \div ? = 28x^2$

83. $\dfrac{a-3}{a^2-9} = \dfrac{1}{?}$

84. $\dfrac{?}{x^2-4} = \dfrac{1}{x-2}$

85. $\dfrac{1}{2} - \dfrac{1}{5} = ?$

86. $\dfrac{1}{4} - \dfrac{1}{5} = ?$

87. $\dfrac{a}{3} + \dfrac{a}{2} = ?$

88. $\dfrac{x}{5} + \dfrac{x}{3} = ?$

89. $\dfrac{1}{a} - \dfrac{1}{b} = ?$

90. $\dfrac{3}{w} - \dfrac{2}{b} = ?$

91. $\dfrac{a}{3} - 1 = ?$

92. $\dfrac{x}{y} - 1 = ?$

93. $2 + \dfrac{1}{a} = ?$

94. $3 - \dfrac{1}{x} = ?$

95. $\dfrac{a}{5} - ? = \dfrac{a-5}{5}$

96. $? - \dfrac{1}{y} = \dfrac{y^2-1}{y}$

97. $(x-1) \div (-1) = ?$

98. $(a-3) \div (3-a) = ?$

99. $(m-2) \div (2-m) = ?$

100. $(x-2) \div (2x-4) = ?$

101. $\dfrac{\dfrac{b}{3a}}{2} = ?$

102. $\dfrac{a-3}{\dfrac{1}{2}} = ?$

103. $\dfrac{a-6}{5} \cdot \dfrac{10}{6-a} = ?$

104. $\dfrac{3}{w-2} + \dfrac{2}{2-w} = ?$

Perform the indicated operations. Variables in exponents represent integers.

105. $\dfrac{2}{x^a - 2} + \dfrac{1}{x^a + 3}$ **106.** $\dfrac{x^k + 1}{x^{k+1} - x} - \dfrac{x^{k-1}}{x^k - 1}$

107. $\dfrac{x^{2k} - 9}{3x^{k+3} + 9x^3} \cdot \dfrac{6x^{k+5}}{x^{k+2} - 3x^2}$

108. $\dfrac{y^{2a} - y^a - 12}{y^{2a} - 4y^a} \div \dfrac{y^{3a} + 3y^{2a}}{y^{6a}}$

Solve each problem.

109. *AIDS.* The ratio of U.S. AIDS deaths in the 40–49 age group to AIDS deaths in the 50–59 age group was 2.875 to 1 for the years 1982 through 1991 (U.S. Bureau of the Census). If there were 19,024 more deaths in the 40–49 age group than in the 50–59 age group, then how many deaths were there in the 50–59 age group?

110. *AIDS.* The ratio of U.S. AIDS deaths in the 30–39 age group to AIDS deaths in the 50–59 age group was 45 to 8 for the years 1982 through 1991 (U.S. Bureau of the Census). If there were 46,926 fewer deaths in the 50–59 age group than in the 30–39 age group, then how many deaths were there in the 30–39 age group? Use the pie chart to estimate the total number of AIDS deaths for the years 1982 through 1991.

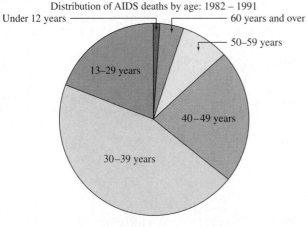

Figure for Exercise 110

111. *Just passing through.* Nikita drove 310 miles on his way to Louisville in the same amount of time that he drove 360 miles after passing through Louisville. If his average speed after passing Louisville was 10 mph more than his average speed on his way to Louisville, then for how many hours did he drive?

Figure for Exercise 111

112. *Pushing a barge.* A tug can push a barge 144 miles down the Mississippi River in the same time that it takes to push the barge 84 miles in the Gulf of Mexico. If the tug's speed is 5 mph greater going down the river, then what is its speed in the Gulf of Mexico?

113. *Quilting bee.* Debbie can make a hand-sewn quilt in 2000 hours, and Rosalina can make an identical quilt in 1000 hours. If Cheryl works just as fast as Rosalina, then how long will it take all three of them working together to make one quilt?

114. *Blood out of a turnip.* A small pump can pump all of the blood out of an average turnip in 30 minutes. A larger pump can pump all of the blood from the same turnip in 20 minutes. If both pumps are hooked to the turnip, then how long would it take to get all of the blood out?

CHAPTER 4 TEST

State the domain of each rational expression.

1. $\dfrac{2x-1}{x^2-9}$

2. $\dfrac{5}{4-3x}$

Reduce to lowest terms.

3. $\dfrac{12a^9b^8}{(2a^2b^3)^3}$

4. $\dfrac{y^2-x^2}{2x^2-4xy+2y^2}$

Perform the indicated operations. Write answers in lowest terms.

5. $\dfrac{5}{12}-\dfrac{4}{9}$

6. $\dfrac{3}{y}+7y$

7. $\dfrac{4}{a-9}-\dfrac{1}{9-a}$

8. $\dfrac{1}{6ab^2}+\dfrac{1}{8a^2b}$

9. $\dfrac{3a^3b}{20ab}\cdot\dfrac{2a^2b}{9ab^3}$

10. $\dfrac{a-b}{7}\div\dfrac{b^2-a^2}{21}$

11. $\dfrac{x-3}{x-1}\div(x^2-2x-3)$

12. $\dfrac{2}{x^2-4}-\dfrac{6}{x^2-3x-10}$

13. $\dfrac{m^3-1}{(m-1)^2}\cdot\dfrac{m^2-1}{3m^2+3m+3}$

Find the solution set to each equation.

14. $\dfrac{3}{x}=\dfrac{7}{4}$

15. $\dfrac{x}{x-2}-\dfrac{5}{x}=\dfrac{3}{4}$

16. $\dfrac{3m}{2}=\dfrac{6}{m}$

Solve each formula for the indicated variable.

17. $W=\dfrac{a^2}{t}$ for t

18. $\dfrac{1}{a}+\dfrac{1}{b}=\dfrac{1}{2}$ for b

Simplify.

19. $\dfrac{\dfrac{1}{x}+\dfrac{1}{3x}}{\dfrac{3}{4x}-\dfrac{1}{2}}$

20. $\dfrac{m^{-2}-w^{-2}}{m^{-2}w^{-1}+m^{-1}w^{-2}}$

21. $\dfrac{\dfrac{a^2b^3}{4a}}{\dfrac{ab^3}{6a^2}}$

Solve each problem.

22. When Jane's wading pool was new, it could be filled in 6 minutes with water from the hose. Now that the pool has several leaks, Jane can play only 8 minutes before the water in the pool is all gone. How long will it take to fill the leaky pool?

23. Milton and Bonnie are hiking the Appalachian Trail together. Milton averages 4 miles per hour, and Bonnie averages 3 miles per hour. If they start out together in the morning, but Milton gets to camp 2 hours and 30 minutes ahead of Bonnie, then how many miles did they hike that day?

24. A group of sailors plans to share equally the cost and use of a $72,000 boat. If they can get three more sailors to join their group, then the cost per person will be reduced by $2000. How many sailors are in the original group?

Chapters 1–4. Find the solution set to each equation.

1. $\dfrac{3}{x} = \dfrac{4}{5}$

2. $\dfrac{2}{x} = \dfrac{x}{8}$

3. $\dfrac{x}{3} = \dfrac{4}{5}$

4. $\dfrac{3}{x} = \dfrac{x+3}{6}$

5. $\dfrac{1}{x} = 4$

6. $\dfrac{2}{3}x = 4$

7. $2x + 3 = 4$

8. $2x + 3 = 4x$

9. $\dfrac{2a}{3} = \dfrac{6}{a}$

10. $\dfrac{12}{x} - \dfrac{14}{x+1} = \dfrac{1}{2}$

11. $|6x - 3| = 1$

12. $\dfrac{x}{2x+9} = \dfrac{3}{x}$

13. $4(6x - 3)(2x + 9) = 0$

14. $\dfrac{x-1}{x+2} - \dfrac{1}{5(x+2)} = 1$

Solve each equation for y. Assume that A, B, and C are constants for which all expressions are defined.

15. $Ax + By = C$

16. $\dfrac{y-3}{x+5} = -\dfrac{1}{3}$

17. $Ay = By + C$

18. $\dfrac{A}{y} = \dfrac{y}{A}$

19. $\dfrac{A}{y} - \dfrac{1}{2} = \dfrac{B}{y}$

20. $\dfrac{A}{y} - \dfrac{1}{2} = \dfrac{B}{C}$

21. $3x - 4y = 6$

22. $y^2 - 2y - Ay + 2A = 0$

23. $A = \dfrac{1}{2}B(C + y)$

24. $y^2 + Cy = BC + By$

Simplify each expression.

25. $3x^5 \cdot 4x^8$

26. $3x^2(x^3 + 5x^6)$

27. $(5x^6)^2$

28. $(3a^3b^2)^3$

29. $\dfrac{12a^9b^4}{-3a^3b^{-2}}$

30. $\left(\dfrac{x^{-2}}{2}\right)^5$

31. $\left(\dfrac{2x^{-4}}{3y^5}\right)^{-3}$

32. $(-2a^{-1}b^3c)^{-2}$

33. $\dfrac{a^{-1} + b^3}{a^{-2} + b^{-1}}$

34. $\dfrac{(a+b)^{-1}}{(a+b)^{-2}}$

Solve each problem.

35. **Negative income tax.** In a negative income tax proposal, the formula $D = 0.75E + 5000$ gives the disposable income D for an earned income of E dollars. (Disposable income is the amount that you have available to spend.) If $E > D$, then the difference is paid in federal taxes. If $D > E$, then the difference is paid to the wage earner by the federal government. Find the amount of tax paid by a person who earns $100,000. Find the amount received from the government by a person who earns $10,000.

Rational Exponents and Radicals

J ust how cold is it in Fargo, North Dakota, in winter? According to local meteorologists, the mercury hit a low of –33°F on January 18, 1994. But air temperature alone is not always a reliable indicator of how cold you feel. On the same date, the average wind velocity was 13.8 miles per hour. This dramatically affected how cold people felt when they stepped outside. High winds along with cold temperatures make exposed skin feel colder because the wind significantly speeds up the loss of body heat. Meteorologists use the terms "wind chill factor," "wind chill index," and "wind chill temperature" to take both air temperature and wind velocity into account.

Through experimentation in Antarctica, Paul A. Siple developed a formula in the 1940s that measures the wind chill from the velocity of the wind and the air temperature. His complex formula involving the square root of the velocity of the wind is still used today to calculate wind chill temperatures. Siple's formula is unlike most scientific formulas in that it is not based on theory. Siple experimented with various formulas involving wind velocity and temperature until he found a formula that seemed to predict how cold the air felt. His formula is stated and used in Exercises 101 and 102 of Section 5.2.

Wind chill temperature (°F) for 25°F air temperature

Wind velocity (mph)

Rational Exponents

In Sections 3.1 and 3.2 you studied integral exponents. In this section we extend the idea of exponent to include rational numbers as exponents.

Roots

If a square has a side of length 7 feet, then it has an area of 7^2, or 49 square feet. If a square has an area of 36 square feet, then the length of its side is 6 feet. If we know the length of a side, then we square it to find the area. If we know the area, then we must reverse the process of squaring to find the length of a side. Reversing the process of squaring is called **taking the square root.**

Because $3^2 = 9$, $2^3 = 8$, and $(-2)^4 = 16$, we say that 3 is a square root of 9, 2 is the cube root of 8, and -2 is a fourth root of 16. In general, reversing an nth power is referred to as **taking an nth root.**

> **nth Roots**
>
> The number a is an **nth root** of b if
>
> $$a^n = b.$$

Both 3 and -3 are **square roots** of 9 because $3^2 = 9$ and $(-3)^2 = 9$. Because $2^4 = 16$ and $(-2)^4 = 16$, there are two real fourth roots of 16: 2 and -2. If n is a positive even integer and a is any positive real number, then there are two real nth roots of a. We call these roots **even roots.** The positive even root of a positive number is called the **principal root.** The principal square root of 9 is 3, and the principal fourth root of 16 is 2. When n is even, the exponent $1/n$ is used to indicate the principal nth root. The principal nth root can also be indicated by the radical symbol $\sqrt[n]{}$, which will be discussed in Section 5.2.

> **Exponent 1/n When n Is Even**
>
> If n is a positive even integer and a is a positive real number, then $a^{1/n}$ denotes the positive real nth root of a and is called the **principal nth root of a.**

CAUTION   An exponent of n indicates nth power, an exponent of $-n$ indicates the reciprocal of the nth power, and an exponent of $1/n$ indicates nth root.   ⊘

We will see later that choosing $1/n$ to indicate nth root fits in nicely with the rules of exponents that we have already studied.

EXAMPLE 1   **Finding even roots**

Evaluate each expression.

a) $4^{1/2}$ b) $16^{1/4}$ c) $-81^{1/4}$ d) $\left(\dfrac{1}{4}\right)^{1/2}$

Solution

a) Because $2^2 = 4$, we have $4^{1/2} = 2$. Note that $4^{1/2} \neq -2$.

b) Because $2^4 = 16$, we have $16^{1/4} = 2$. Note that $16^{1/4} \neq -2$.

c) Following the accepted order of operations from Chapter 1, we find the root first and then take the opposite of it. Because $3^4 = 81$, we have $81^{1/4} = 3$ and $-81^{1/4} = -3$.

d) Because $\left(\dfrac{1}{2}\right)^2 = \dfrac{1}{4}$, we have $\left(\dfrac{1}{4}\right)^{1/2} = \dfrac{1}{2}$.   ■

Note that $2^3 = 8$ but $(-2)^3 = -8$. The **cube root** of 8 is 2, and the cube root of -8 is -2. If n is a positive odd integer and a is any real number, then there is only one real nth root of a. We call this root an **odd root.**

> **Exponent 1/n When n Is Odd**
>
> If n is a positive odd integer and a is any real number, then $a^{1/n}$ denotes the real nth root of a.

EXAMPLE 2   **Finding odd roots**

Evaluate each expression.

a) $8^{1/3}$ b) $(-27)^{1/3}$ c) $-32^{1/5}$

Solution

a) Because $2^3 = 8$, we have $8^{1/3} = 2$.

b) Because $(-3)^3 = -27$, we have $(-27)^{1/3} = -3$.

c) Because $2^5 = 32$, we have $-32^{1/5} = -2$.   ■

We do not allow 0 as the base when we use negative exponents because division by zero is undefined. However, positive powers of zero are defined, and so are roots of zero; for example, $0^4 = 0$, and so $0^{1/4} = 0$.

> **nth Root of Zero**
>
> If n is a positive integer, then $0^{1/n} = 0$.

CAUTION   An expression such as $(-9)^{1/2}$ is not included in the definition of roots because there is no real number whose square is -9. The definition of roots does not include an even root of any negative number because no even power of a real number is negative. However, in Section 5.6 we will define expressions such as $(-9)^{1/2}$ when we study complex numbers.   ⊘

The expression $3^{1/2}$ represents the unique positive real number whose square is 3. Because there is no rational number that has a square equal to 3, the number $3^{1/2}$ is an irrational number. If we use a calculator, we find that $3^{1/2}$ is approximately equal to the rational number 1.732. Because the square root of 3 is not a rational number, the simplest representation for the exact value of the square root of 3 is $3^{1/2}$.

Positive Rational Exponents

When a rational number is used as an exponent, the denominator indicates "root" and the numerator indicates "power." For example, $8^{2/3}$ means $(8^{1/3})^2$. We take the cube root of 8 to get 2, then square 2 to get 4. Thus

$$8^{2/3} = 4.$$

In general, $a^{m/n}$ is the mth power of the nth root of a.

Rational Exponents

If m and n are positive integers, then

$$a^{m/n} = (a^{1/n})^m,$$

provided that $a^{1/n}$ is defined.

EXAMPLE 3   **Evaluating expressions with positive rational exponents**
Evaluate each expression.

a) $16^{3/4}$ b) $27^{4/3}$ c) $(-8)^{2/3}$ d) $-4^{3/2}$

Solution

a) To evaluate $16^{3/4}$, take the fourth root of 16 to get 2 and then cube 2 to get 8. These steps are written as follows:

$$16^{3/4} = (16^{1/4})^3 = 2^3 = 8$$

b) To evaluate $27^{4/3}$, take the cube root of 27 to get 3, and then raise 3 to the fourth power to get 81:

$$27^{4/3} = (27^{1/3})^4 = 3^4 = 81$$

c) $(-8)^{2/3} = [(-8)^{1/3}]^2 = (-2)^2 = 4$
d) $-4^{3/2} = -(4^{1/2})^3 = -2^3 = -8$ ■

CAUTION   We can evaluate $-4^{3/2}$ because the negative sign is saved until last. But $(-4)^{3/2}$ is not a real number because $(-4)^{1/2}$ is not a real number. ⊘

By the definition of rational exponents we find the root and then the power. For example,

$$8^{2/3} = (8^{1/3})^2 = 2^2 = 4.$$

However, we get the same result if we find the power and then the root:

$$8^{2/3} = (8^2)^{1/3} = 64^{1/3} = 4$$

In general, $a^{m/n}$ is also the nth root of the mth power of a.

> **Evaluating $a^{m/n}$ in Either Order**
>
> If m and n are positive integers, then
>
> $$a^{m/n} = (a^{1/n})^m = (a^m)^{1/n},$$
>
> provided that $a^{1/n}$ is defined.

The fact that the power can be found before the root will be used later in this chapter. It is not useful in simply evaluating expressions because if the power is evaluated first, we might have to find the root of a very large number. For example, finding the power first in Example 3(b), we get

$$27^{4/3} = (27^4)^{1/3} = 531{,}441^{1/3} = 81.$$

Negative Rational Exponents

Negative integral exponents were defined by using reciprocals, and so are negative rational exponents. For example, $8^{-2/3}$ is the reciprocal of $8^{2/3}$. So

$$8^{-2/3} = \frac{1}{8^{2/3}} = \frac{1}{4}.$$

> **Negative Rational Exponents**
>
> If m and n are positive integers, then
>
> $$a^{-m/n} = \frac{1}{a^{m/n}},$$
>
> provided that $a^{1/n}$ is defined and nonzero.

Three operations are involved in evaluating the expression $a^{-m/n}$. The operations (root, power, reciprocal) can be performed in any order, but the simplest way to evaluate $a^{-m/n}$ is usually the following order.

> **Evaluating $a^{-m/n}$**
>
>
>
> (1) Take the nth root.
> (2) Raise it to the mth power.
> (3) Find the reciprocal.

EXAMPLE 4   **Evaluating expressions with negative rational exponents**

Evaluate each expression.

a) $4^{-3/2}$ **b)** $(-27)^{-1/3}$ **c)** $(-16)^{-3/4}$

Solution

a) The square root of 4 is 2. The cube of 2 is 8. The reciprocal of 8 is $\frac{1}{8}$. So

$$4^{-3/2} = \frac{1}{8}.$$

b) The cube root of -27 is -3. The first power of -3 is -3. The reciprocal of -3 is $-\frac{1}{3}$. So

$$(-27)^{-1/3} = -\frac{1}{3}.$$

c) The expression $(-16)^{-3/4}$ is not a real number because it involves the fourth root (an even root) of a negative number.   ■

Using the Rules of Exponents

All of the rules for exponents hold for rational exponents as well as integral exponents. Of course, we cannot apply the rules of exponents to expressions that are not real numbers.

Rules for Rational Exponents

The following rules hold for any nonzero real numbers a and b and rational numbers r and s for which the expressions represent real numbers.

1. $a^r a^s = a^{r+s}$ Product rule

2. $\dfrac{a^r}{a^s} = a^{r-s}$ Quotient rule

3. $(a^r)^s = a^{rs}$ Power of a power rule

4. $(ab)^r = a^r b^r$ Power of a product rule

5. $\left(\dfrac{a}{b}\right)^r = \dfrac{a^r}{b^r}$ Power of a quotient rule

---

We can use the product rule to add rational exponents. For example,

$$16^{1/4} \cdot 16^{1/4} = 16^{2/4}.$$

The fourth root of 16 is 2, and 2 squared is 4. So $16^{2/4} = 4$. Because we also have $16^{1/2} = 4$, we see that rational exponents can be reduced to lowest terms. If an exponent can be reduced, it is usually simpler to reduce it before we evaluate the expression. We can simplify $16^{1/4} \cdot 16^{1/4}$ as follows:

$$16^{1/4} \cdot 16^{1/4} = 16^{2/4} = 16^{1/2} = 4$$

EXAMPLE 5   **Using the product and quotient rules with rational exponents**

Simplify each expression.

a) $27^{1/6} \cdot 27^{1/2}$ **b)** $\dfrac{5^{3/4}}{5^{1/4}}$

Solution

a) $27^{1/6} \cdot 27^{1/2} = 27^{1/6 \,+\, 1/2}$ Product rule for exponents

$= 27^{2/3}$

$= 9$

b) $\dfrac{5^{3/4}}{5^{1/4}} = 5^{3/4 \,-\, 1/4} = 5^{2/4} = 5^{1/2}$ We used the quotient rule to subtract the exponents.

EXAMPLE 6   **Using the power rules with rational exponents**

Simplify each expression.

a) $3^{1/2} \cdot 12^{1/2}$ **b)** $(3^{10})^{1/2}$ **c)** $\left(\dfrac{2^6}{3^9}\right)^{-1/3}$

Solution

a) Because the bases 3 and 12 are different, we cannot use the product rule to add the exponents. Instead, we use the power of a product rule to place the 1/2 power outside the parentheses:

$$3^{1/2} \cdot 12^{1/2} = (3 \cdot 12)^{1/2} = 36^{1/2} = 6$$

b) Use the power of a power rule to multiply the exponents:

$$(3^{10})^{1/2} = 3^5$$

c) $\left(\dfrac{2^6}{3^9}\right)^{-1/3} = \dfrac{(2^6)^{-1/3}}{(3^9)^{-1/3}}$ Power of a quotient rule

$= \dfrac{2^{-2}}{3^{-3}}$ Power of a power rule

$= \dfrac{3^3}{2^2}$ Definition of negative exponent

$= \dfrac{27}{4}$

Simplifying Expressions Involving Variables

When simplifying expressions involving rational exponents and variables, we must be careful to write equivalent expressions. For example, in the equation

$$(x^2)^{1/2} = x$$

it looks as if we are correctly applying the power of a power rule. However, this statement is false if x is negative because the 1/2 power on the left side indicates the positive square root of x^2. For example, if $x = -3$, we get

$$[(-3)^2]^{1/2} = 9^{1/2} = 3,$$

which is not equal to -3. To write a simpler equivalent expression for $(x^2)^{1/2}$, we use absolute value as follows.

Square Root of x^2

$$(x^2)^{1/2} = |x| \text{ for any real number } x.$$

Note that the equation $(x^2)^{1/2} = |x|$ is an identity. It is also necessary to use absolute value when writing identities for other even roots of expressions involving variables.

EXAMPLE 7   **Using absolute value symbols with exponents**

Simplify each expression. Assume that the variables represent any real numbers and use absolute value symbols as necessary.

a) $(x^8 y^4)^{1/4}$

b) $\left(\dfrac{x^9}{8}\right)^{1/3}$

Solution

a) Apply the power of a product rule to get the equation $(x^8 y^4)^{1/4} = x^2 y$. The left side is nonnegative for any choices of x and y, but the right side is negative when y is negative. So for any real values of x and y we have

$$(x^8 y^4)^{1/4} = x^2 |y|.$$

b) Using the power of a quotient rule, we get

$$\left(\frac{x^9}{8}\right)^{1/3} = \frac{x^3}{2}.$$

This equation is valid for every real number x, so no absolute value signs are used. ∎

Because there are no real even roots of negative numbers, the expressions

$$a^{1/2}, \quad x^{-3/4}, \quad \text{and} \quad y^{1/6}$$

are not real numbers if the variables have negative values. To simplify matters, we sometimes assume that the variables represent only positive numbers when we are working with expressions involving variables with rational exponents. That way we do not have to be concerned with undefined expressions and absolute value.

EXAMPLE 8   **Expressions involving variables with rational exponents**

Use the rules of exponents to simplify the following. Write your answers with positive exponents. Assume that all variables represent *positive* real numbers.

a) $x^{2/3} x^{4/3}$

b) $\dfrac{a^{1/2}}{a^{1/4}}$

c) $(x^{1/2} y^{-3})^{1/2}$

d) $\left(\dfrac{x^2}{y^{1/3}}\right)^{-1/2}$

Solution

a) $x^{2/3}x^{4/3} = x^{6/3}$ Use the product rule to add the exponents.

$\qquad\qquad = x^2$ Reduce the exponent.

b) $\dfrac{a^{1/2}}{a^{1/4}} = a^{1/2 - 1/4}$ Use the quotient rule to subtract the exponents.

$\qquad\quad = a^{1/4}$ Simplify.

c) $(x^{1/2}y^{-3})^{1/2} = (x^{1/2})^{1/2}(y^{-3})^{1/2}$ Power of a product rule

$\qquad\qquad\quad = x^{1/4}y^{-3/2}$ Power of a power rule

$\qquad\qquad\quad = \dfrac{x^{1/4}}{y^{3/2}}$ Definition of negative exponent

d) Because this expression is a negative power of a quotient, we can first find the reciprocal of the quotient, then apply the power of a power rule:

$$\left(\frac{x^2}{y^{1/3}}\right)^{-1/2} = \left(\frac{y^{1/3}}{x^2}\right)^{1/2} = \frac{y^{1/6}}{x} \qquad \frac{1}{3} \cdot \frac{1}{2} = \frac{1}{6} \qquad \blacksquare$$

Calculator Close-up

Expressions with rational exponents are evaluated on a calculator in the same way as expressions with integral exponents. To evaluate $10^{1/3}$ on a scientific calculator, try the following keystrokes:

$$10 \;\boxed{x^y}\;\boxed{(}\;1\;\boxed{\div}\;3\;\boxed{)}\;\boxed{=}$$

The display will read

$$\boxed{\;2.1544\;}$$

If your calculator has a $\boxed{x^{1/y}}$ key for finding the yth root, try the following:

$$10 \;\boxed{x^{1/y}}\;3\;\boxed{=}$$

Again, the display will read

$$\boxed{\;2.1544\;}$$

On a graphing calculator your display should look like the following:

$$10\wedge(1/3)\;\boxed{\text{ENTER}}$$

$$\boxed{\;2.1544\;}$$

Warm-ups

True or false? Explain.

1. $4^{-1/2} = \dfrac{1}{2}$ 2. $16^{1/2} = 8$ 3. $(3^{2/3})^3 = 9$

4. $8^{-2/3} = -4$ 5. $2^{1/2} \cdot 2^{1/2} = 2$ 6. $\left(\dfrac{1}{4}\right)^{1/2} = \dfrac{1}{2}$

7. $\dfrac{3}{3^{1/2}} = 3^{1/2}$ 8. $(2^9)^{1/2} = 2^3$ 9. $3^{1/3} \cdot 6^{1/3} = 18^{2/3}$

10. $2^{3/4} \cdot 2^{1/4} = 4$

5.1 EXERCISES

Simplify. Some of these expressions are not real numbers. See Examples 1 and 2.

1. $100^{1/2}$
2. $169^{1/2}$
3. $81^{1/4}$
4. $64^{1/6}$
5. $-9^{1/2}$
6. $-4^{1/2}$
7. $\left(\dfrac{1}{64}\right)^{1/6}$
8. $\left(-\dfrac{1}{8}\right)^{1/3}$
9. $(-25)^{1/2}$
10. $(-16)^{1/4}$
11. $1000^{1/3}$
12. $27^{1/3}$
13. $(-64)^{1/3}$
14. $(-32)^{1/5}$
15. $-1^{1/5}$
16. $-125^{1/3}$

Evaluate each expression. Some of these expressions are not real numbers. See Examples 3 and 4.

17. $32^{3/5}$
18. $25^{3/2}$
19. $(-27)^{2/3}$
20. $(-32)^{3/5}$
21. $-25^{3/2}$
22. $-100^{3/2}$
23. $(-25)^{3/2}$
24. $(-64)^{5/6}$
25. $4^{-1/2}$
26. $9^{-1/2}$
27. $8^{-4/3}$
28. $4^{-3/2}$
29. $(-32)^{-3/5}$
30. $(-27)^{-4/3}$
31. $(-9)^{-1/2}$
32. $(-4)^{-1/2}$

Use the rules of exponents to simplify each expression. See Examples 5 and 6.

33. $3^{1/3}3^{1/4}$
34. $2^{1/2}2^{1/3}$
35. $3^{1/3}3^{-1/3}$
36. $5^{1/4}5^{-1/4}$
37. $\dfrac{8^{1/3}}{8^{2/3}}$
38. $\dfrac{27^{-2/3}}{27^{-1/3}}$
39. $4^{3/4} \div 4^{1/4}$
40. $9^{1/4} \div 9^{3/4}$
41. $18^{1/2}2^{1/2}$
42. $8^{1/2}2^{1/2}$
43. $(2^6)^{1/3}$
44. $(3^{10})^{1/5}$
45. $(3^8)^{1/2}$
46. $(3^{-6})^{1/3}$
47. $(2^{-4})^{1/2}$
48. $(5^4)^{1/2}$
49. $\left(\dfrac{3^4}{2^6}\right)^{1/2}$
50. $\left(\dfrac{5^4}{3^6}\right)^{1/2}$
51. $\left(\dfrac{2^{-9}}{3^{-6}}\right)^{-1/3}$
52. $\left(\dfrac{3^{-6}}{5^{-3}}\right)^{-2/3}$

Simplify each expression. Assume that the variables represent any real numbers and use absolute value as necessary. See Example 7.

53. $(x^4)^{1/4}$
54. $(y^6)^{1/6}$
55. $(a^8)^{1/2}$
56. $(b^{10})^{1/2}$
57. $(y^3)^{1/3}$
58. $(w^9)^{1/3}$
59. $(9x^6y^2)^{1/2}$
60. $(16a^8b^4)^{1/4}$
61. $\left(\dfrac{81x^{12}}{y^{20}}\right)^{1/4}$
62. $\left(\dfrac{144a^8}{9y^{18}}\right)^{1/2}$

Simplify. Assume that all variables represent positive numbers. Write answers with positive exponents only. See Example 8.

63. $x^{1/2}x^{1/4}$
64. $y^{1/3}y^{1/3}$
65. $(x^{1/2}y)(x^{-3/4}y^{1/2})$
66. $(a^{1/2}b^{-1/3})(ab)$
67. $\dfrac{w^{1/3}}{w^3}$
68. $\dfrac{a^{1/2}}{a^2}$
69. $\dfrac{x^{1/2}y}{x}$
70. $\dfrac{x^{1/3}y^{-1/2}}{xy^{-1}}$
71. $(144x^{16})^{1/2}$
72. $(125a^8)^{1/3}$
73. $(4x^{-1/2}yz^{1/2})^{-1/2}$
74. $(9x^8y^{-10}z^{12})^{1/2}$
75. $\left(\dfrac{a^{-1/2}}{b^{-1/4}}\right)^{-4}$
76. $\left(\dfrac{2a^{1/2}}{b^{1/3}}\right)^{6}$
77. $\left(\dfrac{9x^{10}}{w^{12}}\right)^{-1/2}$
78. $\left(\dfrac{27a^6b^9}{c^{-12}}\right)^{-2/3}$

Simplify each expression. Write your answers with positive exponents. Assume that all variables represent positive real numbers.

79. $(9^2)^{1/2}$
80. $(4^{16})^{1/2}$
81. $-16^{-3/4}$
82. $-25^{-3/2}$
83. $125^{-4/3}$
84. $27^{-2/3}$
85. $2^{1/2}2^{-1/4}$
86. $9^{-1}9^{1/2}$
87. $3^{0.26}3^{0.74}$
88. $2^{1.5}2^{0.5}$
89. $3^{1/4}27^{1/4}$
90. $3^{2/3}9^{2/3}$
91. $\left(-\dfrac{8}{27}\right)^{2/3}$
92. $\left(-\dfrac{8}{27}\right)^{-1/3}$
93. $\left(-\dfrac{1}{16}\right)^{-3/4}$
94. $\left(\dfrac{9}{16}\right)^{-1/2}$
95. $(9x^9)^{1/2}$
96. $(-27x^9)^{1/3}$
97. $(3a^{-2/3})^{-3}$
98. $(5x^{-1/2})^{-2}$
99. $(a^{1/2}b)^{1/2}(ab^{1/2})$
100. $(m^{1/4}n^{1/2})^2(m^2n^3)^{1/2}$
101. $(km^{1/2})^3(k^3m^5)^{1/2}$
102. $(tv^{1/3})^2(t^2v^{-3})^{-1/2}$
103. $\left(\dfrac{w^{-3/4}a^{-2/3}}{w^2a^{-1/2}}\right)^6\left(\dfrac{w^{1/2}a^{-1/2}}{w^{-1/2}a^3}\right)^{-1/2}$
104. $\left(\dfrac{-2m^{-3/5}p^{-1/4}}{2^{-2}m^{-1/2}p^{1/3}}\right)^2\left(\dfrac{3m^{-2}p^4}{9^{-1}mp^{-2}}\right)^{1/3}$

Use a scientific calculator with a power key (x^y) to find the decimal value of each expression. Round answers to four decimal places.

105. $2^{1/3}$
106. $5^{1/2}$
107. $-2^{1/2}$
108. $(-3)^{1/3}$
109. $1024^{1/10}$
110. $7776^{0.2}$
111. $8^{0.33}$
112. $289^{0.5}$
113. $\left(\dfrac{64}{15,625}\right)^{-1/6}$

114. $\left(\dfrac{32}{243}\right)^{-3/5}$ **115.** $\left(\dfrac{5^{-2/3}}{6^{1/4}}\right)^{-1/2}$ **116.** $\left(\dfrac{5^{1/3}3^{1/2}}{(-6)^{1/3}}\right)^{1/3}$

Simplify each expression. Assume that a and b are positive real numbers and m and n are rational numbers.

117. $a^{m/2} \cdot a^{m/4}$ **118.** $b^{n/2} \cdot b^{-n/3}$

119. $\dfrac{a^{-m/5}}{a^{-m/3}}$ **120.** $\dfrac{h^{-n/4}}{b^{-n/3}}$

121. $(a^{-1/m}b^{-1/n})^{-mn}$ **122.** $(a^{-m/2}b^{-n/3})^{-6}$

123. $\left(\dfrac{a^{-3m}b^{-6n}}{a^{9m}}\right)^{-1/3}$ **124.** $\left(\dfrac{a^{-3/m}b^{6/n}}{a^{-6/m}b^{9/n}}\right)^{-1/3}$

In Exercises 125–132, solve each problem.

125. Diagonal of a box. The length of the diagonal of a box can be found from the formula

$$D = (L^2 + W^2 + H^2)^{1/2},$$

where L, W, and H represent the length, width, and height of the box, respectively. If the box is 12 inches long, 4 inches wide, and 3 inches high, then what is the length of the diagonal?

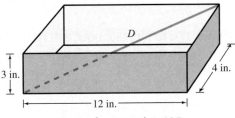

Figure for Exercise 125

126. Radius of a sphere. The radius of a sphere is given in terms of its volume by the formula

$$r = \left(\dfrac{0.75V}{\pi}\right)^{1/3}.$$

Find the radius of a spherical tank that has a volume of $\dfrac{32\pi}{3}$ cubic meters.

Figure for Exercise 126

127. Maximum sail area. According to the new International America's Cup Class Rules, the maximum sail area in square meters for a yacht in the America's Cup race is given by

$$S = (13.0368 + 7.84D^{1/3} - 0.8L)^2$$

where D is the displacement in cubic meters and L is the length in meters. (*Scientific American*, May 1992). Find the maximum sail area for a boat that has a displacement of 18.42 m³ and a length of 21.45 m.

Figure for Exercise 127

128. Orbits of the planets. According to Kepler's third law of planetary motion, the average radius R of the orbit of a planet around the sun is determined by $R = T^{2/3}$, where T is the number of years for one orbit and R is measured in astronomical units (AU). If it takes Mars 1.881 years to make one orbit of the sun, then what is the average radius (in AUs) of the orbit of Mars? (One AU is the distance from the earth to the sun.)

129. Stock fund. The average annual return for an investment is given by the formula

$$r = \left(\dfrac{S}{P}\right)^{1/n} - 1,$$

where P is the initial investment and S is the amount it is worth after n years. An investment of $10,000 in Great-West's Ultra Investors fund in 1983 grew to $41,200 in 1993. Find the ten-year average annual return for this fund.

130. Bond fund. Use the formula from Exercise 129 to find the five-year average annual return for Great-West's Bond Portfolio fund. An investment of $10,000 in this fund in 1988 grew to $15,218 in 1993.

131. Overdue loan payment. In 1777, a wealthy Pennsylvania merchant, Jacob DeHaven, lent $450,000 to the Continental Congress to rescue the troops at Valley Forge. The loan was not repaid. In 1990, DeHaven's descendants filed suit for $141.6 billion (*New York Times*, May 27, 1990). What average annual rate of return are they using to calculate the value of the debt after 213 years? (See Exercise 129.)

132. ***California growin'.*** According to the U.S. Census Bureau, the population of California grew from 19.9 million in 1970 to 30.4 million in 1991. Find the average annual rate of growth for that time period. (Use the formula from Exercise 129 with P being initial population and S being the population n years later.)

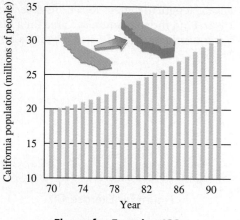

Figure for Exercise 132

Getting More Involved

133. ***Discussion.*** If we use the product rule to simplify $(-1)^{1/2} \cdot (-1)^{1/2}$, we get

$$(-1)^{1/2} \cdot (-1)^{1/2} = (-1)^1 = -1.$$

If we use the power of a product rule, we get

$$(-1)^{1/2} \cdot (-1)^{1/2} = (-1 \cdot -1)^{1/2} = 1^{1/2} = 1.$$

Which of these computations is incorrect? Explain your answer.

134. ***Discussion.*** Determine whether each equation is an identity. Explain.

a) $(w^2 x^2)^{1/2} = |w| \cdot |x|$

b) $(w^2 x^2)^{1/2} = |wx|$

c) $(w^2 x^2)^{1/2} = w|x|$

5.2 Radicals

In this section:

▶ Radical Notation

▶ Product Rule for Radicals

▶ Quotient Rule for Radicals

▶ Rationalizing the Denominator

▶ Simplifying Radicals

▶ Simplifying Radicals Involving Variables

In Section 5.1, we used the exponent $1/n$ to represent the nth root. The symbol $\sqrt[n]{}$ has exactly the same meaning. In this section we learn to use the rules of exponents with this new notation.

Radical Notation

Any expression involving an nth root can be written using radical notation. The symbol $\sqrt{}$ is called the **radical symbol.**

> **Radicals**
>
> If n is a positive integer and a is a real number for which $a^{1/n}$ is defined, then the expression $\sqrt[n]{a}$ is called a **radical,** and
>
> $$\sqrt[n]{a} = a^{1/n}.$$

The number a is called the **radicand.** The number n is called the **index** of the radical. Radicals of index 2 and 3 are referred to as "square roots" and "cube roots," respectively, while the expression $\sqrt[n]{a}$ is "the nth root of a." The index 2 is usually omitted. We write $\sqrt{a}$ rather than $\sqrt[2]{a}$. Remember that $\sqrt[n]{a}$ is simply another notation for $a^{1/n}$. So $\sqrt[n]{a}$ is the *positive* nth root of a when n is even and a is positive, while $\sqrt[n]{a}$ is the real nth root of a when n is odd and a is any real number. The expressions

$$\sqrt{-4}, \quad \sqrt[4]{-16}, \quad \text{and} \quad \sqrt[6]{-2}$$

are not real numbers because there are no even roots of negative numbers in the real number system.

EXAMPLE 1
Changing notations

Write each exponential expression using radical notation and each radical expression using exponential notation. Assume that the variables represent positive numbers. Do not simplify.

a) $\sqrt{36}$ b) $\sqrt[3]{-8}$ c) $\sqrt[3]{y^6}$ d) $x^{3/4}$ e) $x^{-5/2}$

Solution

a) $\sqrt{36} = 36^{1/2}$

b) $\sqrt[3]{-8} = (-8)^{1/3}$

c) $\sqrt[3]{y^6} = y^{6/3}$

d) $x^{3/4} = \sqrt[4]{x^3} = \left(\sqrt[4]{x}\right)^3$ We can write either the root or the power first.

e) $x^{-5/2} = \sqrt{x^{-5}} = \left(\sqrt{x}\right)^{-5}$ ∎

EXAMPLE 2
Simplifying radical expressions

Simplify.

a) $\sqrt{49}$ b) $\sqrt[3]{-125}$ c) $\sqrt[4]{81}$ d) $\sqrt{x^{20}}$ e) $\sqrt[3]{2^{21}}$

Solution

a) $\sqrt{49} = 7$ because $7^2 = 49$.

b) $\sqrt[3]{-125} = -5$ because $(-5)^3 = -125$.

c) $\sqrt[4]{81} = 3$ because $3^4 = 81$.

d) $\sqrt{x^{20}} = x^{20/2} = x^{10}$

e) $\sqrt[3]{2^{21}} = 2^{21/3} = 2^7 = 128$ ∎

Product Rule for Radicals

Recall that the power of a product rule is valid for rational exponents as well as integers. For example, the power of a product rule allows us to write

$$(4y)^{1/2} = 4^{1/2} \cdot y^{1/2} \quad \text{and} \quad (8 \cdot 7)^{1/3} = 8^{1/3} \cdot 7^{1/3}.$$

These equations can be written using radical notation as

$$\sqrt{4y} = \sqrt{4} \cdot \sqrt{y} \quad \text{and} \quad \sqrt[3]{8 \cdot 7} = \sqrt[3]{8} \cdot \sqrt[3]{7}.$$

The power of a product rule (for the power $1/n$) can be stated using radical notation. In this form the rule is called the **product rule for radicals.**

Product Rule for Radicals

The nth root of a product is equal to the product of the nth roots. In symbols,

$$\sqrt[n]{ab} = \sqrt[n]{a} \cdot \sqrt[n]{b},$$

provided that all of the expressions represent real numbers.

The numbers 1, 4, 9, 16, 25, 49, 64, and so on are called **perfect squares** because they are the squares of the positive integers. If the radicand of a square root has a perfect square (other than 1) as a factor, the product rule can be used to simplify the radical expression. For example, the radicand of $\sqrt{50}$ has 25 as a factor, so we can use the product rule to factor $\sqrt{50}$ into a product of two square roots:

$$\sqrt{50} = \sqrt{25} \cdot \sqrt{2} = 5\sqrt{2}.$$

When simplifying a cube root, we check the radicand for factors that are perfect cubes: 8, 27, 64, 125, and so on. In general, when simplifying an nth root, we look for a perfect nth power as a factor of the radicand.

EXAMPLE 3 **Using the product rule to simplify radicals**

Simplify each expression. Assume that all variables represent positive numbers.

a) $\sqrt{4y}$ b) $\sqrt{18}$ c) $\sqrt[3]{56}$ d) $\sqrt[4]{32}$

Solution

a) The radicand $4y$ has the perfect square 4 as a factor. So
$$\sqrt{4y} = \sqrt{4} \cdot \sqrt{y}$$
$$= 2\sqrt{y}.$$

b) The radicand 18 has a factor of 9. So
$$\sqrt{18} = \sqrt{9} \cdot \sqrt{2}$$
$$= 3\sqrt{2}.$$

c) The radicand 56 in this cube root has the perfect cube 8 as a factor. So
$$\sqrt[3]{56} = \sqrt[3]{8} \cdot \sqrt[3]{7}$$
$$= 2\sqrt[3]{7}.$$

d) The radicand in this fourth root has the perfect fourth power 16 as a factor. So
$$\sqrt[4]{32} = \sqrt[4]{16} \cdot \sqrt[4]{2}$$
$$= 2\sqrt[4]{2}. \qquad \blacksquare$$

Quotient Rule for Radicals

The power of a quotient rule is also valid for integral and rational exponents. This rule allows us to write

$$\left(\frac{y}{9}\right)^{1/2} = \frac{y^{1/2}}{9^{1/2}} = \frac{y^{1/2}}{3} \quad \text{and} \quad \left(\frac{3}{32}\right)^{1/5} = \frac{3^{1/5}}{32^{1/5}} = \frac{3^{1/5}}{2}.$$

These equations can be written using radical notation as

$$\sqrt{\frac{y}{9}} = \frac{\sqrt{y}}{\sqrt{9}} = \frac{\sqrt{y}}{3} \quad \text{and} \quad \sqrt[5]{\frac{3}{32}} = \frac{\sqrt[5]{3}}{\sqrt[5]{32}} = \frac{\sqrt[5]{3}}{2}.$$

The power of a quotient rule (for the power $1/n$) can be stated using radical notation. When written with radicals, it is called the **quotient rule for radicals.**

> **Quotient Rule for Radicals**
>
> The nth root of a quotient is equal to the quotient of the nth roots. In symbols,
>
> $$\sqrt[n]{\frac{a}{b}} = \frac{\sqrt[n]{a}}{\sqrt[n]{b}},$$
>
> provided that all of the expressions represent real numbers and $b \neq 0$.

The quotient rule is used to simplify radicals by rewriting the root of a quotient as the quotient of the roots.

EXAMPLE 4 **Using the quotient rule to simplify radicals**

Simplify each expression.

a) $\sqrt{\dfrac{4}{25}}$ b) $\sqrt{\dfrac{8}{9}}$ c) $\sqrt[3]{-\dfrac{8}{27}}$ d) $\sqrt[3]{\dfrac{24y}{125}}$

Solution

a) $\sqrt{\dfrac{4}{25}} = \dfrac{\sqrt{4}}{\sqrt{25}} = \dfrac{2}{5}$

b) $\sqrt{\dfrac{8}{9}} = \dfrac{\sqrt{8}}{\sqrt{9}} = \dfrac{\sqrt{4}\cdot\sqrt{2}}{3} = \dfrac{2\sqrt{2}}{3}$

c) $\sqrt[3]{-\dfrac{8}{27}} = \dfrac{\sqrt[3]{-8}}{\sqrt[3]{27}} = \dfrac{-2}{3} = \dfrac{2}{3}$

d) $\sqrt[3]{\dfrac{24y}{125}} = \dfrac{\sqrt[3]{8}\cdot\sqrt[3]{3y}}{\sqrt[3]{125}} = \dfrac{2\sqrt[3]{3y}}{5}$ ■

MATH AT WORK

Yachtsman

Ernie Godshalk, avid sailor and owner of the sloop Golden Eye, has competed in races all over the world. He has learned that success in a race depends on the winds, a good crew, lots of skill, and knowing what makes his boat go fast.

Because of its shape and hull design, the maximum speed (in knots) of a sailboat can be calculated by finding the value 1.3 times the square root of her waterline (in feet). Thus Godshalk knows the maximum possible speed of the Golden Eye. But decisions made while she is under sail make the difference between attaining her maximum speed and only approaching it. For example, Godshalk knows that the pressure on the sails is what makes the boat move at a certain speed. It is especially important to sail where the wind is strongest. This means that choosing the correct tack, as well as "going where the wind is" can make the difference between coming in first and just finishing the race. In Exercise 105 of this section you will calculate the maximum speed of the Golden Eye.

Rationalizing the Denominator

Square roots such as $\sqrt{2}$, $\sqrt{3}$, and $\sqrt{5}$ are irrational numbers. If roots of this type appear in the denominator of a fraction, it is customary to rewrite the fraction with a rational number in the denominator, or **rationalize** it. We rationalize a denominator by multiplying both numerator and denominator by another radical that makes the denominator rational.

EXAMPLE 5   **Rationalizing the denominator**

Rewrite each expression with a rational denominator.

a) $\dfrac{\sqrt{3}}{\sqrt{5}}$ 　　　　　　　　　　　　　　　　**b)** $\dfrac{3}{\sqrt[3]{2}}$

Solution

a) Because $\sqrt{5} \cdot \sqrt{5} = 5$, multiplying both the numerator and denominator by $\sqrt{5}$ will rationalize the denominator:

$$\frac{\sqrt{3}}{\sqrt{5}} = \frac{\sqrt{3}}{\sqrt{5}} \cdot \frac{\sqrt{5}}{\sqrt{5}} = \frac{\sqrt{15}}{5} \qquad \text{By the product rule, } \sqrt{3} \cdot \sqrt{5} = \sqrt{15}.$$

b) We must build up the denominator to be the cube root of a perfect cube. So we multiply by $\sqrt[3]{4}$ to get $\sqrt[3]{4} \cdot \sqrt[3]{2} = \sqrt[3]{8}$:

$$\frac{3}{\sqrt[3]{2}} = \frac{3}{\sqrt[3]{2}} \cdot \frac{\sqrt[3]{4}}{\sqrt[3]{4}} = \frac{3\sqrt[3]{4}}{\sqrt[3]{8}} = \frac{3\sqrt[3]{4}}{2} \qquad ■$$

CAUTION   To rationalize a denominator with a single square root, you simply multiply by that square root. If the denominator has a cube root, you build the denominator to a cube root of a perfect cube, as in Example 5(b). For a fourth root you build to a fourth root of a perfect fourth power, and so on. ⊘

Simplifying Radicals

When we simplify any expression, we try to make it look "simpler." When we simplify a radical expression, we have three specific conditions to satisfy.

Simplified Radical Form for Radicals of Index n

A radical expression of index n is in **simplified radical form** if it has

1. *no* perfect nth powers as factors of the radicand,
2. *no* fractions inside the radical, and
3. *no* radicals in the denominator.

The radical expressions in the next example do not satisfy the three conditions for simplified radical form. To rewrite an expression in simplified form, we use the product rule, the quotient rule, and rationalizing the denominator.

EXAMPLE 6   **Writing radical expressions in simplified radical form**

Simplify.

a) $\dfrac{\sqrt{10}}{\sqrt{6}}$ 　　　　　　　**b)** $\sqrt[3]{\dfrac{5}{9}}$ 　　　　　　　**c)** $\sqrt[4]{\dfrac{1}{3}}$

Solution

a) To rationalize the denominator, multiply the numerator and denominator by $\sqrt{6}$:

$$\frac{\sqrt{10}}{\sqrt{6}} = \frac{\sqrt{10}}{\sqrt{6}} \cdot \frac{\sqrt{6}}{\sqrt{6}} \qquad \text{Rationalize the denominator.}$$

$$= \frac{\sqrt{60}}{6}$$

$$= \frac{\sqrt{4}\sqrt{15}}{6} \qquad \text{Remove the perfect square from } \sqrt{60}.$$

$$= \frac{2\sqrt{15}}{6}$$

$$= \frac{\sqrt{15}}{3} \qquad \text{Reduce } \tfrac{2}{6} \text{ to } \tfrac{1}{3}. \text{ Note that } \sqrt{15} \div 3 \neq \sqrt{5}.$$

b) To rationalize the denominator, build up the denominator to a cube root of a perfect cube. Because $\sqrt[3]{9} \cdot \sqrt[3]{3} = \sqrt[3]{27} = 3$, we multiply by $\sqrt[3]{3}$:

$$\sqrt[3]{\frac{5}{9}} = \frac{\sqrt[3]{5}}{\sqrt[3]{9}} \qquad \text{Quotient rule for radicals}$$

$$= \frac{\sqrt[3]{5}}{\sqrt[3]{9}} \cdot \frac{\sqrt[3]{3}}{\sqrt[3]{3}} \qquad \text{Rationalize the denominator.}$$

$$= \frac{\sqrt[3]{15}}{\sqrt[3]{27}}$$

$$= \frac{\sqrt[3]{15}}{3}$$

c) To rationalize the denominator, observe that $\sqrt[4]{3} \cdot \sqrt[4]{27} = \sqrt[4]{81} = 3$.

$$\sqrt[4]{\frac{1}{3}} = \frac{\sqrt[4]{1}}{\sqrt[4]{3}} \qquad \text{Quotient rule for radicals}$$

$$= \frac{1}{\sqrt[4]{3}} \cdot \frac{\sqrt[4]{27}}{\sqrt[4]{27}} \qquad \text{Rationalize the denominator.}$$

$$= \frac{\sqrt[4]{27}}{\sqrt[4]{81}} = \frac{\sqrt[4]{27}}{3} \qquad \text{Do not omit the index of the radical in any step.} \quad \blacksquare$$

Simplifying Radicals Involving Variables

To simplify radicals involving variables, we must recognize exponential expressions that are perfect squares, perfect cubes, and so on. The expressions

$$x^2, \quad w^4, \quad y^8, \quad z^{14}, \quad \text{and} \quad x^{50}$$

are **perfect squares** because they are squares of variables with integral powers. *Any even power of a variable is a perfect square.* If we assume that the variables represent positive numbers, we can write

$$\sqrt{x^2} = x, \quad \sqrt{w^4} = w^2, \quad \sqrt{y^8} = y^4, \quad \sqrt{z^{14}} = z^7, \quad \text{and} \quad \sqrt{x^{50}} = x^{25}.$$

Note that when we find the square root, the result has one-half of the original exponent.

EXAMPLE 7 **Radicals with variables**

Simplify each expression. Assume that all variables represent positive real numbers.

a) $\sqrt{12x^6}$ b) $\sqrt{98x^5y^9}$

Solution

a) Use the product rule to place all perfect squares under the first radical symbol and the remaining factors under the second:

$$\sqrt{12x^6} = \sqrt{4x^6 \cdot 3} \qquad \text{Factor out the perfect squares.}$$
$$= \sqrt{4x^6} \cdot \sqrt{3} \qquad \text{Product rule for radicals}$$
$$= 2x^3\sqrt{3}$$

b) $\sqrt{98x^5y^9} = \sqrt{49x^4y^8} \cdot \sqrt{2xy} \qquad \text{Product rule for radicals}$
$$= 7x^2y^4\sqrt{2xy}$$ ■

In the next example we start with a square root of a quotient.

EXAMPLE 8 **Radicals with variables**

Simplify each expression. Assume that all variables represent positive real numbers.

a) $\sqrt{\dfrac{a}{b}}$ b) $\sqrt{\dfrac{x^3}{y^5}}$

Solution

a) $\sqrt{\dfrac{a}{b}} = \dfrac{\sqrt{a}}{\sqrt{b}}$ Quotient rule for radicals

$\phantom{a)\ \sqrt{\dfrac{a}{b}}} = \dfrac{\sqrt{a} \cdot \sqrt{b}}{\sqrt{b} \cdot \sqrt{b}}$ Rationalize the denominator.

$\phantom{a)\ \sqrt{\dfrac{a}{b}}} = \dfrac{\sqrt{ab}}{b}$

b) $\sqrt{\dfrac{x^3}{y^5}} = \dfrac{\sqrt{x^3}}{\sqrt{y^5}}$ Quotient rule for radicals

$\phantom{b)\ \sqrt{\dfrac{x^3}{y^5}}} = \dfrac{\sqrt{x^2} \cdot \sqrt{x}}{\sqrt{y^4} \cdot \sqrt{y}}$ Product rule for radicals

$\phantom{b)\ \sqrt{\dfrac{x^3}{y^5}}} = \dfrac{x\sqrt{x}}{y^2\sqrt{y}}$ Simplify.

$\phantom{b)\ \sqrt{\dfrac{x^3}{y^5}}} = \dfrac{x\sqrt{x} \cdot \sqrt{y}}{y^2\sqrt{y} \cdot \sqrt{y}}$ Rationalize the denominator.

$\phantom{b)\ \sqrt{\dfrac{x^3}{y^5}}} = \dfrac{x\sqrt{xy}}{y^2 \cdot y} = \dfrac{x\sqrt{xy}}{y^3}$ ■

Any variable with an exponent that is a multiple of 3 is a **perfect cube.** For example,

$$a^3, \quad b^6, \quad c^{15}, \quad \text{and} \quad w^{39}$$

are perfect cubes. Each of these expressions is the cube of a variable with an integral exponent. For any values of the variables we can write

$$\sqrt[3]{a^3} = a, \qquad \sqrt[3]{b^6} = b^2, \qquad \sqrt[3]{c^{15}} = c^5, \qquad \text{and} \qquad \sqrt[3]{w^{39}} = w^{13}.$$

Note that when we find the cube root, the result has one-third of the original exponent.

If the exponent on a variable is a multiple of 4, we have a perfect fourth power; if the exponent is a multiple of 5, we have a perfect fifth power; and so on. In the next example we simplify radicals with index higher than 2.

EXAMPLE 9   **Simplifying higher-index radicals with variables**
Simplify. Assume that the variables represent positive numbers.

a) $\sqrt[3]{40x^8}$

b) $\sqrt[4]{x^{12}y^5}$

c) $\sqrt[3]{\dfrac{x}{y}}$

Solution

a) Use the product rule to place the largest perfect cube factors under the first radical and the remaining factors under the second:

$$\sqrt[3]{40x^8} = \sqrt[3]{8x^6} \cdot \sqrt[3]{5x^2} = 2x^2 \sqrt[3]{5x^2}$$

b) Place the largest perfect fourth power factors under the first radical and the remaining factors under the second:

$$\sqrt[4]{x^{12}y^5} = \sqrt[4]{x^{12}y^4} \cdot \sqrt[4]{y} = x^3y \sqrt[4]{y}$$

c) Multiply by $\sqrt[3]{y^2}$ to rationalize the denominator:

$$\sqrt[3]{\dfrac{x}{y}} = \dfrac{\sqrt[3]{x}}{\sqrt[3]{y}} = \dfrac{\sqrt[3]{x}}{\sqrt[3]{y}} \cdot \dfrac{\sqrt[3]{y^2}}{\sqrt[3]{y^2}} = \dfrac{\sqrt[3]{xy^2}}{\sqrt[3]{y^3}} = \dfrac{\sqrt[3]{xy^2}}{y}$$ ∎

Calculator Close-up

Most scientific and graphing calculators have a square root key, ⌐√⌐ On a scientific calculator the square root key gives the square root of the number on the display. With 2 on the display, press ⌐√⌐ to get 1.414 on the display. On a graphing calculator your display should look like the following:

$$\sqrt{\;} \; 2 \; \boxed{\text{ENTER}}$$

| 1.414 |

You can also evaluate a radical expression with a calculator by using its exponential form. For example, to evaluate $\sqrt[3]{20}$ (or $20^{1/3}$) on a scientific calculator, try the following keystrokes:

$$20 \; \boxed{x^y} \; \boxed{(} \; 1 \; \boxed{\div} \; 3 \; \boxed{)} \; \boxed{=}$$

If your calculator has an *n*th root key, ⌐√⌐, try the following:

$$20 \; \boxed{\sqrt{\;}} \; 3 \; \boxed{=}$$

In both cases the display should read

| 2.7144 |

On a graphing calculator your display should look like the following:

$$20^\wedge(1/3) \; \boxed{\text{ENTER}}$$

| 2.7144 |

Warm-ups

True or false? Explain.

1. $2^{1/2} = \sqrt{2}$
2. $3^{1/3} = \sqrt{3}$
3. $2^{2/3} = \sqrt[3]{4}$
4. $\sqrt{81} = \sqrt{9}$
5. $\sqrt{417^2} = 417$
6. $\sqrt[3]{a^{27}} = a^3$
7. $\dfrac{\sqrt{2}}{2} = \dfrac{1}{\sqrt{2}}$
8. $\dfrac{\sqrt{10}}{2} = \sqrt{5}$
9. $\sqrt{2^{-4}} = \dfrac{1}{4}$
10. $\dfrac{\sqrt{6}}{\sqrt{3}} = \sqrt{2}$

5.2 EXERCISES

All variables in the following exercises represent positive numbers.

Write the radical expressions in exponential notation and the exponential expressions in radical notation. Do not simplify. See Example 1.

1. $\sqrt{27}$
2. $\sqrt[3]{-27}$
3. $\sqrt{x^5}$
4. $\sqrt{a^3}$
5. $\sqrt[3]{a^{12}}$
6. $\sqrt[3]{w^{-27}}$
7. $-5^{1/3}$
8. $-7^{1/2}$
9. $2^{2/5}$
10. $3^{2/3}$
11. $x^{-2/3}$
12. $x^{-2/5}$

Simplify each radical expression. See Example 2.

13. $\sqrt{121}$
14. $\sqrt{64}$
15. $\sqrt[3]{-1000}$
16. $\sqrt[5]{-1}$
17. $\sqrt[4]{-81}$
18. $\sqrt[4]{16}$
19. $\sqrt{a^{16}}$
20. $\sqrt{b^{36}}$
21. $\sqrt{4^{16}}$
22. $\sqrt{w^4}$
23. $\sqrt[5]{w^{30}}$
24. $\sqrt[5]{a^{20}}$

Simplify each expression. See Example 3.

25. $\sqrt{9w}$
26. $\sqrt{36m}$
27. $\sqrt{20}$
28. $\sqrt{50}$
29. $\sqrt{45w}$
30. $\sqrt{48t}$
31. $\sqrt{288}$
32. $\sqrt{242}$
33. $\sqrt[3]{54}$
34. $\sqrt[3]{-48}$
35. $\sqrt[4]{32a}$
36. $\sqrt[3]{80xy}$

Simplify each expression. See Example 4.

37. $\sqrt{\dfrac{9}{100}}$
38. $\sqrt{\dfrac{25}{4}}$
39. $\sqrt{\dfrac{50}{9}}$
40. $\sqrt{\dfrac{18}{25}}$
41. $\sqrt[3]{-\dfrac{125}{8}}$
42. $\sqrt[4]{\dfrac{16}{81}}$
43. $\sqrt[3]{\dfrac{16x}{27}}$
44. $\sqrt[3]{\dfrac{-81a}{1000}}$

Rewrite each expression with a rational denominator. See Example 5.

45. $\dfrac{2}{\sqrt{5}}$
46. $\dfrac{5}{\sqrt{3}}$
47. $\dfrac{\sqrt{3}}{\sqrt{7}}$
48. $\dfrac{\sqrt{6}}{\sqrt{5}}$
49. $\dfrac{1}{\sqrt[3]{4}}$
50. $\dfrac{7}{\sqrt[3]{3}}$
51. $\dfrac{\sqrt[3]{6}}{\sqrt[3]{5}}$
52. $\dfrac{\sqrt[4]{2}}{\sqrt[4]{27}}$

Write each radical expression in simplified radical form. See Example 6.

53. $\dfrac{\sqrt{5}}{\sqrt{12}}$
54. $\dfrac{\sqrt{7}}{\sqrt{18}}$
55. $\dfrac{\sqrt{3}}{\sqrt{12}}$
56. $\dfrac{\sqrt{2}}{\sqrt{18}}$
57. $\sqrt{\dfrac{1}{2}}$
58. $\sqrt{\dfrac{3}{8}}$
59. $\sqrt[3]{\dfrac{7}{4}}$
60. $\sqrt[4]{\dfrac{1}{5}}$

Simplify. See Examples 7 and 8.

61. $\sqrt{12x^8}$
62. $\sqrt{72x^{10}}$
63. $\sqrt{60a^9b^3}$
64. $\sqrt{63w^{15}z^7}$
65. $\sqrt{\dfrac{x}{y}}$
66. $\sqrt{\dfrac{x^2}{a}}$
67. $\sqrt{\dfrac{a^3}{b^7}}$
68. $\sqrt{\dfrac{w^5}{y^8}}$

Simplify. Assume that the variables represent positive real numbers. See Example 9.

69. $\sqrt[3]{16x^{13}}$
70. $\sqrt[3]{24x^{17}}$
71. $\sqrt[4]{x^9y^6}$
72. $\sqrt[4]{w^{14}y^7}$
73. $\sqrt[5]{64x^{22}}$
74. $\sqrt[5]{x^{12}y^5z^3}$
75. $\sqrt[3]{\dfrac{a}{b}}$
76. $\sqrt[3]{\dfrac{a}{w^2}}$

Simplify.

77. $\sqrt[4]{3^{12}}$

78. $\sqrt[3]{2^{-9}}$

79. $\sqrt{10^{-2}}$

80. $\sqrt{-10^{-4}}$

81. $\sqrt{\dfrac{8x}{49}}$

82. $\sqrt{\dfrac{12b}{121}}$

83. $\sqrt[4]{\dfrac{32a}{81}}$

84. $\sqrt[4]{\dfrac{162y}{625}}$

85. $\sqrt[3]{-27x^9y^8}$

86. $\sqrt[4]{32y^8z^{11}}$

87. $\dfrac{\sqrt{ab^3}}{\sqrt{a^3b^2}}$

88. $\dfrac{\sqrt{m^3n^5}}{\sqrt{m^5n}}$

89. $\dfrac{\sqrt[3]{a^2b}}{\sqrt[3]{4ab^2}\,\sqrt[3]{3ab^5}}$

90. $\dfrac{\sqrt[3]{5xy^2}}{\sqrt[3]{18x^2y}}$

91. $\dfrac{6m\sqrt[5]{5mn^2}}{\sqrt[5]{12m^3n}}$

92. $\dfrac{3ab\sqrt[5]{2b}}{\sqrt[5]{24a^2}}$

 Use a calculator to find a decimal approximation to each radical expression. Round to three decimal places.

93. $\dfrac{5}{\sqrt{3}}$

94. $\sqrt{\dfrac{2}{27}}$

95. $\sqrt[3]{\dfrac{1}{3}}$

96. $\sqrt[3]{56}$

97. $\dfrac{\sqrt[3]{9}}{\sqrt[3]{4}}$

98. $\dfrac{\sqrt[4]{25}}{\sqrt{5}}$

99. $\dfrac{\sqrt[6]{16}}{\sqrt[3]{4}}$

100. $\sqrt[5]{2.48832}$

In Exercises 101–108, solve each problem.

101. *Factoring in the wind.* Through experimentation in Antarctica, Paul Siple developed the formula

$$W = 91.4 - \dfrac{(10.5 + 6.7\sqrt{v} - 0.45v)(457 - 5t)}{110}$$

to calculate the wind chill temperature W (in degrees Fahrenheit) from the wind velocity v (in miles per hour) and the air temperature t (in degrees Fahrenheit). Find the wind chill temperature when the air temperature is 25°F and the wind velocity is 20 mph. Use the accompanying graph to estimate the wind chill temperature when the air temperature is 25°F and the wind velocity is 30 mph.

Figure for Exercise 101

102. *Comparing wind chills.* Use the formula from Exercise 101 to determine who will feel colder: a person in Minneapolis at 10°F with a 15-mph wind or a person in Chicago at 20°F with a 25-mph wind.

103. *Diving time.* The time t in seconds that it takes for an object to fall a distance of S feet is given by $t = \sqrt{\dfrac{S}{16}}$. Use the properties of radicals to simplify this formula. How long does it take a diver to reach the water when diving from a height of 30 feet?

104. *Sky diving.* The formula in the previous exercise accounts for the effect of gravity only on a falling object. According to that formula, how long would it take a sky diver to reach the earth when jumping from 17,000 feet? (A sky diver can actually get about twice as much falling time by spreading out and using the air to slow the fall.)

105. *Maximum sailing speed.* To find the maximum possible speed in knots (nautical miles per hour) for a sailboat, sailors use the formula $M = 1.3\sqrt{w}$, where w is the length of the waterline in feet. If the waterline for the sloop Golden Eye is 20 feet, then what is the maximum speed of the Golden Eye?

106. *America's Cup.* Since 1988, basic yacht dimensions for the America's Cup competition have satisfied the inequality

$$L + 1.25\sqrt{S} - 9.8\sqrt[3]{D} \leq 16.296,$$

where L is the boat's length in meters, S is the sail area in square meters, and D is the displacement in cubic meters (*Scientific American*, May 1992). A team of naval architects is planning to build a boat with a displacement of 21.44 m³, a sail area of 320.13 m², and a length of 21.22 m. Does this boat satisfy the inequality? If the length and displacement of this boat cannot be changed, then how many square meters of sail area must be removed so that the boat satisfies the inequality?

107. *Landing a Piper Cheyenne.* Aircraft design engineers determine the proper landing speed V (in feet per second) for an airplane from the formula

$$V = \sqrt{\dfrac{841L}{CS}},$$

where L is the gross weight of the aircraft in pounds, C is the coefficient of lift, and S is the wing surface area in square feet. According to Piper Aircraft of Vero Beach, Florida, the Piper Cheyenne has a gross weight of 8700 lb, a coefficient of lift of 2.81, and a wing surface area of 200 ft². Find the proper landing speed for this plane. What is the landing speed in miles per hour?

108. *Landing speed and weight.* Because the gross weight of the Piper Cheyenne depends on how much fuel and cargo are on board, the proper landing speed (from Exercise 107) is not always the same. The formula $V = \sqrt{1.496L}$ gives the landing speed for the Piper Cheyenne in terms of the gross weight only. Find the landing speed if the gross weight is 7,000 lb. What gross weight corresponds to a landing speed of 115 feet per second?

Getting More Involved

109. *Cooperative learning.* Work in a group to determine whether each equation is an identity. Explain your answers.

a) $\sqrt{x^2} = |x|$ **b)** $\sqrt[3]{x^3} = |x|$
c) $\sqrt{x^4} = x^2$ **d)** $\sqrt[4]{x^4} = |x|$

For which values of n is $\sqrt[n]{x^n} = x$ an identity?

110. *Cooperative learning.* Work in a group to determine whether each inequality is correct.

a) $\sqrt{0.9} > 0.9$ **b)** $\sqrt{1.01} > 1.01$
c) $\sqrt[3]{0.99} > 0.99$ **d)** $\sqrt[3]{1.001} > 1.001$

For which values of x and n is $\sqrt[n]{x} > x$?

111. *Discussion.* If your test scores are 80 and 100, then the arithmetic mean of your scores is 90. The geometric mean of the scores is a number h such that

$$\frac{80}{h} = \frac{h}{100}.$$

Are you better off with the arithmetic mean or the geometric mean?

5.3

Operations with Radicals

In this section:

▶ Adding and Subtracting Radicals
▶ Multiplying Radicals
▶ Conjugates

In this section we use the ideas of Section 5.2 in performing arithmetic operations with radical expressions.

Adding and Subtracting Radicals

To find the sum of $\sqrt{2}$ and $\sqrt{3}$, we can use a calculator to get $\sqrt{2} \approx 1.414$ and $\sqrt{3} \approx 1.732$. (The symbol $\approx$ means "is approximately equal to.") We can then add the decimal numbers and get

$$\sqrt{2} + \sqrt{3} \approx 1.414 + 1.732 = 3.146.$$

We cannot write an exact decimal form for $\sqrt{2} + \sqrt{3}$; the number 3.146 is an approximation of $\sqrt{2} + \sqrt{3}$. To represent the exact value of $\sqrt{2} + \sqrt{3}$, we just use the form $\sqrt{2} + \sqrt{3}$. This form cannot be simplified any further. However, a sum of like radicals can be simplified. **Like radicals** are radicals that have the same index and the same radicand.

To simplify the sum $3\sqrt{2} + 5\sqrt{2}$, we can use the fact that $3x + 5x = 8x$ is true for any value of x. Substituting $\sqrt{2}$ for x gives us $3\sqrt{2} + 5\sqrt{2} = 8\sqrt{2}$. So like radicals can be combined just as like terms are combined.

EXAMPLE 1   **Adding and subtracting like radicals**

Simplify the following expressions. Assume that the variables represent positive numbers.

a) $3\sqrt{5} + 4\sqrt{5}$
b) $\sqrt[4]{w} - 6\sqrt[4]{w}$
c) $\sqrt{3} + \sqrt{5} - 4\sqrt{3} + 6\sqrt{5}$
d) $3\sqrt[3]{6x} + 2\sqrt[3]{x} + \sqrt[3]{6x} + \sqrt[3]{x}$

Solution

a) $3\sqrt{5} + 4\sqrt{5} = 7\sqrt{5}$

b) $\sqrt[4]{w} - 6\sqrt[4]{w} = -5\sqrt[4]{w}$

c) $\sqrt{3} + \sqrt{5} - 4\sqrt{3} + 6\sqrt{5} = -3\sqrt{3} + 7\sqrt{5}$ Only like radicals are combined.

d) $3\sqrt[3]{6x} + 2\sqrt[3]{x} + \sqrt[3]{6x} + \sqrt[3]{x} = 4\sqrt[3]{6x} + 3\sqrt[3]{x}$ ∎

We may have to simplify the radicals to determine which ones can be combined.

EXAMPLE 2   **Simplifying radicals before combining**

Perform the indicated operations. Assume that the variables represent positive numbers.

a) $\sqrt{8} + \sqrt{18}$ **b)** $\sqrt{\dfrac{1}{5}} + \sqrt{20}$

c) $\sqrt{2x^3} - \sqrt{4x^2} + 5\sqrt{18x^3}$ **d)** $\sqrt[3]{16x^4y^3} - \sqrt[3]{54x^4y^3}$

Solution

a) $\sqrt{8} + \sqrt{18} = \sqrt{4} \cdot \sqrt{2} + \sqrt{9} \cdot \sqrt{2}$

$\qquad\qquad\quad = 2\sqrt{2} + 3\sqrt{2}$ Simplify each radical.

$\qquad\qquad\quad = 5\sqrt{2}$ Add like radicals .

Note that $\sqrt{8} + \sqrt{18} \neq \sqrt{26}$.

b) $\sqrt{\dfrac{1}{5}} + \sqrt{20} = \dfrac{\sqrt{5}}{5} + 2\sqrt{5}$ Because $\sqrt{\dfrac{1}{5}} = \dfrac{1}{\sqrt{5}} \cdot \dfrac{\sqrt{5}}{\sqrt{5}} = \dfrac{\sqrt{5}}{5}$ and $\sqrt{20} = 2\sqrt{5}$

$\qquad\qquad\quad\; = \dfrac{\sqrt{5}}{5} + \dfrac{10\sqrt{5}}{5}$ Use the LCD of 5.

$\qquad\qquad\quad\; = \dfrac{11\sqrt{5}}{5}$

c) $\sqrt{2x^3} - \sqrt{4x^2} + 5\sqrt{18x^3} = \sqrt{x^2} \cdot \sqrt{2x} - 2x + 5 \cdot \sqrt{9x^2} \cdot \sqrt{2x}$

$\qquad\qquad\qquad\qquad\qquad\quad = x\sqrt{2x} - 2x + 15x\sqrt{2x}$ Simplify each radical.

$\qquad\qquad\qquad\qquad\qquad\quad = 16x\sqrt{2x} - 2x$ Add like radicals only.

d) $\sqrt[3]{16x^4y^3} - \sqrt[3]{54x^4y^3} = \sqrt[3]{8x^3y^3} \cdot \sqrt[3]{2x} - \sqrt[3]{27x^3y^3} \cdot \sqrt[3]{2x}$

$\qquad\qquad\qquad\qquad\qquad\; = 2xy\sqrt[3]{2x} - 3xy\sqrt[3]{2x}$ Simplify each radical.

$\qquad\qquad\qquad\qquad\qquad\; = -xy\sqrt[3]{2x}$ ∎

Multiplying Radicals

We have already multiplied radicals, in Section 5.2 when we rationalized denominators. The product rule for radicals, $\sqrt[n]{a} \cdot \sqrt[n]{b} = \sqrt[n]{ab}$, allows multiplication of radicals with the same index, such as

$$\sqrt{5} \cdot \sqrt{3} = \sqrt{15}, \qquad \sqrt[3]{2} \cdot \sqrt[3]{5} = \sqrt[3]{10}, \qquad \text{and} \qquad \sqrt[5]{x^2} \cdot \sqrt[5]{x} = \sqrt[5]{x^3}.$$

CAUTION   The product rule does not allow multiplication of radicals that have different indices. We cannot use the product rule to multiply $\sqrt{2}$ and $\sqrt[3]{5}$.   ⊘

EXAMPLE 3 **Multiplying radicals with the same index**

Multiply and simplify the following expressions. Assume that the variables represent positive numbers.

a) $5\sqrt{6} \cdot 4\sqrt{3}$ **b)** $\sqrt{3a^2} \cdot \sqrt{6a}$ **c)** $\sqrt[3]{4} \cdot \sqrt[3]{4}$ **d)** $\sqrt[4]{\dfrac{x^3}{2}} \cdot \sqrt[4]{\dfrac{x^2}{4}}$

Solution

a) $\begin{aligned} 5\sqrt{6} \cdot 4\sqrt{3} &= 5 \cdot 4 \cdot \sqrt{6} \cdot \sqrt{3} \\ &= 20\sqrt{18} \qquad \text{Product rule for radicals} \\ &= 20 \cdot 3\sqrt{2} \qquad \sqrt{18} = \sqrt{9} \cdot \sqrt{2} = 3\sqrt{2} \\ &= 60\sqrt{2} \end{aligned}$

b) $\begin{aligned} \sqrt{3a^2} \cdot \sqrt{6a} &= \sqrt{18a^3} \qquad \text{Product rule for radicals} \\ &= \sqrt{9a^2} \cdot \sqrt{2a} \\ &= 3a\sqrt{2a} \qquad \text{Simplify.} \end{aligned}$

c) $\begin{aligned} \sqrt[3]{4} \cdot \sqrt[3]{4} &= \sqrt[3]{16} \\ &= \sqrt[3]{8} \cdot \sqrt[3]{2} \qquad \text{Simplify.} \\ &= 2\sqrt[3]{2} \end{aligned}$

d) $\begin{aligned} \sqrt[4]{\dfrac{x^3}{2}} \cdot \sqrt[4]{\dfrac{x^2}{4}} &= \sqrt[4]{\dfrac{x^5}{8}} \qquad \text{Product rule for radicals} \\ &= \dfrac{\sqrt[4]{x^4} \cdot \sqrt[4]{x}}{\sqrt[4]{8}} \qquad \text{Simplify.} \\ &= \dfrac{x\sqrt[4]{x}}{\sqrt[4]{8}} \\ &= \dfrac{x\sqrt[4]{x} \cdot \sqrt[4]{2}}{\sqrt[4]{8} \cdot \sqrt[4]{2}} \qquad \text{Rationalize the denominator.} \\ &= \dfrac{x\sqrt[4]{2x}}{2} \qquad \sqrt[4]{8} \cdot \sqrt[4]{2} = \sqrt[4]{16} = 2 \end{aligned}$ ■

We find a product such as $3\sqrt{2}(4\sqrt{2} - \sqrt{3})$ by using the distributive property as we do when multiplying a monomial and a binomial. A product such as $(2\sqrt{3} + \sqrt{5})(3\sqrt{3} - 2\sqrt{5})$ can be found by using FOIL as we do for the product of two binomials.

EXAMPLE 4 **Multiplying radicals**

Multiply and simplify.

a) $3\sqrt{2}(4\sqrt{2} - \sqrt{3})$ **b)** $\sqrt[3]{a}(\sqrt[3]{a} - \sqrt[3]{a^2})$
c) $(2\sqrt{3} + \sqrt{5})(3\sqrt{3} - 2\sqrt{5})$ **d)** $(3 + \sqrt{x - 9})^2$

Solution

a) $\begin{aligned} 3\sqrt{2}(4\sqrt{2} - \sqrt{3}) &= 3\sqrt{2} \cdot 4\sqrt{2} - 3\sqrt{2} \cdot \sqrt{3} \qquad \text{Distributive property} \\ &= 12 \cdot 2 - 3\sqrt{6} \qquad \text{Because } \sqrt{2} \cdot \sqrt{2} = 2 \\ &= 24 - 3\sqrt{6} \qquad \text{and } \sqrt{2} \cdot \sqrt{3} = \sqrt{6} \end{aligned}$

b) $\sqrt[3]{a}\,(\sqrt[3]{a} - \sqrt[3]{a^2}) = \sqrt[3]{a^2} - \sqrt[3]{a^3}$ Distributive property

$\qquad\qquad\qquad\quad = \sqrt[3]{a^2} - a$

c) $(2\sqrt{3} + \sqrt{5}\,)(3\sqrt{3} - 2\sqrt{5}\,)$

$$\overbrace{\phantom{= 2\sqrt{3}\cdot 3\sqrt{3}}}^{F}\quad\overbrace{\phantom{- 2\sqrt{3}\cdot 2\sqrt{5}}}^{O}\quad\overbrace{\phantom{+ \sqrt{5}\cdot 3\sqrt{3}}}^{I}\quad\overbrace{\phantom{- \sqrt{5}\cdot 2\sqrt{5}}}^{L}$$

$= 2\sqrt{3}\cdot 3\sqrt{3} - 2\sqrt{3}\cdot 2\sqrt{5} + \sqrt{5}\cdot 3\sqrt{3} - \sqrt{5}\cdot 2\sqrt{5}$

$= 18 - 4\sqrt{15} + 3\sqrt{15} - 10$

$= 8 - \sqrt{15}$ Combine like radicals.

d) To square a sum, we use $(a + b)^2 = a^2 + 2ab + b^2$:

$(3 + \sqrt{x - 9}\,)^2 = 3^2 + 2\cdot 3\sqrt{x - 9} + (\sqrt{x - 9}\,)^2$

$\qquad\qquad\qquad = 9 + 6\sqrt{x - 9} + x - 9$

$\qquad\qquad\qquad = x + 6\sqrt{x - 9}$ ■

In the next example we multiply radicals that have different indices.

EXAMPLE 5   **Multiplying radicals with different indices**

Write each product as a single radical expression.

a) $\sqrt[3]{2}\cdot\sqrt[4]{2}$ \qquad\qquad\qquad\qquad **b)** $\sqrt[3]{2}\cdot\sqrt{3}$

Solution

a) $\sqrt[3]{2}\cdot\sqrt[4]{2} = 2^{1/3}\cdot 2^{1/4}$ Write in exponential notation.

$\qquad\qquad = 2^{7/12}$ Product rule for exponents: $\frac{1}{3} + \frac{1}{4} = \frac{7}{12}$

$\qquad\qquad = \sqrt[12]{2^7}$ Write in radical notation.

$\qquad\qquad = \sqrt[12]{128}$

b) $\sqrt[3]{2}\cdot\sqrt{3} = 2^{1/3}\cdot 3^{1/2}$ Write in exponential notation.

$\qquad\qquad = 2^{2/6}\cdot 3^{3/6}$ Write the exponents with the LCD of 6.

$\qquad\qquad = \sqrt[6]{2^2}\cdot\sqrt[6]{3^3}$ Write in radical notation.

$\qquad\qquad = \sqrt[6]{2^2\cdot 3^3}$ Product rule for radicals

$\qquad\qquad = \sqrt[6]{108}$ $2^2\cdot 3^3 = 4\cdot 27 = 108$ ■

CAUTION   Because the bases in $2^{1/3}\cdot 2^{1/4}$ are identical, we can add the exponents (Example 5a). Because the bases in $2^{2/6}\cdot 3^{3/6}$ are not the same, we cannot add the exponents (Example 5b). Instead, we write each factor as a sixth root and use the product rule for radicals. ⊘

Conjugates

Recall the special product rule $(a + b)(a - b) = a^2 - b^2$. The product of the sum $4 + \sqrt{3}$ and the difference $4 - \sqrt{3}$ can be found by using this rule:

$$(4 + \sqrt{3}\,)(4 - \sqrt{3}\,) = 4^2 - (\sqrt{3}\,)^2 = 16 - 3 = 13$$

The product of the irrational number $4 + \sqrt{3}$ and the irrational number $4 - \sqrt{3}$ is the rational number 13. For this reason the expressions $4 + \sqrt{3}$ and $4 - \sqrt{3}$ are called **conjugates** of one another. We will use conjugates in Section 5.4 to rationalize some denominators.

EXAMPLE 6   **Multiplying conjugates**

Find the products. Assume that the variables represent positive real numbers.

a) $(2 + 3\sqrt{5})(2 - 3\sqrt{5})$ **b)** $(\sqrt{3} - \sqrt{2})(\sqrt{3} + \sqrt{2})$

c) $(\sqrt{2x} - \sqrt{y})(\sqrt{2x} + \sqrt{y})$

Solution

a) $(2 + 3\sqrt{5})(2 - 3\sqrt{5}) = 2^2 - (3\sqrt{5})^2$ $(a + b)(a - b) = a^2 - b^2$

$$= 4 - 45 \qquad (3\sqrt{5})^2 = 9 \cdot 5 = 45$$

$$= -41$$

b) $(\sqrt{3} - \sqrt{2})(\sqrt{3} + \sqrt{2}) = 3 - 2$

$$= 1$$

c) $(\sqrt{2x} - \sqrt{y})(\sqrt{2x} + \sqrt{y}) = 2x - y$ ■

Warm-ups

True or false? Explain.

1. $\sqrt{3} + \sqrt{3} = \sqrt{6}$ 2. $\sqrt{8} + \sqrt{2} = 3\sqrt{2}$

3. $2\sqrt{3} \cdot 3\sqrt{3} = 6\sqrt{3}$ 4. $\sqrt[3]{2} \cdot \sqrt[3]{2} = 2$

5. $2\sqrt{5} \cdot 3\sqrt{2} = 6\sqrt{10}$ 6. $2\sqrt{5} + 3\sqrt{5} = 5\sqrt{10}$

7. $\sqrt{2}(\sqrt{3} - \sqrt{2}) = \sqrt{6} - 2$ 8. $\sqrt{12} = 2\sqrt{6}$

9. $(\sqrt{2} + \sqrt{3})^2 = 2 + 3$ 10. $(\sqrt{3} - \sqrt{2})(\sqrt{3} + \sqrt{2}) = 1$

5.3 EXERCISES

All variables in the following exercises represent positive numbers.

Simplify the sums and differences. Give exact answers. See Example 1.

1. $\sqrt{3} - 2\sqrt{3}$ 2. $\sqrt{5} - 3\sqrt{5}$

3. $5\sqrt{7x} + 4\sqrt{7x}$ 4. $3\sqrt{6a} + 7\sqrt{6a}$

5. $2\sqrt[3]{2} + 3\sqrt[3]{2}$ 6. $\sqrt[3]{4} + 4\sqrt[3]{4}$

7. $\sqrt{3} - \sqrt{5} + 3\sqrt{3} - \sqrt{5}$

8. $\sqrt{2} - 5\sqrt{3} - 7\sqrt{2} + 9\sqrt{3}$

9. $\sqrt[3]{2} + \sqrt[3]{x} - \sqrt[3]{2} + 4\sqrt[3]{x}$

10. $\sqrt[3]{5y} - 4\sqrt[3]{5y} + \sqrt[3]{x} + \sqrt[3]{x}$

11. $\sqrt[3]{x} - \sqrt{2x} + \sqrt[3]{x}$

12. $\sqrt[3]{ab} + \sqrt{a} + 5\sqrt{a} + \sqrt[3]{ab}$

Simplify each expression. Give exact answers. See Example 2.

13. $\sqrt{8} + \sqrt{28}$ 14. $\sqrt{12} + \sqrt{24}$

15. $\sqrt{2} - \sqrt{8}$ 16. $\sqrt{20} - \sqrt{125}$

17. $\dfrac{\sqrt{2}}{2} + \sqrt{2}$ 18. $\dfrac{\sqrt{3}}{3} - \sqrt{3}$

19. $\sqrt{80} + \sqrt{\dfrac{1}{5}}$ 20. $\sqrt{32} + \sqrt{\dfrac{1}{2}}$

21. $\sqrt{45x^3} - \sqrt{18x^2} + \sqrt{50x^2} - \sqrt{20x^3}$

22. $\sqrt{12x^5} - \sqrt{18x} - \sqrt{300x^5} + \sqrt{98x}$

23. $\sqrt[3]{24} + \sqrt[3]{81}$ 24. $\sqrt[3]{24} + \sqrt[3]{375}$

25. $\sqrt[4]{48} - \sqrt[4]{243}$ 26. $\sqrt[5]{64} + \sqrt[5]{2}$

27. $\sqrt[3]{54t^4y^3} - \sqrt[3]{16t^4y^3}$

28. $\sqrt[3]{2000w^2z^5} - \sqrt[3]{16w^2z^5}$

Simplify the products. Give exact answers. See Examples 3 and 4.

29. $\sqrt{3} \cdot \sqrt{5}$ 30. $\sqrt{5} \cdot \sqrt{7}$

31. $2\sqrt{5} \cdot 3\sqrt{10}$ 32. $(3\sqrt{2})(-4\sqrt{10})$

33. $2\sqrt{7a} \cdot 3\sqrt{2a}$

34. $2\sqrt{5c} \cdot 5\sqrt{5}$

35. $\sqrt[4]{9} \cdot \sqrt[4]{27}$

36. $\sqrt[3]{5} \cdot \sqrt[3]{100}$

37. $(2\sqrt{3})^2$

38. $(-4\sqrt{2})^2$

39. $\sqrt[3]{\dfrac{4x^2}{3}} \cdot \sqrt[3]{\dfrac{2x^2}{3}}$

40. $\sqrt[4]{\dfrac{4x^2}{5}} \cdot \sqrt[4]{\dfrac{4x^3}{25}}$

41. $2\sqrt{3}(\sqrt{6} + 3\sqrt{3})$

42. $2\sqrt{5}(\sqrt{3} + 3\sqrt{5})$

43. $\sqrt{5}(\sqrt{10} - 2)$

44. $\sqrt{6}(\sqrt{15} - 1)$

45. $\sqrt[3]{3t}(\sqrt[3]{9t} - \sqrt[3]{t^2})$

46. $\sqrt[3]{2}(\sqrt[3]{12x} - \sqrt[3]{2x})$

47. $(\sqrt{3} + 2)(\sqrt{3} - 5)$

48. $(\sqrt{5} + 2)(\sqrt{5} - 6)$

49. $(\sqrt{11} - 3)(\sqrt{11} + 3)$

50. $(\sqrt{2} + 5)(\sqrt{2} + 5)$

51. $(2\sqrt{5} - 7)(2\sqrt{5} + 4)$

52. $(2\sqrt{6} - 3)(2\sqrt{6} + 4)$

53. $(2\sqrt{3} - \sqrt{6})(\sqrt{3} + 2\sqrt{6})$

54. $(3\sqrt{3} - \sqrt{2})(\sqrt{2} + \sqrt{3})$

55. $(\sqrt{2} + 3)^2$

56. $(\sqrt{2} - \sqrt{3})^2$

57. $(5 + \sqrt{x - 25})^2$

58. $(1 - \sqrt{x - 1})^2$

Write each product as a single radical expression. See Example 5.

59. $\sqrt[3]{3} \cdot \sqrt{3}$

60. $\sqrt{3} \cdot \sqrt[4]{3}$

61. $\sqrt[3]{5} \cdot \sqrt[4]{5}$

62. $\sqrt[3]{2} \cdot \sqrt[5]{2}$

63. $\sqrt[3]{2} \cdot \sqrt{5}$

64. $\sqrt{6} \cdot \sqrt[3]{2}$

65. $\sqrt[3]{2} \cdot \sqrt[4]{3}$

66. $\sqrt[3]{3} \cdot \sqrt[4]{2}$

Find the product of each pair of conjugates. See Example 6.

67. $(\sqrt{3} - 2)(\sqrt{3} + 2)$

68. $(7 - \sqrt{3})(7 + \sqrt{3})$

69. $(\sqrt{5} + \sqrt{2})(\sqrt{5} - \sqrt{2})$

70. $(\sqrt{6} + \sqrt{5})(\sqrt{6} - \sqrt{5})$

71. $(2\sqrt{5} + 1)(2\sqrt{5} - 1)$

72. $(3\sqrt{2} - 4)(3\sqrt{2} + 4)$

73. $(3\sqrt{2} + \sqrt{5})(3\sqrt{2} - \sqrt{5})$

74. $(2\sqrt{3} - \sqrt{7})(2\sqrt{3} + \sqrt{7})$

75. $(5 - 3\sqrt{x})(5 + 3\sqrt{x})$

76. $(4\sqrt{y} + 3\sqrt{z})(4\sqrt{y} - 3\sqrt{z})$

Simplify each expression.

77. $\sqrt{300} + \sqrt{3}$

78. $\sqrt{50} + \sqrt{2}$

79. $2\sqrt{5} \cdot 5\sqrt{6}$

80. $3\sqrt{6} \cdot 5\sqrt{10}$

81. $(3 + 2\sqrt{7})(\sqrt{7} - 2)$

82. $(2 + \sqrt{7})(\sqrt{7} - 2)$

83. $4\sqrt{w} \cdot 4\sqrt{w}$

84. $3\sqrt{m} \cdot 5\sqrt{m}$

85. $\sqrt{3x^3} \cdot \sqrt{6x^2}$

86. $\sqrt{2t^5} \cdot \sqrt{10t^4}$

87. $\dfrac{1}{\sqrt{2}} - \dfrac{1}{\sqrt{8}} + \dfrac{1}{\sqrt{18}}$

88. $\dfrac{1}{\sqrt{3}} + \sqrt{\dfrac{1}{3}} - \sqrt{3}$

89. $(2\sqrt{5} + \sqrt{2})(3\sqrt{5} - \sqrt{2})$

90. $(3\sqrt{2} - \sqrt{3})(2\sqrt{2} + 3\sqrt{3})$

91. $\dfrac{\sqrt{2}}{3} + \dfrac{\sqrt{2}}{5}$

92. $\dfrac{\sqrt{2}}{4} + \dfrac{\sqrt{3}}{5}$

93. $(5 + 2\sqrt{2})(5 - 2\sqrt{2})$

94. $(3 - 2\sqrt{7})(3 + 2\sqrt{7})$

95. $(3 + \sqrt{x})^2$

96. $(1 - \sqrt{x})^2$

97. $(5\sqrt{x} - 3)^2$

98. $(3\sqrt{a} + 2)^2$

99. $(1 + \sqrt{x + 2})^2$

100. $(\sqrt{x - 1} + 1)^2$

101. $\sqrt{4w} - \sqrt{9w}$

102. $10\sqrt{m} - \sqrt{16m}$

103. $2\sqrt{a^3} + 3\sqrt{a^3} - 2a\sqrt{4a}$

104. $5\sqrt{w^2y} - 7\sqrt{w^2y} + 6\sqrt{w^2y}$

105. $\sqrt{50a} + \sqrt{18a} - \sqrt{2a}$

106. $\sqrt{200z} + \sqrt{128z} - \sqrt{8z}$

107. $\sqrt{x^5} + 2x\sqrt{x^3}$

108. $\sqrt{8x^3} + \sqrt{50x^3} - x\sqrt{2x}$

109. $(\sqrt{a} + a^3)(\sqrt{a} - a^3)$

110. $(\sqrt{wz} - 2y^4)(\sqrt{wz} + 2y^4)$

111. $\sqrt[3]{-16x^4} + 5x\sqrt[3]{54x}$

112. $\sqrt[3]{3x^5y^7} - \sqrt[3]{24x^5y^7}$

113. $\sqrt[3]{\dfrac{y^7}{4x}}$

114. $\sqrt[4]{\dfrac{16}{9z^3}}$

115. $\sqrt[3]{\dfrac{x}{5}} \cdot \sqrt[3]{\dfrac{x^5}{5}}$

116. $\sqrt[4]{a^3}(\sqrt[4]{a} - \sqrt[4]{a^5})$

117. $\sqrt[3]{2x} \cdot \sqrt{2x}$

118. $\sqrt[3]{2m} \cdot \sqrt[4]{2n}$

In Exercises 119–122, solve each problem.

119. ***Area of a rectangle.*** Find the exact area of a rectangle that has a length of $\sqrt{6}$ feet and a width of $\sqrt{3}$ feet.

120. ***Volume of a cube.*** Find the exact volume of a cube with sides of length $\sqrt{3}$ meters.

$\sqrt{3}\ m$

$\sqrt{3}\ m$

$\sqrt{3}\ m$

Figure for Exercise 120

121. *Area of a trapezoid.* Find the exact area of a trapezoid with a height of $\sqrt{6}$ feet and bases of $\sqrt{3}$ feet and $\sqrt{12}$ feet.

Figure for Exercise 121

122. *Area of a triangle.* Find the exact area of a triangle with a base of $\sqrt{30}$ meters and a height of $\sqrt{6}$ meters.

Figure for Exercise 122

Getting More Involved

123. *Discussion.* Is $\sqrt{a} + \sqrt{b} = \sqrt{a + b}$ for all values of a and b?

124. *Discussion.* Which of the following equations are identities? Explain your answers.
   a) $\sqrt{9x} = 3\sqrt{x}$
   b) $\sqrt{9 + x} = 3 + \sqrt{x}$
   c) $\sqrt{x - 4} = \sqrt{x} - 2$
   d) $\sqrt{\dfrac{x}{4}} = \dfrac{\sqrt{x}}{2}$

125. *Exploration.* Because 3 is the square of $\sqrt{3}$, a binomial such as $y^2 - 3$ is a difference of two squares.

   a) Factor $y^2 - 3$ and $2a^2 - 7$ using radicals.
   b) Use factoring with radicals to solve the equations $x^2 - 8 = 0$ and $3y^2 - 11 = 0$.
   c) Assuming that a and b are positive real numbers, solve the equations $x^2 - a = 0$ and $ax^2 - b = 0$.

5.4   More Operations with Radicals

In this section:
▶ Dividing Radicals
▶ Rationalizing the Denominator
▶ Powers of Radical Expressions

In this section you continue studying operations with radicals. We learn to rationalize some denominators that are different from those rationalized in Section 5.2.

Dividing Radicals

In Section 5.3 you learned how to add, subtract, and multiply radical expressions. To divide two radical expressions, simply write the quotient as a ratio and then simplify, as we did in Section 5.2. In general, we have

$$\sqrt[n]{a} \div \sqrt[n]{b} = \frac{\sqrt[n]{a}}{\sqrt[n]{b}} = \sqrt[n]{\frac{a}{b}}.$$

Note that the quotient rule is applied only to radicals that have the same index.

EXAMPLE 1   Dividing radicals with the same index

Divide and simplify. Assume that the variables represent positive numbers.
   a) $\sqrt{10} \div \sqrt{5}$ **b)** $(3\sqrt{2}) \div (2\sqrt{3})$ **c)** $\sqrt[3]{10x^2} \div \sqrt[3]{5x}$

Solution

a) $\sqrt{10} \div \sqrt{5} = \dfrac{\sqrt{10}}{\sqrt{5}}$ $a \div b = \frac{a}{b}$, provided that $b \neq 0$.

$\qquad\qquad\quad = \sqrt{\dfrac{10}{5}}$ Quotient rule for radicals

$\qquad\qquad\quad = \sqrt{2}$ Reduce.

b) $(3\sqrt{2}) \div (2\sqrt{3}) = \dfrac{3\sqrt{2}}{2\sqrt{3}}$

$\qquad\qquad\qquad = \dfrac{3\sqrt{2}}{2\sqrt{3}} \cdot \dfrac{\sqrt{3}}{\sqrt{3}}$ Rationalize the denominator.

$\qquad\qquad\qquad = \dfrac{3\sqrt{6}}{2 \cdot 3}$

$\qquad\qquad\qquad = \dfrac{\sqrt{6}}{2}$ Note that $\sqrt{6} \div 2 \neq \sqrt{3}$.

c) $\sqrt[3]{10x^2} \div \sqrt[3]{5x} = \dfrac{\sqrt[3]{10x^2}}{\sqrt[3]{5x}}$

$\qquad\qquad\qquad = \sqrt[3]{\dfrac{10x^2}{5x}}$ Quotient rule for radicals

$\qquad\qquad\qquad = \sqrt[3]{2x}$ Reduce.   ■

Note that in Example 1(a) we applied the quotient rule to get $\sqrt{10} \div \sqrt{5} = \sqrt{2}$. In Example 1(b) we did not use the quotient rule because 2 is not evenly divisible by 3. Instead, we rationalized the denominator to get the result in simplified form.

In Chapter 6 it will be necessary to simplify expressions of the type found in the next example.

EXAMPLE 2   **Simplifying radical expressions**
Simplify.

a) $\dfrac{4 - \sqrt{12}}{4}$ **b)** $\dfrac{-6 + \sqrt{20}}{-2}$

Solution

a) First write $\sqrt{12}$ in simplified form. Then simplify the expression.

$\qquad \dfrac{4 - \sqrt{12}}{4} = \dfrac{4 - 2\sqrt{3}}{4}$ Simplify $\sqrt{12}$.

$\qquad\qquad\quad = \dfrac{2(2 - \sqrt{3})}{2 \cdot 2}$ Factor.

$\qquad\qquad\quad = \dfrac{2 - \sqrt{3}}{2}$ Divide out the common factor.

b) $\dfrac{-6 + \sqrt{20}}{-2} = \dfrac{-6 + 2\sqrt{5}}{-2}$

$\qquad\qquad\quad = \dfrac{-2(3 - \sqrt{5})}{-2}$

$\qquad\qquad\quad = 3 - \sqrt{5}$ ■

CAUTION   To simplify the expression in Example 2, you must simplify the radical, factor the numerator, and then divide out the common factors. You cannot simply "cancel" the 4's in $\dfrac{4 - \sqrt{12}}{4}$ or the 2's in $\dfrac{2 - \sqrt{3}}{2}$ because they are not common factors.   ⊘

Rationalizing the Denominator

In Section 5.2 you learned that a simplified expression involving radicals does not have radicals in the denominator. If an expression such as $4 - \sqrt{3}$ appears in a denominator, we can multiply both numerator and denominator by its conjugate $4 + \sqrt{3}$ to get a rational number in the denominator.

EXAMPLE 3 **Rationalizing the denominator using conjugates.**

Write in simplified form.

a) $\dfrac{2 + \sqrt{3}}{4 - \sqrt{3}}$ b) $\dfrac{\sqrt{5}}{\sqrt{6} + \sqrt{2}}$

Solution

a) $\dfrac{2 + \sqrt{3}}{4 - \sqrt{3}} = \dfrac{(2 + \sqrt{3})(4 + \sqrt{3})}{(4 - \sqrt{3})(4 + \sqrt{3})}$ Multiple numerator and denominator by $4 + \sqrt{3}$.

$= \dfrac{8 + 6\sqrt{3} + 3}{13}$ $(4 - \sqrt{3})(4 + \sqrt{3}) = 16 - 3 = 13$

$= \dfrac{11 + 6\sqrt{3}}{13}$ Simplify.

b) $\dfrac{\sqrt{5}}{\sqrt{6} + \sqrt{2}} = \dfrac{\sqrt{5}(\sqrt{6} - \sqrt{2})}{(\sqrt{6} + \sqrt{2})(\sqrt{6} - \sqrt{2})}$ Multiply numerator and denominator by $\sqrt{6} - \sqrt{2}$.

$= \dfrac{\sqrt{30} - \sqrt{10}}{4}$ $(\sqrt{6} + \sqrt{2})(\sqrt{6} - \sqrt{2}) = 6 - 2 = 4$ ∎

Powers of Radical Expressions

We can use the power of a product rule and the power of a power rule to simplify a radical expression raised to a power. In the next example we also use the fact that a root and a power can be found in either order.

EXAMPLE 4 **Finding powers of rational expressions**

Simplify. Assume that the variables represent positive numbers.

a) $(5\sqrt{2})^3$ b) $(2\sqrt{x^3})^4$ c) $(3w\sqrt[3]{2w})^3$ d) $(2t\sqrt[4]{3t})^3$

Solution

a) $(5\sqrt{2})^3 = 5^3(\sqrt{2})^3$ Power of a product rule

$= 125\sqrt{8}$ $(\sqrt{2})^3 = \sqrt{2^3} = \sqrt{8}$

$= 125 \cdot 2\sqrt{2}$ $\sqrt{8} = \sqrt{4}\sqrt{2} = 2\sqrt{2}$

$= 250\sqrt{2}$

b) $(2\sqrt{x^3})^4 = 2^4(\sqrt{x^3})^4$

$= 16\sqrt{x^{12}}$

$= 16x^6$

c) $(3w\sqrt[3]{2w})^3 = 3^3 w^3 (\sqrt[3]{2w})^3$

$= 27w^3(2w)$

$= 54w^4$

d) $(2t\sqrt[4]{3t})^3 = 2^3 t^3 (\sqrt[4]{3t})^3 = 8t^3 \sqrt[4]{27t^3}$ ∎

Warm-ups

True or false? Explain.

1. $\dfrac{\sqrt{6}}{\sqrt{2}} = \sqrt{3}$

2. $\dfrac{2}{\sqrt{2}} = \sqrt{2}$

3. $\dfrac{4 - \sqrt{10}}{2} = 2 - \sqrt{10}$

4. $\dfrac{1}{\sqrt{3}} = \dfrac{\sqrt{3}}{3}$

5. $\dfrac{8\sqrt{7}}{2\sqrt{7}} = 4\sqrt{7}$

6. $\dfrac{2(2 + \sqrt{3})}{(2 - \sqrt{3})(2 + \sqrt{3})} = 4 + 2\sqrt{3}$

7. $\dfrac{\sqrt{12}}{3} = \sqrt{4}$

8. $\dfrac{\sqrt{20}}{\sqrt{5}} = 2$

9. $(2\sqrt{4})^2 = 16$

10. $(3\sqrt{5})^3 = 27\sqrt{125}$

5.4 EXERCISES

All variables in the following exercises represent positive numbers.

Divide and simplify. See Example 1.

1. $\sqrt{15} \div \sqrt{5}$
2. $\sqrt{14} \div \sqrt{7}$
3. $\sqrt{3} \div \sqrt{5}$
4. $\sqrt{5} \div \sqrt{7}$
5. $(3\sqrt{3}) \div (5\sqrt{6})$
6. $(2\sqrt{2}) \div (4\sqrt{10})$
7. $(2\sqrt{3}) \div (3\sqrt{6})$
8. $(5\sqrt{12}) \div (4\sqrt{6})$
9. $\sqrt[3]{20} \div \sqrt[3]{2}$
10. $\sqrt[4]{48} \div \sqrt[4]{3}$
11. $\sqrt[3]{8x^7} \div \sqrt[3]{2x}$
12. $\sqrt[4]{4a^{10}} \div \sqrt[4]{2a^2}$

Simplify. See Example 2.

13. $\dfrac{6 + \sqrt{45}}{3}$
14. $\dfrac{10 + \sqrt{50}}{5}$
15. $\dfrac{-2 + \sqrt{12}}{-2}$
16. $\dfrac{-6 + \sqrt{72}}{-6}$

Simplify each expression by rationalizing the denominator. See Example 3.

17. $\dfrac{1 + \sqrt{2}}{\sqrt{3} - 1}$
18. $\dfrac{2 - \sqrt{3}}{\sqrt{2} + \sqrt{6}}$
19. $\dfrac{\sqrt{2}}{\sqrt{6} + \sqrt{3}}$
20. $\dfrac{5}{\sqrt{7} - \sqrt{5}}$
21. $\dfrac{2\sqrt{3}}{3\sqrt{2} - \sqrt{5}}$
22. $\dfrac{3\sqrt{5}}{5\sqrt{2} + \sqrt{6}}$
23. $\dfrac{1 + 3\sqrt{2}}{2\sqrt{6} + 3\sqrt{10}}$
24. $\dfrac{3\sqrt{3} + 1}{4 - 5\sqrt{3}}$

Simplify. See Example 4.

25. $(2\sqrt{2})^5$
26. $(3\sqrt{3})^4$
27. $(\sqrt{x})^5$
28. $(2\sqrt{y})^3$
29. $(-3\sqrt{x^3})^3$
30. $(-2\sqrt{x^3})^4$
31. $(2x\sqrt[3]{x^2})^3$
32. $(2y\sqrt[3]{4y})^3$
33. $(-2\sqrt[3]{5})^2$
34. $(-3\sqrt[4]{4})^2$
35. $(\sqrt[3]{x^2})^6$
36. $(2\sqrt[4]{y^3})^3$

In Exercises 37–74, simplify.

37. $\dfrac{\sqrt{3}}{\sqrt{2}} + \dfrac{2}{\sqrt{2}}$
38. $\dfrac{2}{\sqrt{7}} + \dfrac{5}{\sqrt{7}}$
39. $\dfrac{\sqrt{3}}{\sqrt{2}} + \dfrac{3\sqrt{6}}{2}$
40. $\dfrac{\sqrt{3}}{2\sqrt{2}} + \dfrac{\sqrt{5}}{3\sqrt{2}}$
41. $\dfrac{\sqrt{6}}{2} \cdot \dfrac{1}{\sqrt{3}}$
42. $\dfrac{\sqrt{6}}{\sqrt{7}} \cdot \dfrac{\sqrt{14}}{\sqrt{3}}$
43. $(2\sqrt{w}) \div (3\sqrt{w})$
44. $2 \div (3\sqrt{a})$
45. $\dfrac{8 - \sqrt{32}}{20}$
46. $\dfrac{4 - \sqrt{28}}{6}$
47. $\dfrac{5 + \sqrt{75}}{10}$
48. $\dfrac{3 + \sqrt{18}}{6}$
49. $\sqrt{a}(\sqrt{a} - 3)$
50. $3\sqrt{m}(2\sqrt{m} - 6)$
51. $4\sqrt{a}(a + \sqrt{a})$
52. $\sqrt{3ab}(\sqrt{3a} + \sqrt{3})$
53. $(2\sqrt{3m})^2$
54. $(-3\sqrt{4y})^2$
55. $(-2\sqrt{xy^2z})^2$
56. $(5a\sqrt{ab})^2$
57. $\sqrt[3]{m}(\sqrt[3]{m^2} - \sqrt[3]{m^5})$
58. $\sqrt[4]{w}(\sqrt[4]{w^3} - \sqrt[4]{w^7})$
59. $\sqrt[3]{8x^4} + \sqrt[3]{27x^4}$
60. $\sqrt[3]{16a^4} + a\sqrt[3]{2a}$
61. $(2m\sqrt[4]{2m^2})^3$
62. $(-2t\sqrt[6]{2t^2})^5$
63. $\dfrac{4}{2 + \sqrt{8}}$
64. $\dfrac{6}{3 - \sqrt{18}}$

65. $\dfrac{5}{\sqrt{2} - 1} + \dfrac{3}{\sqrt{2} + 1}$ **66.** $\dfrac{\sqrt{3}}{\sqrt{6} - 1} - \dfrac{\sqrt{3}}{\sqrt{6} + 1}$

67. $\dfrac{1}{\sqrt{2}} + \dfrac{1}{\sqrt{3}}$ **68.** $\dfrac{4}{2\sqrt{3}} + \dfrac{1}{\sqrt{5}}$

69. $\dfrac{3}{\sqrt{2} - 1} + \dfrac{4}{\sqrt{2} + 1}$

70. $\dfrac{3}{\sqrt{5} - \sqrt{3}} - \dfrac{2}{\sqrt{5} + \sqrt{3}}$

71. $\dfrac{\sqrt{x}}{\sqrt{x} + 2} + \dfrac{3\sqrt{x}}{\sqrt{x} - 2}$ **72.** $\dfrac{\sqrt{5}}{3 - \sqrt{y}} - \dfrac{\sqrt{5y}}{3 + \sqrt{y}}$

73. $\dfrac{1}{\sqrt{x}} + \dfrac{1}{1 - \sqrt{x}}$ **74.** $\dfrac{\sqrt{x}}{\sqrt{x} - 3} + \dfrac{5}{\sqrt{x}}$

Replace the question mark by an expression that makes the equation correct. Equations involving variables are to be identities.

75. $\dfrac{\sqrt{2}}{\sqrt{3}} = \dfrac{\sqrt{6}}{?}$ **76.** $\dfrac{2}{?} = \sqrt{2}$

77. $\dfrac{1}{\sqrt{2} - 1} = \dfrac{\sqrt{2} + 1}{?}$ **78.** $\dfrac{\sqrt{6}}{\sqrt{6} + 2} = \dfrac{?}{2}$

79. $\dfrac{1}{\sqrt{x} - 1} = \dfrac{?}{x - 1}$ **80.** $\dfrac{5}{3 - \sqrt{x}} = \dfrac{?}{9 - x}$

81. $\dfrac{3}{\sqrt{2} + x} = \dfrac{?}{2 - x^2}$ **82.** $\dfrac{4}{2\sqrt{3} + a} = \dfrac{?}{12 - a^2}$

Use a calculator to find a decimal approximation for each radical expression. Round your answers to three decimal places.

83. $\sqrt{3} + \sqrt{5}$ **84.** $\sqrt{5} + \sqrt{7}$

85. $2\sqrt{3} + 5\sqrt{3}$ **86.** $7\sqrt{3}$

87. $(2\sqrt{3})(3\sqrt{2})$ **88.** $6\sqrt{6}$

89. $\sqrt{5}(\sqrt{5} + \sqrt{3})$ **90.** $5 + \sqrt{15}$

91. $\dfrac{-1 + \sqrt{6}}{2}$ **92.** $\dfrac{-1 - \sqrt{6}}{2}$

93. $\dfrac{4 - \sqrt{10}}{-2}$ **94.** $\dfrac{4 + \sqrt{10}}{-2}$

Getting More Involved

95. *Exploration.* A polynomial is prime if it cannot be factored by using integers, but many prime polynomials can be factored if we use radicals.
  a) Find the product $(x - \sqrt[3]{2})(x^2 + \sqrt[3]{2}\,x + \sqrt[3]{4})$.
  b) Factor $x^3 + 5$ using radicals.
  c) Find the product
    $(\sqrt[3]{5} - \sqrt[3]{2})(\sqrt[3]{25} + \sqrt[3]{10} + \sqrt[3]{4})$.
  d) Use radicals to factor $a + b$ as a sum of two cubes and $a - b$ as a difference of two cubes.

96. *Discussion.* Which one of the following expressions is not equivalent to the others?
  a) $(\sqrt[3]{x})^4$ **b)** $\sqrt[4]{x^3}$ **c)** $\sqrt[3]{x^4}$
  d) $x^{4/3}$ **e)** $(x^{1/3})^4$

5.5   Solving Equations with Radicals and Exponents

In this section:
▶ The Odd-Root Property
▶ The Even-Root Property
▶ Raising Each Side to a Power
▶ Equations Involving Rational Exponents
▶ Summary of Methods
▶ Applications

One of our goals in algebra is to keep increasing our knowledge of solving equations because the solutions to equations can give us the answers to various applied questions. In this section we apply our knowledge of radicals and exponents to solving some new types of equations.

The Odd-Root Property

Because $(-2)^3 = -8$ and $2^3 = 8$, the equation $x^3 = 8$ is equivalent to $x = 2$. The equations $x^3 = -8$ is equivalent to $x = -2$. Because there is only one real odd root of each real number, there is a simple rule for writing an equivalent equation in this situation.

Odd-Root Property

If n is an odd positive integer,

$$x^n = k \quad \text{is equivalent to} \quad x = \sqrt[n]{k}$$

for any real number k.

EXAMPLE 1   **Using the odd-root property**

Solve each equation.

a) $x^3 = 27$ **b)** $x^5 + 32 = 0$ **c)** $(x - 2)^3 = 24$

Solution

a) $x^3 = 27$

$\qquad x = \sqrt[3]{27}$ Odd-root property

$\qquad x = 3$

Check 3 in the original equation. The solution set is {3}.

b) $x^5 + 32 = 0$

$\qquad x^5 = -32$ Isolate the variable.

$\qquad x = \sqrt[5]{-32}$ Odd-root property

$\qquad x = -2$

Check -2 in the original equation. The solution set is {-2}.

c) $(x - 2)^3 = 24$

$\qquad x - 2 = \sqrt[3]{24}$ Odd-root property

$\qquad x = 2 + 2\sqrt[3]{3}$ $\sqrt[3]{24} = \sqrt[3]{8} \cdot \sqrt[3]{3} = 2\sqrt[3]{3}$

Check. The solution set is {$2 + 2\sqrt[3]{3}$}.   ∎

The Even-Root Property

In solving the equation $x^2 = 4$, you might be tempted to write $x = 2$ as an equivalent equation. But $x = 2$ is not equivalent to $x^2 = 4$ because $2^2 = 4$ and $(-2)^2 = 4$. So the solution set to $x^2 = 4$ is {$-2, 2$}. The equation $x^2 = 4$ is equivalent to the compound sentence $x = 2$ or $x = -2$, which we can abbreviate as $x = \pm 2$. The equation $x = \pm 2$ is read "x equals positive or negative 2."

Equations involving other even powers are handled like the squares. Because $2^4 = 16$ and $(-2)^4 = 16$, the equation $x^4 = 16$ is equivalent to $x = \pm 2$. So $x^4 = 16$ has two real solutions. Note that $x^4 = -16$ has no real solutions. The equation $x^6 = 5$ is equivalent to $x = \pm\sqrt[6]{5}$. We can now state a general rule.

Even-Root Property

Suppose n is a positive even integer.

If $k > 0$, then $x^n = k$ is equivalent to $x = \pm\sqrt[n]{k}$.

If $k = 0$, then $x^n = k$ is equivalent to $x = 0$.

If $k < 0$, then $x^n = k$ has no real solution.

EXAMPLE 2   **Using the even-root property**

Solve each equation.

a) $x^2 = 10$ **b)** $w^8 = 0$ **c)** $x^4 = -4$

Solution

a) $x^2 = 10$

       $x = \pm\sqrt{10}$ Even-root property

     The solution set is $\{-\sqrt{10}, \sqrt{10}\}$, or $\{\pm\sqrt{10}\}$.

b) $w^8 = 0$

       $w = 0$ Even-root property

     The solution set is $\{0\}$.

c)   By the even-root property, $x^4 = -4$ has no real solution. (The fourth power of any real number is nonnegative.)   ■

     In the next example the even-root property is used to solve some equations that are a bit more complicated than those of Example 2.

EXAMPLE 3   **Using the even-root property**

Solve each equation.

a) $(x - 3)^2 = 4$ **b)** $2(x - 5)^2 - 7 = 0$ **c)** $x^4 - 1 = 80$

Solution

a) $(x - 3)^2 = 4$

       $x - 3 = 2$ or $x - 3 = -2$ Even-root property

          $x = 5$ or $x = 1$ Add 3 to each side.

     The solution set is $\{1, 5\}$.

b) $2(x - 5)^2 - 7 = 0$

         $2(x - 5)^2 = 7$ Add 7 to each side.

         $(x - 5)^2 = \dfrac{7}{2}$ Divide each side by 2.

       $x - 5 = \sqrt{\dfrac{7}{2}}$ or $x - 5 = -\sqrt{\dfrac{7}{2}}$ Even-root property

         $x = 5 + \dfrac{\sqrt{14}}{2}$ or $x = 5 - \dfrac{\sqrt{14}}{2}$ $\sqrt{\dfrac{7}{2}} = \dfrac{\sqrt{7}\cdot\sqrt{2}}{\sqrt{2}\cdot\sqrt{2}} = \dfrac{\sqrt{14}}{2}$

           $x = \dfrac{10 + \sqrt{14}}{2}$ or $x = \dfrac{10 - \sqrt{14}}{2}$

     The solution set is $\left\{ \dfrac{10 + \sqrt{14}}{2}, \dfrac{10 - \sqrt{14}}{2} \right\}$.

c) $x^4 - 1 = 80$

       $x^4 = 81$

       $x = \pm\sqrt[4]{81} = \pm 3$

     The solution set is $\{-3, 3\}$.   ■

     In Chapter 3 we solved quadratic equations by factoring. The quadratic equations that we encounter in this chapter can be solved by using the even-root property as in parts (a) and (b) of Example 3. In Chapter 6 you will learn general methods for solving any quadratic equation.

Raising Each Side to a Power

If we start with the equation $x = 3$ and square both sides, we get $x^2 = 9$. The solution set to $x^2 = 9$ is $\{-3, 3\}$; the solution set to the original equation is $\{3\}$. Squaring both sides of an equation might produce a *nonequivalent* equation that has more solutions than the original equation. We call these additional solutions **extraneous solutions.** However, any solution of the original must be among the solutions to the new equation.

CAUTION   When you solve an equation by raising each side to a power, you must check your answers. Raising each side to an odd power will always give an equivalent equation; raising each side to an even power might not.   ⊘

EXAMPLE 4   **Raising each side to a power to eliminate radicals**

Solve each equation.

a) $\sqrt{2x - 3} - 5 = 0$ **b)** $\sqrt[3]{3x + 5} = \sqrt[3]{x - 1}$ **c)** $\sqrt{3x + 18} = x$

Solution

a) Eliminate the square root by raising each side to the power 2:

$$\sqrt{2x - 3} - 5 = 0 \qquad \text{Original equation}$$
$$\sqrt{2x - 3} = 5 \qquad \text{Isolate the radical.}$$
$$(\sqrt{2x - 3})^2 = 5^2 \qquad \text{Square both sides.}$$
$$2x - 3 = 25$$
$$2x = 28$$
$$x = 14$$

Check by evaluating $x = 14$ in the original equation:

$$\sqrt{2(14) - 3} - 5 = 0$$
$$\sqrt{28 - 3} - 5 = 0$$
$$\sqrt{25} - 5 = 0$$
$$0 = 0$$

The solution set is $\{14\}$.

b)
$$\sqrt[3]{3x + 5} = \sqrt[3]{x - 1} \qquad \text{Original equation}$$
$$(\sqrt[3]{3x + 5})^3 = (\sqrt[3]{x - 1})^3 \qquad \text{Cube each side.}$$
$$3x + 5 = x - 1$$
$$2x = -6$$
$$x = -3$$

Check $x = -3$ in the original equation:

$$\sqrt[3]{3(-3) + 5} = \sqrt[3]{-3 - 1}$$
$$\sqrt[3]{-4} = \sqrt[3]{-4}$$

Note that $\sqrt[3]{-4}$ is a real number. The solution set is $\{-3\}$. In this example we checked for arithmetic mistakes. There was no possibility of extraneous solutions here because we raised each side to an odd power.

c)

$$\sqrt{3x + 18} = x \qquad \text{Original equation}$$

$$(\sqrt{3x + 18})^2 = x^2 \qquad \text{Square both sides.}$$

$$3x + 18 = x^2 \qquad \text{Simplify.}$$

$$-x^2 + 3x + 18 = 0 \qquad \text{Subtract } x^2 \text{ from each side to get zero on one side.}$$

$$x^2 - 3x - 18 = 0 \qquad \text{Multiply each side by } -1 \text{ for easier factoring.}$$

$$(x - 6)(x + 3) = 0 \qquad \text{Factor.}$$

$$x - 6 = 0 \quad \text{or} \quad x + 3 = 0 \qquad \text{Zero factor property}$$

$$x = 6 \quad \text{or} \quad x = -3$$

Because we squared both sides, we must check for extraneous solutions. If $x = -3$ in the original equation $\sqrt{3x + 18} = x$, we get

$$\sqrt{3(-3) + 18} = -3$$

$$\sqrt{9} = -3$$

$$3 = -3,$$

which is not correct. If $x = 6$ in the original equation, we get

$$\sqrt{3(6) + 18} = 6,$$

which is correct. The solution set is $\{6\}$. ∎

In the next example the radicals are not eliminated after squaring both sides of the equation. In this case we must square both sides a second time. Note that we square the side with two terms the same way we square a binomial.

EXAMPLE 5   **Squaring both sides twice**

Solve $\sqrt{5x - 1} - \sqrt{x + 2} = 1$.

Solution

It is easier to square both sides if the two radicals are not on the same side, so we first rewrite the equation:

$$\sqrt{5x - 1} - \sqrt{x + 2} = 1 \qquad \text{Original equation}$$

$$\sqrt{5x - 1} = 1 + \sqrt{x + 2} \qquad \text{Add } \sqrt{x + 2} \text{ to each side.}$$

$$(\sqrt{5x - 1})^2 = (1 + \sqrt{x + 2})^2 \qquad \text{Square both sides.}$$

$$5x - 1 = 1 + 2\sqrt{x + 2} + x + 2 \qquad \text{Square the right side like a binomial.}$$

$$5x - 1 = 3 + x + 2\sqrt{x + 2} \qquad \text{Combine like terms on the right side.}$$

$$4x - 4 = 2\sqrt{x + 2} \qquad \text{Isolate the square root.}$$

$$2x - 2 = \sqrt{x + 2} \qquad \text{Divide each side by 2.}$$

$$(2x - 2)^2 = (\sqrt{x + 2})^2 \qquad \text{Square both sides.}$$

$$4x^2 - 8x + 4 = x + 2 \qquad \text{Square the binomial on the left side.}$$

$$4x^2 - 9x + 2 = 0$$

$$(4x - 1)(x - 2) = 0$$

$$4x - 1 = 0 \quad \text{or} \quad x - 2 = 0$$

$$x = \frac{1}{4} \quad \text{or} \quad x = 2$$

Check to see whether $\sqrt{5x - 1} - \sqrt{x + 2} = 1$ for $x = \frac{1}{4}$ and for $x = 2$:

$$\sqrt{5 \cdot \frac{1}{4} - 1} - \sqrt{\frac{1}{4} + 2} = \sqrt{\frac{1}{4}} - \sqrt{\frac{9}{4}} = \frac{1}{2} - \frac{3}{2} = -1$$

$$\sqrt{5 \cdot 2 - 1} - \sqrt{2 + 2} = \sqrt{9} - \sqrt{4} = 3 - 2 = 1$$

Because $\frac{1}{4}$ does not satisfy the original equation, the solution set is $\{2\}$. ∎

Equations Involving Rational Exponents

Equations involving rational exponents can be solved by combining the methods that you just learned for eliminating radicals and integral exponents. For equations involving rational exponents, always eliminate the root first and the power second.

EXAMPLE 6   **Eliminating the root, then the power**

Solve each equation.

a) $x^{2/3} = 4$ **b)** $(w - 1)^{-2/5} = 4$

Solution

a) Because the exponent $2/3$ indicates a cube root, raise each side to the power 3:

$x^{2/3} = 4$	Original equation
$(x^{2/3})^3 = 4^3$	Cube each side.
$x^2 = 64$	Multiply the exponents: $\frac{2}{3} \cdot 3 = 2$.
$x = 8$ or $x = -8$	Even-root property

All of the equations are equivalent. Check 8 and -8 in the original equation. The solution set is $\{-8, 8\}$.

b)

$(w - 1)^{-2/5} = 4$	Original equation
$[(w - 1)^{-2/5}]^{-5} = 4^{-5}$	Raise each side to the power -5 to eliminate the negative exponent.
$(w - 1)^2 = \dfrac{1}{1024}$	Multiply the exponents: $-\frac{2}{5}(-5) = 2$.
$w - 1 = \pm\sqrt{\dfrac{1}{1024}}$	Even-root property
$w - 1 = \dfrac{1}{32}$ or $w - 1 = -\dfrac{1}{32}$	
$w = \dfrac{33}{32}$ or $w = \dfrac{31}{32}$	

Check the values in the original equation. The solution set is $\left\{\frac{31}{32}, \frac{33}{32}\right\}$. ∎

An equation with a rational exponent might not have a real solution because all even powers of real numbers are nonnegative.

EXAMPLE 7 **An equation with no solution**

Solve $(2t - 3)^{-2/3} = -1$.

Solution

Raise each side to the power -3 to eliminate the root and the negative sign in the exponent:

$$(2t - 3)^{-2/3} = -1 \qquad \text{Original equation}$$
$$[(2t - 3)^{-2/3}]^{-3} = (-1)^{-3} \qquad \text{Raise each side to the } -3 \text{ power.}$$
$$(2t - 3)^2 = -1 \qquad \text{Multiply the exponents: } -\tfrac{2}{3}(-3) = 2.$$

By the even-root property this equation has no real solution. The square of every real number is nonnegative. ∎

Summary of Methods

The three most important rules for solving equations with exponents and radicals are restated here.

> **Strategy for Solving Equations with Exponents and Radicals**
>
> 1. In raising each side of an equation to an even power, we can create an equation that gives extraneous solutions. We must check all possible solutions in the original equation.
> 2. When applying the even-root property, remember that there is a positive and a negative even root for any positive real number.
> 3. For equations with rational exponents, raise each side to a positive or negative integral power first, then apply the even- or odd-root property. (Positive fraction—raise to a positive power; negative fraction—raise to a negative power.)

Applications

Applications of exponents and roots often involve the Pythagorean theorem, which was introduced in Section 3.9. This theorem says that in any right triangle the sum of the squares of the lengths of the legs is equal to the hypotenuse squared. We will use the Pythagorean theorem to solve the following geometric problem.

EXAMPLE 8 **Diagonal of a baseball diamond**

A baseball diamond is actually a square, 90 feet on each side. What is the distance from third base to first base?

Solution

First make a sketch as in Fig. 5.1. The distance x from third base to first base is the length of the diagonal of the square shown in Fig. 5.1. The Pythagorean theorem can be applied to the right triangle formed from the diagonal and two

Figure 5.1

sides of the square. The sum of the squares of the sides is equal to the diagonal squared:

$$x^2 = 90^2 + 90^2$$
$$x^2 = 8100 + 8100$$
$$x^2 = 16{,}200$$
$$x = \pm\sqrt{16{,}200} = \pm 90\sqrt{2}$$

The length of the diagonal of a square must be positive, so we disregard the negative solution. Checking the answer in the original equation verifies that the *exact* length of the diagonal is $90\sqrt{2}$ feet. ■

Warm-ups

True or false? Explain.

1. The equations $x^2 = 4$ and $x = 2$ are equivalent.
2. The equation $x^2 = -25$ has no real solution.
3. There is no solution to the equation $x^2 = 0$.
4. The equation $x^3 = 8$ is equivalent to $x = \pm 2$.
5. The equation $-\sqrt{x} = 16$ has no solution.
6. To solve $\sqrt{x-3} = \sqrt{2x+5}$, first apply the even-root property.
7. Extraneous solutions are solutions that cannot be found.
8. Squaring both sides of $\sqrt{x} = -7$ yields an equation with an extraneous solution.
9. The equations $x^2 - 6 = 0$ and $x = \pm\sqrt{6}$ are equivalent.
10. Cubing each side of an equation will not produce an extraneous solution.

5.5 EXERCISES

Solve each equation. See Example 1.

1. $x^3 = -1000$
2. $y^3 = 125$
3. $32m^5 - 1 = 0$
4. $243a^5 + 1 = 0$
5. $(y-3)^3 = -8$
6. $(x-1)^3 = -1$
7. $\frac{1}{2}x^3 + 4 = 0$
8. $3(x-9)^7 = 0$

Solve each equation. See Examples 2 and 3.

9. $x^2 = 25$
10. $x^2 = 36$
11. $x^2 - 20 = 0$
12. $a^2 - 40 = 0$
13. $x^2 = -9$
14. $w^2 + 49 = 0$
15. $(x-3)^2 = 16$
16. $(a-2)^2 = 25$
17. $(x+1)^2 - 8 = 0$
18. $(w+3)^2 - 12 = 0$
19. $\frac{1}{2}x^2 = 5$
20. $\frac{1}{3}x^2 = 6$
21. $4(a+1)^2 = 1$
22. $9(m-3)^2 = 1$

23. $(y-3)^4 = 0$
24. $(2x-3)^6 = 0$
25. $2x^6 = 128$
26. $3y^4 = 48$
27. $81a^4 - 16 = 0$
28. $64w^6 - 1 = 0$

Solve each equation and check for extraneous solutions. See Example 4.

29. $\sqrt{x-3} - 7 = 0$
30. $\sqrt{a-1} - 6 = 0$
31. $2\sqrt{w+4} = 5$
32. $3\sqrt{w+1} = 6$
33. $\sqrt[3]{2x+3} = \sqrt[3]{x+12}$
34. $\sqrt[3]{a+3} = \sqrt[3]{2a-7}$
35. $\sqrt{2t+4} = \sqrt{t-1}$
36. $\sqrt{w-3} = \sqrt{4w+15}$
37. $\sqrt{4x^2+x-3} = 2x$
38. $\sqrt{x^2-5x+2} = x$
39. $\sqrt{x^2+2x-6} = 3$
40. $\sqrt{x^2-x-4} = 4$
41. $\sqrt{2x^2-1} = x$
42. $\sqrt{2x^2-3x-10} = x$
43. $\sqrt{2x^2+5x+6} = x$
44. $\sqrt{5x^2-9} = 2x$
45. $\sqrt{2x^2+6x+4} = x+1$
46. $\sqrt{2x^2+2x} = x+1$

Solve each equation and check for extraneous solutions. See Example 5.

47. $\sqrt{x+3} - \sqrt{x-2} = 1$

48. $\sqrt{2x+1} - \sqrt{x} = 1$

49. $\sqrt{2x+2} - \sqrt{x-3} = 2$

50. $\sqrt{3x} - \sqrt{x-2} = 4$

51. $\sqrt{4-x} - \sqrt{x+6} = 2$

52. $\sqrt{6-x} - \sqrt{x-2} = 2$

Solve each equation. See Examples 6 and 7.

53. $x^{2/3} = 3$ **54.** $a^{2/3} = 2$ **55.** $y^{-2/3} = 9$

56. $w^{-2/3} = 4$ **57.** $w^{1/3} = 8$ **58.** $a^{1/3} = 27$

59. $t^{-1/2} = 9$ **60.** $w^{-1/4} = \dfrac{1}{2}$

61. $(3a-1)^{-2/5} = 1$ **62.** $(r-1)^{-2/3} = 1$

63. $(t-1)^{-2/3} = 2$ **64.** $(w+3)^{-1/3} = \dfrac{1}{3}$

65. $(x-3)^{2/3} = -4$ **66.** $(x+2)^{3/2} = -1$

Solve each equation.

67. $2x^2 + 3 = 7$ **68.** $3x^2 - 5 = 16$

69. $\sqrt[3]{2w+3} = \sqrt[3]{w-2}$

70. $\sqrt[3]{2-w} = \sqrt[3]{2w-28}$ **71.** $9x^2 - 1 = 0$

72. $4x^2 - 1 = 0$ **73.** $(w+1)^{2/3} = -3$

74. $(x-2)^{3/4} = 2$ **75.** $(a+1)^{1/3} = -2$

76. $(a-1)^{1/3} = -3$ **77.** $(4y-5)^7 = 0$

78. $(5x)^9 = 0$ **79.** $\sqrt{x^2+5x} = 6$

80. $\sqrt{x^2-8x} = -3$ **81.** $\sqrt{4x^2} = x+2$

82. $\sqrt{9x^2} = x+6$ **83.** $(t+2)^4 = 32$

84. $(w+1)^4 = 48$ **85.** $\sqrt{x^2-3x} = x$

86. $\sqrt[4]{4x^4-48} = -x$ **87.** $x^{-3} = 8$

88. $x^{-2} = 4$ **89.** $a^{-2} = 3$ **90.** $w^{-2} = 18$

91. $\sqrt{x+1} - \sqrt{2x+9} = -2$

92. $\sqrt{x-4} + \sqrt{2x-1} = 4$

Solve each problem by writing an equation and solving it. Find the exact answer and simplify it using the rules for radicals. See Example 8.

93. *Side of a square.* Find the length of the side of a square whose diagonal is 8 feet.

94. *Diagonal of a patio.* Find the length of the diagonal of a square patio with an area of 40 square meters.

95. *Side of a sign.* Find the length of the side of a square sign whose area is 50 square feet.

96. *Side of a cube.* Find the length of the side of a cubic box whose volume is 80 cubic feet.

97. *Diagonal of a rectangle.* If the sides of a rectangle are 30 feet and 40 feet in length, find the length of the diagonal of the rectangle.

98. *Diagonal of a sign.* What is the length of the diagonal of a rectangular billboard whose sides are 5 meters and 12 meters?

99. *Diagonal of a side.* Find the length of the diagonal of a side of a cubic packing crate whose volume is 2 cubic meters.

100. *Volume of a cube.* Find the volume of a cube on which the diagonal of a side measures 2 feet.

101. *Length of a road.* An architect designs a public park in the shape of a trapezoid. Find the length of the diagonal road marked a in the figure.

Figure for Exercises 101 and 102

102. *Length of a boundary.* Find the length of the border of the park marked b in the trapezoid shown in the figure.

103. *Average annual return.* The formula

$$r = \left(\frac{S}{P}\right)^{1/n} - 1$$

was used to find the average annual return on an investment in Exercise 129 in Section 5.1. Solve the formula for S (the amount). Solve it for P (the original principal).

104. *Surface area of a cube.* The formula $A = 6V^{2/3}$ gives the surface area of a cube in terms of its volume V. What is the volume of a cube with surface area 12 square feet?

105. *Kepler's third law.* According to Kepler's third law of planetary motion, the ratio $\dfrac{T^2}{R^3}$ has the same value for every planet in our solar system. R is the average radius of the orbit of the planet measured in astronomical units (AU), and T is the number of years it takes for one complete orbit of the sun. Jupiter orbits the sun in 11.86 years with an average radius of 5.2 AU, while Saturn orbits the sun in

29.46 years. Find the average radius of the orbit of Saturn. (One AU is the distance from the earth to the sun.)

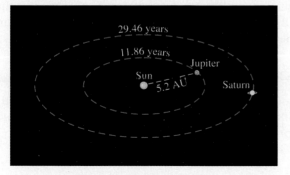

Figure for Exercise 105

106. *Orbit of Venus.* If the average radius of the orbit of Venus is is 0.723 AU, then how many years does it take for Venus to complete one orbit of the sun? Use the information in Exercise 105.

Use a calculator to find approximate solutions to the following equations. Round your answers to three decimal places.

107. $x^2 = 3.24$ **108.** $(x + 4)^3 = 7.51$

109. $\sqrt{x - 2} = 1.73$ **110.** $\sqrt[3]{x - 5} = 3.7$

111. $x^{2/3} = 8.86$ **112.** $(x - 1)^{-3/4} = 7.065$

Getting More Involved

113. *Cooperative learning.* Work in a small group to write a formula that gives the side of a cube in terms of the volume of the cube and explain the formula to the other groups.

114. *Cooperative learning.* Work in a small group to write a formula that gives the side of a square in terms of the diagonal of the square and explain the formula to the other groups.

5.6 Complex Numbers

In this section:

▶ Definition
▶ Addition, Subtraction, and Multiplication
▶ Division of Complex Numbers
▶ Square Roots of Negative Numbers
▶ Imaginary Solutions to Equations

In Chapter 1 we discussed the real numbers and the various subsets of the real numbers. In this section we define a set of numbers that has the real numbers as a subset.

Definition

The equation $2x = 1$ has no solution in the set of integers, but in the set of rational numbers, $2x = 1$ has a solution. The situation is similar for the equation $x^2 = -4$. It has no solution in the set of real numbers because the square of every real number is nonnegative. However, in the set of complex numbers, $x^2 = -4$ has two solutions. The complex numbers were developed so that equations such as $x^2 = -4$ would have solutions.

The complex numbers are based on the symbol $\sqrt{-1}$. In the real number system this symbol has no meaning. In the set of complex numbers this symbol is given meaning. We call it i. We make the definition that

$$i = \sqrt{-1} \quad \text{and} \quad i^2 = -1.$$

Complex Numbers

The set of **complex numbers** is the set of all numbers of the form

$$a + bi,$$

where a and b are real numbers, $i = \sqrt{-1}$, and $i^2 = -1$.

In the complex number $a + bi$, a is called the **real part** and b is called the **imaginary part.** If $b \neq 0$, the number $a + bi$ is called an **imaginary number.**

In dealing with complex numbers, we treat $a + bi$ as if it were a binomial, with i being a variable. Thus we would write $2 + (-3)i$ as $2 - 3i$. We agree that $2 + i3$, $3i + 2$, and $i3 + 2$ are just different ways of writing $2 + 3i$ (the standard form). Some examples of complex numbers are

$$-3 - 5i, \quad \frac{2}{3} - \frac{3}{4}i, \quad 1 + i\sqrt{2}, \quad 9 + 0i, \quad \text{and} \quad 0 + 7i.$$

For simplicity we write only $7i$ for $0 + 7i$. The complex number $9 + 0i$ is the real number 9, and $0 + 0i$ is the real number 0. Any complex number with $b = 0$ is a real number. For any real number a,

$$a + 0i = a.$$

The set of real numbers is a subset of the set of complex numbers. See Fig. 5.2.

Complex numbers

Real numbers	Imaginary numbers
$3, \pi, \frac{5}{2}, 0, -9, \sqrt{2}$	$i, 2 + 3i, \sqrt{-5}, -3 - 8i$

Figure 5.2

Addition, Subtraction, and Multiplication

Addition and subtraction of complex numbers is performed as if the complex numbers were algebraic expressions with i being a variable.

EXAMPLE 1   **Addition and subtraction of complex numbers**

Find the sums and differences.

a) $(2 + 3i) + (6 + i)$ **b)** $(-2 + 3i) + (-2 - 5i)$

c) $(3 + 5i) - (1 + 2i)$ **d)** $(-2 - 3i) - (1 - i)$

Solution

a) $(2 + 3i) + (6 + i) = 8 + 4i$

b) $(-2 + 3i) + (-2 - 5i) = -4 - 2i$

c) $(3 + 5i) - (1 + 2i) = 2 + 3i$

d) $(-2 - 3i) - (1 - i) = -3 - 2i$ ■

We can give a symbolic definition of addition and subtraction as follows.

Addition and Subtraction of Complex Numbers

The sum and difference of $a + bi$ and $c + di$ are defined as follows:

$$(a + bi) + (c + di) = (a + c) + (b + d)i$$
$$(a + bi) - (c + di) = (a - c) + (b - d)i$$

Complex numbers are multiplied as if they were algebraic expressions. Whenever i^2 appears, we replace it by -1.

EXAMPLE 2 **Products of complex numbers**

Find each product.

a) $2i(1 + i)$ b) $(2 + 3i)(4 + 5i)$ c) $(3 + i)(3 - i)$

Solution

a) $2i(1 + i) = 2i + 2i^2$ Distributive property

$\qquad\qquad = 2i + 2(-1)$ $i^2 = -1$

$\qquad\qquad = -2 + 2i$

b) Use the FOIL method to find the product:

$(2 + 3i)(4 + 5i) = 8 + 10i + 12i + 15i^2$

$\qquad\qquad\qquad = 8 + 22i + 15(-1)$ Replace i^2 by -1.

$\qquad\qquad\qquad = 8 + 22i - 15$

$\qquad\qquad\qquad = -7 + 22i$

c) This product is the product of a sum and a difference.

$(3 + i)(3 - i) = 9 - 3i + 3i - i^2$

$\qquad\qquad\qquad = 9 - (-1)$ $i^2 = -1$

$\qquad\qquad\qquad = 10$ ■

We can find powers of i using the fact that $i^2 = -1$. For example,

$$i^3 = i^2 \cdot i = -1 \cdot i = -i.$$

The value of i^4 is found from the value of i^3:

$$i^4 = i^3 \cdot i = -i \cdot i = -i^2 = 1.$$

In the next example we find more powers of imaginary numbers.

EXAMPLE 3 **Powers of imaginary numbers**

Write each expression in the form $a + bi$.

a) $(2i)^2$ b) $(-2i)^2$ c) i^6

Solution

a) $(2i)^2 = 2^2 \cdot i^2 = 4(-1) = -4$

b) $(-2i)^2 = (-2)^2 \cdot i^2 = 4i^2 = 4(-1) = -4$

c) $i^6 = i^2 \cdot i^4 = -1 \cdot 1 = -1$ ■

For completeness we give the following symbolic definition of multiplication of complex numbers. However, it is simpler to find products as we did in Examples 2 and 3 than to use this definition.

Multiplication of Complex Numbers

The complex numbers $a + bi$ and $c + di$ are multiplied as follows:

$$(a + bi)(c + di) = (ac - bd) + (ad + bc)i$$

Division of Complex Numbers

To divide a complex number by a real number, divide each term by the real number, just as we would divide a binomial by a number. For example,

$$\frac{4 + 6i}{2} = \frac{2(2 + 3i)}{2}$$

$$= 2 + 3i.$$

To understand division by a complex number, we first look at imaginary numbers that have a real product. The product of the two imaginary numbers in Example 2(c) is a real number:

$$(3 + i)(3 - i) = 10$$

We say that $3 + i$ and $3 - i$ are complex conjugates of each other.

Complex Conjugates

The complex numbers $a + bi$ and $a - bi$ are called **complex conjugates** of one another. Their product is the real number $a^2 + b^2$.

EXAMPLE 4 **Products of conjugates**

Find the product of the given complex number and its conjugate.

a) $2 + 3i$ **b)** $5 - 4i$

Solution

a) The conjugate of $2 + 3i$ is $2 - 3i$.

$$(2 + 3i)(2 - 3i) = 4 - 9i^2$$

$$= 4 + 9$$

$$= 13$$

b) The conjugate of $5 - 4i$ is $5 + 4i$.

$$(5 - 4i)(5 + 4i) = 25 + 16$$

$$= 41 \qquad \blacksquare$$

We use the idea of complex conjugates to divide complex numbers. The process is similar to rationalizing the denominator. Multiply the numerator and denominator of the quotient by the complex conjugate of the denominator.

EXAMPLE 5 **Dividing complex numbers**

Find each quotient. Write the answer in the form $a + bi$.

a) $\dfrac{5}{3 - 4i}$ **b)** $\dfrac{3 - i}{2 + i}$ **c)** $\dfrac{3 + 2i}{i}$

Solution

a) Multiply the numerator and denominator by $3 + 4i$, the conjugate of $3 - 4i$:

$$\frac{5}{3 - 4i} = \frac{5(3 + 4i)}{(3 - 4i)(3 + 4i)}$$

$$= \frac{15 + 20i}{9 - 16i^2}$$

$$= \frac{15 + 20i}{25} \qquad 9 - 16\,i^2 = 9 - 16(-1) = 25$$

$$= \frac{15}{25} + \frac{20}{25}i$$

$$= \frac{3}{5} + \frac{4}{5}i$$

b) Multiply the numerator and denominator by $2 - i$, the conjugate of $2 + i$:

$$\frac{3 - i}{2 + i} = \frac{(3 - i)(2 - i)}{(2 + i)(2 - i)}$$

$$= \frac{6 - 5i + i^2}{4 - i^2}$$

$$= \frac{6 - 5i - 1}{4 - (-1)}$$

$$= \frac{5 - 5i}{5}$$

$$= 1 - i$$

c) Multiply the numerator and denominator by $-i$, the conjugate of i:

$$\frac{3 + 2i}{i} = \frac{(3 + 2i)(-i)}{i(-i)}$$

$$= \frac{-3i - 2i^2}{-i^2}$$

$$= \frac{-3i + 2}{1}$$

$$= 2 - 3i \qquad\blacksquare$$

The symbolic definition of division of complex numbers follows.

Division of Complex Numbers

We divide the complex number $a + bi$ by the complex number $c + di$ as follows:

$$\frac{a + bi}{c + di} = \frac{(a + bi)(c - di)}{(c + di)(c - di)}$$

Square Roots of Negative Numbers

In Examples 3(a) and 3(b) we saw that both

$$(2i)^2 = -4 \qquad \text{and} \qquad (-2i)^2 = -4.$$

Because the square of each of these complex numbers is -4, both $2i$ and $-2i$ are square roots of -4. We write $\sqrt{-4} = 2i$. In the complex number system the square root of any negative number is an imaginary number.

> ### Square Root of a Negative Number
> For any positive real number b,
> $$\sqrt{-b} = i\sqrt{b}.$$

For example, $\sqrt{-9} = i\sqrt{9} = 3i$ and $\sqrt{-7} = i\sqrt{7}$. Note that the expression $\sqrt{7}\,i$ could easily be mistaken for the expression $\sqrt{7i}$, where i is under the radical. For this reason, when the coefficient of i is a radical, we write i preceding the radical.

EXAMPLE 6 **Square roots of negative numbers**

Write each expression in the form $a + bi$, where a and b are real numbers.

a) $3 + \sqrt{-9}$ **b)** $\sqrt{-12} + \sqrt{-27}$ **c)** $\dfrac{-1 - \sqrt{-18}}{3}$

Solution

a) $\begin{aligned} 3 + \sqrt{-9} &= 3 + i\sqrt{9} \\ &= 3 + 3i \end{aligned}$

b) $\begin{aligned} \sqrt{-12} + \sqrt{-27} &= i\sqrt{12} + i\sqrt{27} \\ &= 2i\sqrt{3} + 3i\sqrt{3} \\ &= 5i\sqrt{3} \end{aligned}$ $\begin{aligned} \sqrt{12} &= \sqrt{4}\sqrt{3} = 2\sqrt{3} \\ \sqrt{27} &= \sqrt{9}\sqrt{3} = 3\sqrt{3} \end{aligned}$

c) $\begin{aligned} \dfrac{-1 - \sqrt{-18}}{3} &= \dfrac{-1 - i\sqrt{18}}{3} \\ &= \dfrac{-1 - 3i\sqrt{2}}{3} \\ &= -\dfrac{1}{3} - i\sqrt{2} \end{aligned}$ ∎

Imaginary Solutions to Equations

In the complex number system the even-root property can be restated so that $x^2 = k$ is equivalent to $x = \pm\sqrt{k}$ for any $k \neq 0$. So an equation such as $x^2 = -9$ that has no real solutions has two imaginary solutions in the complex numbers.

EXAMPLE 7   **Complex solutions to equations**

Find the complex solutions to each equation.

a) $x^2 = -9$ **b)** $3x^2 + 2 = 0$

Solution

a) First apply the even-root property:

$$x^2 = -9$$
$$x = \pm\sqrt{-9} \quad \text{Even-root property}$$
$$= \pm i\sqrt{9}$$
$$= \pm 3i$$

Check these solutions in the original equation:

$$(3i)^2 = 9i^2 = 9(-1) = -9$$
$$(-3i)^2 = 9i^2 = -9$$

The solution set is $\{\pm 3i\}$.

b) First solve the equation for x^2:

$$3x^2 + 2 = 0$$
$$x^2 = -\frac{2}{3}$$
$$x = \pm\sqrt{-\frac{2}{3}}$$
$$= \pm i\sqrt{\frac{2}{3}} = \pm i\frac{\sqrt{6}}{3}$$

Check these solutions in the original equation. The solution set is $\left\{\pm i\dfrac{\sqrt{6}}{3}\right\}$. ∎

The basic facts about complex numbers are listed in the following box.

Complex Numbers

1. Definition of i: $i = \sqrt{-1}$, and $i^2 = -1$.
2. A complex number has the form $a + bi$, where a and b are real numbers.
3. The complex number $a + 0i$ is the real number a.
4. If b is a positive real number, then $\sqrt{-b} = i\sqrt{b}$.
5. The numbers $a + bi$ and $a - bi$ are called complex conjugates of each other. Their product is the real number $a^2 + b^2$.
6. Add, subtract, and multiply complex numbers as if they were algebraic expressions with i being the variable, and replace i^2 by -1.
7. Divide complex numbers by multiplying the numerator and denominator by the conjugate of the denominator.
8. In the complex number system, $x^2 = k$ for any real number k is equivalent to $x = \pm\sqrt{k}$.

Warm-ups

True or false? Explain.

1. The set of real numbers is a subset of the set of complex numbers.
2. $2 - \sqrt{-6} = 2 - 6i$ 3. $\sqrt{-9} = \pm 3i$
4. The solution set to the equation $x^2 = -9$ is $\{\pm 3i\}$.
5. $2 - 3i - (4 - 2i) = -2 - i$
6. $i^4 = 1$ 7. $(2 - i)(2 + i) = 5$
8. $i^3 = i$ 9. $i^{48} = 1$
10. The equation $x^2 = k$ has two complex solutions for any real number k.

5.6 EXERCISES

Find the indicated sums and differences of complex numbers. See Example 1.

1. $(2 + 3i) + (-4 + 5i)$ 2. $(-1 + 6i) + (5 - 4i)$
3. $(2 - 3i) - (6 - 7i)$ 4. $(2 - 3i) - (6 - 2i)$
5. $(-1 + i) + (-1 - i)$ 6. $(-5 + i) + (-5 - i)$
7. $(-2 - 3i) - (6 - i)$ 8. $(-6 + 4i) - (2 - i)$

Find the indicated products of complex numbers. See Example 2.

9. $3(2 + 5i)$ 10. $4(1 - 3i)$ 11. $2i(i - 5)$
12. $3i(2 - 6i)$ 13. $-4i(3 - i)$ 14. $-5i(2 + 3i)$
15. $(2 + 3i)(4 + 6i)$ 16. $(2 + i)(3 + 4i)$
17. $(-1 + i)(2 - i)$ 18. $(3 - 2i)(2 - 5i)$
19. $(-1 - 2i)(2 + i)$ 20. $(1 - 3i)(1 + 3i)$
21. $(5 - 2i)(5 + 2i)$ 22. $(4 + 3i)(4 + 3i)$
23. $(1 - i)(1 + i)$ 24. $(2 + 6i)(2 - 6i)$
25. $(4 + 2i)(4 - 2i)$ 26. $(4 - i)(4 + i)$

Find the indicated powers of complex numbers. See Example 3.

27. $(3i)^2$ 28. $(5i)^2$ 29. $(-5i)^2$
30. $(-9i)^2$ 31. $(2i)^4$ 32. $(-2i)^3$
33. i^9 34. i^{12}

Find the product of the given complex number and its conjugate. See Example 4.

35. $3 + 5i$ 36. $3 + i$ 37. $1 - 2i$
38. $4 - 6i$ 39. $-2 + i$ 40. $-3 - 2i$
41. $2 - i\sqrt{3}$ 42. $\sqrt{5} - 4i$

Find each quotient. See Example 5.

43. $\dfrac{3}{4 + i}$ 44. $\dfrac{6}{7 - 2i}$ 45. $\dfrac{2 + i}{3 - 2i}$
46. $\dfrac{3 + 5i}{2 - i}$ 47. $\dfrac{4 + 3i}{i}$ 48. $\dfrac{5 - 6i}{3i}$
49. $\dfrac{2 + 6i}{2}$ 50. $\dfrac{9 - 3i}{-6}$

Write each expression in the form $a + bi$, where a and b are real numbers. See Example 6.

51. $2 + \sqrt{-4}$ 52. $3 + \sqrt{-9}$ 53. $2\sqrt{-9} + 5$
54. $3\sqrt{-16} + 2$ 55. $7 - \sqrt{-6}$ 56. $\sqrt{-5} + 3$
57. $\sqrt{-8} + \sqrt{-18}$ 58. $2\sqrt{-20} - \sqrt{-45}$
59. $\dfrac{2 + \sqrt{-12}}{2}$ 60. $\dfrac{-6 - \sqrt{-18}}{3}$
61. $\dfrac{-4 - \sqrt{-24}}{4}$ 62. $\dfrac{8 + \sqrt{-20}}{-4}$

Find the complex solutions to each equation. See Example 7.

63. $x^2 = -36$ 64. $x^2 + 4 = 0$ 65. $x^2 = -12$
66. $x^2 = -25$ 67. $2x^2 + 5 = 0$ 68. $3x^2 + 4 = 0$
69. $3x^2 + 6 = 0$ 70. $x^2 + 1 = 0$

Write each expression in the form $a + bi$, where a and b are real numbers.

71. $(2 - 3i)(3 + 4i)$ 72. $(2 - 3i)(2 + 3i)$
73. $(2 - 3i) + (3 + 4i)$ 74. $(3 - 5i) - (2 - 7i)$
75. $\dfrac{2 - 3i}{3 + 4i}$ 76. $\dfrac{-3i}{3 - 6i}$ 77. $i(2 - 3i)$

78. $-3i(4i - 1)$ **79.** $(-3i)^2$ **80.** $(-2i)^6$

81. $\sqrt{-12} + \sqrt{-3}$ **82.** $\sqrt{-49} - \sqrt{-25}$

83. $(2 - 3i)^2$ **84.** $(5 + 3i)^2$

85. $\dfrac{-4 + \sqrt{-32}}{2}$ **86.** $\dfrac{-2 - \sqrt{-27}}{-6}$

Getting More Involved

87. *Writing.* Explain why $2 - i$ is a solution to $x^2 - 4x + 5 = 0$.

88. *Cooperative learning.* Work with a group to verify that $-1 + i\sqrt{3}$ and $-1 - i\sqrt{3}$ satisfy the equation $x^3 - 8 = 0$. In the complex number system there are three cube roots of 8. What are they?

89. *Discussion.* What is wrong with using the product rule for radicals to get

$$\sqrt{-4} \cdot \sqrt{-4} = \sqrt{(-4)(-4)} = \sqrt{16} = 4?$$

What is the correct product?

COLLABORATIVE ACTIVITIES

Laws of Falling Bodies

Grouping: 2 to 4 students per group
Topic: Applications of formulas with square roots

Jaki Sena is a private investigator working on a case. Her client is accused of killing a woman who had lived in the same apartment building as he did. She pulls over the file of notes from across her desk and looks through it yet again.

  "Okay, so my client said he was on the roof of the building just before 4:00 A.M., dropping blocks over into the alleyway. He dropped the last block just as the clock tower began striking the hour. I wonder how high the building is," Jaki says to herself.

  During her initial investigation of the crime scene, Jaki had asked the building manager whether he knew the height of the five-story building. He had been very condescending and had told her nothing. Another tenant of the building, some sort of math instructor at the city college, had told her about an experiment he had done with shadows and a climbing rope. He had given her a drawing.

  Jaki looked through the file until she found the following drawing.

She pulled out a blank piece of paper, and using a calculator, she soon had a number.

  "Now I can find out how long it took the block to fall!" Jaki exclaimed. "Okay, now the witness who saw my client on the top of the building also swears he saw the woman walking past the steps next door as the clock tower started to chime 4:00," Jaki said out loud as she got up from her desk and looked out the window. "I have no idea how fast the woman was walking. That may be a hard one to answer. But I do know the steps begin about $9\frac{1}{2}$ feet from where the block fell." Jaki turned back to the file of notes on her desk. She soon found another piece of paper with measurements on it.

  "And I do know the woman was 5 feet tall!" Jaki exclaimed, taking out her calculator again.

1. Working in your groups, find the height of the building from the information on Jaki's paper. Round your answer to the nearest tenth of a foot.

2. Find out how many seconds it would take the block to fall from the top of the building to hit the woman on the head. Round your answer to the nearest tenth of a second. (*Hint:* You may need to look in other chapters of the book to find the formulas you need here.)

3. How fast would the woman have to walk to have been hit on the head by the block? Round your answer to the nearest tenth of a foot per second.

4. Could Jaki's client have "done it"? Give a reason for your answer.

Wrap-up CHAPTER 5

SUMMARY

Roots and Rational Exponents		Examples
Definition of nth roots	The number a is an nth root of b if $a^n = b$.	$3^4 = 81$, $(-3)^4 = 81$ Both 3 and -3 are fourth roots of 81.
Definition of $1/n$ as an exponent	If n is a positive *even* integer and a is positive, then $a^{1/n}$ denotes the positive real nth root of a. If n is a positive *odd* integer and a is any real number, then $a^{1/n}$ is the real nth root of a.	$81^{1/4} = 3$ $(-8)^{1/3} = -2$ $8^{1/3} = 2$
Definition of nth root of zero	If n is a positive integer, then $0^{1/n} = 0$.	$0^{1/6} = 0$
Definition of rational exponents	If m and n are positive integers, then $a^{m/n} = (a^{1/n})^m$, provided that $a^{1/n}$ is real. We can find the root first or the power first.	$8^{2/3} = (8^{1/3})^2$ $= 2^2 = 4$ $(-16)^{3/4}$ is not real.
Definition of negative rational exponents	If m and n are positive integers, then $$a^{-m/n} = \frac{1}{a^{m/n}},$$ provided that $a^{1/n}$ is defined and nonzero.	$8^{-2/3} = \dfrac{1}{4}$ Root, Power, Reciprocal

Rules for Rational Exponents		Examples

If a and b are nonzero real numbers and r and s are rational numbers, then the following rules hold.

Product rule	$a^r \cdot a^s = a^{r+s}$	$3^{1/4} \cdot 3^{1/2} = 3^{3/4}$
Quotient rule	$\dfrac{a^r}{a^s} = a^{r-s}$	$\dfrac{x^{3/4}}{x^{1/4}} = x^{1/2}$
Power of a power rule	$(a^r)^s = a^{rs}$	$(2^{1/2})^{-1/2} = 2^{-1/4}$ $(x^{3/4})^4 = x^3$
Power of a product rule	$(ab)^r = a^r b^r$	$(a^2 b^6)^{1/2} = ab^3$
Power of a quotient rule	$\left(\dfrac{a}{b}\right)^r = \dfrac{a^r}{b^r}$	$\left(\dfrac{8}{x^6}\right)^{2/3} = \dfrac{4}{x^4}$

Radicals		Examples
Definition of radicals	If n is a positive integer and a is a number for which $a^{1/n}$ is real, then $\sqrt[n]{a} = a^{1/n}$. If $n = 2$, omit the 2 and write $\sqrt[2]{a}$ as $\sqrt{a}$.	$8^{1/3} = \sqrt[3]{8}$ $4^{1/2} = \sqrt{4}$

Product rule for radicals	Provided that all roots are real, $$\sqrt[n]{ab} = \sqrt[n]{a} \cdot \sqrt[n]{b}.$$	$\sqrt{2} \cdot \sqrt{3} = \sqrt{6}$ $\sqrt{4x} = 2\sqrt{x}$
Quotient rule for radicals	Provided that all roots are real and $b \neq 0$, $$\sqrt[n]{\dfrac{a}{b}} = \dfrac{\sqrt[n]{a}}{\sqrt[n]{b}}.$$	$\sqrt{\dfrac{5}{9}} = \dfrac{\sqrt{5}}{3}$ $\sqrt{10} \div \sqrt{5} = \sqrt{2}$
Simplified radical form for radicals of index n	A simplified radical of index n has 1. *no* perfect nth powers as factors of the radicand, 2. *no* fractions inside the radical, and 3. *no* radicals in the denominator.	$\sqrt{20} = \sqrt{4 \cdot 5} = 2\sqrt{5}$ $\sqrt{\dfrac{3}{2}} = \dfrac{\sqrt{3}}{\sqrt{2}}$ $\dfrac{\sqrt{3}}{\sqrt{2}} = \dfrac{\sqrt{3}}{\sqrt{2}} \cdot \dfrac{\sqrt{2}}{\sqrt{2}} = \dfrac{\sqrt{6}}{2}$

Equations

Examples

Equations with radicals and exponents	1. In raising each side of an equation to an even power, we can create an equation that gives extraneous solutions. We must check.	$\sqrt{x} = -3$ $x = 9$
	2. When applying the even-root property, remember that there is a positive and a negative root.	$x^2 = 36$ $x = \pm 6$
	3. For equations with rational exponents, raise each side to a positive or a negative power first, then apply the even- or odd-root property.	$x^{-2/3} = 4$ $(x^{-2/3})^{-3} = 4^{-3}$ $x^2 = \dfrac{1}{64}$ $x = \pm\dfrac{1}{8}$

Complex Numbers

Examples

Complex numbers	Numbers of form $a + bi$, where a and b are real numbers: $i = \sqrt{-1}$, $i^2 = -1$	$2 + 3i$ $-6i$ $\sqrt{2} + i$
Complex conjugates	Complex numbers of the form $a + bi$ and $a - bi$: Their product is the real number $a^2 + b^2$.	$(2 + 3i)(2 - 3i) = 2^2 + 3^2$ $\qquad = 13$
Complex number operations	Add, subtract, and multiply as algebraic expressions with i being the variable. Simplify using $i^2 = -1$.	$(2 + 5i) + (4 - 2i) = 6 + 3i$ $(2 + 5i) - (4 - 2i) = -2 + 7i$ $(2 + 5i)(4 - 2i) = 18 + 16i$
	Divide complex numbers by multiplying numerator and denominator by the conjugate of the denominator.	$(2 + 5i) \div (4 - 2i)$ $= \dfrac{(2 + 5i)(4 + 2i)}{(4 - 2i)(4 + 2i)}$ $= \dfrac{-2 + 24i}{20} = -\dfrac{1}{10} + \dfrac{6}{5}i$
Square root of a negative number	For any positive real number b, $\sqrt{-b} = i\sqrt{b}$.	$\sqrt{-9} = i\sqrt{9} = 3i$
Imaginary solutions to equations	In the complex number system, $x^2 = k$ for any real k is equivalent to $x = \pm\sqrt{k}$.	$x^2 = -25$ $x = \pm\sqrt{-25} = \pm 5i$

REVIEW EXERCISES

5.1 *Simplify the expressions involving rational exponents. Assume that all variables represent positive real numbers. Write your answers with positive exponents.*

1. $(-27)^{-2/3}$ **2.** $-25^{3/2}$ **3.** $(2^6)^{1/3}$

4. $(5^2)^{1/2}$ **5.** $100^{-3/2}$ **6.** $1000^{-2/3}$

7. $\dfrac{3x^{-1/2}}{3^{-2}x^{-1}}$ **8.** $\dfrac{(x^2 y^{-3} z)^{1/2}}{x^{1/2} y z^{-1/2}}$

9. $(a^{1/2} b)^3 (ab^{1/4})^2$ **10.** $(t^{-1/2})^{-2}(t^{-2}v^2)$

11. $(x^{1/2} y^{1/4})(x^{1/4} y)$ **12.** $(a^{1/3} b^{1/6})^2 (a^{1/3} b^{2/3})$

5.2 *Simplify each radical expression. Assume that all variables represent positive real numbers.*

13. $\sqrt{72x^5}$ **14.** $\sqrt{90y^9 z^4}$ **15.** $\sqrt[3]{72x^5}$

16. $\sqrt[3]{81a^8 b^9}$ **17.** $\sqrt{2^6}$ **18.** $\sqrt{2^7}$

19. $\sqrt{\dfrac{2}{5}}$ **20.** $\sqrt{\dfrac{1}{6}}$ **21.** $\sqrt[3]{\dfrac{2}{3}}$

22. $\sqrt[3]{\dfrac{1}{9}}$ **23.** $\dfrac{2}{\sqrt{3x}}$ **24.** $\dfrac{3}{\sqrt{2y}}$

25. $\dfrac{\sqrt{10y^3}}{\sqrt{6}}$ **26.** $\dfrac{\sqrt{5x^5}}{\sqrt{8}}$ **27.** $\dfrac{3}{\sqrt[3]{2a}}$

28. $\dfrac{a}{\sqrt[3]{a^2}}$ **29.** $\dfrac{5}{\sqrt[4]{3x^2}}$ **30.** $\dfrac{b}{\sqrt[4]{a^2 b^3}}$

31. $\sqrt[4]{48x^5 y^{12}}$ **32.** $\sqrt[5]{32x^{10} y^{12}}$

5.3 *Perform the operations and simplify. Assume that the variables represent positive real numbers.*

33. $\sqrt{13} \cdot \sqrt{13}$ **34.** $\sqrt[3]{14} \cdot \sqrt[3]{14} \cdot \sqrt[3]{14}$

35. $\sqrt{27} + \sqrt{45} - \sqrt{75}$ **36.** $\sqrt{12} - \sqrt{50} + \sqrt{72}$

37. $\sqrt{\dfrac{1}{3}} + \sqrt{27}$ **38.** $\sqrt{\dfrac{1}{2}} - \sqrt{\dfrac{1}{8}}$

39. $3\sqrt{2}\,(5\sqrt{2} - 7\sqrt{3}\,)$

40. $-2\sqrt{a}\,(\sqrt{a} - \sqrt{ab^6}\,)$

41. $(2 - \sqrt{3}\,)(3 + \sqrt{2}\,)$

42. $(2\sqrt{x} - \sqrt{y}\,)(\sqrt{x} + \sqrt{y}\,)$

43. $\sqrt[3]{40} - \sqrt[3]{5}$

44. $\sqrt[3]{54x^4} + x\sqrt[3]{16x}$

5.4 *Perform the operations and simplify.*

45. $5 \div \sqrt{2}$ **46.** $(10\sqrt{6}\,) \div (2\sqrt{2}\,)$

47. $(\sqrt{3}\,)^4$ **48.** $(-2\sqrt{x}\,)^9$

49. $\dfrac{2 - \sqrt{8}}{2}$ **50.** $\dfrac{-3 - \sqrt{18}}{-6}$

51. $\dfrac{\sqrt{6}}{1 - \sqrt{3}}$ **52.** $\dfrac{\sqrt{15}}{2 + \sqrt{5}}$

53. $\dfrac{2\sqrt{3}}{3\sqrt{6} - \sqrt{12}}$ **54.** $\dfrac{-\sqrt{xy}}{3\sqrt{x} + \sqrt{xy}}$

55. $\left(2w\sqrt[3]{2w^2}\,\right)^6$

56. $\left(m\sqrt[4]{m^3}\,\right)^8$

5.5 *Find all real solutions to each equation.*

57. $x^2 = 16$ **58.** $w^2 = 100$

59. $(a - 5)^2 = 4$ **60.** $(m - 7)^2 = 25$

61. $(a + 1)^2 = 5$ **62.** $(x + 5)^2 = 3$

63. $(m + 1)^2 = -8$ **64.** $(w + 4)^2 = 16$

65. $\sqrt{m - 1} = 3$ **66.** $3\sqrt{x + 5} = 12$

67. $\sqrt[3]{2x + 9} = 3$ **68.** $\sqrt[4]{2x - 1} = 2$

69. $w^{2/3} = 4$ **70.** $m^{-4/3} = 16$

71. $(m + 1)^{1/3} = 5$ **72.** $(w - 3)^{-2/3} = 4$

73. $\sqrt{x - 3} = \sqrt{x + 2} - 1$

74. $\sqrt{x^2 + 3x + 6} = 4$

75. $\sqrt{5x - x^2} = \sqrt{6}$

76. $\sqrt{x + 4} - 2\sqrt{x - 1} = -1$

77. $\sqrt{x + 7} - 2\sqrt{x} = -2$

78. $\sqrt{x} - \sqrt{x - 1} = 1$

79. $2\sqrt{x} - \sqrt{x - 3} = 3$

80. $1 + \sqrt{x + 7} = \sqrt{2x + 7}$

5.6 *Perform the indicated operations. Write answers in the form a + bi.*

81. $(2 - 3i)(-5 + 5i)$ **82.** $(2 + i)(5 - 2i)$

83. $(2 + i) + (5 - 4i)$ **84.** $(2 + i) + (3 - 6i)$

85. $(1 - i) - (2 - 3i)$ **86.** $(3 - 2i) - (1 - i)$

87. $\dfrac{6 + 3i}{3}$ **88.** $\dfrac{8 + 12i}{4}$

89. $\dfrac{4 - \sqrt{-12}}{2}$ **90.** $\dfrac{6 + \sqrt{-18}}{3}$

91. $\dfrac{2 - 3i}{4 + i}$ **92.** $\dfrac{3 + i}{2 - 3i}$

Find the imaginary solutions to each equation.

93. $x^2 + 100 = 0$ **94.** $25a^2 + 3 = 0$

95. $2b^2 + 9 = 0$ **96.** $3y^2 + 8 = 0$

Miscellaneous

Determine whether each equation is true or false and explain your answer. An equation involving variables should be marked true only if it is an identity. Do not use a calculator.

97. $2^3 \cdot 3^2 = 6^5$

98. $16^{1/4} = 4^{1/2}$

99. $(\sqrt{2})^3 = 2\sqrt{2}$

100. $\sqrt[3]{9} = 3$

101. $8^{200} \cdot 8^{200} = 64^{200}$

102. $\sqrt{295} \cdot \sqrt{295} = 295$

103. $4^{1/2} = \sqrt{2}$

104. $\sqrt{a^2} = |a|$

105. $5^2 \cdot 5^2 = 25^4$

106. $\sqrt{6} \div \sqrt{2} = \sqrt{3}$

107. $\sqrt{w^{10}} = w^5$

108. $\sqrt{a^{16}} = a^4$

109. $\sqrt{x^6} = x^3$

110. $\sqrt[6]{16} = \sqrt[3]{4}$

111. $\sqrt{x^8} = x^4$

112. $\sqrt[9]{2^6} = 2^{2/3}$

113. $\sqrt{16} = 2$

114. $2^{1/2} \cdot 2^{1/4} = 2^{3/4}$

115. $2^{600} = 4^{300}$

116. $\sqrt{2} \cdot \sqrt[4]{2} = \sqrt[6]{2}$

117. $\dfrac{2 + \sqrt{6}}{2} = 1 + \sqrt{6}$

118. $\dfrac{4 + 2\sqrt{3}}{2} = 2 + \sqrt{3}$

119. $\sqrt{\dfrac{4}{6}} = \dfrac{2}{3}$

120. $8^{200} \cdot 8^{200} = 8^{400}$

121. $81^{2/4} = 81^{1/2}$

122. $(-64)^{2/6} = (-64)^{1/3}$

123. $(a^4 b^2)^{1/2} = |a^2 b|$

124. $\left(\dfrac{a^2}{b^6}\right)^{1/2} = \dfrac{|a|}{b^3}$

In Exercises 125–136, solve each problem.

125. *Falling objects.* If we neglect air resistance, the number of feet s that an object falls from rest during t seconds is given by the equation $s = 16t^2$. How long would it take the landing gear of an airplane to reach the earth if it fell off the airplane at 12,000 feet?

126. *Timber.* Anne is pulling on a 60-foot rope attached to the top of a 48-foot tree while Walter is cutting the tree at its base. How far from the base of the tree is Anne standing?

Figure for Exercise 126

127. *Guy wire.* If a guy wire of length 40 feet is attached to an antenna at a height of 30 feet, then how far from the base of the antenna is the wire attached to the ground?

Figure for Exercise 127

128. *Touchdown.* Suppose that at the kickoff of a football game, the receiver catches the football at the left side of the goal line and runs for a touchdown diagonally across the field. How many yards would he run? (A football field is 100 yards long and 160 feet wide.)

129. *Long guy wires.* The manufacturer of an antenna recommends that guy wires from the top of the antenna to the ground be attached to the ground at a distance from the base equal to the height of the antenna. How long would the guy wires be for a 200-foot antenna?

130. *Height of a post.* Betty observed that the lamp post in front of her house casts a shadow of length 8 feet when the angle of inclination of the sun is 60 degrees. How tall is the lamp post? (In a 30-60-90 right triangle, the side opposite 30 is one-half the length of the hypotenuse.)

Figure for Exercise 130

131. ***Manufacturing a box.*** A cubic box has a volume of 40 cubic feet. The amount of recycled cardboard that it takes to make the six-sided box is 10% larger than the surface area of the box. Find the exact amount of recycled cardboard used in manufacturing the box.

132. ***Shipping parts.*** A cubic box with a volume of 32 cubic feet is to be used to ship some machine parts. All of the parts are small except for a long, straight steel connecting rod. What is the maximum length of a connecting rod that will fit into this box?

133. ***Health care reform.*** Total annual expenditures on health in the United States grew from $74.4.billion in 1970 to $666.2 billion in 1990 (*Health Care Financing Review*, 1991). Find the average annual rate r of growth for national health expenditures for that period by solving the equation

$$666.2 = 74.4(1 + r)^{20}.$$

Figure for Exercise 133

134. ***Population growth.*** The formula $P = P_0 (1 + r)^n$ gives the population P at the end of an n-year time period, where P_0 is the population at the beginning of the time period and r is the average annual growth rate. The U.S. population grew from 203.3 million in 1970 to 248.7 million in 1990 (U.S. Census Bureau). Find the average annual rate of growth for the U.S. population for that period.

135. ***Landing speed.*** Aircraft engineers determine the proper landing speed V (in feet per second) for an airplane from the formula

$$V = \sqrt{\frac{841L}{CS}},$$

where L is the gross weight of the aircraft in pounds, C is the coefficient of lift, and S is the wing surface area in square feet. Rewrite the formula so that the expression on the right side is in simplified radical form.

136. ***Spillway capacity.*** Civil engineers use the formula

$$Q = 3.32LH^{3/2}$$

to find the maximum discharge that the dam (a broad-crested weir) shown in the figure can pass before the water breaches its abutments (*Standard Handbook for Civil Engineers*, 1968). In the formula, Q is the discharge in cubic feet per second, L is the length of the spillway in feet, and H is the depth of the water on the spillway. Find Q given that $L = 60$ feet and $H = 5$ feet. Find H given that $Q = 3,000$ cubic feet per second and $L = 70$ feet.

Figure for Exercise 136

CHAPTER 5 TEST

Simplify each expression. Assume that all variables represent positive numbers.

1. $8^{2/3}$

2. $4^{-3/2}$

3. $\sqrt{21} \div \sqrt{7}$

4. $2\sqrt{5} \cdot 3\sqrt{5}$

5. $\sqrt{20} + \sqrt{5}$

6. $\sqrt{5} + \dfrac{1}{\sqrt{5}}$

7. $2^{1/2} \cdot 2^{1/2}$

8. $\sqrt{72}$

9. $\sqrt{\dfrac{5}{12}}$

10. $\dfrac{6 + \sqrt{18}}{6}$

11. $(2\sqrt{3} + 1)(\sqrt{3} - 2)$

12. $\sqrt[4]{32a^5y^8}$

13. $\dfrac{1}{\sqrt[3]{2x^2}}$

14. $\sqrt{\dfrac{8a^9}{b^3}}$

15. $\sqrt[3]{-27x^9}$

16. $\sqrt{20m^3}$

17. $x^{1/2} \cdot x^{1/4}$

18. $(9y^4x^{1/2})^{1/2}$

19. $\sqrt[3]{40x^7}$

20. $(4 + \sqrt{3})^2$

Rationalize the denominator and simplify.

21. $\dfrac{2}{5 - \sqrt{3}}$

22. $\dfrac{\sqrt{6}}{4\sqrt{3} + \sqrt{2}}$

Write each expression in the form a + bi.

23. $(3 - 2i)(4 + 5i)$

24. $i^4 - i^5$

25. $\dfrac{3 - i}{1 + 2i}$

26. $\dfrac{-6 + \sqrt{-12}}{8}$

Find all real or imaginary solutions to each equation.

27. $(x - 2)^2 = 49$

28. $2\sqrt{x + 4} = 3$

29. $w^{2/3} = 4$

30. $9y^2 + 16 = 0$

31. $\sqrt{2x^2 + x - 12} = x$

32. $\sqrt{x - 1} + \sqrt{x + 4} = 5$

Show a complete solution to each problem.

33. Find the exact length of the side of a square whose diagonal is 3 feet.

34. Two positive numbers differ by 11, and their square roots differ by 1. Find the numbers.

35. If the perimeter of a rectangle is 20 feet and the diagonal is $2\sqrt{13}$ feet, then what are the length and width?

36. The average radius R of the orbit of a planet around the sun is determined by $R = T^{2/3}$, where T is the number of years for one orbit and R is measured in astronomical units (AU). If it takes Pluto 248.530 years to make one orbit of the sun, then what is the average radius of the orbit of Pluto? If the average radius of the orbit of Neptune is 30.08 AU, then how many years does it take Neptune to complete one orbit of the sun?

Find all real solutions to each equation or inequality. For the inequalities, also sketch the graph of the solution set.

1. $3(x - 2) + 5 = 7 - 4(x + 3)$

2. $\sqrt{6x + 7} = 4$

3. $|2x + 5| > 1$

4. $8x^3 - 27 = 0$

5. $2x - 3 > 3x - 4$

6. $\sqrt{2x - 3} - \sqrt{3x + 4} = 0$

7. $\dfrac{w}{3} + \dfrac{w - 4}{2} = \dfrac{11}{2}$

8. $2(x + 7) - 4 = x - (10 - x)$

9. $(x + 7)^2 = 25$

10. $a^{-1/2} = 4$

11. $x - 3 > 2$ or $x < 2x + 6$

12. $a^{-2/3} = 16$

13. $3x^2 - 1 = 0$

14. $5 - 2(x - 2) = 3x - 5(x - 2) - 1$

15. $|3x - 4| < 5$

16. $3x - 1 = 0$

17. $\sqrt{y - 1} = 9$

18. $|5(x - 2) + 1| = 3$

19. $0.06x - 0.04(x - 20) = 2.8$

20. $|3x - 1| > -2$

21. $\dfrac{3\sqrt{2}}{x} = \dfrac{\sqrt{3}}{4\sqrt{5}}$

22. $\dfrac{\sqrt{x} - 4}{x} = \dfrac{1}{\sqrt{x} + 5}$

23. $\dfrac{3\sqrt{2} + 4}{\sqrt{2}} = \dfrac{x\sqrt{18}}{3\sqrt{2} + 2}$

24. $\dfrac{x}{2\sqrt{5} - \sqrt{2}} = \dfrac{2\sqrt{5} + \sqrt{2}}{x}$

25. $\dfrac{\sqrt{2x} - 5}{x} = \dfrac{-3}{\sqrt{2x} + 5}$

26. $\dfrac{\sqrt{6} + 2}{x} = \dfrac{2}{\sqrt{6} + 4}$

27. $\dfrac{x - 1}{\sqrt{6}} = \dfrac{\sqrt{6}}{x}$

28. $\dfrac{x + 3}{\sqrt{10}} = \dfrac{\sqrt{10}}{x}$

29. $\dfrac{1}{x} - \dfrac{1}{x - 1} = -\dfrac{1}{6}$

30. $\dfrac{1}{x^2 - 2x} + \dfrac{1}{x} = \dfrac{2}{3}$

The expression $\dfrac{-b + \sqrt{b^2 - 4ac}}{2a}$ *will be used in Chapter 6 to solve quadratic equations. Evaluate it for the given values of a, b, and c.*

31. $a = 1$, $b = 2$, $c = -15$ 32. $a = 1$, $b = 8$, $c = 12$

33. $a = 2$, $b = 5$, $c = -3$ 34. $a = 6$, $b = 7$, $c = -3$

Solve each problem.

35. **Popping corn.** The results of an experiment by D.D. Metzger to determine the relationship between the moisture content of popcorn and the volume of popped corn are shown in the figure (*Cereal Chemistry*, 1989). The formula $v = -94.8 + 21.4x - 0.761x^2$ can be used to model the data shown in the graph. In the formula, v is the number of cubic centimeters of popped corn that result from popping 1 gram of corn with moisture content x% in a hot-air popper. Use the formula to find the volume that results when 1 gram of corn with a moisture content of 11% is popped.

Figure for Exercise 35 and 36

36. **Maximizing the volume of popped corn.** Use the graph to estimate the moisture content that will produce the maximum amount of popped corn. What is the maximum possible volume for popping 1 gram of corn in a hot-air popper?

CHAPTER 6

Quadratic Equations and Inequalities

I s it possible to measure beauty? For thousands of years, artists and philosophers have been challenged to answer this question. The seventeenth century philosopher John Locke said, "Beauty consists of a certain composition of color and figure causing delight in the beholder." Over the centuries, many architects, sculptors, and painters have searched for beauty in their work by exploring numerical patterns in various art forms.

Today, many artists and architects still use the concepts of beauty given to us by the ancient Greeks. One principle, called the Golden Rectangle, concerns the most pleasing proportions of a rectangle. The Golden Rectangle appears in nature as well as in many cultures. Examples of it can be seen in Leonardo da Vinci's *Proportions of the Human Figure* as well as in Indonesian temples and Chinese pagodas. Perhaps one of the best-known examples of the Golden Rectangle is in the facade and floor plan of the Parthenon, built in Athens in the fifth century B.C. In Exercise 65 of Section 6.3 we will see that the principle of the Golden Rectangle is based on a proportion that we can solve using the quadratic formula.

6.1 Factoring and Completing the Square

Factoring and the even-root property were used to solve quadratic equations in Chapters 3, 4, and 5. In this section we first review those methods. Then you will learn the method of completing the square, which can be used to solve any quadratic equation.

Review of Factoring

Recall from Chapter 3 that a quadratic equation is an equation of the form

$$ax^2 + bx + c = 0,$$

where a, b, and c are real numbers with $a \neq 0$. If the quadratic polynomial on the left side can be factored, then we can solve the equation using the following strategy.

> ### Strategy for Solving Quadratic Equations by Factoring
>
> 1. Write the equation with 0 on the right side.
> 2. Factor the left side.
> 3. Use the zero factor property to set each factor equal to 0.
> 4. Solve the simpler equations.
> 5. Check the answers in the original equation.

EXAMPLE 1   Solving a quadratic equation by factoring

Solve $3x^2 - 4x = 15$ by factoring.

Solution

Subtract 15 from each side to get 0 on the right side:

$$3x^2 - 4x - 15 = 0$$

$$(3x + 5)(x - 3) = 0 \qquad \text{Factor the left-hand side.}$$

$$3x + 5 = 0 \qquad \text{or} \qquad x - 3 = 0 \qquad \text{Zero factor property}$$

$$3x = -5 \qquad \text{or} \qquad x = 3$$

$$x = -\frac{5}{3}$$

The solution set is $\left\{-\frac{5}{3}, 3\right\}$. Check the solutions in the original equation. ■

Review of the Even-Root Property

In Chapter 5 we solved quadratic equations by using the even-root property.

EXAMPLE 2   **Solving a quadratic equation by the even-root property**

Solve $(a - 1)^2 = 9$.

Solution

By the even-root property, $x^2 = k$ is equivalent to $x = \pm\sqrt{k}$.

$$(a - 1)^2 = 9$$
$$a - 1 = \pm\sqrt{9} \qquad \text{Even-root property}$$

$a - 1 = 3$	or	$a - 1 = -3$
$a = 4$	or	$a = -2$

Check these solutions in the original equation. The solution set is $\{-2, 4\}$. ∎

Completing the Square

We cannot solve every quadratic by factoring because not all quadratic polynomials can be factored. However, we can write any quadratic equation in the form of Example 2 and then apply the even-root property to solve it. This method is called **completing the square.**

The essential part of completing the square is to recognize a perfect-square trinomial when given its first two terms. For example, the perfect-square trinomial whose first two terms are $x^2 + 6x$ is $x^2 + 6x + 9$. The trinomial $x^2 + 6x + 9$ is a perfect-square trinomial because it is the square of the binomial $x + 3$:

$$x^2 + 6x + 9 = (x + 3)^2$$

If the coefficient of x^2 is 1, there is a simple rule for identifying the last term in a perfect-square trinomial.

Rule for Finding the Last Term

The last term of a perfect-square trinomial is the square of one-half of the coefficient of the middle term. In symbols, the perfect-square trinomial whose first two terms are $x^2 + bx$ is $x^2 + bx + \left(\dfrac{b}{2}\right)^2$.

EXAMPLE 3   **Finding the last term**

Find the perfect-square trinomial whose first two terms are given.

a) $x^2 + 8x$ **b)** $x^2 - 5x$ **c)** $x^2 + \dfrac{4}{7}x$ **d)** $x^2 - \dfrac{3}{2}x$

Solution

a) One-half of 8 is 4, and 4 squared is 16. So the perfect-square trinomial is

$$x^2 + 8x + 16.$$

b) One-half of -5 is $-\dfrac{5}{2}$, and $-\dfrac{5}{2}$ squared is $\dfrac{25}{4}$. So the perfect-square trinomial is $x^2 - 5x + \dfrac{25}{4}$.

c) One-half of $\frac{4}{7}$ is $\frac{2}{7}$, and $\frac{2}{7}$ squared is $\frac{4}{49}$. So the perfect-square trinomial is

$$x^2 + \frac{4}{7}x + \frac{4}{49}.$$

d) One-half of $-\frac{3}{2}$ is $-\frac{3}{4}$, and $\left(-\frac{3}{4}\right)^2 = \frac{9}{16}$. So the perfect-square trinomial is

$$x^2 - \frac{3}{2}x + \frac{9}{16}. \qquad \blacksquare$$

Another essential step in completing the square is to write the perfect square trinomial as the square of a binomial. Recall that

$$a^2 + 2ab + b^2 = (a + b)^2$$

and

$$a^2 - 2ab + b^2 = (a - b)^2.$$

EXAMPLE 4 **Factoring perfect square trinomials**

Factor each trinomial.

a) $x^2 + 12x + 36$ **b)** $y^2 - 7y + \frac{49}{4}$ **c)** $z^2 - \frac{4}{3}z + \frac{4}{9}$

Solution

a) The trinomial $x^2 + 12x + 36$ is of the form $a^2 + 2ab + b^2$ with $a = x$ and $b = 6$. So

$$x^2 + 12x + 36 = (x + 6)^2.$$

Check by squaring $x + 6$.

b) The trinomial $y^2 - 7y + \frac{49}{4}$ is of the form $a^2 - 2ab + b^2$ with $a = y$ and $b = \frac{7}{2}$. So

$$y^2 - 7y + \frac{49}{4} = \left(y - \frac{7}{2}\right)^2.$$

Check by squaring $y - \frac{7}{2}$.

c) The trinomial $z^2 - \frac{4}{3}z + \frac{4}{9}$ is of the form $a^2 - 2ab + b^2$ with $a = z$ and $b = -\frac{2}{3}$. So

$$z^2 - \frac{4}{3}z + \frac{4}{9} = \left(z - \frac{2}{3}\right)^2. \qquad \blacksquare$$

In the next example we use the skills that we practiced in Examples 2, 3, and 4 to solve the quadratic equation $ax^2 + bx + c = 0$ with $a = 1$ by the method of completing the square.

EXAMPLE 5 **Completing the square with $a = 1$**

Solve $x^2 + 6x + 5 = 0$ by completing the square.

Solution

The perfect-square trinomial whose first two terms are $x^2 + 6x$ is

$$x^2 + 6x + 9.$$

So we move 5 to the right side of the equation, then add 9 to each side:

$$x^2 + 6x \quad\;\; = -5 \qquad \text{Subtract 5 from each side.}$$
$$x^2 + 6x + 9 = -5 + 9 \qquad \begin{array}{l}\text{Add 9 to each side to get}\\ \text{a perfect square trinomial.}\end{array}$$
$$(x + 3)^2 = 4 \qquad \text{Factor the left side.}$$
$$x + 3 = \pm\sqrt{4} \qquad \text{Even-root property}$$

$$x + 3 = 2 \qquad \text{or} \qquad x + 3 = -2$$
$$x = -1 \qquad \text{or} \qquad x = -5$$

Check in the original equation:

$$(-1)^2 + 6(-1) + 5 = 0$$

and

$$(-5)^2 + 6(-5) + 5 = 0$$

The solution set is $\{-1, -5\}$. ■

CAUTION All of the perfect-square trinomials that we have used so far had a leading coefficient of 1. If $a \neq 1$, then we must divide each side of the equation by a to get an equation with a leading coefficient of 1.

The strategy for solving a quadratic equation by completing the square is stated in the following box.

> ### Strategy for Solving Quadratic Equations by Completing the Square
>
> 1. The coefficient of x^2 must be 1.
> 2. Get only the x^2 and the x terms on the left side.
> 3. Add to each side the square of $\frac{1}{2}$ the coefficient of x.
> 4. Factor the left side as the square of a binomial.
> 5. Apply the even-root property.
> 6. Solve for x.
> 7. Simplify.

In our procedure for completing the square, the coefficient of x^2 must be 1. We can solve $ax^2 + bx + c = 0$ with $a \neq 1$ by completing the square if we first divide each side of the equation by a.

EXAMPLE 6   **Completing the square with $a \neq 1$**

Solve $2x^2 + 3x - 2 = 0$ by completing the square.

Solution

For completing the square, the coefficient of x^2 must be 1. So we first divide each side of the equation by 2:

$$\frac{2x^2 + 3x - 2}{2} = \frac{0}{2} \qquad \text{Divide each side by 2.}$$

$$x^2 + \frac{3}{2}x - 1 = 0 \qquad \text{Simplify.}$$

$$x^2 + \frac{3}{2}x = 1 \qquad \begin{array}{l}\text{Get only } x^2 \text{ and } x \text{ terms on the}\\ \text{left side.}\end{array}$$

$$x^2 + \frac{3}{2}x + \frac{9}{16} = 1 + \frac{9}{16} \qquad \text{One-half of } \tfrac{3}{2} \text{ is } \tfrac{3}{4}, \text{ and } \left(\tfrac{3}{4}\right)^2 = \tfrac{9}{16}.$$

$$\left(x + \frac{3}{4}\right)^2 = \frac{25}{16} \qquad \text{Factor the left side.}$$

$$x + \frac{3}{4} = \pm\sqrt{\frac{25}{16}} \qquad \text{Even-root property}$$

$$x + \frac{3}{4} = \frac{5}{4} \qquad \text{or} \qquad x + \frac{3}{4} = -\frac{5}{4}$$

$$x = \frac{2}{4} = \frac{1}{2} \qquad \text{or} \qquad x = -\frac{8}{4} = -2$$

Check these values in the original equation. The solution set is $\left\{-2, \frac{1}{2}\right\}$. ∎

In Examples 5 and 6 the solutions were rational numbers, and the equations could have been solved by factoring. In the next example the solutions are irrational numbers, and factoring will not work.

EXAMPLE 7   **A quadratic equation with irrational solutions**

Solve $x^2 - 3x - 6 = 0$ by completing the square.

Solution

Because $a = 1$, we first get the x^2 and x terms on the left side:

$$x^2 - 3x - 6 = 0$$

$$x^2 - 3x = 6 \qquad \text{Add 6 to each side.}$$

$$x^2 - 3x + \frac{9}{4} = 6 + \frac{9}{4} \qquad \text{One-half of } -3 \text{ is } -\tfrac{3}{2}, \text{ and } \left(-\tfrac{3}{2}\right)^2 = \tfrac{9}{4}.$$

$$\left(x - \frac{3}{2}\right)^2 = \frac{33}{4} \qquad 6 + \tfrac{9}{4} = \tfrac{24}{4} + \tfrac{9}{4} = \tfrac{33}{4}$$

$$x - \frac{3}{2} = \pm\sqrt{\frac{33}{4}} \qquad \text{Even-root property}$$

$$x = \frac{3}{2} \pm \frac{\sqrt{33}}{2} \qquad \text{Add } \tfrac{3}{2} \text{ to each side.}$$

$$x = \frac{3 \pm \sqrt{33}}{2}$$

The solution set is $\left\{\dfrac{3 + \sqrt{33}}{2}, \dfrac{3 - \sqrt{33}}{2}\right\}$. ∎

Calculator Close-up

If the solutions to an equation are irrational, as in Example 7, then we can use a calculator to check. For example, to check that $\frac{3 + \sqrt{33}}{2}$ satisfies the equation $x^2 - 3x - 6 = 0$, first use your calculator to get $\frac{3 + \sqrt{33}}{2} \approx 4.372281323$. Store this number in the memory, and use the memory recall key to evaluate the polynomial:

$$\boxed{\text{MR}}\;\boxed{x^2}\;\boxed{-}\;3\;\boxed{\times}\;\boxed{\text{MR}}\;\boxed{-}\;6\;\boxed{=}$$

The display will read

$$\boxed{\qquad\qquad\qquad 0\qquad}$$

Some graphing calculators have an answer key $\boxed{\text{ANS}}$ that has the same value as the last number calculated. In this case you can check as follows:

$$\text{ANS}^2 - 3\text{ANS} - 6\;\boxed{\text{ENTER}}$$

$$\boxed{\qquad\qquad\qquad 0\qquad}$$

Because of round-off errors, an answer like this might not check exactly. Instead of getting 0, your calculator might get an answer such as 1×10^{-10}, which is very close to zero.

The next two examples show equations that are not originally in the form of quadratic equations. However, after simplifying these equations, we get quadratic equations that we can solve.

EXAMPLE 8   **An equation containing a radical**

Solve $x + 3 = \sqrt{153 - x}$.

Solution

Square both sides of the equation to eliminate the radical:

$$
\begin{aligned}
x + 3 &= \sqrt{153 - x} && \text{The original equation}\\
(x + 3)^2 &= (\sqrt{153 - x})^2 && \text{Square each side.}\\
x^2 + 6x + 9 &= 153 - x && \text{Simplify.}\\
x^2 + 7x - 144 &= 0\\
(x - 9)(x + 16) &= 0 && \text{Factor.}
\end{aligned}
$$

$$x - 9 = 0 \quad \text{or} \quad x + 16 = 0 \qquad \text{Zero factor property}$$
$$x = 9 \quad \text{or} \quad x = -16$$

Because we squared each side of the original equation, we must check for extraneous roots. Let $x = 9$ in the original equation:

$$
\begin{aligned}
9 + 3 &= \sqrt{153 - 9}\\
12 &= \sqrt{144} \qquad \text{Correct}
\end{aligned}
$$

Let $x = -16$ in the original equation:

$$
\begin{aligned}
-16 + 3 &= \sqrt{153 - (-16)}\\
-13 &= \sqrt{169} \qquad \text{Incorrect because } \sqrt{169} = 13
\end{aligned}
$$

Because -16 is an extraneous root, the solution set is $\{9\}$.   ■

EXAMPLE 9

An equation containing rational expressions

Solve $\dfrac{1}{x} + \dfrac{3}{x-2} = \dfrac{5}{8}$.

Solution

The LCD for x, $x-2$, and 8 is $8x(x-2)$.

$$\frac{1}{x} + \frac{3}{x-2} = \frac{5}{8}$$

$$8x(x-2)\frac{1}{x} + 8x(x-2)\frac{3}{x-2} = 8x(x-2)\frac{5}{8} \qquad \text{Multiply each side by the LCD.}$$

$$8x - 16 + 24x = 5x^2 - 10x$$

$$32x - 16 = 5x^2 - 10x$$

$$-5x^2 + 42x - 16 = 0$$

$$5x^2 - 42x + 16 = 0 \qquad \text{Multiply each side by } -1 \text{ for easier factoring.}$$

$$(5x - 2)(x - 8) = 0 \qquad \text{Factor.}$$

$$5x - 2 = 0 \qquad \text{or} \qquad x - 8 = 0$$

$$x = \frac{2}{5} \qquad \text{or} \qquad x = 8$$

Check these values in the original equation. The solution set is $\left\{\frac{2}{5}, 8\right\}$. ∎

Imaginary Solutions

In Chapter 5 we found imaginary solutions to quadratic equations using the even-root property. We can get imaginary solutions also by completing the square.

EXAMPLE 10

An equation with imaginary solutions

Find the complex solutions to $x^2 - 4x + 12 = 0$.

Solution

Because the quadratic polynomial cannot be factored, we solve the equation by completing the square.

$$x^2 - 4x + 12 = 0 \qquad \text{The original equation}$$

$$x^2 - 4x \qquad = -12 \qquad \text{Subtract 12 from each side.}$$

$$x^2 - 4x + 4 \ = -12 + 4 \qquad \text{One half of } -4 \text{ is } -2, \text{ and } (-2)^2 = 4.$$

$$(x - 2)^2 = -8$$

$$x - 2 = \pm\sqrt{-8} \qquad \text{Even-root property}$$

$$x = 2 \pm i\sqrt{8}$$

$$= 2 \pm 2i\sqrt{2}$$

Check these values in the original equation. The solution set is $\{2 \pm 2i\sqrt{2}\}$. ∎

Warm-ups

True or false? Explain.

1. Completing the square means drawing the fourth side.

2. The equation $(x - 3)^2 = 12$ is equivalent to $x - 3 = 2\sqrt{3}$.

3. Every quadratic equation can be solved by factoring.

4. The trinomial $x^2 + \frac{4}{3}x + \frac{16}{9}$ is a perfect-square trinomial.

5. Every quadratic equation can be solved by completing the square.

6. To complete the square for $2x^2 + 6x = 4$, add 9 to each side.

7. $(2x - 3)(3x + 5) = 0$ is equivalent to $x = \frac{3}{2}$ or $x = \frac{5}{3}$.

8. In completing the square for $x^2 - 3x = 4$, add $\frac{9}{4}$ to each side.

9. The equation $x^2 = -8$ is equivalent to $x = \pm 2\sqrt{2}$.

10. All quadratic equations have two distinct complex solutions.

6.1 EXERCISES

Solve by factoring. See Example 1.

1. $x^2 - x - 6 = 0$
2. $x^2 + 6x + 8 = 0$
3. $a^2 + 2a = 15$
4. $w^2 - 2w = 15$
5. $2x^2 - x - 3 = 0$
6. $6x^2 - x - 15 = 0$
7. $x^2 = x + 12$
8. $x^2 = x + 20$
9. $y^2 + 14y + 49 = 0$
10. $a^2 - 6a + 9 = 0$
11. $a^2 - 16 = 0$
12. $4w^2 - 25 = 0$

Use the even-root property to solve each equation. See Example 2.

13. $x^2 = 81$
14. $x^2 = \frac{9}{4}$
15. $x^2 = \frac{16}{9}$
16. $a^2 = 32$
17. $(x - 3)^2 = 16$
18. $(x + 5)^2 = 4$
19. $(z + 1)^2 = 5$
20. $(a - 2)^2 = 8$
21. $\left(w - \frac{3}{2}\right)^2 = \frac{7}{4}$
22. $\left(w + \frac{2}{3}\right)^2 = \frac{5}{9}$
23. $\left(y - \frac{1}{2}\right)^2 = \frac{9}{2}$
24. $\left(y - \frac{1}{3}\right)^2 = \frac{1}{3}$

Find the perfect-square trinomial whose first two terms are given. See Example 3.

25. $x^2 + 2x$
26. $m^2 + 14m$
27. $x^2 - 3x$
28. $w^2 - 5w$
29. $y^2 + \frac{1}{4}y$
30. $z^2 + \frac{3}{2}z$
31. $x^2 + \frac{2}{3}x$
32. $p^2 + \frac{6}{5}p$

Factor each perfect-square trinomial. See Example 4.

33. $x^2 + 8x + 16$
34. $x^2 - 10x + 25$
35. $y^2 - 5y + \frac{25}{4}$
36. $w^2 + w + \frac{1}{4}$
37. $z^2 - \frac{4}{7}z + \frac{4}{49}$
38. $m^2 - \frac{6}{5}m + \frac{9}{25}$
39. $t^2 + \frac{3}{5}t + \frac{9}{100}$
40. $h^2 + \frac{3}{2}h + \frac{9}{16}$

Solve by completing the square. See Examples 5–7. Use your calculator to check.

41. $x^2 - 2x - 15 = 0$
42. $x^2 - 6x - 7 = 0$
43. $x^2 + 8x = 20$
44. $x^2 + 10x = -9$
45. $2x^2 - 4x = 70$
46. $3x^2 - 6x = 24$
47. $w^2 - w - 20 = 0$
48. $y^2 - 3y - 10 = 0$
49. $q^2 + 5q = 14$
50. $z^2 + z = 2$
51. $2h^2 - h - 3 = 0$
52. $2m^2 - m - 15 = 0$
53. $x^2 + 4x = 6$
54. $x^2 + 6x - 8 = 0$
55. $x^2 + 8x - 4 = 0$
56. $x^2 + 10x - 3 = 0$
57. $2x^2 + 3x - 4 = 0$
58. $2x^2 + 5x - 1 = 0$

Solve each equation by an appropriate method. See Examples 8 and 9.

59. $\sqrt{2x + 1} = x - 1$

60. $\sqrt{2x - 4} = x - 14$

61. $w = \dfrac{\sqrt{w + 1}}{2}$

62. $y - 1 = \dfrac{\sqrt{y + 1}}{2}$

63. $\dfrac{t}{t - 2} = \dfrac{2t - 3}{t}$

64. $\dfrac{z}{z + 3} = \dfrac{3z}{5z - 1}$

65. $\dfrac{2}{x^2} + \dfrac{4}{x} + 1 = 0$

66. $\dfrac{1}{x^2} + \dfrac{3}{x} + 1 = 0$

Find the complex solutions to each equation. See Example 10.

67. $x^2 + 2x + 5 = 0$

68. $x^2 + 4x + 5 = 0$

69. $x^2 + 12 = 0$

70. $-3x^2 - 21 = 0$

71. $5z^2 - 4z + 1 = 0$

72. $2w^2 - 3w + 2 = 0$

Find all real or imaginary solutions to each equation. Use the method of your choice.

73. $4x^2 + 25 = 0$

74. $5w^2 - 3 = 0$

75. $\left(p + \dfrac{1}{2}\right)^2 = \dfrac{9}{4}$

76. $\left(y - \dfrac{2}{3}\right)^2 = \dfrac{4}{9}$

77. $5t^2 + 4t - 3 = 0$

78. $3v^2 + 4v - 1 = 0$

79. $m^2 + 2m - 24 = 0$

80. $q^2 + 6q - 7 = 0$

81. $\left(a + \dfrac{2}{3}\right)^2 = -\dfrac{32}{9}$

82. $\left(w + \dfrac{1}{2}\right)^2 = -6$

83. $-x^2 + x + 6 = 0$

84. $-x^2 + x + 12 = 0$

85. $x^2 - 6x + 10 = 0$

86. $x^2 - 8x + 17 = 0$

87. $2x - 5 = \sqrt{7x + 7}$

88. $\sqrt{7x + 29} = x + 3$

89. $\dfrac{1}{x} + \dfrac{1}{x - 1} = \dfrac{1}{4}$

90. $\dfrac{1}{x} - \dfrac{2}{1 - x} = \dfrac{1}{2}$

Solve for x.

91. $x^2 + 6x + k = 0$

92. $x^2 - 4x + c = 0$

93. $x^2 + 2kx + w = 0$

94. $x^2 - kx = \dfrac{1}{4}$

95. $\left(x + \dfrac{m}{2}\right)^2 = \dfrac{m^2 - 4}{4}$

96. $2x^2 + mx + n = 0$

If the solution to an equation is imaginary or irrational, it takes a bit more effort to check. Replace x by each given number to verify each statement.

97. Both $2 + \sqrt{3}$ and $2 - \sqrt{3}$ satisfy $x^2 - 4x + 1 = 0$.

98. Both $1 + \sqrt{2}$ and $1 - \sqrt{2}$ satisfy $x^2 - 2x - 1 = 0$.

99. Both $i + 1$ and $1 - i$ satisfy $x^2 - 2x + 2 = 0$.

100. Both $2 + 3i$ and $2 - 3i$ satisfy $x^2 - 4x + 13 = 0$.

Solve each problem.

101. *Approach speed.* The formula $1211.1L = CA^2S$ is used to determine the approach speed for landing an aircraft, where L is the gross weight of the aircraft in pounds, C is the coefficient of lift, S is the surface area of the wings in square feet, and A is approach speed in feet per second. Find A for the Piper Cheyenne, which has a gross weight of 8700 lb, a coefficient of lift of 2.81, and wing surface area of 200 ft^2.

102. *Time to swing.* The period T (time in seconds for one complete cycle) of a simple pendulum is related to the length L (in feet) of the pendulum by the formula $8T^2 = \pi^2 L$. If a child is on a swing with a 10-foot chain, then how long does it take to complete one cycle of the swing?

103. *Time for a swim.* Tropical Pools figures that its monthly revenue in dollars on the sale of x above-ground pools is given by $R = 1500x - 3x^2$, where x is less than 25. What number of pools sold would provide a revenue of $17,568?

104. *Pole vaulting.* In 1981, Vladimir Poliakov (U.S.S.R.) set a world record of 19 ft $\frac{3}{4}$ in. for the pole vault (*Doubleday Almanac*). To reach that height, Poliakov's speed on the runway was approximately 36 feet per second. The formula $h = -16t^2 + 36t$ gives his height t seconds after leaving the ground. For what values of t is his height 18 feet?

Getting More Involved

105. *Discussion.* Which of the following equations is not a quadratic equation?

a) $\pi x^2 - \sqrt{5}\, x - 1 = 0$ b) $3x^2 - 1 = 0$
c) $4x + 5 = 0$ d) $0.009x^2 = 0$

106. *Exploration.* Solve $x^2 - 4x + k = 0$ for $k = 0$, 4, 5, and 10.

a) When does the equation have only one solution?
b) For what values of k are the solutions real?
c) For what values of k are the solutions imaginary?

107. *Cooperative learning.* Write a quadratic equation of each of the following types, then trade your equations with those of a classmate. Solve the equations and verify that they are of the required types.

a) a single rational solution
b) two rational solutions
c) two irrational solutions
d) two imaginary solutions

108. *Exploration.* In the next section we will solve $ax^2 + bx + c = 0$ for x by completing the square. Try it now without looking ahead.

6.2 The Quadratic Formula

Completing the square from Section 6.1 can be used to solve any quadratic equation. Here we apply this method to the general quadratic equation to get a formula for the solutions to any quadratic equation.

Developing the Formula

Start with the general form of the quadratic equation,

$$ax^2 + bx + c = 0.$$

Assume that a is positive for now, and divide each side by a:

$$\frac{ax^2 + bx + c}{a} = \frac{0}{a}$$

$$x^2 + \frac{b}{a}x + \frac{c}{a} = 0$$

$$x^2 + \frac{b}{a}x = -\frac{c}{a} \qquad \text{Subtract } \tfrac{c}{a} \text{ from each side.}$$

One-half of $\dfrac{b}{a}$ is $\dfrac{b}{2a}$, and $\dfrac{b}{2a}$ squared is $\dfrac{b^2}{4a^2}$.

$$x^2 + \frac{b}{a}x + \frac{b^2}{4a^2} = -\frac{c}{a} + \frac{b^2}{4a^2}$$

Factor the left side and get a common denominator for the right side:

$$\left(x + \frac{b}{2a}\right)^2 = \frac{b^2}{4a^2} - \frac{4ac}{4a^2} \qquad \frac{c\,(4a)}{a\,(4a)} = \frac{4ac}{4a^2}$$

$$\left(x + \frac{b}{2a}\right)^2 = \frac{b^2 - 4ac}{4a^2}$$

$$x + \frac{b}{2a} = \pm\sqrt{\frac{b^2 - 4ac}{4a^2}} \qquad \text{Even-root property}$$

$$x = \frac{-b}{2a} \pm \frac{\sqrt{b^2 - 4ac}}{2a} \qquad \text{Because a} > 0,\ \sqrt{4a^2} = 2a.$$

$$x = \frac{-b \pm \sqrt{b^2 - 4ac}}{2a}$$

We assumed that a was positive so that $\sqrt{4a^2} = 2a$ would be correct. If a is negative, then $\sqrt{4a^2} = -2a$, and we get

$$x = \frac{-b}{2a} \pm \frac{\sqrt{b^2 - 4ac}}{-2a}.$$

However, the negative sign can be omitted in $-2a$ because of the $\pm$ symbol preceding it. For example, the results of $5 \pm (-3)$ and 5 ± 3 are the same. So when a is negative, we get the same formula as when a is positive. It is called the **quadratic formula.**

> **The Quadratic Formula**
>
> The solution to $ax^2 + bx + c = 0$, with $a \neq 0$, is given by the formula
>
> $$x = \frac{-b \pm \sqrt{b^2 - 4ac}}{2a}.$$

Using the Formula

The quadratic formula can be used to solve any quadratic equation.

EXAMPLE 1

Two rational solutions

Solve $x^2 + 2x - 15 = 0$ using the quadratic formula.

Solution

To use the formula, we first identify the values of a, b, and c:

$$\underset{\underset{a}{\uparrow}}{1x^2} + \underset{\underset{b}{\uparrow}}{2x} - \underset{\underset{c}{\uparrow}}{15} = 0$$

The coefficient of x^2 is 1, so $a = 1$. The coefficient of $2x$ is 2, so $b = 2$. The constant term is -15, so $c = -15$. Substitute these values into the quadratic formula:

$$x = \frac{-2 \pm \sqrt{2^2 - 4(1)(-15)}}{2(1)}$$

$$= \frac{-2 \pm \sqrt{4 + 60}}{2}$$

$$= \frac{-2 \pm \sqrt{64}}{2}$$

$$= \frac{-2 \pm 8}{2}$$

$$x = \frac{-2 + 8}{2} = 3 \qquad \text{or} \qquad x = \frac{-2 - 8}{2} = -5$$

Check 3 and -5 in the original equation. The solution set is $\{-5, 3\}$. ∎

CAUTION To identify a, b, and c for the quadratic formula, the equation must be in the standard form $ax^2 + bx + c = 0$. If it is not in that form, then you must first rewrite the equation. ⊘

EXAMPLE 2

One rational solution

Solve $4x^2 = 12x - 9$ by using the quadratic formula.

Solution

Rewrite the equation in the form $ax^2 + bx + c = 0$ before identifying a, b, and c:

$$4x^2 - 12x + 9 = 0$$

In this form we get $a = 4$, $b = -12$, and $c = 9$.

$$x = \frac{12 \pm \sqrt{(-12)^2 - 4(4)(9)}}{2(4)}$$ 　　Because $b = -12$, $-b = 12$.

$$= \frac{12 \pm \sqrt{144 - 144}}{8}$$

$$= \frac{12 \pm 0}{8} = \frac{12}{8} = \frac{3}{2}$$

Check $\frac{3}{2}$ in the original equation. The solution set is $\left\{\frac{3}{2}\right\}$. ■

Because the solutions to the equations in Examples 1 and 2 were rational numbers, these equations could have been solved by factoring. In the next example the solutions are irrational.

EXAMPLE 3　**Two irrational solutions**

Solve $2x^2 + 6x + 3 = 0$.

Solution

Let $a = 2$, $b = 6$, and $c = 3$ in the quadratic formula:

$$x = \frac{-6 \pm \sqrt{(6)^2 - 4(2)(3)}}{2(2)}$$

$$= \frac{-6 \pm \sqrt{36 - 24}}{4} = \frac{-6 \pm \sqrt{12}}{4}$$

$$= \frac{-6 \pm 2\sqrt{3}}{4} = \frac{2(-3 \pm \sqrt{3})}{2 \cdot 2}$$

$$= \frac{-3 \pm \sqrt{3}}{2}$$

Check these values in the original equation. The solution set is $\left\{\frac{-3 \pm \sqrt{3}}{2}\right\}$. ■

EXAMPLE 4　**Two imaginary solutions, no real solutions**

Find the complex solutions to $x^2 + x + 5 = 0$.

Solution

Let $a = 1$, $b = 1$, and $c = 5$ in the quadratic formula:

$$x = \frac{-1 \pm \sqrt{(1)^2 - 4(1)(5)}}{2(1)} = \frac{-1 \pm \sqrt{-19}}{2} = \frac{-1 \pm i\sqrt{19}}{2}$$

Check these values in the original equation. The solution set is $\left\{\frac{-1 \pm i\sqrt{19}}{2}\right\}$. There are no real solutions to the equation. ■

You have learned to solve quadratic equations by four different methods: the even-root property, factoring, completing the square, and the quadratic formula. The even-root property and factoring are limited to certain special equations, but you should use those methods when possible. Any quadratic equation can be solved by completing the square or using the quadratic formula. Because the quadratic formula is usually faster, it is used more often than completing the square. However, completing the square is an important skill to learn. It will be used in the study of conic sections in Chapter 11.

Methods for Solving $ax^2 + bx + c = 0$		
Method	**Comments**	**Examples**
Even-root property	Use when $b = 0$.	$(x - 2)^2 = 8$ $x - 2 = \pm\sqrt{8}$
Factoring	Use when the polynomial can be factored.	$x^2 + 5x + 6 = 0$ $(x + 2)(x + 3) = 0$
Quadratic formula	Solves any quadratic equation.	$x^2 + 5x + 3 = 0$ $x = \dfrac{-5 \pm \sqrt{25 - 4(3)}}{2}$
Completing the square	Solves any quadratic equation, but quadratic formula is faster.	$x^2 - 6x + 7 = 0$ $x^2 - 6x + 9 = -7 + 9$ $(x - 3)^2 = 2$

Number of Solutions

The quadratic equations in Examples 1 and 3 had two real solutions each. In each of those examples the value of $b^2 - 4ac$ was positive. In Example 2 the quadratic equation had only one solution because the value of $b^2 - 4ac$ was zero. In Example 4 the quadratic equation had no real solutions because $b^2 - 4ac$ was negative. Because $b^2 - 4ac$ determines the kind and number of solutions to a quadratic equation, it is called the **discriminant.**

MATH AT WORK

Kitchen Designer

Remodeling a kitchen can be an expensive undertaking. Choosing the correct style of cabinets, doors, and floor covering is just a small part of the process. Joe Prendergast, designer for Lee Kimball Kitchens, is involved in every step of creating a new kitchen for a client.

The process begins with customers visiting the store to see what products are available. Once the client has decided on material and style, he or she fills out an extensive questionnaire so that Prendergast can get a sense of the client's lifestyle. Design plans are drawn using a scale of $\frac{1}{2}$ inch to 1 foot. Consideration is given to traffic patterns, doorways, and especially work areas where the cook would want to work unencumbered. Storage areas are a big consideration, as are lighting and color.

A new kitchen can take anywhere from five weeks to a few months or more, depending on how complicated the design and construction is. In Exercise 77 of this section we will find the dimensions of a border around a countertop.

> **Number of Solutions to a Quadratic Equation**
>
> The quadratic equation $ax^2 + bx + c = 0$ with $a \neq 0$ has
>   *two* real solutions if $b^2 - 4ac > 0$,
>   *one* real solution if $b^2 - 4ac = 0$, and
>   *no* real solutions (two imaginary solutions) if $b^2 - 4ac < 0$.

EXAMPLE 5   **Using the discriminant**

Use the discriminant to determine the number of real solutions to each quadratic equation.

a) $x^2 - 3x - 5 = 0$ **b)** $x^2 = 3x - 9$ **c)** $4x^2 - 12x + 9 = 0$

Solution

a)   For $x^2 - 3x - 5 = 0$, use $a = 1$, $b = -3$, and $c = -5$ in $b^2 - 4ac$:

$$b^2 - 4ac = (-3)^2 - 4(1)(-5) = 9 + 20 = 29$$

Because the discriminant is positive, there are two real solutions to this quadratic equation.

b)   For $x^2 - 3x + 9 = 0$, use $a = 1$, $b = -3$, and $c = 9$ in $b^2 - 4ac$:

$$b^2 - 4ac = (-3)^2 - 4(1)(9) = 9 - 36 = -27$$

Because the discriminant is negative, the equation has no real solutions. It has two imaginary solutions.

c)   For $4x^2 - 12x + 9 = 0$, use $a = 4$, $b = -12$, and $c = 9$ in $b^2 - 4ac$:

$$b^2 - 4ac = (-12)^2 - 4(4)(9) = 144 - 144 = 0$$

Because the discriminant is 0, there is only one real solution to this quadratic equation.   ■

Applications

With the quadratic formula we can easily solve problems whose solutions are irrational numbers. When the solutions are irrational numbers, we usually use a calculator to find rational approximations and to check.

EXAMPLE 6   **Area of a table top**

The area of a rectangular table top is 6 square feet. If the width is 2 feet shorter than the length, then what are the dimensions?

Solution

Let x be the length and $x - 2$ be the width, as shown in Fig. 6.1. Because the area is 6 square feet and $A = LW$, we can write the equation

$$x(x - 2) = 6$$

or

$$x^2 - 2x - 6 = 0.$$

Figure 6.1

Because this equation cannot be factored, we use the quadratic formula with $a = 1$, $b = -2$, and $c = -6$:

$$x = \frac{2 \pm \sqrt{(-2)^2 - 4(1)(-6)}}{2(1)}$$

$$= \frac{2 \pm \sqrt{28}}{2} = \frac{2 \pm 2\sqrt{7}}{2} = 1 \pm \sqrt{7}$$

Because $1 - \sqrt{7}$ is a negative number, it cannot be the length of a table top. If $x = 1 + \sqrt{7}$ then $x - 2 = 1 + \sqrt{7} - 2 = \sqrt{7} - 1$. Checking the product of $\sqrt{7} + 1$ and $\sqrt{7} - 1$, we get

$$(\sqrt{7} + 1)(\sqrt{7} - 1) = 7 - 1 = 6.$$

The exact length is $\sqrt{7} + 1$ feet, and the width is $\sqrt{7} - 1$ feet. Using a calculator, we find that the approximate length is 3.65 feet and the approximate width is 1.65 feet.   ■

Warm-ups

True or false? Explain.

1. Completing the square is used to develop the quadratic formula.
2. For the equation $3x^2 = 4x - 7$, we have $a = 3$, $b = 4$, and $c = -7$.
3. If $dx^2 + ex + f = 0$ and $d \neq 0$, then $x = \dfrac{-e \pm \sqrt{e^2 - 4df}}{2d}$.
4. The quadratic formula will not work on the equation $x^2 - 3 = 0$.
5. If $a = 2$, $b = -3$, and $c = -4$, then $b^2 - 4ac = 41$.
6. If the discriminant is zero, then there are no imaginary solutions.
7. If $b^2 - 4ac > 0$, then $ax^2 + bx + c = 0$ has two real solutions.
8. To solve $2x - x^2 = 0$ by the quadratic formula, let $a = -1$, $b = 2$, and $c = 0$.
9. Two numbers that have a sum of 6 can be represented by x and $x + 6$.
10. Some quadratic equations have one real and one imaginary solution.

6.2 EXERCISES

Solve each equation by using the quadratic formula. See Example 1.

1. $x^2 + 5x + 6 = 0$
2. $x^2 - 7x + 12 = 0$
3. $y^2 + y = 6$
4. $m^2 + 2m = 8$
5. $6z^2 - 7z - 3 = 0$
6. $8q^2 + 2q - 1 = 0$

Solve each equation by using the quadratic formula. See Example 2.

7. $4x^2 - 4x + 1 = 0$
8. $4x^2 - 12x + 9 = 0$
9. $9x^2 - 6x + 1 = 0$
10. $9x^2 - 24x + 16 = 0$
11. $9 + 24x + 16x^2 = 0$
12. $4 + 20x = -25x^2$

Solve each equation by using the quadratic formula. See Example 3.

13. $v^2 + 8v + 6 = 0$
14. $p^2 + 6p + 4 = 0$
15. $-x^2 - 5x + 1 = 0$
16. $-x^2 - 3x + 5 = 0$
17. $2t^2 - 6t + 1 = 0$
18. $3z^2 - 8z + 2 = 0$

Solve each equation by using the quadratic formula. See Example 4.

19. $2t^2 - 6t + 5 = 0$
20. $2y^2 + 1 = 2y$
21. $-2x^2 + 3x = 6$
22. $-3x^2 - 2x - 5 = 0$
23. $\dfrac{1}{2}x^2 + 13 = 5x$
24. $\dfrac{1}{4}x^2 + \dfrac{17}{4} = 2x$

Find $b^2 - 4ac$ and the number of real solutions to each equation. See Example 5.

25. $x^2 - 6x + 2 = 0$

26. $x^2 + 6x + 9 = 0$

27. $2x^2 - 5x + 6 = 0$

28. $-x^2 + 3x - 4 = 0$

29. $4m^2 + 25 = 20m$

30. $v^2 = 3v + 5$

31. $y^2 - \frac{1}{2}y + \frac{1}{4} = 0$

32. $\frac{1}{2}w^2 - \frac{1}{3}w + \frac{1}{4} = 0$

33. $-3t^2 + 5t + 6 = 0$

34. $9m^2 + 16 = 24m$

35. $9 - 24z + 16z^2 = 0$

36. $12 - 7x + x^2 = 0$

37. $5x^2 - 7 = 0$

38. $-6x^2 - 5 = 0$

39. $x^2 = x$

40. $-3x^2 + 7x = 0$

Solve each equation by the method of your choice.

41. $\frac{1}{3}x^2 + \frac{1}{2}x = \frac{1}{3}$

42. $\frac{1}{2}x^2 + x = 1$

43. $\frac{w}{w - 2} = \frac{w}{w - 3}$

44. $\frac{y}{3y - 4} = \frac{2}{y + 4}$

45. $\frac{9(3x - 5)^2}{4} = 1$

46. $\frac{25(2x + 1)^2}{9} = 0$

47. $1 + \frac{20}{x^2} = \frac{8}{x}$

48. $\frac{34}{x^2} = \frac{6}{x} - 1$

49. $(x - 8)(x + 4) = -42$

50. $(x - 10)(x - 2) = -20$

51. $y = \frac{3(2y + 5)}{8(y - 1)}$

52. $z = \frac{7z - 4}{12(z - 1)}$

Use the quadratic formula and a calculator to solve each equation. Round answers to three decimal places and check your answers.

53. $x^2 + 3.2x - 5.7 = 0$

54. $x^2 + 7.15x + 3.24 = 0$

55. $x^2 - 7.4x + 13.69 = 0$

56. $1.44x^2 + 5.52x + 5.29 = 0$

57. $1.85x^2 + 6.72x + 3.6 = 0$

58. $3.67x^2 + 4.35x - 2.13 = 0$

59. $3x^2 + 14{,}379x + 243 = 0$

60. $x^2 + 12{,}347x + 6{,}741 = 0$

61. $x^2 + 0.00075x - 0.0062 = 0$

62. $4.3x^2 - 9.86x - 3.75 = 0$

Solve each problem. See Example 6.

63. *Missing numbers.* Find two positive real numbers that differ by 1 and have a product of 16.

64. *Missing numbers.* Find two positive real numbers that differ by 2 and have a product of 10.

65. *More missing numbers.* Find two real numbers that have a sum of 6 and a product of 4.

66. *More missing numbers.* Find two real numbers that have a sum of 8 and a product of 2.

67. *Bulletin board.* The length of a bulletin board is 1 foot more than the width. The diagonal has a length of $\sqrt{3}$ feet. Find the length and width of the bulletin board.

68. *Diagonal brace.* The width of a rectangular gate is 2 meters larger than its height. The diagonal brace measures $\sqrt{6}$ meters. Find the width and height.

Figure for Exercise 68

69. *Area of a rectangle.* The length of a rectangle is 4 feet longer than the width, and its area is 10 square feet. Find the length and width.

70. *Diagonal of a square.* The diagonal of a square is 2 meters longer than a side. Find the length of a side.

If an object is given an initial velocity of v_0 feet per second from a height of s_0 feet, then its height S after t seconds is given by the formula $S = -16t^2 + v_0t + s_0$.

71. *Projected pine cone.* If a pine cone is projected upward at a velocity of 16 feet per second from the top of a 96-foot pine tree, then how long does it take to reach the earth?

72. *Falling pine cone.* If a pine cone falls from the top of a 96-foot pine tree, then how long does it take to reach the earth?

73. *Penny tossing.* If a penny is thrown downward at 30 feet per second from the bridge at Royal Gorge, Colorado, how long does it take to reach the Arkansas River 1,000 feet below?

74. *Foul ball.* If Charlie O'Brian of the Braves hits a baseball upward at 150 feet per second from a height of 5 feet, then how long will it take for the ball to reach the earth?

In Exercises 75–77, solve each problem.

75. *Recovering an investment.* The manager at Cream of the Crop bought a load of watermelons for $200. She priced the melons so that she would make $1.50 profit on each melon. When all but 30 had been sold, she had recovered her initial investment. How many did she buy originally?

76. *Sharing cost.* The members of a flying club plan to share equally the cost of a $200,000 airplane. The members want to find five more people to join the club so that the cost per person will decrease by $2000. How many members are currently in the club?

77. *Kitchen countertop.* A 30-in. by 40-in. countertop for a work island is to be covered with green ceramic tiles, except for a border of uniform width as shown in the figure. If the area covered by the green tiles is 704 in.^{2}, then how wide is the border?

Figure for Exercise 77

Getting More Involved

78. *Discussion.* Find the solutions to $6x^2 + 5x - 4 = 0$. Is the sum of your solutions equal to $-\dfrac{b}{a}$? Explain why the sum of the solutions to any quadratic equation is $-\dfrac{b}{a}$. (*Hint:* Use the quadratic formula.)

79. *Discussion.* Use the result of Exercise 78 to check whether $\left\{\dfrac{2}{3}, \dfrac{1}{3}\right\}$ is the solution set to $9x^2 - 3x - 2 = 0$. If this solution set is not correct, then what is the correct solution set?

80. *Discussion.* What is the product of the two solutions to $6x^2 + 5x - 4 = 0$? Explain why the product of the solutions to any quadratic equation is $\dfrac{c}{a}$.

81. *Discussion.* Use the result of the previous exercise to check whether $\left\{\dfrac{9}{2}, -2\right\}$ is the solution set to $2x^2 - 13x + 18 = 0$. If this solution set is not correct, then what is the correct solution set?

82. *Cooperative learning.* Work in a group to write a quadratic equation that has each given pair of solutions.

a) -4 and 5 b) $2 - \sqrt{3}$ and $2 + \sqrt{3}$
c) $5 + 2i$ and $5 - 2i$

6.3 More on Quadratic Equations

In this section:
▶ Using the Discriminant in Factoring
▶ Writing a Quadratic with Given Solutions
▶ Equations Quadratic in Form
▶ Applications

In this section we use the ideas and methods of the previous sections to explore additional topics involving quadratic equations.

Using the Discriminant in Factoring

Consider $ax^2 + bx + c$, where a, b, and c are integers with a greatest common factor of 1. If $b^2 - 4ac$ is a perfect square, then $\sqrt{b^2 - 4ac}$ is a whole number, and the solutions to $ax^2 + bx + c = 0$ are rational numbers. If the solutions to a quadratic equation are rational numbers, then they could be found by the factoring method. So if $b^2 - 4ac$ is a perfect square, then $ax^2 + bx + c$ factors. It is also true that if $b^2 - 4ac$ is not a perfect square, then $ax^2 + bx + c$ is prime.

EXAMPLE 1   **Using the discriminant**
Use the discriminant to determine whether each polynomial can be factored.

a) $6x^2 + x - 15$ b) $5x^2 - 3x + 2$

Solution

a) Use $a = 6$, $b = 1$, and $c = -15$ to find $b^2 - 4ac$:

$$b^2 - 4ac = 1^2 - 4(6)(-15) = 361$$

Because $\sqrt{361} = 19$, $6x^2 + x - 15$ can be factored. Using the *ac*-method, we get

$$6x^2 + x - 15 = (2x - 3)(3x + 5).$$

b) Use $a = 5$, $b = -3$, and $c = 2$ to find $b^2 - 4ac$:

$$b^2 - 4ac = (-3)^2 - 4(5)(2) = -31$$

Because the discriminant is not a perfect square, $5x^2 - 3x + 2$ is prime. ■

Writing a Quadratic with Given Solutions

Not every quadratic equation can be solved by factoring, but the factoring method can be used (in reverse) to write a quadratic equation with any given solutions. For example, if the solutions to a quadratic equation are 5 and -3, we can reverse the steps in the factoring method as follows:

$$x = 5 \qquad \text{or} \qquad x = -3$$
$$x - 5 = 0 \qquad \text{or} \qquad x + 3 = 0$$
$$(x - 5)(x + 3) = 0 \qquad \text{Zero factor property}$$
$$x^2 - 2x - 15 = 0 \qquad \text{Multiply the factors.}$$

This method will produce the equation even if the solutions are irrational or imaginary.

EXAMPLE 2   **Writing a quadratic given the solutions**
Write a quadratic equation that has each given pair of solutions.

a) 4, -6 **b)** $-\sqrt{2}$, $\sqrt{2}$ **c)** $-3i$, $3i$

Solution

a) Reverse the factoring method using solutions 4 and -6:

$$x = 4 \qquad \text{or} \qquad x = -6$$
$$x - 4 = 0 \qquad \text{or} \qquad x + 6 = 0$$
$$(x - 4)(x + 6) = 0 \qquad \text{Zero factor property}$$
$$x^2 + 2x - 24 = 0 \qquad \text{Multiply the factors.}$$

b) Reverse the factoring method using solutions $-\sqrt{2}$ and $\sqrt{2}$:

$$x = -\sqrt{2} \qquad \text{or} \qquad x = \sqrt{2}$$
$$x + \sqrt{2} = 0 \qquad \text{or} \qquad x - \sqrt{2} = 0$$
$$(x + \sqrt{2})(x - \sqrt{2}) = 0 \qquad \text{Zero factor property}$$
$$x^2 - 2 = 0 \qquad \text{Multiply the factors.}$$

c) Reverse the factoring method using solutions $-3i$ and $3i$:

$$x = -3i \qquad \text{or} \qquad x = 3i$$
$$x + 3i = 0 \qquad \text{or} \qquad x - 3i = 0$$
$$(x + 3i)(x - 3i) = 0 \qquad \text{Zero factor property}$$
$$x^2 - 9i^2 = 0 \qquad \text{Multiply the factors.}$$
$$x^2 + 9 = 0 \qquad \text{Note: } i^2 = -1$$

■

Equations Quadratic in Form

In a quadratic equation we have a variable and its square (x and x^2). An equation that contains an expression and the square of that expression is **quadratic in form** if substituting a single variable for that expression results in a quadratic equation. Equations that are quadratic in form can be solved by using methods for quadratic equations.

EXAMPLE 3 **An equation quadratic in form**

Solve $(x + 15)^2 - 3(x + 15) - 18 = 0$

Solution

Note that $x + 15$ and $(x + 15)^2$ both appear in the equation. Let $a = x + 15$, and substitute a for $x + 15$ in the equation:

$$(x + 15)^2 - 3(x + 15) - 18 = 0$$
$$a^2 - 3a - 18 = 0$$
$$(a - 6)(a + 3) = 0 \qquad \text{Factor.}$$

$$a - 6 = 0 \quad \text{or} \quad a + 3 = 0$$
$$a = 6 \quad \text{or} \quad a = -3$$
$$x + 15 = 6 \quad \text{or} \quad x + 15 = -3 \qquad \text{Replace } a \text{ by } x + 15.$$
$$x = -9 \quad \text{or} \quad x = -18$$

Check in the original equation. The solution set is $\{-18, -9\}$. ∎

In the next example we have a fourth-degree equation that is quadratic in form. Note that the fourth-degree equation has four solutions.

EXAMPLE 4 **A fourth-degree equation**

Solve $x^4 - 6x^2 + 8 = 0$.

Solution

Note that x^4 is the square of x^2. If we let $w = x^2$, then $w^2 = x^4$. Substitute these expressions into the original equation.

$$x^4 - 6x^2 + 8 = 0$$
$$w^2 - 6w + 8 = 0 \qquad \text{Replace } x^4 \text{ by } w^2 \text{ and } x^2 \text{ by } w.$$
$$(w - 2)(w - 4) = 0 \qquad \text{Factor.}$$

$$w - 2 = 0 \quad \text{or} \quad w - 4 = 0$$
$$w = 2 \quad \text{or} \quad w = 4$$
$$x^2 = 2 \quad \text{or} \quad x^2 = 4 \qquad \text{Substitute } x^2 \text{ for } w.$$
$$x = \pm\sqrt{2} \quad \text{or} \quad x = \pm 2 \qquad \text{Even-root property}$$

Check. The solution set is $\{-2, -\sqrt{2}, \sqrt{2}, 2\}$. ∎

CAUTION If you replace x^2 by w, do not quit when you find the values of w. If the variable in the original equation is x, then you must solve for x. ⊘

EXAMPLE 5   **A quadratic within a quadratic**

Solve $(x^2 + 2x)^2 - 11(x^2 + 2x) + 24 = 0$.

Solution

Note that $x^2 + 2x$ and $(x^2 + 2x)^2$ appear in the equation. Let $a = x^2 + 2x$ and substitute.

$$a^2 - 11a + 24 = 0$$
$$(a - 8)(a - 3) = 0 \quad \text{Factor.}$$

$$a - 8 = 0 \quad \text{or} \quad a - 3 = 0$$
$$a = 8 \quad \text{or} \quad a = 3$$
$$x^2 + 2x = 8 \quad \text{or} \quad x^2 + 2x = 3 \quad \text{Replace } a \text{ by } x^2 + 2x.$$
$$x^2 + 2x - 8 = 0 \quad \text{or} \quad x^2 + 2x - 3 = 0$$
$$(x - 2)(x + 4) = 0 \quad \text{or} \quad (x + 3)(x - 1) = 0$$

$$x - 2 = 0 \quad \text{or} \quad x + 4 = 0 \quad \text{or} \quad x + 3 = 0 \quad \text{or} \quad x - 1 = 0$$
$$x = 2 \quad \text{or} \quad x = -4 \quad \text{or} \quad x = -3 \quad \text{or} \quad x = 1$$

Check. The solution set is $\{-4, -3, 1, 2\}$. ∎

The next example involves a fractional exponent. To identify this type of equation as quadratic in form, recall how to square an expression with a fractional exponent. For example, $(x^{1/2})^2 = x$, $(x^{1/4})^2 = x^{1/2}$, and $(x^{1/3})^2 = x^{2/3}$.

EXAMPLE 6   **A fractional exponent**

Solve $x - 9x^{1/2} + 14 = 0$.

Solution

Note that the square of $x^{1/2}$ is x. Let $w = x^{1/2}$; then $w^2 = (x^{1/2})^2 = x$. Now substitute w and w^2 into the original equation:

$$w^2 - 9w + 14 = 0$$
$$(w - 7)(w - 2) = 0$$

$$w - 7 = 0 \quad \text{or} \quad w - 2 = 0$$
$$w = 7 \quad \text{or} \quad w = 2$$
$$x^{1/2} = 7 \quad \text{or} \quad x^{1/2} = 2 \quad \text{Replace } w \text{ by } x^{1/2}.$$
$$x = 49 \quad \text{or} \quad x = 4 \quad \text{Square each side.}$$

Because we squared each side, we must check for extraneous roots. First evaluate $x - 9x^{1/2} + 14$ for $x = 49$:

$$49 - 9 \cdot 49^{1/2} + 14 = 49 - 9 \cdot 7 + 14 = 0$$

Now evaluate $x - 9x^{1/2} + 14$ for $x = 4$:

$$4 - 9 \cdot 4^{1/2} + 14 = 4 - 9 \cdot 2 + 14 = 0$$

Because each solution checks, the solution set is $\{4, 49\}$. ∎

CAUTION   An equation of quadratic form must have a term that is the square of another. Equations such as $x^4 - 5x^3 + 6 = 0$ or $x^{1/2} - 3x^{1/3} - 18 = 0$ are not quadratic in form and cannot be solved by substitution. ⊘

Applications

Applied problems often result in quadratic equations that cannot be factored. For such equations we use the quadratic formula to find exact solutions and a calculator to find decimal approximations for the exact solutions.

EXAMPLE 7 **Changing area**

Marvin's flower bed is rectangular in shape with a length of 10 feet and a width of 5 feet. He wants to increase the length and width by the same amount to obtain a flower bed with an area of 75 square feet. What should the amount of increase be?

Solution

Figure 6.2

Let x be the amount of increase. The length and width of the new flower bed are $x + 10$ feet and $x + 5$ feet, as shown in Fig. 6.2. Because the area is to be 75 square feet, we have

$$(x + 10)(x + 5) = 75.$$

Write this equation in the form $ax^2 + bx + c = 0$:

$$x^2 + 15x + 50 = 75$$
$$x^2 + 15x - 25 = 0 \quad \text{Get 0 on the right.}$$
$$x = \frac{-15 \pm \sqrt{225 - 4(1)(-25)}}{2(1)}$$
$$= \frac{-15 \pm \sqrt{325}}{2} = \frac{-15 \pm 5\sqrt{13}}{2}$$

Because the value of x must be positive, the exact increase is

$$\frac{-15 + 5\sqrt{13}}{2} \text{ feet.}$$

Using a calculator, we can find that x is approximately 1.51 feet. If $x = 1.51$ feet, then the new length is 11.51 feet, and the new width is 6.51 feet. The area of a rectangle with these dimensions is 74.93 square feet. Of course, the approximate dimensions do not give exactly 75 square feet. ∎

EXAMPLE 8 **Mowing the lawn**

It takes Carla one hour longer to mow the lawn than it takes Sharon to mow the lawn. If they can mow the lawn in 5 hours working together, then how long would it take each girl by herself?

Solution

If Sharon can mow the lawn by herself in x hours, then she works at the rate of $\frac{1}{x}$ of the lawn per hour. If Carla can mow the lawn by herself in $x + 1$ hours, then she works at the rate of $\frac{1}{x+1}$ of the lawn per hour. We can use a table to list all of the important quantities.

	Rate	Time	Work
Sharon	$\dfrac{1}{x} \dfrac{\text{lawn}}{\text{hr}}$	5 hr	$\dfrac{5}{x}$ lawn
Carla	$\dfrac{1}{x+1} \dfrac{\text{lawn}}{\text{hr}}$	5 hr	$\dfrac{5}{x+1}$ lawn

Because they complete the lawn in 5 hours, the portion of the lawn done by Sharon and the portion done by Carla have a sum of 1:

$$\frac{5}{x} + \frac{5}{x+1} = 1$$

$$x(x+1)\frac{5}{x} + x(x+1)\frac{5}{x+1} = x(x+1)1 \qquad \text{Multiply by the LCD.}$$

$$5x + 5 + 5x = x^2 + x$$

$$10x + 5 = x^2 + x$$

$$-x^2 + 9x + 5 = 0$$

$$x^2 - 9x - 5 = 0$$

$$x = \frac{9 \pm \sqrt{(-9)^2 - 4(1)(-5)}}{2(1)}$$

$$= \frac{9 \pm \sqrt{101}}{2}$$

Using a calculator, we find that $\frac{9 - \sqrt{101}}{2}$ is negative. So Sharon's time alone is

$$\frac{9 + \sqrt{101}}{2} \text{ hours.}$$

To find Carla's time alone, we add one hour to Sharon's time. So Carla's time alone is

$$\frac{9 + \sqrt{101}}{2} + 1 = \frac{9 + \sqrt{101}}{2} + \frac{2}{2} = \frac{11 + \sqrt{101}}{2} \text{ hours.}$$

Sharon's time alone is approximately 9.525 hours, and Carla's time alone is approximately 10.525 hours. ∎

Warm-ups

True or false? Explain.

1. To solve $x^4 - 5x^2 + 6 = 0$ by substitution, we can let $w = x^2$.
2. We can solve $x^5 - 3x^3 - 10 = 0$ by substitution if we let $w = x^3$.
3. We always use the quadratic formula on equations of quadratic form.
4. If $w = x^{1/6}$, then $w^2 = x^{1/3}$.
5. To solve $x - 7\sqrt{x} + 10 = 0$ by substitution, we let $\sqrt{w} = x$.
6. If $y = 2^{1/2}$, then $y^2 = 2^{1/4}$.
7. If John paints a 100-foot fence in x hours, then his rate is $\frac{100}{x}$ of the fence per hour.
8. If Elvia drives 300 miles in x hours, then her rate is $\frac{300}{x}$ mph.
9. If Ann's boat goes 10 mph in still water, then against a 5-mph current, it will go 2 mph.
10. If squares with sides of length x inches are cut from the corners of an 11-inch by 14-inch rectangular piece of sheet metal and the sides are folded up to form a box, then the dimensions of the bottom will be $11 - x$ by $14 - x$.

6.3 EXERCISES

Use the discriminant to determine whether each quadratic polynomial can be factored, then factor the ones that are not prime. See Example 1.

1. $2x^2 - x + 4$
2. $2x^2 + 3x - 5$
3. $2x^2 + 6x - 5$
4. $3x^2 + 5x - 1$
5. $6x^2 + 19x - 36$
6. $8x^2 + 6x - 27$
7. $4x^2 - 5x - 12$
8. $4x^2 - 27x + 45$
9. $8x^2 - 18x - 45$
10. $6x^2 + 9x - 16$

Write a quadratic equation that has each given pair of solutions. See Example 2.

11. $3, -7$
12. $-8, 2$
13. $4, 1$
14. $3, 2$
15. $\sqrt{5}, -\sqrt{5}$
16. $-\sqrt{7}, \sqrt{7}$
17. $4i, -4i$
18. $-3i, 3i$
19. $i\sqrt{2}, -i\sqrt{2}$
20. $3i\sqrt{2}, -3i\sqrt{2}$
21. $\dfrac{1}{2}, \dfrac{1}{3}$
22. $-\dfrac{1}{5}, -\dfrac{1}{2}$

Find all real solutions to each equation. See Example 3.

23. $(2a - 1)^2 + 2(2a - 1) - 8 = 0$
24. $(3a + 2)^2 - 3(3a + 2) = 10$
25. $(w - 1)^2 + 5(w - 1) + 5 = 0$
26. $(2x - 1)^2 - 4(2x - 1) + 2 = 0$

Find all real solutions to each equation. See Example 4.

27. $x^4 - 14x^2 + 45 = 0$
28. $x^4 + 2x^2 = 15$
29. $x^6 + 7x^3 = 8$
30. $a^6 + 6a^3 = 16$

Find all real solutions to each equation. See Example 5.

31. $(x^2 + 2x)^2 - 7(x^2 + 2x) + 12 = 0$
32. $(x^2 + 3x)^2 + (x^2 + 3x) - 20 = 0$
33. $(y^2 + y)^2 - 8(y^2 + y) + 12 = 0$
34. $(w^2 - 2w)^2 + 24 = 11(w^2 - 2w)$

Find all real solutions to each equation. See Example 6.

35. $x^{1/2} - 5x^{1/4} + 6 = 0$
36. $2x - 5\sqrt{x} + 2 = 0$
37. $2x - 5x^{1/2} - 3 = 0$
38. $x^{1/4} + 2 = x^{1/2}$

Find all real solutions to each equation.

39. $x^{-2} + x^{-1} - 6 = 0$
40. $x^{-2} - 2x^{-1} = 8$
41. $x^{1/6} - x^{1/3} + 2 = 0$
42. $x^{2/3} - x^{1/3} - 20 = 0$
43. $\left(\dfrac{1}{y - 1}\right)^2 + \left(\dfrac{1}{y - 1}\right) = 6$
44. $\left(\dfrac{1}{w + 1}\right)^2 - 2\left(\dfrac{1}{w + 1}\right) - 24 = 0$
45. $2x^2 - 3 - 6\sqrt{2x^2 - 3} + 8 = 0$
46. $x^2 + x + \sqrt{x^2 + x} - 2 = 0$
47. $x^{-2} - 2x^{-1} - 1 = 0$
48. $x^{-2} - 6x^{-1} + 6 = 0$

By definition, a quadratic equation has real coefficients, and the quadratic formula was developed to solve those equations. However, the quadratic formula actually solves $ax^2 + bx + c = 0$ for any complex numbers a, b, and c with $a \neq 0$. Use it to solve the following equations.

49. $x^2 + ix + 2 = 0$
50. $x^2 + 2ix - 3 = 0$
51. $ix^2 + 2x - 3i = 0$
52. $(2 - i)x^2 + 4x + 2 + i = 0$
53. $x^2 + \sqrt{13}\, x - 3 = 0$
54. $\sqrt{2}x^2 + x - \sqrt{2} = 0$

Find the exact solution to each problem. If the exact solution is an irrational number, then also find an approximate decimal solution. See Examples 7 and 8.

55. **Country singers.** Harry and Gary are traveling to Nashville to make their fortunes. Harry leaves on the train at 8:00 A.M. and Gary travels by car, starting at 9:00 A.M. To complete the 300-mile trip and arrive at the same time as Harry, Gary travels 10 miles per hour faster than the train. At what time will they both arrive in Nashville?

56. **Gone fishing.** Debbie traveled by boat 5 miles upstream to fish in her favorite spot. Because of the 4-mph current, it took her 20 minutes longer to get there than to return. How fast will her boat go in still water?

57. **Cross-country cycling.** Erin was traveling across the desert on her bicycle. Before lunch she traveled 60 miles, after lunch she traveled 46 miles. She put in 1 hour more after lunch than before lunch, but her speed was 4 miles per hour slower than before. What was her speed before lunch and after lunch?

58. **Extreme hardship.** Kim starts to walk 3 miles to school at 7:30 A.M. with a temperature of 0°F. Her brother Bryan starts at 7:45 A.M. on his bicycle, traveling 10 miles per hour faster than Kim. If they get to school at the same time, then how fast is each one traveling?

59. **American pie.** John takes 3 hours longer than Andrew to peel 500 pounds of apples. If together they can peel 500 pounds of apples in 8 hours, then how long would it take each one working alone?

60. ***On the half shell.*** It takes Brent 1 hour longer than Calvin to shuck a sack of oysters. If together they shuck a sack of oysters in 45 minutes, then how long would it take each one working alone?

61. ***The growing garden.*** Eric's garden is 20 feet by 30 feet. He wants to increase the length and width by the same amount to have a 1000-square-foot garden. What should the new dimensions of the garden be?

62. ***Open-top box.*** Thomas is going to make an open-top box by cutting equal squares from the four corners of an 11-inch by 14-inch sheet of cardboard and folding up the sides. If the area of the base is to be 80 square inches, then what size square should be cut from each corner?

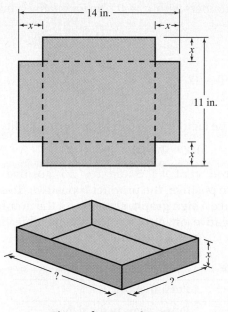

Figure for Exercise 62

63. ***Pumping the pool.*** It takes pump A 2 hours less time than pump B to empty a certain swimming pool. Pump A is started at 8:00 A.M., and pump B is started at 11:00 A.M. If the pool is still half full at 5:00 P.M., then how long would it take pump A working alone?

64. ***Time off for lunch.*** It usually takes Eva 3 hours longer to do the monthly payroll than it takes Cicely. They start working on it together at 9:00 A.M. and at 5:00 P.M. they have 90% of it done. If Eva took a 2 hour-lunch break while Cicely had none, then how much longer will it take for them to finish the payroll working together?

65. ***Golden Rectangle.*** One principle used by the ancient Greeks to get shapes that are pleasing to the eye in art and architecture was the Golden Rectangle. If a square is removed from one end of a Golden Rectangle, as shown in the figure, the sides of the remaining rectangle are proportional to the original rectangle. So the length and width of the original rectangle satisfy

$$\frac{L}{W} = \frac{W}{L - W}.$$

If the length of a Golden Rectangle is 10 meters, then what is its width?

Figure for Exercise 65

Getting More Involved

66. ***Exploration.***
   a) Given that $P(x) = x^4 + 6x^2 - 27$, find $P(3i)$, $P(-3i)$, $P(\sqrt{3})$, and $P(-\sqrt{3})$.
   b) What can you conclude about the values $3i$, $-3i$, $\sqrt{3}$, and $-\sqrt{3}$ and their relationship to each other?

67. ***Cooperative learning.*** Work with a group to write a quadratic equation that has each given pair of solutions.
   a) $3 + \sqrt{5}, 3 - \sqrt{5}$ **b)** $4 - 2i, 4 + 2i$
   c) $\dfrac{1 + i\sqrt{3}}{2}, \dfrac{1 - i\sqrt{3}}{2}$

6.4 Quadratic and Rational Inequalities

We first solved inequalities in Chapter 2. In this section we solve inequalities involving quadratic polynomials. We use a new technique based on the rules for multiplying real numbers.

Solving Quadratic Inequalities with a Sign Graph

An inequality involving a quadratic polynomial is called a **quadratic** inequality.

> **Quadratic Inequality**
>
> A quadratic inequality is an inequality of the form
>
> $$ax^2 + bx + c > 0,$$
>
> where a, b, and c are real numbers with $a \neq 0$. The inequality symbols $<$, $\leq$, and $\geq$ may also be used.

Consider the quadratic inequality

$$x^2 + 3x - 10 > 0.$$

Because the left-hand side can be factored, we can write the inequality as

$$(x + 5)(x - 2) > 0.$$

This inequality says that the product of $x + 5$ and $x - 2$ is positive. If both factors are negative or both are positive, the product is positive. To analyze the signs of each factor, we make a **sign graph,** showing on the number line where the factor is positive, negative, or zero. Consider the possible values of the factor $x + 5$:

Value	Where	On the number line
$x + 5 = 0$	if $x = -5$	Put a 0 above -5.
$x + 5 > 0$	if $x > -5$	Put $+$ signs to the right of -5.
$x + 5 < 0$	if $x < -5$	Put $-$ signs to the left of -5.

The sign graph shown in Fig. 6.3 for the factor $x + 5$ is made from the information in the preceding table.

Figure 6.3

Now consider the possible values of the factor $x - 2$:

Value	Where	On the number line
$x - 2 = 0$	if $x = 2$	Put a 0 above 2.
$x - 2 > 0$	if $x > 2$	Put + signs to the right of 2.
$x - 2 < 0$	if $x < 2$	Put − signs to the left of 2.

We put the information for the factor $x - 2$ on the sign graph for the factor $x + 5$ as shown in Figure 6.4. We can see from Fig. 6.4 that the product is positive if $x < -5$ and the product is positive if $x > 2$. The solution set for the quadratic inequality is shown in Fig. 6.5. Note that -5 and 2 are not included in the graph because for those values of x, the product is 0. The solution set is $(-\infty, -5) \cup (2, \infty)$.

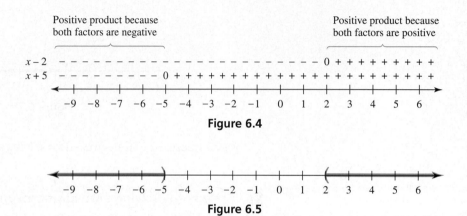

Figure 6.4

Figure 6.5

EXAMPLE 1   **Solving a quadratic inequality**

Solve $2x^2 + 5x \leq 3$ and graph the solution set.

Solution

Rewrite the inequality with 0 on one side:

$$2x^2 + 5x - 3 \leq 0$$

$$(2x - 1)(x + 3) \leq 0 \quad \text{Factor.}$$

Examine the signs of each factor:

$$2x - 1 = 0 \text{ if } x = \frac{1}{2}$$

$$2x - 1 > 0 \text{ if } x > \frac{1}{2}$$

$$2x - 1 < 0 \text{ if } x < \frac{1}{2}$$

$$x + 3 = 0 \text{ if } x = -3$$

$$x + 3 > 0 \text{ if } x > -3$$

$$x + 3 < 0 \text{ if } x < -3$$

Make a sign graph as shown in Fig. 6.6. The product of the factors is negative between -3 and $\frac{1}{2}$, when one factor is negative and the other is positive. The product is 0 at -3 and at $\frac{1}{2}$. So the solution set is the interval $\left[-3, \frac{1}{2}\right]$. The graph of the solution set is shown in Fig. 6.7.

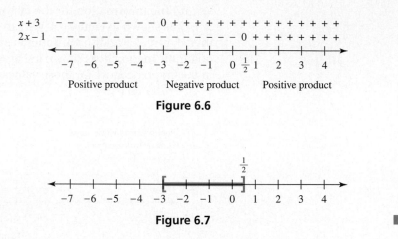

Figure 6.6

Figure 6.7

We summarize the strategy used for solving a quadratic inequality as follows.

▶ Srategy for Solving a Quadratic Inequality with a Sign Graph ◀

1. Write the inequality with 0 on the right.
2. Factor the quadratic polynomial on the left.
3. Make a sign graph showing where each factor is positive, negative, or zero.
4. Use the rules for multiplying signed numbers to determine which regions satisfy the original inequality.

Solving Rational Inequalities with a Sign Graph

The inequalities

$$\frac{x+2}{x-3} \le 2, \qquad \frac{2x-3}{x+5} \le 0 \qquad \text{and} \qquad \frac{2}{x+4} \ge \frac{1}{x+1}$$

are called **rational inequalities.** When we solve *equations* that involve rational expressions, we usually multiply each side by the LCD. However, if we multiply each side of any inequality by a negative number, we must reverse the inequality, and when we multiply by a positive number, we do not. For this reason we generally *do not multiply inequalities by expressions involving variables.* The values of the expressions might be positive or negative. The next two examples show how to use a sign graph to solve rational inequalities that have variables in the denominator.

EXAMPLE 2   **Solving a rational inequality**

Solve $\frac{x+2}{x-3} \le 2$ and graph the solution set.

Solution

We *do not* multiply each side by $x - 3$. Instead, subtract 2 from each side to get 0 on the right:

$$\frac{x+2}{x-3} - 2 \le 0$$

$$\frac{x+2}{x-3} - \frac{2(x-3)}{x-3} \le 0 \qquad \text{Get a common denominator.}$$

$$\frac{x+2}{x-3} - \frac{2x-6}{x-3} \le 0 \qquad \text{Simplify.}$$

$$\frac{x+2-2x+6}{x-3} \le 0 \qquad \text{Subtract the rational expressions.}$$

$$\frac{-x+8}{x-3} \le 0 \qquad \begin{array}{l}\text{The quotient of } -x+8 \text{ and } x-3 \text{ is}\\ \text{less than or equal to } 0.\end{array}$$

Examine the signs of the numerator and denominator.

$$x - 3 = 0 \text{ if } x = 3 \qquad -x + 8 = 0 \text{ if } x = 8$$
$$x - 3 > 0 \text{ if } x > 3 \qquad -x + 8 > 0 \text{ if } x < 8$$
$$x - 3 < 0 \text{ if } x < 3 \qquad -x + 8 < 0 \text{ if } x > 8$$

Make a sign graph as shown in Fig. 6.8. Using the rule for dividing signed numbers and the sign graph, we can identify where the quotient is negative or zero. The solution set is $(-\infty, 3) \cup [8, \infty)$. Note that 3 is not in the solution set because the quotient is undefined if $x = 3$. The graph of the solution set is shown in Figure 6.9.

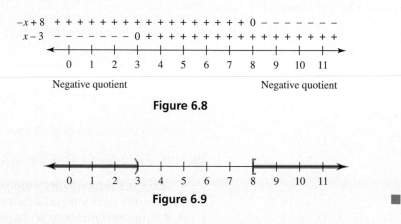

Figure 6.8

Figure 6.9

CAUTION   Remember to reverse the inequality sign when multiplying or dividing by a negative number. For example, $x - 3 > 0$ is equivalent to $x > 3$. But $-x + 8 > 0$ is equivalent to $-x > -8$, or $x < 8$.

EXAMPLE 3   **Solving a rational inequality**

Solve $\dfrac{2}{x+4} \geq \dfrac{1}{x+1}$ and graph the solution set.

Solution

We do not multiply by the LCD as we do in solving equations. Instead, subtract $\dfrac{1}{x+1}$ from each side:

$$\frac{2}{x+4} - \frac{1}{x+1} \geq 0$$

$$\frac{2(x+1)}{(x+4)(x+1)} - \frac{1(x+4)}{(x+1)(x+4)} \geq 0 \quad \text{Get a common denominator.}$$

$$\frac{2x+2-x-4}{(x+1)(x+4)} \geq 0 \quad \text{Simplify.}$$

$$\frac{x-2}{(x+1)(x+4)} \geq 0$$

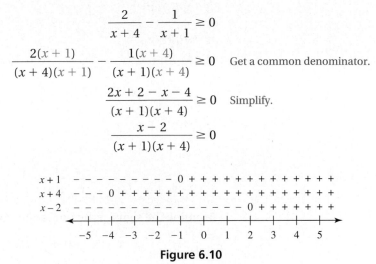

Figure 6.10

Make a sign graph as shown in Fig. 6.10. The computation of

$$\frac{x-2}{(x+1)(x+4)}$$

involves multiplication and division. The result of this computation is positive if all of the three binomials are positive or if only one is positive and the other two are negative. The sign graph shows that this rational expression will have a positive value when x is between -4 and -1 and again when x is larger than 2. The solution set is $(-4, -1) \cup [2, \infty)$. Note that -1 and -4 are not in the solution set because they make the denominator zero. The graph of the solution set is shown in Fig. 6.11.

Figure 6.11

■

Solving rational inequalities with a sign graph is summarized below.

> ▶ **Strategy for Solving a Rational Inequality with a Sign Graph** ◀
>
> 1. Rewrite the inequality with 0 on the right side.
> 2. Use only addition and subtraction to get an equivalent inequality.
> 3. Factor the numerator and denominator if possible.
> 4. Make a sign graph showing where each factor is positive, negative, or zero.
> 5. Use the rules for multiplying and dividing signed numbers to determine the regions that satisfy the original inequality.

Another method for solving quadratic and rational inequalities will be shown in Example 4. This method, called the test point method, can be used instead of the sign graph to solve the inequalities of Examples 1, 2, and 3.

Quadratic Inequalities That Cannot Be Factored

The following example shows how to solve a quadratic inequality that involves a prime polynomial.

EXAMPLE 4   **Solving a quadratic inequality using the quadratic formula**

Solve $x^2 - 4x - 6 > 0$ and graph the solution set.

Solution

The quadratic polynomial is prime, but we can solve $x^2 - 4x - 6 = 0$ by the quadratic formula:

$$x = \frac{4 \pm \sqrt{16 - 4(1)(-6)}}{2(1)} = \frac{4 \pm \sqrt{40}}{2} = \frac{4 \pm 2\sqrt{10}}{2} = 2 \pm \sqrt{10}$$

As in the previous examples, the solutions to the equation divide the number line into intervals on which the quadratic polynomial has either a positive or negative value. To determine which, we select an arbitrary **test point** in each interval. Because $2 + \sqrt{10} \approx 5.2$ and $2 - \sqrt{10} \approx -1.2$, we choose a test point that is less than -1.2, one between -1.2 and 5.2, and one that is greater than 5.2. We have selected -2, 0, and 7 for test points, as shown in Fig. 6.12. Now evaluate $x^2 - 4x - 6$ at each test point.

Figure 6.12

Test point	Value of $x^2 - 4x - 6$ at the test point	Sign of $x^2 - 4x - 6$ in interval of test point
-2	6	Positive
0	-6	Negative
7	15	Positive

Because $x^2 - 4x - 6$ is positive at the test points -2 and 7, it is positive at every point in the intervals containing those test points. So the solution set to the inequality $x^2 - 4x - 6 > 0$ is

$$(-\infty, 2 - \sqrt{10}) \cup (2 + \sqrt{10}, \infty),$$

and its graph is shown in Fig. 6.13.

Figure 6.13

The test-point method used in Example 4 can be used also on inequalities that do factor. We summarize the strategy for solving inequalities using test points in the following box.

▶ Strategy for Solving Quadratic Inequalities Using Test Points ◀

1. Rewrite the inequality with 0 on the right.
2. Solve the quadratic equation that results from replacing the inequality symbol with the equals symbol.
3. Locate the solutions to the quadratic equation on a number line.
4. Select a test point in each interval determined by the solutions to the quadratic equation.
5. Test each point in the original quadratic inequality to determine which intervals satisfy the inequality.

Applications

The following example shows how a quadratic inequality can be used to solve a problem.

EXAMPLE 5 **Making a profit**

Charlene's daily profit P (in dollars) for selling x magazine subscriptions is determined by the formula

$$P = -x^2 + 80x - 1500.$$

For what values of x is her profit positive?

Solution

We can find the values of x for which $P > 0$ by solving a quadratic inequality:

$$-x^2 + 80x - 1500 > 0$$
$$x^2 - 80x + 1500 < 0 \quad \text{Multiply each side by } -1.$$
$$(x - 30)(x - 50) < 0 \quad \text{Factor.}$$

Make a sign graph as shown in Fig. 6.14. The product of the two factors is negative for x between 30 and 50. Because the last inequality is equivalent to the first, the profit is positive when the number of magazine subscriptions sold is greater than 30 and less than 50.

Figure 6.14

Warm-ups

True or false? Explain.

1. The solution set to $x^2 > 4$ is $(2, \infty)$.

2. The inequality $\frac{x}{x-3} > 2$ is equivalent to $x > 2x - 6$.

3. The inequality $(x - 1)(x + 2) < 0$ is equivalent to $x - 1 < 0$ or $x + 2 < 0$.

4. We cannot solve quadratic inequalities that do not factor.

5. One technique for solving quadratic inequalities is based on the rules for multiplying signed numbers.

6. Multiplying each side of an inequality by a variable should be avoided.

7. In solving quadratic or rational inequalities, we always get 0 on one side.

8. The inequality $\frac{x}{2} > 3$ is equivalent to $x > 6$.

9. The inequality $\frac{x-3}{x+2} < 1$ is equivalent to $\frac{x-3}{x+2} - 1 < 0$.

10. The solution set to $\frac{x+2}{x-4} \geq 0$ is $(-\infty, -2] \cup [4, \infty)$.

6.4 EXERCISES

Solve each inequality. State the solution set using interval notation and graph the solution set. See Example 1.

1. $x^2 + x - 6 < 0$
2. $x^2 - 3x - 4 \geq 0$
3. $y^2 - 4 > 0$
4. $z^2 - 16 < 0$
5. $2u^2 + 5u \geq 12$
6. $2v^2 + 7v < 4$
7. $4x^2 - 8x \geq 0$
8. $x^2 + x > 0$
9. $5x - 10x^2 < 0$
10. $3x - x^2 > 0$
11. $x^2 + 6x + 9 \geq 0$
12. $x^2 + 25 < 10x$

Solve each rational inequality. State and graph the solution set. See Examples 2 and 3.

13. $\frac{x}{x-3} > 0$
14. $\frac{a}{a+2} > 0$
15. $\frac{x+2}{x} \leq 0$
16. $\frac{w-6}{w} \leq 0$
17. $\frac{t-3}{t+6} > 0$
18. $\frac{x-2}{2x+5} < 0$
19. $\frac{x}{x+2} > -1$
20. $\frac{x+3}{x} \leq -2$
21. $\frac{2}{x-5} > \frac{1}{x+4}$
22. $\frac{3}{x+2} > \frac{2}{x-1}$

23. $\frac{m}{m-5} + \frac{3}{m-1} > 0$
24. $\frac{p}{p-16} + \frac{2}{p-6} \leq 0$
25. $\frac{x}{x-3} \leq \frac{-8}{x-6}$
26. $\frac{x}{x+20} > \frac{2}{x+8}$

Solve each inequality. State and graph the solution set. See Example 4.

27. $x^2 - 2x - 4 > 0$
28. $x^2 - 2x - 5 \leq 0$
29. $2x^2 - 6x + 3 \geq 0$
30. $2x^2 - 8x + 3 < 0$
31. $y^2 - 3y - 9 \leq 0$
32. $z^2 - 5z - 7 < 0$

In Exercises 33–56, solve each inequality. State the solution set using interval notation.

33. $x^2 \leq 9$
34. $x^2 \geq 36$
35. $16 - x^2 > 0$
36. $9 - x^2 < 0$
37. $x^2 - 4x \geq 0$
38. $4x^2 - 9 > 0$
39. $3(2w^2 - 5) < w$
40. $6(y^2 - 2) + y < 0$
41. $z^2 \geq 4(z + 3)$
42. $t^2 < 3(2t - 3)$
43. $(q + 4)^2 > 10q + 31$
44. $(2p + 4)(p - 1) < (p + 2)^2$
45. $\frac{1}{2}x^2 \geq 4 - x$
46. $\frac{1}{2}x^2 \leq x + 12$

47. $\dfrac{x-4}{x+3} \le 0$

48. $\dfrac{2x-1}{x+5} \ge 0$

49. $(x-2)(x+1)(x-5) \ge 0$

50. $(x-1)(x+2)(2x-5) < 0$

51. $x^3 + 3x^2 - x - 3 < 0$

52. $x^3 + 5x^2 - 4x - 20 \ge 0$

53. $0.23x^2 + 6.5x + 4.3 < 0$

54. $0.65x^2 + 3.2x + 5.1 > 0$

55. $\dfrac{x}{x-2} > \dfrac{-1}{x+3}$

56. $\dfrac{x}{3-x} > \dfrac{2}{x+5}$

Solve each problem by using a quadratic inequality. See Example 5.

57. *Positive profit.* The monthly profit P (in dollars) that Big Jim makes on the sale of x mobile homes is determined by the formula $P = x^2 + 5x - 50$. For what values of x is his profit positive?

58. *Profitable fruitcakes.* Sharon's revenue R (in dollars) on the sale of x fruitcakes is determined by the formula $R = 50x - x^2$. Her cost C (in dollars) for producing x fruit cakes is given by the formula $C = 2x + 40$. For what values of x is Sharon's profit positive? (Profit = revenue − cost.)

If an object is given an initial velocity straight upward of v_0 feet per second from a height of s_0 feet, then its altitude S after t seconds is given by the formula

$$S = -16t^2 + v_0 t + s_0.$$

59. *Flying high.* An arrow is shot straight upward with a velocity of 96 feet per second from an altitude of 6 feet. For how many seconds is this arrow more than 86 feet high?

60. *Putting the shot.* In 1978, Udo Beyer (East Germany) set a world record in the shot-put of 72 ft 8 in. If Beyer had projected the shot straight upward with a velocity of 30 feet per second from a height of 5 feet, then for what values of t would the shot be under 15 feet high?

If a projectile is fired at a 45° angle from a height of s_0 feet with initial velocity v_0 ft/sec, then its altitude S in feet after t seconds is given by

$$S = -16t^2 + \dfrac{v_0}{\sqrt{2}}\, t + s_0.$$

61. *Siege and garrison artillery.* An 8-inch mortar used in the Civil War fired a 44.5-lb projectile a distance of 3600 feet when aimed at a 45° angle (Harold R. Peterson, *Notes on Ordinance of the American Civil War*). If the 8-inch mortar was fired from the ground level with an initial velocity of $240\sqrt{2}$ ft/sec, then for how long did the projectile have an altitude of more than 864 feet?

62. *Seacoast artillery.* The 13-inch mortar used in the Civil War fired a 220-lb projectile a distance of 12,975 feet when aimed at a 45° angle. If the 13-inch mortar was fired from a hill 100 feet above sea level with an initial velocity of 644 feet per second, then for how long was the projectile more than 800 feet above sea level?

Figure for Exercise 62

Getting More Involved

63. *Cooperative learning.* Work in a small group to solve each inequality for x, given that h and k are real numbers with $h < k$.

a) $(x-h)(x-k) < 0$ **b)** $(x-h)(x-k) > 0$

c) $(x+h)(x+k) < 0$ **d)** $(x+h)(x+k) \ge 0$

e) $\dfrac{x-h}{x-k} \ge 0$ **f)** $\dfrac{x+h}{x+k} \le 0$

64. *Cooperative learning.* Work in a small group to solve $ax^2 + bx + c > 0$ for x in each case.

a) $b^2 - 4ac = 0$ and $a > 0$

b) $b^2 - 4ac = 0$ and $a < 0$

c) $b^2 - 4ac < 0$ and $a > 0$

d) $b^2 - 4ac < 0$ and $a < 0$

e) $b^2 - 4ac > 0$ and $a > 0$

f) $b^2 - 4ac > 0$ and $a < 0$

COLLABORATIVE ACTIVITIES

Building a Room Addition

Grouping: 2 to 4 students per group
Topic: Applications of the quadratic formula

Leslie decides to build a rectangular studio on the south side of his house. The studio is attached so that the north side of the studio will be a portion of the current south side of the house. Because Leslie lives in the Southwest, he decides to make the room from rammed earth, thus making the walls of the studio 2 feet thick. Because of the cost of building materials and labor, Leslie decides to limit the total area of the studio. He also decides to make the studio's inside south wall twice as long as its inside west wall.

Leslie would also like to build a semicircular patio around the studio (sliding glass doors will open from the studio out onto the patio). The studio will be circumscribed by the semicircle. A large tree, which he would like to keep for the summer shade is 20 feet from the south side of the house. Below is a sketch Leslie made of the studio and patio.

1. Working in your groups, finish the drawing of the house and studio, adding the information above. Define your variable(s) and find the internal dimensions of the rectangular studio if Leslie wants the *external* area of the studio to be 400 square feet. Round your answers to the nearest foot.

2. Determine how far the semicircular patio extends past the house at the farthest point. Round your answer to the nearest tenth of a foot.

3. Will Leslie be able to build the patio without removing the tree? Give a reason for your answer.

Wrap-up   CHAPTER 6

SUMMARY

Quadratic Equations		Examples
Quadratic equation	An equation of the form $ax^2 + bx + c = 0,$ where a, b, and c are real numbers, with $a \neq 0$	$x^2 = 11$ $(x - 5)^2 = 99$ $x^2 + 3x - 20 = 0$
Methods for solving quadratic equations	Factoring: Factor the quadratic polynomial, then set each factor equal to 0.	$x^2 + x - 6 = 0$ $(x + 3)(x - 2) = 0$ $x + 3 = 0$ or $x - 2 = 0$
	The even-root property: If $x^2 = k (k > 0)$, then $x = \pm\sqrt{k}$. If $x^2 = 0$, then $x = 0$. There are no real solutions to $x^2 = k$ for $k < 0$.	$(x - 5)^2 = 10$ $x - 5 = \pm\sqrt{10}$

	Completing the square: Take one-half of middle term, square it, then add it to each side.	$x^2 + 6x = -4$ $x^2 + 6x + 9 = -4 + 9$ $(x + 3)^2 = 5$
	Quadratic formula: Solves $ax^2 + bx + c = 0$ with $a \neq 0$: $$x = \frac{-b \pm \sqrt{b^2 - 4ac}}{2a}$$	$2x^2 + 3x - 5 = 0$ $$x = \frac{-3 \pm \sqrt{3^2 - 4(2)(-5)}}{2(2)}$$
Number of solutions	Determined by the discriminant $b^2 - 4ac$: $\quad b^2 - 4ac > 0 \quad$ 2 real solutions $\quad b^2 - 4ac = 0 \quad$ 1 real solution $\quad b^2 - 4ac < 0 \quad$ no real solutions, $\qquad\qquad\qquad\quad$ 2 imaginary solutions	$x^2 + 6x - 12 = 0$ $6^2 - 4(1)(-12) > 0$ $x^2 + 10x + 25 = 0$ $10^2 - 4(1)(25) = 0$ $x^2 + 2x + 20 = 0$ $2^2 - 4(1)(20) < 0$
Factoring	The quadratic polynomial $ax^2 + bx + c$ (with integral coefficients) can be factored if and only if $b^2 - 4ac$ is a perfect square.	$2x^2 - 11x + 12$ $b^2 - 4ac = 25$ $(2x - 3)(x - 4)$
Writing equations	To write an equation with given solutions, reverse the steps in solving an equation by factoring.	$x = 2$ or $x = -3$ $(x - 2)(x + 3) = 0$ $x^2 + x - 6 = 0$
Equations quadratic in form	Use substitution to convert to a quadratic.	$x^4 + 3x^2 - 10 = 0$ Let $a = x^2$ $a^2 + 3a - 10 = 0$

Quadratic and Rational Inequalities

		Examples
Quadratic inequality	An inequality involving a quadratic polynomial	$2x^2 - 7x + 6 \geq 0$ $x^2 - 4x - 5 < 0$
Rational inequality	An inequality involving a rational expression	$$\frac{1}{x - 1} < \frac{3}{x - 2}$$
Solving quadratic and rational inequalities	Get 0 on one side and express the other side as a product and/or quotient of linear factors. Make a sign graph showing the signs of the factors. Use test points if the quadratic polynomial is prime.	$(x - 5)(x + 1) < 0$

REVIEW EXERCISES

6.1 *Solve by factoring.*

1. $x^2 - 2x - 15 = 0$

2. $x^2 - 2x - 24 = 0$

3. $2x^2 + x = 15$

4. $2x^2 + 7x = 4$

5. $w^2 - 25 = 0$

6. $a^2 - 121 = 0$

7. $4x^2 - 12x + 9 = 0$

8. $x^2 - 12x + 36 = 0$

Solve by using the even-root property.

9. $x^2 = 12$

10. $x^2 = 20$

11. $(x - 1)^2 = 9$

12. $(x + 4)^2 = 4$

13. $(x - 2)^2 = \dfrac{3}{4}$

14. $(x - 3)^2 = \dfrac{1}{4}$

15. $4x^2 = 9$

16. $2x^2 = 3$

Solve by completing the square.

17. $x^2 - 6x + 8 = 0$

18. $x^2 + 4x + 3 = 0$

19. $x^2 - 5x + 6 = 0$

20. $x^2 - x - 6 = 0$

21. $2x^2 - 7x + 3 = 0$

22. $2x^2 - x = 6$

23. $x^2 + 4x + 1 = 0$

24. $x^2 + 2x - 2 = 0$

6.2 *Solve by the quadratic formula.*

25. $x^2 - 3x - 10 = 0$

26. $x^2 - 5x - 6 = 0$

27. $6x^2 - 7x = 3$

28. $6x^2 = x + 2$

29. $x^2 + 4x + 2 = 0$

30. $x^2 + 6x = 2$

31. $3x^2 + 1 = 5x$

32. $2x^2 + 3x - 1 = 0$

Find the value of the discriminant and the number of real solutions to each equation.

33. $25x^2 - 20x + 4 = 0$

34. $16x^2 + 1 = 8x$

35. $x^2 - 3x + 7 = 0$

36. $3x^2 - x + 8 = 0$

37. $2x^2 + 1 = 5x$

38. $-3x^2 + 6x - 2 = 0$

Find the complex solutions to the quadratic equations.

39. $2x^2 - 4x + 3 = 0$

40. $2x^2 - 6x + 5 = 0$

41. $2x^2 + 3 = 3x$

42. $x^2 + x + 1 = 0$

43. $3x^2 + 2x + 2 = 0$

44. $x^2 + 2 = 2x$

45. $\dfrac{1}{2}x^2 + 3x + 8 = 0$

46. $\dfrac{1}{2}x^2 - 5x + 13 = 0$

6.3 *Use the discriminant to determine whether each quadratic polynomial can be factored, then factor the ones that are not prime.*

47. $8x^2 - 10x - 3$

48. $18x^2 + 9x - 2$

49. $4x^2 - 5x + 2$

50. $6x^2 - 7x - 4$

51. $8y^2 + 10y - 25$

52. $25z^2 - 15z - 18$

Write a quadratic equation that has each given pair of solutions.

53. $-3, -6$

54. $4, -9$

55. $-5\sqrt{2}, 5\sqrt{2}$

56. $-2i\sqrt{3}, 2i\sqrt{3}$

Find all real solutions to each equation.

57. $x^6 + 7x^3 - 8 = 0$

58. $8x^6 + 63x^3 - 8 = 0$

59. $x^4 - 13x^2 + 36 = 0$

60. $x^4 + 7x^2 + 12 = 0$

61. $(x^2 + 3x)^2 - 28(x^2 + 3x) + 180 = 0$

62. $(x^2 + 1)^2 - 8(x^2 + 1) + 15 = 0$

63. $x^2 - 6x + 6\sqrt{x^2 - 6x} - 40 = 0$

64. $x^2 - 3x - 3\sqrt{x^2 - 3x} + 2 = 0$

65. $t^{-2} + 5t^{-1} - 36 = 0$

66. $a^{-2} + a^{-1} - 6 = 0$

67. $w - 13\sqrt{w} + 36 = 0$

68. $4a - 5\sqrt{a} + 1 = 0$

6.4 *Solve each inequality. State the solution set using interval notation and graph it.*

69. $a^2 + a > 6$

70. $x^2 - 5x + 6 > 0$

71. $x^2 - x - 20 \le 0$

72. $a^2 + 2a \le 15$

73. $w^2 - w < 0$

74. $x - x^2 \le 0$

75. $\dfrac{x - 4}{x + 2} \ge 0$

76. $\dfrac{x - 3}{x + 5} < 0$

77. $\dfrac{x - 2}{x + 3} < 1$

78. $\dfrac{x - 3}{x + 4} > 2$

79. $\dfrac{3}{x + 2} > \dfrac{1}{x + 1}$

80. $\dfrac{1}{x + 1} < \dfrac{1}{x - 1}$

Miscellaneous

In Exercises 81–92, find all real or imaginary solutions to each equation.

81. $144x^2 - 120x + 25 = 0$

82. $49x^2 + 9 = 42x$

83. $(2x + 3)^2 + 7 = 12$

84. $6x = -\dfrac{19x + 25}{x + 1}$

85. $1 + \dfrac{20}{9x^2} = \dfrac{8}{3x}$

86. $\dfrac{x - 1}{x + 2} = \dfrac{2x - 3}{x + 4}$

87. $\sqrt{3x^2 + 7x - 30} = x$ **88.** $\dfrac{x^4}{3} = x^2 + 6$

89. $2(2x + 1)^2 + 5(2x + 1) = 3$

90. $(w^2 - 1)^2 + 2(w^2 - 1) = 15$

91. $x^{1/2} - 15x^{1/4} + 50 = 0$

92. $x^{-2} - 9x^{-1} + 18 = 0$

Find exact and approximate solutions to each problem.

93. *Missing numbers.* Find two positive real numbers that differ by 4 and have a product of 4.

94. *One on one.* Find two positive real numbers that differ by 1 and have product of 1.

95. *Big screen TV.* On a 19-inch diagonal measure television picture screen, the height is 4 inches less than the width. Find the height and width.

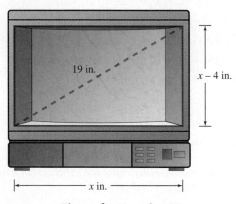

Figure for Exercise 95

96. *Boxing match.* A boxing ring is in the shape of a square, 20 feet on each side. How far apart are the fighters when they are in opposite corners of the ring?

97. *Students for a Clean Environment.* A group of environmentalists plans to print a message on an 8-in. by 10-in. paper. If the typed message requires 24 square inches of paper and the group wants an equal border on all sides, then how wide should the border be?

Figure for Exercise 97

98. *Winston works faster.* Winston can mow his dad's lawn in 1 hour less than it takes his brother Willie. If they take 2 hours to mow it when working together, then how long would it take Winston working alone?

99. *Ping pong.* The table used for table tennis is 4 feet longer than it is wide and has an area of 45 square feet. What are the dimensions of the table?

100. *Swimming pool design.* An architect has designed a motel pool within a rectangular area that is fenced on three sides as shown in the figure. If she uses 60 yards of fencing to enclose an area of 352 square yards, then what are the dimensions marked L and W in the figure? Assume that L is greater than W.

Figure for Exercise 100

101. *Decathlon champion.* For 1989 and 1990, Dave Johnson had the highest decathlon score in the world. When Johnson reached a speed of 32 ft per second on the pole vault runway, his height above the ground t seconds after leaving the ground was given by $h = -16t^2 + 32t$. (The elasticity of the pole converts the horizontal speed into vertical speed.) Find the value of t for which his height was 12 feet.

102. *Time of flight.* Use the information from Exercise 101 to determine how long Johnson was in the air. For how long was he more then 14 feet in the air?

CHAPTER 6 TEST

Calculate the value of $b^2 - 4ac$, and state how many real solutions each equation has.

1. $2x^2 - 3x + 2 = 0$

2. $-3x^2 + 5x - 1 = 0$

3. $4x^2 - 4x + 1 = 0$

Solve by using the quadratic formula.

4. $2x^2 + 5x - 3 = 0$

5. $x^2 + 6x + 6 = 0$

Solve by completing the square.

6. $x^2 + 10x + 25 = 0$

7. $2x^2 + x - 6 = 0$

Solve by any method.

8. $x(x + 1) = 12$

9. $a^4 - 5a^2 + 4 = 0$

10. $x - 2 - 8\sqrt{x - 2} + 15 = 0$

Find the complex solutions to the quadratic equations.

11. $x^2 + 36 = 0$

12. $x^2 + 6x + 10 = 0$

13. $3x^2 - x + 1 = 0$

Write a quadratic equation that has each given pair of solutions.

14. $-4, 6$

15. $-5i, 5i$

Solve each inequality. State and graph the solution set.

16. $w^2 + 3w < 18$

17. $\dfrac{2}{x - 2} < \dfrac{3}{x + 1}$

Find the exact solution to each problem.

18. The length of a rectangle is 2 feet longer than the width. If the area is 16 square feet, then what are the length and width?

19. A new computer can process a company's monthly payroll in 1 hour less time than the old computer. To really save time, the manager used both computers and finished the payroll in 3 hours. How long would it take the new computer to do the payroll by itself?

Tying It All Together

Solve each equation.

1. $2x - 15 = 0$
2. $2x^2 - 15 = 0$
3. $2x^2 + x - 15 = 0$
4. $2x^2 + 4x - 15 = 0$
5. $|4x + 11| = 3$
6. $|4x^2 + 11x| = 3$
7. $\sqrt{x} = x - 6$
8. $(2x - 5)^{2/3} = 4$

Solve each inequality.

9. $1 - 2x < 5 - x$
10. $(1 - 2x)(5 - x) \leq 0$
11. $\dfrac{1 - 2x}{5 - x} \leq 0$
12. $|5 - x| < 3$
13. $3x - 1 < 5$ and $-3 \leq x$
14. $x - 3 < 1$ or $2x \geq 8$

Solve each equation for y.

15. $2x - 3y = 9$
16. $\dfrac{y - 3}{x + 2} = -\dfrac{1}{2}$
17. $3y^2 + cy + d = 0$
18. $my^2 - ny = w$
19. $\dfrac{1}{3}x - \dfrac{2}{5}y = \dfrac{5}{6}$
20. $y - 3 = -\dfrac{2}{3}(x - 4)$

Let $m = \dfrac{y_2 - y_1}{x_2 - x_1}$. Find the value of m for each of the following choices of x_1, x_2, y_1, and y_2.

21. $x_1 = 2$, $x_2 = 5$, $y_1 = 3$, $y_2 = 7$
22. $x_1 = -3$, $x_2 = 4$, $y_1 = 5$, $y_2 = -6$
23. $x_1 = 0.3$, $x_2 = 0.5$, $y_1 = 0.8$, $y_2 = 0.4$
24. $x_1 = \dfrac{1}{2}$, $x_2 = \dfrac{1}{3}$, $y_1 = \dfrac{3}{5}$, $y_2 = -\dfrac{4}{3}$

Solve each problem.

25. **Ticket prices.** In the summer of 1994 the rock group Pearl Jam testified before a congressional committee that Ticketmaster was unfairly raising the prices of the group's concert tickets. One member of the group stated that fans should not have to pay more than $20 to see Pearl Jam. Of course, for any concert, as ticket prices rise, the number of tickets sold decreases, as shown in the figure. If you use the formula $n = 48,000 - 400p$ to predict the number sold depending on the price p, then how many will be sold at $20 per ticket? How many will be sold at $25 per ticket? Use the bar graph to estimate the price if 35,000 tickets were sold.

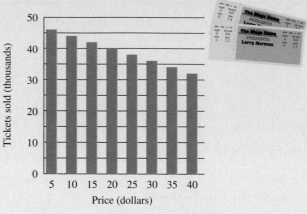

Figure for Exercise 25

26. **Increasing revenue.** Even though the number of tickets sold for a concert decreases with increasing price, the revenue generated does not necessarily decrease. Use the formula $R = p(48,000 - 400p)$ to determine the revenue when the price is $20 and when the price is $25. What price would produce a revenue of $1.28 million? Use the graph to find the price that determines the maximum revenue.

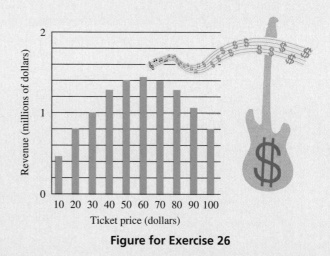

Figure for Exercise 26

Linear Equations in Two Variables

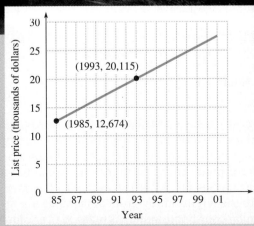

The first self-propelled automobile to carry passengers was built in 1801 by the British inventor Richard Trevithick. By 1911, about 600,000 automobiles were being operated in the United States alone. Some were powered by steam and some by electricity, but most were powered by gasoline. In 1913, to meet the ever growing demand, Henry Ford increased production by introducing a moving assembly line to carry automobile parts. Today, the United States is a nation of cars. Over 11 million automobiles are produced here annually, and total car registrations number over 114 million.

Henry Ford's early model T sold for $850. In 1994 the average price of a new car was $18,000, with some selling for more than $50,000. Prices for new cars rise every year, but, unfortunately for the buyer, the moment a new automobile is bought, its value begins to decrease. Much of the behavior of automobile prices can be modeled by linear equations. In Exercises 83 and 84 of Section 7.1 you will use linear equations to find increasing new car prices and depreciating used car prices.

7.1 Graphing in the Coordinate Plane

In Chapter 1 we graphed numbers on a number line. In Chapters 2 and 6, number lines were used to illustrate the solution sets to inequalities. In this section we graph pairs of numbers in a coordinate system made from two number lines to illustrate solution sets.

Ordered Pairs

An equation in two variables, such as $y = 2x + 3$, is satisfied only if we find a value for x and a value for y that make it true. For example, if $x = 4$ and $y = 11$, then the equation becomes $11 = 2 \cdot 4 + 3$, which is a true statement. We write $x = 4$ and $y = 11$ as the **ordered pair** (4, 11). The order of the numbers in an ordered pair is important. For example, (5, 13) also satisfies $y = 2x + 3$ because using 5 for x and 13 for y gives $13 = 2 \cdot 5 + 3$. However, (13, 5) does not satisfy this equation because $5 \neq 2 \cdot 13 + 3$. In an ordered pair, the value for x is always written first and the value of y second: (x, y).

To graph ordered pairs of real numbers, we need a new coordinate system. The **rectangular** or **Cartesian coordinate system** consists of a horizontal number line, the **x-axis,** and a vertical number line, the **y-axis,** as shown in Fig. 7.1. The intersection of the axes is the **origin.** The axes divide the coordinate plane, or the **xy-plane,** into four regions called **quadrants.** The quadrants are numbered as shown in Fig. 7.1, and they do not include any points on the axes.

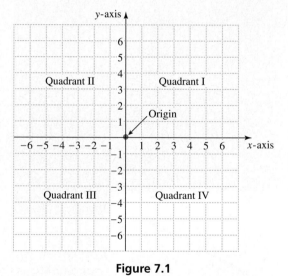

Figure 7.1

Plotting Points

Just as every real number corresponds to a point on the number line, every pair of real numbers corresponds to a point in the rectangular coordinate system. For example, the pair (2, 5) corresponds to the point that lies two units to the right of the origin and five units up. See Fig. 7.2. To locate $(-3, 4)$, start at the origin and move three units to the left and four units upward, as in Fig. 7.2. Locating a point in the rectangular coordinate system is referred to as **plotting** or **graphing** the point.

Figure 7.2

Because ordered pairs of real numbers correspond to points in the plane, we frequently refer to an ordered pair as a point.

EXAMPLE 1

Graphing ordered pairs

Plot the points $(2, 3)$, $(-2, -3)$, $(-1, 3)$, $(0, -4)$, and $(4, -2)$.

Solution

To plot $(2, 3)$, start at the origin, move two units to the right, then up three units. To graph $(-2, -3)$, start at the origin, move two units to the left, then down three units. All five points are shown in Fig. 7.3. ∎

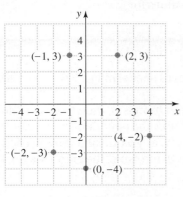

Figure 7.3

Graphing a Linear Equation

Consider again the equation $y = 2x + 3$. To find points that satisfy this equation, we can choose any number for x and then use $y = 2x + 3$ to find y. Because the y-value depends on the x-value, x is the **independent variable** and y is the **dependent variable.** For example, if we choose $x = -4$, then we can find y by replacing x by -4 in the equation:

$$y = 2x + 3$$
$$y = 2(-4) + 3 = -5$$

So $(-4, -5)$ satisfies the equation. The following table contains ordered pairs that satisfy $y = 2x + 3$:

x	-4	-3	-2	-1	0	1
$y = 2x + 3$	-5	-3	-1	1	3	5

The values for x are arbitrary but are chosen so that the ordered pair will fit within the range selected for the graph. For example, $(500, 1003)$ satisfies the equation but does not fit in with the other points. The points listed in the table appear to lie on a straight line in Fig. 7.4.

In fact, any ordered pair satisfying $y = 2x + 3$ lies on this line. So the complete line shown in Fig. 7.5 is the graph of $\{(x, y) \mid y = 2x + 3\}$, the solution set to the equation. Because the graph of $y = 2x + 3$ is a line, the equation is called a **linear equation.** The arrows on the ends of the line indicate that it extends infinitely in each direction.

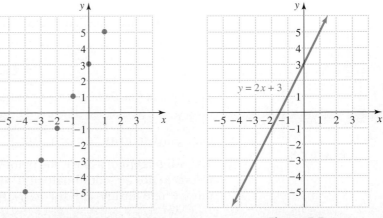

Figure 7.4   **Figure 7.5**

EXAMPLE 2   **Graphing a linear equation**

Graph $y + 2x = 1$. Plot at least four points.

Solution

If we solve the equation for y, we get $y = -2x + 1$. Now arbitrarily select four values for x and calculate the corresponding values for y:

x	-1	0	1	2
$y = -2x + 1$	3	1	-1	-3

Plot these points and draw a line through them as shown in Fig. 7.6. ■

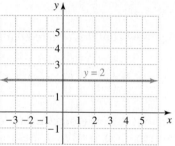

Figure 7.6

If the coefficient of a variable in a linear equation is 0, then that variable is usually omitted from the equation. For example, the equation $y = 0 \cdot x + 2$ is written as $y = 2$. Because x is multiplied by 0, any value of x can be used as long as y is 2. Because the y-coordinates are all the same, the graph is a horizontal line.

EXAMPLE 3   **Graphing a horizontal line**

Graph $y = 2$. Plot at least four points.

Solution

The following table gives four points that satisfy $y = 2$, or $y = 0 \cdot x + 2$:

x	-2	-1	0	1
$y = 0 \cdot x + 2$	2	2	2	2

The horizontal line through these points is shown in Fig. 7.7. ■

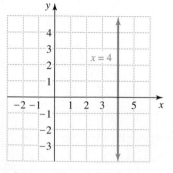

Figure 7.7

If the coefficient of y is 0 in a linear equation, then the graph is a vertical line.

EXAMPLE 4   **Graphing a vertical line**

Graph $x = 4$. Plot at least four points.

Solution

We can think of the equation $x = 4$ as $x = 4 + 0 \cdot y$. Because y is multiplied by 0, the equation is satisfied by any ordered pair with an x-coordinate of 4.

$x = 4 + 0 \cdot y$	4	4	4	4
y	-2	-1	0	1

The vertical line through these points is shown in Fig. 7.8. ■

Figure 7.8

Using Intercepts for Graphing

The **x-intercept** of a line is the point where the line crosses the x-axis. The x-intercept has a y-coordinate of 0. Similarly, the **y-intercept** of a line is the point where the line crosses the y-axis. The y-intercept has an x-coordinate of 0. If a line has distinct x- and y-intercepts, then these intercepts can be used as two points that determine the location of the line. (Horizontal lines, vertical lines, and lines through the origin do not have two distinct intercepts.)

EXAMPLE 5

Using intercepts to graph

Use the intercepts to graph the line $3x - 4y = 6$.

Solution

Let $x = 0$ in $3x - 4y = 6$ to find the y-intercept:

$$3(0) - 4y = 6$$

$$-4y = 6$$

$$y = -\frac{3}{2}$$

Let $y = 0$ in $3x - 4y = 6$ to find the x-intercept:

$$3x - 4(0) = 6$$

$$3x = 6$$

$$x = 2$$

The y-intercept is $\left(0, -\frac{3}{2}\right)$, and the x-intercept is $(2, 0)$. The line through the intercepts is shown in Fig. 7.9. To check, find another point that satisfies the equation. The point $(-2, -3)$ satisfies the equation and is on the line in Fig. 7.9. ∎

Figure 7.9

CAUTION Even though two points determine the location of a line, finding at least three points will help you avoid errors. ⊘

The Midpoint Formula

We now turn our attention from lines to line segments. **A line segment** consists of a pair of **endpoints** together with all points between the endpoints that lie on the line through the endpoints. A line segment has a midpoint and a length; a line has neither.

If you score 6 and 10 on two quizzes, then your average is $\frac{6 + 10}{2}$, or 8. The number 8 is midway between 6 and 10. In the same manner we can find a point in the rectangular coordinate system that is the midpoint of a line segment with two given endpoints.

Midpoint Formula

The midpoint of the line segment with endpoints (x_1, y_1) and (x_2, y_2) is

$$\left(\frac{x_1 + x_2}{2}, \frac{y_1 + y_2}{2}\right).$$

EXAMPLE 6 **Finding a midpoint**

Find the midpoint of the line segment with endpoints $(-5, -2)$ and $(3, 4)$, then graph the line segment and the midpoint.

Solution

Let $(x_1, y_1) = (-5, -2)$ and $(x_2, y_2) = (3, 4)$ and use the expression for midpoint:

$$\left(\frac{x_1 + x_2}{2}, \frac{y_1 + y_2}{2} \right) = \left(\frac{-5 + 3}{2}, \frac{-2 + 4}{2} \right) = (-1, 1)$$

The line segment and its midpoint $(-1, 1)$ are shown in Fig. 7.10.

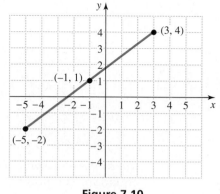

Figure 7.10 ■

The Distance Formula

In Example 6 the distance from the midpoint $(-1, 1)$ to either endpoint of the line segment should be the same. We can use the Pythagorean theorem to find the distance between $(-1, 1)$ and $(3, 4)$. As shown in Fig. 7.11, the distance between these points, d, is the length of the hypotenuse of a right triangle. We find the length of the two legs by subtracting coordinates. The length of side a is $4 - 1$, or 3, and the length of side b is $3 - (-1)$, or 4.

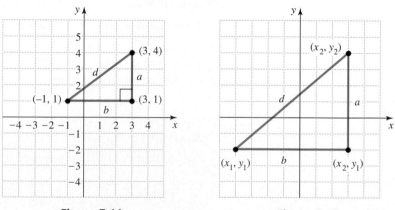

Figure 7.11   **Figure 7.12**

By the Pythagorean theorem,

$$d^2 = 3^2 + 4^2$$
$$d^2 = 9 + 16$$
$$d^2 = 25 \quad \text{Use only the positive square root of 25 because distance is always positive.}$$
$$d = 5.$$

Repeating this computation for $(-5, -2)$ and $(-1, 1)$ also gives a distance of 5.

We can also use the Pythagorean theorem to find a formula for the distance between any two points (x_1, y_1) and (x_2, y_2). As shown in Fig. 7.12, the distance is again the length of the hypotenuse of a right triangle. The length of side a is $y_2 - y_1$, and the length of side b is $x_2 - x_1$. Using the Pythagorean theorem, we can write

$$d^2 = (x_2 - x_1)^2 + (y_2 - y_1)^2.$$

If we apply the even-root property and omit the negative square root, we can express this formula as follows.

Distance Formula

The distance d between (x_1, y_1) and (x_2, y_2) is given by the formula

$$d = \sqrt{(x_2 - x_1)^2 + (y_2 - y_1)^2}.$$

EXAMPLE 7 **Using the distance formula**

Find the length of the line segment with endpoints $(-8, -10)$ and $(6, -4)$.

Solution

Let $(x_1, y_1) = (-8, -10)$ and $(x_2, y_2) = (6, -4)$. Now substitute the appropriate values into the distance formula:

$$d = \sqrt{[6 - (-8)]^2 + [-4 - (-10)]^2}$$
$$= \sqrt{(14)^2 + (6)^2}$$
$$= \sqrt{196 + 36}$$
$$= \sqrt{232}$$
$$= \sqrt{4 \cdot 58}$$
$$= 2\sqrt{58} \quad \text{Simplified form}$$

The exact length of the segment is $2\sqrt{58}$. ■

Note that it does not matter which point is called (x_1, y_1) and which is called (x_2, y_2) for finding midpoint or distance. To see this, repeat Example 7 using $(x_1, y_1) = (6, -4)$ and $(x_2, y_2) = (-8, -10)$.

The distance formula and the midpoint formula can be used to establish geometric facts about figures in a rectangular coordinate system.

EXAMPLE 8 **Using the distance formula in geometry**

Determine whether the triangle with vertices $A(-1, -1)$, $B(2, 0)$, and $C(1, 3)$ is a right triangle.

Solution

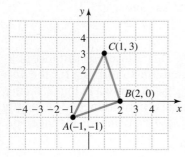

Figure 7.13

A graph of this triangle is shown in Fig. 7.13. We use the distance formula to find the length of each side of the triangle. The notation AB is used for the length of a line segment with endpoints A and B.

$$AB = \sqrt{(-1-2)^2 + (-1-0)^2}$$
$$= \sqrt{9+1} = \sqrt{10}$$
$$AC = \sqrt{(-1-1)^2 + (-1-3)^2}$$
$$= \sqrt{4+16} = \sqrt{20}$$
$$BC = \sqrt{(2-1)^2 + (0-3)^2}$$
$$= \sqrt{1+9} = \sqrt{10}$$

If the sum of the squares of the two shortest sides is equal to the square of the longest side, then the triangle is a right triangle (the Pythagorean theorem). Because

$$(\sqrt{10})^2 + (\sqrt{10})^2 = 10 + 10 = 20$$

and

$$(\sqrt{20})^2 = 20,$$

the triangle is a right triangle. ■

Applications

All of the equations that we graphed in this section are linear equations. Linear equations occur in many applications. Usually, the equations are not as simple as our previous examples. We may have to rename the axes or adjust the scale to graph equations involving large numbers. In graphing large numbers, the axes are often numbered so that the intersection of the two axes is not $(0, 0)$.

EXAMPLE 9 **Graphing with large numbers**

The cost per week, C, in dollars of producing n pairs of shoes for the Reebop Shoe Company is given by the equation $C = 2n + 8000$. Graph the equation for n between 0 and 800 inclusive ($0 \le n \le 800$).

Solution

Figure 7.14

We let the horizontal axis be the n-axis and the vertical axis be the C-axis. The scales are adjusted to accommodate the large numbers. Let each unit on the n-axis represent 200 pairs, and let each unit on the C-axis represent 1000 dollars. Find some ordered pairs the satisfy $C = 2n + 8000$.

n	0	200	400	600	800
$C = 2n + 8000$	8000	8400	8800	9200	9600

Figure 7.14 shows the graph of the equation. ■

Warm-ups

True or false? Explain your answer.

1. The point $(2, 5)$ satisfies the equation $3y - 2x = -4$.
2. The vertical axis is usually called the x-axis.
3. The point $(0, 0)$ is in quadrant I.
4. The point $(0, 1)$ is on the y-axis.
5. The graph of $x = 7$ is a vertical line.
6. The graph of $8 - y = 0$ is a horizontal line.
7. The distance between (a, b) and (c, d) is $\sqrt{(a - c) + (b - d)}$.
8. The distance between $(3, -2)$ and $(3, 5)$ is -7.
9. The distance between $(-2, 6)$ and $(3, 6)$ is 5.
10. The y-intercept for the line $y = 2x - 3$ is $(0, -3)$.

7.1 EXERCISES

Plot the following points in a rectangular coordinate system. For each point, name the quadrant in which it lies or the axis on which it lies. See Example 1.

1. $(2, 5)$
2. $(-5, 1)$
3. $\left(-3, -\dfrac{1}{2}\right)$
4. $(-2, -6)$
5. $(0, 4)$
6. $(0, 2)$
7. $(\pi, -1)$
8. $\left(\dfrac{\pi}{3}, 0\right)$
9. $(-4, 3)$
10. $(0, -3)$
11. $\left(\dfrac{\pi}{2}, 0\right)$
12. $(\pi, 2)$
13. $(0, -1)$
14. $(4, -3)$
15. $(0, 0)$
16. $\left(-2, -\dfrac{1}{2}\right)$

Graph each linear equation. Plot four points for each line. See Examples 2–4.

17. $y = x + 1$
18. $y = x - 1$
19. $y = -2x + 3$
20. $y = 2x - 3$
21. $y = x$
22. $y = -x$
23. $y = 3$
24. $y = -2$
25. $y = 1 - x$
26. $y = 2 - x$
27. $x = 2$
28. $x = -3$
29. $y = \dfrac{1}{2}x - 1$
30. $y = \dfrac{1}{3}x - 2$
31. $x - 4 = 0$
32. $y + 3 = 0$
33. $3x + y = 5$
34. $x + 2y = 4$
35. $2x + y = 1$
36. $x - y = 2$
37. $x - y = -3$
38. $x - 2y = 1$
39. $y = 1.35x - 4.27$
40. $y = -0.26x + 3.86$

Find the x- and y-intercepts for each line and use them to graph the line. See Example 5.

41. $4x - 3y = 12$
42. $2x + 5y = 20$
43. $x - y + 5 = 0$
44. $x + y + 7 = 0$
45. $2x + 3y = 5$
46. $3x - 4y = 7$
47. $y = \dfrac{3}{5}x + \dfrac{2}{3}$
48. $y = -\dfrac{2}{3}x - \dfrac{5}{4}$

Complete the given ordered pairs so that each ordered pair satisfies the given equation.

49. $(2, \ \ ), (\ \ , -3), \quad y = -3x + 6$
50. $(-1, \ \ ), (\ \ , 4), \quad y = \dfrac{1}{2}x + 2$
51. $(-4, \ \ ), (\ \ , 6), \quad \dfrac{1}{2}x - \dfrac{1}{3}y = 9$
52. $(3, \ \ ), (\ \ , -1), \quad 2x - 3y = 5$

In Exercises 53–70, find the midpoint and length of the line segment with the given endpoints. See Examples 6 and 7.

53. $(6, 5), (4, 2)$
54. $(7, 3), (5, 1)$
55. $(3, 5), (1, -3)$
56. $(6, 2), (3, -5)$
57. $(4, -2), (-3, -6)$
58. $(-2, 3), (1, -4)$
59. $(2, -7), (-5, -7)$
60. $(-3, 6), (-3, -1)$
61. $(0, -3), (4, 0)$
62. $(4, 0), (0, 3)$
63. $(0, 0), (-6, 8)$
64. $(0, 0), (-9, -12)$

65. $\left(\dfrac{1}{2}, \dfrac{1}{4}\right)$, $\left(\dfrac{1}{3}, \dfrac{1}{2}\right)$

66. $(3\sqrt{2}, \sqrt{3})$, $(5\sqrt{2}, -\sqrt{3})$

67. $(3.5, -6.2)$, $(-1.4, 3.7)$

68. $(-2.4, 1.5)$, $(5.3, -1.6)$

69. $\left(\dfrac{\pi}{2}, 0\right)$, $(\pi, 1)$ **70.** $\left(-\dfrac{\pi}{2}, 0\right)$, $(\pi, -1)$

Solve each geometric problem. See Example 8 and Appendix A for a review of some geometric terms.

71. Determine whether the triangle shown in the accompanying figure is an isosceles triangle.

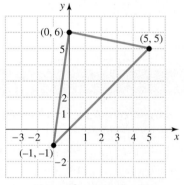

Figure for Exercise 71

72. Determine whether the points $(-2, 10)$, $(4, 3)$, and $(2, -6)$ are the vertices of an isosceles triangle.

73. If two line segments have identical midpoints, then the segments bisect each other. Do the diagonals of the parallelogram with vertices $(0, 0)$, $(4, 0)$, $(6, 2)$, and $(2, 2)$ bisect each other?

74. Do the the diagonals of the rectangle with vertices $(2, 1)$, $(6, 3)$, $(5, 5)$, and $(1, 3)$ bisect each other? See Exercise 73.

75. Is the triangle with vertices $(4, 4)$, $(0, 8)$, and $(-2, -2)$ a right triangle?

76. Is the triangle with vertices $(-1, 7)$, $(2, 1)$, and $(-4, -2)$ a right triangle?

77. Determine whether the points $(-2, -13)$, $(1, -4)$, and $(3, 2)$ lie on the same straight line. If the sum of the lengths of the two shortest line segments is equal to the length of the longest line segment, then the three points lie on a line.

78. Determine whether the points $(-2, 13)$, $(1, 1)$, and $(3, -6)$ lie on the same straight line. See Exercise 77.

79. If the opposite sides of a quadrilateral are equal in length, then the quadrilateral is a parallelogram. Determine whether the quadrilateral with vertices $(1, 5)$, $(2, 3)$, $(3, 9)$, and $(4, 7)$ is a parallelogram.

80. If all sides of a quadrilateral are the same length, then the quadrilateral is a rhombus. Determine whether the points $(-2, -2)$, $(-1, 4)$, $(4, -1)$, and $(5, 5)$ are the vertices of a rhombus.

81. If a parallelogram has at least one right angle, then it is a rectangle. Determine whether the points $(-2, 4)$, $(-1, 6)$, $(4, 1)$, and $(5, 3)$ are the vertices of a rectangle.

82. If a rhombus has at least one right angle, then it is a square. Determine whether the points $(-7, -1)$, $(-4, 5)$, $(-1, -4)$, and $(2, 2)$ are the vertices of a square.

Solve the following. See Example 9.

83. *Camaro Z28 inflation.* The rising list price P (in dollars) for a new Chevrolet Camaro Z28 can be modeled by the formula $P = 12{,}463 + 720n$, where n is the number of years since 1985 (*Edmund's 1994 Used Car Prices*). Sketch a graph of this equation for $0 \leq n \leq 15$. What was the list price for a new Z28 in 1985 and in 1990? What will the list price be in the year 2000 according to this formula?

84. *Camaro Z28 depreciation.* The 1994 average retail price P (in dollars) for an n-year-old Chevrolet Camaro Z28 can be modeled by the formula $P = 17{,}683 - 1787n$, where $2 \leq n \leq 8$ (*Edmund's 1994 Used Car Prices*). What was the average retail price of a four-year-old Z28 in 1994? Sketch a graph of this equation for $2 \leq n \leq 8$.

85. *Rental cost.* For a one-day car rental the X-press Car Company charges C dollars, where C is determined by the formula $C = 0.26m + 42$ and m is the number of miles driven. What is the charge for a car driven 400 miles? Sketch a graph of the equation for m ranging from 0 to 1000.

86. *Measuring risk.* The Friendly Bob Loan Company gives each applicant a rating, t, from 0 to 10 according to the applicant's ability to repay, a higher rating indicating higher risk. The interest rate, r, is then determined by the formula $r = 0.02t + 0.15$. If your rating were 8, then what would be your interest rate? Sketch the graph of the equation for t ranging from 0 to 10.

Getting More Involved

87. *Writing.* In the coordinate plane, why is the graph of $x = 4$ a straight line and not a single point?

88. *Discussion.* Find an example outside of a mathematics class in which a coordinate system is used to locate points in a plane.

7.2 Slope of a Line

In this section:

▶ Slope
▶ Using Coordinates to Find Slope
▶ Perpendicular Lines
▶ Parallel Lines
▶ Geometric Applications of Slope

In Section 7.1 we saw that the graph of a linear equation is a straight line. In this section we look at graphs of straight lines in more detail and develop the concept of slope of a line.

9-12 pitch

Slope

If a highway has a 6% grade, then in 100 feet (measured horizontally) the road rises 6 feet (measured vertically). The ratio of 6 to 100 is 6%. If a roof rises 9 feet in a horizontal distance (or run) of 12 feet, then the roof has a 9–12 pitch. A roof with a 9–12 pitch is steeper than a roof with a 6–12 pitch. The grade of a road and the pitch of a roof are measurements of steepness. In each case the measurement is a ratio of rise (vertical change) to run (horizontal change).

We measure the steepness of a line in the same way that we measure steepness of a road or a roof. The slope of a line is the ratio of the change in y-coordinate, or the **rise,** to the change in x-coordinate, or the **run,** between two points on the line.

Slope

$$\text{Slope} = \frac{\text{change in } y\text{-coordinate}}{\text{change in } x\text{-coordinate}} = \frac{\text{rise}}{\text{run}}$$

Figure 7.15

Consider the line in Fig. 7.15(a). In going from $(0, 1)$ to $(1, 3)$, there is a change of $+1$ in the x-coordinate and a change of $+2$ in the y-coordinate, or a run of 1 and a rise of 2. So the slope is $\frac{2}{1}$ or 2. If we move from $(1, 3)$ to $(0, 1)$ as in Fig. 7.15(b), the rise is -2 and the run is -1. So the slope is $\frac{-2}{-1}$ or 2. If we start at either point and move to the other point, we get the same slope.

EXAMPLE 1   **Finding the slope from a graph**

Find the slope of each line by going from point A to point B.

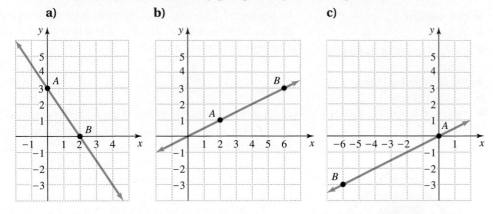

a)   **b)**   **c)**

Solution

a) A is located at $(0, 3)$ and B at $(2, 0)$. In going from A to B, the change in y is -3 and the change in x is 2. So

$$m = -\frac{3}{2}.$$

b) In going from $A(2, 1)$ to $B(6, 3)$, we must rise 2 and run 4. So

$$m = \frac{2}{4} = \frac{1}{2}.$$

c) In going from $A(0, 0)$ to $B(-6, -3)$, the rise is -3 and the run is -6. So

$$m = \frac{-3}{-6} = \frac{1}{2}. \qquad \blacksquare$$

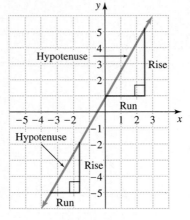

Figure 7.16

Note that in Example 1(c) we found the slope of the line of Example 1(b) by using two different points. The slope is the ratio of the lengths of the two legs of a right triangle whose hypotenuse is on the line. See Fig. 7.16. As long as one leg is vertical and the other leg is horizontal, all such triangles for a given line have the same shape: They are similar triangles. Because ratios of corresponding sides in similar triangles are equal, the slope has the same value no matter which two points of the line are used to find it.

Using Coordinates to Find Slope

We can obtain the rise and run from a graph, or we can get them without a graph by subtracting the y-coordinates to get the rise and the x-coordinates to get the run for two points on the line. See Figure 7.17 on the next page.

Slope Using Coordinates

The slope m of the line containing the points (x_1, y_1) and (x_2, y_2) is given by

$$m = \frac{y_2 - y_1}{x_2 - x_1}, \qquad \text{provided that } x_2 - x_1 \neq 0.$$

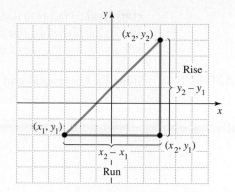

Figure 7.17

EXAMPLE 2   **Finding slope from coordinates**

Find the slope of each line.

a)   The line through $(2, 5)$ and $(6, 3)$

b)   The line through $(-2, 3)$ and $(-5, -1)$

c)   The line through $(-6, 4)$ and the origin

Solution

a)   Let $(x_1, y_1) = (2, 5)$ and $(x_2, y_2) = (6, 3)$. The assignment of (x_1, y_1) and (x_2, y_2) is arbitrary.

$$m = \frac{y_2 - y_1}{x_2 - x_1}$$

$$= \frac{3 - 5}{6 - 2}$$

$$= \frac{-2}{4} = -\frac{1}{2}$$

b)   Let $(x_1, y_1) = (-5, -1)$ and $(x_2, y_2) = (-2, 3)$.

$$m = \frac{y_2 - y_1}{x_2 - x_1}$$

$$= \frac{3 - (-1)}{-2 - (-5)}$$

$$= \frac{4}{3}$$

c)   Let $(x_1, y_1) = (0, 0)$ and $(x_2, y_2) = (-6, 4)$.

$$m = \frac{4 - 0}{-6 - 0}$$

$$= \frac{4}{-6} = -\frac{2}{3}$$

CAUTION Do not reverse the order of subtraction from numerator to denominator when finding the slope. If you divide $y_2 - y_1$ by $x_1 - x_2$, you will get the wrong sign for the slope. ⊘

EXAMPLE 3 **Slope for horizontal and vertical lines**

Find the slope of each line.

a)

b)

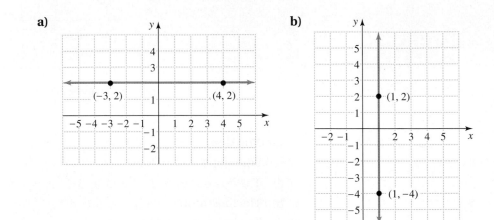

Solution

a) Using $(-3, 2)$ and $(4, 2)$ to find the slope of the horizontal line, we get

$$m = \frac{2 - 2}{-3 - 4}$$

$$= \frac{0}{-7} = 0.$$

b) Using $(1, -4)$ and $(1, 2)$ to find the slope of the vertical line, we get $x_2 - x_1 = 0$. Because the definition of slope using coordinates says that $x_2 - x_1$ must be nonzero, the slope is undefined for this line. ■

For any two points on a horizontal line, the y-coordinates are equal and $y_2 - y_1 = 0$. Because the x-coordinates are all identical on any vertical line, $x_2 - x_1 = 0$ for any two points. So we can make the following general statements.

Horizontal and Vertical Lines

The slope of any horizontal line is 0.
Slope is undefined for any vertical line.

CAUTION Do not say that a vertical line has no slope, because "no slope" could be confused with 0 slope, the slope of a horizontal line. ⊘

As you move the tip of your pencil from left to right along a line with positive slope, the y-coordinates are increasing. As you move the tip of your pencil from left to right along a line with negative slope, the y-coordinates are decreasing. See Fig. 7.18.

Figure 7.18

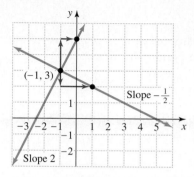

Figure 7.19

Perpendicular Lines

The lines shown in Fig. 7.19 have slopes 2 and $-\frac{1}{2}$. These two lines appear to be perpendicular to each other. It can be shown that *a line is perpendicular to another line if its slope is the negative of the reciprocal of the slope of the other.*

> **Perpendicular Lines**
>
> Two lines with slopes m_1 and m_2 are perpendicular if and only if
>
> $$m_1 = -\frac{1}{m_2}.$$

EXAMPLE 4 **Perpendicular lines**

Line l contains the point $(1, 6)$ and is perpendicular to the line through $(-4, 1)$ and $(3, -2)$. Find the slope of line l.

Solution

The line through $(-4, 1)$ and $(3, -2)$ has slope

$$m = \frac{1 - (-2)}{-4 - 3} = \frac{3}{-7} = -\frac{3}{7}.$$

Because line l is perpendicular to a line with slope $-\frac{3}{7}$, the slope of line l is $\frac{7}{3}$. ∎

Parallel Lines

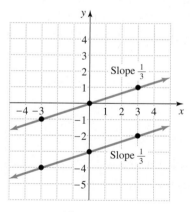

Figure 7.20

Consider the two lines shown in Fig. 7.20. Each of these lines has a slope of $\frac{1}{3}$, and these lines are parallel. In general, we have the following fact.

> **Parallel Lines**
>
> Nonvertical parallel lines have equal slopes.

EXAMPLE 5 **Parallel lines**

Line l goes through the origin and is parallel to the line through $(-2, 3)$ and $(4, -5)$. Find the slope of line l.

Solution

The line through $(-2, 3)$ and $(4, -5)$ has slope

$$m = \frac{-5 - 3}{4 - (-2)} = \frac{-8}{6} = -\frac{4}{3}.$$

Because line l is parallel to a line with slope $-\frac{4}{3}$, the slope of line l is $-\frac{4}{3}$ also. ∎

Geometric Applications of Slope

When a geometric figure is located in a coordinate system, we can use slope to determine whether it has any parallel or perpendicular sides.

EXAMPLE 6 **Using slope with geometric figures**

Determine whether $(-3, 2)$, $(-2, -1)$, $(4, 1)$, and $(3, 4)$ are the vertices of a rectangle.

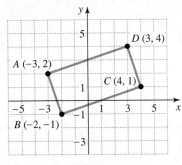

Figure 7.21

Solution

Figure 7.21 shows the quadrilateral determined by these points. If a parallelogram has at least one right angle, then it is a rectangle. Calculate the slope of each side.

$$m_{AB} = \frac{2 - (-1)}{-3 - (-2)} \qquad\qquad m_{BC} = \frac{-1 - 1}{-2 - 4}$$

$$= \frac{3}{-1} = -3 \qquad\qquad = \frac{-2}{-6} = \frac{1}{3}$$

$$m_{CD} = \frac{1 - 4}{4 - 3} \qquad\qquad m_{AD} = \frac{2 - 4}{-3 - 3}$$

$$= \frac{-3}{1} = -3 \qquad\qquad = \frac{-2}{-6} = \frac{1}{3}$$

Because the opposite sides have the same slope, they are parallel, and the figure is a parallelogram. Because $\frac{1}{3}$ is the opposite of the reciprocal of -3, the intersecting sides are perpendicular. Therefore the figure is a rectangle. ■

Warm-ups

True or false? Explain your answer.

1. Slope is a measurement of the steepness of a line.
2. Slope is run divided by rise.
3. The line through $(4, 5)$ and $(-3, 5)$ has undefined slope.
4. The line through $(-2, 6)$ and $(-2, -5)$ has undefined slope.
5. Slope cannot be negative.
6. The slope of the line through $(0, -2)$ and $(5, 0)$ is $-\frac{2}{5}$.
7. The line through $(4, 4)$ and $(5, 5)$ has slope $\frac{5}{4}$.
8. If a line contains points in quadrants I and III, then its slope is positive.
9. Lines with slope $\frac{2}{3}$ and $-\frac{2}{3}$ are perpendicular to each other.
10. Any two parallel lines have equal slopes.

7.2 EXERCISES

Determine the slope of each line. See Example 1.

1.

2.

3.

4.

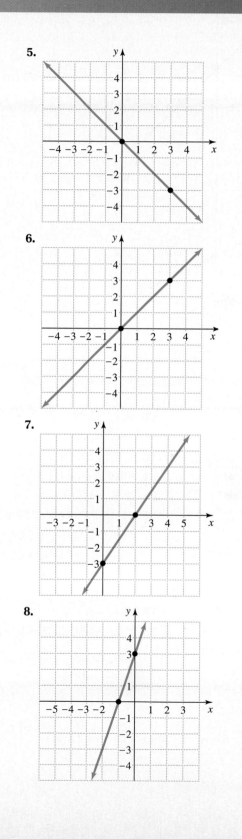

5.

6.

7.

8.

9.

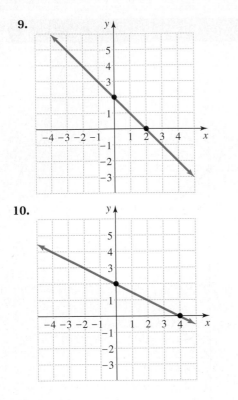

10.

In each case, make a sketch and find the slope of line l. See Examples 4 and 5.

35. Line l contains the point $(3, 4)$ and is perpendicular to the line through $(-5, 1)$ and $(3, -2)$.

36. Line l goes through $(-3, -5)$ and is perpendicular to the line through $(-2, 6)$ and $(5, 3)$.

37. Line l goes through $(2, 5)$ and is parallel to the line through $(-3, -2)$ and $(4, 1)$.

38. Line l goes through the origin and is parallel to the line through $(-3, -5)$ and $(4, -1)$.

39. Line l is perpendicular to a line with slope $\frac{4}{5}$. Both lines contain the origin.

40. Line l is perpendicular to a line with slope -5. Both lines contain the origin.

Solve each geometric figure problem. See Example 6.

41. If the opposite sides of a quadrilateral are parallel, then it is a parallelogram. Use slope to determine whether the points $(-6, 1)$, $(-2, -1)$, $(0, 3)$, and $(4, 1)$ are the vertices of a parallelogram.

42. Use slope to determine whether the points $(-7, 0)$, $(-1, 6)$, $(-1, -2)$, and $(6, 5)$ are the vertices of a parallelogram. See Exercise 41.

43. A trapezoid is a quadrilateral with one pair of parallel sides. Use slope to determine whether the points $(-3, 2)$, $(-1, -1)$, $(3, 6)$, and $(6, 4)$ are the vertices of a trapezoid.

44. Use slope to determine whether the points $(0, 0)$, $(0, 4)$, $(3, 6)$, and $(6, 4)$ are the vertices of a trapezoid. See Exercise 43.

45. A parallelogram with at least one right angle is a rectangle. Determine whether the points $(-4, 4)$, $(-1, -2)$, $(0, 6)$, and $(3, 0)$ are the vertices of a rectangle.

46. A parallelogram with four equal sides and at least one right angle is a square. Determine whether the points $(-3, 2)$, $(1, -2)$, $(5, 2)$, and $(0, 6)$ are the vertices of a square.

47. If a triangle has one right angle, then it is a right triangle. Use slope to determine whether the points $(-3, 3)$, $(-1, 6)$, and $(0, 0)$ are the vertices of a right triangle.

48. Use slope to determine whether the points $(0, -1)$, $(2, 5)$, and $(5, 4)$ are the vertices of a right triangle. See Exercise 47.

Find the slope of the line that contains each of the following pairs of points. See Examples 2 and 3.

11. $(2, 6)$, $(5, 1)$

12. $(3, 4)$, $(6, 10)$

13. $(-3, -1)$, $(4, 3)$

14. $(-2, -3)$, $(1, 3)$

15. $(-2, 2)$, $(-1, 7)$

16. $(-3, 5)$, $(1, -6)$

17. $(3, -5)$, $(0, 0)$

18. $(0, 0)$, $(-2, -1)$

19. $(0, 3)$, $(5, 0)$

20. $(3, 0)$, $(0, 10)$

21. $\left(\frac{3}{4}, -1\right)$, $\left(-\frac{1}{2}, -\frac{1}{2}\right)$

22. $\left(\frac{1}{2}, 2\right)$, $\left(\frac{1}{4}, \frac{1}{2}\right)$

23. $(6, 212)$, $(7, 209)$

24. $(1988, 306)$, $(1990, 315)$

25. $(4, 7)$, $(-12, 7)$

26. $(5, -3)$, $(9, -3)$

27. $(2, 6)$, $(2, -6)$

28. $(-3, 2)$, $(-3, 0)$

29. $(24.3, 11.9)$, $(3.57, 8.4)$

30. $(-2.7, 19.3)$, $(5.46, -3.28)$

31. $(\sqrt{2}, 3)$, $\left(\frac{\sqrt{2}}{2}, 5\right)$

32. $(\sqrt{3}, -1)$, $(-\sqrt{5}, \sqrt{2})$

33. $\left(\frac{\pi}{4}, 1\right)$, $\left(\frac{\pi}{2}, 0\right)$

34. $\left(\frac{\pi}{3}, -1\right)$, $\left(\frac{\pi}{6}, 0\right)$

Solve each problem.

49. ***Crown Victoria.*** The list price of a new Ford Crown Victoria four-door sedan was $12,674 in 1985 and $20,115 in 1993 (*Edmund's 1994 Used Car Prices*). Use the accompanying graph to estimate the list price in the year 2000. Find the slope of the line through (1985, 12,674) and (1993, 20,115). What does the slope tell you in this case? Use the slope to predict the list price of a new Crown Victoria in the year 2000.

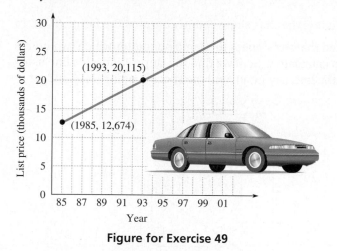

Figure for Exercise 49

50. ***Lumina Euro Coupe.*** In 1994 the average retail price of a one-year-old Chevrolet Lumina Euro Coupe was $13,550, while the average price for a four-year-old Chevrolet Lumina Euro Coupe was $6425. (*Edmund's 1994 Used Car Prices*). Use the accompanying graph to estimate the average retail price of a three-year-old Lumina Euro Coupe in 1994. Find the slope of the line through (1, 13,550) and (4, 6425). What does the slope tell you in this case? Use the slope to predict the average retail price of a three-year-old Lumina Euro Coupe in 1994.

Figure for Exercise 50

51. What is the slope of a line that is perpendicular to a line with slope 0.247?

52. What is the slope of a line that is perpendicular to the line through (3.27, −1.46) and (−5.48, 3.61)?

53. Find k so that the line through (2, k) and (−3, −5) has slope $\frac{1}{2}$.

54. Find k so that the line through (k, 3) and (−2, 0) has slope k.

Getting More Involved

55. ***Writing.*** What is the difference between zero slope and undefined slope?

56. ***Writing.*** Is it possible for a line to be in only one quadrant? Two quadrants? Write a rule for determining whether a line has positive, negative, zero, or undefined slope from knowing in which quadrants the line is found.

57. ***Exploration.*** Draw a rhombus with vertices (−3, −1), (0, 3), (2, −1), and (5, 3). Find the slopes of the diagonals of the rhombus. What can you conclude about the diagonals of this rhombus?

58. ***Exploration.*** Draw a square with vertices (−5, 3), (−3, −3), (1, 5), and (3, −1). Find the slopes of the diagonals of the square. What can you conclude about the diagonals of this square?

7.3 Equations of Lines and Their Graphs

In Section 7.1 you learned how to graph a straight line corresponding to a linear equation. The line contains all of the points that satisfy the equation. In this section we start with a line or a description of a line and write an equation corresponding to the line.

Point-Slope Form

Figure 7.22 shows the line that has slope $\frac{2}{3}$ and contains the point $(3, 5)$. In Section 7.2 you learned that the slope is the same no matter which two points of the line are used to calculate it. So if we find the slope m for this line using an arbitrary point of the line, say (x, y), and the specific point $(3, 5)$, we get

$$m = \frac{y - 5}{x - 3}.$$

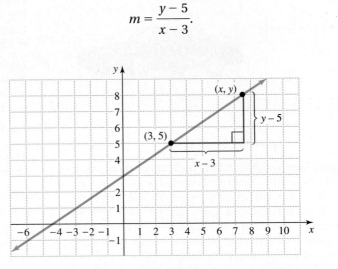

Figure 7.22

Because the slope of this line is $\frac{2}{3}$, we can write

$$\frac{y - 5}{x - 3} = \frac{2}{3}.$$

Multiplying each side by $x - 3$, we get

$$y - 5 = \frac{2}{3}(x - 3).$$

Because (x, y) was an arbitrary point on the line, this equation is satisfied by every point on the line.

If we use (x_1, y_1) as the specific point and (x, y) as an arbitrary point on a line with slope m, we can write

$$\frac{y - y_1}{x - x_1} = m.$$

Multiplying each side of this equation by $x - x_1$ gives us the **point-slope form** of the equation of the line.

> **Point-Slope Form**
>
> The equation of the line through (x_1, y_1) with slope m in point-slope form is
>
> $$y - y_1 = m(x - x_1).$$

EXAMPLE 1 **Using point-slope form to write the equation of a line**

Find the equation of line l and solve it for y.

a) Line l goes through $(-2, 5)$ and has slope -3.

b) Line l goes through $(3, -2)$ and $(-1, 1)$.

Solution

a) Use $x_1 = -2$, $y_1 = 5$, and $m = -3$ in the point-slope form:

$$y - 5 = -3[x - (-2)]$$

Now solve the equation for y:

$$y - 5 = -3[x + 2]$$
$$y - 5 = -3x - 6$$
$$y = -3x - 1$$

b) We are not given the slope, but we can find it because the points $(3, -2)$ and $(-1, 1)$ are on the line:

$$m = \frac{1 - (-2)}{-1 - 3}$$

$$= \frac{3}{-4} = -\frac{3}{4}$$

Now use this slope and one of the points, say $(3, -2)$, to write the equation of line l in point-slope form:

$$y - (-2) = -\frac{3}{4}(x - 3) \quad \text{Point-slope form}$$

$$y + 2 = -\frac{3}{4}x + \frac{9}{4} \quad \text{Distributive property}$$

$$y = -\frac{3}{4}x + \frac{1}{4} \quad \text{Solve for } y: \frac{9}{4} - 2 = \frac{9}{4} - \frac{8}{4} = \frac{1}{4}$$

Note that we would get the same equation if we had used slope $-\frac{3}{4}$ and the other point $(-1, 1)$. Try it. ∎

For the next example, recall that if a line has slope m, then the slope of any line perpendicular to it is $-\dfrac{1}{m}$, provided that $m \neq 0$.

EXAMPLE 2 **The equation of a line perpendicular to another line**

Line l goes through $(2, 0)$ and is perpendicular to the line through $(5, -1)$ and $(-1, 3)$. Find the equation of line l and then solve it for y.

Solution

First find the slope of the line through $(5, -1)$ and $(-1, 3)$:

$$m = \frac{3 - (-1)}{-1 - 5} = \frac{4}{-6} = -\frac{2}{3}$$

Because line l is perpendicular to this line, line l has slope $\frac{3}{2}$. Now use $(2, 0)$ and the slope $\frac{3}{2}$ in the point-slope formula to get the equation of line l:

$$y - 0 = \frac{3}{2}(x - 2)$$

$$y = \frac{3}{2}x - 3 \qquad \text{Distributive property} \qquad \blacksquare$$

Slope-Intercept Form

The line $y = -3x - 1$ in Example 1(a) has slope -3. To find the y-intercept of this line, let $x = 0$ in $y = -3x - 1$: $y = -3(0) - 1 = -1$. The y-intercept is $(0, -1)$. Its y-coordinate appears in the equation:

$$y = -3x - 1$$

Slope y-intercept $(0, -1)$

Because the slope and y-intercept can be read from the equation when it is solved for y, this form of the equation of the line is called slope-intercept form.

> **Slope-Intercept Form**
>
> The equation of a line in **slope-intercept form** is
>
> $$y = mx + b,$$
>
> where m is the slope and $(0, b)$ is the y-intercept.

EXAMPLE 3 **Writing an equation given its slope and y-intercept**

Write the slope-intercept form of the equation of the line shown in Figure 7.23.

Solution

From Fig. 7.23 we see that the y-intercept is $(0, 3)$. If we start at the y-intercept and move down 2 and 3 to the right, we get to another point on the line. So the slope is $-\frac{2}{3}$. The equation of this line in slope-intercept form is

$$y = -\frac{2}{3}x + 3. \qquad \blacksquare$$

Figure 7.23

Standard Form

If x students paid \$5 each and y adults paid \$7 each to attend a play for which the ticket sales totaled \$1900, then we can write the equation $5x + 7y = 1900$. This form of a linear equation is common in applications. It is called standard form.

> **Standard Form**
>
> The equation of a line in **standard form** is
>
> $$Ax + By = C,$$
>
> where A, B, and C are real numbers with A and B not both zero.

The numbers A, B, and C in standard form can be any real numbers, but it is a common practice to write standard form using only integers and a positive coefficient for x.

EXAMPLE 4 **Changing to standard form**

Write the equation $y = \frac{1}{2}x - \frac{3}{4}$ in standard form using only integers and a positive coefficient for x.

Solution

Use the properties of equality to get the equation in the form $Ax + By = C$:

$$y = \frac{1}{2}x - \frac{3}{4} \qquad \text{Original equation}$$

$$-\frac{1}{2}x + y = -\frac{3}{4} \qquad \text{Subtract } \tfrac{1}{2}x \text{ from each side.}$$

$$4\left(-\frac{1}{2}x + y\right) = 4\left(-\frac{3}{4}\right) \qquad \begin{array}{l}\text{Multiply each side by 4 to}\\ \text{get integral coefficients.}\end{array}$$

$$-2x + 4y = -3 \qquad \text{Distributive property}$$

$$2x - 4y = 3 \qquad \begin{array}{l}\text{Multiply by } -1 \text{ to make the}\\ \text{coefficient of } x \text{ positive.}\end{array} \qquad \blacksquare$$

To find the slope and y-intercept of a line written in standard form, we convert the equation to slope-intercept form.

EXAMPLE 5 **Changing to slope-intercept form**

Find the slope and y-intercept of the line $3x - 2y = 5$.

Solution

Solve for y to get slope-intercept form:

$$3x - 2y = 5$$

$$-2y = -3x + 5$$

$$y = \frac{3}{2}x - \frac{5}{2} \qquad \text{Slope-intercept form}$$

The slope is $\frac{3}{2}$, and the y-intercept is $\left(0, -\frac{5}{2}\right)$. $\qquad \blacksquare$

You learned in Section 7.1 that the graph of the equation $x = 4$ is a vertical line. Because slope is undefined for vertical lines, the equation of this line cannot be written in slope-intercept form or point-slope form. Only nonvertical lines can be written in those forms. However, a vertical line can be written in standard form. For example,

$$x = 4$$

can be written as

$$1 \cdot x + 0 \cdot y = 4.$$

Every line has an equation in standard form.

EXAMPLE 6 **Finding the equation of a line**

Write an equation in standard form with integral coefficients for the line l through $(2, 5)$ that is perpendicular to the line $2x + 3y = 1$.

Solution

First solve the equation $2x + 3y = 1$ for y to find its slope:

$$2x + 3y = 1$$
$$3y = -2x + 1$$
$$y = -\frac{2}{3}x + \frac{1}{3} \quad \text{The slope is } -\frac{2}{3}.$$

The slope of line l is the opposite of the reciprocal of $-\frac{2}{3}$. So line l has slope $\frac{3}{2}$ and goes through $(2, 5)$. Now use the point-slope form to write the equation:

$$y - 5 = \frac{3}{2}(x - 2) \quad \text{Point-slope form}$$

$$y - 5 = \frac{3}{2}x - 3 \quad \text{Distributive property}$$

$$y = \frac{3}{2}x + 2$$

$$-\frac{3}{2}x + y = 2$$

$$3x - 2y = -4 \quad \text{Multiply each side by } -2.$$

So $3x - 2y = -4$ is the standard form of the equation of the line through $(2, 5)$ that is perpendicular to $2x + 3y = 1$. ■

Using Slope-Intercept Form for Graphing

In the slope-intercept form, a point on the line (the y-intercept) and the slope are readily available. To graph a line, we can start at the y-intercept and count off the rise and run to get a second point on the line.

EXAMPLE 7 **Using slope and y-intercept to graph**

Graph the line $2x - 3y = -3$.

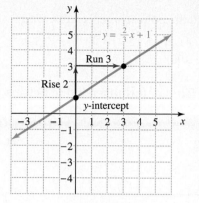

Figure 7.24

Solution

First write the equation in slope-intercept form:

$$2x - 3y = -3$$
$$-3y = -2x - 3$$
$$y = \frac{2}{3}x + 1$$

The slope is $\frac{2}{3}$, and the y-intercept is $(0, 1)$. Start at $(0, 1)$ on the y-axis, then rise 2 and run 3 to locate a second point on the line. Because there is only one line containing any two given points, these two points determine the line. See Fig. 7.24. ∎

The three methods that we used for graphing linear equations are summarized as follows.

> **Methods for Graphing a Linear Equation**
>
> 1. Arbitrarily select some points that satisfy the equation, and draw a line through them.
> 2. Find the x- and y-intercepts (provided that they are not the origin), and draw a line through them.
> 3. Start at the y-intercept and use the slope to locate a second point, then draw a line through the two points.

If the y-coordinate of the y-intercept is an integer and the slope is a rational number, then it is usually easiest to use the y-intercept and slope.

Linear Functions

The linear equation $y = mx + b$ with $m \neq 0$ is a formula that shows how to determine a value of y from a value of x. We say that y is a **linear function** of x. Functions in general will be discussed in Chapter 8. In the next example we use the point-slope formula to write Fahrenheit temperature as a linear function of Celsius temperature.

EXAMPLE 8 **Writing a linear function given two points**

Fahrenheit temperature F is a linear function of Celsius temperature C. Water freezes at 0°C or 32°F and boils at 100°C or 212°F. Find the linear equation that expresses F as a linear function of C.

Solution

We want the equation of the line that contains the points $(0, 32)$ and $(100, 212)$ as shown in Fig. 7.25. Use C as the independent variable (x) and F as the dependent variable (y). The slope of the line is

Figure 7.25

$$m = \frac{F_2 - F_1}{C_2 - C_1} = \frac{212 - 32}{100 - 0} = \frac{180}{100} = \frac{9}{5}.$$

Using a slope of $\frac{9}{5}$ and the point $(100, 212)$ in the point-slope formula, we get

$$F - 212 = \frac{9}{5}(C - 100).$$

We can solve this equation for F to get the familiar formula relating Celsius and Fahrenheit temperature:

$$F = \frac{9}{5}C + 32 \qquad \blacksquare$$

Calculator Close-up

Xmin = -10   Ymin = -10
Xmax = 10   Ymax = 10

We can perform computations with either a scientific or a graphing calculator, but only a graphing calculator can graph an equation in its display. To graph $y = 2x - 3$, enter the equation into the calculator using the $\boxed{\text{Y=}}$ key and then press the $\boxed{\text{GRAPH}}$ key. (Because there are many differences in how different brands of graphing calculators graph equations, you should consult your manual at this point.) Your calculator might show a graph of this straight line in its viewing window as in Fig. 7.26. This viewing window has x-values ranging from a minimum of -10 to a maximum of 10 and y-values ranging from a minimum of -10 to a maximum of 10, the **standard viewing window.** If the line $y = 2x - 3$ does not appear, it is probably because your viewing window is set so that it does not contain any part of the line. Consult your manual to see how to adjust your viewing window. Exercises for graphing calculators are included in this section and in many other sections for the remainder of the text.

Warm-ups

True or false? Explain your answer.

1. There is exactly one line through a given point with a given slope.
2. The line $y - a = m(x - b)$ goes through (a, b) and has slope m.
3. The equation of the line through (a, b) with slope m is $y = mx + b$.
4. The x-coordinate of the y-intercept of a nonvertical line is 0.
5. The y-coordinate of the x-intercept of a nonhorizontal line is 0.
6. Every line in the xy-plane has an equation in slope-intercept form.
7. The line $2y + 3x = 7$ has slope $-\frac{3}{2}$.
8. The line $y = 3x - 1$ is perpendicular to the line $y = \frac{1}{3}x - 1$.
9. The line $2y = 3x + 5$ has a y-intercept of $(0, 5)$.
10. Every line in the xy-plane has an equation in standard form.

7.3 EXERCISES

Find the equation of line l in each case and solve it for y. See Examples 1 and 2.

1. Line l goes through $(2, -3)$ and has slope 2.

2. Line l goes through $(-2, 5)$ and has slope 6.

3. Line l goes through $(-2, 3)$ and has slope $-\frac{1}{2}$.

4. Line l goes through $(3, 5)$ and has slope $\frac{2}{3}$.

5. Line l goes through $(2, 3)$ and $(-5, 6)$.

6. Line l goes through $(-2, 1)$ and $(3, -4)$.

7. Line l goes through $(3, 4)$ and is perpendicular to the line through $(-3, 1)$ and $(5, -1)$.

8. Line l goes through $(0, 0)$ and is perpendicular to the line through $(0, 6)$ and $(-5, 0)$.

9. Line l goes through $(0, 0)$ and is parallel to the line through $(9, -3)$ and $(-3, 6)$.

10. Line l goes through $(2, -4)$ and is parallel to the line through $(6, 2)$ and $(-2, 6)$.

In Exercises 11–20, write an equation in slope-intercept form (if possible) for each of the lines shown. See Example 3.

20.

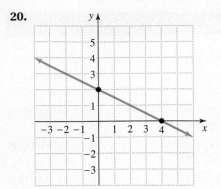

Write each equation in standard form using only integers and a positive coefficient for x. See Example 4.

21. $y = \frac{1}{3}x - 2$ **22.** $y = \frac{1}{2}x + 7$

23. $y - 5 = \frac{1}{2}(x + 3)$

24. $y - 1 = \frac{1}{4}(x - 6)$

25. $y + \frac{1}{2} = \frac{1}{3}(x - 4)$

26. $y + \frac{1}{3} = \frac{1}{4}(x - 3)$

27. $0.05x + 0.06y - 8.9 = 0$

28. $0.03x - 0.07y = 2$

29. $y + 55 = -0.02(x - 600)$

30. $y - 60 = -0.08(x + 200)$

Write each equation in slope-intercept form, and identify the slope and y-intercept. See Example 5.

31. $2x + 5y = 1$ **32.** $3x - 3y = 2$

33. $3x - y - 2 = 0$ **34.** $5 - x - 2y = 0$

35. $y + 3 = 5$ **36.** $y - 9 = 0$

37. $y - 2 = 3(x - 1)$ **38.** $y + 4 = -2(x - 5)$

39. $\dfrac{y - 5}{x + 4} = \dfrac{3}{2}$ **40.** $\dfrac{y - 6}{x - 2} = -\dfrac{3}{5}$

41. $y - \frac{1}{2} = \frac{1}{3}\left(x + \frac{1}{4}\right)$

42. $y - \frac{1}{3} = -\frac{1}{2}\left(x - \frac{1}{4}\right)$

43. $y - 6000 = 0.01(x + 5700)$

44. $y - 5000 = 0.05(x - 1990)$

45. $0.3x - 0.2y = 10$

46. $0.1y - 0.2x = 3.6$

Find the equation of line l in each case and then write it in standard form with integral coefficients. See Example 6.

47. Line l has slope $\frac{1}{2}$ and goes through $(0, 5)$.

48. Line l has slope 5 and goes through $\left(0, \frac{1}{2}\right)$.

49. Line l has x-intercept $(2, 0)$ and y-intercept $(0, 4)$.

50. Line l has y-intercept $(0, 5)$ and x-intercept $(4, 0)$.

51. Line l goes through $(-2, 1)$ and is parallel to $y = 2x + 6$.

52. Line l goes through $(1, -3)$ and is parallel to $y = -3x - 5$.

53. Line l is parallel to $2x + 4y = 1$ and goes through $(-3, 5)$.

54. Line l is parallel to $3x - 5y = -7$ and goes through $(2, 4)$.

55. Line l goes through $(1, 1)$ and is perpendicular to $y = \frac{1}{2}x - 3$.

56. Line l goes through $(-1, -2)$ and is perpendicular to $y = -3x + 7$.

57. Line l goes through $(-2, 3)$ and is perpendicular to $x + 3y = 4$.

58. Line l is perpendicular to $2y + 5 - 3x = 0$ and goes through $(2, 7)$.

59. Line l goes through $(2, 5)$ and is parallel to the x-axis.

60. Line l goes through $(-1, 6)$ and is parallel to the y-axis.

61. Line l is parallel to $x = 3$ and goes through the point $(-3, -2)$.

62. Line l is parallel to $y = 4$ and goes through the point $(-7, 3)$.

Graph each line. Use the slope and y-intercept when possible. See Example 7.

63. $y = \frac{1}{2}x$ **64.** $y = -\frac{2}{3}x$

65. $y = 2x - 3$ **66.** $y = -x + 1$

67. $y = -\frac{2}{3}x + 2$ **68.** $y = 3x - 4$

69. $3y + x = 0$ **70.** $4y - x = 0$

71. $y - x + 3 = 0$ **72.** $x - y = 4$

73. $3x - 2y = 6$ **74.** $3x + 5y = 10$

75. $y + 2 = 0$ **76.** $y - 3 = 0$

77. $x + 3 = 0$ **78.** $x - 5 = 0$

Determine whether each pair of lines is parallel or perpendicular.

79. $y = 3x - 8$, $x + 3y = 7$

80. $y = \dfrac{1}{2}x - 4$, $\dfrac{1}{2}x + \dfrac{1}{4}y = 1$

81. $2x - 4y = 9$, $\dfrac{1}{3}x = \dfrac{2}{3}y - 8$

82. $\dfrac{1}{4}x - \dfrac{1}{6}y = \dfrac{1}{3}$, $\dfrac{1}{3}y = \dfrac{1}{2}x - 2$

83. $x - 6 = 9$, $y - 4 = 12$ **84.** $9 - x = 3$, $\dfrac{1}{2}x = 8$

Solve each problem. See Example 8.

85. ***Heating water.*** Suppose the temperature, t, of a cup of water is a linear function of the number of seconds, s, that it is in the microwave. If the temperature at $s = 0$ seconds is $t = 60°F$ and the temperature at $s = 120$ seconds is $200°F$, find the linear equation that expresses t as a function of s. What should the temperature be after 30 seconds? (*Hint:* Write the equation of the line containing the points $(0, 60)$ and $(120, 200)$ in the form $t = ms + b$.) Draw a graph of this linear function.

86. ***Making circuit boards.*** The accountant at Apollo Manufacturing has determined that the cost, C, per week in dollars for making circuit boards is a linear function of the number, n, of circuit boards produced in a week. If $C = \$1500$ when $n = 1000$, and $C = \$2000$ when $n = 2000$, find the linear equation that expresses C in terms of n. What is the cost if they produce only one circuit board in a week? Draw a graph of this linear function.

87. ***Depth and flow.*** On September 1, 1993, the depth of the water in the Tangipahoa River at Robert, Louisiana, was 8.86 feet and the flow was 1204 cubic feet per second. On September 23, 1993, the depth was 7.07 feet, and the flow was 659 cubic feet per second (U.S. Geological Survey, *Water Resources Data for Louisiana*, 1993). For depths between 7 and 9 feet, the flow w is a linear function of the depth d. Write the equation of the line through $(8.86, 1204)$ and $(7.07, 659)$ and express w as a linear function of d. What is the flow when the depth is 8.24 feet?

88. ***Buying stock.*** On January 11, 1994, Ford Motor stock sold for $68.25 per share, and General Motors sold for $59.50 per share (*The Times-Picayune*, January 12, 1994). An investor bought x shares of Ford and y shares of GM on that day for a total of $64,974. Write a linear equation that models this situation. If 476 shares of Ford were purchased, then how many shares of GM were purchased?

Getting More Involved

89. ***Exploration.*** Plot the points $(1, 1)$, $(2, 3)$, $(3, 4)$, $(4, 6)$, and $(5, 7)$ on graph paper. Use a ruler to draw a straight line that "best fits" the five points. The line drawn does not necessarily have to go through any of the five points.

a) Estimate the slope and y-intercept for the line drawn and write an equation for the line in slope-intercept form.

b) For each x-coordinate from 1 through 5, find the difference between the given y-coordinate and the y-coordinate on your line.

c) To determine how well you have done, square each difference that you found in part (b) and then find the sum of those squares. Compare your sum with your classmates' sums. The person with the smallest sum has done the best job of fitting a line to the five given points.

Graphing Calculator Exercises

90. Graph the equation $y = 0.5x - 1$ using the standard viewing window. Adjust the range of y-values so that the line goes from the lower left corner of your viewing window to the upper right corner.

91. Graph $y = x - 3000$, using a viewing window that shows both the x-intercept and the y-intercept.

92. Graph $y = 2x - 400$ and $y = -0.5x + 1$ on the same screen, using the viewing window $-500 \le x \le 500$ and $-1000 \le y \le 1000$. Should these lines be perpendicular? Explain.

93. The lines $y = 2x - 3$ and $y = 1.9x + 2$ are not parallel. Find a viewing window in which the lines intersect. Estimate the point of intersection.

7.4 Linear Inequalities and Their Graphs

In the first three sections of this chapter you studied linear equations. We now turn our attention to linear inequalities.

Definition

A linear inequality is a linear equation with the equal sign replaced by an inequality symbol.

> **Linear Inequality**
>
> If A, B, and C are real numbers with A and B not both zero, then
>
> $$Ax + By \leq C$$
>
> is called a **linear inequality.** In place of $\leq$, we can also use $\geq$, $<$, or $>$.

Graphing Linear Inequalities

Consider the inequality $-x + y > 1$. If we solve the inequality for y, we get

$$y > x + 1.$$

Which points in the xy-plane satisfy this inequality? We want the points where the y-coordinate is larger than the x-coordinate plus 1. If we locate a point on the line $y = x + 1$, say $(2, 3)$, then the y-coordinate is equal to the x-coordinate plus 1. If we move upward from that point, to say $(2, 4)$, the y-coordinate is larger than the x-coordinate plus 1. Because this argument can be made at every point on the line, all points above the line satisfy $y > x + 1$. Likewise, points below the line satisfy $y < x + 1$. The solution sets, or graphs, for the inequality $y > x + 1$ and the inequality $y < x + 1$ are the shaded regions shown in Figs. 7.27(a) and 7.27(b). In each case the line $y = x + 1$ is dashed to indicate that points on the line do not satisfy the inequality and so are not in the solution set. If the inequality symbol is $\leq$ or $\geq$, then points on the boundary line also satisfy the inequality, and the line is drawn solid.

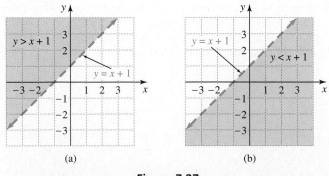

(a)   (b)

Figure 7.27

Every nonvertical line divides the xy-plane into two regions. One region is above the line, and the other is below the line. A vertical line also divides the plane into two regions, but one is on the left side of the line and the other is on the right side of the line. An inequality involving only x has a vertical boundary line, and its graph is one of those regions.

► Graphing a Linear Inequality ◄

1. Solve the inequality for y, then graph $y = mx + b$.

   $y > mx + b$ is satisfied above the line.

   $y = mx + b$ is satisfied on the line itself.

   $y < mx + b$ is satisfied below the line.

2. If the inequality involves x and not y, then graph the vertical line $x = k$.

   $x > k$ is satisfied to the right of the line.

   $x = k$ is satisfied on the line itself.

   $x < k$ is satisfied to the left of the line.

EXAMPLE 1   **Graphing linear inequalities**

Graph each inequality.

a) $y < \dfrac{1}{2}x - 1$ **b)** $y \geq -2x + 1$ **c)** $3x - 2y < 6$

Solution

a) The set of points satisfying this inequality is the region below the line $y = \dfrac{1}{2}x - 1$. To show this region, we first graph the boundary line $y = \dfrac{1}{2}x - 1$. The slope of the line is $\dfrac{1}{2}$, and the y-intercept is $(0, -1)$. Start at $(0, -1)$ on the y-axis, then rise 1 and run 2 to get a second point of the line. We draw the line dashed because points on the line do not satisfy this inequality. The solution set to the inequality is the shaded region shown in Fig. 7.28.

b) Because the inequality symbol is $\geq$, every point on or above the line satisfies this inequality. To show that the line $y = -2x + 1$ is included, we make it a solid line. See Fig. 7.29.

Figure 7.28   **Figure 7.29**

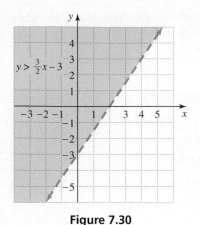

Figure 7.30

c) First solve for y:

$$3x - 2y < 6$$

$$-2y < -3x + 6$$

$$y > \frac{3}{2}x - 3 \quad \text{Divide by } -2 \text{ and reverse the inequality.}$$

To graph this inequality, use a dashed line for the boundary $y = \frac{3}{2}x - 3$ and shade the region above the line. See Fig. 7.30 for the graph. ∎

CAUTION In Example 1(c) we solved the inequality for y before graphing the line. We did that because $<$ corresponds to the region below the line and $>$ corresponds to the region above the line only when the inequality is solved for y. ⊘

EXAMPLE 2 **Inequalities with horizontal and vertical boundaries**

Graph the inequalities.

a) $y \le 5$ **b)** $x > 4$

Solution

a) The line $y = 5$ is the horizontal line with y-intercept $(0, 5)$. Draw a solid horizontal line and shade below it as in Fig. 7.31.

b) The points that satisfy $x > 4$ lie to the right of the vertical line $x = 4$. The solution set is shown in Fig. 7.32.

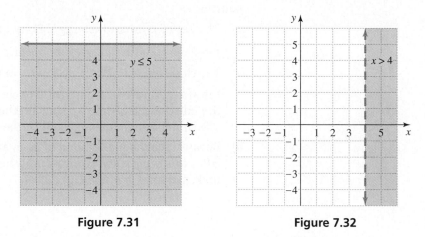

Figure 7.31   **Figure 7.32** ∎

The Test Point Method

The graph of any line $Ax + By = C$ separates the xy-plane into two regions. Every point on one side of the line satisfies the inequality $Ax + By < C$, and every point on the other side satisfies the inequality $Ax + By > C$. We can use these facts to graph an inequality by the **test point method:**

1. Graph the corresponding equation.
2. Choose any point *not* on the line.
3. Test to see whether the point satisfies the inequality.

If the point satisfies the inequality, then the solution set is the region containing the test point. If not, then the solution set is the other region. With this method it is not necessary to solve the inequality for y.

EXAMPLE 3 **Using the test point method**

Graph the inequality $3x - 4y > 7$.

Solution

First graph the equation $3x - 4y = 7$ using the x-intercept and the y-intercept. If $x = 0$, then $y = -\frac{7}{4}$. If $y = 0$, then $x = \frac{7}{3}$. Use the x-intercept $\left(\frac{7}{3}, 0\right)$ and the y-intercept $\left(0, -\frac{7}{4}\right)$ to graph the line as shown in Fig. 7.33(a). Select a point on one side of the line, say $(0, 1)$, to test in the inequality. Because

$$3(0) - 4(1) > 7$$

is false, the region on the other side of the line satisfies the inequality. The graph of $3x - 4y > 7$ is shown in Fig. 7.33(b).

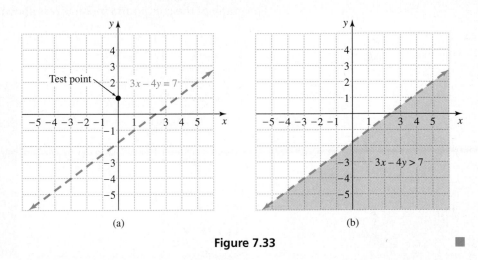

(a)   (b)

Figure 7.33   ■

Graphing Compound Inequalities

We can write compound inequalities with two variables just as we do for one variable. For example,

$$y > x - 3 \text{ and } y < -\frac{1}{2}x + 2$$

is a compound inequality. Because the inequalities are connected by the word *and*, a point is in the solution set to the compound inequality if and only if it is in the solution sets to *both* of the individual inequalities. So the graph of this compound inequality is the intersection of the solution sets to the individual inequalities.

EXAMPLE 4 **Graphing a compound inequality with *and***

Graph the compound inequality $y > x - 3$ and $y < -\frac{1}{2}x + 2$.

Solution

We first graph the equations $y = x - 3$ and $y = -\frac{1}{2}x + 2$. These lines divide the plane into four regions as shown in Fig. 7.34(a). Now test one point of each region to determine which region satisfies the compound inequality. Test the points $(3, 3)$, $(0, 0)$, $(4, -5)$, and $(5, 0)$:

$$3 > 3 - 3 \text{ and } \quad 3 < -\frac{1}{2} \cdot 3 + 2 \qquad \text{Second inequality is incorrect.}$$

$$0 > 0 - 3 \text{ and } \quad 0 < -\frac{1}{2} \cdot 0 + 2 \qquad \text{Both inequalities are correct.}$$

$$-5 > 4 - 3 \text{ and } -5 < -\frac{1}{2} \cdot 4 + 2 \qquad \text{First inequality is incorrect.}$$

$$0 > 5 - 3 \text{ and } \quad 5 < -\frac{1}{2} \cdot 0 + 2 \qquad \text{Both inequalities are incorrect.}$$

The only point that satisfies both inequalities is $(0, 0)$. So the solution set to the compound inequality consists of all points in the region containing $(0, 0)$. The graph of the compound inequality is shown in Fig. 7.34(b).

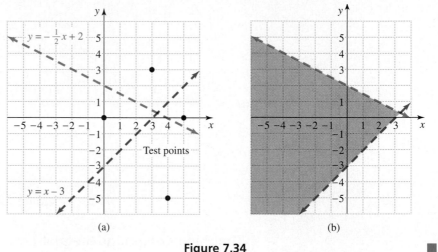

(a)   (b)

Figure 7.34   ■

Compound inequalities are also formed by connecting individual inequalities with the word *or*. A point satisfies a compound inequality connected by *or* if and only if it satisfies one or the other or both of the individual inequalities. The graph is the union of the graphs of the individual inequalities.

EXAMPLE 5   **Graphing a compound inequality with *or***

Graph the compound inequality

$$2x - 3y \leq -6 \text{ or } x + 2y \geq 4.$$

Solution

First graph the lines $2x - 3y = -6$ and $x + 2y = 4$. If we graph the lines using x- and y-intercepts, then we do not have to solve the equations for y. The lines are shown in Fig. 7.35(a). The graph of the compound inequality is the set of all points that satisfy either one inequality or the other (or both). Test the points $(0, 0)$, $(4, 2)$, $(0, 5)$, and $(-3, 2)$. You should verify that only $(0, 0)$ fails to satisfy at least one of the inequalities. So only the region containing the origin is left unshaded. The graph of the compound inequality is shown in Fig. 7.35(b).

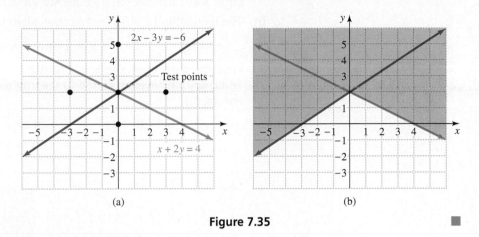

(a)   (b)

Figure 7.35

In the next example we graph absolute value inequalities by writing equivalent compound inequalities.

EXAMPLE 6   **Graphing absolute value inequalities**

Graph each absolute value inequality.

a) $|y - 2x| \leq 3$ **b)** $|x - y| > 1$

Solution

a) The inequality $|y - 2x| \leq 3$ is equivalent to $-3 \leq y - 2x \leq 3$, which is equivalent to the compound inequality

$$y - 2x \leq 3 \text{ and } y - 2x \geq -3.$$

First graph the lines $y - 2x = 3$ and $y - 2x = -3$ as shown in Fig. 7.36(a). These lines divide the plane into three regions. Test a point from each region in the original inequality, say $(-5, 0)$, $(0, 1)$, and $(5, 0)$:

$$|0 - 2(-5)| \leq 3 \qquad |1 - 2 \cdot 0| \leq 3 \qquad |0 - 2 \cdot 5| \leq 3$$
$$10 \leq 3 \qquad\qquad\qquad 1 \leq 3 \qquad\qquad\qquad 10 \leq 3$$

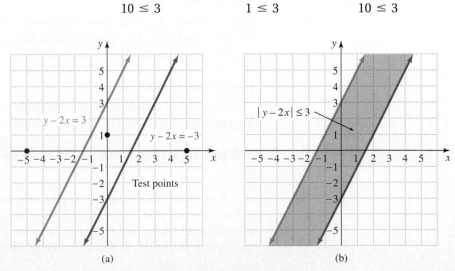

(a)   (b)

Figure 7.36

Only $(0, 1)$ satisfies the original inequality. So the region satisfying the absolute value inequality is the shaded region containing $(0, 1)$ shown in Fig. 7.36(b). The boundary lines are solid because of the $\leq$ symbol.

b) The inequality $|x - y| > 1$ is equivalent to

$$x - y > 1 \text{ or } x - y < -1.$$

First graph the lines $x - y = 1$ and $x - y = -1$ as shown in Fig. 7.37(a). Test a point from each region in the original inequality, say $(-4, 0)$, $(0, 0)$, and $(4, 0)$:

$$|-4 - 0| > 1 \qquad |0 - 0| > 1 \qquad |4 - 0| > 1$$
$$4 > 1 \qquad\qquad 0 > 1 \qquad\qquad 4 > 1$$

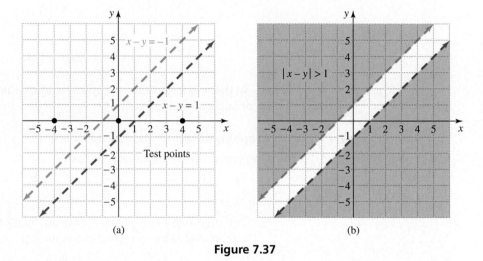

(a)   (b)

Figure 7.37

Because $(-4, 0)$ and $(4, 0)$ satisfy the inequality, we shade those regions as shown in Fig. 7.37(b). The boundary lines are dashed because of the $>$ symbol. ∎

Applications

In real situations, x and y often represent quantities or amounts, which cannot be negative. In this case our graphs are restricted to the first quadrant, where x and y are both nonnegative.

EXAMPLE 7 **Inequalities in business**

The manager of a furniture store can spend a maximum of $3000 on advertising per week. It costs $50 to run a 30-second ad on an AM radio station and $75 to run the ad on an FM station. Graph the region that shows the possible numbers of AM and FM ads that can be purchased.

Solution

If x represents the number of AM ads and y represents the number of FM ads, then x and y must satisfy the inequality $50x + 75y \leq 3000$. Because the number of ads cannot be negative, we also have $x \geq 0$ and $y \geq 0$. So we graph only points in the first quadrant that satisfy $50x + 75y \leq 3000$. The line $50x + 75y = 3000$ goes through $(0, 40)$ and $(60, 0)$. The inequality is satisfied below this line. The region showing the possible numbers of AM ads and FM ads is shown in Fig. 7.38. We shade the entire region in Fig. 7.38, but only points in the shaded region in which both coordinates are whole numbers actually satisfy the given condition.

Figure 7.38

Warm-ups

True or false? Explain your answer.

1. The point $(2, -3)$ satisfies the inequality $y > -3x + 2$.
2. The graph of $3x - y > 2$ is the region above the line $3x - y = 2$.
3. The graph of $3x + y < 5$ is the region below the line $y = -3x + 5$.
4. The graph of $x < -3$ is the region to the left of the vertical line $x = 3$.
5. The graph of $y > x + 3$ and $y < 2x - 6$ is the intersection of two regions.
6. The graph of $y \leq 2x - 3$ or $y \geq 3x + 5$ is the union of two regions.
7. The ordered pair $(2, -5)$ satisfies $y > -3x + 5$ and $y < 2x - 3$.
8. The ordered pair $(-3, 2)$ satisfies $y \leq 3x - 6$ or $y \leq x + 5$.
9. The inequality $|2x - y| \leq 4$ is equivalent to $2x - y \leq 4$ and $2x + y \leq 4$.
10. The inequality $|x - y| > 3$ is equivalent to $x - y > 3$ or $x - y < -3$.

7.4 EXERCISES

Graph each linear inequality. See Examples 1 and 2.

1. $y < x + 2$
2. $y < x - 1$
3. $y \leq -2x + 1$
4. $y \geq -3x + 4$
5. $x + y > 3$
6. $x + y \leq -1$
7. $2x + 3y < 9$
8. $-3x + 2y > 6$
9. $3x - 4y \leq 8$
10. $4x - 5y > 10$
11. $x - y > 0$
12. $2x - y < 0$
13. $x \geq 1$
14. $x < 0$
15. $y < 3$
16. $y > -1$

Graph each linear inequality by using a test point. See Example 3.

17. $2x - 3y < 5$
18. $5x - 4y > 3$
19. $x + y + 3 \geq 0$
20. $x - y - 6 \leq 0$
21. $\frac{1}{2}x + \frac{1}{3}y < 1$
22. $2 - \frac{2}{5}y > \frac{1}{2}x$

Graph each compound inequality. See Examples 4 and 5.

23. $y > x$ and $y > -2x + 3$
24. $y < x$ and $y < -3x + 2$
25. $y < x + 3$ or $y > -x + 2$
26. $y \geq x - 5$ or $y \leq -2x + 1$
27. $x + y \leq 5$ and $x - y \leq 3$
28. $2x - y < 3$ and $3x - y > 0$
29. $x - 2y \leq 4$ or $2x - 3y \leq 6$
30. $4x - 3y \leq 3$ or $2x + y \geq 2$
31. $y > 2$ and $x < 3$

32. $x \leq 5$ and $y \geq -1$
33. $y \geq x$ and $x \leq 2$
34. $y < x$ and $y > 0$
35. $2x < y + 3$ or $y > 2 - x$
36. $3 - x < y + 2$ or $x > y + 5$
37. $x - 1 < y < x + 3$
38. $x - 1 < y < 2x + 5$
39. $0 \leq y \leq x$ and $x \leq 1$
40. $x \leq y \leq 1$ and $x \geq 0$
41. $1 \leq x \leq 3$ and $2 \leq y \leq 5$
42. $-1 < x < 1$ and $-1 < y < 1$

Graph the absolute value inequalities. See Example 6.

43. $|x + y| < 2$
44. $|2x + y| < 1$
45. $|2x + y| \geq 1$
46. $|x + 2y| \geq 6$
47. $|x - y - 3| > 5$
48. $|x - 2y + 4| > 2$
49. $|x - 2y| \leq 4$
50. $|x - 3y| \leq 6$
51. $|x| > 2$
52. $|x| \leq 3$
53. $|y| < 1$
54. $|y| \geq 2$
55. $y < |x|$
56. $y < |x + 3|$
57. $|x| < 2$ and $|y| < 3$
58. $|x| \geq 3$ or $|y| \geq 1$
59. $|x - 3| < 1$ and $|y - 2| < 1$
60. $|x - 2| \geq 3$ or $|y - 5| \geq 2$

Solve each problem. See Example 7.

61. *Budget planning.* The Highway Patrol can spend a maximum of $120,000 on new vehicles this year. They can get a fully equipped compact car for $15,000 or a fully equipped full-size car for $20,000. Graph the region that shows the number of cars of each type that could be purchased.

62. *Allocating resources.* A furniture maker has a shop that can employ 12 workers for 40 hours per week at its maximum capacity. The shop makes tables and chairs. It takes 16 hours of labor to make a table and 8 hours of labor to make a chair. Graph the region that shows the possibilities for the number of tables and chairs that could be made in one week.

63. *More restrictions.* In Exercise 61, add the condition that the number of full-size cars must be greater than the number of compact cars. Graph the region showing the possibilities for the number of cars of each type that could be purchased.

64. *Chairs per table.* In Exercise 62, add the condition that the number of chairs must be at least four times the number of tables and at most six times the number of tables. Graph the region showing the possibilities for the number of tables and chairs that could be made in one week.

65. *Building fitness.* To achieve cardiovascular fitness, you should exercise so that your target heart rate is between 70% and 85% of its maximum rate. Your target heart rate h depends on your age a. For building fitness you should have $h \leq 187 - 0.85a$ and $h \geq 154 - 0.70a$ (NordicTrack brochure). Graph this compound inequality for $20 \leq a \leq 75$ to see the heart rate target zone for building fitness.

66. *Waist-to-hip ratio.* A study by Dr. Aaron R. Folsom concluded that waist-to-hip ratios are a better predictor of five-year survival than more traditional height-to-weight ratios (*New York Times*, January 27, 1993). Dr. Folsom concluded that for good health the waist size of a woman aged 50–69 should be less than 80% of her hip size, $w \leq 0.80h$. Make a graph showing possible waist and hip sizes for good health for women in this group for which hip size is not more than 50 inches.

67. *Advertising dollars.* A restaurant manager can spend at most $9000 on advertising per month and has two choices for advertising. The manager can purchase an ad in the *Daily Chronicle* (a seven-day-per-week newspaper) for $300 per day or a 30-second ad on WBTU television for $1000 each time the ad is aired. Graph the region that shows the possible number of days that an ad can be run in the newspaper and the possible number of times that an ad can be aired on television.

68. *More restrictions.* How many solutions are there to Exercise 67 if you assume that the number of days that an ad appears in the Daily Chronicle is a whole number and the number of times an ad is aired on WBTU is a whole number? (The manager does not have to spend the entire $9000.) How many solutions are there if you also know that the manager is advertising in February 1997?

Getting More Involved

69. *Writing.* Explain the difference between a compound inequality using the word *and* and a compound inequality using the word *or*.

70. *Discussion.* Explain how to write an absolute value inequality as a compound inequality.

COLLABORATIVE ACTIVITIES

Parallel and Perpendicular Explorations

Grouping: 2 students per group
Topic: Parallel and perpendicular lines

Part I

One person in your pair will do the graphing (**G**—Grapher) while the other one is answering the questions (**A**—Answerer). Discuss your answers so that you both know how to find them. Use either graph paper or a graphing calculator so that your graphs will be accurate.

1. **G:** Graph the line through the points (2, 3) and (−1, 4). We will call this line S. **A:** Find the equation for line S.

2. **G:** Plot points three units up (positive y-direction) from each of the points given on line S and draw a line through the new points. We will call this line P. **A:** What are the new points? Find the equation for line P.

3. **G:** On the same graph, plot points five units to the left (negative x-direction) from each of the points given on line S and draw a line through the new points. We will call this line L. **A:** What are the new points? Find the equation for line L.

4. **G, A:** Looking only at your graph, state how these three lines are the same and how they are different. Make sure your equations are in slope-intercept form. State how the equations are the same and how they are different.

5. **G, A:** Find another line parallel to line S. Graph it and state its equation.

Part II

Switch roles and get a new sheet of graph paper or clear the graph on your calculator.

6. **G:** Graph the line through the points (3, 1) and (−1, −2). We will call this line T. **A:** Find the equation for line T.

7. **G:** Plot one point four units down (negative y-direction) and three units to the right (positive x-direction) of the point (3, 1). Draw the line connecting (3, 1) and your new point. Call this line Q. **A:** What is this point? Find the equation for line Q. Explain how the direction and number of units you moved from (3, 1) relate to the slope of your new equation.

8. **G:** Plot the point (−4, 2) and draw a line from the point (−1, −2) to this new point. Call this line M. **A:** Find the equation for line M.

9. **G, A:** Looking at the graph, state how the three lines are related. State how their equations compare.

10. **G, A:** Starting from any point on line Q, how many units left or right (negative or positive x-direction) and how many units up or down (positive or negative y-direction) would you need to move to get to a point that would be on a line perpendicular to line Q? Find a line perpendicular to line Q. Graph it and state its equation. How does it compare to line T?

Wrap-up CHAPTER 7

SUMMARY

Rectangular Coordinate System		Examples
Midpoint formula	The midpoint of the line segment with endpoints (x_1, y_1) and (x_2, y_2) is $$\left(\frac{x_1 + x_2}{2}, \frac{y_1 + y_2}{2} \right).$$	Line segment with endpoints $(1, -2)$ and $(3, -4)$ has midpoint $(2, -3)$.
Distance formula	The distance between (x_1, y_1) and (x_2, y_2) is $$\sqrt{(x_2 - x_1)^2 + (y_2 - y_1)^2}$$	Distance between $(1, -2)$ and $(3, -4)$ is $\sqrt{4 + 4}$, or $2\sqrt{2}$.

| x-intercept | The point where a nonhorizontal line intersects the x-axis | For the line $2x + y = 6$, the x-intercept is $(3, 0)$ and the y-intercept is $(0, 6)$. |
| y-intercept | The point where a nonvertical line intersects the y-axis | |

Slope

Examples

Slope of a line	$$\text{Slope} = \frac{\text{change in } y\text{-coordinate}}{\text{change in } x\text{-coordinate}}$$ $$= \frac{\text{rise}}{\text{run}}$$	
Slope using coordinates	Slope of line through (x_1, y_1) and (x_2, y_2) is $$m = \frac{y_2 - y_1}{x_2 - x_1}, \qquad \text{provided that } x_2 - x_1 \neq 0.$$	If $(x_1, y_1) = (4, -2)$ and $(x_2, y_2) = (3, -6)$, then $$m = \frac{-6 - (-2)}{3 - 4} = 4.$$
Types of slope	Positive slope · Negative slope · Zero slope · Undefined slope	
Perpendicular lines	The slope of one line is the negative of the reciprocal of the slope of the other line.	The lines $y = -\frac{1}{3}x + 5$ and $y = 3x - 9$ are perpendicular.
Parallel lines	Nonvertical parallel lines have equal slopes.	The lines $y = 2x - 3$ and $y = 2x + 7$ are parallel.

Forms of Linear Equations

Examples

Point-slope form	$y - y_1 = m(x - x_1)$ (x_1, y_1) is a point on the line, and m is the slope	Line through $(5, -3)$ with slope 2: $y + 3 = 2(x - 5)$
Slope-intercept form	$y = mx + b$ m is the slope, and $(0, b)$ is the y-intercept	Line through $(0, -3)$ with slope 2: $y = 2x - 3$
Standard form	$Ax + By = C$ A and B are not both 0.	$3x - 2y = 12$
Vertical line	$x = k$, where k is any real number. Slope is undefined for vertical lines.	$x = 5$
Horizontal line	$y = k$, where k is any real number. Slope is zero for horizontal lines.	$y = -2$

Graphing Linear Equations

		Examples
Point-plotting	Arbitrarily select some points that satisfy the the equation, and draw a line through them.	For $y = 2x + 1$, draw a line through $(0, 1)$, $(1, 3)$, and $(2, 5)$.
Intercepts	Find the x- and y-intercepts (provided that they are not the origin), and draw a line through them.	For $x + y = 4$, the intercepts are $(0, 4)$ and $(4, 0)$.
y-intercept and slope	Start at the y-intercept and use the slope to locate a second point, then draw a line through the two points.	For $y = 3x - 2$, start at $(0, -2)$, rise 3 and run 1 to get to $(1, 1)$. Draw a line through the two points.

Linear Inequalities

		Examples
Linear inequality	$Ax + By \leq C$, where A and B are not both zero. The symbols $<$, $>$, and $\geq$ are also used.	$2x - 3y \leq 7$ $x - y > 6$
Graphing linear inequalities	Solve for y, then graph the line $y = mx + b$. $\quad y > mx + b$ is the region above the line. $\quad y < mx + b$ is the region below the line.	Graph of $y = x + 2$ is a line. $y > x + 2$ above $y = x + 2$. $y < x + 2$ below $y = x + 2$.
	For inequalities without y, graph $x = k$. $\quad x > k$ is the region to the right of $x = k$. $\quad x < k$ is the region to the left of $x = k$.	The graph of $x > 5$ is to the right of the vertical line $x = 5$, and the graph of $x < 5$ is to left of $x = 5$.
Test points	A linear inequality may also be graphed by graphing the corresponding line and then testing a point to determine which region satisfies the inequality.	
Compound inequalities	Simple inequalities connected by *and* or *or* For *and*, find the intersection of the regions. For *or*, find the union of the regions.	$x < 3$ and $y > x - 4$ $x + y \geq 1$ or $y \leq x - 1$

REVIEW EXERCISES

7.1 *For each point, name the quadrant in which it lies or the axis on which it lies.*

1. $(-3, -2)$ **2.** $(0, \pi)$ **3.** $(\pi, 0)$

4. $(-5, 4)$ **5.** $(0, -1)$ **6.** $\left(\dfrac{\pi}{2}, 1\right)$

7. $(\sqrt{2}, -3)$ **8.** $(6, -3)$

Find the midpoint and length of the line segment with the given endpoints.

9. $(2, 5), (1, 3)$ **10.** $(3, -1), (-5, 2)$

11. $(-6, 0), (0, -8)$ **12.** $(-2, -4), (4, -2)$

Complete the given ordered pairs so that each ordered pair satisfies the given equation.

13. $(0, \ \ ), (\ \ , 0), (4, \ \ ), (\ \ , -3),\quad y = -3x + 2$

14. $(0, \ \ ), (\ \ , 0), (-6, \ \ ), (\ \ , 5),\quad 2x + 3y = 5$

7.2 *Find the slope of the line through each pair of points.*

15. $(-5, 6), (-2, 9)$ **16.** $(-2, 7), (3, -4)$

17. $(4, 1), (-3, -2)$ **18.** $(6, 0), (0, -3)$

Solve each problem.

19. What is the slope of any line that is parallel to the line through $(-3, -4)$ and $(5, -1)$?

20. What is the slope of the line through $(4, 6)$ that is parallel to the line through $(-2, 1)$ and $(7, 1)$?

21. What is the slope of any line that is perpendicular to the line through $(-3, 5)$ and $(4, -6)$?

22. What is the slope of the line through $(1, 2)$ that is perpendicular to the line through $(5, 4)$ and $(5, -2)$?

7.3 *Find the slope and y-intercept for each line.*

23. $y = -3x + 4$

24. $2y - 3x + 1 = 0$

25. $y - 3 = \frac{2}{3}(x - 1)$

26. $y - 3 = 5$

Write each equation in standard form with integral coefficients.

27. $y = \frac{2}{3}x - 4$

28. $y = -0.05x + 0.26$

29. $y - 1 = \frac{1}{2}(x + 3)$

30. $\frac{1}{2}x - \frac{1}{3}y = \frac{1}{4}$

Write the equation of the line containing the given point and having the given slope. Rewrite each equation in standard form with integral coefficients.

31. $(1, -3)$, $m = \frac{1}{2}$

32. $(0, 2)$, $m = 3$

33. $(-2, 6)$, $m = -\frac{3}{4}$

34. $\left(2, \frac{1}{2}\right)$, $m = \frac{1}{4}$

35. $(3, 5)$, $m = 0$

36. $(0, 0)$, $m = -1$

Graph each equation.

37. $y = 2x - 3$

38. $y = \frac{2}{3}x + 1$

39. $3x - 2y = -6$

40. $4x + 5y = 10$

41. $y - 3 = 10$

42. $2x = 8$

43. $5x - 3y = 7$

44. $3x + 4y = -1$

45. $5x + 4y = 100$

46. $2x - y = 120$

47. $x - 80y = 400$

48. $75x + y = 300$

7.4 *Graph each linear inequality.*

49. $y > 3x - 2$

50. $y \leq 2x + 3$

51. $x - y \leq 5$

52. $2x + y > 1$

53. $3x > 2$

54. $x + 2 \leq 0$

55. $4y \leq 0$

56. $4y - 4 > 0$

57. $4x - 2y \geq 6$

58. $-5x - 3y > 6$

59. $5x - 2y < 9$

60. $3x + 4y \leq -1$

Graph each compound or absolute value inequality.

61. $y > 3$ and $y - x < 5$

62. $x + y \leq 1$ or $y \leq 4$

63. $3x + 2y \geq 8$ or $3x - 2y \leq 6$

64. $x + 8y > 8$ and $x - 2y < 10$

65. $|x + 2y| < 10$

66. $|x - 3y| \geq 9$

67. $|x| \leq 5$

68. $|y| > 6$

69. $|y - x| > 2$

70. $|x - y| \leq 1$

Miscellaneous

Write an equation in standard form with integral coefficients for each line described.

71. The line that crosses the x-axis at $(2, 0)$ and the y-axis at $(0, -6)$

72. The line with an x-intercept of $(4, 0)$ and slope $-\frac{1}{2}$

73. The line through $(-1, 4)$ with slope $-\frac{1}{2}$

74. The line through $(2, -3)$ with slope 0

75. The line through $(2, -6)$ and $(2, 5)$

76. The line through $(-3, 6)$ and $(4, 2)$

77. The line through $(0, 0)$ perpendicular to $x = 5$

78. The line through $(2, -3)$ perpendicular to $y = -3x + 5$

79. The line through $(-1, 4)$ parallel to $y = 2x + 1$

80. The line through $(2, 1)$ perpendicular to $y = 10$

For Exercises 81–84, write an equation in standard form with integral coefficients for each line.

81.

82.

83.

84.

Use slope or distance to solve each geometric problem.

85. Show that the points $(-5, -5)$, $(-3, -1)$, $(6, 2)$, and $(4, -2)$ are the vertices of a parallelogram.

86. Show that the points $(-5, -5)$, $(4, -2)$, and $(3, 1)$ are the vertices of a right triangle.

87. Show that the points $(-4, -3)$, $(4, -1)$, and $(2, 7)$ are the vertices of an isosceles triangle.

88. Show that the points $(-2, 2)$, $(-1, -1)$, $(1, 3)$, and $(2, 0)$ are the vertices of a square.

89. Show that the points $(-1, 0)$, $(3, 0)$, and $\left(1, 2\sqrt{3}\,\right)$ are the vertices of an equilateral triangle.

90. Show that the points $(-2, 2)$, $(0, 0)$, $(2, 6)$, and $(4, 4)$ are the vertices of a rectangle.

91. Find the perimeter of the triangle with vertices $(-4, -3)$, $(4, -1)$, and $(2, 7)$.

92. Determine whether the points $(2, 1)$, $(4, 7)$, and $(-3, -14)$ lie on a straight line.

Solve the following problems.

93. *Maximum heart rate.* The maximum heart rate during exercise for a 20-year old is 200, and the maximum heart rate for a 70-year old is 150 (NordicTrack brochure). Assuming that the maximum heart rate h is a linear function of age a, write a linear equation expressing h in terms of a. What is the maximum heart rate for a 40-year-old?

94. *Resting heart rate.* A subject is given 3 mg of an experimental drug, and a resting heart rate of 82 is recorded. Another subject is given 5 mg of the same drug, and a resting heart rate of 89 is recorded. If we assume that the heart rate, h, is a linear function of the dosage, d, find the linear equation expressing h in terms of d. If a subject is given 10 mg of the drug, what would the expected heart rate be?

95. *Rental costs.* The charge, C, in dollars, for renting an air hammer from the Tools Is Us Rental Company is determined from the formula $C = 26 + 17d$, where d is the number of days in the rental period. Graph this function for d from 1 to 30. If the air hammer is worth \$1080, then in how many days would the rental charge equal the value of the air hammer?

96. *Waist-to-hip ratio.* Dr. Aaron R. Folsom, from the University of Minnesota School of Public Health, has concluded that for a man aged 50–69 to be in good health, his maximum waist size w should be 95% of his hip size h (*New York Times*, January 27, 1993). Write w as a linear function of h. If a man in this group has a waist of 38 inches, then what is his maximum hip size for good health?

CHAPTER 7 TEST

Complete each ordered pair so that it satisfies the given equation.

1. $(0, \quad)$, $(\quad, 0)$, $(4, \quad)$, $(\quad, -8)$, $2x + y = 5$

Find the midpoint and length of each line segment with the given endpoints.

2. $(3, -2)$, $(5, -6)$ **3.** $(0, 3)$, $(-4, 0)$

Find the slope of the line that contains each pair of points.

4. $(-3, 7)$, $(2, 1)$ **5.** $(0, -6)$, $(4, 2)$

Determine the slope and y-intercept for each line.

6. $y = -\dfrac{1}{2}x - 2$ **7.** $8x - 5y = -10$

For each line described below, write its equation in standard form with integral coefficients.

8. The line with y-intercept $(0, 3)$ and slope $-\dfrac{1}{2}$

9. The line through $(-3, 5)$ with slope -4

10. The line through $(4, 1)$ and $(-2, -5)$

11. The line through $(2, 3)$ that is perpendicular to $3x - 5y = 7$

12. The line through $(-4, 6)$ parallel to $y = -1$

13. The line shown in the graph:

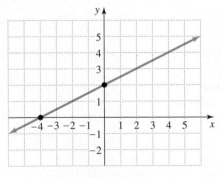

Figure for Exercise 13

Use the slope or distance formula to solve each problem.

14. Show that $(-1, -2)$, $(2, -1)$, and $(1, 2)$ are the vertices of a right triangle.

15. Show that $(-1, -2)$, $(0, 0)$, $(6, 2)$, and $(5, 0)$ are the vertices of a parallelogram.

Sketch the graph of each line.

16. $y = \dfrac{1}{3}x - 2$ **17.** $5x - 2y = 7$

Sketch the graph of each inequality.

18. $y > -\dfrac{1}{2}x + 3$ **19.** $3x - 2y - 6 \le 0$

Sketch the graph of each compound or absolute value inequality.

20. $x > 2$ and $x + y > 0$ **21.** $|2x + y| \ge 3$

22. Suppose the value, V, in dollars of a boat is a linear function of its age, a, in years. If a boat was valued at $22,000 brand new and is worth $16,000 when it is three years old, find the linear equation that expresses V in terms of a.

Simplify each expression.

1. -5^4

2. $125^{2/3}$

3. $\dfrac{4 + \sqrt{8}}{4}$

4. $\dfrac{8x - 2}{4}$

5. $(x + h)^2$

6. $(a + 1)^2 - (a + 1)$

7. $\dfrac{(x + h)^2 - x^2}{h}$

8. $\dfrac{3(x + h)^2 - 2(x + h) - (3x^2 - 2x)}{h}$

Sketch the graph of each equation.

9. $y = \dfrac{1}{3}x$

10. $y = 3x$

11. $y = 2x + 5$

12. $y = \dfrac{x - 5}{2}$

Find all real solutions to each equation.

13. $2x - 5 = 9$

14. $3(y - 2) + 4 = -5$

15. $|2x - 3| = 0$

16. $\sqrt{x - 4} = 0$

17. $x^2 + 9 = 0$

18. $x^2 - 8x + 12 = 0$

Graph the solution set to each inequality or compound inequality in one variable on the number line.

19. $2x - 1 > 7$

20. $5 - 3x \le -1$

21. $x - 5 \le 4$ and $3x - 1 < 8$

22. $2x \le -6$ or $5 - 2x < -7$

23. $|x - 3| < 2$

24. $|1 - 2x| \ge 7$

Find the domain of each expression.

25. $\sqrt{x}$

26. $\sqrt{6 - 2x}$

27. $\dfrac{5x - 3}{x^2 + 1}$

28. $\dfrac{x - 3}{x^2 - 10x + 9}$

Solve the problem.

29. *Equipment replacement.* To decide when to replace company cars, an accountant looks at two cost components: capital cost and operating cost. The capital cost (the difference between the original cost and the salvage value) for a certain car is $3000 plus $0.12 for each mile that the car is driven.

a) Write the capital cost C as a linear function of x, the number of miles that the car has been driven.

b) The operating cost is $0.15 per mile initially and increases to $0.25 per mile when the car reaches 100,000 miles. Assuming that the operating cost P is a linear function of x, write P in terms of x.

c) The total cost per mile is given by $T = \dfrac{C}{x} + P$. Find T for $x = 20{,}000$, $30{,}000$, and $90{,}000$.

d) T decreases at first and then increases as x increases. The accountant has decided to replace the car when T reaches $0.38 for the second time. At what mileage will the car be replaced?

e) For what values of x is T less than or equal to $0.38?

Functions and Graphs

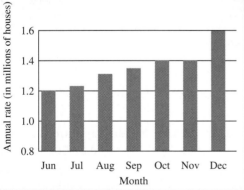

The actions of the Federal Reserve System, also known as "the Fed," affect each of us. But just what is the Federal Reserve System?

The Federal Reserve Act that created the Federal Reserve System was signed into law on December 23, 1913, by President Woodrow Wilson. Congress passed this act to provide a safe, flexible banking system for the United States. The seven members of the Board of Governors are appointed for a 14-year term by the President. The President also appoints the chairman and vice chairman with the advice and consent of the Senate. The chairman usually represents the Federal Reserve System in policy discussions after conferring with the Secretary of the Treasury and other economic advisors.

The Fed's responsibility for the nation's money supply is perhaps its most important goal. The Fed can increase or decrease the money supply by changing the prime rate, the rate at which the Reserve Banks lend money to other banks. These changes are initiated by a director of one of the 12 individual Reserve Banks and approved by the Board of Governors.

Before making a decision on rates, the Fed studies economic indicators such as the inflation rate, unemployment rate, and housing starts. When the Fed raises the prime rate, the inflation rate decreases because the inflation rate is a function of the prime rate. When the Fed lowers the prime rate, housing starts increase because the number of housing starts is also a function of the prime rate. In Exercise 92 of Section 8.1 and Exercise 77 of Section 8.2 you will work with functions that involve housing starts.

8.1 Relations and Functions

In Section 7.3 you studied linear functions. In this section we introduce the general concepts of relations and functions.

Definitions

Apple Imagewriter ribbons are sold in boxes of six in the 1994 K-LOG Catalog and are priced as follows.

Number of boxes	1	2–3	4
Cost per ribbon	$4.85	$4.60	$4.35

We can write this data as a set of ordered pairs in which the first coordinate is the number of boxes purchased and the second is the cost per ribbon in dollars:

$$\{(1, 4.85), (2, 4.60), (3, 4.60), (4, 4.35)\}$$

Because the number of boxes determines the cost per ribbon, we say that *the cost is a function of the number of boxes purchased.*

Suppose the following table appeared in the K-LOG Catalog:

Number of boxes	1	2	2	4
Cost per ribbon	$4.85	$4.60	$4.45	$4.35

Something is wrong with this table. The cost per ribbon when you buy two boxes is not clear because the ordered pairs $(2, 4.60)$ and $(2, 4.45)$ have the same first coordinate and different second coordinates. In this case the cost per ribbon is *not* a function of the number of boxes.

These examples illustrate the definition of function.

Function

A **function** is a set of ordered pairs of real numbers in which no two ordered pairs have the same first coordinate and different second coordinates.

Any set of ordered pairs is called a **relation.** So every function is a relation, but a relation in which two ordered pairs have the same first coordinate and different second coordinates is not a function.

EXAMPLE 1   **Functions defined from a list of ordered pairs**

Determine whether each relation is a function.

a) $\{(1, 2), (1, 5), (3, 7)\}$ **b)** $\{(4, 5), (3, 5), (2, 6), (1, 7)\}$

Solution

a) This relation is not a function because $(1, 2)$ and $(1, 5)$ have the same first coordinates but different second coordinates.

b) This relation is a function. Note that the same second coordinate with different first coordinates is permitted in a function. ■

If variables are used to represent the numbers of the ordered pairs, then the variable for the first coordinate is the **independent variable** and the variable for the second coordinate is the **dependent variable.** When we use variables x and y, we always assume that x is the independent variable and y is the dependent variable. If we state that B is a function of A, then A is the independent variable (first coordinate) and B is the dependent variable (second coordinate), and we write (A, B).

EXAMPLE 2   **Functions defined with set notation**

Determine whether each relation is a function.

a) $\{(x, y) \mid y = 5x^2 - 7x + 2\}$ b) $\{(s, t) \mid t^2 = s\}$ c) $\{(u, v) \mid 2u + 3v = 6\}$

Solution

a) This relation is a function, or y is a function of x, because for any x there is only one value for y determined by the equation $y = 5x^2 - 7x + 2$.

b) Because the dependent variable is squared in $t^2 = s$, a positive value for s corresponds to both a positive and a negative value for t. For example, if $s = 4$, then $t^2 = 4$, or $t = \pm 2$. The ordered pairs $(4, -2)$ and $(4, 2)$ belong to this set. So this relation is not a function; that is, t is not a function of s.

c) If we solve $2u + 3v = 6$ for v, we get $v = -\frac{2}{3}u + 2$. Because each value of u determines only one value for v, this relation is a function. In this case, v is a *linear function* of u. ■

We often omit the set notation and refer to an equation as a relation or a function. There are many well-known formulas that express the value of one variable as a function of another variable. For example, $A = \pi r^2$ gives the area of a circle as a function of its radius. The formula $F = \frac{9}{5}C + 32$ expresses the Fahrenheit temperature as a linear function of the Celsius temperature.

CAUTION   Any equation with two variables is a relation because it determines a set of ordered pairs. However, an equation is a function only if the set of ordered pairs that satisfy the equation is a function. ⊘

EXAMPLE 3   **Functions defined by equations**

Determine whether each relation defines y as a function of x.

a) $x = y^4$ b) $y = x^2$ c) $x^2 + y^2 = 16$

Solution

a) Because the dependent variable appears to the fourth power in $x = y^4$, a positive value for x corresponds to both a positive and a negative value for y. For example, both $(1, -1)$ and $(1, 1)$ satisfy $x = y^4$, and these ordered pairs have the same first coordinate and different second coordinates. So $x = y^4$ does not define y as a function of x. The equation $x = y^4$ is not a function.

b) For each value of x the equation $y = x^2$ determines only one value for y. So $y = x^2$ defines y as a function of x. We also say that $y = x^2$ is a function. Note that $(1, 1)$ and $(-1, 1)$ satisfy $y = x^2$, but ordered pairs with the same second coordinate and different first coordinates are allowed in a function.

c) The even power of the dependent variable in $x^2 + y^2 = 16$ indicates that we can use a number or its opposite for y and get the same value for y^2. For example, if $y = \pm 4$, then both $(0, -4)$ and $(0, 4)$ satisfy $x^2 + y^2 = 16$. So the equation is not a function. The equation does not define y as a function of x. ∎

Domain and Range

The **domain** of a relation (or function) is the set of first coordinates, and the **range** is the set of second coordinates of the ordered pairs. If the ordered pairs of a relation are listed, then the domain and range can be read from the list. Relations are often defined by equations with no domain stated. *If the domain is not stated, we agree that the domain consists of all real numbers that, when substituted for the independent variable, produce real numbers for the dependent variable.*

EXAMPLE 4 **Identifying domain and range**

State the domain and range of each relation.

a) $\{(2, 5), (2, 7), (4, 3)\}$ b) $y = \sqrt{x}$ c) $\{(v, w) \mid w = v + 1 \text{ and } v > 2\}$

Solution

a) The domain is the set of numbers used as first coordinates, $\{2, 4\}$. The range is the set of second coordinates, $\{3, 5, 7\}$.

b) Because $\sqrt{x}$ is a real number only for $x \geq 0$, the domain is $[0, \infty)$, the nonnegative real numbers. The range is the set of numbers that result from taking the principal square root of every nonnegative real number. Thus the range is also the set of nonnegative real numbers, $[0, \infty)$.

c) Because v is the independent variable, the condition $v > 2$ specifies the domain of the function. So the domain is the interval $(2, \infty)$. The range is the set of values of w. If $v > 2$, then $w > 3$, and the range is the interval $(3, \infty)$. ∎

For some relations it is easier to determine the domain than the range. In Section 8.2 we will see that the graph of a relation can be helpful for finding the range.

The *f*-Notation

We often name functions using a single letter. For example, C might represent a cost function, and P might represent a profit function. When we have no specific application in mind, we might name a function f, g, or h. If we are interested in the cost of n concert tickets at \$30 each, then the cost function C pairs the number of tickets with the cost. We use $C(n)$, read "C of n," to represent the cost of n tickets. If tickets are \$30 each, then $C(n) = 30n$. If $n = 5$, we write $C(5) = \$150$.

In general, if x is the independent variable, then $f(x)$, read "f of x," is the dependent variable, and f is thought of as the name of the function. This nota-tion is called *f*-notation. Because $f(x)$ is used in place of y, we have $y = f(x)$. For example, if $y = x^2 + 2$, we could also write $f(x) = x^2 + 2$. The advan-tage of using $f(x)$ as the name of the second coordinate is that it also con-tains the name of the corresponding first coordinate. For example, $f(3)$ is the value of the dependent variable when the independent variable is 3, and

$$f(3) = 3^2 + 2 = 11.$$

CAUTION   The notation $f(3)$ does not mean f times 3. Remember that f is the name of the function and not a variable. By itself, f does not have a numer-ical value, but $f(3)$ has a value.   ⊘

EXAMPLE 5   **Using *f*-notation**

Let $f(x) = 3x - 2$ and $g(x) = x^2 - x$. Find the following:

a) $f(-5)$ **b)** $g(-3)$ **c)** x, if $g(x) = 0$

Solution

a) Replace x by -5 in the equation defining the function f:

$$f(x) = 3x - 2$$
$$f(-5) = 3(-5) - 2$$
$$= -17$$

So $f(-5) = -17$.

b) Replace x by -3 in the equation defining the function g:

$$g(x) = x^2 - x$$
$$g(-3) = (-3)^2 - (-3)$$
$$= 12$$

So $g(-3) = 12$.

c) Because $g(x) = x^2 - x$ and we are given that $g(x) = 0$, we can find x by solving an equation:

$$x^2 - x = 0$$
$$x(x - 1) = 0$$
$$x = 0 \quad \text{or} \quad x = 1 \qquad ■$$

The letter used for the independent variable in $f(x) = x^2 + 2$ is not important. The notation $f(t) = t^2 + 2$ or even

$$f(\text{first coordinate}) = (\text{first coordinate})^2 + 2$$

could be used to define the same set of ordered pairs. The second coordinate of an ordered pair of this function is determined by squaring the first coordi-nate and adding 2.

The first coordinate of an ordered pair can be an expression rather than a real number. For example, if the first coordinate is the expression $a + 4$ and the function is $f(x) = x^2 + 2$, then

$$f(a + 4) = (a + 4)^2 + 2 \quad \text{Replace } x \text{ by } a + 4.$$
$$= a^2 + 8a + 18.$$

The first coordinate can even be an expression involving x. For example, if the first coordinate is $x + 5$ and the function is $f(x) = x^2 + 2$, then

$$f(x + 5) = (x + 5)^2 + 2 \quad \text{Replace } x \text{ by } x + 5.$$
$$= x^2 + 10x + 27.$$

Note that $x + 5$ replaces x and the binomial $x + 5$ is squared, not just x.

EXAMPLE 6   **Using an expression for the independent variable**

Let $f(x) = 3x - 2$ and $g(x) = x^2 - x$. Find the following:

a) $f(a - 3)$ **b)** $g(x + h)$

Solution

a) Replace x by $a - 3$ in the equation defining the function f:

$$f(x) = 3x - 2$$
$$f(a - 3) = 3(a - 3) - 2$$
$$= 3a - 11$$

So $f(a - 3) = 3a - 11$.

b) Replace x by $x + h$ in the equation defining the function g:

$$g(x) = x^2 - x$$
$$g(x + h) = (x + h)^2 - (x + h)$$
$$= x^2 + 2xh + h^2 - x - h \qquad ■$$

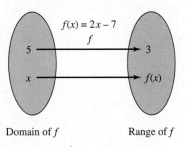

Domain of f Range of f

Figure 8.1

A function is a rule for pairing numbers in one set (the domain) with numbers in another set (the range). Figure 8.1 shows the domain and range of a function. The arrow represents the function and illustrates the pairing of values in the domain with values in the range.

Average Rate of Change

The U.S. Bureau of Labor Statistics reports on the number of people (in millions) in the civilian labor force of the United States. The size of the labor force is a function of the year. Two ordered pairs of this function are $(1985, 88.4)$ and $(1989, 97.3)$. To see how the function is changing in time, we can find the slope of the line through these two points:

$$\frac{97.3 - 88.4}{1989 - 1985} \approx 2.3 \text{ million people per year}$$

During this four-year period the civilian labor force increased on the average by 2.3 million people per year. Note that 2.3 million is not necessarily the increase for any particular year, but it is the average rate at which the labor force was changing during those four years.

In general, the average rate of change of a function over an interval in its domain is defined as follows.

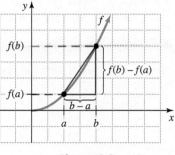

y

$f(b)$

f

$f(b) - f(a)$

$f(a)$

$b - a$

a b x

Figure 8.2

Average Rate of Change of a Function

If a and b are in the domain of the function f and $a < b$, then the **average rate of change** of f over the interval $[a, b]$ is

$$\frac{f(b) - f(a)}{b - a}.$$

Note that the average rate of change of f over $[a, b]$ is the slope of the line through $(a, f(a))$ and $(b, f(b))$, as shown in Fig. 8.2.

EXAMPLE 7 **Average rate of change of a function**

If a stunt man jumps from a bridge that is 144 ft above the water, then his height above the water in feet at time t in seconds is given by the function $h(t) = -16t^2 + 144$. Find the average rate of change of his height above the water over the time interval $[1, 3]$. See Fig. 8.3.

Solution

$h(t)$

150 — (1, 128)
120
90
60
30
0 — (3, 0)
1 2 3 t

Figure 8.3

Because $h(1) = 128$ and $h(3) = 0$, we have

$$\frac{h(3) - h(1)}{3 - 1} = \frac{0 - 128}{2}$$

$$= -64 \text{ ft/sec.}$$

So the average rate of change of the height is -64 ft/sec on the interval $[1, 3]$. ∎

Warm-ups

True or false? Explain your answer.

1. Any set of ordered pairs is a function.
2. The circumference of a circle is a function of the diameter.
3. The set $\{(1, 2), (3, 2), (5, 2)\}$ is a function.
4. Every relation is a function.
5. The set $\{(1, 5), (3, 6), (1, 7)\}$ is a function.
6. The domain of $f(x) = \sqrt{x}$ is the set of positive real numbers.
7. The range of $g(x) = |x|$ is the set of nonnegative real numbers.
8. The set $\{(x, y) \mid x = 4y\}$ is a function.
9. The set $\{(u, v) \mid u = v^2\}$ is a function.
10. If $h(x) = x^2 - 3$, then $h(-2) = 1$.

8.1 EXERCISES

Determine whether each relation is a function. See Example 1.

1. $\{(2, 4), (3, 4), (4, 5)\}$ 2. $\{(2, -5), (2, 5), (3, 10)\}$

3. $\{(-2, 4), (-2, 6), (3, 6)\}$ 4. $\{(3, 6), (6, 3)\}$

5. $\{(\pi, -1), (\pi, 1)\}$

6. $\{(-0.3, -0.3), (-0.2, 0), (-0.3, 1)\}$

7. $\left\{\left(\frac{1}{2}, \frac{1}{2}\right)\right\}$

8. $\left\{\left(\frac{1}{3}, 7\right), \left(-\frac{1}{3}, 7\right), \left(\frac{1}{6}, 7\right)\right\}$

Determine whether each relation is a function. See Example 2.

9. $\{(x, y) \mid y = (x - 1)^2\}$

10. $\{(x, y) \mid y = x^2 - 12x + 1\}$

11. $\{(x, y) \mid x = |y|\}$ 12. $\{(x, y) \mid x = y^2 + 2\}$

13. $\{(s, t) \mid t = s\}$ 14. $\left\{(u, v) \mid v = \dfrac{1}{u}\right\}$

15. $\{(x, y) \mid x = 5y + 2\}$ 16. $\{(x, y) \mid x = 3y\}$

Determine whether each relation defines y as a function of x. See Example 3.

17. $x = 2y^2$ 18. $y = 3x^4$ 19. $y = 3x - 4$

20. $y = 2x + 9$ 21. $x^2 + y^2 = 1$ 22. $x^4 - y^4 = 0$

23. $y = \sqrt{x}$ 24. $x = \sqrt{y}$ 25. $x = 2|y|$

26. $y = |x - 1|$ 27. $y = x^{1/3}$

28. $y = x^{2/3} + 3x^{1/3} + 5$

Determine the domain and range of each relation. See Example 4.

29. $\{(2, 3), (2, 5), (2, 7)\}$ 30. $\{(3, 1), (5, 1), (4, 1)\}$

31. $y = |x|$ 32. $y = 2x + 1$

33. $x = \sqrt{y}$ 34. $y = \sqrt{x} + 1$

35. $\{(s, t) \mid t = s^{1/3}$ and $s \geq 8\}$

36. $\{(s, t) \mid s = t^2$ and $0 \leq s \leq 9\}$

Let $f(x) = 3x - 2$, $g(x) = x^2 - 3x + 2$, and $h(x) = |x + 2|$. Find the following. See Example 5.

37. $f(4)$ 38. $f(100)$ 39. $g(-2)$

40. $g(6)$ 41. $h(-3)$ 42. $h(-19)$

43. x, if $f(x) = 5$ 44. x, if $f(x) = 49$

45. x, if $g(x) = 0$ 46. x, if $g(x) = 2$

47. x, if $h(x) = 3$ 48. x, if $h(x) = 7$

Let $f(x) = 4x - 1$ and $g(x) = x^2 - 2x + 4$. Find the following. See Example 6.

49. $f(a)$ 50. $f(a + 1)$ 51. $f(x + 2)$

52. $f(x + h)$ 53. $g(x + 3)$ 54. $g(x - 2)$

55. $g(x + h)$ 56. $g(a - 2)$

Find the average rate of change of the given function over the given interval. See Example 7.

57. $f(x) = x^2$, $[1, 3]$ 58. $g(x) = 2x - 5$, $[3, 9]$

59. $h(x) = 2x^2 - 3x$, $[4, 8]$ 60. $f(x) = \sqrt{x}$, $[4, 9]$

61. $f(x) = x^2 - x + 2$, $[a, a + 1]$

62. $h(x) = x^2 - 3$, $[a, a + 4]$

63. $g(x) = 3x - 5$, $[x, x + h]$

64. $f(x) = 2x^2$, $[x, x + h]$

Let $f(x) = \sqrt{x + 2}$ and $g(x) = 3x^2 - 8x + 2$. Use a calculator to find the following. Round answers to three decimal places.

65. $f(3.46)$ 66. $g(-1.37)$

67. $g(-3.5)$ 68. $f(-1.2)$

69. x, if $f(x) = 5.6$ 70. x, if $f(x) = 8.251$

71. x, if $g(x) = 0$ 72. x, if $g(x) = 1$

Let $f(x) = 3x - 2$ and $g(x) = x^2 - x$. Find and simplify each expression.

73. $f(a - 5) - f(a)$ 74. $f(x + h) - f(x)$

75. $g(a + 2) - g(a)$ 76. $g(x + h) - g(x)$

77. $\dfrac{f(x + h) - f(x)}{h}$ 78. $\dfrac{f(n + 3) - f(n)}{3}$

79. $\dfrac{g(x + h) - g(x)}{(x + h) - x}$

80. $\dfrac{g(a - 5) - g(a)}{(a - 5) - a}$

In Exercises 81–92, solve each problem.

81. **Area of a square.** Express the area of a square, A, as a function of the length of a side, s.

82. **Perimeter of a square.** Express the perimeter of a square, P, as a function of the length of a side, s.

83. **Cost of fabric.** If a certain fabric is priced at $3.98 per yard, express the cost of a purchase, C, as a function of the number of yards purchased, y.

84. **Earned income.** If Mildred earns $14.50 per hour, express her total pay, P, as a function of the number of hours worked, h.

85. *Cost of pizza.* A pizza parlor charges $14.95 for a pizza plus $0.50 for each topping. Express the cost of a pizza, C, as a function of the number of toppings, n.

86. *Dealing in gravel.* A gravel dealer charges $120 for a minimum load of 9 cubic yards and $10 more for each additional cubic yard. Express the total charge, C, as a function of the number of yards sold, n, where $n \geq 9$.

87. *Staying fit.* Suppose that the heart rate of a certain individual is a linear function of the number of minutes she spends on a treadmill. A heart rate of 78 was measured after 2 minutes, and a heart rate of 86 was measured after 4 minutes. Write the heart rate, h, as a linear function of the number of minutes, t, on the treadmill, for $0 \leq t \leq 8$.

88. *Printing costs.* To determine the cost of printing a book, a printer uses a linear function of the number of pages. If the cost is $8.60 for a 400-page book and $12.20 for a 580-page book, then what is the linear function that is used?

89. *Chevrolet Beretta.* According to *Edmund's 1994 Used Car Prices*, a 1990 Chevrolet Beretta two-door coupe that sold for $11,640 in 1990 sells for $5675 in 1994. Find the average rate of change of the price of this car over the time interval [1990, 1994].

90. *Porsche 928S.* According to *Edmund's 1994 Used Car Prices*, a 1988 Porsche 928S that sold for $69,680 in 1988 sells for $25,250 in 1994. Find the average rate of change of the price of this car over the time interval [1988, 1994].

91. *Mazda RX-7.* The 1993 Mazda RX-7 had a list price of $34,790 and could go from 0 to 60 mph in 5.4 seconds. (*Motor Trend*, March 1994). Find the average rate of change of the velocity of this car over the time interval [0, 5.4].

92. *Housing starts.* Housing starts in the United States went from a 1.2 million annual rate in the sixth month of 1993 to a 1.6 million annual rate in the twelfth month of 1993 (*USA Today*, July 6, 1994). Find the average rate of change of housing starts over the time interval [6, 12]. (This rapid increase in housing starts was one of the reasons the Fed began a series of interest rate hikes in 1994.)

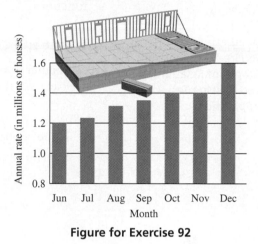

Figure for Exercise 92

Getting More Involved

93. *Exploration.* In Example 7 we found that the stunt man traveled 128 feet in 2 seconds for an average velocity of −64 feet per second over the interval [1, 3]. His velocity actually starts out at 0 ft/sec and keeps increasing as he falls. Find the average rate of change of his height, or his average velocity, over the time intervals [2.8, 3], [2.9, 3], and [2.99, 3].

94. *Exploration.* In Exercise 93 we found the stunt man's average velocity for some time intervals right before he hits the water. His velocity at the moment he hits the water is his *instantaneous velocity* at $t = 3$. Find the average velocity over the time intervals [2.999, 3] and [2.9999, 3]. What do you think is his instantaneous velocity at time $t = 3$? Explain your answer.

8.2 Graphs of Functions

In Chapter 7 you learned that graphs are a useful visual aid in the study of linear equations. In this section we will again use graphs, this time to study functions.

Linear and Constant Functions

In Chapter 7 you learned that if $y = mx + b$, then y is a linear function of x. We can state this definition using f-notation as follows.

> **Linear Function**
>
> A **linear function** is a function of the form
>
> $$f(x) = mx + b,$$
>
> where m and b are real numbers with $m \neq 0$.

The graph of the linear function $f(x) = mx + b$ is exactly the same as the graph of the linear equation $y = mx + b$. If $m = 0$, then we get $f(x) = b$, which is called a **constant function.** If $m = 1$ and $b = 0$, then we get the function $f(x) = x$, which is called the **identity function.** When we graph a function given in f-notation, we usually label the vertical axis as $f(x)$ rather than y.

EXAMPLE 1 **Graphing a constant function**

Graph $f(x) = 3$ and state the domain and range.

Solution

The graph of the function $f(x) = 3$ is the same as the graph of the horizontal line $y = 3$. See Figure 8.4. The domain is the set of all real numbers. We can see from the graph that the only y-coordinate used is 3. So the range is {3}. ■

Figure 8.4   **Figure 8.5**

EXAMPLE 2   **Graphing a linear function**

Graph the function $f(x) = 3x - 4$ and state the domain and range.

Solution

The y-intercept is $(0, -4)$, and the slope of the line is 3. The slope and intercept determine the position of the line, just as they did in Chapter 7. See Figure 8.5. The domain is R. We can see from the graph that every real number is used as a y-coordinate. So the range is also R. ■

Absolute Value Functions

The equation $y = |x|$ defines a function because every value of x determines a unique value of y. We call this function the absolute value function.

Absolute Value Function

The **absolute value function** is the function defined by

$$f(x) = |x|.$$

To graph the absolute value function, we simply plot enough ordered pairs of the function to see what the graph looks like.

EXAMPLE 3   **The absolute value function**

Graph $f(x) = |x|$ and state the domain and range.

Solution

To graph this function, we find points that satisfy the equation $f(x) = |x|$.

x	−2	−1	0	1	2
$f(x) = |x|$	2	1	0	1	2

These points are graphed in Fig. 8.6(a). If we plotted all of the points satisfying $f(x) = |x|$, we would get the V-shaped graph shown in Fig. 8.6(b). The domain is $(-\infty, \infty)$. Because the graph does not go below the x-axis, the range is the set of nonnegative real numbers, $[0, \infty)$.

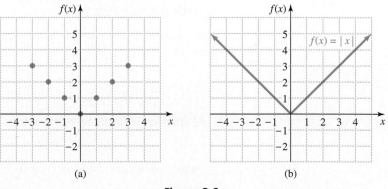

(a)   (b)

Figure 8.6 ■

Many functions involving absolute value have graphs that are V-shaped, as in Fig. 8.6(b). To graph functions involving absolute value, we must choose points that determine the correct location of the V-shaped graph.

EXAMPLE 4 **Other functions involving absolute value**

Graph each function and state the domain and range.

a) $f(x) = |x| - 2$ **b)** $g(x) = |2x - 6|$

Solution

a) Choose values for x and find $f(x)$.

x	-2	-1	0	1	2
$f(x) = |x| - 2$	0	-1	-2	-1	0

Plot these points and draw a V-shaped graph through them, as shown in Fig. 8.7. The domain is $(-\infty, \infty)$, and the range is $[-2, \infty)$.

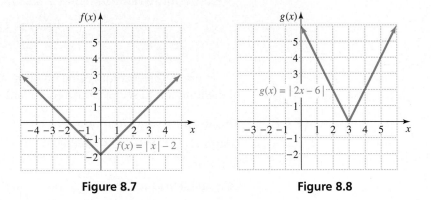

Figure 8.7   Figure 8.8

b) Make a table of values for x and $g(x)$.

x	1	2	3	4	5
$g(x) = |2x - 6|$	4	2	0	2	4

Draw the graph as shown in Fig. 8.8. The domain is $(-\infty, \infty)$, and the range is $[0, \infty)$. ∎

Quadratic Functions

The general quadratic polynomial $ax^2 + bx + c$ can be used to define a function.

> **Quadratic Function**
>
> A **quadratic function** is a function of the form
>
> $$f(x) = ax^2 + bx + c,$$
>
> where a, b, and c are real numbers, with $a \neq 0$.

Without the term ax^2 this function would be a linear function. That is why we specify $a \neq 0$.

EXAMPLE 5   **The simplest quadratic function**

Graph the function $h(x) = x^2$ and state the domain and range.

Solution

Make a table of values for x and $h(x)$:

x	-2	-1	0	1	2
$h(x) = x^2$	4	1	0	1	4

See Fig. 8.9 for the graph. The domain is $(-\infty, \infty)$. From the graph we see that the smallest y-coordinate of the function is 0. So the range is $[0, \infty)$.

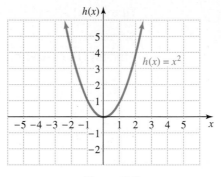

Figure 8.9

EXAMPLE 6   **A quadratic function**

Graph the function $g(x) = 4 - x^2$ and state the domain and range.

Solution

We plot enough points to get the correct shape of the graph.

x	-2	-1	0	1	2
$g(x) = 4 - x^2$	0	3	4	3	0

See Fig. 8.10 for the graph. The domain is $(-\infty, \infty)$. From the graph we see that the largest y-coordinate is 4. So the range is $(-\infty, 4]$.

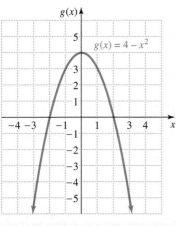

Figure 8.10

The graph of any quadratic function is called a **parabola.** All parabolas are similar to the ones in Figs. 8.9 and 8.10. The parabola in Fig. 8.9 **opens upward,** and the parabola in Fig. 8.10 **opens downward.** The lowest point on a parabola that opens upward or the highest point on a parabola that opens downward is called the **vertex.** The y-coordinate of the vertex is the **minimum value** of the function if the parabola opens upward, and it is the **maximum value** of the function if the parabola opens downward. For $h(x) = x^2$ the vertex is $(0, 0)$, and 0 is the minimum value of the function. For $g(x) = 4 - x^2$ the vertex is $(0, 4)$, and 4 is the maximum value of the function. It will be shown in Chapter 11 that the x-coordinate of the vertex for the graph of any quadratic function is $\dfrac{-b}{2a}$.

Vertex of a Parabola

The x-coordinate of the vertex of the parabola $f(x) = ax^2 + bx + c$ $(a \neq 0)$ is given by

$$x = \frac{-b}{2a}.$$

When you graph a parabola, you should always locate the vertex. With the vertex and several nearby points, you can see the correct shape of the parabola.

EXAMPLE 7 **Using the vertex to graph a quadratic function**

Graph $f(x) = -x^2 - x + 2$, and state the domain and range.

Solution

First find the x-coordinate of the vertex:

$$x = \frac{-b}{2a} = \frac{-(-1)}{2(-1)} = \frac{1}{-2} = -\frac{1}{2}$$

Now find $f(x)$ for $x = -\frac{1}{2}$:

$$f\left(-\frac{1}{2}\right) = -\left(-\frac{1}{2}\right)^2 - \left(-\frac{1}{2}\right) + 2$$

$$= -\frac{1}{4} + \frac{1}{2} + 2$$

$$= \frac{9}{4}$$

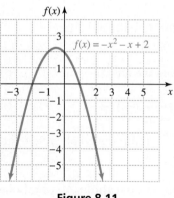

Figure 8.11

The vertex is $\left(-\frac{1}{2}, \frac{9}{4}\right)$. Now find a few points on either side of the vertex.

x	-2	-1	$-\dfrac{1}{2}$	0	1
$f(x) = -x^2 - x + 2$	0	2	$\dfrac{9}{4}$	2	0

Sketch a parabola through these points, as in Fig. 8.11. The domain is $(-\infty, \infty)$. Because the graph goes no higher than $\frac{9}{4}$, the range is $\left(-\infty, \frac{9}{4}\right]$. ∎

Square-Root Functions

Functions involving square roots typically have graphs that look like half of a parabola.

Square-Root Function

The **square-root function** is the function defined by

$$f(x) = \sqrt{x}.$$

EXAMPLE 8   **The square-root function**
Graph the equation $y = \sqrt{x}$, and state the domain and range.

Solution

The graph of the equation $y = \sqrt{x}$ and the graph of the function $f(x) = \sqrt{x}$ are the same. Because $\sqrt{x}$ is a real number only if $x \geq 0$, the domain of this function is the set of nonnegative real numbers. The following ordered pairs are on the graph:

x	0	1	4	9
$y = \sqrt{x}$	0	1	2	3

The graph goes through these ordered pairs, as shown in Fig. 8.12. Note that x is chosen from the nonnegative numbers. The domain is $[0, \infty)$, and the range is $[0, \infty)$.

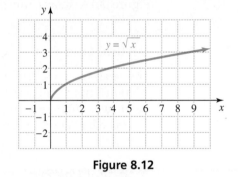

Figure 8.12

■

Calculator Close-up

There are infinitely many points on each of the graphs in the previous examples. Of course, it is impossible to plot every point. We try to plot enough points to see the general shape of the graph. We can get the correct graph of a function from only a few points if we recall the graphs of similar functions. Even computers and graphing calculators plot only a finite number of points of a function. Most graphing calculators have a **dot mode** and a **connected mode**. In dot mode the calculator shows only the points (about 96 of them) that it calculates for the function. In connected mode the calculator connects the points with line segments, making the graph look more like a smooth curve. Fig. 8.13 shows the graph of $y = x^2$ in both dot and connected mode. Use your graphing calculator to graph each of the functions in the previous examples.

Xmin = –2   Ymin = –1
Xmax = 2   Ymax = 4

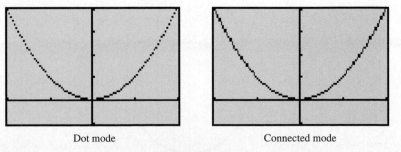

Dot mode   Connected mode

Figure 8.13

Graphs of Relations

In the next example we graph a relation by simply finding enough points to see the shape of the graph. Relations can also be graphed on a graphing calculator. See Exercises 85 and 86.

EXAMPLE 9 **The graph of a relation**

Graph $x = y^2$ and state the domain and range.

Solution

The set of ordered pairs that satisfy $x = y^2$ is not a function. However, we can still make a table of values and sketch the graph. For convenience of calculation we choose the y-coordinate first and then find the x-coordinate.

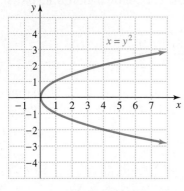

$x = y^2$	4	1	0	1	4
y	-2	-1	0	1	2

Figure 8.14 shows the graph. The domain is $[0, \infty)$, and the range is $(-\infty, \infty)$. ∎

Figure 8.14

Vertical-Line Test

The relation $x = y^2$ from Example 9 does not define y as a function of x. A relation is not a function if there are two ordered pairs with the same first coordinate and different second coordinates. Whether a relation has such ordered pairs can be determined by a simple visual test called the **vertical-line test.**

> **Vertical-Line Test**
>
> If it is possible to draw a vertical line that crosses the graph of a relation two or more times, then the graph is not the graph of a function.

If there is a vertical line that crosses a graph twice (or more), then we have two points (or more) with the same x-coordinate and different y-coordinates, and so the graph is not the graph of a function. If you mentally consider every possible vertical line and none of them cross the graph more than once, then you can conclude that the graph is the graph of a function.

EXAMPLE 10 **Using the vertical-line test**

Which of the following graphs are graphs of functions?

a)   b)   c)

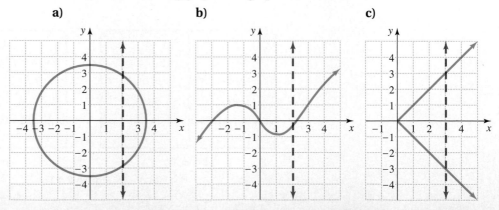

Solution

Neither (a) nor (c) is the graph of a function, since we can draw vertical lines that cross these graphs twice. The graph (b) is the graph of a function, since no vertical line crosses it twice. ■

Applications

The maximum or minimum value of a function is important in applications. Many applications involve functions that determine quantities such as profit or cost. We may be interested in the maximum profit or the minimum cost. In the next example we find the maximum altitude reached by a projectile.

EXAMPLE 11

Maximum altitude

A flare is fired upward from the surface of the earth with an initial velocity of 96 feet per second. The altitude of the flare in feet for any time, t, in seconds is given by the function $h(t) = -16t^2 + 96t$. Graph this function for $0 \leq t \leq 6$, and determine the maximum height reached by the flare.

Solution

The function $h(t)$ is a quadratic function. The t-coordinate of the vertex of the parabola is

$$t = \frac{-b}{2a} = \frac{-96}{2(-16)} = 3.$$

Because the maximum or minimum value of any quadratic function is reached at the vertex, the maximum value of this function is the second coordinate of the vertex:

$$h(3) = -16 \cdot 3^2 + 96 \cdot 3 = 144$$

The maximum height reached by the flare is 144 feet. The graph of this function is shown in Figure 8.15. ■

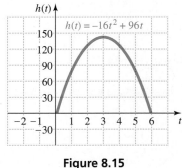

Figure 8.15

Warm-ups

True or false? Explain your answer.

1. The graph of a function is a picture of all ordered pairs of the function.

2. The graph of every linear function is a straight line.

3. The absolute value function has a V-shaped graph.

4. The domain of $f(x) = \dfrac{1}{x}$ is $(-\infty, \infty)$.

5. The vertex of a parabola is determined by $x = \dfrac{b}{2a}$.

6. The range of any quadratic function is $(-\infty, \infty)$.

7. The y-axis and the $f(x)$-axis are the same.

8. The vertical line test is used to determine whether a relation is a function.

9. The domain of $f(x) = \sqrt{x-1}$ is $(1, \infty)$.

10. The domain of any quadratic function is $(-\infty, \infty)$.

8.2 EXERCISES

Graph each linear function and state its domain and range. See Examples 1 and 2.

1. $h(x) = -2$

2. $f(x) = 4$

3. $f(x) = 2x - 1$

4. $g(x) = x + 2$

5. $g(x) = \dfrac{1}{2}x + 2$

6. $h(x) = \dfrac{2}{3}x - 4$

7. $y = -\dfrac{2}{3}x + 3$

8. $y = -\dfrac{3}{4}x + 4$

9. $y = -0.3x + 6.5$

10. $y = 0.25x - 0.5$

Graph each absolute value function and state its domain and range. See Examples 3 and 4.

11. $f(x) = |x| + 1$

12. $g(x) = |x| - 3$

13. $h(x) = |x + 1|$

14. $f(x) = |x - 2|$

15. $g(x) = |3x|$

16. $h(x) = |-2x|$

17. $f(x) = |2x - 1|$

18. $y = |2x - 3|$

19. $f(x) = |x - 2| + 1$

20. $y = |x - 1| + 2$

Graph each quadratic function and state its domain and range. See Examples 5 and 6.

21. $g(x) = x^2 + 2$

22. $f(x) = x^2 - 4$

23. $f(x) = 2x^2$

24. $h(x) = -3x^2$

25. $y = 6 - x^2$

26. $y = -2x^2 + 3$

Graph each guadratic function and state its vertex, domain, and range. See Example 7.

27. $g(x) = x^2 - x - 2$

28. $f(x) = x^2 + 2x - 3$

29. $h(x) = 2x^2 + 9x + 4$

30. $g(x) = 2x^2 - x - 3$

31. $y = -x^2 + 2x + 8$

32. $y = -x^2 + 2x$

33. $y = -2x^2 - 3x + 2$

34. $y = -x^2 + 4x - 3$

Graph each square-root function and state its domain and range. See Example 8.

35. $g(x) = 2\sqrt{x}$

36. $g(x) = \sqrt{x} - 1$

37. $f(x) = \sqrt{x - 1}$

38. $f(x) = \sqrt{x + 1}$

39. $h(x) = -\sqrt{x}$

40. $h(x) = -\sqrt{x - 1}$

41. $y = \sqrt{x} + 2$

42. $y = 2\sqrt{x} + 1$

Graph each relation and state its domain and range. See Example 9.

43. $x = |y|$

44. $x = -|y|$

45. $x = -y^2$

46. $x = 1 - y^2$

47. $x = 5$

48. $x = -3$

49. $x + 9 = y^2$

50. $x + 3 = |y|$

51. $x = \sqrt{y}$

52. $x = -\sqrt{y}$

53. $x = (y - 1)^2$

54. $x = (y + 2)^2$

Use the vertical line test to determine which of the graphs are graphs of functions. See Example 10.

55.

56.

57.

58.

59.

60.

Graph each function and state the domain and range.

61. $f(x) = 1 - |x|$

62. $h(x) = \sqrt{x - 3}$

63. $y = (x - 3)^2 - 1$

64. $y = x^2 - 2x - 3$

65. $y = |x + 3| + 1$

66. $f(x) = -2x + 4$

67. $y = \sqrt{x} - 3$

68. $y = 2|x|$

69. $y = 3x - 5$

70. $g(x) = (x + 2)^2$

71. $y = -x^2 + 5x - 4$

72. $y = -2|x - 1| + 4$

Solve each problem. See Example 11.

73. *Maximum height.* If an object is projected upward with an initial velocity of 64 feet per second, then its height is a function of time, given by $h(t) = -16t^2 + 64t$. Graph this function for $0 \le t \le 4$. What is the maximum height reached by this object?

74. *Maximum height.* If a soccer ball is kicked straight up with an initial velocity of 32 feet per second, then its height above the earth is a function of time given by $h(t) = -16t^2 + 32t$. Graph this function for $0 \le t \le 2$. What is the maximum height reached by this ball?

75. *Minimum cost.* The sheriff has determined that the cost in dollars per day per prisoner for housing prisoners in the Macon County jail is a function of the number of prisoners n, where

$$C(n) = 0.015n^2 - 4.5n + 400.$$

Find the minimum cost per prisoner?

76. *Air pollution.* The amount of nitrogen dioxide present in the air in the city of Springfield on a certain day in July is shown in the accompanying figure.
   a) Use the graph to estimate the time of day when the nitrogen dioxide level was at its maximum.
   b) The amount of nitrogen dioxide in parts per million (ppm) is approximated by the function

$$A(t) = -2t^2 + 32t + 42,$$

where t is the number of hours after 6:00 A.M. Use this function to find the time at which the nitrogen dioxide level was at its maximum.

Figure for Exercise 76

Solve each problem.

77. *Amarillo's recovery.* Since 1990 the number of housing permits issued annually in Amarillo, Texas, has been increasing steadily (American Economic Development Corporation). If $h(t)$ is the number of permits issued in the year $1990 + t$, then the function $h(t) = 103t + 93$ can be used to model this growth. If this trend continues, then how many housing permits will be issued in 1998? In what year were 402 housing permits issued?

78. *Radius of a circle.* The function $r(A) = \sqrt{\dfrac{A}{\pi}}$ gives the radius of a circle as a function of its area A. Find the radius of a circle that has area 500 cm². Round your answer to the nearest tenth of a centimeter. Find the area of a circle that has radius 30 ft. Round the answer to the nearest tenth of a square foot.

Getting More Involved

79. *Exploration.* Write the function $f(x) = 3(x - 2)^2 + 6$ in the form $f(x) = ax^2 + bx + c$ and find the vertex of the parabola. How is the vertex related to the original form of the function? What is the vertex of $f(x) = -4(x - 5)^2 + 9$? What is the vertex of $f(x) = 3(x + 2)^2 - 6$?

80. *Exploration.* Write the function $f(x) = a(x - h)^2 + k$ in the form $f(x) = ax^2 + bx + c$ and find the vertex of the parabola.

⚙ Graphing Calculator Exercises

81. Graph the function $f(x) = \sqrt{x^2}$ and explain what this graph illustrates.

82. Graph the function $f(x) = \dfrac{1}{x}$ and state the domain and range.

83. Graph $y = x^2$, $y = \frac{1}{2}x^2$, and $y = 2x^2$ on the same coordinate system. What can you say about the graph of $y = kx^2$?

84. Graph $y = (x - 3)^2$, $y = |x - 3|$, and $y = \sqrt{x - 3}$ on the same coordinate system. How does the graph of $y = f(x - k)$ compare to the graph of $y = f(x)$?

85. You can graph a relation such as $x = y^2$ by solving it for y to get $y = \pm\sqrt{x}$ and graphing the two functions $y = \sqrt{x}$ and $y = -\sqrt{x}$. Try it.

86. Graph each of the following relations by solving for y. See Exercise 85.

    a) $x = y^2 - 1$ **b)** $x = -y^2$ **c)** $x^2 + y^2 = 4$

8.3 Combining Functions

In this section:

▶ Basic Operations with Functions

▶ Composition

In Sections 8.1 and 8.2 you learned the fundamental ideas about functions and their graphs. In this section you will learn how to combine functions to obtain new functions.

Basic Operations with Functions

An entrepreneur plans to rent a stand at a farmers market for $25 per day to sell strawberries. If she buys x flats of berries for $5 per flat and sells them for $9 per flat, then her daily cost in dollars can be written as a function of x:

$$C(x) = 5x + 25$$

Assuming that she sells as many flats as she buys, her revenue in dollars is also a function of x:

$$R(x) = 9x$$

Because profit is revenue minus cost, we can find a function for the profit by subtracting the functions for cost and revenue:

$$P(x) = R(x) - C(x)$$
$$= 9x - (5x + 25)$$
$$= 4x - 25$$

The function $P(x) = 4x - 25$ expresses the daily profit as a function of x.

In the example of the entrepreneur we subtracted two functions to find a new function. In other cases we may use addition, multiplication, or division to combine two functions. For any two given functions we can define the sum, difference, product, and quotient functions as follows.

Sum, Difference, Product, and Quotient Functions

Given two functions f and g, the functions $f + g$, $f - g$, $f \cdot g$, and $\frac{f}{g}$ are defined as follows:

Sum function: $\qquad (f + g)(x) = f(x) + g(x)$

Difference function: $\qquad (f - g)(x) = f(x) - g(x)$

Product function: $\qquad (f \cdot g)(x) = f(x) \cdot g(x)$

Quotient function: $\qquad \left(\dfrac{f}{g}\right)(x) = \dfrac{f(x)}{g(x)} \qquad$ provided that $g(x) \neq 0$

The domain of the function $f + g$, $f - g$, $f \cdot g$, or $\frac{f}{g}$ is the intersection of the domain of f and the domain of g. For the function $\frac{f}{g}$ we also rule out any values of x for which $g(x) = 0$.

EXAMPLE 1   **Operations with functions**

Let $f(x) = 4x - 12$ and $g(x) = x - 3$. Find the following.

a) $(f + g)(x)$ **b)** $(f - g)(x)$ **c)** $(f \cdot g)(x)$ **d)** $\left(\dfrac{f}{g}\right)(x)$

Solution

a) $(f + g)(x) = f(x) + g(x)$
$$= 4x - 12 + x - 3$$
$$= 5x - 15$$

b) $(f - g)(x) = f(x) - g(x)$
$$= 4x - 12 - (x - 3)$$
$$= 3x - 9$$

c) $(f \cdot g)(x) = f(x) \cdot g(x)$
$$= (4x - 12)(x - 3)$$
$$= 4x^2 - 24x + 36$$

d) $\left(\dfrac{f}{g}\right)(x) = \dfrac{f(x)}{g(x)} = \dfrac{4x - 12}{x - 3} = \dfrac{4(x - 3)}{x - 3} = 4 \qquad$ for $x \neq 3$.   ∎

EXAMPLE 2   **Evaluating a sum function**

Let $f(x) = 4x - 12$ and $g(x) = x - 3$. Find $(f + g)(2)$.

Solution

In Example l(a) we found a general formula for the function $f + g$, namely, $(f + g)(x) = 5x - 15$. If we replace x by 2, we get

$$(f + g)(2) = 5(2) - 15$$
$$= -5.$$

We can also find $(f + g)(2)$ by evaluating each function separately and then adding the results. Because $f(2) = -4$ and $g(2) = -1$, we get

$$(f + g)(2) = f(2) + g(2)$$
$$= -4 + (-1)$$
$$= -5.$$ ∎

Composition

A salesperson's monthly salary is a function of the number of cars he sells: $1000 plus $50 for each car sold. If we let S be his salary and n be the number of cars sold, then S in dollars is a function of n:

$$S = 1000 + 50n$$

Each month the dealer contributes $100 plus 5% of his salary to a profit-sharing plan. If P represents the amount put into profit sharing, then P (in dollars) is a function of S:

$$P = 100 + 0.05S$$

Now P is a function of S, and S is a function of n. Is P a function of n? The value of n certainly determines the value of P. In fact, we can write a formula for P in terms of n by substituting one formula into the other:

$$P = 100 + 0.05S$$
$$= 100 + 0.05(1000 + 50n) \quad \text{Substitute } S = 1000 + 50n.$$
$$= 100 + 50 + 2.5n \quad \text{Distributive property}$$
$$= 150 + 2.5n$$

Now P is written as a function of n, bypassing S. We call this idea **composition of functions.**

EXAMPLE 3 **The composition of two functions**

Given that $y = x^2 - 2x + 3$ and $z = 2y - 5$, write z as a function of x.

Solution

Replace y in $z = 2y - 5$ by $x^2 - 2x + 3$:

$$z = 2y - 5$$
$$= 2(x^2 - 2x + 3) - 5 \quad \text{Replace } y \text{ by } x^2 - 2x + 3.$$
$$= 2x^2 - 4x + 1$$

The equation $z = 2x^2 - 4x + 1$ expresses z as a function of x. ∎

The composition of two functions using f-notation is defined as follows.

Composition of Functions

The **composition** of f and g is denoted $f \circ g$ and is defined by the equation

$$(f \circ g)(x) = f(g(x)),$$

provided that $g(x)$ is in the domain of f.

The notation $f \circ g$ is read as "the composition of f and g" or "f compose g." The diagram in Fig. 8.16 shows a function g pairing numbers in its domain with numbers in its range. If the range of g is contained in or equal to the domain of f, then f pairs the second coordinates of g with numbers in the range of f. The composition function $f \circ g$ is a rule for pairing numbers in the domain of g directly with numbers in the range of f, bypassing the middle set. The domain of the function $f \circ g$ is the domain of g (or a subset of it) and the range of $f \circ g$ is the range of f (or a subset of it).

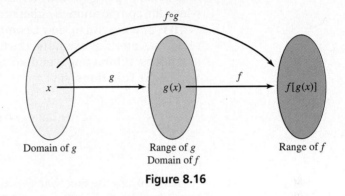

Figure 8.16

CAUTION The order in which functions are written is important in composition. For the function $f \circ g$ the function f is applied to $g(x)$. For the function $g \circ f$ the function g is applied to $f(x)$. The function closest to the variable x is applied first. ⊘

EXAMPLE 4 **Composition of functions**

Let $f(x) = 3x - 2$ and $g(x) = x^2 + 2x$. Find the following.

a) $(g \circ f)(2)$　　b) $(f \circ g)(2)$　　c) $(g \circ f)(x)$　　d) $(f \circ g)(x)$

Solution

a) Because $(g \circ f)(2) = g(f(2))$, we first find $f(2)$:

$$f(2) = 3 \cdot 2 - 2 = 4$$

Because $f(2) = 4$, we have

$$(g \circ f)(2) = g(f(2)) = g(4) = 4^2 + 2 \cdot 4 = 24.$$

So $(g \circ f)(2) = 24$.

b) Because $(f \circ g)(2) = f(g(2))$, we first find $g(2)$:

$$g(2) = 2^2 + 2 \cdot 2 = 8$$

Because $g(2) = 8$, we have

$$(f \circ g)(2) = f(g(2)) = f(8) = 3 \cdot 8 - 2 = 22.$$

Thus $(f \circ g)(2) = 22$.

c) $(g \circ f)(x) = g(f(x))$

$\qquad = g(3x - 2)$

$\qquad = (3x - 2)^2 + 2(3x - 2)$

$\qquad = 9x^2 - 12x + 4 + 6x - 4 = 9x^2 - 6x$

So $(g \circ f)(x) = 9x^2 - 6x$.

d) $(f \circ g)(x) = f(g(x))$

$\qquad = f(x^2 + 2x)$

$\qquad = 3(x^2 + 2x) - 2 = 3x^2 + 6x - 2$

So $(f \circ g)(x) = 3x^2 + 6x - 2.$ ∎

Notice that in Example 4(a) and (b), $(g \circ f)(2) \neq (f \circ g)(2)$. In Example 4(c) and (d) we see that $(g \circ f)(x)$ and $(f \circ g)(x)$ have different formulas defining them. In general, $f \circ g \neq g \circ f$. However, in Section 8.4 we will see some functions for which the composition in either order results in the same function.

It is often useful to view a complicated function as a composition of simpler functions. For example, the function $Q(x) = (x - 3)^2$ consists of two operations, subtracting 3 and squaring. So Q can be described as a composition of the functions $f(x) = x - 3$ and $g(x) = x^2$. To check this, we find $(g \circ f)(x)$:

$$(g \circ f)(x) = g(f(x))$$
$$= g(x - 3)$$
$$= (x - 3)^2$$

We can express the fact that Q is the same as the composition function $g \circ f$ by writing $Q = g \circ f$ or $Q(x) = (g \circ f)(x)$.

EXAMPLE 5 **Expressing a function as a composition of simpler functions**

Let $f(x) = x - 2$, $g(x) = 3x$, and $h(x) = \sqrt{x}$. Write each of the following functions as a composition, using f, g, and h.

a) $F(x) = \sqrt{x - 2}$ **b)** $H(x) = x - 4$ **c)** $K(x) = 3x - 6$

Solution

a) The function F consists of first subtracting 2 from x and then taking the square root of that result. So $F = h \circ f$. Check this result by finding $(h \circ f)(x)$:

$$(h \circ f)(x) = h(f(x)) = h(x - 2) = \sqrt{x - 2}$$

b) Subtracting 4 from x can be accomplished by subtracting 2 from x and then subtracting 2 from that result. So $H = f \circ f$. Check by finding $(f \circ f)(x)$:

$$(f \circ f)(x) = f(f(x)) = f(x - 2) = x - 2 - 2 = x - 4$$

c) Notice that $K(x) = 3(x - 2)$. The function K consists of subtracting 2 from x and then multiplying the result by 3. So $K = g \circ f$. Check by finding $(g \circ f)(x)$:

$$(g \circ f)(x) = g(f(x)) = g(x - 2) = 3(x - 2) = 3x - 6$$ ∎

CAUTION In Example 5(a) we have $F = h \circ f$ because in F we subtract 2 before taking the square root. If we had the function $G(x) = \sqrt{x} - 2$, we would take the square root before subtracting 2. So $G = f \circ h$. Notice how important the order of operations is here. ⊘

In the next example we see functions for which the composition is the identity function. Each function undoes what the other function does. We will study functions of this type further in Section 8.4.

EXAMPLE 6   **Composition of functions**

Show that $(f \circ g)(x) = x$ for each pair of functions.

a) $f(x) = 2x - 1$ and $g(x) = \dfrac{x+1}{2}$ **b)** $f(x) = x^3 + 5$ and $g(x) = (x-5)^{1/3}$

Solution

a) $(f \circ g)(x) = f(g(x)) = f\left(\dfrac{x+1}{2}\right)$

$$= 2\left(\dfrac{x+1}{2}\right) - 1$$
$$= x + 1 - 1$$
$$= x$$

b) $(f \circ g)(x) = f(g(x)) = f((x-5)^{1/3})$
$$= ((x-5)^{1/3})^3 + 5$$
$$= x - 5 + 5$$
$$= x$$

Warm-ups

True or false? Explain your answer.

1. If $f(x) = x - 2$ and $g(x) = x + 3$, then $(f - g)(x) = -5$.
2. If $f(x) = x + 4$ and $g(x) = 3x$, then $\left(\dfrac{f}{g}\right)(2) = 1$.
3. The functions $f \circ g$ and $g \circ f$ are always the same.
4. If $f(x) = x^2$ and $g(x) = x + 2$, then $(f \circ g)(x) = x^2 + 2$.
5. The functions $f \circ g$ and $f \cdot g$ are always the same.
6. If $f(x) = \sqrt{x}$ and $g(x) = x - 9$, then $g(f(x)) = f(g(x))$ for every x.
7. If $f(x) = 3x$ and $g(x) = \dfrac{x}{3}$, then $(f \circ g)(x) = x$.
8. If $a = 3b^2 - 7b$, and $c = a^2 + 3a$, then c is a function of b.
9. The function $F(x) = \sqrt{x-5}$ is a composition of two functions.
10. If $F(x) = (x-1)^2$, $h(x) = x - 1$, and $g(x) = x^2$, then $F = g \circ h$.

8.3 EXERCISES

Let $f(x) = 4x - 3$, and $g(x) = x^2 - 2x$. Find the following. See Examples 1 and 2.

1. $(f + g)(x)$
2. $(f - g)(x)$
3. $(f \cdot g)(x)$
4. $\left(\dfrac{f}{g}\right)(x)$
5. $(f + g)(3)$
6. $(f + g)(2)$
7. $(f - g)(-3)$
8. $(f - g)(-2)$
9. $(f \cdot g)(-1)$
10. $(f \cdot g)(-2)$
11. $\left(\dfrac{f}{g}\right)(4)$
12. $\left(\dfrac{f}{g}\right)(-2)$

For Exercises 13–20, use the two functions to write y as a function of x. See Example 3.

13. $y = 3a - 2$, $a = 2x - 6$
14. $y = 2c + 3$, $c = -3x + 4$
15. $y = 2d + 1$, $d = \dfrac{x+1}{2}$
16. $y = -3d + 2$, $d = \dfrac{2-x}{3}$

17. $y = m^2 - 1$, $m = x + 1$

18. $y = n^2 - 3n + 1$, $n = x + 2$

19. $y = \dfrac{a - 3}{a + 2}$, $a = \dfrac{2x + 3}{1 - x}$

20. $y = \dfrac{w + 2}{w - 5}$, $w = \dfrac{5x + 2}{x - 1}$

Let $f(x) = 2x - 3$, $g(x) = x^2 + 3x$, and $h(x) = \dfrac{x + 3}{2}$. Find the following. See Example 4.

21. $(g \circ f)(1)$ **22.** $(f \circ g)(-2)$

23. $(f \circ g)(1)$ **24.** $(g \circ f)(-2)$

25. $(f \circ f)(4)$ **26.** $(h \circ h)(3)$

27. $(h \circ f)(5)$ **28.** $(f \circ h)(0)$

29. $(f \circ h)(5)$ **30.** $(h \circ f)(0)$

31. $(g \circ h)(-1)$ **32.** $(h \circ g)(-1)$

33. $(f \circ g)(2.36)$ **34.** $(h \circ f)(23.761)$

35. $(g \circ f)(x)$ **36.** $(g \circ h)(x)$

37. $(f \circ g)(x)$ **38.** $(h \circ g)(x)$

39. $(h \circ f)(x)$ **40.** $(f \circ h)(x)$

41. $(f \circ f)(x)$ **42.** $(g \circ g)(x)$

43. $(h \circ h)(x)$ **44.** $(f \circ f \circ f)(x)$

Let $f(x) = \sqrt{x}$, $g(x) = x^2$, and $h(x) = x - 3$. Write each of the following functions as a composition using f, g, or h. See Example 5.

45. $F(x) = \sqrt{x - 3}$ **46.** $N(x) = \sqrt{x} - 3$

47. $G(x) = x^2 - 6x + 9$ **48.** $P(x) = x$

49. $H(x) = x^2 - 3$ **50.** $M(x) = x^{1/4}$

51. $J(x) = x - 6$ **52.** $R(x) = \sqrt{x^2 - 3}$

53. $K(x) = x^4$ **54.** $Q(x) = \sqrt{x^2 - 6x + 9}$

Show that $(f \circ g)(x) = x$ and $(g \circ f)(x) = x$ for each given pair of functions. See Example 6.

55. $f(x) = 3x + 5$, $g(x) = \dfrac{x - 5}{3}$

56. $f(x) = 3x - 7$, $g(x) = \dfrac{x + 7}{3}$

57. $f(x) = x^3 - 9$, $g(x) = \sqrt[3]{x + 9}$

58. $f(x) = x^3 + 1$, $g(x) = \sqrt[3]{x - 1}$

59. $f(x) = \dfrac{x - 1}{x + 1}$, $g(x) = \dfrac{x + 1}{1 - x}$

60. $f(x) = \dfrac{x + 1}{x - 3}$, $g(x) = \dfrac{3x + 1}{x - 1}$

61. $f(x) = \dfrac{1}{x}$, $g(x) = \dfrac{1}{x}$

62. $f(x) = 2x^3$, $g(x) = \left(\dfrac{x}{2}\right)^{1/3}$

Solve each problem.

63. *Color monitor.* The CTX CMS-1561 color monitor has a square viewing area that has a diagonal measure of 15 inches (Midwest Micro Catalog). Find the area of the viewing area in square inches. Write a formula for the area of a square as a function of the length of its diagonal.

64. *Perimeter.* Write a formula for the perimeter of a square as a function of its area.

65. *Profit function.* A plastic-bag manufacturer has determined that the company can sell as many bags as it can produce each month. If it produces x thousand bags in a month, the revenue is $R(x) = x^2 - 10x + 30$ dollars, and the cost is $C(x) = 2x^2 - 30x + 200$ dollars. Use the fact that profit is revenue minus cost to write the profit as a function of x.

66. *Area of a sign.* A sign is in the shape of a square with a semicircle of radius x adjoining one side, and a semicircle of diameter x removed from the opposite side. If the sides of the square are length $2x$, then write the area of the sign as a function of x.

Figure for Exercise 66

67. *Junk food expenditures.* Suppose that the average family spends 25% of its income on food, $F = 0.25I$, and 10% of each food dollar on junk food, $J = 0.10F$. Write J as a function of I.

68. *Area of an inscribed circle.* A pipe of radius r must pass through a square hole of area M as shown in the figure. Write the cross-sectional area of the pipe A as a function of M.

Figure for Exercise 68

Getting More Involved

69. *Discussion.* Let $f(x) = \sqrt{x} - 4$ and $g(x) = \sqrt{x}$. Find the domains of f, g, and $g \circ f$.

70. *Discussion.* Let $f(x) = \sqrt{x - 4}$ and $g(x) = \sqrt{x - 8}$. Find the domains of f, g, and $f + g$.

Graphing Calculator Exercises

71. Graph $y_1 = x$, $y_2 = \sqrt{x}$, and $y_3 = x + \sqrt{x}$ in the same screen. Find the domain and range of $y_3 = x + \sqrt{x}$ by examining its graph. (On some graphing calculators you can enter y_3 as $y_3 = y_1 + y_2$.)

72. Graph $y_1 = |x|$, $y_2 = |x - 3|$, and $y_3 = |x| + |x - 3|$. Find the domain and range of $y_3 = |x| + |x - 3|$ by examining its graph.

8.4   Inverse Functions

In this section:

▶ Inverse of a Function
▶ Identifying Inverse Functions
▶ Switch-and-Solve Strategy
▶ Even Roots or Even Powers
▶ Graphs of f and f^{-1}

In Section 8.3 we introduced the idea of a pair of functions such that $(f \circ g)(x) = x$ and $(g \circ f)(x) = x$. Each function reverses what the other function does. In this section we explore that idea further.

Inverse of a Function

You can buy a 6-, 7-, or 8-foot conference table in the 1994 K-LOG Catalog for \$299, \$329, or \$349, respectively. The set

$$f = \{(6, 299), (7, 329), (8, 349)\}$$

gives the price as a function of the length. We use the letter f as a name for this set or function, just as we use the letter f as a name for a function in the f-notation. In the function f, lengths in the domain $\{6, 7, 8\}$ are paired with prices in the range $\{299, 329, 349\}$. The **inverse** of the function f, denoted f^{-1}, is a function whose ordered pairs are obtained from f by interchanging the x- and y-coordinates:

$$f^{-1} = \{(299, 6), (329, 7), (349, 8)\}$$

We read f^{-1} as "f inverse." The domain of f^{-1} is $\{299, 329, 349\}$, and the range of f^{-1} is $\{6, 7, 8\}$. The inverse function f^{-1} pairs prices in the range of f with lengths in the domain of f. It reverses what the function f does. In general, *the domain of f^{-1} is the range of f, and the range of f^{-1} is the domain of f.* See Fig. 8.17.

Domain of f Range of f

6 f 299
7 f^{-1} 329
8 349

Range of f^{-1} Domain of f^{-1}

Figure 8.17

CAUTION   The -1 in f^{-1} is not read as an exponent. It does not mean $\dfrac{1}{f}$. ⊘

In Section 8.1 we saw that the cost per ribbon for Apple Imagewriter ribbons is a function of the number of boxes purchased:

$$g = \{(1, 4.85), (2, 4.60), (3, 4.60), (4, 4.35)\}$$

If we interchange the first and second coordinates in the ordered pairs of this function, we get

$$\{(4.85, 1),\ (4.60, 2),\ (4.60, 3),\ (4.35, 4)\}.$$

This set of ordered pairs is not a function because it contains ordered pairs with the same first coordinates and different second coordinates. So g does not have an inverse function. We say that f is **invertible** and g is **not invertible.** The reason g is not invertible is that the definition of function allows more than one number of the domain to be paired with the same number in the range. Of course, when this pairing is reversed, the definition of function is violated.

One-to-One Function

If a function is such that no two ordered pairs have different x-coordinates and the same y-coordinate, then the function is called a **one-to-one** function.

In a one-to-one function, each member of the domain corresponds to just one member of the range, and each member of the range corresponds to just one member of the domain. *Functions that are one-to-one are invertible functions.*

Inverse Function

The inverse of a one-to-one function f is the function f^{-1}, which is obtained from f by interchanging the coordinates in each ordered pair of f.

EXAMPLE 1 **Identifying invertible functions**

Determine whether each function is invertible. If it is invertible, then find the inverse function.

a) $f = \{(2, 4),\ (-2, 4),\ (3, 9)\}$

b) $g = \left\{ \left(2, \dfrac{1}{2}\right),\ \left(5, \dfrac{1}{5}\right),\ \left(7, \dfrac{1}{7}\right) \right\}$

c) $h = \{(3, 5),\ (7, 9)\}$

Solution

a) Since $(2, 4)$ and $(-2, 4)$ have the same y-coordinate, this function is not one-to-one, and it is not invertible.

b) This function is one-to-one, and so it is invertible.

$$g^{-1} = \left\{ \left(\dfrac{1}{2}, 2\right),\ \left(\dfrac{1}{5}, 5\right),\ \left(\dfrac{1}{7}, 7\right) \right\}$$

c) This function is invertible, and $h^{-1} = \{(5, 3),\ (9, 7)\}$. ■

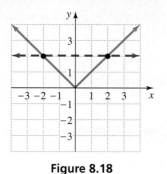

Figure 8.18

You learned to use the vertical-line test in Section 8.2 to determine whether a graph is the graph of a function. The **horizontal-line test** is a similar visual test for determining whether a function is invertible. If a horizontal line crosses a graph two (or more) times, as in Fig. 8.18, then there are two points on the graph, say (x_1, y) and (x_2, y), that have different x-coordinates and the same y-coordinate. So the function is not one-to-one, and the function is not invertible.

> **Horizontal-Line Test**
>
> A function is invertible if and only if no horizontal line crosses its graph more than once.

EXAMPLE 2 **Using the horizontal-line test**

Determine whether each function is invertible by examining its graph.

a) **b)**

Solution

a) This function is not invertible, because a horizontal line can be drawn so that it crosses the graph at $(2, 4)$ and $(-2, 4)$.

b) This function is invertible, because every horizontal line that crosses the graph crosses it only once. ∎

Identifying Inverse Functions

Consider the one-to-one function $f(x) = 3x$. The inverse function must reverse the ordered pairs of the function. Because division by 3 undoes multiplication by 3, we could guess that $g(x) = \dfrac{x}{3}$ is the inverse function. To verify our guess, we can use the following rule for determining whether two given functions are inverses of each other.

> **Identifying Inverse Functions**
>
> Functions f and g are inverses of each other if and only if
>
> $(g \circ f)(x) = x$ for every number x in the domain of f and
> $(f \circ g)(x) = x$ for every number x in the domain of g.

In the next example we verify that $f(x) = 3x$ and $g(x) = \dfrac{x}{3}$ are inverses.

EXAMPLE 3   **Identifying inverse functions**
Determine whether the functions f and g are inverses of each other.

a) $f(x) = 3x$ and $g(x) = \dfrac{x}{3}$

b) $f(x) = 2x - 1$ and $g(x) = \dfrac{1}{2}x + 1$

c) $f(x) = x^2$ and $g(x) = \sqrt{x}$

Solution

a) Find $g \circ f$ and $f \circ g$:

$$(g \circ f)(x) = g(f(x)) = g(3x) = \frac{3x}{3} = x$$

$$(f \circ g)(x) = f(g(x)) = f\left(\frac{x}{3}\right) = 3 \cdot \frac{x}{3} = x$$

Because each of these equations is true for any real number x, f and g are inverses of each other. We write $g = f^{-1}$ or $f^{-1}(x) = \dfrac{x}{3}$.

b) Find the composition of g and f:

$$(g \circ f)(x) = g(f(x))$$
$$= g(2x - 1) = \frac{1}{2}(2x - 1) + 1 = x + \frac{1}{2}$$

So f and g are not inverses of each other.

c) If x is any real number, we can write

$$(g \circ f)(x) = g(f(x))$$
$$= g(x^2) = \sqrt{x^2} = |x|.$$

The domain of f is $(-\infty, \infty)$, and $|x| \neq x$ if x is negative. So g and f are not inverses of each other. Note that $f(x) = x^2$ is not a one-to-one function, since both $(3, 9)$ and $(-3, 9)$ are ordered pairs of this function. Thus $f(x) = x^2$ does not have an inverse. ∎

Switch-and-Solve Strategy

If an invertible function is defined by a list of ordered pairs, as in Example 1, then the inverse function is found by simply interchanging the coordinates in the ordered pairs. If an invertible function is defined by a formula, then the inverse function must reverse or undo what the function does. Because the inverse function interchanges the roles of x and y, we interchange x and y in the formula and then solve the new formula for y to undo what the original function did. This **switch-and-solve** strategy is illustrated in the next two examples.

EXAMPLE 4   **The switch-and-solve strategy**

Find the inverse of $h(x) = 2x + 1$.

Solution

First write the function as $y = 2x + 1$, then interchange x and y:

$$y = 2x + 1$$

$$x = 2y + 1 \qquad \text{Interchange } x \text{ and } y.$$

$$x - 1 = 2y \qquad \text{Solve for } y.$$

$$\frac{x - 1}{2} = y$$

$$h^{-1}(x) = \frac{x - 1}{2} \qquad \text{Replace } y \text{ by } h^{-1}(x).$$

We can verify that h and h^{-1} are inverses by using composition:

$$(h^{-1} \circ h)(x) = h^{-1}(h(x)) = h^{-1}(2x + 1)$$

$$= \frac{2x + 1 - 1}{2} = \frac{2x}{2} = x$$

$$(h \circ h^{-1})(x) = h(h^{-1}(x)) = h\left(\frac{x - 1}{2}\right)$$

$$= 2 \cdot \frac{x - 1}{2} + 1 = x - 1 + 1 = x \qquad \blacksquare$$

EXAMPLE 5   **The switch-and-solve strategy**

If $f(x) = \dfrac{x + 1}{x - 3}$, find $f^{-1}(x)$.

Solution

Replace $f(x)$ by y, interchange x and y, then solve for y:

$$y = \frac{x + 1}{x - 3} \qquad \text{Use } y \text{ in place of } f(x).$$

$$x = \frac{y + 1}{y - 3} \qquad \text{Switch } x \text{ and } y.$$

$$x(y - 3) = y + 1 \qquad \text{Multiply each side by } y - 3.$$

$$xy - 3x = y + 1 \qquad \text{Distributive property}$$

$$xy - y = 3x + 1$$

$$y(x - 1) = 3x + 1 \qquad \text{Factor out } y.$$

$$y = \frac{3x + 1}{x - 1} \qquad \text{Divide each side by } x - 1.$$

$$f^{-1}(x) = \frac{3x + 1}{x - 1} \qquad \text{Replace } y \text{ by } f^{-1}(x).$$

You should check that $(f \circ f^{-1})(x) = x$ and $(f^{-1} \circ f)(x) = x$. $\qquad \blacksquare$

The strategy for finding the inverse of a function $f(x)$ is summarized as follows.

>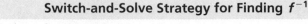
> ### Switch-and-Solve Strategy for Finding f^{-1}
> 1. Replace $f(x)$ by y.
> 2. Interchange x and y.
> 3. Solve the equation for y.
> 4. Replace y by $f^{-1}(x)$.

If we use the switch-and-solve strategy to find the inverse of $f(x) = x^3$, then we get $f^{-1}(x) = x^{1/3}$. For $h(x) = 6x$ we have $h^{-1}(x) = \dfrac{x}{6}$. The inverse of $k(x) = x - 9$ is $k^{-1}(x) = x + 9$. For each of these functions there is an appropriate operation of arithmetic that undoes what the function does.

If a function involves two operations, the inverse function undoes those operations in the opposite order from which the function does them. For example, the function $g(x) = 3x - 5$ multiplies x by 3 and then subtracts 5 from that result. To undo these operations, we add 5 and then divide the result by 3. So

$$g^{-1}(x) = \frac{x + 5}{3}.$$

Note that $g^{-1}(x) \neq \dfrac{x}{3} + 5$.

MATH AT WORK

State Police Officer

When a serious automobile accident occurs in Massachusetts, Stephen Benanti, the Commanding Officer of the Accident Reconstruction Section of the State Police, may be called to examine the physical evidence at the scene. Physical evidence can consist of debris, scrapes, and gouges on the road; damage to fixed objects such as utility poles, trees, or guardrails; and the final resting position of and damage to the vehicles.

One critical type of evidence that is sometimes found are skid marks on the road. Sergeant Benanti can use the lengths of the skid marks to calculate the speeds of accident vehicles. Using a sophisticated laser measuring device, Benanti can store data that can later be downloaded into a computer to reconstruct the accident scene. He must also conduct tests on the road surface to calculate the drag factor, which is the resistance between the tire and the road surface. A smooth, icy surface yields a much lower drag factor than a dry asphalt surface. The minimum speed formula that state troopers use to determine vehicles' speeds is a function of the drag factor and the skid distance: $S = \sqrt{30DF}$, where $S =$ speed, $D =$ distance skidded, and $F =$ drag factor.

In Exercise 65 of this section you will use this minimum speed formula with a given length of skid marks to determine the speed of a vehicle.

Even Roots or Even Powers

We need to use special care in finding inverses for functions that involve even roots or even powers. We saw in Example 3(c) that $f(x) = x^2$ is not the inverse of $g(x) = \sqrt{x}$. However, because $g(x) = \sqrt{x}$ is a one-to-one function, it has an inverse. The domain of g is $[0, \infty)$, and the range is $[0, \infty)$. So the inverse of g must have domain $[0, \infty)$ and range $[0, \infty)$. See Fig. 8.19. The only reason that $f(x) = x^2$ is not the inverse of g is that it has the wrong domain. So to write the inverse function, we must use the appropriate domain:

$$g^{-1}(x) = x^2 \qquad \text{for } x \geq 0$$

Note that by restricting the domain of g^{-1} to $[0, \infty)$, g^{-1} is one-to-one. With this restriction it is true that $(g \circ g^{-1})(x) = x$ and $(g^{-1} \circ g)(x) = x$ for every nonnegative number x.

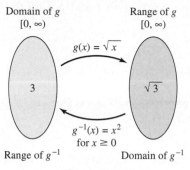

Domain of g
$[0, \infty)$

Range of g
$[0, \infty)$

$g(x) = \sqrt{x}$

3

$\sqrt{3}$

$g^{-1}(x) = x^2$
for $x \geq 0$

Range of g^{-1} Domain of g^{-1}

Figure 8.19

EXAMPLE 6   **Inverse of a function with an even exponent**

Find the inverse of the function $f(x) = (x - 3)^2$ for $x \geq 3$.

Solution

Because of the restriction $x \geq 3$, f is a one-to-one function with domain $[3, \infty)$ and range $[0, \infty)$. The domain of the inverse function is $[0, \infty)$, and its range is $[3, \infty)$. Use the switch-and-solve strategy to find the formula for the inverse:

$$y = (x - 3)^2$$
$$x = (y - 3)^2$$
$$y - 3 = \pm\sqrt{x}$$
$$y = 3 \pm \sqrt{x}$$

To get the range $[3, \infty)$, we use the formula $f^{-1}(x) = 3 + \sqrt{x}$. Because the domain of f^{-1} is assumed to be $[0, \infty)$, no restriction is required on x. ■

Graphs of f and f^{-1}

Consider $f(x) = x^2$ for $x \geq 0$ and $f^{-1}(x) = \sqrt{x}$. Their graphs are shown in Fig. 8.20. Notice the symmetry. If we folded the paper along the line $y = x$, the two graphs would coincide.

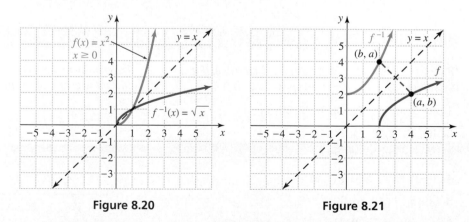

Figure 8.20   **Figure 8.21**

If a point (a, b) is on the graph of the function f, then (b, a) must be on the graph of f^{-1}. See Fig. 8.21. The points (a, b) and (b, a) lie on opposite sides of the diagonal line $y = x$ and are the same distance from it. For this reason the graphs of f and f^{-1} are symmetric with respect to the line $y = x$.

Figure 8.22

EXAMPLE 7

Inverses and their graphs

Find the inverse of the function $f(x) = \sqrt{x-1}$ and graph f and f^{-1} on the same pair of axes.

Solution

To find f^{-1}, first switch x and y in the formula $y = \sqrt{x-1}$:

$$x = \sqrt{y-1}$$
$$x^2 = y - 1 \quad \text{Square both sides.}$$
$$x^2 + 1 = y$$

Because the range of f is the set of nonnegative real numbers $[0, \infty)$, we must restrict the domain of f^{-1} to be $[0, \infty)$. Thus $f^{-1}(x) = x^2 + 1$ for $x \geq 0$. The two graphs are shown in Fig. 8.22. ∎

Warm-ups

True or false? Explain your answer.

1. The inverse of $\{(1, 3), (2, 5)\}$ is $\{(3, 1), (2, 5)\}$

2. The function $f(x) = 3$ is a one-to-one function.

3. If $g(x) = 2x$, then $g^{-1}(x) = \dfrac{1}{2x}$.

4. Only one-to-one functions are invertible.

5. The domain of g is the same as the range of g^{-1}.

6. The function $f(x) = x^4$ is invertible.

7. If $f(x) = -x$, then $f^{-1}(x) = -x$.

8. If h is invertible and $h(7) = -95$, then $h^{-1}(-95) = 7$.

9. If $k(x) = 3x - 6$, then $k^{-1}(x) = \frac{1}{3}x + 2$.

10. If $f(x) = 3x - 4$, then $f^{-1}(x) = x + 4$.

8.4 EXERCISES

Determine whether each function is invertible. If it is invertible, then find the inverse. See Example 1.

1. $\{(-3, 3), (-2, 2), (0, 0), (2, 2)\}$

2. $\{(1, 1), (2, 8), (3, 27)\}$

3. $\{(16, 4), (9, 3), (0, 0)\}$

4. $\{(-1, 1), (-3, 81), (3, 81)\}$

5. $\{(0, 5), (5, 0), (6, 0)\}$

6. $\{(3, -3), (-2, 2), (1, -1)\}$

7. $\{(0, 0), (2, 2), (9, 9)\}$

8. $\{(9, 1), (2, 1), (7, 1), (0, 1)\}$

Determine whether each function is invertible by examining the graph of the function. See Example 2.

9.

10.

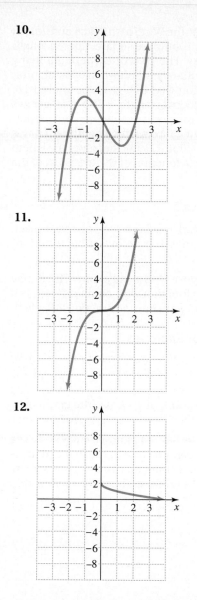

11.

12.

Determine whether each pair of functions f and g are inverses of each other. See Example 3.

13. $f(x) = 2x$ and $g(x) = 0.5x$

14. $f(x) = 3x$ and $g(x) = 0.33x$

15. $f(x) = 2x - 10$ and $g(x) = \frac{1}{2}x + 5$

16. $f(x) = 3x + 7$ and $g(x) = \frac{x - 7}{3}$

17. $f(x) = -x$ and $g(x) = -x$

18. $f(x) = \frac{1}{x}$ and $g(x) = \frac{1}{x}$

19. $f(x) = x^4$ and $g(x) = x^{1/4}$

20. $f(x) = |2x|$ and $g(x) = \left|\frac{x}{2}\right|$

Determine f^{-1} for each function by using the switch-and-solve strategy. Check that $(f \circ f^{-1})(x) = x$ and $(f^{-1} \circ f)(x) = x$. See Examples 4 and 5.

21. $f(x) = 5x$

22. $h(x) = -3x$

23. $g(x) = x - 9$

24. $j(x) = x + 7$

25. $k(x) = 5x - 9$

26. $r(x) = 2x - 8$

27. $m(x) = \frac{2}{x}$

28. $s(x) = \frac{-1}{x}$

29. $f(x) = \sqrt[3]{x - 4}$

30. $f(x) = \sqrt[3]{x + 2}$

31. $f(x) = \frac{3}{x - 4}$

32. $f(x) = \frac{2}{x + 1}$

33. $f(x) = \sqrt[3]{3x + 7}$

34. $f(x) = \sqrt[3]{7 - 5x}$

35. $f(x) = \frac{x + 1}{x - 2}$

36. $f(x) = \frac{1 - x}{x + 3}$

37. $f(x) = \frac{x + 1}{3x - 4}$

38. $g(x) = \frac{3x + 5}{2x - 3}$

Find the inverse of each function. See Example 6.

39. $p(x) = \sqrt[4]{x}$

40. $v(x) = \sqrt[6]{x}$

41. $f(x) = (x - 2)^2$ for $x \geq 2$

42. $g(x) = (x + 5)^2$ for $x \geq -5$

43. $f(x) = x^2 + 3$ for $x \geq 0$

44. $f(x) = x^2 - 5$ for $x \geq 0$

45. $f(x) = \sqrt{x + 2}$

46. $f(x) = \sqrt{x - 4}$

In Exercises 47–64, find the inverse of each function and graph f and f^{-1} on the same pair of axes. See Example 7.

47. $f(x) = 2x + 3$

48. $f(x) = -3x + 2$

49. $f(x) = x^2 - 1$ for $x \geq 0$

50. $f(x) = x^2 + 3$ for $x \geq 0$

51. $f(x) = 5x$

52. $f(x) = \frac{x}{4}$

53. $f(x) = x^3$

54. $f(x) = 2x^3$

55. $f(x) = \sqrt{x - 2}$

56. $f(x) = \sqrt{x + 3}$

For each pair of functions, find $(f^{-1} \circ f)(x)$.

57. $f(x) = x^3 - 1$ and $f^{-1}(x) = \sqrt[3]{x + 1}$

58. $f(x) = 2x^3 + 1$ and $f^{-1}(x) = \sqrt[3]{\frac{x - 1}{2}}$

59. $f(x) = \frac{1}{2}x - 3$ and $f^{-1}(x) = 2x + 6$

60. $f(x) = 3x - 9$ and $f^{-1}(x) = \frac{1}{3}x + 3$

61. $f(x) = \frac{1}{x} + 2$ and $f^{-1}(x) = \frac{1}{x - 2}$

62. $f(x) = 4 - \frac{1}{x}$ and $f^{-1}(x) = \frac{1}{4 - x}$

63. $f(x) = \frac{x + 1}{x - 2}$ and $f^{-1}(x) = \frac{2x + 1}{x - 1}$

64. $f(x) = \frac{3x - 2}{x + 2}$ and $f^{-1}(x) = \frac{2x + 2}{3 - x}$

Solve each problem.

65. *Accident reconstruction.*

   a) State troopers know that the distance it takes a car to stop depends on the speed of the car and the drag factor. The drag factor is the resistance between the tire and the road surface. For a surface with drag factor F, troopers use the formula $S = \sqrt{30DF}$ to find the minimum speed S (in miles per hour) for a car that has left skid marks of length D feet. Find the minimum speed for a car that has left skid marks 150 ft long, where the drag factor is 0.75.

   b) Suppose that for a certain road surface the drag factor is 1. The formula for finding minimum speed becomes $S = \sqrt{30D}$. Write D as a function of S.

66. *Area of a circle.* Let x be the radius of a circle and $h(x)$ be the area of the circle. Write a formula for $h(x)$ in terms of x. What does x represent in the notation $h^{-1}(x)$? Write a formula for $h^{-1}(x)$.

67. *Vehicle cost.* At Bill Hood Ford in Hammond, a sales tax of 9% of the selling price x and a $125 title and license fee are added to the selling price to get the total cost of a vehicle. Find the function $T(x)$ that the dealer uses to get the total cost as a function of the selling price x. Citizens National Bank will not include sales tax or fees in a loan. Find the function $T^{-1}(x)$ that the bank can use to get the selling price as a function of the total cost x.

68. *Carpeting cost.* At the Windrush Trace apartment complex, all living rooms are square, but the length of x feet may vary. The cost of carpeting a living room is $18 per square yard plus a $50 installation fee. Find the function $C(x)$ that gives the total cost of carpeting a living room of length x. The manager has an invoice for the total cost of a living room carpeting job but does not know in which apartment it was done. Find the function $C^{-1}(x)$ that gives the length of a living room as a function of the total cost of the carpeting job x.

Getting More Involved

69. *Discussion.* Let $f(x) = x^n$ for n a positive integer. For which values of n is f an invertible function? Explain.

70. *Discussion.* Suppose f is a function with range $(-\infty, \infty)$ and g is a function with domain $(0, \infty)$. Is it possible that g and f are inverse functions? Explain.

Graphing Calculator Exercises

71. Most graphing calculators can form compositions of functions. Let $f(x) = x^2$ and $g(x) = \sqrt{x}$. To graph the composition $g \circ f$, let $y_1 = x^2$ and $y_2 = \sqrt{y_1}$. The graph of y_2 is the graph of $g \circ f$. Use the graph of y_2 to determine whether f and g are inverse functions.

72. Let $y_1 = x^3 - 4$, $y_2 = \sqrt[3]{x + 4}$, and $y_3 = \sqrt[3]{y_1 + 4}$. The function y_3 is the composition of the first two functions. Graph all three functions on the same screen. What do the graphs indicate about the relationship between y_1 and y_2?

8.5   Variation

If $y = 3x$, then as x varies, so does y. Certain functions are customarily expressed in terms of variation. In this section you will learn to write formulas for those functions from verbal descriptions of the functions.

Direct Variation

In a community with an 8% sales tax rate, the amount of tax, t (in dollars), is a function of the amount of the purchase, a (in dollars). This function is expressed by the formula

$$t = 0.08a.$$

If the amount increases, then the tax increases. If a decreases, then t decreases. In this situation we say that t *varies directly with a, or t is directly proportional to a.* The constant tax rate, 0.08, is called the **variation constant** or **proportionality constant.** Notice that t is just a simple linear function of a. We are merely introducing some new terms to express an old idea.

> **Direct Variation**
>
> The statement **y varies directly as x,** or **y is directly proportional to x,** means that
>
> $$y = kx$$
>
> for some constant, k. The constant, k, is a fixed nonzero real number.

Finding the Proportionality Constant

If y varies directly as x and we know corresponding values for x and y, then we can find the proportionality constant.

EXAMPLE 1   **Finding the proportionality constant**

Joyce is traveling by car, and the distance she travels, d, varies directly with the amount of time, t, that she drives. In 3 hours she drove 120 miles. Find the proportionality constant and write d as a function of t.

Solution

Because d varies directly as t, we must have a constant k such that

$$d = kt.$$

Because $d = 120$ when $t = 3$, we can write

$$120 = k \cdot 3,$$

or

$$40 = k.$$

So the proportionality constant is 40 mph, and $d = 40t$.   ∎

EXAMPLE 2   **Direct variation**

In a downtown office building, the monthly rent for an office is directly proportional to the size of the office. If a 420-square-foot office rents for $1260 per month, then what is the rent for a 900-square-foot office?

Solution

Because the rent, R, varies directly with the area of the office, A, we have

$$R = kA.$$

Because a 420-square-foot office rents for $1260, We can substitute to find k:

$$1260 = k \cdot 420$$
$$3 = k$$

Now that we know the value of k, we can write

$$R = 3A.$$

To get the rent for a 900-square-foot office, insert 900 into this formula:

$$R = 3 \cdot 900$$
$$= 2700$$

So a 900-square-foot office rents for $2700 per month.   ■

Inverse Variation

In making a 500-mile trip by car, the time it takes is a function of the speed of the car. The greater the speed, the less time it will take. If you decrease the speed, the time increases. We say that the time is *inversely proportional* to the speed. Using the formula $D = RT$ or $T = \dfrac{D}{R}$, we can write

$$T = \frac{500}{R}.$$

In general, we make the following definition.

Inverse Variation

The statement **y varies inversely as x**, or **y is inversely proportional to x,** means that

$$y = \frac{k}{x}$$

for some nonzero constant, k.

CAUTION   Be sure to understand the difference between direct and inverse variation. If y varies directly as x (with $k > 0$), then as x increases, y increases. If y varies inversely as x (with $k > 0$), then as x increases, y decreases.   ⊘

EXAMPLE 3   **Inverse variation**

Suppose a is inversely proportional to b, and when $b = 5$, $a = \frac{1}{2}$. Find a when $b = 12$.

Solution

Because a is inversely proportional to b, we have

$$a = \frac{k}{b}$$

for some constant, k. Because $a = \frac{1}{2}$ when $b = 5$, we can find k by substituting these values into the formula:

$$\frac{1}{2} = \frac{k}{5}$$

$$\frac{5}{2} = k \qquad \text{Multiply each side by 5.}$$

Now to find a when $b = 12$, we use the formula with k replaced by $\frac{5}{2}$:

$$a = \frac{\frac{5}{2}}{b}$$

$$a = \frac{\frac{5}{2}}{12} = \frac{5}{2} \cdot \frac{1}{12} = \frac{5}{24} \qquad \blacksquare$$

Joint Variation

On a deposit of \$5000 in a savings account, the interest earned, I, depends on the rate, r, and the time, t. Assuming that the interest is simple interest, we can use the formula $I = Prt$ to write

$$I = 5000rt.$$

The variable I is a function of two independent variables, r and t. In this case we say that I *varies jointly* as r and t.

Joint Variation

The statement y **varies jointly as** x **and** z, or y **is jointly proportional to** x **and** z, means that

$$y = kxz$$

for some nonzero constant, k.

EXAMPLE 4   **Joint variation**

Suppose that y varies jointly with x and z, and $y = 12$ when $x = 5$ and $z = 2$. Find y when $x = 10$ and $z = -3$.

Solution

Because y varies jointly with x and z, we can write

$$y = kxz$$

for some constant, k. Now substitute $y = 12$, $x = 5$, and $z = 2$, and solve for k:

$$12 = k \cdot 5 \cdot 2$$
$$12 = 10k$$
$$\frac{6}{5} = k$$

Now that we know the value of k, we can rewrite the equation as

$$y = \frac{6}{5}xz.$$

To find y when $x = 10$ and $z = -3$, substitute into the equation:

$$y = \frac{6}{5}(10)(-3)$$
$$y = -36$$ ■

More Variation

We frequently combine the ideas of direct, inverse, and joint variation with powers and roots. A combination of direct and inverse variation is referred to as **combined variation.** Study the examples that follow.

More Variation Examples	
Statement	**Formula**
y varies directly as the square root of x.	$y = k\sqrt{x}$
y is directly proportional to the cube of x.	$y = kx^3$
y is inversely proportional to x^2.	$y = \dfrac{k}{x^2}$
y varies inversely as the square root of x.	$y = \dfrac{k}{\sqrt{x}}$
y varies jointly as x and the square of z.	$y = kxz^2$
y varies directly with x and inversely with the square root of z (combined variation).	$y = \dfrac{kx}{\sqrt{z}}$

CAUTION   The variation terms never signify addition or subtraction. We always use multiplication unless we see the word "inversely." In that case we divide.   ⊘

EXAMPLE 5   **Newton's law of gravity**

According to Newton's law of gravity, the gravitational attraction F between two objects with masses m_1 and m_2 is directly proportional to the product of their masses and inversely proportional to the square of the distance r between their centers. Write a formula for Newton's law of gravity.

Solution

Letting k be the constant of proportionality, we have

$$F = \frac{km_1m_2}{r^2}.$$ ∎

EXAMPLE 6   **House framing**

The time t that it takes to frame a house varies directly with the size of the house s in square feet and inversely with the number of framers n working on the job. If three framers can complete a 2500-square-foot house in six days, then how long will it take six framers to complete a 4500-square-foot house?

Solution

Because t varies directly with s and inversely with n, we have

$$t = \frac{ks}{n}.$$

Substitute $t = 6$, $s = 2500$, and $n = 3$ into this equation to find k:

$$6 = \frac{k \cdot 2500}{3}$$

$$18 = 2500k$$

$$0.0072 = k$$

Now use $k = 0.0072$, $s = 4500$, and $n = 6$ to find t:

$$t = \frac{0.0072 \cdot 4500}{6}$$

$$t = 5.4$$

So six framers can frame a 4500-square-foot house in 5.4 days.   ∎

Warm-ups

True or false? Explain your answer.

1. If a varies directly as b, then $a = kb$.
2. If a is inversely proportional to b, then $a = bk$.
3. If a is jointly proportional to b and c, then $a = bc$.
4. If a is directly proportional to the square root of c, then $a = k\sqrt{c}$.
5. If b is directly proportional to a, then $b = ka^2$.
6. If a varies directly as b and inversely as c, then $a = \dfrac{kb}{c}$.
7. If a is jointly proportional to c and the square of b, then $a = \dfrac{kc}{b^2}$.
8. If a varies directly as c and inversely as the square root of b, then $a = \dfrac{kc}{b}$.
9. If b varies directly as a and inversely as the square of c, then $b = ka\sqrt{c}$.
10. If b varies inversely with the square of c, then $b = \dfrac{k}{c^2}$.

8.5 EXERCISES

Write a formula that expresses the relationship described by each statement. Use k as a constant of variation. See Examples 1–6.

1. a varies directly as m.
2. w varies directly with P.
3. d varies inversely with e.
4. y varies inversely as x.
5. I varies jointly as r and t.
6. q varies jointly as w and v.
7. m is directly proportional to the square of p.
8. g is directly proportional to the cube of r.
9. B is directly proportional to the cube root of w.
10. F is directly proportional to the square of m.
11. t is inversely proportional to the square of x.
12. y is inversely proportional to the square root of z.
13. v varies directly as m and inversely as n.
14. b varies directly as the square of n and inversely as the square root of v.

Find the proportionality constant and write a formula that expresses the indicated variation. See Example 1.

15. y varies directly as x, and $y = 6$ when $x = 4$.
16. m varies directly as w, and $m = \frac{1}{3}$ when $w = \frac{1}{4}$.
17. A varies inversely as B, and $A = 10$ when $B = 3$.
18. c varies inversely as d, and $c = 0.31$ when $d = 2$.
19. m varies inversely as the square root of p, and $m = 12$ when $p = 9$.
20. s varies inversely as the square root of v, and $s = 6$ when $v = \frac{3}{2}$.
21. A varies jointly as t and u, and $A = 6$ when $t = 5$ and $u = 3$.
22. N varies jointly as the square of p and the cube of q, and $N = 72$ when $p = 3$ and $q = 2$.
23. y varies directly as x and inversely as z, and $y = 2.37$ when $x = \pi$ and $z = \sqrt{2}$.
24. a varies directly as the square root of m and inversely as the square of n, and $a = 5.47$ when $m = 3$ and $n = 1.625$.

Solve each variation problem. See Examples 2–6.

25. If y varies directly as x, and $y = 7$ when $x = 5$, find y when $x = -3$.

26. If n varies directly as p, and $n = 0.6$ when $p = 0.2$, find n when $p = \sqrt{2}$.

27. If w varies inversely as z, and $w = 6$ when $z = 2$, find w when $z = -8$.

28. If p varies inversely as q, and $p = 5$ when $q = \sqrt{3}$, find p when $q = 5$.

29. If A varies jointly as F and T, and $A = 6$ when $F = 3\sqrt{2}$ and $T = 4$, find A when $F = 2\sqrt{2}$ and $T = \frac{1}{2}$.

30. If j varies jointly as the square of r and the cube of v, and $j = -3$ when $r = 2\sqrt{3}$ and $v = \frac{1}{2}$, find j when $r = 3\sqrt{5}$ and $v = 2$.

31. If D varies directly with t and inversely with the square of s, and $D = 12.35$ when $t = 2.8$ and $s = 2.48$, find D when $t = 5.63$ and $s = 6.81$.

32. If M varies jointly with x and the square of v, and $M = 39.5$ when $x = \sqrt{10}$ and $v = 3.87$, find M when $x = \sqrt{30}$ and $v = 7.21$.

Determine whether each equation represents direct, inverse, joint, or combined variation.

33. $y = \dfrac{78}{x}$

34. $y = \dfrac{\pi}{x}$

35. $y = \dfrac{1}{2}x$

36. $y = \dfrac{x}{4}$

37. $y = \dfrac{3x}{w}$

38. $y = \dfrac{4t^2}{\sqrt{x}}$

39. $y = \dfrac{1}{3}xz$

40. $y = 99qv$

In Exercises 41–54, solve each problem.

41. *Lawn maintenance.* At Larry's Lawn Service the cost of lawn maintenance varies directly with the size of the lawn. If the monthly maintenance on a 4,000-square-foot lawn is $280, then what is the maintenance fee for a 6,000-square-foot lawn?

42. *Weight of the iguana.* The weight of an iguana is directly proportional to its length. If a 4-foot iguana weighs 30 pounds, then how much should a 5-foot iguana weigh?

43. *Gas laws.* The volume of a gas in a cylinder at a fixed temperature is inversely proportional to the weight on the piston. If the gas has a volume of 6 cubic centimeters for a weight of 30 kilograms, then what would the volume be for a weight of 20 kilograms?

44. *Selling software.* A software vendor sells a software package at a price that is inversely proportional to the number of packages sold per month. When they are selling 900 packages per month, the price is $80 each. If they sell 1000 packages per month, then what should the new price be?

45. *Costly culvert.* The price of an aluminum culvert is jointly proportional to its radius and length. If a 12-foot culvert with a 6-inch radius costs $324, then what is the price of a 10-foot culvert with an 8-inch radius?

46. *Pricing plastic.* The cost of a piece of PVC water pipe varies jointly as its diameter and length. If a 20-foot pipe with a diameter of 1 inch costs $6.80, then what will be the cost of a 10-foot pipe with a $\frac{3}{4}$-inch diameter?

47. *Reinforcing rods.* The price of a steel rod varies jointly as the length and the square of the diameter. If an 18-foot rod with a 2-inch diameter costs $12.60, then what is the cost of a 12-foot rod with a 3-inch diameter?

48. *Pea soup.* The weight of a cylindrical can of pea soup varies jointly with the height and the square of the radius. If a 4-inch-high can with a 1.5-inch radius weighs 16 ounces, then what is the weight of a 5-inch-high can with a radius of 3 inches?

49. *Falling objects.* The distance an object falls in a vacuum varies directly as the square of the time it is falling. If an object falls 0.16 foot in 0.1 second, then how far does it fall in 0.5 second?

50. *Making Frisbees.* The cost of material used in making a Frisbee varies directly with the square of the diameter. If it costs the manufacturer $0.45 for the material in a Frisbee with a 9-inch diameter, then what is the cost for the material in a 12-inch-diameter Frisbee?

51. *Using leverage.* The basic law of leverage is that the force required to lift an object is inversely proportional to the length of the lever. If a force of 2000 pounds applied 2 feet from the pivot point would lift a car, then what force would be required at 10 feet to lift the car?

52. *Resistance.* The resistance of a wire varies directly with the length and inversely as the square of the diameter. If a wire of length 20 feet and diameter 0.1 inch has a resistance of 2 ohms, then what is the resistance of a 30-foot wire with a diameter of 0.2 inch?

53. ***Computer programming.*** The time t required to complete a programming job varies directly with the complexity of the job and inversely with the number n of programmers working on the job. The complexity c is an arbitrarily assigned number between 1 and 10, with 10 being the most complex. It takes 8 days for a team of three programmers to complete a job with complexity 6. How long will it take five programmers to complete a job with complexity 9?

54. ***Shock absorbers.*** The volume of gas in a gas shock absorber varies directly with the temperature and inversely with the pressure. The volume is 10 cubic centimeters when the temperature is 20°C and the pressure is 40 kg. What is the volume when the temperature is 30°C and the pressure is 25 kg?

Graphing Calculator Exercises

55. To see the difference between direct and inverse variation, graph $y_1 = 2x$ and $y_2 = \dfrac{2}{x}$ using $0 \le x \le 5$ and $0 \le y \le 10$. Which of these functions is increasing and which is decreasing?

56. Graph $y_1 = 2\sqrt{x}$ and $y_2 = \dfrac{2}{\sqrt{x}}$ by using $0 \le x \le 5$ and $0 \le y \le 10$. At what point in the first quadrant do the curves cross? Which function is increasing and which is decreasing? Which represents direct variation and which represents inverse variation?

8.6 The Factor Theorem

In this section:

▶ The Factor Theorem
▶ Solving Polynomial Equations

In Chapter 3 you learned to add, subtract, multiply, divide, and factor polynomials. In this section we study functions defined by polynomials and learn to solve some higher-degree polynomial equations.

The Factor Theorem

Consider the polynomial function

$$P(x) = x^2 + 2x - 15.$$

The values of x for which $P(x) = 0$ are called the **zeros** of the function. We can find the zeros of the function by solving the equation $P(x) = 0$:

$$x^2 + 2x - 15 = 0$$
$$(x + 5)(x - 3) = 0$$
$$x + 5 = 0 \quad \text{or} \quad x - 3 = 0$$
$$x = -5 \quad \text{or} \quad x = 3$$

Because $x + 5$ is a factor of $x^2 + 2x - 15$, -5 is a solution to the equation $x^2 + 2x - 15 = 0$ and a zero of the function $P(x) = x^2 + 2x - 15$. We can check that -5 is a zero of $P(x) = x^2 + 2x - 15$ as follows:

$$P(-5) = (-5)^2 + 2(-5) - 15$$
$$= 25 - 10 - 15$$
$$= 0$$

Because $x - 3$ is a factor of the polynomial, 3 is also a solution to the equation $x^2 + 2x - 15 = 0$ and a zero of the polynomial function. Check that $P(3) = 0$:

$$P(3) = 3^2 + 2 \cdot 3 - 15$$
$$= 9 + 6 - 15$$
$$= 0$$

Every linear factor of the polynomial corresponds to a zero of the polynomial function, and every zero of the polynomial function corresponds to a linear factor.

Now suppose that $P(x)$ represents an arbitrary polynomial. If $x - c$ is a factor of the polynomial $P(x)$, then c is a solution to the equation $P(x) = 0$, and so $P(c) = 0$. If we divide $P(x)$ by $x - c$ and the remainder is 0, we must have

$$P(x) = (x - c)(\text{quotient}).$$ Dividend equals the divisor times the quotient.

If the remainder is 0, then $x - c$ is a factor of $P(x)$.

The **factor theorem** summarizes these ideas.

> **The Factor Theorem**
>
> The following statements are equivalent for any polynomial $P(x)$.
>
> 1. The remainder is zero when $P(x)$ is divided by $x - c$.
> 2. $x - c$ is a factor of $P(x)$.
> 3. c is a solution to $P(x) = 0$.
> 4. c is a zero of the function $P(x)$, or $P(c) = 0$.

To say that statements are equivalent means that the truth of any one of them implies that the others are true.

According to the factor theorem, if we want to determine whether a given number c is a zero of a polynomial function, we can divide the polynomial by $x - c$. The remainder is zero if and only if c is a zero of the polynomial function. The quickest way to divide by $x - c$ is to use synthetic division from Section 3.5.

EXAMPLE 1   **Using the factor theorem**

Use synthetic division to determine whether 2 is a zero of $P(x) = x^3 - 3x^2 + 5x - 2$.

Solution

By the factor theorem, 2 is a zero of the function if and only if the remainder is 0 when $P(x)$ is divided by $x - 2$. We can use synthetic division to determine the remainder. If we divide by $x - 2$, we use 2 on the left in synthetic division along with the coefficients 1, -3, 5, -2 from the polynomial:

$$
\begin{array}{r|rrrr}
2 & 1 & -3 & 5 & -2 \\
  &   & 2 & -2 & 6 \\
\hline
  & 1 & -1 & 3 & 4
\end{array}
$$

Because the remainder is 4, 2 is not a zero of the function. ∎

EXAMPLE 2   **Using the factor theorem**

Use synthetic division to determine whether -4 is a solution to the equation $2x^4 - 28x^2 + 14x - 8 = 0$.

Solution

By the factor theorem, -4 is a solution to the equation if and only if the remainder is 0 when $P(x)$ is divided by $x + 4$. When dividing by $x + 4$, we use -4 in the synthetic division:

$$
\begin{array}{r|rrrrr}
-4 & 2 & 0 & -28 & 14 & -8 \\
   &   & -8 & 32 & -16 & 8 \\
\hline
   & 2 & -8 & 4 & -2 & 0
\end{array}
$$

Because the remainder is 0, -4 is a solution to $2x^4 - 28x^2 + 14x - 8 = 0$. ∎

In the next example we use the factor theorem to determine whether a given binomial is a factor of a polynomial.

EXAMPLE 3   **Using the factor theorem**

Use synthetic division to determine whether $x + 4$ is a factor of $x^3 + 3x^2 + 16$.

Solution

According to the factor theorem, $x + 4$ is a factor of $x^3 + 3x^2 + 16$ if and only if the remainder is zero when the polynomial is divided by $x + 4$. Use synthetic division to determine the remainder:

$$
\begin{array}{r|rrrr}
-4 & 1 & 3 & 0 & 16 \\
   &   & -4 & 4 & -16 \\
\hline
   & 1 & -1 & 4 & 0
\end{array}
$$

Because the remainder is 0, $x + 4$ is a factor, and the polynomial can be written as

$$x^3 + 3x^2 + 16 = (x + 4)(x^2 - x + 4).$$

Because $x^2 - x + 4$ is a prime polynomial, the factoring is complete. ∎

Solving Polynomial Equations

The techniques used to solve polynomial equations of degree 3 or higher are not as straightforward as those used to solve linear equations and quadratic equations. The next example shows how the factor theorem can be used to solve a third-degree polynomial equation.

EXAMPLE 4   **Solving a third-degree equation**

Suppose the equation $x^3 - 4x^2 - 17x + 60 = 0$ is known to have a solution that is an integer between -3 and 3 inclusive. Find the solution set.

Solution

Because one of the numbers -3, -2, -1, 0, 1, 2, and 3 is a solution to the equation, we can use synthetic division with these numbers until we discover which one is a solution. We arbitrarily select 1 to try first:

$$\begin{array}{r|rrrr} 1 & 1 & -4 & -17 & 60 \\ & & 1 & -3 & -20 \\ \hline & 1 & -3 & -20 & 40 \end{array}$$

Because the remainder is 40, 1 is not a solution to the equation. Next try 2:

$$\begin{array}{r|rrrr} 2 & 1 & -4 & -17 & 60 \\ & & 2 & 4 & -42 \\ \hline & 1 & -2 & -21 & 18 \end{array}$$

Because the remainder is not zero, 2 is not a solution to the equation. Next try 3:

$$\begin{array}{r|rrrr} 3 & 1 & -4 & -17 & 60 \\ & & 3 & -3 & -60 \\ \hline & 1 & -1 & -20 & 0 \end{array}$$

The remainder is 0, so 3 is a solution to the equation, and $x - 3$ is a factor of the polynomial. (If 3 had not produced a remainder of zero, then we would have tried -3, -2, -1, and 0.) The other factor is the quotient, $x^2 - x - 20$.

$$x^3 - 4x^2 - 17x + 60 = 0$$
$$(x - 3)(x^2 - x - 20) = 0 \quad \text{Use the results of synthetic division to factor.}$$
$$(x - 3)(x - 5)(x + 4) = 0 \quad \text{Factor completely.}$$
$$x - 3 = 0 \quad \text{or} \quad x - 5 = 0 \quad \text{or} \quad x + 4 = 0$$
$$x = 3 \quad \text{or} \quad x = 5 \quad \text{or} \quad x = -4$$

Check each of these solutions in the original equation. The solution set is $\{3, 5, -4\}$. ■

Warm-ups

True or false? Explain your answers.

1. To divide $x^3 - 4x^2 - 3$ by $x - 5$, use 5 in the synthetic division.
2. To divide $5x^4 - x^3 + x - 2$ by $x + 7$, use -7 in the synthetic division.
3. The number 2 is a zero of $P(x) = 3x^3 - 5x^2 - 2x + 2$.
4. If $x^3 - 8$ is divided by $x - 2$, then $R = 0$.
5. If $R = 0$ when $x^4 - 1$ is divided by $x - a$, then $x - a$ is a factor of $x^4 - 1$.
6. If -2 satisfies $x^4 + 8x = 0$, then $x + 2$ is a factor of $x^4 + 8x$.
7. The binomial $x - 1$ is a factor of $x^{35} - 3x^{24} + 2x^{18}$.
8. The binomial $x + 1$ is a factor of $x^3 - 3x^2 + x + 5$.
9. If $x^3 - 5x + 4$ is divided by $x - 1$, then $R = 0$.
10. If $R = 0$ when $P(x) = x^3 - 5x - 2$ is divided by $x + 2$, then $P(-2) = 0$.

8.6 EXERCISES

Determine whether each given value of x is a zero of the given function. See Example 1.

1. $x = 1, \quad P(x) = x^3 - x^2 + x - 1$
2. $x = -2, \quad P(x) = -2x^3 - 5x^2 + 3x + 10$
3. $x = -3, \quad P(x) = -x^4 - 3x^3 - 2x^2 + 18$
4. $x = 4, \quad P(x) = x^4 - x^2 - 8x - 16$
5. $x = 2, \quad P(x) = 2x^3 - 4x^2 - 5x + 9$
6. $x = -3, \quad P(x) = x^3 + 5x^2 + 2x + 1$

Use synthetic division to determine whether each given value of x is a solution to the given equation. See Example 2.

7. $x = -3, \quad x^3 + 5x^2 + 2x - 12 = 0$
8. $x = -5, \quad x^2 - 3x - 40 = 0$
9. $x = -2, \quad x^4 + 3x^3 - 5x^2 - 10x + 5 = 0$
10. $x = -3, \quad -x^3 - 4x^2 + x + 12 = 0$
11. $x = 4, \quad -2x^4 + 30x^2 + 5x + 12 = 0$
12. $x = 6, \quad x^4 + x^3 - 40x^2 - 72 = 0$
13. $x = 3, \quad 0.8x^2 - 0.3x - 6.3 = 0$
14. $x = 5, \quad 6.2x^2 - 28.2x - 41.7 = 0$

Use synthetic division to determine whether the first polynomial is a factor of the second. If it is, then factor the polynomial completely. See Example 3.

15. $x - 3, \quad x^3 - 6x - 9$
16. $x + 2, \quad x^3 - 6x - 4$
17. $x + 5, \quad x^3 + 9x^2 + 23x + 15$
18. $x - 3, \quad x^4 - 9x^2 + x - 7$
19. $x - 2, \quad x^3 - 8x^2 + 4x - 6$
20. $x + 5, \quad x^3 + 125$
21. $x + 1, \quad x^4 + x^3 - 8x - 8$
22. $x - 2, \quad x^3 - 6x^2 + 12x - 8$
23. $x - 0.5, \quad 2x^3 - 3x^2 - 11x + 6$
24. $x - \dfrac{1}{3}, \quad 3x^3 - 10x^2 - 27x + 10$

Solve each equation, given that at least one of the solutions to each equation is an integer between −5 and 5. See Example 4.

25. $x^3 - 13x + 12 = 0$
26. $x^3 + 2x^2 - 5x - 6 = 0$
27. $2x^3 - 9x^2 + 7x + 6 = 0$
28. $6x^3 + 13x^2 - 4 = 0$
29. $2x^3 - 3x^2 - 50x - 24 = 0$
30. $x^3 - 7x^2 + 2x + 40 = 0$
31. $x^3 + 5x^2 + 3x - 9 = 0$
32. $x^3 + 6x^2 + 12x + 8 = 0$
33. $x^4 - 4x^3 + 3x^2 + 4x - 4 = 0$
34. $x^4 + x^3 - 7x^2 - x + 6 = 0$

Getting More Involved

35. **Exploration.** We can find the zeros of a polynomial function by solving a polynomial equation. We can also work backwards to find a polynomial function that has given zeros.
   a) Write a first-degree polynomial function whose zero is −2.
   b) Write a second-degree polynomial function whose zeros are 5 and −5.
   c) Write a third-degree polynomial function whose zeros are 1, −3, and 4.
   d) Is there a polynomial function with any given number of zeros? What is its degree?

Graphing Calculator Exercises

36. The x-coordinate of each x-intercept on the graph of a polynomial function is a zero of the polynomial function. Find the zeros of each function from its graph. Use synthetic division to check that the zeros found on your calculator really are zeros of the function.
   a) $P(x) = x^3 - 2x^2 - 5x + 6$
   b) $P(x) = 12x^3 - 20x^2 + x + 3$

37. With a graphing calculator an equation can be solved without the kind of hint that was given for Exercises 25–34. Solve each of the following equations by examining the graph of a corresponding function. Use synthetic division to check.
   a) $x^3 - 4x^2 - 7x + 10 = 0$
   b) $8x^3 - 20x^2 - 18x + 45 = 0$

———————COLLABORATIVE ACTIVITIES———————

Betting on Rockets

Grouping: 3 students per group
Topic: Applications of maximum and minimum of a parabola

Gretchen and Rafael are launching model rockets in the park. They have a bet to see whose rocket can go the highest. Both have taken this math course and have some clues on how to use algebra to find out whose went the highest. Below is the formula that they will use to determine whose rocket has gone the highest:

$$h = -16t^2 + v_0\, t,$$

where h is the distance (feet) of the rocket above the ground, t is the time passed (seconds) since the rocket was launched, and v_0 is the speed of the rocket when launched (initial velocity).

   They have decided to do three trials and see whose is the highest two out of the three times.

   From the following list, have each member of your group choose a different trial.

- **First Trial:** On the first launch, Gretchen's rocket has a time in the air of 14 seconds, and Rafael's rocket has a time of 10 seconds.

- **Second Trial:** On the second launch, Gretchen's rocket was in the air 11.5 seconds, and Rafael's rocket was in the air 12 seconds.

- **Third Trial:** On the third launch, Gretchen's rocket was in the air 11.25 seconds, and Rafael's rocket was in the air 11.3 seconds.

   Individually, complete the following for the trial you selected.

1. Find the initial velocity for each rocket using the formula $h = -16t^2 + v_0\, t$.

2. Graph the two quadratic functions. What is the $h =$ value for each vertex? Whose rocket went the highest?

   Together in your groups, compare your results and decide who wins the bet.

Wrap-up   CHAPTER 8

SUMMARY

Relations and Functions		Examples
Relation	Any set of ordered pairs of real numbers	$\{(1, 2),\ (1, 3)\}$
Function	A relation in which no two ordered pairs have the same first coordinate and different second coordinates	$\{(1, 2),\ (3, 5),\ (4, 5)\}$ $\{(x, y) \mid y = x^2\}$
	If y is a function of x, then y is uniquely determined by x. A function may be defined by a table, a listing of ordered pairs in set notation, or an equation.	
Domain	The set of first coordinates of the ordered pairs	Function: $y = x^2$ Domain: $(-\infty, \infty)$
Range	The set of second coordinates of the ordered pairs	Range: $[0, \infty)$
f-notation	If y is a function of x, we write $y = f(x)$.	$y = 2x + 3$ $f(x) = 2x + 3$

Average rate of change of f over $[a, b]$	$\dfrac{f(b) - f(a)}{b - a}$, where a and b are in the domain of f	Average rate of change of $f(x) = x^2$ over $[1, 3]$ is $\dfrac{3^2 - 1^2}{3 - 1} = 4.$

Types of Functions | | Examples

Linear function	$y = mx + b$ or $f(x) = mx + b$ for $m \neq 0$ Domain $(-\infty, \infty)$, range $(-\infty, \infty)$ If $m = 0$, $y = b$ is a constant function. Domain $(-\infty, \infty)$, range $\{b\}$	$f(x) = 2x - 3$
Absolute value function	$y = |x|$ or $f(x) = |x|$ Domain $(-\infty, \infty)$, range $[0, \infty)$	$f(x) = |x + 5|$
Quadratic function	$f(x) = ax^2 + bx + c$ for $a \neq 0$ Minimum or maximum value of f occurs at $x = \dfrac{-b}{2a}$.	$f(x) = x^2 - 4x + 3$
Square-root function	$f(x) = \sqrt{x}$ Domain $[0, \infty)$, range $[0, \infty)$	$f(x) = \sqrt{x - 4}$
Vertical-line test	If a graph can be crossed more than once by a vertical line, then it is not the graph of a function.	

Combining Functions | | Examples

Sum	$(f + g)(x) = f(x) + g(x)$	For $f(x) = x^2$ and $g(x) = x + 1$ $(f + g)(x) = x^2 + x + 1$
Difference	$(f - g)(x) = f(x) - g(x)$	$(f - g)(x) = x^2 - x - 1$
Product	$(f \cdot g)(x) = f(x) \cdot g(x)$	$(f \cdot g)(x) = x^3 + x^2$
Quotient	$\left(\dfrac{f}{g}\right)(x) = \dfrac{f(x)}{g(x)}$	$\left(\dfrac{f}{g}\right)(x) = \dfrac{x^2}{x + 1}$
Composition of functions	$(g \circ f)(x) = g(f(x))$ $(f \circ g)(x) = f(g(x))$	$(g \circ f)(x) = g(x^2) = x^2 + 1$ $(f \circ g)(x) = f(x + 1)$ $\qquad = x^2 + 2x + 1$

Inverse Functions | | Examples

One-to-one function	A function in which no two ordered pairs have different x-coordinates and the same y-coordinate.	$f = \{(2, 20), (3, 30)\}$
Inverse function	The inverse of a one-to-one function f is the function f^{-1}, which is obtained from f by interchanging the coordinates in each ordered pair of f. The domain of f^{-1} is the range of f, and the range of f^{-1} is the domain of f.	$f^{-1} = \{(20, 2), (30, 3)\}$
Horizontal-line test	If there is a horizontal line that crosses the graph of a function more than once, then the function is not invertible.	

f-Notation for inverse	Two functions f and g are inverses of each other if and only if both of the following conditions are met.	$f(x) = x^3 + 1$
	1. $(g \circ f)(x) = x$ for every number x in the domain of f.	
	2. $(f \circ g)(x) = x$ for every number x in the domain of g.	$f^{-1}(x) = \sqrt[3]{x-1}$
Switch-and-solve strategy for finding f^{-1}	1. Replace $f(x)$ by y.	$y = x^3 + 1$
	2. Interchange x and y.	$x = y^3 + 1$
	3. Solve for y.	$x - 1 = y^3$
	4. Replace y by $f^{-1}(x)$.	$y = \sqrt[3]{x-1}$
		$f^{-1}(x) = \sqrt[3]{x-1}$
Graphs of f and f^{-1}	Graphs of inverse functions are symmetric with respect to the line $y = x$.	

The Language of Variation

		Examples
Direct	y varies directly as x, $y = kx$	$z = 5m$
Inverse	y varies inversely as x, $y = \dfrac{k}{x}$	$a = \dfrac{1}{c}$
Joint	y varies jointly as x and z, $y = kxz$	$V = 6LW$
Combined	y varies directly as x and inversely as z, $y = \dfrac{kx}{z}$	$S = \dfrac{3A}{B}$

The Factor Theorem

		Examples
Factor theorem	The following are equivalent for $P(x)$ a polynomial in x.	$P(x) = x^2 - x - 2$
	1. The remainder is zero when $P(x)$ is divided by $x - c$.	$P(x) = (x - 2)(x + 1)$
	2. $x - c$ is a factor of $P(x)$.	
	3. c is a solution to $P(x) = 0$.	$P(-1) = 0$, $P(2) = 0$
	4. c is a zero of the function $P(x)$, or $P(c) = 0$.	-1 and 2 both satisfy $x^2 - x - 2 = 0$

REVIEW EXERCISES

8.1 *Determine whether each relation is a function.*

1. $\{(5, 7), (5, 10)\}$

2. $\{(10, 7), (6, 7)\}$

3. $\{(x, y) \mid y = x^2\}$

4. $x^2 = 1 + y^2$

5. $\{(x, y) \mid x = y^4\}$

6. $y = \sqrt{x - 1}$

Determine the domain and range of each relation.

7. $\{(1, 3), (2, 3)\}$

8. $\{(u, v) \mid v = -u^2\}$

9. $f(t) = 2t - 7$

10. $x = y^{1/2}$

11. $y = |x|$

12. $y = \sqrt{x} - 1$

Let $f(x) = 2x - 5$ and $g(x) = x^2 + x - 6$. Find the following.

13. $f(0)$

14. $f(-3)$

15. $g(0)$

16. $g(-2)$

17. $g\left(\dfrac{1}{2}\right)$

18. $g\left(-\dfrac{1}{2}\right)$

19. $f(a)$

20. $f(x + 3)$

21. $g(a - 1)$

22. $g(x + h)$

23. a, if $f(a) = 1$

24. x, if $g(x) = 0$

Find the average rate of change of the given function on the given interval.

25. $f(x) = 20x^2$, $[2, 8]$

26. $g(x) = \sqrt{x - 8}$, $[9, 89]$

27. $g(x) = x^2 + 2x$, $[x, x + h]$

28. $f(x) = 2x^2 - x$, $[x, x + h]$

8.2 *Graph each function and state the domain and range.*

29. $f(x) = 3x - 4$

30. $y = 0.3x$

31. $h(x) = |x| - 2$

32. $y = |x - 2|$

33. $y = x^2 - 2x + 1$

34. $g(x) = x^2 - 2x - 15$

35. $k(x) = \sqrt{x} + 2$

36. $y = \sqrt{x - 2}$

37. $y = \dfrac{2}{x}$

38. $y = 4 - x^2$

Graph each relation and state its domain and range.

39. $x = 2$

40. $x = y^2 - 1$

41. $x = |y| + 1$

42. $x = \sqrt{y - 1}$

Find the vertex of each parabola and determine whether the parabola opens upward or downward.

43. $f(x) = x^2 - 4x + 3$

44. $f(x) = x^2 - 3x + 2$

45. $y = -2x^2 + 8x + 9$

46. $y = -3x^2 + 6x - 7$

8.3 *Let $f(x) = 3x + 5$, $g(x) = x^2 - 2x$, and $h(x) = \dfrac{x - 5}{3}$. Find the following.*

47. $f(-3)$

48. $h(-4)$

49. $(h \circ f)(\sqrt{2})$

50. $(f \circ h)(\pi)$

51. $(g \circ f)(2)$

52. $(g \circ f)(x)$

53. $(f + g)(3)$

54. $(f - g)(x)$

55. $(f \cdot g)(x)$

56. $\left(\dfrac{f}{g}\right)(1)$

57. $(f \circ f)(0)$

58. $(f \circ f)(x)$

Let $f(x) = |x|$, $g(x) = x + 2$, and $h(x) = x^2$. Write each of the following functions as a composition of functions, using f, g, or h.

59. $F(x) = |x + 2|$

60. $G(x) = |x| + 2$

61. $H(x) = x^2 + 2$

62. $K(x) = x^2 + 4x + 4$

63. $I(x) = x + 4$

64. $J(x) = x^4 + 2$

8.4 *Determine whether each function is invertible. If it is invertible, find the inverse.*

65. $\{(-2, 4), (2, 4)\}$

66. $\{(1, 1), (3, 3)\}$

67. $f(x) = 8x$

68. $i(x) = -\dfrac{x}{3}$

69. $g(x) = 13x - 6$

70. $h(x) = \sqrt[3]{x - 6}$

71. $j(x) = \dfrac{x + 1}{x - 1}$

72. $k(x) = |x| + 7$

73. $m(x) = (x - 1)^2$

74. $n(x) = \dfrac{3}{x}$

Find the inverse of each function, and graph f and f^{-1} on the same pair of axes.

75. $f(x) = 3x - 1$

76. $f(x) = 2 - x^2$ for $x \geq 0$

77. $f(x) = \dfrac{x^3}{2}$

78. $f(x) = -\dfrac{1}{4}x$

8.5 *Solve each variation problem.*

79. If y varies directly as m and $y = -3$ when $m = \dfrac{1}{4}$, find y when $m = -2$.

80. If a varies inversely as b and $a = 6$ when $b = -3$, find a when $b = 4$.

81. If c varies directly as m and inversely as n, and $c = 20$ when $m = 10$ and $n = 4$, find c when $m = 6$ and $n = -3$.

82. If V varies jointly as h and the square of r, and $V = 32$ when $h = 6$ and $r = 3$, find V when $h = 3$ and $r = 4$.

8.6 *Given that either 2 or -2 is a solution to each of the following equations, solve each equation.*

83. $x^3 - 4x^2 - 11x + 30 = 0$

84. $x^3 - 2x^2 - 6x + 12 = 0$

85. $x^3 - 5x^2 - 2x + 24 = 0$

86. $x^3 + 7x^2 + 4x - 12 = 0$

Determine whether the first polynomial is a factor of the second. If it is, then factor the polynomial completely.

87. $x - 3$, $\quad x^3 + 4x^2 - 11x - 30$

88. $x + 4$, $\quad x^3 + x^2 - 10x + 8$

89. $x + 2$, $\quad x^3 - 5x - 6$

90. $x - 1$, $\quad 2x^3 - 5x^2 + 2x - 9$

Miscellaneous

91. *Crop dusting.* The cost, C, of crop dusting is a linear function of the number of pounds of chemical, n, distributed over the fields. Betty paid $1800 for 400 pounds and then paid $2300 for 600 pounds. Write C as a linear function of n.

92. *Drug concentration.* The concentration of a certain drug in a patient's blood is found to be 0.3 milligram per cubic centimeter 1 hour after the patient is injected, and 0.25 milligram per cubic centimeter 3 hours after the patient is injected. Assuming that the concentration C is a linear function of the time t, find the function. In how many hours will the drug be eliminated from the patient's blood?

93. *Falling object.* If a ball is dropped from a tall building, then the distance traveled by the ball in t seconds varies directly as the square of the time t. If the ball travels 144 feet in 3 seconds, then how far does it travel in 4 seconds?

94. *Studying or partying.* Evelyn's grade on a math test varies directly with the number of hours spent studying and inversely with the number of hours spent partying during the 24 hours preceding the test. If she scored a 90 on a test after she studied 10 hours and partied 2 hours, then what should she score after studying 4 hours and partying 6 hours?

95. *Inscribed square.* Given that B is the area of a square inscribed in a circle of radius r and area A, write B as a function of A.

96. *Area of a window.* A window is in the shape of a square of side s, with a semicircle of diameter s above it. Write a function that expresses the total area of the window as a function of s.

97. *Composition of functions.* Given that $a = 3k + 2$ and $k = 5w - 6$, write a as a function of w.

98. *Volume of a cylinder.* The volume of a cylinder with a fixed height of 10 cm is given by $V = 10\pi r^2$, where r is the radius of the circular base. Write the volume as a function of the area of the base, A.

99. *Magazine sales.* The profit that Luther makes on the sale of magazine subscriptions is determined by the formula

$$P(x) = 50x - x^2 - 400,$$

where x is the number of subscriptions he sells and $P(x)$ is in dollars. What value of x will maximize his profit? What is his maximum profit?

100. *Gross national product.* The gross national product (GNP) for a certain developing country is given by

$$G(t) = 0.1t^2 - 1.8t + 52.1,$$

where t is the number of years since 1980 and $G(t)$ is in billions of dollars. What was the smallest GNP for this country since 1980?

101. *Thunderbird LX.* The 1994 Thunderbird LX V8 goes from 0 to 60 mph in 8.5 seconds (*Motor Trend*, February 1994). Find the average rate of change of its velocity over the time interval [0, 8.5].

102. *Camry SE.* The 1994 Toyota Camry SE Coupe can travel a distance of 1320 feet from a standing start in 16.1 seconds (*Motor Trend*, February 1994). Find the average rate of change of its distance traveled over the time interval [0, 16.1].

Figure for Exercise 96

CHAPTER 8 TEST

Determine whether each relation is a function.

1. $\{(-2, 5), (3, 4), (-2, 3)\}$

2. $\{(0, 0.5), (9, 0.5), (4, 0.5)\}$

3. $\{(x, y) \mid y = -\sqrt{x}\}$

4. $\{(x, y) \mid x = |y| + 6\}$

Determine the domain and range of each relation.

5. $f(x) = |x|$

6. $x = \sqrt{y}$

Let $f(x) = -2x + 5$ and $g(x) = x^2 + 4$. Find the following.

7. $f(-3)$

8. $g(a + 1)$

9. $(g \circ f)(x)$

10. $f^{-1}(x)$

11. $(g + f)(2)$

12. $(f \cdot g)(1)$

13. $(f^{-1} \circ f)(-1782)$

14. x, if $g(x) = 9$

Sketch the graph of each function or relation.

15. $g(x) = -\dfrac{2}{3}x + 1$

16. $y = |x| - 4$

17. $y = 5 - x^2$

18. $f(x) = x^2 + 2x - 8$

19. $h(x) = 2\sqrt{x} - 1$

20. $x = y^2$

Solve each problem.

21. Find the average rate of change of the function $f(x) = 3x^2 - 4x$ over the interval $[1, 4]$.

22. Find the inverse of the function $f(x) = \dfrac{2x + 1}{x - 1}$.

23. The volume of a sphere varies directly as the cube of the radius. If a sphere with radius 3 feet has a volume of 36π cubic feet, then what is the volume of a sphere with a radius of 2 feet?

24. A mail-order firm charges its customers a shipping and handling fee of $3.00 plus $0.50 per pound for each order shipped. Express the shipping and handling fee as function of the weight of the order, n.

25. Suppose y varies directly as x and inversely as the square root of z. If $y = 12$ when $x = 7$ and $z = 9$, then what is the proportionality constant?

26. If a ball is tossed into the air from a height of 6 feet with a velocity of 40 feet per second, then its altitude at time t can be described by the quadratic function $A(t) = -16t^2 + 40t + 6$. For what value of t is the altitude at its maximum?

27. Is $x + 1$ a factor of $2x^4 - 5x^3 + 3x^2 + 6x - 4$?

28. Given that all of the solutions to the equation $x^3 - 12x^2 + 47x - 60 = 0$ are positive integers smaller than 10, find the solution set.

Tying It All Together

Simplify each expression.

1. $125^{-2/3}$

2. $\left(\dfrac{8}{27}\right)^{-1/3}$

3. $\sqrt{18} - \sqrt{8}$

4. $x^5 \cdot x^3$

5. $16^{1/4}$

6. $\dfrac{x^{12}}{x^3}$

Find the solution set to each equation.

7. $x^2 = 9$

8. $x^2 = 8$

9. $x^2 = x$

10. $x^2 - 4x - 6 = 0$

11. $x^{1/4} = 3$

12. $x^{1/6} = -2$

13. $|x| = 8$

14. $|5x - 4| = 21$

15. $x^3 = 8$

16. $(3x - 2)^3 = 27$

17. $\sqrt{2x - 3} = 9$

18. $\sqrt{x - 2} = x - 8$

Sketch the graph of each set.

19. $\{(x, y) \mid y = 5\}$

20. $\{(x, y) \mid y = 2x - 5\}$

21. $\{(x, y) \mid x = 5\}$

22. $\{(x, y) \mid 3y = x\}$

23. $\{(x, y) \mid y = 5x^2\}$

24. $\{(x, y) \mid y = -2x^2\}$

Find the missing coordinates in each ordered pair so that the ordered pair satisfies the given equation.

25. $(2, \ )$, $(3, \ )$, $(\ , 2)$, $(\ , 16)$, $2^x = y$

26. $\left(\dfrac{1}{2}, \ \right)$, $(-1, \ )$, $(\ , 16)$, $(\ , 1)$, $4^x = y$

Solve.

27. ***Basic energy requirement.*** Clinical dietitians must design diets that meet patients' basic energy requirements and are suitable for the condition of their health (*Snapshots of Applications in Mathematics*). The basic energy requirement B (in calories) for a male is a function of three variables, determined by the formula

$$B = 655 + 9.56W + 1.85H - 4.68A,$$

where W is the patient's weight in kilograms, H is the height in centimeters, and A is the age in years.

a) Find the basic energy requirement for a 30-year-old male with a height of 6 ft 1 in. and a weight of 180 lb. (1 in. = 2.54 cm, 1 kg = 2.2 lb.)

b) If a male with a height of 6 ft maintains a weight of 170 lb from ages 20 through 70, then his basic energy requirement is a function of his age. For this male, write B as a function of A and graph this function for $20 \le A \le 70$.

Exponential and Logarithmic Functions

Water is one of the essentials of life, yet it is something that most of us take for granted. Among other things, the U.S. Geological Survey (U.S.G.S.) studies fresh water. For over 50 years the Water Resources Division of the U.S.G.S. has been gathering basic data about the flow of both fresh and salt water from streams and groundwater surfaces. This division collects, compiles, analyzes, verifies, organizes, and publishes data gathered from groundwater data collection networks in each of the 50 states, Puerto Rico, and the Trust Territories. Records of stream flow, groundwater levels, and water quality provide hydrological information needed by local, state, and federal agencies as well as the private sector.

There are many instances of the importance of the data collected by the U.S.G.S. For example, before 1987 the Tangipahoa River in Louisiana was used extensively for swimming and boating. In 1987, data gathered by the U.S.G.S. showed that fecal coliform levels in the river exceeded safe levels. Consequently, Louisiana banned recreational use of the river. Other studies by the Water Resources Division include the results of pollutants on salt marsh environments and the effect that salting highways in winter has on our drinking water supply.

In Exercises 79 and 80 of Section 9.2 you will see how data from the U.S.G.S. is used in a logarithmic function to measure water quality.

9.1 Exponential Functions

In Chapter 8 we studied functions such as

$$f(x) = x^2, \qquad g(x) = x^3, \qquad \text{and} \qquad h(x) = x^{1/2}.$$

For these functions the variable is the base. In this section we discuss functions that have a variable as an exponent. These functions are called *exponential functions*.

Definition

Some examples of exponential functions are

$$f(x) = 2^x, \qquad f(x) = \left(\frac{1}{2}\right)^x, \qquad \text{and} \qquad f(x) = 3^x.$$

> **Exponential Function**
> An **exponential function** is a function of the form
> $$f(x) = a^x,$$
> where $a > 0$ and $a \neq 1$.

We rule out the base 1 in the definition because $f(x) = 1^x$ is the same as the constant function $f(x) = 1$. Zero is not used as a base because $0^x = 0$ for any positive x and nonpositive powers of 0 are undefined. Negative numbers are not used as bases because an expression such as $(-4)^x$ is not a real number if $x = \frac{1}{2}$.

EXAMPLE 1 **Evaluating exponential functions**

Let $f(x) = 2^x$, $g(x) = \left(\frac{1}{4}\right)^{1-x}$, and $h(x) = -3^x$. Find the following.

a) $f\left(\frac{3}{2}\right)$ **b)** $f(-3)$ **c)** $g(3)$ **d)** $h(2)$

Solution

a) $f\left(\frac{3}{2}\right) = 2^{3/2} = \sqrt{2^3} = \sqrt{8} = 2\sqrt{2}$

b) $f(-3) = 2^{-3} = \frac{1}{2^3} = \frac{1}{8}$

c) $g(3) = \left(\frac{1}{4}\right)^{1-3} = \left(\frac{1}{4}\right)^{-2} = 4^2 = 16$

d) $h(2) = -3^2 = -9$ Note that $-3^2 \neq (-3)^2$. ∎

For many applications of exponential functions we use base 10 or another base called e. The number e is an irrational number that is approximately 2.718. We will see how e is used in compound interest in Example 10 of this section. Base 10 will be used in the next section. Base 10 is called the **common base,** and base e is called the **natural base.**

Calculator Close-up

Scientific and graphing calculators both have function keys for 10^x and e^x. On most calculators they are second or inverse functions. You can find a better approximation for e than 2.718 that was given previously by finding e^1 on your calculator. On a scientific calculator, try the following keystrokes:

$$1 \;\; \boxed{\text{2nd}} \;\; \boxed{e^x}$$

The display will read

$$\boxed{2.718281828}$$

Note that when e^1 is rounded to nine decimal places, the digits appear to repeat, but since e is irrational, it is actually a nonrepeating decimal number. To find e^1 on a graphing calculator, use the e^x key. The display should appear as follows:

$$e\,\wedge 1 \;\; \boxed{\text{ENTER}}$$

$$\boxed{2.718281828}$$

Evaluating 10^x is similar to evaluating e^x. For example, to evaluate $10^{0.5}$ on a scientific calculator, try the following keystrokes:

$$0.5 \;\; \boxed{\text{2nd}} \;\; \boxed{10^x}$$

The display will read

$$\boxed{3.1623}$$

EXAMPLE 2   **Base 10 and base e**

Let $f(x) = 10^x$ and $g(x) = e^x$. Find the following and round approximate answers to four decimal places.

a) $f(3)$ **b)** $f(1.51)$ **c)** $g(0)$ **d)** $g(2)$

Solution

a) $f(3) = 10^3 = 1000$

b) $f(1.51) = 10^{1.51} \approx 32.3594$ Use the 10^x key on a calculator.

c) $g(0) = e^0 = 1$

d) $g(2) = e^2 \approx 7.3891$ Use the e^x key on a calculator.   ■

Domain

In the definition of an exponential function, no restrictions were placed on the exponent x because the domain of an exponential function is the set of all real numbers. So both rational and irrational numbers can be used as the exponent. We have been using rational numbers for exponents since Chapter 5, but we have not yet seen an irrational number as an exponent. Even though we do not formally define irrational exponents in this text, an irrational number such as π can be used as an exponent, and you can evaluate an expression such as 2^π by using a calculator. Try it:

$$2^\pi \approx 8.824977827$$

Graphing Exponential Functions

Even though the domain of an exponential function is the set of all real numbers, we can graph an exponential function by evaluating it for just a few integers.

EXAMPLE 3 **An exponential function with $a > 1$**

Sketch the graph of each function.

a) $f(x) = 2^x$ **b)** $g(x) = 3^x$

Solution

a) We first make a table of ordered pairs that satisfy $f(x) = 2^x$:

x	-2	-1	0	1	2	3
$f(x) = 2^x$	$\dfrac{1}{4}$	$\dfrac{1}{2}$	1	2	4	8

As x increases, 2^x increases and 2^x is always positive. Because the domain of the function is $(-\infty, \infty)$, we draw the graph in Fig. 9.1 as a smooth curve through these points. From the graph we can see that the range is $(0, \infty)$.

b) Make a table of ordered pairs that satisfy $g(x) = 3^x$:

x	-2	-1	0	1	2	3
$g(x) = 3^x$	$\dfrac{1}{9}$	$\dfrac{1}{3}$	1	3	9	27

As x increases, 3^x increases and 3^x is always positive. The graph is shown in Fig. 9.2. From the graph we see that the range is $(0, \infty)$.

Figure 9.1   Figure 9.2

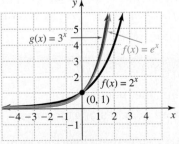

Figure 9.3

Because $e \approx 2.718$, the graph of $f(x) = e^x$ lies between the graphs of $f(x) = 2^x$ and $g(x) = 3^x$, as shown in Fig. 9.3. Note that all three functions have the same domain and range and the same y-intercept. In general, the function $f(x) = a^x$ for $a > 1$ has the following characteristics:

1. The y-intercept of the curve is $(0, 1)$.
2. The domain is $(-\infty, \infty)$, and the range is $(0, \infty)$.
3. The curve approaches the negative x-axis but does not touch it.
4. The y-values are increasing as we go from left to right along the curve.

EXAMPLE 4 **Exponential functions with $0 < a < 1$**
Graph each function.

a) $f(x) = \left(\frac{1}{2}\right)^x$　　　　　　　　　　**b)** $f(x) = 4^{-x}$

Solution

a) First make a table of ordered pairs that satisfy $f(x) = \left(\frac{1}{2}\right)^x$:

x	-2	-1	0	1	2	3
$f(x) = \left(\frac{1}{2}\right)^x$	4	2	1	$\frac{1}{2}$	$\frac{1}{4}$	$\frac{1}{8}$

As x increases, $\left(\frac{1}{2}\right)^x$ decreases, getting closer and closer to 0. Draw a smooth curve through these points as shown in Fig. 9.4.

b) Because $4^{-x} = \left(\frac{1}{4}\right)^x$, we make a table for $f(x) = \left(\frac{1}{4}\right)^x$:

x	-2	-1	0	1	2	3
$f(x) = \left(\frac{1}{4}\right)^x$	16	4	1	$\frac{1}{4}$	$\frac{1}{16}$	$\frac{1}{64}$

As x increases, $\left(\frac{1}{4}\right)^x$, or 4^{-x}, decreases, getting closer and closer to 0. Draw a smooth curve through these points, as shown in Fig. 9.5.

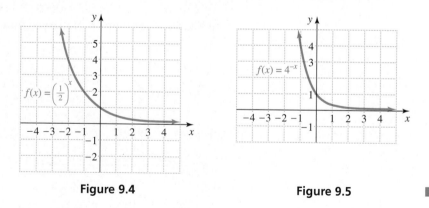

Figure 9.4　　　　　　　　　　　　　　Figure 9.5

Notice the similarities and differences between the exponential function with $a > 1$ and with $0 < a < 1$. The function $f(x) = a^x$ for $0 < a < 1$ has the following characteristics:

1. The y-intercept of the curve is $(0, 1)$.
2. The domain is $(-\infty, \infty)$, and the range is $(0, \infty)$.
3. The curve approaches the positive x-axis but does not touch it.
4. The y-values are decreasing as we go from left to right along the curve.

CAUTION　An exponential function can be written in more than one form. For example, $f(x) = \left(\frac{1}{2}\right)^x$ is the same as $f(x) = \frac{1}{2^x}$, or $f(x) = 2^{-x}$.

Although exponential functions have the form $f(x) = a^x$, other functions that have similar forms are also called exponential functions. Notice how changing the form $f(x) = a^x$ in the next two examples changes the shape and location of the graph.

EXAMPLE 5   **Changing the shape and location**

Sketch the graph of $f(x) = 3^{2x-1}$.

Solution

Make a table of ordered pairs:

x	-1	0	$\frac{1}{2}$	1	2
$f(x) = 3^{2x-1}$	$\frac{1}{27}$	$\frac{1}{3}$	1	3	27

The graph through these points is shown in Fig. 9.6. ∎

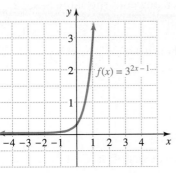

Figure 9.6

EXAMPLE 6   **Changing the shape and location**

Sketch the graph of $y = -2^{-x}$.

Solution

Because $-2^{-x} = -(2^{-x})$, all y-coordinates are negative. Make a table of ordered pairs:

x	-2	-1	0	1	2
$f(x) = -2^{-x}$	-4	-2	-1	$-\frac{1}{2}$	$-\frac{1}{4}$

The graph through these points is shown in Fig. 9.7. ∎

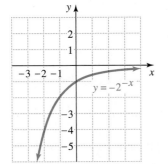

Figure 9.7

Exponential Equations

In Chapter 8 we used the horizontal line test to determine whether a function is one-to-one. Because no horizontal line can cross the graph of an exponential function more than once, exponential functions are one-to-one functions. For an exponential function, one-to-one means that *if two powers of the same base are equal, then the exponents are equal.*

> **One-to-One Property of Exponential Functions**
>
> For $a > 0$ and $a \neq 1$,
>
> $$\text{if} \quad a^m = a^n, \quad \text{then} \quad m = n.$$

In the next example we use the one-to-one property to solve equations involving exponential functions.

EXAMPLE 7 **Using the one-to-one property**

Solve each equation.

a) $2^{2x-1} = 8$ **b)** $9^{|x|} = 3$ **c)** $\dfrac{1}{8} = 4^x$

Solution

a) Because 8 is 2^3, we can write each side as a power of the same base, 2:

$$2^{2x-1} = 8 \qquad \text{Original equation}$$
$$2^{2x-1} = 2^3 \qquad \text{Write each side as a power of the same base.}$$
$$2x - 1 = 3 \qquad \text{One-to-one property}$$
$$2x = 4$$
$$x = 2$$

Check: $2^{2 \cdot 2 - 1} = 2^3 = 8$. The solution set is $\{2\}$.

b) Because $9 = 3^2$, we can write each side as a power of 3:

$$9^{|x|} = 3 \qquad \text{Original equation}$$
$$(3^2)^{|x|} = 3^1$$
$$3^{2|x|} = 3^1 \qquad \text{Power of a power rule}$$
$$2|x| = 1 \qquad \text{One-to-one property}$$
$$|x| = \frac{1}{2}$$
$$x = \pm\frac{1}{2}$$

Check $x = \pm\frac{1}{2}$ in the original equation. The solution set is $\left\{-\frac{1}{2}, \frac{1}{2}\right\}$.

c) Because $\frac{1}{8} = 2^{-3}$ and $4 = 2^2$, we can write each side as a power of 2:

$$\frac{1}{8} = 4^x \qquad \text{Original equation}$$
$$2^{-3} = (2^2)^x \qquad \text{Write each side as a power of 2.}$$
$$2^{-3} = 2^{2x} \qquad \text{Power of a power rule}$$
$$2x = -3 \qquad \text{One-to-one property}$$
$$x = -\frac{3}{2}$$

Check $x = -\frac{3}{2}$ in the original equation. The solution set is $\left\{-\frac{3}{2}\right\}$. ■

The one-to-one property is also used to find the first coordinate when given the second coordinate of an exponential function.

EXAMPLE 8 **Finding the x-coordinate in an exponential function**

Let $f(x) = 2^x$ and $g(x) = \left(\frac{1}{2}\right)^{1-x}$. Find x if:

a) $f(x) = 32$ **b)** $g(x) = 8$

Solution

a) Because $f(x) = 2^x$ and $f(x) = 32$, we can find x by solving $2^x = 32$:

$$2^x = 32$$
$$2^x = 2^5 \quad \text{Write both sides as a power of the same base.}$$
$$x = 5 \quad \text{One-to-one property}$$

b) Because $g(x) = \left(\frac{1}{2}\right)^{1-x}$ and $g(x) = 8$, we can find x by solving $\left(\frac{1}{2}\right)^{1-x} = 8$:

$$\left(\frac{1}{2}\right)^{1-x} = 8$$
$$(2^{-1})^{1-x} = 2^3 \quad \text{Because } \tfrac{1}{2} = 2^{-1} \text{ and } 8 = 2^3$$
$$2^{x-1} = 2^3 \quad \text{Power of a power rule}$$
$$x - 1 = 3 \quad \text{One-to-one property}$$
$$x = 4$$

∎

Applications

Exponential functions are used to describe phenomena such as population growth, radioactive decay, and compound interest. Here we discuss compound interest. If an investment is earning **compound interest,** then interest is periodically paid into the account and the interest that is paid also earns interest. If a bank pays 6% compounded quarterly on an account, then the interest is computed four times per year (every three months) at 1.5% (one-quarter of 6%). Suppose an account has $5000 in it at the beginning of a quarter. We can apply the simple-interest formula $A = P + Prt$, with $r = 6\%$ and $t = \frac{1}{4}$, to find how much is in the account at the end of the first quarter.

$$A = P + Prt$$
$$= P(1 + rt) \quad \text{Factor.}$$
$$= 5000\left(1 + 0.06 \cdot \frac{1}{4}\right) \quad \text{Substitute.}$$
$$= 5000(1.015)$$
$$= \$5075$$

To repeat this computation for another quarter, we multiply $5075 by 1.015. If A represents the amount in the account at the end of n quarters, we can write A as an exponential function of n:

$$A = \$5000(1.015)^n$$

In general, the amount A is given by the following formula.

Compound-Interest Formula

If P represents the principal, i the interest rate per period, n the number of periods, and A the amount at the end of n periods, then

$$A = P(1 + i)^n.$$

EXAMPLE 9 **Compound-interest formula**

If $350 is deposited in an account paying 12% compounded monthly, then how much is in the account at the end of 6 years and 6 months?

Solution

Interest is paid 12 times per year, so the account earns $\frac{1}{12}$ of 12%, or 1% each month, for 78 months. So $i = 0.01$, $n = 78$, and $P = \$350$:

$$A = P(1 + i)^n$$
$$A = \$350(1.01)^{78}$$
$$= \$760.56 \qquad \blacksquare$$

If we shorten the length of the time period (yearly, quarterly, monthly, daily, hourly, etc.), the number of periods n increases while the interest rate for the period decreases. As n increases, the amount A also increases but will not exceed a certain amount. That certain amount is the amount obtained from *continuous compounding* of the interest. It is shown in more advanced courses that the following formula gives the amount when interest is compounded continuously.

Continuous Compounding Formula

If P is the principal or beginning balance, r is the annual percentage rate compounded continuously, t is the time in years, and A is the amount or ending balance, then

$$A = Pe^{rt}.$$

CAUTION The value of t in the continuous-compounding formula must be in years. For example, if the time is 1 year and 3 months, then $t = 1.25$ years. If the time is 3 years and 145 days, then $t = 3 + \frac{145}{365} \approx 3.3973$ years. ⊘

EXAMPLE 10 **Continuous-compounding formula**

If $350 is deposited in an account paying 12% compounded continuously, then how much is in the account after 6 years and 6 months?

Solution

Use $r = 12\%$, $t = 6.5$ years, and $P = \$350$ in the formula for compounding interest continuously:

$$A = Pe^{rt}$$
$$= 350e^{(0.12)(6.5)}$$
$$= 350e^{0.78}$$
$$= \$763.52 \qquad \text{Use the } e^x \text{ key on a scientific calculator.}$$

Note that compounding continuously amounts to a few dollars more than compounding monthly did in Example 9. $\blacksquare$

Calculator Close-up

(50, 141,200)

(42.1, 54,752)

Xmin = 0   Ymin = 0
Xmax = 50   Ymax = 150,000

Figure 9.8

In Example 10 the investment of $350 more than doubled in 6.5 years. Let's look at what happens to the $350 over a longer period of time by graphing the function $A = 350e^{0.12t}$ for t ranging from 0 to 50 years. If we enter the function

$$y_1 = 350e\,\widehat{}\,(0.12x)$$

into a graphing calculator for $0 \le x \le 50$ and $0 \le y \le 150,000$ we get the graph shown in Fig. 9.8. Using the trace feature of the calculator we can see that the curve goes through points such as (42.1, 54,752) and (50, 141,200). So after 42 years the investment is worth over $54,000, and after 50 years it is worth over $141,000. The value of the investment increases slowly at first and then increases very rapidly in the later years.

MATH AT WORK

Geophysicist

Neal Driscoll, a geophysicist at the Lamont-Doherty Earth Observatory of Columbia University, explores both the ocean and the continents to understand the processes that shape the earth. What he finds fascinating is the interaction between the ocean and the land—not just at the shoreline, but underneath the sea as well.

To get a preliminary picture of the ocean floor, Dr. Driscoll spent part of the past summer working with the U.S.G.S., studying the effects of storms on beaches and underwater landscapes. The results of these studies can be used as a baseline to provide help to coastal planners building waterfront homes. Other information obtained can be used to direct transporters of dredged material to places where the material is least likely to affect plant, fish, and human life. The most recent study found many different types of ocean floor, ranging from sand and mud to large tracts of algae.

Imaging the sea floor is a difficult problem. It can be a costly venture, and there are numerous logistical problems. Recently developed technology, such as towable undersea cameras and satellite position systems, has made the task easier. Dr. Driscoll and his team use this new technology and sound reflection to gather data about how the sediment on the ocean floor changes in response to storm events. This research is funded by the Office of Naval Research (ONR).

In Exercise 28 of the Tying It All Together exercises you will see how a geophysicist uses sound to measure the depth of the ocean.

Warm-ups

True or false? Explain your answer.

1. If $f(x) = 4^x$, then $f\left(-\frac{1}{2}\right) = -2$.

2. If $f(x) = \left(\frac{1}{3}\right)^x$, then $f(-1) = 3$.

3. The function $f(x) = x^4$ is an exponential function.

4. The functions $f(x) = \left(\frac{1}{2}\right)^x$ and $g(x) = 2^{-x}$ have the same graph.

5. The function $f(x) = 2^x$ is invertible.

6. The graph of $y = \left(\frac{1}{3}\right)^x$ has an x-intercept.

7. The y-intercept for $f(x) = e^x$ is $(0, 1)$.
8. The expression $2^{\sqrt{2}}$ is undefined.

9. The functions $f(x) = 2^{-x}$ and $g(x) = \frac{1}{2^x}$ have the same graph.

10. If \$500 earns 6% compounded monthly, then at the end of 3 years the investment is worth $500(1.005)^3$ dollars.

9.1 EXERCISES

Let $f(x) = 4^x$, $g(x) = \left(\frac{1}{3}\right)^{x+1}$, and $h(x) = -2^x$. Find the following. See Example 1.

1. $f(2)$
2. $f(-1)$
3. $f\left(\frac{1}{2}\right)$

4. $f\left(-\frac{3}{2}\right)$
5. $g(-2)$
6. $g(1)$

7. $g(0)$
8. $g(-3)$
9. $h(0)$

10. $h(3)$
11. $h(-2)$
12. $h(-4)$

Let $h(x) = 10^x$ and $j(x) = e^x$. Find the following. Use a calculator as necessary and round approximate answers to four decimal places. See Example 2.

13. $h(0)$
14. $h(-1)$
15. $h(2)$

16. $h(3.4)$
17. $j(1)$
18. $j(3.5)$

19. $j(-2)$
20. $j(0)$

Sketch the graph of each function. See Examples 3 and 4.

21. $f(x) = 4^x$
22. $g(x) = 5^x$

23. $h(x) = \left(\frac{1}{3}\right)^x$
24. $i(x) = \left(\frac{1}{5}\right)^x$

25. $y = 10^x$
26. $y = (0.1)^x$

Sketch the graph of each function. See Examples 5 and 6.

27. $y = 10^{x+2}$
28. $y = 3^{2x+1}$

29. $f(x) = -2^x$
30. $k(x) = -2^{x-2}$

31. $g(x) = 2^{-x}$
32. $A(x) = 10^{1-x}$

33. $f(x) = -e^x$
34. $g(x) = e^{-x}$

35. $H(x) = 10^{|x|}$
36. $s(x) = 2^{(x^2)}$

37. $P = 5000(1.05)^t$
38. $d = 800 \cdot 10^{-4t}$

Solve each equation. See Example 7.

39. $2^x = 64$
40. $3^x = 9$

41. $10^x = 0.001$
42. $10^{2x} = 0.1$

43. $2^x = \frac{1}{4}$
44. $3^x = \frac{1}{9}$

45. $\left(\frac{2}{3}\right)^{x-1} = \frac{9}{4}$
46. $\left(\frac{1}{4}\right)^{3x} = 16$

47. $5^{-x} = 25$
48. $10^{-x} = 0.01$

49. $-2^{1-x} = -8$
50. $-3^{2-x} = -81$

51. $10^{|x|} = 1000$
52. $3^{|2x-5|} = 81$

Let $f(x) = 2^x$, $g(x) = \left(\frac{1}{3}\right)^x$, and $h(x) = 4^{2x-1}$. Find x in each case. See Example 8.

53. $f(x) = 4$

54. $f(x) = \dfrac{1}{4}$

55. $f(x) = 4^{2/3}$

56. $f(x) = 1$

57. $g(x) = 9$

58. $g(x) = \dfrac{1}{9}$

59. $g(x) = 1$

60. $g(x) = \sqrt{3}$

61. $h(x) = 16$

62. $h(x) = \dfrac{1}{2}$

63. $h(x) = 1$

64. $h(x) = \sqrt{2}$

Solve each problem. See Example 9.

65. ***Compounding quarterly.*** If $6000 is deposited in an account paying 5% compounded quarterly, then what amount will be in the account after 10 years?

66. ***Compounding quarterly.*** If $400 is deposited in an account paying 10% compounded quarterly, then what amount will be in the account after 7 years?

67. ***Outstanding performance.*** The Janus mutual fund is among the top performing mutual funds because it has returned an average of 16.24% annually for the last 20 years (*Money*, February 1994). How much would an investment of $10,000 in 1974 in the Janus fund be worth in 1994? Use the accompanying graph to estimate the year in which the $10,000 investment was worth $100,000.

Figure for Exercise 67

68. ***High-risk investment.*** The high-risk Prudential Speculator mutual fund had an average annual return of only 2% for the past five years. (*Money*, February 1994) How much would an investment of $10,000 in 1989 in this fund be worth in 1994?

69. ***Depreciating knowledge.*** The value of a certain textbook seems to decrease according to the formula $V = 45 \cdot 2^{-0.9t}$, where V is the value in dollars and t is the age of the book in years. What is the book worth when it is new? What is it worth when it is 2 years old?

70. ***Mosquito abatement.*** In a Minnesota swamp in the springtime, the number of mosquitoes per acre appears to grow according to the formula $N = 10^{0.1t+2}$, where t is the number of days since the last frost. What is the size of the mosquito population at times $t = 10$, $t = 20$, and $t = 30$?

In Exercises 71–76, solve each problem. See Example 10.

71. ***Compounding continuously.*** If $500 is deposited in an account paying 7% compounded continuously, then how much will be in the account after 3 years?

72. ***Compounding continuously.*** If $7000 is deposited in an account paying 8% compounded continuously, then what will it amount to after 4 years?

73. ***One year's interest.*** How much interest will be earned the first year on $80,000 on deposit in an account paying 7.5% compounded continuously?

74. ***Partial year.*** If $7500 is deposited in an account paying 6.75% compounded continuously, then how much will be in the account after 5 years and 215 days?

75. ***Radioactive decay.*** The number of grams of a certain radioactive substance present at time t is given by the formula $A = 300 \cdot e^{-0.06t}$, where t is the number of years. Find the amount present at time $t = 0$. Find the amount present after 20 years. Use the accompanying graph to estimate the number of years that it takes for one-half of the substance to decay. Will the substance ever decay completely?

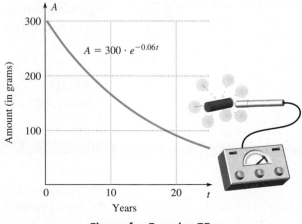

Figure for Exercise 75

76. *Population growth.* The population of a certain country appears to be growing according to the formula $P = 20 \cdot e^{0.1t}$, where P is the population in millions and t is the number of years since 1980. What was the population in 1980? What will the population be in the year 2000?

Getting More Involved

77. *Exploration.* An approximate value for e can be found by adding the terms in the following infinite sum:

$$1 + \frac{1}{1} + \frac{1}{2 \cdot 1} + \frac{1}{3 \cdot 2 \cdot 1} + \frac{1}{4 \cdot 3 \cdot 2 \cdot 1} + \cdots$$

Use a calculator to find the sum of the first four terms. Find the difference between the sum of the first four terms and e. (For e, use all of the digits that your calculator gives for e^1.) What is the difference between e and the sum of the first eight terms?

Graphing Calculator Exercises

78. Graph $y_1 = 2^x$, $y_2 = e^x$, and $y_3 = 3^x$ on the same coordinate system. Which point do all three graphs have in common?

79. Graph $y_1 = 3^x$, $y_2 = 3^{x-1}$, and $y_3 = 3^{x-2}$ on the same coordinate system. What can you say about the graph of $y = 3^{x-k}$ for any real number k?

9.2 Logarithms

In Section 9.1 you learned that exponential functions are one-to-one functions. Because they are one-to-one functions, they have inverse functions. In this section we study the inverses of the exponential functions.

In this section:
▶ Definition
▶ Domain and Range
▶ Graphing Logarithmic Functions
▶ Logarithmic Equations
▶ Applications

Definition

Consider the base-2 exponential function $f(x) = 2^x$. Because any exponential function is one-to-one, f has an inverse function. The inverse function is called the **base-2 logarithm function.** We write $f^{-1}(x) = \log_2(x)$ and read $\log_2(x)$ as "the base-2 logarithm of x." The inverse function undoes what the function does. For example, because $f(5) = 2^5 = 32$, we have $\log_2(32) = 5$. See Fig. 9.9. So $\log_2(32)$ is the exponent that is used on the base 2 to obtain 32. In general, the inverse of the base-a exponential function is called the **base-a logarithm function** and $\log_a(x)$ *is the exponent that is used on the base a to obtain x.* The expression $\log_a(x)$ is called a **logarithm.**

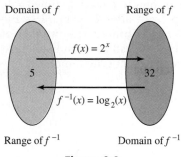

Domain of f Range of f

$f(x) = 2^x$

5   32

$f^{-1}(x) = \log_2(x)$

Range of f^{-1} Domain of f^{-1}

Figure 9.9

$\log_a(x)$

For any $a > 0$ and $a \neq 1$,

$$y = \log_a(x) \qquad \text{if and only if} \qquad a^y = x.$$

The definition of base-a logarithm says that the logarithmic equation $y = \log_a(x)$ and the exponential equation $a^y = x$ are equivalent.

EXAMPLE 1   **Using the definition of logarithm**

Write each logarithmic equation as an exponential equation and each exponential equation as a logarithmic equation.

a) $\log_5 (125) = 3$ b) $6 = \log_{1/4} (x)$

c) $\left(\dfrac{1}{2}\right)^m = 8$ d) $7 = 3^z$

Solution

a) "The base 5 logarithm of 125 equals 3" means that 3 is the exponent on 5 that produces 125. So $5^3 = 125$.

b) The equation $6 = \log_{1/4} (x)$ is equivalent to $\left(\dfrac{1}{4}\right)^6 = x$ by the definition of logarithm.

c) The equation $\left(\dfrac{1}{2}\right)^m = 8$ is equivalent to $\log_{1/2} (8) = m$.

d) The equation $7 = 3^z$ is equivalent to $\log_3 (7) = z$.   ■

The definition of logarithm is also used to evaluate logarithmic functions.

EXAMPLE 2   **Finding logarithms**

Evaluate each logarithm.

a) $\log_5 (25)$ b) $\log_2 \left(\dfrac{1}{8}\right)$

c) $\log_{1/2} (4)$ d) $\log_{10} (0.001)$

Solution

a) The number $\log_5 (25)$ is the exponent that is used on the base 5 to obtain 25. Because $25 = 5^2$, we have $\log_5 (25) = 2$.

b) The number $\log_2 \left(\dfrac{1}{8}\right)$ is the power of 2 that gives us $\dfrac{1}{8}$. Because $\dfrac{1}{8} = 2^{-3}$, we have $\log_2 \left(\dfrac{1}{8}\right) = -3$.

c) The number $\log_{1/2} (4)$ is the power of $\dfrac{1}{2}$ that produces 4. Because $4 = \left(\dfrac{1}{2}\right)^{-2}$, we have $\log_{1/2} (4) = -2$.

d) Because $0.001 = 10^{-3}$, we have $\log_{10} (0.001) = -3$.   ■

There are two bases for logarithms that are used more frequently than the others: They are 10 and e. The base-10 logarithm is called the **common logarithm** and is usually written as $\log (x)$. The base-e logarithm is called the **natural logarithm,** and it is usually written as $\ln (x)$. Most scientific calculators have function keys for $\log (x)$ and $\ln (x)$. The simplest way to obtain a common or natural logarithm is to use a scientific calculator. However, a table of common logarithms can be found in Appendix C of this text.

Calculator Close-up

To find the natural logarithm ln (5.44) on a scientific calculator, enter the following:

$$5.44 \boxed{\text{ln}}$$

The display will read

$$\boxed{ 1.6938}$$

Finding ln (5.44) on a graphing calculator, the display should appear as follows:

$$\text{ln } 5.44 \boxed{\text{ENTER}}$$

$$\boxed{ 1.6938}$$

You can use your calculator to check this result by finding $e^{1.6938}$.

Common logarithms are found by using the log key. For example, to find log (25.3) on a scientific calculator, enter the following:

$$25.3 \boxed{\text{LOG}}$$

The display will read

$$\boxed{ 1.4031}$$

Finding log (25.3) on a graphing calculator, the display should appear as follows:

$$\text{log } 25.3 \boxed{\text{ENTER}}$$

$$\boxed{ 1.4031}$$

You can check this result by finding $10^{1.4031}$.

In the next example we find natural and common logarithms of certain numbers without a calculator or a table.

EXAMPLE 3 **Finding common and natural logarithms**

Evaluate each logarithm.

a) log (1000) **b)** ln (e) **c)** $\log\left(\dfrac{1}{10}\right)$

Solution

a) Because $10^3 = 1000$, we have log (1000) = 3.

b) Because $e^1 = e$, we have ln (e) = 1.

c) Because $10^{-1} = \dfrac{1}{10}$, we have $\log\left(\dfrac{1}{10}\right) = -1$. ∎

Domain and Range

The domain of the exponential function $y = 2^x$ is $(-\infty, \infty)$, and its range is $(0, \infty)$. Because the logarithmic function $y = \log_2 (x)$ is the inverse of $y = 2^x$, the domain of $y = \log_2 (x)$ is $(0, \infty)$, and its range is $(-\infty, \infty)$.

CAUTION   Because the domain of $y = \log_a (x)$ is $(0, \infty)$ for any $a > 0$ and $a \neq 1$, expressions such as $\log_2 (-4)$, $\log_{1/3} (0)$, and $\ln (-1)$ are undefined.⊘

Graphing Logarithmic Functions

In Chapter 8 we saw that the graphs of a function and its inverse function are symmetric about the line $y = x$. Because the logarithm functions are inverses of exponential functions, their graphs are also symmetric about $y = x$.

EXAMPLE 4   **A logarithmic function with base greater than 1**

Sketch the graph of $g(x) = \log_2 (x)$ and compare it to the graph of $y = 2^x$.

Solution

Make a table of ordered pairs for $g(x) = \log_2 (x)$ using positive numbers for x:

x	$\frac{1}{4}$	$\frac{1}{2}$	1	2	4	8
$g(x) = \log_2 (x)$	-2	-1	0	1	2	3

Draw a curve through these points as shown in Fig. 9.10. The graph of the inverse function $y = 2^x$ is also shown in Fig. 9.10 for comparison. Note the symmetry of the two curves about the line $y = x$.   ■

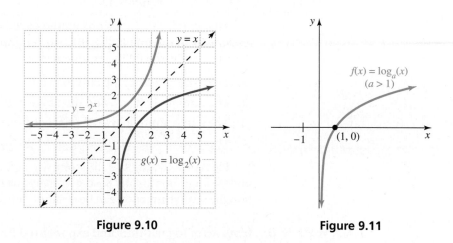

Figure 9.10   Figure 9.11

All logarithmic functions with base greater than 1 have graphs that are similar to the one in Fig. 9.10. In general, the graph of $f(x) = \log_a (x)$ for $a > 1$ has the following characteristics (see Fig. 9.11):

1. The x-intercept of the curve is $(1, 0)$.
2. The domain is $(0, \infty)$, and the range is $(-\infty, \infty)$.
3. The curve approaches the negative y-axis, but does not touch it.
4. The y-values are increasing as we go from left to right along the curve.

EXAMPLE 5   **A logarithmic function with base less than 1**

Sketch the graph of $f(x) = \log_{1/2}(x)$ and compare it to the graph of $y = \left(\frac{1}{2}\right)^x$.

Solution

Make a table of ordered pairs for $f(x) = \log_{1/2}(x)$ using positive numbers for x:

x	$\frac{1}{4}$	$\frac{1}{2}$	1	2	4	8
$f(x) = \log_{1/2}(x)$	2	1	0	−1	−2	−3

The curve through these points is shown in Fig. 9.12. The graph of the inverse function $y = \left(\frac{1}{2}\right)^x$ is also shown in Fig. 9.12 for comparison.   ■

Figure 9.12   Figure 9.13

All logarithmic functions with base between 0 and 1 have graphs that are similar to the one in Fig. 9.12. In general, the graph of $f(x) = \log_a(x)$ for $0 < a < 1$ has the following characteristics (see Fig. 9.13):

1. The x-intercept of the curve is $(1, 0)$.
2. The domain is $(0, \infty)$, and the range is $(-\infty, \infty)$.
3. The curve approaches the positive y-axis but does not touch it.
4. The y-values are decreasing as we go from left to right along the curve.

Figures 9.10 and 9.12 illustrate the fact that $y = \log_a(x)$ and $y = a^x$ are inverse functions for any base a. For any given exponential or logarithmic function the inverse function can be easily obtained from the definition of logarithm.

EXAMPLE 6   **Inverses of logarithmic and exponential functions**

Find the inverse of each function.

a) $f(x) = 10^x$ **b)** $g(x) = \log_3(x)$

Solution

a) The inverse of $f(x)$ is $f^{-1}(x) = \log_{10}(x)$ or $y = \log(x)$.

b) The inverse of $g(x) = \log_3(x)$ is $g^{-1}(x) = 3^x$ or $y = 3^x$.   ■

Logarithmic Equations

In Section 9.1 we learned that the exponential functions are one-to-one functions. Because logarithmic functions are inverses of exponential functions, they are one-to-one functions also. For a base-*a* logarithmic function, *one-to-one means that if the base-a logarithms of two numbers are equal, then the numbers are equal.*

One-to-One Property of Logarithms

For $a > 0$ and $a \neq 1$,

$$\text{if} \quad \log_a (m) = \log_a (n), \quad \text{then} \quad m = n.$$

The one-to-one property of logarithms and the definition of logarithms are the two basic tools that we use to solve equations involving logarithms. We use these tools in the next example.

EXAMPLE 7   **Logarithmic equations**

Solve each equation.

a) $\log_3 (x) = -2$ **b)** $\log_x (8) = -3$ **c)** $\log (x^2) = \log (4)$

Solution

a) Use the definition of logarithms to rewrite the logarithmic equation as an equivalent exponential equation:

$$\log_3 (x) = -2$$
$$3^{-2} = x \quad \text{Definition of logarithm}$$
$$\frac{1}{9} = x$$

The solution set is $\left\{\frac{1}{9}\right\}$.

b) Use the definition of logarithms to rewrite the logarithmic equation as an equivalent exponential equation:

$$\log_x (8) = -3$$
$$x^{-3} = 8 \qquad \text{Definition of logarithm}$$
$$(x^{-3})^{-1} = 8^{-1} \qquad \text{Raise each side to the } -1 \text{ power.}$$
$$x^3 = \frac{1}{8}$$
$$x = \sqrt[3]{\frac{1}{8}} = \frac{1}{2} \qquad \text{Odd-root property}$$

The solution set is $\left\{\frac{1}{2}\right\}$.

c) To write an equation equivalent to $\log (x^2) = \log (4)$, we use the one-to-one property of logarithms:

$$\log (x^2) = \log (4)$$
$$x^2 = 4 \qquad \text{One-to-one property of logarithms}$$
$$x = \pm 2 \qquad \text{Even-root property}$$

The solution set is $\{-2, 2\}$. ■

CAUTION If we have equality of two logarithms with the same base, we use the one-to-one property to eliminate the logarithms. If we have an equation with only one logarithm, such as $\log_a (x) = y$, we use the definition of logarithm to write $a^y = x$ and eliminate the logarithm. ⊘

Applications

When money earns interest compounded continuously, the formula

$$t = \frac{1}{r} \ln\left(\frac{A}{P}\right)$$

expresses the relationship between the time in years t, the annual interest rate r, the principal P, and the amount A. This formula is used to determine how long it takes for a deposit to grow to a specific amount.

EXAMPLE 8 **Finding the time for a specified growth**

How long does it take $80 to grow to $240 at 12% compounded continuously?

Solution

Use $r = 0.12$, $P = \$80$, and $A = \$240$ in the formula, and use a calculator to evaluate the logarithm:

$$t = \frac{1}{0.12} \ln\left(\frac{240}{80}\right)$$

$$= \frac{\ln(3)}{0.12}$$

$$\approx 9.155$$

It takes approximately 9.155 years, or 9 years and 57 days. ■

Warm-ups

True or false? Explain.

1. The equation $a^3 = 2$ is equivalent to $\log_a (2) = 3$.
2. If (a, b) satisfies $y = 8^x$, then (a, b) satisfies $y = \log_8 (x)$.
3. If $f(x) = a^x$ for $a > 0$ and $a \neq 1$, then $f^{-1}(x) = \log_a (x)$.
4. If $f(x) = \ln(x)$, then $f^{-1}(x) = e^x$.
5. The domain of $f(x) = \log_6 (x)$ is $(-\infty, \infty)$.
6. $\log_{25} (5) = 2$
7. $\log(-10) = 1$
8. $\log(0) = 0$
9. $5^{\log_5 (125)} = 125$
10. $\log_{1/2} (32) = -5$

9.2 EXERCISES

Write each exponential equation as a logarithmic equation and each logarithmic equation as an exponential equation. See Example 1.

1. $\log_2 (8) = 3$ **2.** $\log_{10} (10) = 1$

3. $10^2 = 100$ **4.** $5^3 = 125$

5. $y = \log_5 (x)$ **6.** $m = \log_b (N)$

7. $2^a = b$ **8.** $a^3 = c$

9. $\log_3 (x) = 10$ **10.** $\log_c (t) = 4$

11. $e^3 = x$ **12.** $m = e^x$

Evaluate each logarithm. See Examples 2 and 3.

13. $\log_2 (4)$ **14.** $\log_2 (1)$

15. $\log_2 (16)$ **16.** $\log_4 (16)$

17. $\log_2 (64)$ **18.** $\log_8 (64)$

19. $\log_4 (64)$ **20.** $\log_{64} (64)$

21. $\log_2 \left(\dfrac{1}{4}\right)$ **22.** $\log_2 \left(\dfrac{1}{8}\right)$

23. $\log (100)$ **24.** $\log (1)$

25. $\log (0.01)$ **26.** $\log (10{,}000)$

27. $\log_{1/3} \left(\dfrac{1}{3}\right)$ **28.** $\log_{1/3} \left(\dfrac{1}{9}\right)$

29. $\log_{1/3} (27)$ **30.** $\log_{1/3} (1)$

31. $\ln (e^3)$ **32.** $\ln (1)$

33. $\ln (e^2)$ **34.** $\ln \left(\dfrac{1}{e}\right)$

Use a calculator to evaluate each logarithm. Round answers to four decimal places.

35. $\log (5)$ **36.** $\log (0.03)$

37. $\ln (6.238)$ **38.** $\ln (0.23)$

Sketch the graph of each function. See Examples 4 and 5.

39. $f(x) = \log_3 (x)$ **40.** $g(x) = \log_{10} (x)$

41. $y = \log_4 (x)$ **42.** $y = \log_5 (x)$

43. $h(x) = \log_{1/4} (x)$ **44.** $y = \log_{1/3} (x)$

45. $y = \log_{1/5} (x)$ **46.** $y = \log_{1/6} (x)$

Find the inverse of each function. See Example 6.

47. $f(x) = 6^x$ **48.** $f(x) = 4^x$

49. $f(x) = \ln (x)$ **50.** $f(x) = \log (x)$

51. $f(x) = \log_{1/2} (x)$ **52.** $f(x) = \log_{1/4} (x)$

Solve each equation. See Example 7.

53. $x = \left(\dfrac{1}{2}\right)^{-2}$ **54.** $x = 3^2$

55. $5 = 25^x$ **56.** $0.1 = 10^x$

57. $\log (x) = -3$ **58.** $\log (x) = 5$

59. $\log_x (36) = 2$ **60.** $\log_x (100) = 2$

61. $\log_x (5) = -1$ **62.** $\log_x (16) = -2$

63. $\log (x^2) = \log (9)$ **64.** $\ln (2x - 3) = \ln (x + 1)$

Use a calculator to solve each equation. Round answers to four decimal places.

65. $3 = 10^x$ **66.** $10^x = 0.03$ **67.** $10^x = \dfrac{1}{2}$

68. $75 = 10^x$ **69.** $e^x = 7.2$ **70.** $e^{3x} = 0.4$

Solve each problem. See Example 8. Use a calculator as necessary.

71. **Double your money.** How long does it take $5000 to grow to $10,000 at 12% compounded continuously?

72. **Half the rate.** How long does it take $5000 to grow to $10,000 at 6% compounded continuously?

73. **Earning interest.** How long does it take to earn $1000 in interest on a deposit of $6000 at 8% compounded continuously?

74. **Lottery winnings.** How long does it take to earn $1000 interest on a deposit of one million dollars at 9% compounded continuously?

The annual growth rate for an investment that is growing continuously is given by

$$r = \frac{1}{t} \ln \left(\frac{A}{P}\right),$$

where P is the principal and A is the amount after t years.

75. **Close shave.** An investment of $10,000 in 1980 in Gillette stock was worth $302,000 in 1994 (*Money*, February 1994). Assuming that the investment grew continuously, what was the annual growth rate? If the stock continues to grow at this rate, what will the $10,000 investment be worth in the year 2000?

76. **Chocolate bars.** An investment of $10,000 in 1980 in Hershey stock was worth $179,900 in 1994 (*Money*, February 1994). Assuming that the investment grew continuously, what was the annual growth rate?

In chemistry, the pH of a solution is defined by

$$pH = -\log_{10} [H+],$$

where H+ is the hydrogen ion concentration of the solution in moles per liter. Distilled water has a pH of approximately 7. A solution with a pH under 7 is called an acid, and one with a pH over 7 is called a base.

77. *Tomato juice.* Tomato juice has a hydrogen ion concentration of $10^{-4.1}$ mole per liter. Find the pH of tomato juice.

78. *Stomach acid.* The gastric juices in your stomach have a hydrogen ion concentration of 10^{-1} mole per liter. Find the pH of your gastric juices.

79. *Red River pH.* The pH of a water sample is one of the many measurements of water quality done by the U.S. Geological Survey. The hydrogen ion concentration of the water in the Red River above Shreveport, Louisiana, was 9.77×10^{-9} mole per liter on October 25, 1989 (U.S.G.S., *Water Resources Data for Louisiana 1990*). What was the pH of the water in the Red River at that time?

80. *Changing pH.* On February 28, 1990, the hydrogen ion concentration of the water in the Red River above Shreveport, Louisiana, was 4.07×10^{-8} mole per liter (U.S.G.S., *Water Resources Data for Louisiana 1990*). What was the pH of the water at that time?

Solve each problem.

81. *Sound level.* The level of sound in decibels (db) is given by the formula

$$L = 10 \cdot \log (I \times 10^{12})$$

where I is the intensity of the sound in watts per square meter. If the intensity of the sound at a rock concert is 0.001 watt per square meter at a distance of 75 meters from the stage, then what is the level of the sound at this point in the audience?

82. *Logistic growth.* If a rancher has one cow with a contagious disease in a herd of 1000, then the time in days t for n of the cows to become infected is modeled by

$$t = -5 \cdot \ln \left(\frac{1000 - n}{999n} \right).$$

Find the number of days that it takes for the disease to spread to 100, 200, 998, and 999 cows. This model, called a *logistic growth model*, describes how a disease can spread very rapidly at first and then very slowly as nearly all of the population has become infected.

Figure for Exercise 82

Getting More Involved

83. *Discussion.* Use the switch-and-solve method from Chapter 8 to find the inverse of the function $f(x) = 5 + \log_2 (x - 3)$. State the domain and range of the inverse function.

84. *Discussion.* Find the inverse of the function $f(x) = 2 + e^{x+4}$. State the domain and range of the inverse function.

⌁ Graphing Calculator Exercises

85. *Composition of inverses.* Graph the functions $y = \ln (e^x)$ and $y = e^{\ln (x)}$. Explain the similarities and differences between the graphs.

86. *The population bomb.* The population of the earth is growing continuously with an annual rate of about 1.6%. If the present population is 6 billion, then the function $y = 6e^{0.016x}$ gives the population in billions x years from now. Graph this function for $0 \le x \le 200$. What will the population be in 100 years and in 200 years?

9.3 Properties

The properties of logarithms are very similar to the properties of exponents because *logarithms are exponents*. In this section we use the properties of exponents to write some properties of logarithms. The properties will be used in solving logarithmic equations in Section 9.4.

Product Rule for Logarithms

If $M = a^x$ and $N = a^y$, we can use the product rule for exponents to write

$$MN = a^x \cdot a^y = a^{x+y}.$$

The equation $MN = a^{x+y}$ is equivalent to

$$\log_a (MN) = x + y.$$

Because $M = a^x$ and $N = a^y$ are equivalent to $x = \log_a (M)$ and $y = \log_a (N)$, we can replace x and y in $\log_a (MN) = x + y$ to get

$$\log_a (MN) = \log_a (M) + \log_a (N).$$

So *the logarithm of a product is the sum of the logarithms*, provided that all of the logarithms are defined. This rule is called the **product rule for logarithms.**

> **Product Rule for Logarithms**
>
> $$\log_a (MN) = \log_a (M) + \log_a (N)$$

EXAMPLE 1   **Using the product rule for logarithms**
Write each expression as a single logarithm.
a) $\log_2 (7) + \log_2 (5)$ **b)** $\ln (\sqrt{2}) + \ln (\sqrt{3})$

Solution

a) $\log_2 (7) + \log_2 (5) = \log_2 (35)$ Product rule for logarithms
b) $\ln (\sqrt{2}) + \ln (\sqrt{3}) = \ln (\sqrt{6})$ Product rule for logarithms   ■

Quotient Rule for Logarithms

If $M = a^x$ and $N = a^y$, we can use the quotient rule for exponents to write

$$\frac{M}{N} = \frac{a^x}{a^y} = a^{x-y}.$$

By the definition of logarithm, $\dfrac{M}{N} = a^{x-y}$ is equivalent to

$$\log_a \left(\frac{M}{N} \right) = x - y.$$

Because $x = \log_a (M)$ and $y = \log_a (N)$, we have

$$\log_a \left(\frac{M}{N} \right) = \log_a (M) - \log_a (N).$$

So *the logarithm of a quotient is equal to the difference of the logarithms*, provided that all logarithms are defined. This rule is called the **quotient rule for logarithms.**

Quotient Rule for Logarithms

$$\log_a \left(\frac{M}{N} \right) = \log_a (M) - \log_a (N)$$

EXAMPLE 2   **Using the quotient rule for logarithms**
Write each expression as a single logarithm.
a) $\log_2 (3) - \log_2 (7)$
b) $\ln (w^8) - \ln (w^2)$

Solution

a) $\log_2 (3) - \log_2 (7) = \log_2 \left(\dfrac{3}{7} \right)$ Quotient rule for logarithms

b) $\ln (w^8) - \ln (w^2) = \ln \left(\dfrac{w^8}{w^2} \right)$ Quotient rule for logarithms

$\qquad\qquad\qquad\quad\;\; = \ln (w^6)$ Quotient rule for exponents   ■

Power Rule for Logarithms

If $M = a^x$, we can use the power rule for exponents to write

$$M^N = (a^x)^N = a^{Nx}.$$

By the definition of logarithms, $M^N = a^{Nx}$ is equivalent to

$$\log_a (M^N) = Nx.$$

Because $x = \log_a (M)$, we have

$$\log_a (M^N) = N \cdot \log_a (M).$$

So *the logarithm of a power of a number is the power times the logarithm of the number*, provided that all logarithms are defined. This rule is called the **power rule for logarithms.**

> ### Power Rule for Logarithms
>
> $$\log_a (M^N) = N \cdot \log_a (M)$$

EXAMPLE 3 **Using the power rule for logarithms**

Rewrite each logarithm in terms of $\log (2)$.

a) $\log (2^{10})$ **b)** $\log (\sqrt{2})$ **c)** $\log \left(\dfrac{1}{2}\right)$

Solution

a) $\log (2^{10}) = 10 \cdot \log (2)$ Power rule for logarithms

b) $\log (\sqrt{2}) = \log (2^{1/2})$ Write $\sqrt{2}$ as a power of 2.

$\qquad\qquad = \dfrac{1}{2} \log (2)$ Power rule for logarithms

c) $\log \left(\dfrac{1}{2}\right) = \log (2^{-1})$ Write $\frac{1}{2}$ as a power of 2.

$\qquad\qquad = -1 \cdot \log (2)$ Power rule for logarithms

$\qquad\qquad = -\log (2)$ ■

Inverse Properties

The base-a exponential function and the base-a logarithmic function are inverses of each other. We can express this fact with the following two inverse properties.

> ### Inverse Properties
>
> 1. $\log_a (a^M) = M$ 2. $a^{\log_a (M)} = M$

EXAMPLE 4 **Using the inverse properties**

Simplify each expression.

a) $\ln (e^5)$ **b)** $2^{\log_2 (8)}$

Solution

a) Using the first inverse property, we get $\ln (e^5) = 5$.

b) Using the second inverse property, we get $2^{\log_2 (8)} = 8$. ■

Note that there is more than one way to simplify the expressions in Example 4. Using the power rule for logarithms and the fact that $\ln (e) = 1$, we have $\ln (e^5) = 5 \cdot \ln (e) = 5$. Using $\log_2 (8) = 3$, we have $2^{\log_2 (8)} = 2^3 = 8$.

Using the Properties

The most important properties of logarithms are summarized as follows.

> **Properties of Logarithms**
>
> If M, N, and a are positive numbers, $a \neq 1$, then
>
> 1. $\log_a (a) = 1$
> 2. $\log_a (1) = 0$
> 3. $\log_a (a^M) = M$
> 4. $a^{\log_a (M)} = M$
> 5. $\log_a (MN) = \log_a (M) + \log_a (N)$ Product rule
> 6. $\log_a \left(\dfrac{M}{N}\right) = \log_a (M) - \log_a (N)$ Quotient rule
> 7. $\log_a \left(\dfrac{1}{N}\right) = -\log_a (N)$
> 8. $\log_a (M^N) = N \cdot \log_a (M)$ Power rule

Note that Property 7 is derived from Property 6 with $M = 1$.

    We have already seen several ways in which to use the properties of logarithms. In the next three examples we see more uses of the properties. First we use the rules of logarithms to write the logarithm of a complicated expression in terms of logarithms of simpler expressions.

EXAMPLE 5   **Using the properties of logarithms**

Rewrite each expression in terms of $\log (2)$ and/or $\log (3)$.

a) $\log (6)$ **b)** $\log (16)$

c) $\log \left(\dfrac{9}{2}\right)$ **d)** $\log \left(\dfrac{1}{3}\right)$

Solution

a) $\log (6) = \log (2 \cdot 3)$

            $= \log (2) + \log (3)$ Product rule

b) $\log (16) = \log (2^4)$

             $= 4 \cdot \log (2)$ Power rule

c) $\log \left(\dfrac{9}{2}\right) = \log (9) - \log (2)$ Quotient rule

            $= \log (3^2) - \log (2)$

            $= 2 \cdot \log (3) - \log (2)$ Power rule

d) $\log \left(\dfrac{1}{3}\right) = -\log (3)$ Property 8   ■

CAUTION   Do not confuse $\frac{\log{(9)}}{\log{(2)}}$ with $\log\left(\frac{9}{2}\right)$. We can use the quotient rule to write $\log\left(\frac{9}{2}\right) = \log{(9)} - \log{(2)}$, but $\frac{\log{(9)}}{\log{(2)}} \neq \log{(9)} - \log{(2)}$. The expression $\frac{\log{(9)}}{\log{(2)}}$ means $\log{(9)} \div \log{(2)}$. Use your calculator to verify these two statements.   ⊘

The properties of logarithms can be used to combine several logarithms into a single logarithm (as in Examples 1 and 2) or to write a logarithm of a complicated expression in terms of logarithms of simpler expressions.

EXAMPLE 6   **Using the properties of logarithms**

Rewrite each expression as a sum or difference of multiples of logarithms.

a) $\log\left(\dfrac{xz}{y}\right)$ **b)** $\log_3\left(\dfrac{(x-3)^{2/3}}{\sqrt{x}}\right)$

Solution

a) $\log\left(\dfrac{xz}{y}\right) = \log{(xz)} - \log{(y)}$ Quotient rule

$\phantom{\log\left(\dfrac{xz}{y}\right)} = \log{(x)} + \log{(z)} - \log{(y)}$ Product rule

b) $\log_3\left(\dfrac{(x-3)^{2/3}}{\sqrt{x}}\right) = \log_3{((x-3)^{2/3})} - \log_3{(x^{1/2})}$ Quotient rule

$\phantom{\log_3\left(\dfrac{(x-3)^{2/3}}{\sqrt{x}}\right)} = \dfrac{2}{3}\log_3{(x-3)} - \dfrac{1}{2}\log_3{(x)}$ Power rule   ■

In the next example we use the properties of logarithms to convert expressions involving several logarithms into a single logarithm. The skills we are learning here will be used to solve logarithmic equations in Section 9.4.

EXAMPLE 7   **Combining logarithms**

Rewrite each expression as a single logarithm.

a) $\dfrac{1}{2}\log{(x)} - 2 \cdot \log{(x+1)}$ **b)** $3 \cdot \log{(y)} + \dfrac{1}{2}\log{(z)} - \log{(x)}$

Solution

a) $\dfrac{1}{2}\log{(x)} - 2 \cdot \log{(x+1)} = \log{(x^{1/2})} - \log{((x+1)^2)}$ Power rule

$\phantom{\dfrac{1}{2}\log{(x)} - 2 \cdot} = \log\left(\dfrac{\sqrt{x}}{(x+1)^2}\right)$ Quotient rule

b) $3 \cdot \log{(y)} + \dfrac{1}{2}\log{(z)} - \log{(x)} = \log{(y^3)} + \log{(\sqrt{z})} - \log{(x)}$ Power rule

$\phantom{3 \cdot \log{(y)} + \dfrac{1}{2}} = \log{(y^3 \cdot \sqrt{z})} - \log{(x)}$ Product rule

$\phantom{3 \cdot \log{(y)} + \dfrac{1}{2}} = \log\left(\dfrac{y^3 \cdot \sqrt{z}}{x}\right)$ Quotient rule   ■

Warm-ups

True or false? Explain.

1. $\log_2\left(\dfrac{x^2}{8}\right) = \log_2(x^2) - 3$

2. $\dfrac{\log(100)}{\log(10)} = \log(100) - \log(10)$

3. $\ln(\sqrt{2}) = \dfrac{\ln(2)}{2}$

4. $3^{\log_3(17)} = 17$

5. $\log_2\left(\dfrac{1}{8}\right) = \dfrac{1}{\log_2(8)}$

6. $\ln(8) = 3\cdot\ln(2)$

7. $\ln(1) = e$

8. $\dfrac{\log(100)}{10} = \log(10)$

9. $\dfrac{\log_2(8)}{\log_2(2)} = \log_2(4)$

10. $\ln(2) + \ln(3) - \ln(7) = \ln\left(\dfrac{6}{7}\right)$

9.3 EXERCISES

Assume that all variables involved in logarithms represent numbers for which the logarithms are defined.

Write each expression as a single logarithm. See Example 1.

1. $\log(3) + \log(7)$
2. $\ln(5) + \ln(4)$
3. $\log_3(\sqrt{5}) + \log_3(\sqrt{x})$
4. $\ln(\sqrt{x}) + \ln(\sqrt{y})$
5. $\log(x^2) + \log(x^3)$
6. $\ln(a^3) + \ln(a^5)$
7. $\ln(2) + \ln(3) + \ln(5)$
8. $\log_2(x) + \log_2(y) + \log_2(z)$
9. $\log(x) + \log(x + 3)$
10. $\ln(x - 1) + \ln(x + 1)$
11. $\log_2(x - 3) + \log_2(x + 2)$
12. $\log_3(x - 5) + \log_3(x - 4)$

Write each expression as a single logarithm. See Example 2.

13. $\log(8) - \log(2)$
14. $\ln(3) - \ln(6)$
15. $\log_2(x^6) - \log_2(x^2)$
16. $\ln(w^9) - \ln(w^3)$
17. $\log(\sqrt{10}) - \log(\sqrt{2})$
18. $\log_3(\sqrt{6}) - \log_3(\sqrt{3})$
19. $\ln(4h - 8) - \ln(4)$
20. $\log(3x - 6) - \log(3)$
21. $\log_2(w^2 - 4) - \log_2(w + 2)$
22. $\log_3(k^2 - 9) - \log_3(k - 3)$
23. $\ln(x^2 + x - 6) - \ln(x + 3)$
24. $\ln(t^2 - t - 12) - \ln(t - 4)$

Write each expression in terms of log (3). See Example 3.

25. $\log(27)$
26. $\log\left(\dfrac{1}{9}\right)$
27. $\log(\sqrt{3})$
28. $\log(\sqrt[4]{3})$
29. $\log(3^x)$
30. $\log(3^{-99})$

Simplify each expression. See Example 4.

31. $\log_2(2^{10})$
32. $\ln(e^9)$
33. $5^{\log_5(19)}$
34. $10^{\log(2.3)}$
35. $\log(10^8)$
36. $\log_4(4^5)$
37. $e^{\ln(4.3)}$
38. $3^{\log_3(5.5)}$

Rewrite each expression in terms of log (3) and/or log (5). See Example 5.

39. $\log(15)$
40. $\log(9)$
41. $\log\left(\dfrac{5}{3}\right)$
42. $\log\left(\dfrac{3}{5}\right)$
43. $\log(25)$
44. $\log\left(\dfrac{1}{27}\right)$
45. $\log(75)$
46. $\log(0.6)$
47. $\log\left(\dfrac{1}{3}\right)$
48. $\log(45)$
49. $\log(0.2)$
50. $\log\left(\dfrac{9}{25}\right)$

Rewrite each expression as a sum or difference of multiples of logarithms. See Example 6.

51. $\log(xyz)$
52. $\log(3y)$
53. $\log_2(8x)$
54. $\log_2(16y)$
55. $\ln\left(\dfrac{x}{y}\right)$
56. $\ln\left(\dfrac{z}{3}\right)$
57. $\log(10x^2)$
58. $\log(100\sqrt{x})$

59. $\log_5\left(\dfrac{(x-3)^2}{\sqrt{w}}\right)$ **60.** $\log_3\left(\dfrac{(y+6)^3}{y-5}\right)$

61. $\ln\left(\dfrac{yz\sqrt{x}}{w}\right)$ **62.** $\ln\left(\dfrac{(x-1)\sqrt{w}}{x^3}\right)$

Rewrite each expression as a single logarithm. See Example 7.

63. $\log(x) + \log(x-1)$

64. $\log_2(x-2) + \log_2(5)$

65. $\ln(3x-6) - \ln(x-2)$

66. $\log_3(x^2-1) - \log_3(x-1)$

67. $\ln(x) - \ln(w) + \ln(z)$

68. $\ln(x) - \ln(3) - \ln(7)$

69. $3 \cdot \ln(y) + 2 \cdot \ln(x) - \ln(w)$

70. $5 \cdot \ln(r) + 3 \cdot \ln(t) - 4 \cdot \ln(s)$

71. $\dfrac{1}{2}\log(x-3) - \dfrac{2}{3}\log(x+1)$

72. $\dfrac{1}{2}\log(y-4) + \dfrac{1}{2}\log(y+4)$

73. $\dfrac{2}{3}\log_2(x-1) - \dfrac{1}{4}\log_2(x+2)$

74. $\dfrac{1}{2}\log_3(y+3) + 6 \cdot \log_3(y)$

Determine whether each equation is true or false.

75. $\log(56) = \log(7) \cdot \log(8)$ **76.** $\log\left(\dfrac{5}{9}\right) = \dfrac{\log(5)}{\log(9)}$

77. $\log_2(4^2) = (\log_2(4))^2$ **78.** $\ln(4^2) = (\ln(4))^2$

79. $\ln(25) = 2 \cdot \ln(5)$ **80.** $\ln(3e) = 1 + \ln(3)$

81. $\dfrac{\log_2(64)}{\log_2(8)} = \log_2(8)$ **82.** $\dfrac{\log_2(16)}{\log_2(4)} = \log_2(4)$

83. $\log\left(\dfrac{1}{3}\right) = -\log(3)$ **84.** $\log_2(8 \cdot 2^{59}) = 62$

85. $\log_2(16^5) = 20$

86. $(\log_2(5))(\log_5(2)) = 1$

87. $\log(10^3) = 3$ **88.** $\log_3(3^7) = 7$

89. $\log(100 + 3) = 2 + \log(3)$ **90.** $\dfrac{\log_7(32)}{\log_7(8)} = \dfrac{5}{3}$

Solve each problem.

91. *Growth rate.* The annual growth rate for continuous growth is given by

$$r = \dfrac{\ln(A) - \ln(P)}{t},$$

where P is the principal and A is the amount after t years. Rewrite the formula using a single logarithm. If A is three times as large as P after five years, then what is the annual growth rate r?

92. *Diversity index.* The U.S.G.S. measures the quality of a water sample by using the diversity index d, given by

$$d = -[p_1 \cdot \log_2(p_1) + p_2 \cdot \log_2(p_2) + \cdots + p_n \cdot \log_2(p_n)],$$

where n is the number of different taxons (biological classifications) represented in the sample and p_1 through p_n are the percentages of organisms in each of the n taxons. The value of d ranges from 0 when all organisms in the water sample are the same to some positive number when all organisms in the sample are different. If $\frac{2}{3}$ of the organisms in a water sample are in one taxon and $\frac{1}{3}$ of the organisms are in a second taxon, then $n = 2$, and

$$d = -\left[\dfrac{2}{3}\log_2\left(\dfrac{2}{3}\right) + \dfrac{1}{3}\log_2\left(\dfrac{1}{3}\right)\right].$$

Use the properties of logarithms to write the expression on the right as $\log_2\left(\dfrac{3\sqrt[3]{2}}{2}\right)$. (In Section 9.4 you will learn how to evaluate a base-2 logarithm using a calculator.)

Getting More Involved

93. *Discussion.* Which of the following equations is an identity? Explain.

a) $\ln(3x) = \ln(3) \cdot \ln(x)$
b) $\ln(3x) = \ln(3) + \ln(x)$
c) $\ln(3x) = 3 \cdot \ln(x)$
d) $\ln(3x) = \ln(x^3)$

94. *Discussion.* Which of the following expressions is not equal to $\log(5^{2/3})$? Explain.

a) $\dfrac{2}{3}\log(5)$ **b)** $\dfrac{\log(5) + \log(5)}{3}$

c) $(\log(5))^{2/3}$ **d)** $\dfrac{1}{3}\log(25)$

Graphing Calculator Exercises

95. Graph the functions $y_1 = \ln(\sqrt{x})$ and $y_2 = 0.5 \cdot \ln(x)$ on the same screen. Explain your results.

96. Graph the functions $y_1 = \log(x)$, $y_2 = \log(10x)$, $y_3 = \log(100x)$, and $y_4 = \log(1000x)$ using the viewing window $-2 \le x \le 5$ and $-2 \le y \le 5$. Why do these curves appear as they do?

97. Graph the function $y = \log(e^x)$. Explain why the graph is a straight line. What is its slope?

$\underline{9.4}$ Solving Equations

▶ Logarithmic Equations
▶ Exponential Equations
▶ Changing the Base
▶ Strategy for Solving Equations
▶ Applications

We solved some equations involving exponents and logarithms in Sections 9.1 and 9.2. In this section we use the properties of exponents and logarithms to solve more complex equations.

Logarithmic Equations

The main tool that we have for solving logarithmic equations is the definition of logarithms: $y = \log_a(x)$ if and only if $a^y = x$. We can use the definition to rewrite any equation that has only one logarithm as an equivalent exponential equation.

EXAMPLE 1 **A logarithmic equation with only one logarithm**
Solve $\log(x + 3) = 2$.

Solution

Write the equivalent exponential equation:

$$
\begin{aligned}
\log(x + 3) &= 2 &&\text{Original equation} \\
10^2 &= x + 3 &&\text{Definition of logarithm} \\
100 &= x + 3 \\
97 &= x
\end{aligned}
$$

Check: $\log(97 + 3) = \log(100) = 2$. The solution set is $\{97\}$. ∎

In the next example we use the product rule for logarithms to write a sum of two logarithms as a single logarithm.

EXAMPLE 2 **Using the product rule to solve an equation**
Solve $\log_2(x + 3) + \log_2(x - 3) = 4$.

Solution

Rewrite the sum of the logarithms as the logarithm of a product:

$$
\begin{aligned}
\log_2(x + 3) + \log_2(x - 3) &= 4 &&\text{Original equation} \\
\log_2[(x + 3)(x - 3)] &= 4 &&\text{Product rule} \\
\log_2[x^2 - 9] &= 4 &&\text{Multiply the binomials.} \\
x^2 - 9 &= 2^4 &&\text{Definition of logarithm} \\
x^2 - 9 &= 16 \\
x^2 &= 25 \\
x &= \pm 5 &&\text{Even-root property}
\end{aligned}
$$

To check, first let $x = -5$ in the original equation:

$$
\begin{aligned}
\log_2(-5 + 3) + \log_2(-5 - 3) &= 4 \\
\log_2(-2) + \log_2(-8) &= 4 &&\text{Incorrect.}
\end{aligned}
$$

Because the domain of any logarithm function is the set of positive real numbers, these logarithms are undefined. Now check $x = 5$ in the original equation:

$$\log_2 (5 + 3) + \log_2 (5 - 3) = 4$$
$$\log_2 (8) + \log_2 (2) = 4$$
$$3 + 1 = 4 \quad \text{Correct.}$$

The solution set is {5}. ∎

CAUTION Always check that your solutions to a logarithmic equation do not produce undefined logarithms in the original equation. ⊘

EXAMPLE 3 **Using the one-to-one property of logarithms**

Solve $\log (x) + \log (x - 1) = \log (8x - 12) - \log (2)$.

Solution

Apply the product rule to the left side and the quotient rule to the right side to get a single logarithm on each side:

$$\log (x) + \log (x - 1) = \log (8x - 12) - \log (2).$$

$$\log [x(x - 1)] = \log \left(\frac{8x - 12}{2} \right) \quad \text{Product rule; quotient rule}$$

$$\log (x^2 - x) = \log (4x - 6) \quad \text{Simplify.}$$

$$x^2 - x = 4x - 6 \quad \text{One-to-one property of logarithms.}$$

$$x^2 - 5x + 6 = 0$$

$$(x - 2)(x - 3) = 0$$

$$x - 2 = 0 \quad \text{or} \quad x - 3 = 0$$

$$x = 2 \quad \text{or} \quad x = 3$$

Neither $x = 2$ nor $x = 3$ produces undefined terms in the original equation. Use a calculator to check that they both satisfy the original equation. The solution set is {2, 3}. ∎

CAUTION The product rule, quotient rule, and power rule do not eliminate logarithms from equations. To do so, we use the definition to change $y = \log_a (x)$ into $a^y = x$ or the one-to-one property to change $\log_a (m) = \log_a (n)$ into $m = n$. ⊘

Exponential Equations

If an equation has a single exponential expression, we can write the equivalent logarithmic equation.

EXAMPLE 4   **A single exponential expression**

Find the exact solution to $2^x = 10$.

Solution

The equivalent logarithmic equation is

$$x = \log_2 (10).$$

The solution set is $\{\log_2 (10)\}$. The number $\log_2 (10)$ is the exact solution to the equation. Later in this section you will learn how to use the base-change formula to find an approximate value for an expression of this type. ∎

In Section 9.1 we solved some exponential equations by writing each side as a power of the same base and then applying the one-to-one property of exponential functions. We review that method in the next example.

EXAMPLE 5   **Powers of the same base**

Solve $2^{(x^2)} = 4^{3x-4}$.

Solution

We can write each side as a power of the same base:

$$
\begin{aligned}
2^{(x^2)} &= (2^2)^{3x-4} && \text{Because } 4 = 2^2 \\
2^{(x^2)} &= 2^{6x-8} && \text{Power of a power rule} \\
x^2 &= 6x - 8 && \text{One-to-one property of} \\
& && \text{exponential functions} \\
x^2 - 6x + 8 &= 0 \\
(x - 4)(x - 2) &= 0 \\
x - 4 = 0 \quad &\text{or} \quad x - 2 = 0 \\
x = 4 \quad &\text{or} \quad x = 2
\end{aligned}
$$

Check $x = 2$ and $x = 4$ in the original equation. The solution set is $\{2, 4\}$. ∎

For some exponential equations we cannot write each side as a power of the same base as we did in Example 5. In this case we take a logarithm of each side and simplify, using the rules for logarithms.

EXAMPLE 6   **Exponential equation with two different bases**

Find the exact and approximate solution to $2^{x-1} = 3^x$.

Solution

We first take the base 10 logarithm of each side:

$$
\begin{aligned}
2^{x-1} &= 3^x && \text{Original equation} \\
\log (2^{x-1}) &= \log (3^x) && \text{Take log of each side.} \\
(x - 1) \log (2) &= x \cdot \log (3) && \text{Power rule} \\
x \cdot \log (2) - \log (2) &= x \cdot \log (3) && \text{Distributive property} \\
x \cdot \log (2) - x \cdot \log (3) &= \log (2) && \text{Get all } x\text{-terms on one side.} \\
x[\log (2) - \log (3)] &= \log (2) && \text{Factor out } x. \\
x &= \frac{\log (2)}{\log (2) - \log (3)} && \text{Exact solution} \\
x &\approx -1.7095 && \text{Approximate solution}
\end{aligned}
$$

You can use a calculator to check -1.7095 in the original equation. As the first step of the solution, we could have taken the logarithm of each side using any base. We chose base 10 so that we could use a calculator to find an approximate solution from the exact solution. ∎

Changing the Base

Scientific calculators have an x^y key for computing any power of any base, in addition to the function keys for computing 10^x and e^x. For logarithms we have the keys ln and log, but there are no function keys for logarithms using other bases. To solve this problem, we develop a formula for expressing a base-a logarithm in terms of base-b logarithms.

If $y = \log_a (M)$, then $a^y = M$. Now we solve $a^y = M$ for y, using base-b logarithms:

$$a^y = M$$

$$\log_b (a^y) = \log_b (M) \qquad \text{Take the base-}b\text{ logarithm of each side.}$$

$$y \cdot \log_b (a) = \log_b (M) \qquad \text{Power rule}$$

$$y = \frac{\log_b (M)}{\log_b (a)} \qquad \text{Divide each side by } \log_b (a).$$

Because $y = \log_a (M)$, we can write $\log_a (M)$ in terms of base-b logarithms.

Base-Change Formula

If a and b are positive numbers not equal to 1 and M is positive, then

$$\log_a (M) = \frac{\log_b (M)}{\log_b (a)}.$$

In words, we take the logarithm with the new base and divide by the logarithm of the old base. The most important use of the base-change formula is to find base-a logarithms using a calculator. If the new base is 10 or e, then

$$\log_a (M) = \frac{\log (M)}{\log (a)} = \frac{\ln (M)}{\ln (a)}.$$

EXAMPLE 7 **Using the base-change formula**

Find $\log_7 (99)$ to four decimal places.

Solution

Use the base-change formula with $a = 7$ and $b = 10$:

$$\log_7 (99) = \frac{\log (99)}{\log (7)} \approx 2.3614$$

Check by finding $7^{2.3614}$ with your calculator. Note that we also have

$$\log_7 (99) = \frac{\ln (99)}{\ln (7)} \approx 2.3614.$$ ∎

Strategy for Solving Equations

There is no formula that will solve every equation in this section. However, we have a strategy for solving exponential and logarithmic equations. The following list summarizes the ideas that we need for solving these equations.

> ### Solving Exponential and Logarithmic Equations
>
> 1. If the equation has a single logarithm or a single exponential expression, rewrite the equation using the definition $y = \log_a(x)$ if and only if $a^y = x$.
> 2. Use the properties of logarithms to combine logarithms as much as possible.
> 3. Use the one-to-one properties:
>   a) If $\log_a(m) = \log_a(n)$, then $m = n$.
>   b) If $a^m = a^n$, then $m = n$.
> 4. To get an approximate solution of an exponential equation, take the common or natural logarithm of each side of the equation.

Applications

In compound-interest problems, logarithms are used to find the time it takes for money to grow to a specified amount.

EXAMPLE 8   **Finding the time**

If $500 is deposited into an account paying 8% compounded quarterly, then in how many quarters will the account have $1000 in it?

Solution

We use the compound-interest formula $A = P(1 + i)^n$ with a principal of $500, an amount of $1000, and an interest rate of 2% each quarter:

$$A = P(1 + i)^n$$

$$1000 = 500(1.02)^n \qquad \text{Substitute.}$$

$$2 = (1.02)^n \qquad \text{Divide each side by 500.}$$

$$n = \log_{1.02}(2) \qquad \text{Definition of logarithm}$$

$$= \frac{\ln(2)}{\ln(1.02)} \qquad \text{Base-change formula}$$

$$\approx 35.0028 \qquad \text{Use a calculator.}$$

It takes approximately 35 quarters, or 8 years and 9 months, for the initial investment to be worth $1000. Note that we could also solve $2 = (1.02)^n$ by taking the common or natural logarithm of each side. Try it. ∎

Calculator Close-up

A graphing calculator can be used to find approximate solutions to equations that cannot be solved by the methods shown in this chapter. For example, to solve

$$e^x = \ln (x + 3),$$

graph $y_1 = e^x$ and $y_2 = \ln (x + 3)$ on the same screen of your graphing calculator, as shown in Fig. 9.14. Use the trace feature to find the points of intersection of these curves. Because the y-coordinates on each curve are the same at a point of intersection, the x-coordinate of an intersection point satisfies $e^x = \ln (x + 3)$. The solutions to this equation are $x \approx 0.133$ and $x \approx -1.825$.

Xmin = −4   Ymin = −4
Xmax = 4   Ymax = 4

Figure 9.14

Warm-ups

True or false? Explain.

1. If $\log (x - 2) + \log (x + 2) = 7$, then $\log (x^2 - 4) = 7$.
2. If $\log (3x + 7) = \log (5x - 8)$, then $3x + 7 = 5x - 8$.
3. If $e^{x-6} = e^{x^2 - 5x}$, then $x - 6 = x^2 - 5x$.
4. If $2^{3x-1} = 3^{5x-4}$, then $3x - 1 = 5x - 4$.
5. If $\log_2 (x^2 - 3x + 5) = 3$, then $x^2 - 3x + 5 = 8$.
6. If $2^{2x-1} = 3$, then $2x - 1 = \log_2 (3)$.
7. If $5^x = 23$, then $x \cdot \ln (5) = \ln (23)$.

8. $\log_3 (5) = \dfrac{\ln (3)}{\ln (5)}$ 9. $\dfrac{\ln (2)}{\ln (6)} = \dfrac{\log (2)}{\log (6)}$ 10. $\log (5) = \ln (5)$

9.4 EXERCISES

Solve each equation. See Examples 1 and 2.

1. $\log_2 (x + 1) = 3$
2. $\log_3 (x^2) = 4$
3. $\log (x) + \log (5) = 1$
4. $\ln (x) + \ln (3) = 0$
5. $\log_2 (x - 1) + \log_2 (x + 1) = 3$

6. $\log_3 (x - 4) + \log_3 (x + 4) = 2$
7. $\log_2 (x - 1) - \log_2 (x + 2) = 2$
8. $\log_4 (8x) - \log_4 (x - 1) = 2$
9. $\log_2 (x - 4) + \log_2 (x + 2) = 4$
10. $\log_6 (x + 6) + \log_6 (x - 3) = 2$

Solve each equation. See Example 3.

11. $\ln(x) + \ln(x+5) = \ln(x+1) + \ln(x+3)$

12. $\log(x) + \log(x+5) = 2 \cdot \log(x+2)$

13. $\log(x+3) + \log(x+4) = \log(x^3 + 13x^2) - \log(x)$

14. $\log(x^2 - 1) - \log(x-1) = \log(6)$

15. $2 \cdot \log(x) = \log(20 - x)$

16. $2 \cdot \log(x) + \log(3) = \log(2 - 5x)$

Solve each equation. See Examples 4 and 5.

17. $3^x = 7$ **18.** $2^{x-1} = 5$

19. $2^{3x+4} = 4^{x-1}$ **20.** $9^{2x-1} = 27^{1/2}$

21. $\left(\dfrac{1}{3}\right)^x = 3^{1+x}$ **22.** $4^{3x} = \left(\dfrac{1}{2}\right)^{1-x}$

Find the exact solution and approximate solution to each equation. See Example 6.

23. $2^x = 3^{x+5}$ **24.** $e^x = 10^x$

25. $5^{x+2} = 10^{x-4}$ **26.** $3^{2x} = 6^{x+1}$

27. $8^x = 9^{x-1}$ **28.** $5^{x+1} = 8^{x-1}$

🖩 *Use the base-change formula to find each logarithm to four decimal places. See Example 7.*

29. $\log_2(3)$ **30.** $\log_3(5)$

31. $\log_3\left(\dfrac{1}{2}\right)$ **32.** $\log_5(2.56)$

33. $\log_{1/2}(4.6)$ **34.** $\log_{1/3}(3.5)$

35. $\log_{0.1}(0.03)$ **36.** $\log_{0.2}(1.06)$

For each equation, find the exact solution and an approximate solution when appropriate. Round approximate answers to four decimal places.

37. $x \cdot \ln(2) = \ln(7)$

38. $x \cdot \log(3) = \log(5)$

39. $3x - x \cdot \ln(2) = 1$

40. $2x + x \cdot \log(5) = \log(7)$

41. $3^x = 5$

42. $2^x = \dfrac{1}{3}$

43. $2^{x-1} = 9$

44. $10^{x-2} = 6$

45. $3^x = 20$

46. $2^x = 128$

47. $\log_3(x) + \log_3(5) = 1$

48. $\log(x) - \log(3) = \log(6)$

49. $8^x = 2^{x+1}$

50. $2^x = 5^{x+1}$

🖩 *In Exercises 51–62, solve each problem. See Example 8.*

51. *Finding the time.* How many months does it take for $1000 to grow to $1500 in an account paying 12% compounded monthly?

52. *Finding the time.* How many years does it take for $25 to grow to $100 in an account paying 8% compounded annually?

53. *Going with the flow.* The flow (in cubic feet per second) of the Tangipahoa River at Robert, Louisiana, is a linear function of its depth (in feet) when the river is at normal levels (Exercise 83, Section 7.3). When flood stages are included, the flow y is better modeled by the exponential function $y = 114.308e^{0.265x}$, where x is the depth. Find the flow when the depth is 15.8 feet.

Figure for Exercises 53 and 54

54. *Flood waters run deep.* Use the formula of Exercise 53 to find the depth of the Tangipahoa River at Robert, Louisiana, on March 7, 1993, when the flow was 11,700 ft³/sec (Data from U.S. Geological Survey, Baton Rouge, Louisiana).

55. *Above the poverty level.* In a certain country the number of people above the poverty level is currently 28 million and growing 5% annually. Assuming that the population is growing continuously, the population P (in millions), t years from now, is determined by the formula $P = 28e^{0.05t}$. In how many years will there be 40 million people above the poverty level?

56. *Below the poverty level.* In the same country as in Exercise 55, the number of people below the poverty level is currently 20 million and growing 7% annually. This population (in millions), t years from now, is determined by the formula $P = 20e^{0.07t}$. In how many years will there be 40 million people below the poverty level?

57. *Fifty-fifty.* For this exercise, use the information given in Exercises 55 and 56. In how many years will the number of people above the poverty level equal the number of people below the poverty level?

58. *Golden years.* In a certain country there are currently 100 million workers and 40 million retired people. The population of workers is decreasing according to the formula $W = 100e^{-0.01t}$, where t is in years and W is in millions. The population of retired people is increasing according to the formula $R = 40e^{0.09t}$, where t is in years and R is in millions. In how many years will the number of workers equal the number of retired people?

59. *Ions for breakfast.* Orange Juice has a pH of 3.7. What is the hydrogen ion concentration of orange juice? (See Exercises 77–80 of Section 9.2.)

60. *Ions in your veins.* Normal human blood has a pH of 7.4. What is the hydrogen ion concentration of normal human blood?

61. *Diversity index.* In Exercise 92 of Section 9.3 we expressed the diversity index d for a certain water sample as

$$d = \log_2\left(\frac{3\sqrt[3]{2}}{2}\right).$$

Use the base-change formula and a calculator to calculate the value of d. Round the answer to four decimal places.

62. *Quality water.* In a certain water sample, 5% of the organisms are in one taxon, 10% are in a second taxon, 20% are in a third taxon, 15% are in a fourth taxon, 23% are in a fifth taxon, and the rest are in a sixth taxon. Use the formula given in Exercise 92 of Section 9.3 with $n = 6$ to find the diversity index of the water sample.

Getting More Involved

63. *Exploration.* Logarithms were designed to solve equations that have variables in the exponents, but logarithms can be used to solve certain polynomial equations. Consider the following example:

$$x^5 = 88$$
$$5 \cdot \ln(x) = \ln(88)$$
$$\ln(x) = \frac{\ln(88)}{5} \approx 0.895467$$
$$x = e^{0.895467} \approx 2.4485$$

Solve $x^3 = 12$ by taking the natural logarithm of each side. Round the approximate solution to four decimal places. Solve $x^3 = 12$ without using logarithms and compare with your previous answer.

64. *Discussion.* Determine whether each logarithm is positive or negative without using a calculator. Explain your answers.

a) $\log_2(0.45)$ **b)** $\ln(1.01)$
c) $\log_{1/2}(4.3)$ **d)** $\log_{1/3}(0.44)$

Graphing Calculator Exercises

65. Graph $y_1 = 2^x$ and $y_2 = 3^{x-1}$ on the same coordinate system. Use the trace feature of your calculator to find the point of intersection of the two curves. Round to two decimal places.

66. Bob invested $1000 at 6% compounded continuously. At the same time, Paula invested $1200 at 5% compounded monthly. Write two functions that give the amounts of Bob's and Paula's investments after x years. Graph these functions on a graphing calculator. Use the trace feature of your graphing calculator to find the approximate value of x for which the investments are equal in value?

67. Graph the functions $y_1 = \log_2(x)$ and $y_2 = 3^{x-4}$ on the same coordinate system and use the trace feature to find the points of intersection of the curves. Round to two decimal places. (*Hint:* To graph $y = \log_2(x)$, use the base-change formula to write the function as $y = \ln(x)/\ln(2)$.)

──────────────COLLABORATIVE ACTIVITIES──────────────

In How Much Space Could We Live?

Grouping: 2 students per group
Topic: Exponential and logarithmic functions

The formula for population growth is

$$P(t) = P_0 e^{kt},$$

where $P(t)$ is the population after t years, P_0 is the population initially, k is the growth rate per year, and t is the number of years elapsed. We will find out how long it would take to completely cover the earth if the human population were growing exponentially. For all of your calculations, round your answers to one decimal place unless directed otherwise.

1. The population of the world in 1994 was approximately 5.5×10^9 people. If the population was 2.75×10^9 people in 1964, then what is the current growth rate (express as a percent)? Round your answer to the nearest tenth of a percent.

    The earth has a total surface area of approximately 5.1×10^{14} square meters. Seventy percent of this surface area is rock, ice, sand, and open ocean. Another 8% of the total surface area is tundra, lakes and streams, continental shelves, algal beds and reefs, and estuaries. We will consider the remaining area to be suitable for growing food and for living space.

2. Determine the surface area available for growing food and for living space.

3. If each person needs 100 square meters of the earth's surface for living space and growing food, then in how many years after 1994 will the livable surface of the earth be used up? (Use the rate and the 1994 population from Question 1 and the surface area from Question 2.)

4. With your partner, think about the following questions and report your conclusions.

    • Does 100 square meters per person for living space and growing food seem reasonable? Take into account that many people live in tall apartment buildings and how that translates into surface area used per person.

    • How much room do you think it takes to grow animals for food (cows, chickens, pigs, etc.)? What about grains, vegetables, nuts, and fruit? Would food grow as well in desert areas, mountainous areas, or jungle areas?

    • Would there be any space left for wild animals or natural plant life? Would there be any space left for shopping malls, movie theaters, concert halls, factories, office buildings, or parking lots?

Wrap-up CHAPTER 9

SUMMARY

Exponential and Logarithmic Functions		Examples
Exponential function	A function of the form $f(x) = a^x$ for $a > 0$ and $a \neq 1$	$f(x) = 3^x$
Logarithm function	A function of the form $f(x) = \log_a(x)$ for $a > 0$ and $a \neq 1$ $$y = \log_a(x) \text{ if and only if } a^y = x.$$	$f(x) = \log_2(x)$ $\log_3(8) = x \leftrightarrow 3^x = 8$
Common logarithm	Base-10: $f(x) = \log(x)$	$\log(100) = 2$ because $100 = 10^2$.
Natural logarithm	Base-e: $f(x) = \ln(x)$ $e \approx 2.718$	$\ln(e) = 1$ because $e^1 = e$.

| Inverse functions | $f(x) = a^x$ and $g(x) = \log_a (x)$ are inverse functions. | If $f(x) = e^x$, then $f^{-1}(x) = \ln (x)$. |

Properties

Examples

M, N, and *a* are positive numbers with $a \neq 1$.

	$\log_a (a) = 1 \qquad \log_a (1) = 0$	$\log_5 (5) = 1$, $\log_5 (1) = 0$
Inverse properties	$\log_a (a^M) = M \qquad a^{\log_a (M)} = M$	$\log (10^7) = 7$, $e^{\ln (3.4)} = 3.4$
Product rule	$\log_a (MN) = \log_a (M) + \log_a (N)$	$\ln (3x) = \ln (3) + \ln (x)$
Quotient rule	$\log_a \left(\dfrac{M}{N} \right) = \log_a (M) - \log_a (N)$	$\ln \left(\dfrac{1}{3} \right) = \ln (1) - \ln (3)$
	$\log_a \left(\dfrac{1}{N} \right) = -\log_a (N)$	$\ln \left(\dfrac{1}{3} \right) = -\ln (3)$
Power rule	$\log_a (M^N) = N \cdot \log_a (M)$	$\log (x^3) = 3 \cdot \log (x)$
Base-change formula	$\log_a (M) = \dfrac{\log_b (M)}{\log_b (a)}$	$\log_3 (5) = \dfrac{\ln (5)}{\ln (3)}$

Equations Involving Logarithms and Exponents

Examples

Strategy	1. If there is a single logarithm or a single exponential expression, rewrite the equation using $y = \log_a (x)$ if and only if $a^y = x$.	$2^x = 3$ and $x = \log_2 (3)$ are equivalent.
	2. Use the properties of logarithms to combine logarithms as much as possible.	$\log (x) + \log (x - 3) = 1$ $\log (x^2 - 3x) = 1$
	3. Use the one-to-one properties: a) If $\log_a (m) = \log_a (n)$, then $m = n$.	$\ln (x) = \ln (5 - x)$, $x = 5 - x$
	b) If $a^m = a^n$, then $m = n$.	$2^{3x} = 2^{5x - 7}$, $3x = 5x - 7$
	4. To get an approximate solution, take the common or natural logarithm of each side of an exponential equation.	$2^x = 3$, $\ln (2^x) = \ln (3)$ $x \cdot \ln (2) = \ln (3)$ $x = \dfrac{\ln (3)}{\ln (2)}$

REVIEW EXERCISES

9.1 *Use $f(x) = 5^x$, $g(x) = 10^{x-1}$, and $h(x) = \left(\frac{1}{4} \right)^x$ for Exercises 1–28. Find the following.*

Find x in each case.

1. $f(-2)$ **2.** $f(0)$ **3.** $f(3)$

4. $f(4)$ **5.** $g(1)$ **6.** $g(-1)$

7. $g(0)$ **8.** $g(3)$ **9.** $h(-1)$

10. $h(2)$ **11.** $h\left(\dfrac{1}{2} \right)$ **12.** $h\left(-\dfrac{1}{2} \right)$

13. $f(x) = 25$ **14.** $f(x) = -\dfrac{1}{125}$

15. $g(x) = 1000$ **16.** $g(x) = 0.001$

17. $h(x) = 32$ **18.** $h(x) = 8$

19. $h(x) = \dfrac{1}{16}$ **20.** $h(x) = 1$

▦ *Find the following.*

21. $f(1.34)$ **22.** $f(-3.6)$

23. $g(3.25)$ **24.** $g(4.87)$

25. $h(2.82)$ **26.** $h(\pi)$

27. $h(\sqrt{2})$ **28.** $h\left(\dfrac{1}{3}\right)$

Sketch the graph of each function.

29. $f(x) = 5^x$ **30.** $g(x) = e^x$

31. $y = \left(\dfrac{1}{5}\right)^x$ **32.** $y = e^{-x}$

33. $f(x) = 3^{-x}$ **34.** $f(x) = -3^{x-1}$

35. $y = 1 + 2^x$ **36.** $y = 1 - 2^x$

9.2 *Write each exponential equation as a logarithmic equation and each logarithmic equation as an exponential equation.*

37. $10^m = n$ **38.** $b = a^5$

39. $h = \log_k (t)$ **40.** $\log_v (5) = u$

Let $f(x) = \log_2 (x)$, $g(x) = \log (x)$, and $h(x) = \log_{1/2} (x)$. Find the following.

41. $f\left(\dfrac{1}{8}\right)$ **42.** $f(64)$

43. $g(0.1)$ **44.** $g(1)$

45. $g(100)$ **46.** $h\left(\dfrac{1}{8}\right)$

47. $h(1)$ **48.** $h(4)$

49. x, if $f(x) = 8$ **50.** x, if $g(x) = 3$

▦ **51.** $f(77)$ ▦ **52.** $g(88.4)$

▦ **53.** $h(33.9)$ ▦ **54.** $h(0.05)$

▦ **55.** x, if $f(x) = 2.475$ ▦ **56.** x, if $g(x) = 1.426$

For each function f, find f^{-1} and sketch the graphs of f and f^{-1} on the same set of axes.

57. $f(x) = 10^x$ **58.** $f(x) = \log_8 (x)$

59. $f(x) = e^x$ **60.** $f(x) = \log_3 (x)$

9.3 *Rewrite each expression as a sum or difference of multiples of logarithms.*

61. $\log (x^2 y)$ **62.** $\log_3 (x^2 + 2x)$

63. $\ln (16)$ **64.** $\log\left(\dfrac{y}{\sqrt{x}}\right)$

65. $\log_5\left(\dfrac{1}{x}\right)$ **66.** $\ln\left(\dfrac{xy}{z}\right)$

Rewrite each expression as a single logarithm.

67. $\dfrac{1}{2}\log (x + 2) - 2 \cdot \log (x - 1)$

68. $3 \cdot \ln (x) + 2 \cdot \ln (y) - \dfrac{1}{3}\ln (z)$

9.4 *Find the exact solution to each equation.*

69. $\log_2 (x) = 8$ **70.** $\log_3 (x) = 0.5$

71. $\log_2 (8) = x$ **72.** $3^x = 8$

73. $x^3 = 8$ **74.** $3^2 = x$

75. $\log_x (27) = 3$ **76.** $\log_x (9) = -\dfrac{1}{3}$

77. $x \cdot \ln (3) - x = \ln (7)$

78. $x \cdot \log (8) = x \cdot \log (4) + \log (9)$

79. $3^x = 5^{x-1}$ **80.** $5^{(2x^2)} = 5^{3-5x}$

81. $4^{2x} = 2^{x+1}$

82. $\log (12) = \log (x) + \log (7 - x)$

83. $\ln (x + 2) - \ln (x - 10) = \ln (2)$

84. $2 \cdot \ln (x + 3) = 3 \cdot \ln (4)$

85. $\log (x) - \log (x - 2) = 2$

86. $\log_2 (x) = \log_2 (x + 16) - 1$

▦ *Use a calculator to find an approximate solution to each of the following. Round your answers to four decimal places.*

87. $6^x = 12$ **88.** $5^x = 8^{3x+2}$

89. $3^{x+1} = 5$ **90.** $\log_3 (x) = 2.634$

Miscellaneous

▦ *Solve each problem.*

91. **Compounding annually.** What does $10,000 invested at 11.5% compounded annually amount to after 15 years?

92. **Doubling time.** How many years does it take for an investment to double at 6.5% compounded annually?

93. **Decaying substance.** The amount, A, of a certain radioactive substance remaining after t years, is given by the formula $A = A_0 e^{-0.0003t}$, where A_0 is the initial amount. If we have 218 grams of this substance today, then how much of it will be left 1000 years from now?

94. **Wildlife management.** The population of white-tailed deer in a certain national forest is believed to be growing according to the formula $P = 517 + 10 \cdot \ln (8t + 1)$, where t is the time in years from the present. What is the current population? In how many years will there be 600 deer?

95. *Comparing investments.* Melissa deposited $1000 into an account paying 5% annually; on the same day, Frank deposited $900 into an account paying 7% compounded continuously. Find the number of years that it will take for the amounts in the accounts to be equal.

96. *Imports and exports.* The value of imported goods for a small South American country is believed to be growing according to the formula $I = 15 \cdot \log(16t + 33)$, and the value of exported goods appears to be growing according to the formula $E = 30 \cdot \log(t + 3)$, where I and E are in millions of dollars and t is the time in years from the present. What is the current value of imports and exports? In how many years will the value of imports equal the value of exports?

97. *Finding river flow.* The U.S.G.S. measures the water height h (in feet above sea level) for the Tangipahoa River at Robert, Louisiana, and then finds the flow y (in cubic feet per second), using the formula

$$y = 114.308e^{0.265(h - 6.87)}.$$

Find the flow when the river at Robert is 20.6 ft above sea level.

98. *Finding the height.* Rewrite the formula in Exercise 97 to express h as a function of y. Use the new formula to find the water height above sea level when the flow is 10,000 ft^3/sec.

CHAPTER 9 TEST

Let $f(x) = 5^x$ and $g(x) = \log_5(x)$. Find the following.

1. $f(2)$ **2.** $f(-1)$ **3.** $f(0)$

4. $g(125)$ **5.** $g(1)$ **6.** $g\left(\dfrac{1}{5}\right)$

Sketch the graph of each function.

7. $y = 2^x$ **8.** $f(x) = \log_2(x)$

9. $y = \left(\dfrac{1}{3}\right)^x$ **10.** $g(x) = \log_{1/3}(x)$

Suppose $\log_a(M) = 6$ and $\log_a(N) = 4$. Find the following.

11. $\log_a(MN)$

12. $\log_a\left(\dfrac{M^2}{N}\right)$

13. $\dfrac{\log_a(M)}{\log_a(N)}$

14. $\log_a(a^3M^2)$

15. $\log_a\left(\dfrac{1}{N}\right)$

Find the exact solution to each equation.

16. $3^x = 12$

17. $\log_3(x) = \dfrac{1}{2}$

18. $5^x = 8^{x-1}$

19. $\log(x) + \log(x + 15) = 2$

20. $2 \cdot \ln(x) = \ln(3) + \ln(6 - x)$

Use a scientific calculator to find an approximate solution to each of the following. Round your answers to four decimal places.

21. Solve $20^x = 5$.

22. Solve $\log_3(x) = 2.75$.

23. The number of bacteria present in a culture at time t is given by the formula $N = 10e^{0.4t}$, where t is in hours. How many bacteria are present initially? How many are present after 24 hours?

24. How many hours does it take for the bacteria population of Problem 23 to double?

Find the exact solution to each equation.

1. $(x - 3)^2 = 8$

2. $\log_2 (x - 3) = 8$

3. $2^{x-3} = 8$

4. $2x - 3 = 8$

5. $|x - 3| = 8$

6. $\sqrt{x - 3} = 8$

7. $\log_2 (x - 3) + \log_2 (x) = \log_2 (18)$

8. $2 \cdot \log_2 (x - 3) = \log_2 (5 - x)$

9. $\dfrac{1}{2}x - \dfrac{2}{3} = \dfrac{3}{4}x + \dfrac{1}{5}$

10. $3x^2 - 6x + 2 = 0$

Find the inverse of each function.

11. $f(x) = \dfrac{1}{3}x$ 12. $g(x) = \log_3 (x)$

13. $f(x) = 2x - 4$ 14. $h(x) = \sqrt{x}$

15. $j(x) = \dfrac{1}{x}$ 16. $k(x) = 5^x$

17. $m(x) = e^{x-1}$ 18. $n(x) = \ln (x)$

Sketch the graph of each equation.

19. $y = 2x$ 20. $y = 2^x$

21. $y = x^2$ 22. $y = \log_2 (x)$

23. $y = \dfrac{1}{2}x - 4$ 24. $y = |2 - x|$

25. $y = 2 - x^2$ 26. $y = e^2$

Solve each problem.

27. **Civilian labor force.** Using data from the U.S. Bureau of Labor statistics for 1985 through 1989, one economist declared that the number of workers in the civilian labor force could be predicted by the linear model

$$n = 2.13t + 88.38,$$

where t is the number of years since 1985 and n is in millions of workers. Using the same data, a second economist declared that the exponential model

$$n = 88.43e^{0.023t}$$

would better predict the size of the labor force.

a) Graph both functions on the same coordinate system for $0 \le t \le 15$.

b) What does each model predict for the value of n in 1992?

c) Consult your library to find the actual size of the civilian labor force in 1992.

d) Which model's prediction is closer to the actual number?

e) Answer parts (b), (c), and (d) for the present year.

28. **Measuring ocean depths.** In this exercise you will see how a geophysicist uses sound reflection to measure the depth of the ocean. Let v be the speed of sound through the water and d_1 be the depth of the ocean below the ship, as shown in the accompanying figure.

a) The time it takes for sound to travel from the ship at point S straight down to the ocean floor at point B_1 and back to point S is 0.270 sec. Write d_1 as a function of v.

b) It takes 0.432 sec for sound to travel from point S to point B_2 and then to a receiver at R, which is towed 500 meters behind the ship. Assuming that $d_2 = d_3$, write d_2 as a function of v.

c) Use the Pythagorean theorem to find v. Then find the ocean depth d_1.

Figure for Exercise 28

CHAPTER 10

Systems of Equations

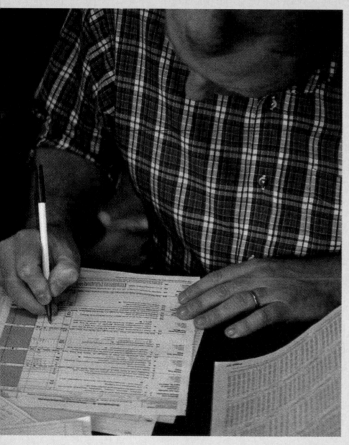

In his letter to M. Leroy in 1789, Benjamin Franklin said, "in this world nothing is certain but death and taxes." Since that time, taxes have become not only inevitable, but intricate and complex.

Each year, the U.S. Congress revises parts of the Federal Income Tax Code. To help clarify these revisions, the Internal Revenue Service issues frequent revenue rulings. In addition there are seven tax courts that further interpret changes and revisions, sometimes in entirely different ways. Is it any wonder that tax preparation has become complicated and few individuals actually prepare their own taxes? Both corporate and individual tax preparation is a growing business, and there are over 500,000 tax counselors helping more than 60 million taxpayers file their returns correctly.

Everyone knows that doing taxes involves a lot of arithmetic, but not everyone knows that computing taxes can also involve algebra. In fact, to find state and federal taxes for certain corporations, you must solve a system of equations. You will see an example of using algebra to find amounts of income taxes in Exercises 45 and 46 of Section 10.1.

10.1 Systems of Linear Equations in Two Variables

In previous chapters you studied many different types of equations, but we have usually considered only one equation at a time. In this chapter we will see problems that involve more than one equation. Any collection of two or more equations is called a **system** of equations. If the equations of a system involve two variables, then the set of ordered pairs that satisfy all of the equations is the **solution set of the system.** In this section we solve systems of linear equations in two variables and use systems to solve problems.

Solving a System by Graphing

Because the graph of each linear equation is a line, points that satisfy both equations lie on both lines. For some systems these points can be found by graphing.

EXAMPLE 1 **A system with only one solution**

Solve the system by graphing:

$$y = x + 2$$
$$x + y = 4$$

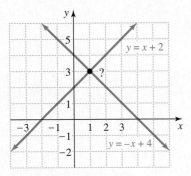

Figure 10.1

Solution

First write the equations in slope-intercept form:

$$y = x + 2$$
$$y = -x + 4$$

Use the y-intercept and the slope to graph each line. The graph of the system is shown in Fig. 10.1. From the graph it appears that these lines intersect at $(1, 3)$. To be certain, we can check that $(1, 3)$ satisfies both equations. Let $x = 1$ and $y = 3$ in $y = x + 2$ to get

$$3 = 1 + 2.$$

Let $x = 1$ and $y = 3$ in $x + y = 4$ to get

$$1 + 3 = 4.$$

Because $(1, 3)$ satisfies both equations, the solution set to the system is $\{(1, 3)\}$. ■

The graphs of the equations in the next example are parallel lines, and there is no point of intersection.

EXAMPLE 2 **A system with no solution**

Solve the system by graphing:

$$2x - 3y = 6$$
$$3y - 2x = 3$$

Solution

First write each equation in slope-intercept form.

$$2x - 3y = 6 \qquad\qquad 3y - 2x = 3$$
$$-3y = -2x + 6 \qquad\qquad 3y = 2x + 3$$
$$y = \frac{2}{3}x - 2 \qquad\qquad y = \frac{2}{3}x + 1$$

The graph of the system is shown in Fig. 10.2. Because the two lines in Fig. 10.2 are parallel, there is no ordered pair that satisfies both equations. The solution set to the system is the empty set, $\varnothing$.

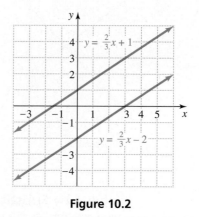

Figure 10.2

The equations in the next example are two different-looking equations for the same straight line.

EXAMPLE 3   **A system with infinitely many solutions**

Solve the system by graphing:

$$2(y + 2) = x$$
$$x - 2y = 4$$

Solution

Write each equation in slope-intercept form:

$$2(y + 2) = x \qquad\qquad x - 2y = 4$$
$$2y + 4 = x \qquad\qquad -2y = -x + 4$$
$$y = \frac{1}{2}x - 2 \qquad\qquad y = \frac{1}{2}x - 2$$

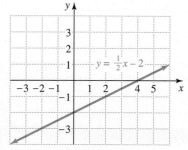

Figure 10.3

Because the equations have the same slope-intercept form, the original equations are equivalent. Their graphs are the same straight line, as shown in Fig. 10.3. Every point on the line satisfies both equations of the system. There are infinitely many points in the solution set. The solution set is $\{(x, y) \mid x - 2y = 4\}$.

Independent, Inconsistent, and Dependent Equations

Our first three examples illustrate the three possible ways in which two lines can be positioned in a plane. In Example 1 the lines intersect in a single point. In this case we say that the equations are **independent** or the system is independent. If the two lines are parallel, as in Example 2, then there is no solution to the system, and the equations are **inconsistent** or the system is inconsistent. If the two equations of a system are equivalent, as in Example 3, the equations are **dependent** or the system is dependent. Figure 10.4 shows the types of graphs that correspond to independent, inconsistent, and dependent systems.

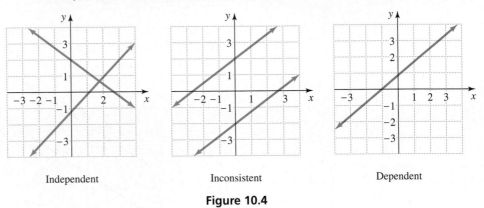

Independent   Inconsistent   Dependent

Figure 10.4

Solving by Substitution

Solving a system by graphing is certainly limited by the accuracy of the graph. If the lines intersect at a point whose coordinates are not integers, then it is difficult to determine those coordinates from the graph. The method of solving a system by **substitution** does not depend on a graph and is totally accurate. For substitution we replace a variable in one equation with an equivalent expression obtained from the other equation.

EXAMPLE 4   **An independent system solved by substitution**

Solve the system by substitution:

$$2x + 3y = 8$$
$$y + 2x = 6$$

Solution

We can easily solve $y + 2x = 6$ for y to get $y = -2x + 6$. Now replace y in the first equation by $-2x + 6$:

$$2x + 3y = 8$$
$$2x + 3(-2x + 6) = 8 \quad \text{Substitute } -2x + 6 \text{ for } y.$$
$$2x - 6x + 18 = 8$$
$$-4x = -10$$
$$x = \frac{5}{2}$$

To find y, we let $x = \frac{5}{2}$ in the equation $y = -2x + 6$:

$$y = -2\left(\frac{5}{2}\right) + 6 = -5 + 6 = 1$$

The next step is to check $x = \frac{5}{2}$ and $y = 1$ in each equation. If $x = \frac{5}{2}$ and $y = 1$ in $2x + 3y = 8$, we get

$$2\left(\frac{5}{2}\right) + 3(1) = 8.$$

If $x = \frac{5}{2}$ and $y = 1$ in $y + 2x = 6$, we get

$$1 + 2\left(\frac{5}{2}\right) = 6.$$

Because both of these equations are true, the solution set to the system is $\left\{\left(\frac{5}{2}, 1\right)\right\}$. The equations of this system are independent.   ■

EXAMPLE 5   **An inconsistent system solved by substitution**
Solve by substitution:

$$x - 2y = 3$$
$$2x - 4y = 7$$

Solution

Solve the first equation for x to get $x = 2y + 3$. Substitute $2y + 3$ for x in the second equation:

$$2x - 4y = 7$$
$$2(2y + 3) - 4y = 7$$
$$4y + 6 - 4y = 7$$
$$6 = 7$$

Because $6 = 7$ is incorrect no matter what values are chosen for x and y, there is no solution to this system of equations. The equations are inconsistent. To check, we write each equation in slope-intercept form:

$$x - 2y = 3 \qquad\qquad 2x - 4y = 7$$
$$-2y = -x + 3 \qquad\qquad -4y = -2x + 7$$
$$y = \frac{1}{2}x - \frac{3}{2} \qquad\qquad y = \frac{1}{2}x - \frac{7}{4}$$

The graphs of these equations are parallel lines with different y-intercepts. The solution set to the system is the empty set, $\varnothing$.   ■

EXAMPLE 6   **A dependent system solved by substitution**
Solve by substitution:

$$2x + 3y = 5 + x + 4y$$
$$y = x - 5$$

Solution
Substitute $y = x - 5$ into the first equation:

$$2x + 3(x - 5) = 5 + x + 4(x - 5)$$
$$2x + 3x - 15 = 5 + x + 4x - 20$$
$$5x - 15 = 5x - 15$$

Because the last equation is an identity, any ordered pair that satisfies $y = x - 5$ will also satisfy $2x + 3y = 5 + x + 4y$. The equations of this system are dependent. The solution set to the system is the set of all points that satisfy $y = x - 5$. We write the solution set in set notation as

$$\{(x, y) \mid y = x - 5\}.$$

We can verify this result by writing $2x + 3y = 5 + x + 4y$ in slope-intercept form:

$$2x + 3y = 5 + x + 4y$$
$$3y = -x + 5 + 4y$$
$$-y = -x + 5$$
$$y = x - 5$$

Because this slope-intercept form is identical to the slope-intercept form of the other equation, they are two different-looking equations for the same straight line.   ■

   In solving a dependent system by substitution, an identity will result after the substitution. If the system is inconsistent, then an inconsistent equation will result after the substitution. The strategy for solving an independent system by substitution can be summarized as follows.

> ### The Substitution Method
>
> 1. Solve one of the equations for one variable in terms of the others.
> 2. Substitute into the other equation to get an equation in one variable.
> 3. Solve for the remaining variable (if possible).
> 4. Insert the value just found into one of the original equations to find the value of the other variable.
> 5. Check the two values in both equations.

Applications

Many of the problems that we solved in previous chapters involved more than one unknown quantity. To solve them, we wrote expressions for all of the unknowns in terms of one variable. Now we can solve problems involving two unknowns by using two variables and writing a system of equations.

EXAMPLE 7   **Tale of two investments**

Belinda had $20,000 to invest. She invested part of it at 10% and the remainder at 12%. If her income from the two investments was $2160, then how much did she invest at each rate?

Solution

Let x be the amount invested at 10% and y be the amount invested at 12%. We can summarize all of the given information in a table:

	Amount	Rate	Interest
First investment	x	10%	$0.10x$
Second investment	y	12%	$0.12y$

We can write one equation about the amounts invested and another about the interest from the investments:

$$x + y = 20{,}000 \qquad \text{Total amount invested}$$
$$0.10x + 0.12y = 2160 \qquad \text{Total interest}$$

Solve the first equation for x to get $x = 20{,}000 - y$. Substitute $20{,}000 - y$ for x in the second equation:

$$0.10x + 0.12y = 2160$$
$$0.10(20{,}000 - y) + 0.12y = 2160 \qquad \text{Replace } x \text{ by } 20{,}000 - y.$$
$$2000 - 0.10y + 0.12y = 2160 \qquad \text{Solve for } y.$$
$$0.02y = 160$$
$$y = 8{,}000$$
$$x = 12{,}000 \qquad \text{Because } x = 20{,}000 - y$$

To check this answer, find 10% of $12,000 and 12% of $8,000:

$$0.10(12{,}000) = 1{,}200$$
$$0.12(8{,}000) = 960$$

Because $1,200 + $960 = $2,160 and $8,000 + $12,000 = $20,000, we can be certain that Belinda invested $12,000 at 10% and $8,000 at 12%.   ∎

MATH AT WORK

Accountant

Accountants work with both people and numbers. When Maria L. Manning, an auditor at Deloitte & Touche LLP, is preparing for an audit, she studies the numbers on the balance sheet and income statement, comparing the current fiscal year to the prior one. The purpose of an independent audit is to give investors a realistic view of a company's finances. To determine that a company's financial statements are fairly stated in accordance with the General Accepted Accounting Procedures (GAAP), she first interviews the comptroller to get the story behind the numbers. Typical questions she could ask are: How productive was your year? Are there any new products? What was behind the big stories in the newspapers? Then Ms. Manning and members of the audit team test the financial statement in detail, closely examining accounts relating to unusual losses or profits.

Ms. Manning is responsible for both manufacturing and mutual fund companies. At a manufacturing company, accounts receivable and inventory are two key components of an audit. For example, to test inventory, Ms. Manning visits a company's warehouse and physically counts all the items for sale to verify a company's assets. For a mutual fund company the audit team pays close attention to current events, for they indirectly affect the financial industry.

In Exercises 47 and 48 of this section you will work problems that involve one aspect of cost accounting: calculating the amount of taxes and bonuses paid by a company.

Warm-ups

True or false? Explain your answer.

1. The ordered pair $(1, 2)$ is in the solution set to the equation $2x + y = 4$.

2. The ordered pair $(1, 2)$ satisfies $2x + y = 4$ and $3x - y = 6$.

3. The ordered pair $(2, 3)$ satisfies $4x - y = 5$ and $4x - y = -5$.

4. If two distinct straight lines in the coordinate plane are not parallel, then they intersect in exactly one point.

5. The substitution method is used to eliminate a variable.

6. No ordered pair satisfies $y = 3x - 5$ and $y = 3x + 1$.

7. The equations $y = 3x - 6$ and $y = 2x + 4$ are independent.

8. The equations $y = 2x + 7$ and $y = 2x + 8$ are inconsistent.

9. The graphs of dependent equations are the same.

10. The graphs of independent linear equations intersect at exactly one point.

10.1 EXERCISES

Solve each system by graphing. See Examples 1–3.

1. $y = 2x$
$y = -x + 3$

2. $y = x - 3$
$y = -x + 1$

3. $y = 2x - 1$
$2y = x - 2$

4. $y = 2x + 1$
$x + y = -2$

5. $y = x - 3$
$x - 2y = 4$

6. $y = -3x$
$x + y = 2$

7. $2y - 2x = 2$
$2y - 2x = 6$

8. $3y - 3x = 9$
$x - y = 1$

9. $y = -\dfrac{1}{2}x + 4$
$x + 2y = 8$

10. $2x - 3y = 6$
$y = \dfrac{2}{3}x - 2$

The graphs of the following systems are given in (a) through (d). Match each system with the correct graph.

11. $5x + 4y = 7$
$x - 3y = 9$

12. $3x - 5y = -9$
$5x - 6y = -8$

13. $4x - 5y = -2$
$3y - x = -3$

14. $4x + 5y = -2$
$4y - x = 11$

a)

b)

c)

d)

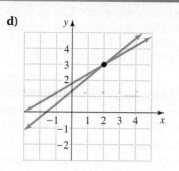

Solve each system by the substitution method. Determine whether the equations are independent, dependent, or inconsistent. See Examples 4–6.

15. $y = x - 5$
$2x - 5y = 1$

16. $y = x + 4$
$3y - 5x = 6$

17. $x = 2y - 7$
$3x + 2y = -5$

18. $x = y + 3$
$3x - 2y = 4$

19. $x - y = 5$
$2x = 2y + 14$

20. $2x - y = 3$
$2y = 4x - 6$

21. $y = 2x - 5$
$y + 1 = 2(x - 2)$

22. $3x - 6y = 5$
$2y = 4x - 6$

23. $2x + y = 9$
$2x - 5y = 15$

24. $3y - x = 0$
$x - 4y = -2$

25. $x - y = 0$
$2x + 3y = 35$

26. $2y = x + 6$
$-3x + 2y = -2$

27. $x + y = 40$
$0.1x + 0.08y = 3.5$

28. $x - y = 10$
$0.2x + 0.05y = 7$

29. $y = 2x - 30$
$\dfrac{1}{5}x - \dfrac{1}{2}y = -1$

30. $3x - 5y = 4$
$y = \dfrac{3}{4}x - 2$

31. $x + y = 4$
$x - y = 5$

32. $y = 2x - 3$
$y = 3x - 3$

33. $2x - y = 4$
$2x - y = 3$

34. $y = 3(x - 4)$
$3x - y = 12$

35. $3(y - 1) = 2(x - 3)$
$3y - 2x = -3$

36. $y = 3x$
$y = 3x + 1$

37. $x - y = -0.375$
$1.5x - 3y = -2.25$

38. $y - 2x = 1.875$
$2.5y - 3.5x = 11.8125$

In Exercises 39–50, write a system of two equations in two unknowns for each problem. Solve each system by substitution. See Example 7.

39. *Different interest rates.* Mrs. Brighton invested $30,000 and received a total of $2,300 in interest. If she invested part of the money at 10% and the remainder at 5%, then how much did she invest at each rate?

40. *Different growth rates.* The combined population of Marysville and Springfield was 25,000 in 1990. By 1995 the population of Marysville had increased by 10%, while Springfield had increased by 9%. If the total population increased by 2,380 people, then what was the population of each city in 1990?

41. *Finding numbers.* The sum of two numbers is 2, and their difference is 26. Find the numbers.

42. *Finding more numbers.* The sum of two numbers is −16, and their difference is 8. Find the numbers.

43. *Toasters and vacations.* During one week, a land developer gave away Florida vacation coupons or toasters to 100 potential customers who listened to a sales presentation. It costs the developer $6 for a toaster and $24 for a Florida vacation coupon. If his bill for prizes that week was $708, then how many of each prize did he give away?

44. *Ticket sales.* Tickets for a concert were sold to adults for $3 and to students for $2. If the total receipts were $824 and twice as many adult tickets as student tickets were sold, then how many of each were sold?

45. *Corporate taxes.* According to Bruce Harrell, CPA, the amount of federal income tax for a class C corporation is deductible on the Louisiana state tax return, and the amount of state income tax for a class C corporation is deductible on the federal tax return. So for a state tax rate of 5% and a federal tax rate of 30%, we have

state tax = 0.05(taxable income − federal tax)

and

federal tax = 0.30(taxable income − state tax).

Find the amounts of state and federal income tax for a class C corporation that has a taxable income of $100,000.

46. *More taxes.* Use information given in Exercise 45 to find the amounts of state and federal income tax for a class C corporation that has a taxable income of $300,000. Use a state tax rate of 6% and a federal tax rate of 40%.

47. *Cost accounting.* The problems presented in this exercise and the next are encountered in cost accounting. A company has agreed to distribute 20% of its net income N to its employees as a bonus; $B = 0.20N$. If the company has income of $120,000 before the bonus, the bonus B is deducted from the $120,000 as an expense to determine net income; $N = 120,000 − B$. Solve the system of two equations in N and B to find the amount of the bonus.

48. *Bonus and taxes.* A company has an income of $100,000 before paying taxes and a bonus. The bonus B is to be 20% of the income after deducting income taxes T but before deducting the bonus.

So

$$B = 0.20(100,000 − T).$$

Because the bonus is a deductible expense, the amount of income tax T at a 40% tax rate is 40% of the income after deducting the bonus. So

$$T = 0.40(100,000 − B).$$

Solve the system to find the amount of the bonus and the amount of income tax.

49. *Textbook case.* The accompanying graph shows the cost of producing textbooks and the revenue from the sale of those textbooks.

a) What is the cost of producing 10,000 textbooks?
b) What is the revenue when 10,000 textbooks are sold?
c) For what number of textbooks is the cost equal to the revenue?
d) The cost of producing zero textbooks is called the *fixed cost*. Find the fixed cost.

Figure for Exercise 49

50. *Free market.* The function $S = 5000 + 200x$ and $D = 9500 − 100x$ express the supply S and the demand D, respectively, for a popular compact disk brand as a function of its price x (in dollars).

a) Graph the functions on the same coordinate system.
b) What happens to the supply as the price increases?
c) What happens to the demand as the price increases?
d) The price at which supply and demand are equal is called the *equilibrium price*. What is the equilibrium price?

Getting More Involved

51. *Discussion.* Which of the following equations is not equivalent to $2x − 3y = 6$?

a) $3y − 2x = 6$ b) $y = \dfrac{2}{3}x − 2$

c) $x = \dfrac{3}{2}y + 3$ d) $2(x − 5) = 3y − 4$

52. *Discussion.* Which of the following equations is inconsistent with the equation $3x + 4y = 8$?

a) $y = \dfrac{3}{4}x + 2$ **b)** $6x + 8y = 16$

c) $y = -\dfrac{3}{4}x + 8$ **d)** $3x - 4y = 8$

Graphing Calculator Exercises

53. Solve each system by graphing each pair of equations on a graphing calculator and using the trace feature to estimate the point of intersection. Find the coordinates of the intersection to the nearest tenth.

a) $y = 3.5x - 7.2$
  $y = -2.3x + 9.1$

b) $2.3x - 4.1y = 3.3$
  $3.4x + 9.2y = 1.3$

10.2 The Addition Method

In this section:

▶ The Addition Method
▶ Equations Involving Fractions or Decimals
▶ Applications

In Section 10.1 you used substitution to eliminate a variable in a system of equations. In this section we see another method for eliminating a variable in a system of equations.

The Addition Method

In the **addition method** we eliminate a variable by adding the equations.

EXAMPLE 1 **An independent system solved by addition**

Solve the system by the addition method: $3x - 5y = 9$ ← TYPO
  $4x + 5y = 23$

Solution

The addition property of equality allows us to add the same number to each side of an equation. We can also use the addition property of equality to add the two left sides and add the two right sides:

$$3x - 5y = -9$$
$$\underline{4x + 5y = 23}$$
$$7x \qquad = 14 \quad \text{Add.}$$
$$x = 2$$

The y-term was eliminated when we added the equations because the coefficients of the y-terms were opposites. Now use $x = 2$ in one of the original equations to find y. It does not matter which original equation we use. In this example we will use both equations to see that we get the same y in either case.

$$3x - 5y = -9 \qquad\qquad 4x + 5y = 23$$
$$3(2) - 5y = -9 \quad \text{Replace } x \text{ by 2.} \qquad 4(2) + 5y = 23$$
$$6 - 5y = -9 \quad \text{Solve for } y. \qquad 8 + 5y = 23$$
$$-5y = -15 \qquad\qquad 5y = 15$$
$$y = 3 \qquad\qquad y = 3$$

Because $3(2) - 5(3) = -9$ and $4(2) + 5(3) = 23$ are both true, $(2, 3)$ satisfies both equations. The solution set is $\{(2, 3)\}$. ∎

Actually the addition method can be used to eliminate any variable whose coefficients are opposites. If neither variable has coefficients that are opposites, then we use the multiplication property of equality to change the coefficients of the variables, as shown in Examples 2 and 3.

EXAMPLE 2 **Using multiplication and addition**

Solve the system by the addition method:
$$2x - 3y = -13$$
$$5x - 12y = -46$$

Solution

If we multiply both sides of the first equation by -4, the coefficients of y will be 12 and -12, and y will be eliminated by addition.

$$(-4)(2x - 3y) = (-4)(-13) \quad \text{Multiply each side by } -4.$$
$$5x - 12y = -46$$

$$\begin{array}{rl} -8x + 12y = 52 & \\ \underline{5x - 12y = -46} & \quad \text{Add.} \\ -3x = 6 & \\ x = -2 & \end{array}$$

Replace x by -2 in one of the original equations to find y:

$$2x - 3y = -13$$
$$2(-2) - 3y = -13$$
$$-4 - 3y = -13$$
$$-3y = -9$$
$$y = 3$$

Because $2(-2) - 3(3) = -13$ and $5(-2) - 12(3) = -46$ are both true, the solution set is $\{(-2, 3)\}$. ∎

EXAMPLE 3 **Multiplying both equations before adding**

Solve the system by the addition method:
$$-2x + 3y = 6$$
$$3x - 5y = -11$$

Solution

To eliminate x, we multiply the first equation by 3 and the second by 2:

$$3(-2x + 3y) = 3(6) \quad \text{Multiply each side by 3.}$$
$$2(3x - 5y) = 2(-11) \quad \text{Multiply each side by 2.}$$

$$\begin{array}{rl} -6x + 9y = 18 & \\ \underline{6x - 10y = -22} & \quad \text{Add.} \\ -y = -4 & \\ y = 4 & \end{array}$$

Note that we could have eliminated y by multiplying by 5 and 3. Now insert $y = 4$ into one of the original equations to find x:

$$-2x + 3(4) = 6 \quad \text{Let } y = 4 \text{ in } -2x + 3y = 6.$$
$$-2x + 12 = 6$$
$$-2x = -6$$
$$x = 3$$

Check that $(3, 4)$ satisfies both equations. The solution set is $\{(3, 4)\}$. ∎

We can always use the addition method as long as the equations in a system are in the same form.

EXAMPLE 4   **Using the addition method for an inconsistent system**

Solve the system: $-4y = 5x + 7$

$ 4y = -5x + 12$

Solution

If these equations are added, both variables are eliminated:

$$-4y = 5x + 7$$
$$\underline{4y = -5x + 12}$$
$$0 = 19$$

Because this equation is inconsistent, the original equations are inconsistent. The solution set to the system is the empty set, $\varnothing$.   ■

Equations Involving Fractions or Decimals

When a system of equations involves fractions or decimals, we can use the multiplication property of equality to eliminate the fractions or decimals.

EXAMPLE 5   **A system with fractions**

Solve the system: $\dfrac{1}{2}x - \dfrac{2}{3}y = 7$

$ \dfrac{2}{3}x - \dfrac{3}{4}y = 11$

Solution

Multiply the first equation by 6 and the second equation by 12:

$$6\left(\frac{1}{2}x - \frac{2}{3}y\right) = 6(7) \qquad \rightarrow \qquad 3x - 4y = 42$$

$$12\left(\frac{2}{3}x - \frac{3}{4}y\right) = 12(11) \qquad \rightarrow \qquad 8x - 9y = 132$$

To eliminate x, multiply the first equation by -8 and the second by 3:

$$-8(3x - 4y) = -8(42) \qquad \rightarrow \qquad -24x + 32y = -336$$
$$3(8x - 9y) = 3(132) \qquad \rightarrow \qquad \underline{24x - 27y = 396}$$
$$ 5y = 60$$
$$ y = 12$$

Substitute $y = 12$ into the first of the original equations:

$$\frac{1}{2}x - \frac{2}{3}(12) = 7$$

$$\frac{1}{2}x - 8 = 7$$

$$\frac{1}{2}x = 15$$

$$x = 30$$

Check $(30, 12)$ in the original system. The solution set is $\{(30, 12)\}$.   ■

The strategy for solving a system by addition is summarized as follows.

The Addition Method

1. Write both equations in the same form (usually $Ax + By = C$).
2. Multiply one or both of the equations by appropriate numbers (if necessary) so that one of the variables will be eliminated by addition.
3. Add the equations to get an equation in one variable.
4. Solve the equation in one variable.
5. Substitute the value obtained for one variable into one of the original equations to obtain the value of the other variable.
6. Check the two values in both of the original equations.

Applications

Any system of two linear equations in two variables can be solved by either the addition method or substitution. In applications we use whichever method appears to be the simpler for the problem at hand.

EXAMPLE 6   **Fajitas and burritos**

At the Cactus Cafe the total price for 4 fajita dinners and 3 burrito dinners is $48, and the total price for 3 fajita dinners and 2 burrito dinners is $34. What is the price of each type of dinner?

Solution

Let x represent the price (in dollars) of a fajita dinner, and let y represent the price (in dollars) of a burrito dinner. We can write two equations to describe the given information:

$$4x + 3y = 48$$
$$3x + 2y = 34$$

To solve this system by addition, we multiply the first equation by -3 and the second by 4:

$$-3(4x + 3y) = -3(48) \qquad \text{Multiply each side by } -3.$$
$$4(3x + 2y) = 4(34) \qquad \text{Multiply each side by } 4.$$

$$
\begin{aligned}
-12x - 9y &= -144 \\
\underline{12x + 8y} &= \underline{136} \qquad \text{Add.} \\
-y &= -8 \\
y &= 8
\end{aligned}
$$

To find x, use $y = 8$ in the first equation $4x + 3y = 48$:

$$
\begin{aligned}
4x + 3(8) &= 48 \\
4x + 24 &= 48 \\
4x &= 24 \\
x &= 6
\end{aligned}
$$

So the fajita dinners are $6 each, and the burrito dinners are $8 each. Check this solution in the original problem. ∎

EXAMPLE 7 **Mixing cooking oil**

Canola oil is 7% saturated fat, and corn oil is 14% saturated fat. Crisco sells a blend, Crisco Canola and Corn Oil, which is 11% saturated fat. How many gallons of each type of oil must be mixed to get 280 gallons of this blend?

Solution

Let x represent the number of gallons of canola oil, and let y represent the number of gallons of corn oil. Make a table to summarize all facts:

	Amount (gallons)	% fat	Amount of fat (gallons)
Canola oil	x	7%	$0.07x$
Corn oil	y	14%	$0.14y$
Canola and Corn Oil	280	11%	0.11(280) or 30.8

We can write two equations to express the following facts: (1) the total amount of oil is 280 gallons and (2) the total amount of fat is 30.8 gallons. Then we can use multiplication and addition to solve the system.

(1) $\qquad x + y = 280 \qquad$ Multiply by -0.07.

(2) $\quad 0.07x + 0.14y = 30.80$

$$-0.07x - 0.07y = -19.6$$
$$\underline{0.07x + 0.14y = 30.8}$$
$$0.07y = 11.2$$
$$y = \frac{11.2}{0.07} = 160$$

If $y = 160$ and $x + y = 280$, then $x = 120$. Check that $0.07(120) + 0.14(160) = 30.8$. So it takes 120 gallons of canola oil and 160 gallons of corn oil to make 280 gallons of Crisco Canola and Corn Oil. ∎

Warm-ups

True or false? Explain your answer.

Exercises 1–6 refer to the following systems.

a) $3x - y = 9$
$\ 2x + y = 6$

b) $4x - 2y = 20$
$\ -2x + y = -10$

c) $x - y = 6$
$\ x - y = 7$

1. To solve System (a) by addition, we simply add the equations.

2. To solve System (a) by addition, we can multiply the first equation by 2 and the second by 3 and then add.

3. To solve System (b) by addition, we can multiply the second equation by 2 and then add.

4. Both $(0, -10)$ and $(5, 0)$ are in the solution set to System (b).

5. The solution set to System (b) is the set of all real numbers.

6. System (c) has no solution.

7. Both the addition method and substitution are used to eliminate a variable from a system of two linear equations in two variables.

8. For the addition method, both equations must be in standard form.

9. To eliminate fractions in an equation, we multiply each side by the least common denominator of all fractions involved.

10. We can eliminate either variable by using the addition method.

10.2 EXERCISES

Solve each system by the addition method. See Examples 1–3.

1. $x + y = 7$
   $x - y = 9$

2. $3x - 4y = 11$
   $-3x + 2y = -7$

3. $x - y = 12$
   $2x + y = 3$

4. $x - 2y = -1$
   $-x + 5y = 4$

5. $2x - y = -5$
   $3x + 2y = 3$

6. $3x + 5y = -11$
   $x - 2y = 11$

7. $2x - 5y = 13$
   $3x + 4y = -15$

8. $3x + 4y = -5$
   $5x + 6y = -7$

9. $2x = 3y + 11$
   $7x - 4y = 6$

10. $2x = 2 - y$
    $3x + y = -1$

11. $x + y = 48$
    $12x + 14y = 628$

12. $x + y = 13$
    $22x + 36y = 356$

Solve each system by the addition method. Determine whether the equations are independent, dependent, or inconsistent. See Example 4.

13. $3x - 4y = 9$
    $-3x + 4y = 12$

14. $x - y = 3$
    $-6x + 6y = 17$

15. $5x - y = 1$
    $10x - 2y = 2$

16. $4x + 3y = 2$
    $-12x - 9y = -6$

17. $2x - y = 5$
    $2x + y = 5$

18. $-3x + 2y = 8$
    $3x + 2y = 8$

Solve each system by the addition method. See Example 5.

19. $\frac{1}{4}x + \frac{1}{3}y = 5$
    $x - y = 6$

20. $\frac{3x}{2} - \frac{2y}{3} = 10$
    $\frac{1}{2}x + \frac{1}{2}y = -1$

21. $\frac{x}{4} - \frac{y}{3} = -4$
    $\frac{x}{8} + \frac{y}{6} = 0$

22. $\frac{x}{3} - \frac{y}{2} = -\frac{5}{6}$
    $\frac{x}{5} - \frac{y}{3} = -\frac{3}{5}$

23. $\frac{1}{8}x + \frac{1}{4}y = 5$
    $\frac{1}{16}x + \frac{1}{2}y = 7$

24. $\frac{3}{7}x + \frac{5}{9}y = 27$
    $\frac{1}{9}x + \frac{2}{7}y = 7$

25. $0.05x + 0.10y = 1.30$
    $x + y = 19$

26. $0.1x + 0.06y = 9$
    $0.09x + 0.5y = 52.7$

27. $x + y = 1200$
    $0.12x + 0.09y = 120$

28. $x - y = 100$
    $0.20x + 0.06y = 150$

29. $1.5x - 2y = -0.25$
    $3x + 1.5y = 6.375$

30. $3x - 2.5y = 7.125$
    $2.5x - 3y = 7.3125$

Write a system of two equations in two unknowns for each problem. Solve each system by the method of your choice. See Example 6.

31. Coffee and doughnuts. On Monday, Archie paid $2.54 for 3 doughnuts and 2 coffees. On Tuesday he paid $2.46 for 2 doughnuts and 3 coffees. On Wednesday he was tired of paying the tab and went out for coffee by himself. What was his bill for 1 doughnut and 1 coffee?

32. Books and magazines. At Gwen's garage sale, all books were one price, and all magazines were another price. Harriet bought 4 books and 3 magazines for $1.45, and June bought 2 books and 5 magazines for $1.25. What was the price of a book and what was the price of a magazine?

33. Boys and girls. One-half of the boys and one-third of the girls of Freemont High attended the homecoming game, while one-third of the boys and one-half of the girls attended the homecoming dance. If there were 570 students at the game and 580 at the dance, then how many students are there at Freemont High?

34. Girls and boys. There are 385 surfers in Surf City. Two-thirds of the boys are surfers and one-twelfth of the girls are surfers. If there are two girls for every boy, then how many boys and how many girls are there in Surf City?

35. Nickels and dimes. Winborne has 35 coins consisting of dimes and nickels. If the value of his coins is $3.30, then how many of each type does he have?

36. Pennies and nickels. Wendy has 52 coins consisting of nickels and pennies. If the value of the coins is $1.20, then how many of each type does she have?

37. Blending fudge. The Chocolate Factory blends its Double-Dark-Chocolate fudge, which is 35% fat, with its Peanut Butter fudge, which is 25% fat, to obtain Double-Dark-Peanut fudge, which is 29% fat. How many pounds of Double-Dark-Chocolate and how many pounds of Peanut Butter fudge must be mixed together to obtain 50 pounds of Double-Dark-Peanut fudge?

38. Low-fat yogurt. Ziggy's Famous Yogurt blends regular yogurt that is 3% fat with its no-fat yogurt to obtain low-fat yogurt that is 1% fat. How many pounds of regular yogurt and how many pounds of no-fat yogurt should be mixed to obtain 60 pounds of low-fat yogurt?

39. *Keystone state.* Judy averaged 42 mph driving from Allentown to Harrisburg and 51 mph driving from Harrisburg to Pittsburgh. If she drove a total of 288 miles in 6 hours, then how long did it take her to drive from Harrisburg to Pittsburgh?

Figure for Exercise 39

40. *Empire state.* Spike averaged 45 mph driving from Rochester to Syracuse and 49 mph driving from Syracuse to Albany. If he drove a total of 237 miles in 5 hours, then how far is it from Syracuse to Albany?

41. *Probability of rain.* If Valerie Voss states that the probability of rain tomorrow is four times the probability that it doesn't rain, then what is the probability of rain tomorrow? (*Hint:* The probability that it rains plus the probability that it doesn't rain is 1.)

42. *Super Bowl contender.* A Las Vegas odds-maker believes that the probability that San Francisco plays in the next Super Bowl is nine times the probability that they do not play in the next Super Bowl. What is the odds-maker's probability that San Francisco plays in the next Super Bowl?

43. *Rectangular lot.* The width of a rectangular lot is 75% of its length. If the perimeter is 700 meters, then what are the length and width?

44. *Fence painting.* Darren and Douglas must paint the 792-foot fence that encircles their family home. Because Darren is older, he has agreed to paint 20% more than Douglas. How much of the fence will each boy paint?

Getting More Involved

45. *Discussion.* Explain how you decide whether it is easier to solve a system by substitution or addition.

46. *Exploration.* a) Write a linear equation in two variables that is satisfied by $(-3, 5)$. **b)** Write another linear equation in two variables that is satisfied by $(-3, 5)$. **c)** Are your equations independent or dependent? **d)** Explain how to select the second equation so that it will be independent of the first.

47. *Exploration.* a) Make up a system of two linear equations in two variables such that both $(-1, 2)$ and $(4, 5)$ are in the solution set. **b)** Are your equations independent or dependent? **c)** Is it possible to find an independent system that is satisfied by both ordered pairs? Explain.

10.3 Systems of Linear Equations in Three Variables

In this section:
▶ Definition
▶ Solving a System by Elimination
▶ Graphs of Equations in Three Variables
▶ Applications

The techniques that you learned in Section 10.2 can be extended to systems of equations in more than two variables. In this section we use elimination of variables to solve systems of equations in three variables.

Definition

The equation $5x - 4y = 7$ is called a linear equation in two variables because its graph is a straight line. The equation $2x + 3y - 4z = 12$ is similar in form, and so it is a linear equation in three variables. An equation in three variables is graphed in a three-dimensional coordinate system. The graph of a linear equation in three variables is a plane, not a line. We will not graph equations in three variables in this text, but we can solve systems without graphing. In general, we make the following definition.

> **Linear Equation in Three Variables**
>
> If A, B, C, and D are real numbers, with A, B, and C not all zero, then
>
> $$Ax + By + Cz = D$$
>
> is called a **linear equation in three variables.**

Solving a System by Elimination

A solution to an equation in three variables is an **ordered triple** such as $(-2, 1, 5)$, where the first coordinate is the value of x, the second coordinate is the value of y, and the third coordinate is the value of z. There are infinitely many solutions to a linear equation in three variables.

The solution to a system of equations in three variables is the set of all ordered triples that satisfy all of the equations of the system. The techniques for solving a system of linear equations in three variables are similar to those used on systems of linear equations in two variables. We eliminate variables by either substitution or addition.

EXAMPLE 1   **A linear system with a single solution**

Solve the system:

$$
\begin{aligned}
(1) \qquad & x + y - z = -1 \\
(2) \qquad & 2x - 2y + 3z = 8 \\
(3) \qquad & 2x - y + 2z = 9
\end{aligned}
$$

Solution

We can eliminate z from Eqs. (1) and (2) by multiplying Eq. (1) by 3 and adding it to Eq. (2):

$$
\begin{array}{ll}
3x + 3y - 3z = -3 & \text{Eq. (1) multiplied by 3} \\
\underline{2x - 2y + 3z = 8} & \text{Eq. (2)} \\
(4) \qquad 5x + y \qquad\quad = 5 &
\end{array}
$$

Now we must eliminate the same variable, z, from another pair of equations. Eliminate z from (1) and (3):

$$2x + 2y - 2z = -2 \quad \text{Eq. (1) multiplied by 2}$$
$$\underline{2x - y + 2z = 9} \quad \text{Eq. (3)}$$
$$(5) \qquad 4x + y = 7$$

Equations (4) and (5) give us a system with two variables. We now solve this system. Eliminate y by multiplying Eq. (5) by -1 and adding the equations:

$$5x + y = 5 \quad \text{Eq. (4)}$$
$$\underline{-4x - y = -7} \quad \text{Eq. (5) multiplied by } -1$$
$$x = -2$$

Now that we have x, we can replace x by -2 in Eq. (5) to find y:

$$4x + y = 7$$
$$4(-2) + y = 7$$
$$-8 + y = 7$$
$$y = 15$$

Now replace x by -2 and y by 15 in Eq. (1) to find z:

$$x + y - z = -1$$
$$-2 + 15 - z = -1$$
$$13 - z = -1$$
$$-z = -14$$
$$z = 14$$

Check that $(-2, 15, 14)$ satisfies all three of the original equations. The solution set is $\{(-2, 15, 14)\}$. ∎

The strategy that we follow for solving a system of three linear equations in three variables is stated as follows.

▶ **Solving a System in Three Variables** ◀

1. Use substitution or addition to eliminate any one of the variables from a pair of equations of the system. Look for the easiest variable to eliminate.
2. Eliminate the same variable from another pair of equations of the system.
3. Solve the resulting system of two equations in two unknowns.
4. After you have found the values of two of the variables, substitute into one of the original equations to find the value of the third variable.
5. Check the three values in all of the original equations.

In the next example we use a combination of addition and substitution.

EXAMPLE 2   **Using addition and substitution**

Solve the system:

$$
\begin{array}{lrcl}
(1) & x + y & = & 4 \\
(2) & 2x - 3z & = & 14 \\
(3) & 2y + z & = & 2
\end{array}
$$

Solution

From Eq. (1) we get $y = 4 - x$. If we substitute $y = 4 - x$ into Eq. (3), then Eqs. (2) and (3) will be equations involving x and z only.

$$
\begin{array}{ll}
(3) & 2y + z = 2 \\
 & 2(4 - x) + z = 2 \quad \text{Replace } y \text{ by } 4 - x. \\
 & 8 - 2x + z = 2 \quad \text{Simplify.} \\
(4) & -2x + z = -6
\end{array}
$$

Now solve the system consisting of Eqs. (2) and (4) by addition:

$$
\begin{array}{ll}
2x - 3z = 14 & \text{Eq. (2)} \\
\underline{-2x + z = -6} & \text{Eq. (4)} \\
-2z = 8 & \\
z = -4 &
\end{array}
$$

Use Eq. (3) to find y:

$$
\begin{array}{ll}
2y + z = 2 & \text{Eq. (3)} \\
2y + (-4) = 2 & \text{Let } z = -4. \\
2y = 6 & \\
y = 3 &
\end{array}
$$

Use Eq. (1) to find x:

$$
\begin{array}{ll}
x + y = 4 & \text{Eq. (1)} \\
x + 3 = 4 & \text{Let } y = 3. \\
x = 1 &
\end{array}
$$

Check that $(1, 3, -4)$ satisfies all three of the original equations. The solution set is $\{(1, 3, -4)\}$. ∎

CAUTION   In solving a system in three variables it is essential to keep your work organized and neat. Writing short notes that explain your steps (as was done in the examples) will allow you to go back and check your work.   ⊘

Graphs of Equations in Three Variables

The graph of any equation in three variables can be drawn on a three-dimensional coordinate system. The graph of a linear equation in three variables is a plane. To solve a system of three linear equations in three variables by graphing, we would have to draw the three planes and then identify the points that lie on all three of them. This method would be difficult even when the points have simple coordinates. So we will not attempt to solve these systems by graphing.

By considering how three planes might intersect, we can better under-stand the different types of solutions to a system of three equations in three variables. Figure 10.5 shows some of the possibilities for the positioning of three planes in three-dimensional space. In most of the problems we will solve, the planes intersect at a single point, as in Figure 10.5(a). The solution set consists of one ordered triple. However, the system may include two equations corresponding to parallel planes that have no intersection. In this case the equations are said to be **inconsistent.** If the system has at least two incon-sistent equations, then the solution set is the empty set (see Figs. 10.5(b) and 10.5 (c)).

There are two ways in which the intersection of three planes can consist of infinitely many points. The intersection could be a line or a plane. To get a line, we can have either three different planes intersecting along a line, as in Fig. 10.5(d) or two equations for the same plane, with the third plane intersecting that plane. If all three equations are equations of the same plane, we get that plane for the intersection. We will not solve systems corresponding to all of the possible configurations described. The following examples illustrate two of these cases.

(a)   (b)   (c)   (d)

Figure 10.5

EXAMPLE 3   **An inconsistent system of three linear equations**

Solve the system:

$$
\begin{aligned}
(1) \qquad x + \ y - \ z &= 5 \\
(2) \qquad 3x - 2y + \ z &= 8 \\
(3) \qquad 2x + 2y - 2z &= 7
\end{aligned}
$$

Solution

We can eliminate the variable z from Eqs. (1) and (2) by adding them:

$$
\begin{aligned}
(1) \qquad x + \ y - z &= 5 \\
(2) \qquad \underline{3x - 2y + z} &= \underline{8} \\
4x - \ y \quad\ \ &= 13
\end{aligned}
$$

To eliminate z from Eqs. (1) and (3), multiply Eq. (1) by -2 and add the resulting equation to Eq. (3):

$$
\begin{aligned}
-2x - 2y + 2z &= -10 \qquad \text{Eq. (1) multiplied by } -2 \\
\underline{2x + 2y - 2z} &= \underline{7} \qquad\ \ \text{Eq. (3)} \\
0 &= -3
\end{aligned}
$$

Because the last equation is false, there are two inconsistent equations in the system. Therefore the solution set is the empty set. ∎

EXAMPLE 4 **A dependent system of three equations**

Solve the system:

$$(1) \qquad 2x - 3y - z = 4$$
$$(2) \qquad -6x + 9y + 3z = -12$$
$$(3) \qquad 4x - 6y - 2z = 8$$

Solution

We will first eliminate x from Eqs. (1) and (2). Multiply Eq. (1) by 3 and add the resulting equation to Eq. (2):

$$\begin{array}{ll} 6x - 9y - 3z = 12 & \text{Eq. (1) multiplied by 3} \\ \underline{-6x + 9y + 3z = -12} & \text{Eq. (2)} \\ 0 = 0 & \end{array}$$

The last statement is an identity. The identity occurred because Eq. (2) is a multiple of Eq. (1). In fact, Eq. (3) is also a multiple of Eq. (1). These equations are dependent. They are all equations for the same plane. The solution set is the set of all points on that plane,

$$\{(x,\, y,\, z) \mid 2x - 3y - z = 4\}. \qquad \blacksquare$$

Applications

Problems involving three unknown quantities can often be solved by using a system of three equations in three variables.

EXAMPLE 5 **Finding three unknown rents**

Theresa took in a total of $1240 last week from the rental of three condominiums. She had to pay 10% of the rent from the one-bedroom condo for repairs, 20% of the rent from the two-bedroom condo for repairs, and 30% of the rent from the three-bedroom condo for repairs. If the three-bedroom condo rents for twice as much as the one-bedroom condo and her total repair bill was $276, then what is the rent for each condo?

Solution

Let x, y, and z represent the rent on the one-bedroom, two-bedroom, and three-bedroom condos, respectively. We can write one equation for the total rent, another equation for the total repairs, and a third equation expressing the fact that the rent for the three-bedroom condo is twice that for the one-bedroom condo:

$$x + y + z = 1240$$
$$0.1x + 0.2y + 0.3z = 276$$
$$z = 2x$$

Substitute $z = 2x$ into both of the other equations to eliminate z:

$$x + y + 2x = 1240$$
$$0.1x + 0.2y + 0.3(2x) = 276$$

$$3x + y = 1240$$
$$0.7x + 0.2y = 276$$

$$-2(3x + y) = -2(1240) \quad \text{Multiply each side by } -2.$$
$$10(0.7x + 0.2y) = 10(276) \quad \text{Multiply each side by } 10.$$

$$-6x - 2y = -2480$$
$$\underline{7x + 2y = 2760} \qquad \text{Add.}$$
$$x \qquad = 280$$

$$z = 2(280) = 560 \qquad \text{Because } z = 2x$$
$$280 + y + 560 = 1240 \qquad \text{Because } x + y + z = 1240$$
$$y = 400$$

Check that $(280, 400, 560)$ satisfies all three of the original equations. The condos rent for $280, $400, and $560 per week. ∎

Warm-ups

True or false? Explain your answer.

1. The point $(1, -2, 3)$ is in the solution set to the equation $x + y - z = 4$.

2. The point $(4, 1, 1)$ is the only solution to the equation $x + y - z = 4$.

3. The ordered triple $(1, -1, 2)$ satisfies $x + y + z = 2$, $x - y - z = 0$, and $2x + y - z = -1$.

4. Substitution cannot be used on three equations in three variables.

5. Two distinct planes either are parallel or intersect in a single point.

6. The equations $x - y + 2z = 6$ and $x - y + 2z = 4$ are inconsistent.

7. The equations $3x + 2y - 6z = 4$ and $-6x - 4y + 12z = -8$ are dependent.

8. The graph of $y = 2x - 3z + 4$ is a straight line.

9. The value of x nickels, y dimes, and z quarters is $0.05x + 0.10y + 0.25z$ cents.

10. If $x = -2$, $z = 3$, and $x + y + z = 6$, then $y = 7$.

10.3 EXERCISES

Solve each system of equations. See Examples 1 and 2.

1. $x + y + z = 2$
$x + 2y - z = 6$
$2x + y - z = 5$

2. $2x - y + 3z = 14$
$x + y - 2z = -5$
$3x + y - z = 2$

3. $x - 2y + 4z = 3$
$x + 3y - 2z = 6$
$x - 4y + 3z = -5$

4. $2x + 3y + z = 13$
$-3x + 2y + z = -4$
$4x - 4y + z = 5$

5. $2x - y + z = 10$
$3x - 2y - 2z = 7$
$x - 3y - 2z = 10$

6. $x - 3y + 2z = -11$
$2x - 4y + 3z = -15$
$3x - 5y - 4z = 5$

7. $2x - 3y + z = -9$
$-2x + y - 3z = 7$
$x - y + 2z = -5$

8. $3x - 4y + z = 19$
$2x + 4y + z = 0$
$x - 2y + 5z = 17$

9. $2x - 5y + 2z = 16$
$3x + 2y - 3z = -19$
$4x - 3y + 4z = 18$

10. $-2x + 3y - 4z = 3$
$3x - 5y + 2z = 4$
$-4x + 2y - 3z = 0$

11. $x + y = 4$
$y - z = -2$
$x + y + z = 9$

12. $x + y - z = 0$
$x - y = -2$
$y + z = 10$

13. $x + y = 7$
$y - z = -1$
$x + 3z = 18$

14. $2x - y = -8$
$y + 3z = 22$
$x - z = -8$

Solve each system. See Examples 3 and 4.

15. $x - y + 2z = 3$
$2x + y - z = 5$
$3x - 3y + 6z = 4$

16. $2x - 4y + 6z = 12$
$6x - 12y + 18z = 36$
$-x + 2y - 3z = -6$

17. $3x - y + z = 5$
$9x - 3y + 3z = 15$
$-12x + 4y - 4z = -20$

18. $4x - 2y - 2z = 5$
$2x - y - z = 7$
$-4x + 2y + 2z = 6$

19. $x - y = 3$
$y + z = 8$
$2x + 2z = 7$

20. $2x - y = 6$
$2y + z = -4$
$8x + 2z = 3$

21. $0.10x + 0.08y - 0.04z = 3$
$5x + 4y - 2z = 150$
$0.3x + 0.24y - 0.12z = 9$

22. $0.06x - 0.04y + z = 6$
$3x - 2y + 50z = 300$
$0.03x - 0.02y + 0.5z = 3$

⌨ *Use a calculator to solve each system.*

23. $3x + 2y - 0.4z = 0.1$
$3.7x - 0.2y + 0.05z = 0.41$
$-2x + 3.8y - 2.1z = -3.26$

24. $3x - 0.4y + 9z = 1.668$
$0.3x + 5y - 8z = -0.972$
$5x - 4y - 8z = 1.8$

Solve each problem by using a system of three equations in three unknowns. See Example 5.

25. *Diversification.* Ann invested a total of $12,000 in stocks, bonds, and a mutual fund. She received a 10% return on her stock investment, an 8% return on her bond investment, and a 12% return on her mutual fund. Her total return was $1230. If the total invest-ment in stocks and bonds equaled her mutual fund investment, then how much did she invest in each?

26. *Paranoia.* Fearful of a bank failure, Norman split his life savings of $60,000 among three banks. He received 5%, 6%, and 7% on the three deposits. In the account earning 7% interest, he deposited twice as much as in the account earning 5% interest. If his total earnings were $3,760, then how much did he deposit in each account?

27. *Big tipper.* On Monday, Headley paid $1.70 for 2 cups of coffee and 1 doughnut, including the tip. On Tuesday, he paid $1.65 for 2 doughnuts and a cup of coffee, including the tip. On Wednesday, he paid $1.30 for 1 coffee and 1 doughnut, including the tip. If he always tips the same amount, then what is the amount of each item?

28. *Weighing in.* Anna, Bob, and Chris will not disclose their weights but agree to be weighed in pairs. Anna and Bob together weigh 226 pounds. Bob and Chris together weigh 210 pounds. Anna and Chris together weigh 200 pounds. How much does each student weigh?

29. *Lunch-box special.* Salvador's Fruit Mart sells vari-ety packs. The small pack contains 3 bananas, 2 apples, and 1 orange, for $1.80. The medium pack contains 4 bananas, 3 apples, and 3 oranges, for $3.05. The family size contains 6 bananas, 5 apples, and 4 oranges, for $4.65. What price should Salvador charge for his lunch-box special that consists of 1 banana, 1 apple, and 1 orange?

30. *Three generations.* Edwin, his father, and his grand-father have an average age of 53. One-half of his grandfather's age, plus one-third of his father's age, plus one-fourth of Edwin's age is 65. If four years ago, Edwin's grandfather was 4 times as old as Edwin, then how old are they all now?

31. *Error in the scale.* Alex is using a scale that is known to have a constant error. A can of soup and a can of tuna are placed on this scale, and it reads 24 ounces. Now 4 identical cans of soup and 3 identical cans of tuna are placed on an accurate scale, and a weight of 80 ounces is recorded. If 2 cans of tuna weigh 18 ounces on the bad scale, then what is the amount of error in the scale and what is the correct weight of each type of can?

32. *Three-digit number.* The sum of the digits of a three-digit number is 11. If the digits are reversed, the new number is 46 more than 5 times the old number. If the hundreds digit plus twice the tens digit is equal to the units digit, then what is the number?

33. *Working overtime.* To make ends meet, Ms. Farnsby works three jobs. Her total income last year was $48,000. Her income from teaching was just $6,000 more than her income from house painting. Royalties from her textbook sales were one-seventh of the total money she received from teaching and house painting. How much did she make from each source last year?

34. *Pocket change.* Harry has $2.25 in nickels, dimes, and quarters. If he had twice as many nickels, half as many dimes, and the same number of quarters, he would have $2.50. If he has 27 coins altogether, then how many of each does he have?

Getting More Involved

35. *Exploration.* Draw diagrams showing the possible ways to position three-planes in three-dimensional space.

36. *Discussion.* Make up a system of three linear equations in three variables for which the solution set is $\{(0, 0, 0)\}$. A system with this solution set is called a *homogeneous* system. Why do you think it is given that name?

37. *Cooperative learning.* Working in groups, do parts (a)–(d) below. Then write a report on your findings.
a) Find values of a, b, and c so that the graph of $y = ax^2 + bx + c$ goes through the points $(-1, -2)$, $(1, 0)$, and $(2, 7)$.
b) Arbitrarily select three ordered pairs and find the equation of the parabola that goes through the three points.
c) Could more than one parabola pass through three given points? Give reasons for your answer.
d) Explain how to pick three points for which no parabola passes through all of them.

10.4 Nonlinear Systems of Equations

In this section:
▶ Solving by Elimination
▶ Applications

You have been studying systems of linear equations in this chapter. In this section we turn our attention to nonlinear systems of equations.

Solving by Elimination

Equations such as

$$y = x^2, \quad y = \sqrt{x}, \quad y = |x|, \quad y = 2^x, \quad \text{and} \quad y = \log_2(x)$$

are **nonlinear equations** because their graphs are not straight lines. We say that a system of equations is nonlinear if at least one equation in the system is nonlinear. We solve a nonlinear system just like a linear system, by elimination of variables. However, because the graphs of nonlinear equations may intersect at more than one point, there may be more than one ordered pair in the solution set to the system.

EXAMPLE 1 **A parabola and a line**

Solve the system of equations and draw the graph of each equation on the same coordinate system:

$$y = x^2 - 1$$
$$x + y = 1$$

Solution

We can eliminate y by substituting $y = x^2 - 1$ into $x + y = 1$:

$$x + y = 1$$
$$x + (x^2 - 1) = 1 \quad \text{Substitute } x^2 - 1 \text{ for } y.$$
$$x^2 + x - 2 = 0$$
$$(x - 1)(x + 2) = 0$$
$$x - 1 = 0 \quad \text{or} \quad x + 2 = 0$$
$$x = 1 \quad \text{or} \quad x = -2$$

Replace x by 1 and -2 in $y = x^2 - 1$ to find the corresponding values of y:

$$y = (1)^2 - 1 \qquad y = (-2)^2 - 1$$
$$y = 0 \qquad\qquad y = 3$$

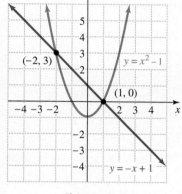

Figure 10.6

Check that each of the points $(1, 0)$ and $(-2, 3)$ satisfies both of the original equations. The solution set is $\{(1, 0), (-2, 3)\}$. If we solve $x + y = 1$ for y, we get $y = -x + 1$. The line of $y = -x + 1$ has y-intercept $(0, 1)$ and slope -1. The graph of $y = x^2 - 1$ is a parabola with vertex $(0, -1)$. Of course, $(1, 0)$ and $(-2, 3)$ are on both graphs. The two graphs are shown in Fig. 10.6. ∎

Graphing is not an accurate method for solving any system of equations. However, the graphs of the equations in a nonlinear system help us to understand how many solutions we should have for the system. It is not necessary to graph a system to solve it. Even when the graphs are too difficult to sketch, we can solve the system.

EXAMPLE 2 **No graph needed**

Solve the system:
$$x^2 + y^2 + 2y = 3$$
$$x^2 - y = 5$$

Solution

If we substitute $y = x^2 - 5$ into the first equation to eliminate y, we will get a fourth-degree equation to solve. Instead, we can eliminate the variable x by writing $x^2 - y = 5$ as $x^2 = y + 5$. Now replace x^2 by $y + 5$ in the first equation:

$$x^2 + y^2 + 2y = 3$$
$$(y + 5) + y^2 + 2y = 3$$
$$y^2 + 3y + 5 = 3$$
$$y^2 + 3y + 2 = 0$$
$$(y + 2)(y + 1) = 0 \quad \text{Solve by factoring.}$$
$$y + 2 = 0 \quad \text{or} \quad y + 1 = 0$$
$$y = -2 \quad \text{or} \quad y = -1$$

Let $y = -2$ in the equation $x^2 = y + 5$ to find the corresponding x:

$$x^2 = -2 + 5$$
$$x^2 = 3$$
$$x = \pm\sqrt{3}$$

Now let $y = -1$ in the equation $x^2 = y + 5$ to find the corresponding x:

$$x^2 = -1 + 5$$
$$x^2 = 4$$
$$x = \pm 2$$

Check these values in the original equations. The solution set is

$$\{(\sqrt{3}, -2), (-\sqrt{3}, -2), (2, -1), (-2, -1)\}.$$

The graphs of these two equations intersect at four points. ■

EXAMPLE 3 **No graph needed**

Solve the system: (1) $\dfrac{2}{x} + \dfrac{1}{y} = \dfrac{1}{5}$

(2) $\dfrac{1}{x} - \dfrac{3}{y} = \dfrac{1}{3}$

Solution

Usually with equations involving rational expressions we first multiply by the LCD, but this would make the given system more complicated. So we will just use the addition method to eliminate y:

$$\dfrac{6}{x} + \dfrac{3}{y} = \dfrac{3}{5} \qquad \text{Eq. (1) multiplied by 3}$$

$$\dfrac{1}{x} - \dfrac{3}{y} = \dfrac{1}{3} \qquad \text{Eq. (2)}$$

$$\overline{}$$

$$\dfrac{7}{x} = \dfrac{14}{15} \qquad \tfrac{3}{5} + \tfrac{1}{3} = \tfrac{14}{15}$$

$$14x = 7 \cdot 15$$

$$x = \dfrac{7 \cdot 15}{14} = \dfrac{15}{2}$$

To find y, substitute $x = \dfrac{15}{2}$ into Eq. (1):

$$\dfrac{2}{\dfrac{15}{2}} + \dfrac{1}{y} = \dfrac{1}{5}$$

$$\dfrac{4}{15} + \dfrac{1}{y} = \dfrac{1}{5} \qquad \dfrac{2}{\frac{15}{2}} = 2 \cdot \tfrac{2}{15} = \tfrac{4}{15}$$

$$15y \cdot \dfrac{4}{15} + 15y \cdot \dfrac{1}{y} = 15y \cdot \dfrac{1}{5} \qquad \text{Multiply each side by the LCD, } 15y.$$

$$4y + 15 = 3y$$

$$y = -15$$

Check that $x = \dfrac{15}{2}$ and $y = -15$ satisfy both original equations. The solution set is $\left\{\left(\dfrac{15}{2}, -15\right)\right\}$. ■

Applications

The next example shows a geometric problem that can be solved with a system of nonlinear equations.

EXAMPLE 4

Figure 10.7

Nonlinear equations in applications

A 15-foot ladder is leaning against a wall so that the distance from the bottom of the ladder to the wall is one-half of the distance from the top of the ladder to the ground. Find the distance from the top of the ladder to the ground.

Solution

Let x be the distance from the bottom of the ladder to the wall and y be the distance from the top of the ladder to the ground (see Figure 10.7). We can write two equations involving x and y:

$$x^2 + y^2 = 15^2 \quad \text{Pythagorean theorem}$$
$$y = 2x$$

Solve by substitution:

$$x^2 + (2x)^2 = 225 \quad \text{Replace } y \text{ by } 2x.$$
$$x^2 + 4x^2 = 225$$
$$5x^2 = 225$$
$$x^2 = 45$$
$$x = \pm\sqrt{45} = \pm3\sqrt{5}$$

Because x represent distance, x must be positive. So $x = 3\sqrt{5}$ feet. Because $y = 2x$, we get $y = 6\sqrt{5}$ feet. ∎

The next example shows how a nonlinear system can be used to solve a problem involving work.

EXAMPLE 5

Nonlinear equations in applications

A large fish tank at the Gulf Aquarium can usually be filled in 10 minutes using pumps A and B. However, pump B can pump water in or out at the same rate. If pump B is inadvertently run in reverse, then the tank will be filled in 30 minutes. How long would it take each pump to fill the tank by itself?

Solution

Let a represent the number of minutes that it takes pump A to fill the tank alone and b represent the number of minutes it takes pump B to fill the tank alone. The rate at which pump A fills the tank is $\frac{1}{a}$ of the tank per minute, and the rate at which pump B fills the tank is $\frac{1}{b}$ of the tank per minute. Because the work completed is the product of the rate and time, we can make the following table when the pumps work together to fill the tank:

	Rate	Time	Work
Pump A	$\frac{1}{a}$ $\frac{\text{tank}}{\text{min}}$	10 min	$\frac{10}{a}$ tank
Pump B	$\frac{1}{b}$ $\frac{\text{tank}}{\text{min}}$	10 min	$\frac{10}{b}$ tank

Note that each pump fills a fraction of the tank and those fractions have a sum of 1:

$$(1) \qquad \frac{10}{a} + \frac{10}{b} = 1$$

A similar equation describes what happens when pump B is working against pump A for 30 minutes:

$$(2) \qquad \frac{30}{a} - \frac{30}{b} = 1$$

Multiply Eq. (1) by 3 and add the result to Eq. (2) to eliminate b:

$$\frac{30}{a} + \frac{30}{b} = 3 \qquad \text{Eq. (1) multiplied by 3}$$

$$\frac{30}{a} - \frac{30}{b} = 1 \qquad \text{Eq. (2)}$$

$$\overline{\frac{60}{a} \qquad\quad = 4}$$

$$4a = 60$$

$$a = 15$$

Use $a = 15$ in Eq. (1) to find b:

$$\frac{10}{15} + \frac{10}{b} = 1$$

$$\frac{10}{b} = \frac{1}{3} \qquad \text{Subtract } \tfrac{10}{15} \text{ from each side.}$$

$$b = 30$$

So pump A fills the tank in 15 minutes working alone, and pump B fills the tank in 30 minutes working alone. ∎

Warm-ups

True or false? Explain your answer.

1. The graph of $y = x^2$ is a parabola.
2. The graph of $y = |x|$ is a straight line.
3. The point $(3, -4)$ satisfies both $x^2 + y^2 = 25$ and $y = \sqrt{5x + 1}$.
4. The graphs of $y = \sqrt{x}$ and $y = -x - 2$ do not intersect.
5. For nonlinear systems we can use only substitution to eliminate a variable.
6. If Bob paints a fence in x hours, then he paints $\frac{1}{x}$ of the fence per hour.
7. In a triangle whose angles are 30°, 60°, and 90°, the length of the side opposite the 30° angle is one-half the length of the hypotenuse.
8. The formula $V = LWH$ gives the volume of a rectangular box in which the sides have lengths L, W, and H.
9. The surface area of a rectangular box is $2LW + 2WH + 2LH$.
10. The area of a right triangle is one-half the product of the lengths of its legs.

10.4 EXERCISES

Solve each system and graph both equations on the same set of axes. See Example 1.

1. $y = x^2$
$x + y = 6$

2. $y = x^2 - 1$
$x + y = 11$

3. $y = |x|$
$2y - x = 6$

4. $y = |x|$
$3y = x + 6$

5. $y = \sqrt{2x}$
$x - y = 4$

6. $y = \sqrt{x}$
$x - y = 6$

7. $4x - 9y = 9$
$xy = 1$

8. $2x + 2y = 3$
$xy = -1$

9. $y = -x^2 + 1$
$y = x^2$

10. $y = x^2$
$y = \sqrt{x}$

Solve each system. See Examples 2 and 3.

11. $x^2 + y^2 = 25$
$y = x^2 - 5$

12. $x^2 + y^2 = 25$
$y = x + 1$

13. $xy - 3x = 8$
$y = x + 1$

14. $xy + 2x = 9$
$x - y = 2$

15. $xy - x = 8$
$xy + 3x = -4$

16. $2xy - 3x = -1$
$xy + 5x = -7$

17. $\dfrac{1}{x} - \dfrac{1}{y} = 5$
$\dfrac{2}{x} + \dfrac{1}{y} = -3$

18. $\dfrac{2}{x} - \dfrac{3}{y} = \dfrac{1}{2}$
$\dfrac{3}{x} + \dfrac{1}{y} = \dfrac{1}{2}$

19. $x^2 y = 20$
$xy + 2 = 6x$

20. $y^2 x = 3$
$xy + 1 = 6x$

21. $x^2 + y^2 = 8$
$x^2 - y^2 = 2$

22. $x^2 + 2y^2 = 8$
$2x^2 - y^2 = 1$

23. $x^2 + xy - y^2 = -11$
$x + y = 7$

24. $x^2 + xy + y^2 = 3$
$y = 2x - 5$

25. $3y - 2 = x^4$
$y = x^2$

26. $y - 3 = 2x^4$
$y = 7x^2$

Solve the following systems involving logarithmic and exponential functions.

27. $y = \log_2 (x - 1)$
$y = 3 - \log_2 (x + 1)$

28. $y = \log_3 (x - 4)$
$y = 2 - \log_3 (x + 4)$

29. $y = \log_2 (x - 1)$
$y = 2 + \log_2 (x + 2)$

30. $y = \log_4 (8x)$
$y = 2 + \log_4 (x - 1)$

31. $y = 2^{3x + 4}$
$y = 4^{x - 1}$

32. $y = 4^{3x}$
$y = \left(\dfrac{1}{2}\right)^{1 - x}$

The following systems require two steps to eliminate a variable. Solve each system.

33. $x^2 + xy + y^2 = 19$
$x^2 + y^2 = 13$

34. $x^2 - 2xy + y^2 = 25$
$x^2 + y^2 = 37$

35. $x^2 + xy + 2y^2 = 11$
$x^2 - 2xy + 2y^2 = 5$

36. $3x^2 - 2xy + y^2 = 18$
$3x^2 + xy + y^2 = 9$

Solve each problem by using a system of two equations in two unknowns. See Examples 4 and 5.

37. *Known hypotenuse.* Find the lengths of the legs of a right triangle whose hypotenuse is $\sqrt{15}$ feet and whose area is 3 square feet.

38. *Known diagonal.* A small television is advertised to have a picture with a diagonal measure of 5 inches and a viewing area of 12 square inches. What are the length and width of the screen?

Figure for Exercise 38

39. *House of seven gables.* Vincent has plans to build a house with seven gables. The plans call for an attic vent in the shape of an isosceles triangle in each gable. Because of the slope of the roof, the ratio of the height to the base of each triangle must be 1 to 4. If the vents are to provide a total ventilating area of 3,500 square inches, then what should be the height and base of each triangle?

Figure for Exercise 39

40. *Known perimeter.* Find the lengths of the sides of a triangle whose perimeter is 6 feet and whose angles are 30°, 60°, and 90° (see Appendix A).

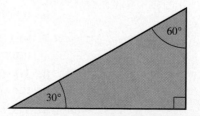

Figure for Exercise 40

41. Filling a tank. Pump A can either fill a tank or empty it in the same amount of time. If pump A and pump B are working together, the tank can be filled in 6 hours. When pump A was inadvertently left in the drain position while pump B was trying to fill the tank, it took 12 hours to fill the tank. How long would it take either pump working alone to fill the tank?

42. Cleaning a house. Roxanne either cleans the house or messes it up at the same rate. When Roxanne is cleaning with her mother, they can clean up a completely messed up house in 6 hours. If Roxanne is not cooperating, it takes her mother 9 hours to clean the house, with Roxanne continually messing it up. How long would it take her mother to clean the entire house if Roxanne were sent to her grandmother's house?

43. Cleaning fish. Jan and Beth work in a seafood market that processes 200 pounds of catfish every morning. On Monday, Jan started cleaning catfish at 8:00 A.M. and finished cleaning 100 pounds just as Beth arrived. Beth then took over and finished the job at 8:50 A.M. On Tuesday they both started at 8 A.M. and worked together to finish the job at 8:24 A.M. On Wednesday, Beth was sick. If Jan is the faster worker, then how long did it take Jan to complete all of the catfish by herself?

44. Building a patio. Richard has already formed a rectangular area for a flagstone patio, but his wife Susan is unsure of the size of the patio they want. If the width is increased by 2 feet, then the area is increased by 30 square feet. If the width is increased by 1 foot and the length by 3 feet, then the area is increased by 54 square feet. What are the dimensions of the rectangle that Richard has already formed?

Figure for Exercise 44

45. Fencing a rectangle. If 34 feet of fencing are used to enclose a rectangular area of 72 square feet, then what are the dimensions of the area.

46. Real numbers. Find two numbers that have a sum of 8 and a product of 10.

47. Imaginary numbers. Find two complex numbers whose sum is 8 and whose product is 20.

48. Imaginary numbers. Find two complex numbers whose sum is -6 and whose product is 10.

49. Making a sign. Rico's Sign Shop has a contract to make a sign in the shape of a square with an isosceles triangle on top of it, as shown in the figure. The contract calls for a total height of 10 feet with an area of 72 square feet. How long should Rico make the side of the square and what should be the height of the triangle?

Figure for Exercise 49

50. Designing a box. Angelina is designing a rectangular box of 120 cubic inches that is to contain new Eaties breakfast cereal. The box must be 2 inches thick so that it is easy to hold. It must have 184 square inches of surface area to provide enough space for all of the special offers and coupons. What should be the dimensions of the box?

Graphing Calculator Exercises

51. Solve each system by graphing each pair of equations on a graphing calculator and using the trace feature to estimate the point of intersection. Find the coordinates of each intersection to the nearest hundredth.

a) $y = e^x - 4$
$y = \ln(x + 3)$

b) $3^{y-1} = x$
$y = x^2$

c) $x^2 + y^2 = 4$
$y = x^3$

10.5 Solving Linear Systems Using Matrices

In this section:
▶ Matrices
▶ The Augmented Matrix
▶ The Gauss-Jordan Method
▶ Inconsistent and Dependent Equations

You solved linear systems in two variables by substitution and addition in Sections 10.1 and 10.2. Those methods are done differently on each system. In this section you will learn the Gauss-Jordan method, which is related to the addition method. The Gauss-Jordan method is performed in the same way on every system. We first need to introduce some new terminology.

Matrices

A **matrix** is a rectangular array of numbers. The **rows** of a matrix run horizontally, and the **columns** of a matrix run vertically. A matrix with m rows and n columns has **order** $m \times n$ (read "m by n"). Each number in a matrix is called an **element** or **entry** of the matrix.

EXAMPLE 1 **Order of a matrix**

Determine the order of each matrix.

a) $\begin{bmatrix} -1 & 2 \\ 5 & \sqrt{2} \\ 0 & 3 \end{bmatrix}$ **b)** $\begin{bmatrix} 2 & 3 \\ -1 & 5 \end{bmatrix}$ **c)** $\begin{bmatrix} 1 & 2 & 3 \\ 4 & 5 & 6 \\ -1 & 0 & 2 \end{bmatrix}$ **d)** $[1 \ 3 \ 6]$

Solution

Because matrix (a) has 3 rows and 2 columns, its order is 3×2. Matrix (b) is a 2×2 matrix, matrix (c) is a 3×3 matrix, and matrix (d) is a 1×3 matrix. ∎

The Augmented Matrix

The solution to a system of linear equations such as

$$x - 2y = -5$$
$$3x + y = 6$$

depends on the coefficients of x and y and the constants on the right-hand side of the equation. The matrix of coefficients for this system is the 2×2 matrix

$$\begin{bmatrix} 1 & -2 \\ 3 & 1 \end{bmatrix}.$$

If we insert the constants from the right hand-side of the system into the matrix of coefficients, we get the 2×3 matrix

$$\begin{bmatrix} 1 & -2 & -5 \\ 3 & 1 & 6 \end{bmatrix}.$$

We use a vertical line between the coefficients and the constants to represent the equal signs. This matrix is the **augmented matrix** of the system. Two systems of linear equations are **equivalent** if they have the same solution set. Two augmented matrices are **equivalent** if the systems they represent are equivalent.

EXAMPLE 2   **Writing the augmented matrix**

Write the augmented matrix for each system of equations.

a) $3x - 5y = 7$
  $x + y = 4$

b) $x + y - z = 5$
  $2x + z = 3$
  $2x - y + 4z = 0$

c) $x + y = 1$
  $y + z = 6$
  $z = -5$

Solution

a) $\begin{bmatrix} 3 & -5 & 7 \\ 1 & 1 & 4 \end{bmatrix}$

b) $\begin{bmatrix} 1 & 1 & -1 & 5 \\ 2 & 0 & 1 & 3 \\ 2 & -1 & 4 & 0 \end{bmatrix}$

c) $\begin{bmatrix} 1 & 1 & 0 & 1 \\ 0 & 1 & 1 & 6 \\ 0 & 0 & 1 & -5 \end{bmatrix}$ ■

EXAMPLE 3   **Writing the system**

Write the system of equations represented by each augmented matrix.

a) $\begin{bmatrix} 1 & 4 & -2 \\ 1 & -1 & 3 \end{bmatrix}$

b) $\begin{bmatrix} 1 & 0 & 5 \\ 0 & 1 & 1 \end{bmatrix}$

c) $\begin{bmatrix} 2 & 3 & 4 & 6 \\ -1 & 0 & 5 & -2 \\ 1 & -2 & 3 & 1 \end{bmatrix}$

Solution

a) Use the first two numbers in each row as the coefficients of x and y and the last number as the constant to get the following system:

$$x + 4y = -2$$
$$x - y = 3$$

b) Use the first two numbers in each row as the coefficients of x and y and the last number as the constant to get the following system:

$$x = 5$$
$$y = 1$$

c) Use the first three numbers in each row as the coefficients of x, y, and z and the last number as the constant to get the following system:

$$2x + 3y + 4z = 6$$
$$-x + 5z = -2$$
$$x - 2y + 3z = 1$$

■

The Gauss-Jordan Method

When we solve a single equation, we write simpler and simpler equivalent equations to get an equation whose solution is obvious. In the **Gauss-Jordan method** we write simpler and simpler equivalent augmented matrices until we get an augmented matrix (like the one in Example 3(b)) in which the solution to the corresponding system is obvious.

Because each row of an augmented matrix represents an equation, we can perform the **row operations** on the augmented matrix. These row operations, given on the next page, correspond to the usual operations with equations used in the addition method.

> ### Row Operations
>
> The following row operations on an augmented matrix give an equivalent augmented matrix:
>
> 1. Interchange two rows of the matrix.
> 2. Multiply every element in a row by a nonzero real number.
> 3. Add to a row a multiple of another row.

In the Gauss-Jordan method our goal is to use row operations to obtain an augmented matrix that has ones on the **diagonal** in its matrix of coefficients and zeros elsewhere:

$$\left[\begin{array}{cc|c} 1 & 0 & a \\ 0 & 1 & b \end{array}\right]$$

The system corresponding to this augmented matrix is $x = a$ and $y = b$. So the solution set to the system is $\{(a, b)\}$.

EXAMPLE 4 **Gauss-Jordan with two equations in two variables**

Use the Gauss-Jordan method to solve the system:

$$x - 3y = 11$$
$$2x + y = 1$$

Solution

Start with the augmented matrix:

$$\left[\begin{array}{cc|c} 1 & -3 & 11 \\ 2 & 1 & 1 \end{array}\right]$$

Multiply row 1 by -2 and add the result to row 2 (in symbols, $-2R_1 + R_2 \rightarrow R_2$). Because $-2R_1 = [-2, 6, -22]$ and $R_2 = [2, 1, 1]$, $-2R_1 + R_2 = [0, 7, -21]$. Note that the coefficient of x in the second equation is now 0. We get the following matrix:

$$\left[\begin{array}{cc|c} 1 & -3 & 11 \\ 0 & 7 & -21 \end{array}\right] \quad -2R_1 + R_2 \rightarrow R_2$$

Multiply each element of row 2 by $\frac{1}{7}$ (in symbols, $\frac{1}{7}R_2 \rightarrow R_2$):

$$\left[\begin{array}{cc|c} 1 & -3 & 11 \\ 0 & 1 & -3 \end{array}\right] \quad \frac{1}{7}R_2 \rightarrow R_2$$

Multiply row 2 by 3 and add the result to row 1. Because $3R_2 = [0, 3, -9]$ and $R_1 = [1, -3, 11]$, $3R_2 + R_1 = [1, 0, 2]$. Note that the coefficient of y in the first equation is now 0. We get the following matrix:

$$\left[\begin{array}{cc|c} 1 & 0 & 2 \\ 0 & 1 & -3 \end{array}\right] \quad 3R_2 + R_1 \rightarrow R_1$$

This augmented matrix represents the system $x = 2$ and $y = -3$. So the solution set to the system is $\{(2, -3)\}$. Check in the original system. ∎

In the next example we use the row operations on the augmented matrix of a system of three linear equations in three variables.

EXAMPLE 5   **Gauss-Jordan with three equations in three variables**

Use the Gauss-Jordan method to solve the following system:

$$2x - y + z = -3$$
$$x + y - z = 6$$
$$3x - y - z = 4$$

Solution

Start with the augmented matrix and interchange the first and second row to get a 1 in the upper left position in the matrix:

$$\begin{bmatrix} 2 & -1 & 1 & | & -3 \\ 1 & 1 & -1 & | & 6 \\ 3 & -1 & -1 & | & 4 \end{bmatrix} \quad \text{The augmented matrix}$$

$$\begin{bmatrix} 1 & 1 & -1 & | & 6 \\ 2 & -1 & 1 & | & -3 \\ 3 & -1 & -1 & | & 4 \end{bmatrix} \quad R_1 \leftrightarrow R_2$$

Now multiply the first row by -2 and add the result onto the second row. Multiply the first row by -3 and add the result onto the third row. These two steps eliminate the variable x from the second and third row:

$$\begin{bmatrix} 1 & 1 & -1 & | & 6 \\ 0 & -3 & 3 & | & -15 \\ 0 & -4 & 2 & | & -14 \end{bmatrix} \quad \begin{array}{l} -2R_1 + R_2 \to R_2 \\ -3R_1 + R_3 \to R_3 \end{array}$$

Multiply the second row by $-\frac{1}{3}$ to get 1 in the second position on the diagonal:

$$\begin{bmatrix} 1 & 1 & -1 & | & 6 \\ 0 & 1 & -1 & | & 5 \\ 0 & -4 & 2 & | & -14 \end{bmatrix} \quad -\frac{1}{3}R_2 \to R_2$$

Use the second row to eliminate the variable y from the first and third rows:

$$\begin{bmatrix} 1 & 0 & 0 & | & 1 \\ 0 & 1 & -1 & | & 5 \\ 0 & 0 & -2 & | & 6 \end{bmatrix} \quad \begin{array}{l} -1R_2 + R_1 \to R_1 \\ \\ 4R_2 + R_3 \to R_3 \end{array}$$

Multiply the third row by $-\frac{1}{2}$ to get a 1 in the third position on the diagonal:

$$\begin{bmatrix} 1 & 0 & 0 & | & 1 \\ 0 & 1 & -1 & | & 5 \\ 0 & 0 & 1 & | & -3 \end{bmatrix} \quad -\frac{1}{2}R_3 \to R_3$$

Use the third row to eliminate the variable z from the second row:

$$\begin{bmatrix} 1 & 0 & 0 & | & 1 \\ 0 & 1 & 0 & | & 2 \\ 0 & 0 & 1 & | & -3 \end{bmatrix} \quad R_3 + R_2 \to R_2$$

This last augmented matrix represents the system $x = 1$, $y = 2$, and $z = -3$. So the solution set to the system is $\{(1, 2, -3)\}$. ∎

Inconsistent and Dependent Equations

Inconsistent and dependent equations are easily recognized in using the Gauss-Jordan method.

EXAMPLE 6 **Gauss-Jordan with an inconsistent system**

Solve the system:

$$x - y = 1$$
$$-3x + 3y = 4$$

Solution

Start with the augmented matrix:

$$\left[\begin{array}{rr|r} 1 & -1 & 1 \\ -3 & 3 & 4 \end{array}\right]$$

Multiply row 1 by 3 and add the result to row 2. We get the following matrix:

$$\left[\begin{array}{rr|r} 1 & -1 & 1 \\ 0 & 0 & 7 \end{array}\right] \quad 3R_1 + R_2 \to R_2$$

The second row of the augmented matrix corresponds to the equation $0 = 7$. So the equations are inconsistent, and there is no solution to the system. ■

EXAMPLE 7 **Gauss-Jordan with a dependent system**

Solve the system:

$$3x + y = 1$$
$$6x + 2y = 2$$

Solution

Start with the augmented matrix:

$$\left[\begin{array}{rr|r} 3 & 1 & 1 \\ 6 & 2 & 2 \end{array}\right]$$

Multiply row 1 by -2 and add the result to row 2. We get the following matrix:

$$\left[\begin{array}{rr|r} 3 & 1 & 1 \\ 0 & 0 & 0 \end{array}\right] \quad -2R_1 + R_2 \to R_2$$

In the second row of the augmented matrix we have the equation $0 = 0$. So the equations are dependent. Every ordered pair that satisfies the first equation satisfies both equations. The solution set is $\{(x, y) \mid 3x + y = 1\}$. ■

The Gauss-Jordan method may be applied to a system of n linear equations in n unknowns, where $n \geq 2$. However, it is a rather tedious method to perform when n is greater than 2, especially when fractions are involved. Computers are programmed to work with matrices, and the Gauss-Jordan method is a popular method for computers.

Warm-ups

True or false? Explain your answer.

Statements 1–7 refer to the following matrices:

a) $\begin{bmatrix} 1 & 3 & 5 \\ -1 & -3 & 2 \end{bmatrix}$ **b)** $\begin{bmatrix} 1 & 3 & 5 \\ 0 & 0 & 7 \end{bmatrix}$ **c)** $\begin{bmatrix} -1 & 2 & -3 \\ 2 & -4 & 3 \end{bmatrix}$ **d)** $\begin{bmatrix} 1 & 3 & 5 \\ 0 & 0 & 0 \end{bmatrix}$

1. The augmented matrix for $x + 3y = 5$ and $-x - 3y = 2$ is matrix (a).
2. The augmented matrix for $2y - x = -3$ and $2x - 4y = 3$ is matrix (c).
3. Matrix (a) is equivalent to matrix (b).
4. Matrix (c) is equivalent to matrix (d).
5. The system corresponding to matrix (b) is inconsistent.
6. The system corresponding to matrix (c) is dependent.
7. The system corresponding to matrix (d) is independent.
8. The augmented matrix for a system of two linear equations in two unknowns is a 2×2 matrix.
9. The notation $2R_1 + R_3 \to R_3$ means to replace R_3 by $2R_1 + R_3$.
10. The notation $R_1 \leftrightarrow R_2$ means to replace R_2 by R_1.

10.5 EXERCISES

Determine the order of each matrix. See Example 1.

1. $\begin{bmatrix} 5 & 0 \\ -2 & 3 \end{bmatrix}$ 2. $\begin{bmatrix} 1 & 3 & 6 \\ -7 & 0 & 2 \end{bmatrix}$ 3. $\begin{bmatrix} a & c \\ 0 & d \\ 3 & w \end{bmatrix}$

4. $\begin{bmatrix} 0 & a & b \\ 5 & 7 & -8 \\ a & b & 2 \end{bmatrix}$ 5. $\begin{bmatrix} -\sqrt{3} \\ \pi \\ \frac{1}{2} \end{bmatrix}$ 6. $[3 \ 0 \ 4]$

Write the augmented matrix for each system of equations. See Example 2.

7. $2x - 3y = 9$
   $-3x + y = -1$

8. $x - y = 4$
   $2x + y = 3$

9. $x - y + z = 1$
   $x + y - 2z = 3$
   $y - 3z = 4$

10. $x + y = 2$
    $y - 3z = 5$
    $-3x + 2z = 8$

Write the system of equations represented by each augmented matrix. See Example 3.

11. $\begin{bmatrix} 5 & 1 & -1 \\ 2 & -3 & 0 \end{bmatrix}$ 12. $\begin{bmatrix} 1 & 0 & 4 \\ 0 & 1 & -3 \end{bmatrix}$

13. $\begin{bmatrix} 1 & 0 & 0 & 6 \\ -1 & 0 & 1 & -3 \\ 1 & 1 & 0 & 1 \end{bmatrix}$ 14. $\begin{bmatrix} 1 & 0 & 4 & 3 \\ 0 & 2 & 1 & -1 \\ 1 & 1 & 1 & 1 \end{bmatrix}$

Determine the row operation that was used to convert each given augmented matrix into the equivalent augmented matrix that follows it. See Example 4.

15. $\begin{bmatrix} 3 & 2 & 12 \\ 1 & -1 & -1 \end{bmatrix}$, $\begin{bmatrix} 1 & -1 & -1 \\ 3 & 2 & 12 \end{bmatrix}$

16. $\begin{bmatrix} 1 & -1 & -1 \\ 3 & 2 & 12 \end{bmatrix}$, $\begin{bmatrix} 1 & -1 & -1 \\ 0 & 5 & 15 \end{bmatrix}$

17. $\begin{bmatrix} 1 & -1 & -1 \\ 0 & 5 & 15 \end{bmatrix}$, $\begin{bmatrix} 1 & -1 & -1 \\ 0 & 1 & 3 \end{bmatrix}$

18. $\begin{bmatrix} 1 & -1 & -1 \\ 0 & 1 & 3 \end{bmatrix}$, $\begin{bmatrix} 1 & 0 & 2 \\ 0 & 1 & 3 \end{bmatrix}$

Solve each system using the Gauss-Jordan method. See Examples 4–7.

19. $x + y = 3$
    $-3x + y = -1$

20. $x - y = -1$
    $2x - y = 2$

21. $2x - y = 3$
    $x + y = 9$

22. $3x - 4y = -1$
    $x - y = 0$

23. $3x - y = 4$
    $2x + y = 1$

24. $2x - y = -3$
    $3x + y = -2$

25. $6x - 7y = 0$
    $2x + y = 20$

26. $2x + y = 11$
    $2x - y = 1$

27. $2x - 3y = 4$
    $-2x + 3y = 5$

28. $x - 3y = 8$
    $2x - 6y = 1$

29. $x + 2y = 1$
$3x + 6y = 3$

30. $2x - 3y = 1$
$-6x + 9y = -3$

31. $x + y + z = 6$
$x - y + z = 2$
$2y - z = 1$

32. $x - y - z = 0$
$-x - y + z = -4$
$-x + y - z = -2$

33. $2x + y + z = 4$
$x + y - z = 1$
$x - y + 2z = 2$

34. $3x - y = 1$
$x + y + z = 4$
$x + 2z = 3$

35. $2x - y + z = 0$
$x + y - 3z = 3$
$x - y + z = -1$

36. $x - y - z = 0$
$-x - y + 2z = -1$
$-x + y - 2z = -3$

37. $-x + 3y + z = 0$
$x - y - 4z = -3$
$x + y + 2z = 3$

38. $-x + z = -2$
$2x - y = 5$
$y + 3z = 9$

39. $x - y + z = 1$
$2x - 2y + 2z = 2$
$-3x + 3y - 3z = -3$

40. $4x - 2y + 2z = 2$
$2x - y + z = 1$
$-2x + y - z = -1$

41. $x + y - z = 2$
$2x - y + z = 1$
$3x + 3y - 3z = 8$

42. $x + y + z = 5$
$x - y - z = 8$
$-x + y + z = 2$

Getting More Involved

43. *Cooperative learning.* Write a step-by-step procedure for solving any system of two linear equations in two variables by the Gauss-Jordan method. Have a classmate evaluate your procedure by using it to solve a system.

44. *Cooperative learning.* Repeat Exercise 43 for a system of three linear equations in three variables.

10.6 Cramer's Rule for Systems in Two Variables

In this section:

▶ Determinants

▶ Cramer's Rule

The Gauss-Jordan method of Section 10.5 can be performed the same way on every system. Another method that is applied the same way for every system is Cramer's rule, which we study in this section. Before you learn Cramer's rule, we need to introduce a new number associated with a matrix, called a *determinant*.

Determinants

The determinant of a square matrix is a real number corresponding to the matrix. For a 2×2 matrix the determinant is defined as follows.

Determinant of a 2 × 2 Matrix

The **determinant** of the matrix $\begin{bmatrix} a & b \\ c & d \end{bmatrix}$ is defined to be the real number $ad - bc$. We write $\begin{vmatrix} a & b \\ c & d \end{vmatrix} = ad - bc$.

Note that the symbol for determinant is a pair of vertical lines similar to the absolute value symbol, while a matrix is enclosed in brackets.

EXAMPLE 1   **Using the definition of determinant**

Find the determinant of each matrix.

a) $\begin{bmatrix} 1 & 3 \\ -2 & 5 \end{bmatrix}$

b) $\begin{bmatrix} 2 & 4 \\ 6 & 12 \end{bmatrix}$

Solution

a) $\begin{vmatrix} 1 & 3 \\ -2 & 5 \end{vmatrix} = 1 \cdot 5 - 3(-2)$

$= 5 + 6$

$= 11$

b) $\begin{vmatrix} 2 & 4 \\ 6 & 12 \end{vmatrix} = 2 \cdot 12 - 4 \cdot 6$

$= 24 - 24$

$= 0$ ∎

Cramer's Rule

To understand Cramer's rule, we first solve a general system of two linear equations in two variables. Consider the system

$$(1) \qquad a_1x + b_1y = c_1$$
$$(2) \qquad a_2x + b_2y = c_2$$

where a_1, b_1, c_1, a_2, b_2, and c_2 represent real numbers. To eliminate y, we multiply Eq. (1) by b_2 and Eq. (2) by $-b_1$:

$$a_1b_2x + b_1b_2y = c_1b_2 \qquad \text{Eq. (1) multiplied by } b_2$$
$$\underline{-a_2b_1x - b_1b_2y = -c_2b_1} \qquad \text{Eq. (2) multiplied by } -b_1$$
$$a_1b_2x - a_2b_1x \qquad\quad = c_1b_2 - c_2b_1 \qquad \text{Add.}$$

$$(a_1b_2 - a_2b_1)x = c_1b_2 - c_2b_1$$

$$x = \frac{c_1b_2 - c_2b_1}{a_1b_2 - a_2b_1} \qquad \text{Provided that } a_1b_2 - a_2b_1 \neq 0$$

Using similar steps to eliminate x from the system, we get

$$y = \frac{a_1c_2 - a_2c_1}{a_1b_2 - a_2b_1},$$

provided that $a_1b_2 - a_2b_1 \neq 0$. These formulas for x and y can be written by using determinants. In the determinant form, they are known as **Cramer's rule.**

Cramer's Rule

The solution to the system

$$a_1x + b_1y = c_1$$
$$a_2x + b_2y = c_2$$

is given by $x = \dfrac{D_x}{D}$ and $y = \dfrac{D_y}{D}$, where

$$D = \begin{vmatrix} a_1 & b_1 \\ a_2 & b_2 \end{vmatrix}, \qquad D_x = \begin{vmatrix} c_1 & b_1 \\ c_2 & b_2 \end{vmatrix}, \qquad \text{and} \qquad D_y = \begin{vmatrix} a_1 & c_1 \\ a_2 & c_2 \end{vmatrix},$$

provided that $D \neq 0$.

Note that D is the determinant made up of the original coefficients of x and y. D is used in the denominator for both x and y. D_x is obtained by replacing the first (or x) column of D by the constants c_1 and c_2. D_y is found by replacing the second (or y) column of D by the constants c_1 and c_2.

EXAMPLE 2   **An independent system**

Use Cramer's rule to solve the system:

$$3x - 2y = 4$$
$$2x + y = -3$$

Solution

First find the determinants D, D_x, and D_y:

$$D = \begin{vmatrix} 3 & -2 \\ 2 & 1 \end{vmatrix} = 3 - (-4) = 7$$

$$D_x = \begin{vmatrix} 4 & -2 \\ -3 & 1 \end{vmatrix} = 4 - 6 = -2, \qquad D_y = \begin{vmatrix} 3 & 4 \\ 2 & -3 \end{vmatrix} = -9 - 8 = -17$$

By Cramer's rule, we have

$$x = \frac{D_x}{D} = -\frac{2}{7} \qquad \text{and} \qquad y = \frac{D_y}{D} = -\frac{17}{7}.$$

Check in the original equations. The solution set is $\left\{ \left(-\frac{2}{7}, -\frac{17}{7} \right) \right\}$. ■

CAUTION   Cramer's rule works *only* when the determinant D is *not* equal to zero. Cramer's rule solves only those systems that have a single point in their solution set. If $D = 0$, we use elimination to determine whether the solution set is empty or contains all points of a line. ⊘

EXAMPLE 3   **An inconsistent system**

Solve the system:

$$2x + 3y = 9$$
$$-2x - 3y = 5$$

Solution

Cramer's rule does not work because

$$D = \begin{vmatrix} 2 & 3 \\ -2 & -3 \end{vmatrix} = -6 - (-6) = 0.$$

Because Cramer's rule fails to solve the system, we apply the addition method:

$$\begin{array}{r} 2x + 3y = 9 \\ -2x - 3y = 5 \\ \hline 0 = 14 \end{array}$$

Because this last statement is false, the solution set is empty. The original equations are inconsistent. ■

EXAMPLE 4   **A dependent system**

Solve the system:

$$(1) \qquad 3x - 5y = 7$$
$$(2) \qquad 6x - 10y = 14$$

Solution

Cramer's rule does not apply because

$$D = \begin{vmatrix} 3 & -5 \\ 6 & -10 \end{vmatrix} = -30 - (-30) = 0.$$

Multiply Eq. (1) by -2 and add it to Eq. (2):

$$\begin{array}{ll} -6x + 10y = -14 & \text{Eq. (1) times } -2 \\ \underline{6x - 10y = 14} & \text{Eq. (2)} \\ 0 = 0 & \text{Add.} \end{array}$$

Because the last statement is an identity, the equations are dependent. The solution set is $\{(x, y) \,|\, 3x - 5y = 7\}$. ◼

To apply Cramer's rule, the equations must be in standard form.

EXAMPLE 5   **Standard form for Cramer's rule**

Use Cramer's rule to solve the system:

$$\begin{array}{ll} (1) & 2x - 3(y + 1) = -3 \\ (2) & 2y = 3x - 5 \end{array}$$

Solution

First write Eq. (1) in standard form, $Ax + By = C$:

$$2x - 3(y + 1) = -3$$
$$2x - 3y - 3 = -3$$
$$2x - 3y = 0$$

In standard form, Eq. (2) is $-3x + 2y = -5$. Now solve the system:

$$2x - 3y = 0$$
$$-3x + 2y = -5$$

Find D, D_x, and D_y:

$$D = \begin{vmatrix} 2 & -3 \\ -3 & 2 \end{vmatrix} = 4 - 9 = -5$$

$$D_x = \begin{vmatrix} 0 & -3 \\ -5 & 2 \end{vmatrix} = 0 - 15 = -15, \qquad D_y = \begin{vmatrix} 2 & 0 \\ -3 & -5 \end{vmatrix} = -10 - 0 = -10$$

Using Cramer's rule, we get

$$x = \frac{D_x}{D} = \frac{-15}{-5} = 3 \qquad \text{and} \qquad y = \frac{D_y}{D} = \frac{-10}{-5} = 2.$$

Because $(3, 2)$ satisfies both of the original equations, the solution set is $\{(3, 2)\}$. ◼

Calculator Close-up

Most graphing calculators can work with matrices and determinants. A matrix is entered into the calculator one element at a time. The elements of a matrix are identified by their row and column. Consider the following 2 × 2 matrix:

$$A = \begin{bmatrix} 2 & 3 \\ -4 & 5 \end{bmatrix}$$

```
[A] 2x2
1,1=2
1,2=3
2,1=-4
2,2=5
```

Figure 10.8

The 1-1 element is 2, the 1-2 element is 3, the 2-1 element is −4, and the 2-2 element is 5. Figure 10.8 shows how the display looks when matrix A is entered into a graphing calculator. See your calculator manual for details. Once the matrix is entered into the calculator, it is a simple matter to find the determinant:

det [A] ENTER

```
                    22
```

Warm-ups

True or false? Explain your answer.

1. $\begin{vmatrix} -1 & 2 \\ 3 & -5 \end{vmatrix} = -1$

2. $\begin{vmatrix} 2 & 4 \\ -4 & 8 \end{vmatrix} = 0$

3. Cramer's rule solves any system of two linear equations in two variables.

4. The determinant of a 2 × 2 matrix is a real number.

5. If $D = 0$, then there might be no solution to the system.

6. Cramer's rule is used to solve systems of linear equations only.

7. If the graphs of a pair of linear equations intersect at exactly one point, then this point can be found by using Cramer's rule.

8. If x and y represent the digits of a two-digit number, then the value of the number is either $10x + y$ or $10y + x$.

9. If x represents the length of a side of a square and y represents the length of a side of a separate equilateral triangle, then $4x + y$ represents the total of their perimeters.

10. If a and b are the digits of a two-digit number whose value is $10a + b$, then $b + 10a$ is the value of the number when the digits are reversed.

10.6 EXERCISES

Find the value of each determinant. See Example 1.

1. $\begin{vmatrix} 2 & 5 \\ 3 & 7 \end{vmatrix}$

2. $\begin{vmatrix} -1 & 0 \\ 1 & 1 \end{vmatrix}$

3. $\begin{vmatrix} 0 & 3 \\ 1 & 5 \end{vmatrix}$

4. $\begin{vmatrix} 2 & 4 \\ 6 & 12 \end{vmatrix}$

5. $\begin{vmatrix} -3 & -2 \\ -4 & 2 \end{vmatrix}$

6. $\begin{vmatrix} -2 & 2 \\ -3 & -5 \end{vmatrix}$

7. $\begin{vmatrix} 0.05 & 0.06 \\ 10 & 20 \end{vmatrix}$

8. $\begin{vmatrix} 0.02 & -0.5 \\ 30 & 50 \end{vmatrix}$

Solve each system using Cramer's rule. See Example 2.

9. $2x - y = 5$
   $3x + 2y = -3$

10. $3x + y = -1$
    $x + 2y = 8$

11. $3x - 5y = -2$
    $2x + 3y = 5$

12. $x - y = 1$
    $3x - 2y = 0$

13. $4x - 3y = 5$
    $2x + 5y = 7$

14. $2x - y = 2$
    $3x - 2y = 1$

15. $0.5x + 0.2y = 8$
$0.4x - 0.6y = -5$

16. $0.6x + 0.5y = 18$
$0.5x - 0.25y = 7$

17. $\frac{1}{2}x + \frac{1}{4}y = 5$
$\frac{1}{3}x - \frac{1}{2}y = -1$

18. $\frac{1}{2}x + \frac{2}{3}y = 4$
$\frac{3}{4}x + \frac{1}{3}y = -2$

Solve each system. Use Cramer's rule when possible. See Examples 3 and 4.

19. $2x - 3y = 5$
$4x - 6y = 8$

20. $-x + 3y = 6$
$3x - 9y = -18$

21. $x - y = 4$
$x + 2y = 6$

22. $2x - y = 7$
$3x + 2y = -7$

23. $4x - y = 6$
$-8x + 2y = -12$

24. $-x + 2y = 3$
$3x - 6y = 10$

Solve each system using Cramer's rule. See Example 5.

25. $y = 3x - 12$
$3(x + 1) - 11 = y + 4$

26. $y = x - 7$
$y = -2x + 5$

27. $x - 6 = y + 1$
$y = -3x + 1$

28. $1 - x = 6 + y$
$5 - 2x = 6 - y$

29. $y = 0.05x$
$x + y = 504$

30. $0.05x + 0.10y = 6$
$y = 98 - x$

31. $x\sqrt{2} = y\sqrt{2} - 2$
$x + y = \sqrt{18}$

32. $x\sqrt{2} + y\sqrt{3} = 5$
$y = x\sqrt{6} - \sqrt{3}$

Solve each problem by using two equations in two variables and Cramer's rule.

33. *Peas and beets.* One serving of canned peas contains 3 grams of protein and 11 grams of carbohydrates. One serving of canned beets contains 1 gram of protein and 8 grams of carbohydrates. How many servings of each would provide exactly 38 grams of protein and 187 grams of carbohydrates?

34. *Protein and carbohydrates.* One serving of Cornies breakfast cereal contains 2 grams of protein and 25 grams of carbohydrates. One serving of Oaties breakfast cereal contains 4 grams of protein and 20 grams of carbohydrates. How many servings of each would provide exactly 24 grams of protein and 210 grams of carbohydrates?

35. *Two-digit number.* The sum of the digits of a two-digit number is 10. If the digits are reversed, the new number is 1 less than twice the original number. Find the original number.

36. *Another two-digit number.* The units digit of a two-digit number is twice the tens digit. If the digits are reversed, the new number is 36 more than the original number. Find the original number.

37. *Milk and a magazine.* Althia bought a gallon of milk and a magazine for a total of $4.65, excluding tax.

Including the tax, the bill was $4.95. If there is a 5% sales tax on milk and an 8% sales tax on magazines, then what was the price of each item?

38. *Washing machines and refrigerators.* A truck carrying 3600 cubic feet of cargo consisting of washing machines and refrigerators was hijacked. The washing machines are worth $300 each and are shipped in 36-cubic foot cartons. The refrigerators are worth $900 each and are shipped in 45-cubic-foot cartons. If the total value of the cargo was $51,000, then how many of each were there on the truck?

39. *Singles and doubles.* Windy's Hamburger Palace sells singles and doubles. Toward the end of the evening, Windy himself noticed that he had on hand only 32 patties and 34 slices of tomatoes. If a single takes l patty and 2 slices, and a double takes 2 patties and 1 slice, then how many more singles and doubles must Windy sell to use up all of his patties and tomato slices?

40. *Valuable wrenches.* Carmen has a total of 28 wrenches, all of which are either box wrenches or open-end wrenches. For insurance purposes she values the box wrenches at $3.00 each and the open-end wrenches at $2.50 each. If the value of her wrench collection is $78, then how many of each type does she have?

41. *Gary and Harry.* Gary is 5 years older than Harry. Twenty-nine years ago, Gary was twice as old as Harry. How old are they now?

42. *Acute angles.* One acute angle of a right triangle is 3° more than twice the other acute angle. What are the sizes of the acute angles?

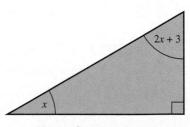

Figure for Exercise 42

43. *Equal perimeters.* A rope of length 80 feet is to be cut into two pieces. One piece will be used to form a square, and the other will be used to form an equilateral triangle. If the figures are to have equal perimeters, then what should be the length of a side of each?

44. *Coffee and doughnuts.* For a cup of coffee and a doughnut, Thurrel spent $2.25, including a tip. Later, he spent $4.00 for 2 coffees and 3 doughnuts, including a tip. If he always tips $1.00, then what is the price of a cup of coffee?

45. *Chlorine mixture.* A 10% chlorine solution is to be mixed with a 25% chlorine solution to obtain 30 gallons of 20% solution. How many gallons of each must be used?

46. *Safe drivers.* Emily and Camille started from the same city and drove in opposite directions on the freeway. After 3 hours they were 354 miles apart. If they had gone in the same direction, they would have been only 18 miles apart. How fast did each woman drive?

Getting More Involved

47. *Writing.* Explain what to do when you are trying to use Cramer's rule and $D = 0$.

48. *Exploration.* For what value of a does the system

$$ax - y = 3$$
$$x + 2y = 1$$

have a single solution?

49. *Exploration.* Can Cramer's rule be used to solve the following system? Explain.

$$2x^2 - y = 3$$
$$3x^2 + 2y = 22$$

Graphing Calculator Exercises

50. Use the determinant feature on your graphing calculator to find the determinants in Exercises 1–8 of this section.

51. Solve the systems in Exercises 9–18 of this section by using your graphing calculator to find the necessary determinants.

10.7 Cramer's Rule for Systems in Three Variables

In this section:

▶ Minors
▶ Evaluating a 3 × 3 Determinant
▶ Cramer's Rule

The solution of linear systems involving three variables using determinants is very similar to the solution of linear systems in two variables using determinants. However, you first need to learn to find the determinant of a 3 × 3 matrix.

Minors

To each element of a 3 × 3 matrix there corresponds a 2 × 2 matrix called the **minor** of that element. The minor of an element of a 3 × 3 matrix is the 2 × 2 matrix that remains after deleting the row and column of that element.

EXAMPLE 1   **Finding minors**

Find the minors for the elements 2, 3, and −6 of the 3 × 3 matrix

$$\begin{bmatrix} 2 & -1 & -8 \\ 0 & -2 & 3 \\ 4 & -6 & 7 \end{bmatrix}.$$

Solution

To find the minor for 2, delete the first row and first column of the matrix:

$$\begin{bmatrix} 2 & -1 & -8 \\ 0 & -2 & 3 \\ 4 & -6 & 7 \end{bmatrix} \qquad \text{The minor for 2 is} \quad \begin{bmatrix} -2 & 3 \\ -6 & 7 \end{bmatrix}.$$

To find the minor for 3, delete the second row and third column of the matrix:

$$\begin{bmatrix} 2 & -1 & -8 \\ 0 & 2 & 3 \\ 4 & -6 & 7 \end{bmatrix}$$ The minor for 3 is $\begin{bmatrix} 2 & -1 \\ 4 & -6 \end{bmatrix}$.

To find the minor for -6, delete the third row and second column of the matrix:

$$\begin{bmatrix} 2 & -1 & -8 \\ 0 & -2 & 3 \\ 4 & -6 & 7 \end{bmatrix}$$ The minor for -6 is $\begin{bmatrix} 2 & -8 \\ 0 & 3 \end{bmatrix}$.   ∎

Evaluating a 3 × 3 Determinant

The determinant of a 3×3 matrix is defined in terms of the determinants of minors.

Determinant of a 3 × 3 Matrix

The determinant of a 3×3 matrix is defined as follows:

$$\begin{vmatrix} a_1 & b_1 & c_1 \\ a_2 & b_2 & c_2 \\ a_3 & b_3 & c_3 \end{vmatrix} = a_1 \cdot \begin{vmatrix} b_2 & c_2 \\ b_3 & c_3 \end{vmatrix} - a_2 \cdot \begin{vmatrix} b_1 & c_1 \\ b_3 & c_3 \end{vmatrix} + a_3 \cdot \begin{vmatrix} b_1 & c_1 \\ b_2 & c_2 \end{vmatrix}$$

Note that the determinants following a_1, a_2, and a_3 are determinants of the minors for a_1, a_2, and a_3, respectively. Writing the determinant of a 3×3 matrix in terms of determinants of minors is called **expansion by minors.** In the definition we expanded by minors about the first column. Later we will see how to expand by minors using any row or column and get the same value for the determinant.

EXAMPLE 2   **Determinant of a 3 × 3 matrix**

Find the determinant of the matrix by expansion by minors about the first column.

$$\begin{bmatrix} 1 & 3 & -5 \\ -2 & 4 & 6 \\ 0 & -7 & 9 \end{bmatrix}$$

Solution

$$\begin{vmatrix} 1 & 3 & -5 \\ -2 & 4 & 6 \\ 0 & -7 & 9 \end{vmatrix} = 1 \cdot \begin{vmatrix} 4 & 6 \\ -7 & 9 \end{vmatrix} - (-2) \cdot \begin{vmatrix} 3 & -5 \\ -7 & 9 \end{vmatrix} + 0 \cdot \begin{vmatrix} 3 & -5 \\ 4 & 6 \end{vmatrix}$$

$$= 1 \cdot [36 - (-42)] + 2 \cdot (27 - 35) + 0 \cdot [18 - (-20)]$$

$$= 1 \cdot 78 + 2 \cdot (-8) + 0$$

$$= 78 - 16$$

$$= 62$$ ∎

In the next example we evaluate a determinant using expansion by minors about the second row. In expanding about any row or column, the signs of the coefficients of the minors alternate according to the **sign array** that follows:

$$\begin{bmatrix} + & - & + \\ - & + & - \\ + & - & + \end{bmatrix}$$

The sign array is easily remembered by observing that there is a "+" sign in the upper left position and then alternating signs for all of the remaining positions.

EXAMPLE 3   **Determinant of a 3 × 3 matrix**

Evaluate the determinant of the matrix by expanding by minors about the second row.

$$\begin{bmatrix} 1 & 3 & -5 \\ -2 & 4 & 6 \\ 0 & -7 & 9 \end{bmatrix}$$

Solution

For expansion using the second row, we prefix the signs "$- + -$" from the second row of the sign array to the corresponding numbers in the second row of the matrix, $-2, 4$, and 6. Note that the signs from the sign array are used in addition to any signs that occur on the numbers in the second row.

From the sign array, second row

$$\begin{vmatrix} 1 & 3 & -5 \\ -2 & 4 & 6 \\ 0 & -7 & 9 \end{vmatrix} = -(-2) \cdot \begin{vmatrix} 3 & -5 \\ -7 & 9 \end{vmatrix} + 4 \cdot \begin{vmatrix} 1 & -5 \\ 0 & 9 \end{vmatrix} - 6 \cdot \begin{vmatrix} 1 & 3 \\ 0 & -7 \end{vmatrix}$$

$$= 2(27 - 35) + 4(9 - 0) - 6(-7 - 0)$$

$$= 2(-8) + 4(9) - 6(-7)$$

$$= -16 + 36 + 42$$

$$= 62$$

Note that 62 is the same value that was obtained for this determinant in Example 2.   ∎

It can be shown that expanding by minors using any row or column prefixed by the corresponding signs from the sign array yields the same value for the determinant. Because we can use any row or column to evaluate a determinant of a 3 × 3 matrix, we can choose a row or column that makes the work easier. We can shorten the work considerably by picking a row or column with zeros in it.

EXAMPLE 4   **Choosing the simplest row or column**

Find the determinant of the matrix

$$\begin{bmatrix} 3 & -5 & 0 \\ 4 & -6 & 0 \\ 7 & 9 & 2 \end{bmatrix}$$

Solution

We choose to expand by minors about the third column of the matrix because the third column contains two zeros. Prefix the third-column entries 0, 0, 2 by the signs "+ − +" from the third column of the sign array:

$$\begin{vmatrix} 3 & -5 & 0 \\ 4 & -6 & 0 \\ 7 & 9 & 2 \end{vmatrix} = 0 \cdot \begin{vmatrix} 4 & -6 \\ 7 & 9 \end{vmatrix} - 0 \cdot \begin{vmatrix} 3 & -5 \\ 7 & 9 \end{vmatrix} + 2 \cdot \begin{vmatrix} 3 & -5 \\ 4 & -6 \end{vmatrix}$$

$$= 0 - 0 + 2[-18 - (-20)]$$

$$= 4$$

∎

Cramer's Rule

A system of three linear equations in three variables can be solved by using determinants in a manner similar to that of the previous section. This rule is also called Cramer's rule.

Cramer's Rule for Three Equations in Three Unknowns

The solution to the system

$$a_1 x + b_1 y + c_1 z = d_1$$
$$a_2 x + b_2 y + c_2 z = d_2$$
$$a_3 x + b_3 y + c_3 z = d_3$$

is given by $x = \dfrac{D_x}{D}$, $y = \dfrac{D_y}{D}$, and $z = \dfrac{D_z}{D}$, where

$$D = \begin{vmatrix} a_1 & b_1 & c_1 \\ a_2 & b_2 & c_2 \\ a_3 & b_3 & c_3 \end{vmatrix}, \qquad D_x = \begin{vmatrix} d_1 & b_1 & c_1 \\ d_2 & b_2 & c_2 \\ d_3 & b_3 & c_3 \end{vmatrix},$$

$$D_y = \begin{vmatrix} a_1 & d_1 & c_1 \\ a_2 & d_2 & c_2 \\ a_3 & d_3 & c_3 \end{vmatrix}, \qquad D_z = \begin{vmatrix} a_1 & b_1 & d_1 \\ a_2 & b_2 & d_2 \\ a_3 & b_3 & d_3 \end{vmatrix},$$

provided that $D \neq 0$.

Note that D_x, D_y, and D_z are obtained from D by replacing the x-, y-, or z-column with the constants d_1, d_2, and d_3.

EXAMPLE 5   **Solving an independent system**

Use Cramer's rule to solve the system:

$$x + y + z = 4$$
$$x - y = -3$$
$$x + 2y - z = 0$$

Solution

We first calculate D, D_x, D_y, and D_z. To calculate D, expand by minors about the third column, because the third column has a zero in it:

$$D = \begin{vmatrix} 1 & 1 & 1 \\ 1 & -1 & 0 \\ 1 & 2 & -1 \end{vmatrix} = 1 \cdot \begin{vmatrix} 1 & -1 \\ 1 & 2 \end{vmatrix} - 0 \cdot \begin{vmatrix} 1 & 1 \\ 1 & 2 \end{vmatrix} + (-1) \cdot \begin{vmatrix} 1 & 1 \\ 1 & -1 \end{vmatrix}$$

$$= 1 \cdot [2 - (-1)] - 0 + (-1)[-1 - 1]$$
$$= 3 - 0 + 2$$
$$= 5$$

For D_x, expand by minors about the first column:

$$D_x = \begin{vmatrix} 4 & 1 & 1 \\ -3 & -1 & 0 \\ 0 & 2 & -1 \end{vmatrix} = 4 \cdot \begin{vmatrix} -1 & 0 \\ 2 & -1 \end{vmatrix} - (-3) \cdot \begin{vmatrix} 1 & 1 \\ 2 & -1 \end{vmatrix} + 0 \cdot \begin{vmatrix} 1 & 1 \\ -1 & 0 \end{vmatrix}$$

$$= 4 \cdot (1 - 0) + 3 \cdot (-1 - 2) + 0$$
$$= 4 - 9 + 0$$
$$= -5$$

For D_y, expand by minors about the third row:

$$D_y = \begin{vmatrix} 1 & 4 & 1 \\ 1 & -3 & 0 \\ 1 & 0 & -1 \end{vmatrix} = 1 \cdot \begin{vmatrix} 4 & 1 \\ -3 & 0 \end{vmatrix} - 0 \cdot \begin{vmatrix} 1 & 1 \\ 1 & 0 \end{vmatrix} + (-1) \cdot \begin{vmatrix} 1 & 4 \\ 1 & -3 \end{vmatrix}$$

$$= 1 \cdot 3 - 0 + (-1)(-7)$$
$$= 10$$

To get D_z, expand by minors about the third row:

$$D_z = \begin{vmatrix} 1 & 1 & 4 \\ 1 & -1 & -3 \\ 1 & 2 & 0 \end{vmatrix} = 1 \cdot \begin{vmatrix} 1 & 4 \\ -1 & -3 \end{vmatrix} - 2 \cdot \begin{vmatrix} 1 & 4 \\ 1 & -3 \end{vmatrix} + 0 \cdot \begin{vmatrix} 1 & 1 \\ 1 & -1 \end{vmatrix}$$

$$= 1 \cdot 1 - 2(-7) + 0$$
$$= 15$$

Now, by Cramer's rule,

$$x = \frac{D_x}{D} = \frac{-5}{5} = -1, \qquad y = \frac{D_y}{D} = \frac{10}{5} = 2, \qquad \text{and} \qquad z = \frac{D_z}{D} = \frac{15}{5} = 3.$$

Check $(-1, 2, 3)$ in the original equations. The solution set is $\{(-1, 2, 3)\}$. ∎

If $D = 0$, Cramer's rule does not apply. Cramer's rule provides the solution only to a system of three equations with three variables that has a single point in the solution set. If $D = 0$, then the solution set either is empty or consists of infinitely many points, and we use elimination of variables to find the solution.

EXAMPLE 6   **Solving a dependent system**

Solve the system:

$$
\begin{aligned}
(1) \quad & x + y - z = 2 \\
(2) \quad & 2x + 2y - 2z = 4 \\
(3) \quad & -3x - 3y + 3z = -6
\end{aligned}
$$

Solution

Calculate D by expanding about the first column:

$$
D = \begin{vmatrix} 1 & 1 & -1 \\ 2 & 2 & -2 \\ -3 & -3 & 3 \end{vmatrix} = 1 \cdot \begin{vmatrix} 2 & -2 \\ -3 & 3 \end{vmatrix} - 2 \cdot \begin{vmatrix} 1 & -1 \\ -3 & 3 \end{vmatrix} + (-3) \cdot \begin{vmatrix} 1 & -1 \\ 2 & -2 \end{vmatrix}
$$

$$
= 1 \cdot 0 - 2 \cdot 0 + (-3) \cdot 0
$$

$$
= 0
$$

Because $D = 0$, Cramer's rule does not apply to this system. If we multiply Eq. (1) by 2, we get Eq. (2). If we multiply Eq. (1) by -3, we get Eq. (3). Thus all three equations are equivalent, and they are dependent. The solution set to the system is $\{(x, y, z) \mid x + y - z = 2\}$.   ■

Calculator Close-up

Most graphing calculators can find determinants of matrices as large as 6×6. Enter the 3×3 matrix

$$
A = \begin{bmatrix} 3 & -2 & 1 \\ 1 & -1 & 5 \\ 4 & 2 & 7 \end{bmatrix}
$$

into your graphing calculator. The elements of a matrix are located by their row and column. For example, the 2-3 element in matrix A is 5. See your calculator manual for details. To find the determinant of A, try the following:

det [A] ENTER

$^{-}71$

Warm-ups

True or false? Explain your answer.

1. A minor is a 2×2 matrix.

2. The minor for an element is found by deleting the element from the matrix.

3. The determinant of a 3×3 matrix is found by using determinants of minors.

4. Expansion by minors converts a 3×3 matrix into a 4×4 matrix.

5. Using Cramer's rule, we use $\dfrac{D}{D_x}$ to get the value of x.

6. Expansion by minors about any row or any column gives the same value for the determinant of a 3×3 matrix.

7. The sign array is used in evaluating the determinant of a 3×3 matrix.

8. It is easier to find the determinant of a 3×3 matrix with several zero elements than one with no zero elements.

9. If $D = 0$, then x, y, and z are all zero.

10. Cramer's rule solves nonlinear systems of three equations in three unknowns.

10.7 EXERCISES

Find the indicated minors using the following matrix See Example 1.

$$\begin{bmatrix} 3 & -2 & 5 \\ 4 & -3 & 7 \\ 0 & 1 & -6 \end{bmatrix}$$

1. Minor for 3
2. Minor for -2
3. Minor for 5
4. Minor for -3
5. Minor for 7
6. Minor for 0
7. Minor for 1
8. Minor for -6

Find the determinant of each 3×3 matrix by using expansion by minors about the first column. See Example 2.

9. $\begin{bmatrix} 1 & 1 & 2 \\ 2 & 3 & 1 \\ 3 & 1 & 5 \end{bmatrix}$

10. $\begin{bmatrix} 2 & 1 & 3 \\ 1 & 1 & 2 \\ 3 & 4 & 6 \end{bmatrix}$

11. $\begin{bmatrix} 2 & 1 & 0 \\ 1 & 0 & 1 \\ 3 & 1 & 2 \end{bmatrix}$

12. $\begin{bmatrix} 1 & 0 & 2 \\ 2 & 1 & 3 \\ 4 & 3 & 0 \end{bmatrix}$

13. $\begin{bmatrix} -2 & 1 & 2 \\ -3 & 3 & 1 \\ -5 & 4 & 0 \end{bmatrix}$

14. $\begin{bmatrix} -2 & 1 & 3 \\ -1 & 4 & 2 \\ 2 & 1 & 1 \end{bmatrix}$

15. $\begin{bmatrix} 1 & 1 & 5 \\ 0 & 3 & 2 \\ 0 & 2 & 3 \end{bmatrix}$

16. $\begin{bmatrix} 1 & 0 & 6 \\ 0 & 1 & 4 \\ 0 & 0 & 9 \end{bmatrix}$

Evaluate the determinant of each 3×3 matrix using expansion by minors about the row or column of your choice. See Examples 3 and 4.

17. $\begin{bmatrix} 3 & 1 & 5 \\ 2 & 0 & 6 \\ 4 & 0 & 1 \end{bmatrix}$

18. $\begin{bmatrix} 2 & 1 & 2 \\ 1 & 2 & 5 \\ 3 & 0 & 0 \end{bmatrix}$

19. $\begin{bmatrix} -2 & 1 & 3 \\ 0 & 1 & -1 \\ 2 & -4 & -3 \end{bmatrix}$

20. $\begin{bmatrix} -2 & 0 & 1 \\ -3 & 2 & -5 \\ 4 & -2 & 6 \end{bmatrix}$

21. $\begin{bmatrix} -2 & -3 & 0 \\ 4 & -1 & 0 \\ 0 & 3 & 5 \end{bmatrix}$

22. $\begin{bmatrix} -2 & 6 & 3 \\ 0 & 4 & 0 \\ -1 & -4 & 5 \end{bmatrix}$

23. $\begin{bmatrix} 2 & 1 & 1 \\ 0 & 0 & 5 \\ 5 & 0 & 4 \end{bmatrix}$

24. $\begin{bmatrix} 2 & 3 & 0 \\ 6 & 4 & 1 \\ 1 & 2 & 0 \end{bmatrix}$

Use Cramer's rule to solve each system. See Example 5.

25. $x + y + z = 6$
$x - y + z = 2$
$2x + y + z = 7$

26. $x + y + z = 2$
$x - y - 2z = -3$
$2x - y + z = 7$

27. $x - 3y + 2z = 0$
$x + y + z = 2$
$x - y + z = 0$

28. $3x + 2y + 2z = 0$
$x - y + z = 1$
$x + y - z = 3$

29. $x + y = -1$
$2y - z = 3$
$x + y + z = 0$

30. $x - y = 8$
$x - 2z = 0$
$x + y - z = 1$

31. $x + y - z = 0$
$2x + 2y + z = 6$
$x - 3y = 0$

32. $x + y + z = 1$
$5x - y = 0$
$3x + y + 2z = 0$

33. $x + y + z = 0$
$2y + 2z = 0$
$3x - y = -1$

34. $x + z = 0$
$x - 3y = 1$
$4y - 3z = 3$

Solve each system. Use Cramer's rule if possible. See Example 6.

35. $2x - y + z = 1$
$-6x + 3y - 3z = -3$
$4x - 2y + 2z = 2$

36. $x - y + z = 4$
$2x - 2y + 2z = 3$
$4x - y - z = 1$

37. $x + y = 1$
$y + 2z = 3$
$x + 2y + 2z = 5$

38. $x - y + z = 5$
$2x - 2y + 2z = 10$
$3x - 3y + 3z = 15$

39. $x + y = 4$
$y + z = -3$
$x + z = -5$

40. $x + y = 0$
$x - z = -1$
$y + z = 3$

Write a system of three equations in three variables for each word problem. Use Cramer's rule to solve each system.

41. *Weighing dogs.* Cassandra wants to determine the weights of her two dogs, Mimi and Mitzi. However, neither dog will sit on the scale by herself. Cassandra, Mimi, and Mitzi all together weigh 175 pounds. Cassandra and Mimi together weigh 143 pounds. Cassandra and Mitzi together weigh 139 pounds. How much does each weigh individually?

42. *Nickels, dimes, and quarters.* Bernard has 41 coins consisting of nickels, dimes, and quarters, and they are worth a total of $4.00. If the number of dimes plus the number of quarters is one more than the number of nickels, then how many of each does he have?

43. *Finding three angles.* If the two acute angles of a right triangle differ by 12°, then what are the measures of the three angles of this triangle?

44. *Two acute and one obtuse.* The obtuse angle of a triangle is twice as large as the sum of the two acute angles. If the smallest angle is only one-eighth as large as the sum of the other two, then what is the measure of each angle?

Getting More Involved

45. *Writing.* For what values of a, b, c, and d is the determinant of the matrix

$$\begin{bmatrix} a & b & 0 \\ c & d & 0 \\ b & a & 0 \end{bmatrix}$$

equal to zero? Explain your answer.

46. *Exploration.* The determinant of a 4×4 matrix is found by expanding by 3×3 minors and using a 4×4 sign array. Find the determinant of a 4×4 matrix of your choice. Use the definition of the determinant of a 3×3 matrix as a guide.

Graphing Calculator Exercises

47. Use the determinant feature of your graphing calculator to find the determinant of each matrix in Exercises 9–16 of this section.

48. Solve the systems in Exercises 25–34 by using Cramer's rule and your graphing calculator to find the determinants.

49. The solution to an independent system of four linear equations in four variables can be found by using determinants of 4×4 matrices in the same manner as Cramer's rule for three variables. Write Cramer's rule for four variables. Make up a system of four linear equations in four variables and solve it using your new rule and a graphing calculator to evaluate the determinants.

—————————————COLLABORATIVE ACTIVITIES—————————————

Types of Systems

Assign roles in your groups. Set up the system of equations for each scenario below and attempt to solve them. Analyze your results.

Part I

Talia's Tallitot: Talia plans to make tallitot (prayer shawls) and matching kipot (prayer hats) from a bolt of material. The fabric she plans to use is 250 yards long. She will need 2 yards of material for each tallit (prayer shawl) and $\frac{1}{6}$ of a yard for each kipah (prayer hat). She also plans to put tzitzit (fringes) on the tallit and will need 1 yard of cord for each tallit. The kipot will not have any fringes. She has 100 yards of cord to use for the tzitzit.

1. Write two equations to use to solve for how many kipot and tallitot she can make from the supplies on hand. Let t = the number of tallitot made and k = number of kipot made.

2. Try to solve this system of equations. If you get a solution, state what it is. If not, describe what your result means.

Part II

Karif's Kerchiefs: Karif plans to make red and green kerchiefs in two sizes, medium and large. He has 60 yards of each color of material. He will need $\frac{1}{4}$ of a yard of the appropriate colored fabric for each medium kerchief and $\frac{1}{2}$ of a yard for each large kerchief. He also plans to trim each kerchief with the alternate color of ribbon. He has 240 yards of each color of ribbon. He will need 1 yard of ribbon for each medium kerchief and 2 yards for each large kerchief.

Grouping: 2 to 4 students per group
Topic: Types of solutions to systems of equations

3. Let m = the number of medium kerchiefs and l = the number of large kerchiefs. Set up two equations to use to solve for the number of each color of kerchiefs Karif can make from the supplies on hand.

4. Try to solve this system of equations. If you get a solution, state what it is. If not, describe what your result means.

Part III

Maria's Mantillas: Maria plans to make two different kinds of mantillas (veils), some out of lace and some out of a sheer cotton fabric. She also plans to make two sizes of mantillas, one size for women and one size for young girls. She has 150 yards of the lace fabric and 200 yards of the cotton fabric. She will need 2 yards of either fabric for the smaller mantillas and $3\frac{1}{2}$ yards of either fabric for the larger mantillas.

5. Let w = the number of mantillas for the women and y = the number of mantillas for the young girls. Set up two equations to use to solve for the number of each mantilla Maria can make from each type of fabric.

6. Try to solve this system of equations. If you get a solution, state what it is. If not, describe what your result means.

Extension: Explain what you would need to do for the systems of equations you could not solve to make them solvable.

Wrap-up CHAPTER 10

SUMMARY

Systems of Linear Equations		Examples
Methods for solving systems in two variables	Graphing: Sketch the graphs to see the solution.	The graphs of $y = x - 1$ and $x + y = 3$ intersect at $(2, 1)$.
	Substitution: Solve one equation for one variable in terms of the other, then substitute into the other equation.	Substitution: $x + (x - 1) = 3$
	Addition: Multiply each equation as necessary to eliminate a variable upon addition of the equations.	$\begin{array}{r} -x + y = -1 \\ x + y = 3 \\ \hline 2y = 2 \end{array}$
Types of linear systems in two variables	Independent: One point in solution set — The lines intersect at one point.	$y = x - 5$ $y = 2x + 3$
	Inconsistent: Empty solution set — The lines are parallel.	$2x + y = 1$ $2x + y = 5$
	Dependent: Infinite solution set — The lines are the same.	$2x + 3y = 4$ $4x + 6y = 8$
Linear equation in three variables	$Ax + By + Cz = D$ In a three-dimensional coordinate system the graph is a plane.	$2x - y + 3z = 5$
Linear systems in three variables	Use substitution or addition to eliminate variables in the system. The solution set may be a single point, the empty set, or an infinite set of points.	$\begin{array}{r} x + y - z = 3 \\ 2x - 3y + z = 2 \\ x - y - 4z = 14 \end{array}$
Nonlinear systems in two variables	Use substitution or addition to eliminate variables. Nonlinear systems may have several points in the solution set.	$y = x^2$ $x^2 + y^2 = 4$ Substitution: $y + y^2 = 4$

Matrices and Determinants		Examples
Matrix	A rectangular array of real numbers. An $n \times m$ matrix has n rows and m columns.	$\begin{bmatrix} 1 & -3 \\ 2 & 5 \end{bmatrix}, \begin{bmatrix} 1 & 0 & 1 \\ 2 & 1 & 4 \end{bmatrix}$
Augmented matrix	The matrix of coefficients and constants from a system of linear equations	$x - 3y = -7$ $2x + 5y = 19$ Augmented matrix: $\begin{bmatrix} 1 & -3 & \vline & -7 \\ 2 & 5 & \vline & 19 \end{bmatrix}$

Gauss-Jordan method	Use the row operations to get ones on the diagonal and zeros elsewhere for the coefficients in the augmented matrix.	$\begin{bmatrix} 1 & 0 & \vert & 2 \\ 0 & 1 & \vert & 3 \end{bmatrix}$ $x = 2$ and $y = 3$
Determinant	A real number corresponding to a square matrix	
Determinant of a 2×2 matrix	$\begin{vmatrix} a_1 & b_1 \\ a_2 & b_2 \end{vmatrix} = a_1 b_2 - a_2 b_1$	$\begin{vmatrix} 1 & -3 \\ 2 & 5 \end{vmatrix} = 5 - (-6)$ $= 11$
Determinant of a 3×3 matrix	Expand by minors about any row or column, using signs from the sign array. $\begin{vmatrix} a_1 & b_1 & c_1 \\ a_2 & b_2 & c_2 \\ a_3 & b_3 & c_3 \end{vmatrix} = a_1 \cdot \begin{vmatrix} b_2 & c_2 \\ b_3 & c_3 \end{vmatrix} - a_2 \cdot \begin{vmatrix} b_1 & c_1 \\ b_3 & c_3 \end{vmatrix} + a_3 \cdot \begin{vmatrix} b_1 & c_1 \\ b_2 & c_2 \end{vmatrix}$	Sign array: $\begin{bmatrix} + & - & + \\ - & + & - \\ + & - & + \end{bmatrix}$

Cramer's Rules

Two linear equations in two variables	The solution to the system $$a_1 x + b_1 y = c_1$$ $$a_2 x + b_2 y = c_2$$ is given by $x = \dfrac{D_x}{D}$ and $y = \dfrac{D_y}{D}$, where $D = \begin{vmatrix} a_1 & b_1 \\ a_2 & b_2 \end{vmatrix}$, $\qquad D_x = \begin{vmatrix} c_1 & b_1 \\ c_2 & b_2 \end{vmatrix}$, and $\qquad D_y = \begin{vmatrix} a_1 & c_1 \\ a_2 & c_2 \end{vmatrix}$ provided that $D \neq 0$.
Three linear equations in three variables	The solution to the system $$a_1 x + b_1 y + c_1 z = d_1$$ $$a_2 x + b_2 y + c_2 z = d_2$$ $$a_3 x + b_3 y + c_3 z = d_3$$ is given by $x = \dfrac{D_x}{D}$, $y = \dfrac{D_y}{D}$, and $z = \dfrac{D_z}{D}$, where $D = \begin{vmatrix} a_1 & b_1 & c_1 \\ a_2 & b_2 & c_2 \\ a_3 & b_3 & c_3 \end{vmatrix}$, $\qquad D_x = \begin{vmatrix} d_1 & b_1 & c_1 \\ d_2 & b_2 & c_2 \\ d_3 & b_3 & c_3 \end{vmatrix}$, $D_y = \begin{vmatrix} a_1 & d_1 & c_1 \\ a_2 & d_2 & c_2 \\ a_3 & d_3 & c_3 \end{vmatrix}$, $\qquad D_z = \begin{vmatrix} a_1 & b_1 & d_1 \\ a_2 & b_2 & d_2 \\ a_3 & b_3 & d_3 \end{vmatrix}$, provided that $D \neq 0$.

REVIEW EXERCISES

10.1 *Solve by graphing. Indicate whether each system is independent, inconsistent, or dependent.*

1. $y = 2x - 1$
  $x + y = 2$

2. $y = -x$
  $y = -x + 3$

3. $y - 3x \quad 4$
  $y = -2x + 1$

4. $x + y = 5$
  $x - y = -1$

Solve each system by the substitution method. Indicate whether each system is independent, inconsistent, or dependent.

5. $y = 3x + 11$
  $2x + 3y = 0$

6. $x - y = 3$
  $3x - 2y = 3$

7. $x = y + 5$
  $2x - 2y = 12$

8. $2x - y = 3$
  $6x - 9 = 3y$

10.2 *Solve each system by the addition method. Indicate whether each system is independent, inconsistent, or dependent.*

9. $5x - 3y = -20$
  $3x + 2y = 7$

10. $-3x + y = 3$
  $2x - 3y = 5$

11. $2(y - 5) + 4 = 3(x - 6)$
  $3x - 2y = 12$

12. $3x - 4(y - 5) = x + 2$
  $2y - x = 7$

10.3 *Solve each system by elimination of variables.*

13. $2x - y - z = 3$
  $3x + y + 2z = 4$
  $4x + 2y - z = -4$

14. $2x + 3y - 2z = -11$
  $3x - 2y + 3z = 7$
  $x - 4y + 4z = 14$

15. $x - 3y + z = 5$
  $2x - 4y - z = 7$
  $2x - 6y + 2z = 6$

16. $x - y + z = 1$
  $2x - 2y + 2z = 2$
  $-3x + 3y - 3z = -3$

10.4 *Graph both equations on the same set of axes, then determine the points of intersection of the graphs by solving the system.*

17. $y = x^2$
  $y = -2x + 15$

18. $y = \sqrt{x}$
  $y = \dfrac{1}{3}x$

19. $y = 3x$
  $y = \dfrac{1}{x}$

20. $y = |x|$
  $y = -3x + 5$

Solve each system.

21. $x^2 + y^2 = 4$
  $y = \dfrac{1}{3}x^2$

22. $12y^2 - 4x^2 = 9$
  $x = y^2$

23. $x^2 + y^2 = 34$
  $y = x + 2$

24. $y = 2x + 1$
  $xy - y = 5$

25. $y = \log(x - 3)$
  $y = 1 - \log(x)$

26. $y = \left(\dfrac{1}{2}\right)^x$
  $y = 2^{x-1}$

27. $x^4 = 2(12 - y)$
  $y = x^2$

28. $x^2 - xy + 2y^2 = 7$
  $x^2 + 2y^2 = 9$

10.5 *Solve each system by using the Gauss-Jordan method.*

29. $2x + y = 0$
  $x - 3y = 14$

30. $2x - y = 8$
  $3x + 2y = -2$

31. $x + y - z = 0$
  $x - y + 2z = 4$
  $2x + y - z = 1$

32. $2x - y + 2z = 9$
  $x + 3y = 5$
  $3x + z = 9$

10.6 *Evaluate each determinant.*

33. $\begin{vmatrix} 1 & 3 \\ 0 & 2 \end{vmatrix}$

34. $\begin{vmatrix} -1 & 2 \\ -3 & 5 \end{vmatrix}$

35. $\begin{vmatrix} 0.01 & 0.02 \\ 50 & 80 \end{vmatrix}$

36. $\begin{vmatrix} \dfrac{1}{2} & \dfrac{1}{3} \\ \dfrac{1}{4} & \dfrac{1}{5} \end{vmatrix}$

Solve each system. Use Cramer's rule if possible.

37. $2x - y = 0$
  $3x + y = -5$

38. $3x - 2y = 14$
  $2x + 3y = -8$

39. $y = 2x - 3$
  $3x - 2y = 4$

40. $2x - 5y = -1$
  $10y - 4x = 2$

41. $3x - y = -1$
  $2y - 6x = 5$

42. $y = 2x - 5$
  $y = 3x - 3y$

10.7 *Evaluate each determinant.*

43. $\begin{vmatrix} 2 & 3 & 1 \\ -1 & 2 & 4 \\ 6 & 1 & 1 \end{vmatrix}$

44. $\begin{vmatrix} 1 & -1 & 0 \\ -2 & 0 & 0 \\ 3 & 1 & 5 \end{vmatrix}$

45. $\begin{vmatrix} 2 & 3 & -2 \\ 2 & 0 & 4 \\ -1 & 0 & 3 \end{vmatrix}$

46. $\begin{vmatrix} 3 & -1 & 4 \\ 2 & -1 & 1 \\ -2 & 0 & 1 \end{vmatrix}$

Solve each system. Use Cramer's rule if possible.

47. $x + y = 3$
  $x + y + z = 0$
  $x - y - z = 2$

48. $2x - 4y + 2z = 6$
  $x - y - z = 1$
  $x - 2y + z = 4$

49. $2x - y + z = 0$
  $4x + 6y - 2z = 0$
  $x - 2y - z = -9$

50. $3x - 3y - 3z = 3$
  $2x - 2y - 2z = 2$
  $x - y - z = 1$

Miscellaneous

Use a system of equations in two or three variables to solve each word problem. Solve by the method of your choice.

51. *Two-digit number.* The sum of the digits in a two-digit number is 15. When the digits are reversed, the new number is 9 more than the original number. What is the original number?

52. *Two-digit number.* The sum of the digits in a two-digit number is 8. When the digits are reversed, the new number is 18 less than the original number. What is the original number?

53. *Perimeter of a rectangle.* A rectangle has a perimeter of 16 feet and an area of 12 square feet. Find its length and width.

54. *Traveling by boat.* Alonzo can travel from his camp downstream to the mouth of the river in 30 minutes. If it takes him 45 minutes to come back, then how long would it take him to go that same distance in the lake with no current?

Time with current = 30 min
Time against current = 45 min

Figure for Exercise 54

55. *Tale of two circles.* Find the radii of two circles such that the difference in areas of the two is 10π square inches and the difference in radii of the two is 2 inches.

56. *Two lost numbers.* Find two numbers that have a sum of 10 and a product of 22.

57. *Driving and dating.* In four years, Gasper will be old enough to drive. His parents said that he must have a driver's license for two years before he can date. Three years ago, Gasper's age was only one-half of the age necessary to date. How old must Gasper be to drive, and how old is he now?

58. *Three solutions.* A chemist has three solutions of acid that must be mixed to obtain 20 liters of a solution that is 38% acid. Solution A is 30% acid, solution B is 20% acid, and solution C is 60% acid. Because of another chemical in these solutions, the chemist must keep the ratio of solution C to solution A at 2 to 1. How many liters of each should she mix together?

59. *Two investments.* Darlene had $20,000 to invest. She invested part of it at 10% and the remainder at 8%. If her income from the two investments was $1920, then how much did she invest at each rate?

60. *Beets and beans.* One serving of canned beets contains 1 gram of protein and 6 grams of carbohydrates. One serving of canned red beans contains 6 grams of protein and 20 grams of carbohydrates. How many servings of each would it take to get exactly 21 grams of protein and 78 grams of carbohydrates?

CHAPTER 10 TEST

Solve the system by graphing.

1. $x + y = 4$
   $y = 2x + 1$

Solve each system by substitution.

2. $y = 2x - 8$
   $4x + 3y = 1$

3. $y = x - 5$
   $3x - 4(y - 2) = 28 - x$

Solve each system by the addition method.

4. $3x + 2y = 3$
   $4x - 3y = -13$

5. $3x - y = 5$
   $-6x + 2y = 1$

Determine whether each system is independent, inconsistent, or dependent.

6. $y = 3x - 5$
   $y = 3x + 2$

7. $2x + 2y = 8$
   $x + y = 4$

8. $y = 2x - 3$
   $y = 5x - 14$

Solve the nonlinear system.

9. $y^2 - x^2 = 7$
   $y = -x^2 + 13$

Solve the following system by elimination of variables.

10. $x + y - z = 2$
    $2x - y + 3z = -5$
    $x - 3y + z = 4$

Solve by the Gauss-Jordan method.

11. $3x - y = 1$
    $x + 2y = 12$

12. $x - y - z = 1$
    $-x - y + 2z = -2$
    $-x - 3y + z = -5$

Evaluate each determinant.

13. $\begin{vmatrix} 2 & 3 \\ 4 & -3 \end{vmatrix}$

14. $\begin{vmatrix} 1 & -2 & -1 \\ 2 & 3 & 1 \\ 1 & 1 & 0 \end{vmatrix}$

Solve each system by using Cramer's rule.

15. $2x - y = -4$
    $3x + y = -1$

16. $x + y = 0$
    $x - y + 2z = 6$
    $2x + y - z = 1$

For each word problem, write a system of equations in two or three variables. Use the method of your choice to solve each system.

17. Find the length and width of a rectangular room that has an area of 108 square feet and a perimeter of 42 feet.

18. One night the manager of the Sea Breeze Motel rented 5 singles and 12 doubles for a total of $390. The next night he rented 9 singles and 10 doubles for a total of $412. What is the rental charge for each type of room?

19. Jill, Karen, and Betsy studied a total of 93 hours last week. Jill's and Karen's study time totaled only one-half as much as Betsy's. If Jill studied 3 hours more than Karen, then how many hours did each one of the girls spend studying?

Tying It All Together

Solve each equation.

1. $2(x - 3) + 5x = 8$

2. $|2x - 3| = 6$

3. $\sqrt{2x + 1} = 3$

4. $4x - 3 = 0$

5. $\dfrac{x^2}{9} = 1$

6. $4x^2 - 3 = 0$

7. $x^{2/3} = 4$

8. $x^2 + 6x + 8 = 0$

9. $x^2 + 8x - 2 = 0$

10. $\dfrac{1}{2} + \dfrac{1}{x} - \dfrac{4}{x^2} = 0$

11. $(x + 3)^2 = x^2$

12. $x^4 - 10x^2 + 9 = 0$

13. $\dfrac{x + 1}{x - 2} = \dfrac{5}{4}$

14. $2^{x-1} = 6$

15. $\ln(x) + \ln(x + 2) = 3 \cdot \ln(2)$

16. $\log_2(x) = 5$

Solve each inequality in one variable and graph the solution set on the number line.

17. $3x - 5 \geq 10$

18. $3 - x < 6$ and $3x < 6$

19. $|2x - 1| < 7$

20. $x^2 + x - 6 > 0$

Sketch the graph of each equation or function.

21. $y = 3x - 4$

22. $f(x) = 3 - x$

23. $h(x) = |x| + 1$

24. $y = \sqrt{x} - 1$

25. $y = \dfrac{1}{x}$

26. $y = 1 - x^2$

27. $y = 2^x$

28. $f(x) = \log_3(x)$

29. $f(x) = 3^{-x}$

30. $g(x) = \ln(x)$

Perform the operations and simplify.

31. $\sqrt{18} + \sqrt{50}$

32. $\sqrt{242} - \sqrt{98}$

33. $(3 + 2\sqrt{2})(4 - 3\sqrt{2})$

34. $(5 - \sqrt{3})(5 + \sqrt{3})$

35. $(3 + \sqrt{2})^2$

36. $\dfrac{6 - \sqrt{12}}{2}$

37. $\dfrac{\sqrt{2}}{\sqrt{3} - 2}$

38. $\dfrac{3}{\sqrt{2}} - \dfrac{1}{2} + \sqrt{12}$

Solve.

39. *Comparing copiers.* A self-employed consultant has prepared the accompanying graph to compare the total cost of purchasing and using two different copy machines.

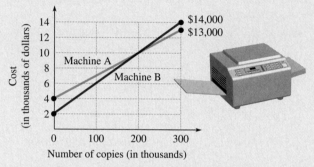

Figure for Exercise 39

a) Which machine has the larger purchase price?

b) What is the per-copy cost for operating each machine, not including the purchase price?

c) Find the slope of each line and interpret your findings.

d) Find the equation of each line.

e) Find the number of copies for which the total cost is the same for both machines.

Conic Sections

Courtesy, British Airways Plc

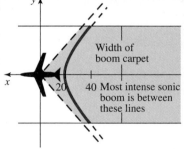

Width of boom carpet

Most intense sonic boom is between these lines

At a cruising speed of 1540 miles per hour, the Concorde can fly from London to New York in about three hours. So why isn't the same aircraft used for a fast flight from New York to Los Angeles? Concerns about cost efficiency and pollution are two reasons. However, most people agree that the biggest problem is noise. Traveling at Mach 2, the supersonic jet is flying faster than the speed of sound. At this speed the Concorde creates a cone-shaped wave in the air, on which there is a momentary change in air pressure. This change in air pressure causes a thunderlike sonic boom. When the jet is traveling parallel to the ground, the cone-shaped wave intersects the ground along one branch of a hyperbola. People on the ground hear the boom as the hyperbola passes them.

Sonic booms not only are noisy, but have also been known to cause physical destruction such as broken windows and cracked plaster. For this reason, supersonic jets are restricted from flying over land areas in the United States and much of the world. Some engineers believe that changing the silhouette of the plane can lessen the sonic boom, but most agree that it is impossible to eliminate the noise altogether.

In this chapter we discuss curves, including the hyperbola, that occur when a geometric plane intersects a cone. In Exercise 46 of Section 11.3 you will see how the altitude of the aircraft is related to the width of the area where the sonic boom is heard.

11.1 The Parabola

The parabola is one of four different curves that can be obtained by intersecting a cone and a plane as in Fig. 11.1. These curves, called **conic sections,** are the parabola, circle, ellipse, and hyperbola. You studied parabolas in Section 8.2, but in this section you will study parabolas in more detail. You will see how the parabola and the other conic sections are used in applications such as satellite dishes, telescopes, spotlights, and navigation.

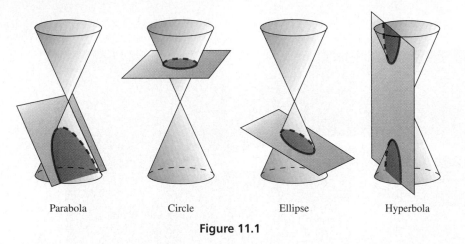

| Parabola | Circle | Ellipse | Hyperbola |

Figure 11.1

Developing the Equation

In Section 8.2 we called the graph of $y = ax^2 + bx + c$ a parabola. This equation is the **standard equation** of a parabola. In this section you will see that the following definition describes the same curve as the equation.

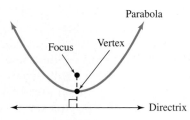

Figure 11.2

> **Parabola**
>
> Given a line (the **directrix**) and a point not on the line (the **focus**), the set of all points in the plane that are equidistant from the point and the line is called a **parabola.**

The **vertex** of the parabola is the midpoint of the line segment joining the focus and the directrix, perpendicular to the directrix. See Fig. 11.2.

The focus of a parabola is important in applications. When parallel rays of light travel into a parabolic reflector, they are reflected toward the focus as in Fig. 11.3. This property is used in telescopes to magnify the light from distant stars. If the light source is at the focus, as in a searchlight, the light is reflected off the parabola and projected outward in a narrow beam. This

Figure 11.3

542

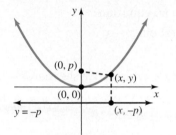

Figure 11.4

reflecting property is also used in camera lenses, satellite dishes, and eaves-dropping devices.

To develop an equation for a parabola, given the focus and directrix, choose the point $(0, p)$, where $p > 0$ as the focus and the line $y = -p$ as the directrix, as shown in Fig. 11.4. The vertex of this parabola is $(0, 0)$. For an arbitrary point (x, y) on the parabola the distance to the directrix is the distance from (x, y) to $(x, -p)$. The distance to the focus is the distance between (x, y) and $(0, p)$. We use the fact that these distances are equal to write the equation of the parabola:

$$\sqrt{(x - 0)^2 + (y - p)^2} = \sqrt{(x - x)^2 + (y - (-p))^2}$$

To simplify the equation, first remove the parentheses inside the radicals:

$$\sqrt{x^2 + y^2 - 2py + p^2} = \sqrt{y^2 + 2py + p^2}$$

$x^2 + y^2 - 2py + p^2 = y^2 + 2py + p^2$ Square each side.

$x^2 = 4py$ Subtract y^2 and p^2 from each side.

$$y = \frac{1}{4p}x^2$$

$y = ax^2$ Let $a = \frac{1}{4p}$.

So the equation of the parabola in Fig. 11.4 is of the form $y = ax^2$, which is a special case of the standard equation $y = ax^2 + bx + c$.

If a parabola has vertex $(0, 0)$ and opens up or down, then its equation is of the form $y = ax^2$. In general, if (h, k) is the vertex, $(h, k + p)$ is the focus, and $y = k - p$ is the directrix, then we can develop the equation $y = a(x - h)^2 + k$ for the parabola just as we developed the equation $y = ax^2$.

> **The Equation of a Parabola**
>
> The graph of the equation $y = a(x - h)^2 + k(a \neq 0)$ is a parabola with vertex (h, k), focus $(h, k + p)$, and directrix $y = k - p$, where $a = \frac{1}{4p}$.
>
> If $a > 0$, the parabola opens upward; if $a < 0$, the parabola opens downward.

Note that the locations of the focus and directrix determine the value of a and the shape and opening of the parabola.

CAUTION For a parabola that opens upward, $p > 0$, and the focus $(h, k + p)$ is above the vertex (h, k). For a parabola that opens downward, $p < 0$, and the focus $(h, k + p)$ is below the vertex (h, k). In either case the distance from the vertex to the focus and the vertex to the directrix is $|p|$. ⊘

EXAMPLE 1

Finding the vertex, focus, and directrix, given an equation

Find the vertex, focus, and directrix for the parabola $y = x^2$.

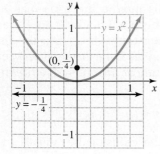

Solution

Compare $y = x^2$ to the general formula $y = a(x - h)^2 + k$. We see that $h = 0$, $k = 0$, and $a = 1$. So the vertex is $(0, 0)$. Because $a = 1$, we can use $a = \frac{1}{4p}$ to get

$$1 = \frac{1}{4p},$$

or $p = \frac{1}{4}$. Use $(h, k + p)$ to get the focus $\left(0, \frac{1}{4}\right)$. Use the equation $y = k - p$ to get $y = -\frac{1}{4}$ as the equation of the directrix. See Fig. 11.5. ∎

Figure 11.5

EXAMPLE 2

Finding an equation, given a focus and directrix

Find the equation of the parabola with focus $(-1, 4)$ and directrix $y = 3$.

Solution

Because the vertex is halfway between the focus and directrix, the vertex is $\left(-1, \frac{7}{2}\right)$. See Fig. 11.6. The distance from the vertex to the focus is $\frac{1}{2}$. Because the focus is above the vertex, p is positive. So $p = \frac{1}{2}$, and $a = \frac{1}{4p} = \frac{1}{2}$. The equation is

$$y = \frac{1}{2}(x - (-1))^2 + \frac{7}{2}.$$

Figure 11.6

Simplify to get the equation $y = \frac{1}{2}x^2 + x + 4$. ∎

The equation of a parabola can be written in two different forms. To change the form $y = a(x - h)^2 + k$ to the form $y = ax^2 + bx + c$, we square the binomial and combine like terms, as in Example 2. To change from $y = ax^2 + bx + c$ to $y = a(x - h)^2 + k$, we complete the square.

EXAMPLE 3

Converting $y = ax^2 + bx + c$ to $y = a(x - h)^2 + k$

Write $y = 2x^2 - 4x + 5$ in the form $y = a(x - h)^2 + k$ and identify the vertex, focus, and directrix of the parabola.

Solution

Use completing the square to rewrite the equation:

$$y = 2(x^2 - 2x) + 5$$
$$y = 2(x^2 - 2x + 1 - 1) + 5 \qquad \text{Complete the square.}$$
$$y = 2(x^2 - 2x + 1) - 2 + 5 \qquad \text{Move } 2(-1) \text{ outside the parentheses.}$$
$$y = 2(x - 1)^2 + 3$$

The vertex is $(1, 3)$. Because $a = \dfrac{1}{4p}$, we have

$$\frac{1}{4p} = 2,$$

and $p = \frac{1}{8}$. Because the parabola opens upward, the focus is $\frac{1}{8}$ unit above the vertex at $\left(1, 3\frac{1}{8}\right)$, or $\left(1, \frac{25}{8}\right)$, and the directrix is the horizontal line $\frac{1}{8}$ unit below the vertex, $y = 2\frac{7}{8}$ or $y = \frac{23}{8}$.   ■

CAUTION   Be careful when you complete a square within parentheses as in Example 3. For another example, consider the equivalent equations

$$y = -3(x^2 + 4x),$$
$$y = -3(x^2 + 4x + 4 - 4), \qquad \text{and}$$
$$y = -3(x + 2)^2 + 12.$$ ⊘

The Standard Equation of a Parabola

In Chapter 8 we stated that the x-coordinate of the vertex for the parabola $y = ax^2 + bx + c$ is $\dfrac{-b}{2a}$. We will now complete the square in the standard equation to verify this fact:

$$y = a\left(x^2 + \frac{bx}{a}\right) + c \qquad \text{Factor } a \text{ out of the first two terms.}$$

$$y = a\left(x^2 + \frac{bx}{a} + \frac{b^2}{4a^2} - \frac{b^2}{4a^2}\right) + c \qquad \frac{1}{2} \cdot \frac{b}{a} = \frac{b}{2a} \text{ and } \left(\frac{b}{2a}\right)^2 = \frac{b^2}{4a^2}$$

$$y = a\left(x^2 + \frac{bx}{a} + \frac{b^2}{4a^2}\right) - \frac{b^2}{4a} + c \qquad \text{Move } a\left(-\frac{b^2}{4a^2}\right) \text{ outside the parentheses.}$$

$$y = a\left(x^2 + \frac{bx}{a} + \frac{b^2}{4a^2}\right) - \frac{b^2}{4a} + \frac{c \cdot 4a}{4a} \qquad \text{Build up the denominator.}$$

$$y = a\left(x + \frac{b}{2a}\right)^2 + \frac{4ac - b^2}{4a} \qquad \text{Simplify.}$$

The equation is now in the form $y = a(x - h)^2 + k$, and the vertex is

$$\left(\frac{-b}{2a}, \frac{4ac - b^2}{4a}\right).$$

We summarize these results as follows.

> **Parabolas in the Form $y = ax^2 + bx + c$**
>
> The graph of $y = ax^2 + bx + c$ (for $a \neq 0$) is a parabola opening upward if $a > 0$ and downward if $a < 0$. The x-coordinate of the vertex is $x = \dfrac{-b}{2a}$.

You can use $\dfrac{4ac - b^2}{4a}$ to get the y-coordinate of the vertex, but it is usually easier to get it from $y = ax^2 + bx + c$ for $x = \dfrac{-b}{2a}$. We use the second approach in Example 4.

EXAMPLE 4   **Finding the features of a parabola from standard form**

Find the vertex, focus, and directrix of the parabola $y = -3x^2 + 9x - 5$, and determine whether the parabola opens upward or downward.

Solution

The x-coordinate of the vertex is

$$x = \frac{-b}{2a} = \frac{-9}{2(-3)} = \frac{-9}{-6} = \frac{3}{2}.$$

To find the y-coordinate of the vertex, let $x = \frac{3}{2}$ in $y = -3x^2 + 9x - 5$:

$$y = -3\left(\frac{3}{2}\right)^2 + 9\left(\frac{3}{2}\right) - 5 = -\frac{27}{4} + \frac{27}{2} - 5 = \frac{7}{4}$$

The vertex is $\left(\frac{3}{2}, \frac{7}{4}\right)$. Because $a = -3$, the parabola opens downward. To find the focus, use $-3 = \dfrac{1}{4p}$ to get $p = -\dfrac{1}{12}$. The focus is $\dfrac{1}{12}$ of a unit below the vertex at $\left(\frac{3}{2}, \frac{7}{4} - \frac{1}{12}\right)$ or $\left(\frac{3}{2}, \frac{5}{3}\right)$. The directrix is the horizontal line $\dfrac{1}{12}$ of a unit above the vertex, $y = \frac{7}{4} + \frac{1}{12}$ or $y = \frac{11}{6}$. ∎

Symmetry and Intercepts

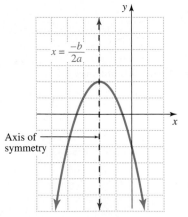

Figure 11.7

The graph of $y = x^2$ shown in Fig. 11.5 is said to be **symmetric about the y-axis** because the two halves of the parabola would coincide if the paper were folded on the y-axis. In general, the vertical line through the vertex is called the **axis of symmetry** for the parabola. Because the x-coordinate of the vertex for $y = ax^2 + bx + c$ is $\dfrac{-b}{2a}$, the axis of symmetry is $x = \dfrac{-b}{2a}$. See Fig. 11.7.

The parabola $y = ax^2 + bx + c$ has exactly one y-intercept. If $x = 0$ in the equation, then $y = a(0)^2 + b(0) + c = c$. So the y-intercept is $(0, c)$.

To find the x-intercepts, we let $y = 0$ in the equation $y = ax^2 + bx + c$ and solve the quadratic equation $ax^2 + bx + c = 0$. The number of x-intercepts may be 0, 1, or 2, depending on the number of solutions to the quadratic equation. See Fig. 11.8.

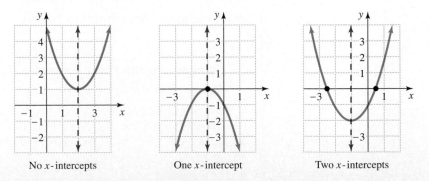

No x-intercepts   One x-intercept   Two x-intercepts

Figure 11.8

Graphing a Parabola

When graphing a parabola, we can use the features that we have discussed to improve accuracy and understanding. These features are summarized in the following box as a strategy for graphing parabolas.

Graphing the Parabola $y = ax^2 + bx + c$

To graph the parabola $y = ax^2 + bx + c$, use the following facts:

1. The parabola opens upward if $a > 0$ and opens downward if $a < 0$.
2. The x-coordinate of the vertex is $\frac{-b}{2a}$.
3. The graph is symmetric about the vertical line $x = \frac{-b}{2a}$.
4. The y-intercept is $(0, c)$.
5. The x-intercepts are found by solving $ax^2 + bx + c = 0$.

CAUTION The focus and directrix are important features of a parabola, but they are not part of the curve itself. So we do not usually find the focus and directrix when graphing a parabola from an equation.

EXAMPLE 5   **Graphing a parabola**

Determine whether the parabola $y = -x^2 - x + 6$ opens upward or downward, and find the vertex, axis of symmetry, x-intercepts, and y-intercept. Find several additional points on the parabola, and sketch the graph.

Solution

Because $a = -1$, this parabola opens downward. The x-coordinate of the vertex is

$$x = \frac{-b}{2a} = \frac{-(-1)}{2(-1)} = -\frac{1}{2}.$$

If $x = -\frac{1}{2}$, then

$$y = -\left(-\frac{1}{2}\right)^2 - \left(-\frac{1}{2}\right) + 6 = -\frac{1}{4} + \frac{1}{2} + 6 = \frac{25}{4}.$$

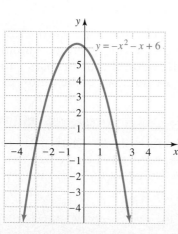

So the vertex is $\left(-\frac{1}{2}, \frac{25}{4}\right)$. The axis of symmetry is the vertical line $x = -\frac{1}{2}$. To find the x-intercepts, let $y = 0$ in the original equation:

$$-x^2 - x + 6 = 0$$
$$x^2 + x - 6 = 0$$
$$(x + 3)(x - 2) = 0$$
$$x + 3 = 0 \quad \text{or} \quad x - 2 = 0$$
$$x = -3 \quad \text{or} \quad x = 2$$

The x-intercepts are $(-3, 0)$ and $(2, 0)$. The y-intercept is $(0, 6)$. Using all of this information and the additional points $(1, 4)$, $(-2, 4)$, and $(-1, 6)$, we get the graph shown in Fig. 11.9.

Figure 11.9

Maximum or Minimum Value of y

On a parabola that opens upward, the minimum value that y attains is the y-coordinate of the vertex. On a parabola that opens downward, the maximum value of y is again the y-coordinate of the vertex. We use the fact that the x-coordinate of the vertex is $\dfrac{-b}{2a}$ to obtain the maximum or minimum value of y.

EXAMPLE 6 **Maximizing a product**

Find two numbers that have the maximum product, subject to the condition that their sum is 1.

Solution

Because their sum is 1, we can let x represent one number and $1 - x$ represent the other. If y represents their product, we can write the equation

$$y = x(1 - x)$$

or

$$y = -x^2 + x.$$

This equation is the equation of a parabola that opens downward. Its highest point, which is the maximum value of y, occurs when

$$x = \frac{-b}{2a}$$

$$= \frac{-1}{2(-1)}$$

$$= \frac{1}{2}.$$

If $x = \frac{1}{2}$, then $1 - x = \frac{1}{2}$. The two numbers are $\frac{1}{2}$ and $\frac{1}{2}$. Among the numbers with a sum of 1, they give the maximum product. So no two numbers with a sum of 1 can have a product larger than $\frac{1}{4}$. ■

CAUTION Be sure to answer the question that is asked in a maximum-minimum problem. If $y = ax^2 + bx + c$, then the maximum (or minimum) value of y is the y-coordinate of the vertex. The value of x that causes y to reach its maximum (or minimum) value is the x-coordinate of the vertex. ⊘

EXAMPLE 7 **Maximizing revenue**

A manufacturer of in-line roller skates uses the formula $R = 360x - 2x^2$ to predict the weekly revenue in dollars that will be produced when the skates are priced at x dollars per pair. Find the price that will produce the maximum revenue, and find the maximum possible revenue.

Solution

The graph of $R = 360x - 2x^2$ is a parabola that opens downward as shown in Fig. 11.10. To find the vertex of the parabola, use $a = -2$ and $b = 360$ in $x = \dfrac{-b}{2a}$:

$$x = \frac{-360}{2(-2)} = 90$$

Now use $x = 90$ in $R = 360x - 2x^2$ to find R:

$$R = 360(90) - 2 \cdot 90^2 = 16{,}200$$

So if the skates are priced at $90 per pair, then the manufacturer will receive the maximum weekly revenue $16,200.

Figure 11.10

Warm-ups

True or false? Explain your answer.

1. There is a parabola with focus $(2, 3)$, directrix $y = 1$, and vertex $(0, 0)$.

2. The focus for the parabola $y = \frac{1}{4}x^2 + 1$ is $(0, 2)$.

3. The graph of $y - 3 = 5(x - 4)^2$ is a parabola with vertex $(4, 3)$.

4. The graph of $y = 6x + 3x + 2$ is a parabola.

5. The graph of $y = 2x - x^2 + 9$ is a parabola opening upward.

6. For $y = x^2$ the vertex and y-intercept are the same point.

7. A parabola with vertex $(2, 3)$ and focus $(2, 4)$ has no x-intercepts.

8. The parabola $y = 3x^2 + 9x$ has no y-intercept.

9. If $(1, 5)$ satisfies $y = a(x - 2)^2 + k$, then so does $(3, 5)$.

10. The parabola $y = (x - 5)^2$ has only one x-intercept.

11.1 EXERCISES

Find the vertex, focus, and directrix for each parabola. See Example 1.

1. $y = 2x^2$

2. $y = \dfrac{1}{2}x^2$

3. $y = -\dfrac{1}{4}x^2$

4. $y = -\dfrac{1}{12}x^2$

5. $y = \dfrac{1}{2}(x - 3)^2 + 2$

6. $y = \dfrac{1}{4}(x + 2)^2 - 5$

7. $y = -(x + 1)^2 + 6$

8. $y = -3(x - 4)^2 + 1$

Find the equation of the parabola with the given focus and directrix. See Example 2.

9. Focus $(0, 2)$, directrix $y = -2$

10. Focus $(0, -3)$, directrix $y = 3$

11. Focus $\left(0, -\dfrac{1}{2}\right)$, directrix $y = \dfrac{1}{2}$

12. Focus $\left(0, \dfrac{1}{8}\right)$, directrix $y = -\dfrac{1}{8}$

13. Focus $(3, 2)$, directrix $y = 1$

14. Focus $(-4, 5)$, directrix $y = 4$

15. Focus $(1, -2)$, directrix $y = 2$

16. Focus $(2, -3)$, directrix $y = 1$

17. Focus $(-3, 1.25)$, directrix $y = 0.75$

18. Focus $\left(5, \dfrac{17}{8}\right)$, directrix $y = \dfrac{15}{8}$

Write each equation in the form $y = a(x - h)^2 + k$. Identify the vertex, focus, and directrix of each parabola. See Example 3.

19. $y = x^2 - 6x + 1$

20. $y = x^2 + 4x - 7$

21. $y = 2x^2 + 12x + 5$

22. $y = 3x^2 + 6x - 7$

23. $y = -2x^2 + 16x + 1$

24. $y = -3x^2 - 6x + 7$

25. $y = 5x^2 + 40x$

26. $y = -2x^2 + 10x$

Find the vertex, focus, and directrix of each parabola (without completing the square), and determine whether the parabola opens upward or downward. See Example 4.

27. $y = x^2 - 4x + 1$

28. $y = x^2 - 6x - 7$

29. $y = -x^2 + 2x - 3$

30. $y = -x^2 + 4x + 9$

31. $y = 3x^2 - 6x + 1$

32. $y = 2x^2 + 4x - 3$

33. $y = -x^2 - 3x + 2$

34. $y = -x^2 + 3x - 1$

35. $y = 3x^2 + 5$

36. $y = -2x^2 - 6$

Find the vertex, axis of symmetry, x-intercepts, and y-intercept for each parabola. Find several additional points on the parabola, and then sketch its graph. See Example 5.

37. $y = x^2 - 3x + 2$

38. $y = x^2 + 6x + 8$

39. $y = -x^2 - 2x + 8$

40. $y = -x^2 - 2x + 15$

41. $y = (x + 2)^2 + 1$

42. $y = -2(x + 1)^2 + 3$

43. $y = x^2 + 2x + 1$

44. $y = x^2 - 6x + 9$

45. $y = -4x^2 + 4x - 1$

46. $y = -4x^2 + 12x - 9$

47. $y = x^2 - 5x$

48. $y = 3x^2 - 9x$

49. $y = 3x^2 + 5$

50. $y = -2x^2 + 3$

51. $y = x^2 - 2x - 1$

52. $y = x^2 - 4x + 1$

53. $y = (x - 5)^2$

54. $y = 3(x - 1)^2 - 4$

Solve each problem. See Examples 6 and 7.

55. **Maximum product.** Find two numbers that have the maximum possible product among numbers that have a sum of 8.

56. **Maximum product.** What is the maximum product that can be obtained by two numbers that a have a sum of -6?

57. **Maximum area.** A gardener plans to enclose a rectangular area with 160 feet of fencing. What dimensions for the length and width would maximize the area?

58. **Maximum area.** Another gardener has a 4-foot-wide gate for her rectangular garden in place. If she has 100 feet of fencing in addition to the gate, then what dimensions for the length and width would maximize the area?

59. **Minimum hypotenuse.** If the total length of the legs of a right triangle is 6 feet, then what lengths for the legs would minimize the square of the hypotenuse?

60. **Maximum revenue.** A concert promoter uses the formula $R = 7500p - 250p^2$ to calculate the total revenue for a concert if the tickets sell for p dollars each. What value of p will maximize the revenue? See the accompanying figure on the next page.

Figure for Exercise 60

61. *Maximum height.* If a punter kicks a football straight up at a velocity of 128 feet per second from a height of 6 feet, then the football's distance above the earth after t seconds is given by the formula $s = -16t^2 + 128t + 6$. What is the maximum height reached by the ball? Could the punter hit the roof of the New Orleans Superdome, which is 273 feet above the playing field?

Figure for Exercise 61

62. *Maximum height.* If a baseball is hit straight upward at 150 feet per second from a height of 5 feet, then its distance above the earth after t seconds is given by the formula $s = -16t^2 + 150t + 5$. What is the maximum height attained by the ball?

Figure for Exercise 62

63. *Minimum cost.* It costs Acme Manufacturing C dollars per hour to operate its golf ball division. An analyst has determined that C is related to the number of golf balls produced per hour, x, by the equation $C = 0.009x^2 - 1.8x + 100$. What number of balls per hour should Acme produce to minimize the cost per hour of manufacturing these golf balls?

64. *Maximum profit.* A chain-store manager has been told by the main office that daily profit, P, is related to the number of clerks working that day, x, according to the equation $P = -25x^2 + 300x$. What number of clerks will maximize the profit, and what is the maximum possible profit?

65. *World's largest telescope.* The largest reflecting telescope in the world is the 6-m reflector on Mount Pastukhov in Russia. The accompanying figure shows a cross section of a parabolic mirror 6 m in diameter with the vertex at the origin and the focus at $(0, 15)$. Find the equation of the parabola.

Figure for Exercise 65

66. *Arecibo Observatory.* The largest radio telescope in the world uses a 1000-ft parabolic dish, suspended in a valley in Arecibo, Puerto Rico. The antenna hangs above the vertex of the dish on cables stretching from two towers. The accompanying figure shows a cross section of the parabolic dish and the towers. Assuming that the vertex is at $(0, 0)$, find the equation for the parabola. Find the distance from the vertex to the antenna located at the focus.

Figure for Exercise 66

Graph both equations of each system on the same coordinate axes. Use elimination of variables (from Section 10.4) to find all points of intersection.

67. $y = -x^2 + 3$
$y = x^2 + 1$

68. $y = x^2 - 3$
$y = -x^2 + 5$

69. $y = x^2 - 2$
$y = 2x - 3$

70. $y = x^2 + x - 6$
$y = 7x - 15$

71. $y = x^2 + 3x - 4$
$y = -x^2 - 2x + 8$

72. $y = x^2 + 2x - 8$
$y = -x^2 - x + 12$

73. $y = x^2 + 3x - 4$
$y = 2x + 2$

74. $y = x^2 + 5x + 6$
$y = x + 11$

Write an equation in the form $y = ax^2 + bx + c$ for each given parabola.

75.

76.

77.

78.

Getting More Involved

79. *Exploration.* Consider the parabola with focus $(p, 0)$ and directrix $x = -p$ for $p > 0$. Let (x, y) be an arbitrary point on the parabola. Write an equation expressing the fact that the distance from (x, y) to the focus is equal to the distance from (x, y) to the directrix. Rewrite the equation in the form $x = ay^2$, where $a = \dfrac{1}{4p}$.

80. *Exploration.* In general, the graph of $x = a(y - h)^2 + k$ for $a \ne 0$ is a parabola opening left or right with vertex at (k, h).

a) For which values of a does the parabola open to the right, and for which values of a does it open to the left?

b) What is the equation of its axis of symmetry?

c) Sketch the graphs $x = 2(y - 3)^2 + 1$ and $x = -(y + 1)^2 + 2$.

Graphing Calculator Exercises

81. Graph $y = x^2$ using the viewing window with $-1 \le x \le 1$ and $0 \le y \le 1$. Next graph $y = 2x^2 - 1$ using the viewing window $-2 \le x \le 2$ and $-1 \le y \le 7$. Explain what you see.

82. Graph $y = x^2$ and $y = 6x - 9$ in the viewing window $-5 \le x \le 5$ and $-5 \le y \le 20$. Does the line appear to be tangent to the parabola? Solve the system $y = x^2$ and $y = 6x - 9$ to find all points of intersection for the parabola and the line.

11.2 The Circle

In this section:

▶ Developing the Equation
▶ Equations Not in Standard Form
▶ Systems of Equations

In this section we continue the study of the conic sections with a discussion of the circle.

Developing the Equation

A circle is obtained by cutting a cone, as was shown in Figure 11.1. We can also define a circle using points and distance, as we did for the parabola.

> **Circle**
>
> A **circle** is the set of all points in a plane that lie a fixed distance from a given point in the plane. The fixed distance is called the **radius,** and the given point is called the **center.**

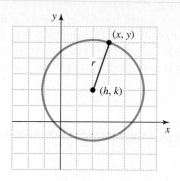

Figure 11.11

We can use the distance formula of Section 7.1 to write an equation for the circle with center (h, k) and radius r, shown in Fig. 11.11. If (x, y) is a point on the circle, its distance from the center is r. So

$$\sqrt{(x - h)^2 + (y - k)^2} = r.$$

We square both sides of this equation to get the **standard form** for the equation of a circle.

Standard Equation for a Circle

The graph of the equation

$$(x - h)^2 + (y - k)^2 = r^2$$

with $r > 0$, is a circle with center (h, k) and radius r.

Note that a circle centered at the origin with radius $r (r > 0)$ has the standard equation

$$x^2 + y^2 = r^2.$$

EXAMPLE 1   **Finding the equation, given the center and radius**

Write the standard equation for the circle with the given center and radius.

a)   Center $(0, 0)$, radius 2   b)   Center $(-1, 2)$, radius 4

Solution

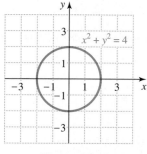

Figure 11.12

a)   The center at $(0, 0)$ means that $h = 0$ and $k = 0$ in the standard equation. So the equation is $(x - 0)^2 + (y - 0)^2 = 2^2$, or $x^2 + y^2 = 4$. The circle with radius 2 centered at the origin is shown in Fig. 11.12.

b)   The center at $(-1, 2)$ means that $h = -1$ and $k = 2$. So

$$[x - (-1)]^2 + [y - 2]^2 = 4^2.$$

Simplify this equation to get

$$(x + 1)^2 + (y - 2)^2 = 16.$$

The circle with center $(-1, 2)$ and radius 4 is shown in Fig. 11.13.

Figure 11.13

CAUTION The equations $(x-1)^2 + (y+3)^2 = -9$ and $(x-1)^2 + (y+3)^2 = 0$ might look like equations of circles, but they are not. The first equation is not satisfied by any ordered pair of real numbers because the left side is nonnegative for any x and y. The second equation is satisfied only by the point $(1, -3)$. ⊘

EXAMPLE 2 **Finding the center and radius, given the equation**

Determine the center and radius of the circle $x^2 + (y+5)^2 = 2$.

Solution

We can write this equation as

$$(x-0)^2 + [y - (-5)]^2 = (\sqrt{2})^2.$$

In this form, we see that the center is $(0, -5)$ and the radius is $\sqrt{2}$. ∎

EXAMPLE 3 **Graphing a circle**

Find the center and radius of $(x-1)^2 + (y+2)^2 = 9$, and sketch the graph.

Solution

The graph of this equation is a circle with center at $(1, -2)$ and radius 3. See Fig. 11.14 for the graph.

Figure 11.14 ∎

Xmin = –2   Ymin = –2
Xmax = 2   Ymax = 2

Figure 11.15

A graphing calculator will usually plot points only for equations in which y is written as a function of x. So to graph a circle such as $x^2 + y^2 = 1$, we solve for y to get $y = \pm\sqrt{1-x^2}$ and then graph both $y = \sqrt{1-x^2}$ and $y = -\sqrt{1-x^2}$. Note that on a graphing calculator the expression $1 - x^2$ must be in parentheses because the radical must apply to the entire expression. These two functions should appear on your calculator's display as

$$y_1 = \sqrt{\ }(1-x^2) \qquad \text{and} \qquad y_2 = -\sqrt{\ }(1-x^2).$$

If the unit length on the x-axis is not equal to the unit length on the y-axis, then a circle will not look round. See Fig. 11.15. You can adjust the viewing window or use the "square" feature to get the unit lengths the same.

Equations Not in Standard Form

It is not easy to recognize that $x^2 - 6x + y^2 + 10y = -30$ is the equation of a circle, but it is. In the next example we convert this equation into the standard form for a circle by completing the squares for the variables x and y.

EXAMPLE 4   **Converting to standard form**

Find the center and radius of the circle given by the equation

$$x^2 - 6x + y^2 + 10y = -30.$$

Solution

To complete the square for $x^2 - 6x$, we add 9, and for $y^2 + 10y$, we add 25. To get an equivalent equation, we must add on both sides:

$$x^2 - 6x + \quad y^2 + 10y \quad = -30$$
$$x^2 - 6x + 9 + y^2 + 10y + 25 = -30 + 9 + 25 \qquad \text{Add 9 and 25 to both sides.}$$
$$(x - 3)^2 + (y + 5)^2 = 4 \qquad \text{Factor the trinomials on the left side.}$$

We see that in standard form the center is $(3, -5)$ and the radius is 2.   ■

Systems of Equations

We first solved systems of nonlinear equations in two variables in Section 10.4. We found the points of intersection of two graphs without drawing the graphs. Here we will solve systems involving circles, parabolas, and lines. In the next example we find the points of intersection of a line and a circle.

EXAMPLE 5   **Intersection of a line and a circle**

Graph both equations of the system

$$(x - 3)^2 + (y + 1)^2 = 9$$
$$y = x - 1$$

on the same coordinate axes, and solve the system by elimination of variables.

Solution

The graph of the first equation is a circle with center at $(3, -1)$ and radius 3. The graph of the second equation is a straight line with slope 1 and y-intercept $(0, -1)$. Both graphs are shown in Figure 11.16. To solve the system by elimination, we substitute $y = x - 1$ into the equation of the circle:

$$(x - 3)^2 + (x - 1 + 1)^2 = 9$$
$$(x - 3)^2 + x^2 = 9$$
$$x^2 - 6x + 9 + x^2 = 9$$
$$2x^2 - 6x = 0$$
$$x^2 - 3x = 0$$
$$x(x - 3) = 0$$

$$x = 0 \qquad \text{or} \qquad x = 3$$
$$y = -1 \qquad\qquad\qquad y = 2 \quad \text{Because } y = x - 1$$

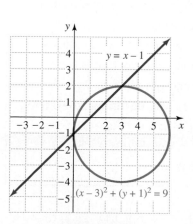

Figure 11.16

Check $(0, -1)$ and $(3, 2)$ in the original system and with the graphs in Fig. 11.16. The solution set is $\{(0, -1), (3, 2)\}$.   ■

MATH
AT
WORK

**Early Musical
Instrument Maker**

THE
EARLY MUSIC SHOP
OF
NEW ENGLAND

Friedrich von Huene, a flautist and recorder player, has been crafting woodwind instruments in his family business for over 30 years. Because it is best to play music of earlier centuries on the instruments of their time, von Huene is using many originals as models for his flutes, recorders, and oboes of different sizes.

Because museum instruments have many different pitch standards, their dimensions frequently have to be changed to accommodate pitch standards that musicians use today. For a lower pitch, the length of the instrument as well as the inside diameter must be increased. For a higher pitch, the length has to be shortened and the diameter decreased. However, the factor for changing the length will be different from the factor for changing the diameter. A row of organ pipes demonstrates that the larger and longer pipes are proportionately more slender than the shorter high-pitched pipes. A pipe an octave higher in pitch is about half as long as the pipe an octave lower, but its diameter will be about 0.6 as large as the diameter of the lower pipe.

When making a very large recorder, von Huene carefully chooses the length, the position of the tone holes, and the bore to get the proper volume of air inside the instrument. In Exercises 55 and 56 of this section you will make the kinds of calculations von Huene makes when he crafts a modern reproduction of a Renaissance flute.

Warm-ups

True or false? Explain your answer.

1. The radius of a circle can be any nonzero real number.
2. The coordinates of the center must satisfy the equation of the circle.
3. The circle $x^2 + y^2 = 4$ has its center at the origin.
4. The graph of $x^2 + y^2 = 9$ is a circle centered at $(0, 0)$ with radius 9.
5. The graph of $(x - 2)^2 + (y - 3)^2 + 4 = 0$ is a circle of radius 2.
6. The graph of $(x - 3) + (y + 5) = 9$ is a circle of radius 3.
7. There is only one circle centered at $(-3, -1)$ passing through the origin.
8. The center of the circle $(x - 3)^2 + (y - 4)^2 = 10$ is $(-3, -4)$.
9. The center of the circle $x^2 + y^2 + 6y - 4 = 0$ is on the y-axis.
10. The radius of the circle $x^2 - 3x + y^2 = 4$ is 2.

11.2 EXERCISES

Write the standard equation for each circle with the given center and radius. See Example 1.

1. Center $(0, 3)$, radius 5

2. Center $(2, 0)$, radius 3

3. Center $(1, -2)$, radius 9

4. Center $(-3, 5)$, radius 4

5. Center $(0, 0)$, radius $\sqrt{3}$

6. Center $(0, 0)$, radius $\sqrt{2}$

7. Center $(-6, -3)$, radius $\frac{1}{2}$

8. Center $(-3, -5)$, radius $\frac{1}{4}$

9. Center $\left(\frac{1}{2}, \frac{1}{3}\right)$, radius 0.1

10. Center $\left(-\frac{1}{2}, 3\right)$, radius 0.2

Find the center and radius for each circle. See Example 2.

11. $(x - 3)^2 + (y - 5)^2 = 2$

12. $(x + 3)^2 + (y - 7)^2 = 6$

13. $x^2 + \left(y - \frac{1}{2}\right)^2 = \frac{1}{2}$

14. $5x^2 + 5y^2 = 5$

15. $4x^2 + 4y^2 = 9$

16. $9x^2 + 9y^2 = 49$

17. $3 - y^2 = (x - 2)^2$

18. $9 - x^2 = (y + 1)^2$

Sketch the graph of each equation. See Example 3.

19. $x^2 + y^2 = 9$

20. $x^2 + y^2 = 16$

21. $x^2 + (y - 3)^2 = 9$

22. $(x - 4)^2 + y^2 = 16$

23. $(x + 1)^2 + (y - 1)^2 = 2$

24. $(x - 2)^2 + (y + 2)^2 = 8$

25. $(x - 4)^2 + (y + 3)^2 = 16$

26. $(x - 3)^2 + (y - 7)^2 = 25$

27. $\left(x - \frac{1}{2}\right)^2 + \left(y + \frac{1}{2}\right)^2 = \frac{1}{4}$

28. $\left(x + \frac{1}{3}\right)^2 + y^2 = \frac{1}{9}$

Rewrite each equation in the standard form for the equation of a circle, and identify its center and radius. See Example 4.

29. $x^2 + 4x + y^2 + 6y = 0$

30. $x^2 - 10x + y^2 + 8y = 0$

31. $x^2 - 2x + y^2 - 4y - 3 = 0$

32. $x^2 - 6x + y^2 - 2y + 9 = 0$

33. $x^2 + y^2 = 8y + 10x - 32$

34. $x^2 + y^2 = 8x - 10y$

35. $x^2 - x + y^2 + y = 0$

36. $x^2 - 3x + y^2 = 0$

37. $x^2 - 3x + y^2 - y = 1$

38. $x^2 - 5x + y^2 + 3y = 2$

39. $x^2 - \frac{2}{3}x + y^2 + \frac{3}{2}y = 0$

40. $x^2 + \frac{1}{3}x + y^2 - \frac{2}{3}y = \frac{1}{9}$

Graph both equations of each system on the same coordinate axes. Solve the system by elimination of variables to find all points of intersection of the graphs. See Example 5.

41. $x^2 + y^2 = 10$
    $y = 3x$

42. $x^2 + y^2 = 4$
    $y = x - 2$

43. $x^2 + y^2 = 9$
    $y = x^2 - 3$

44. $x^2 + y^2 = 4$
    $y = x^2 - 2$

45. $(x - 2)^2 + (y + 3)^2 = 4$
    $y = x - 3$

46. $(x + 1)^2 + (y - 4)^2 = 17$
    $y = x + 2$

In Exercises 47–56, solve each problem.

47. Determine all points of intersection of the circle $(x - 1)^2 + (y - 2)^2 = 4$ with the y-axis.

48. Determine the points of intersection of the circle $x^2 + (y - 3)^2 = 25$ with the x-axis.

49. Find the radius of the circle that has center $(2, -5)$ and passes through the origin.

50. Find the radius of the circle that has center $(-2, 3)$ and passes through $(3, -1)$.

51. Determine the equation of the circle that is centered at $(2, 3)$ and passes through $(-2, -1)$.

52. Determine the equation of the circle that is centered at $(3, 4)$ and passes through the origin.

53. Find all points of intersection of the circles $x^2 + y^2 = 9$ and $(x - 5)^2 + y^2 = 9$.

54. A donkey is tied at the point $(2, -3)$ on a rope of length 12. Turnips are growing at the point $(6, 7)$. Can the donkey reach them?

55. *Volume of a flute.* The volume of air in a flute is a critical factor in determining its pitch. A cross section of a Renaissance flute in C is shown in the accompanying figure. If the length of the flute is 2874 millimeters, then what is the volume of air in the flute (to the nearest cubic millimeter)? (*Hint:* Use the formula for the volume of a cylinder.)

Figure for Exercises 55 and 56

56. *Flute reproduction.* To make the smaller C# flute, Friedrich von Huene multiplies the length and cross-sectional area of the flute of Exercise 55 by 0.943. Find the equation for the bore hole (centered at the origin) and the volume of air in the C# flute.

Graph each equation.

57. $x^2 + y^2 = 0$

58. $x^2 - y^2 = 0$

59. $y = \sqrt{1 - x^2}$

60. $y = -\sqrt{1 - x^2}$

Getting More Involved

61. *Cooperative learning.* The equation of a circle is a special case of the general equation $Ax^2 + Bx + Cy^2 + Dy = E$, where A, B, C, D, and E are real numbers. Working in small groups, find restrictions that must be placed on A, B, C, D, and E so that the graph of this equation is a circle. What does the graph of $x^2 + y^2 = -9$ look like?

62. *Discussion.* Suppose lighthouse A is located at the origin and lighthouse B is located at coordinates $(0, 6)$. The captain of a ship has determined that the ship's distance from lighthouse A is 2 and its distance from lighthouse B is 5. What are the possible coordinates for the location of the ship?

Graphing Calculator Exercises

Graph each relation on a graphing calculator by solving for y and graphing two functions.

63. $x^2 + y^2 = 4$

64. $(x - 1)^2 + (y + 2)^2 = 1$

65. $x = y^2$

66. $x = (y + 2)^2 - 1$

67. $x = y^2 + 2y + 1$

68. $x = 4y^2 + 4y + 1$

11.3 The Ellipse and Hyperbola

In this section:

▶ The Ellipse

▶ The Hyperbola

In this section we study the remaining two conic sections: the ellipse and the hyperbola.

The Ellipse

An ellipse can be obtained by intersecting a plane and a cone, as was shown in Fig. 11.1. We can also give a definition of an ellipse in terms of points and distance.

Figure 11.17

> **Ellipse**
>
> An **ellipse** is the set of all points in a plane such that the sum of their distances from two fixed points is a constant. Each fixed point is called a **focus** (plural: foci).

An easy way to draw an ellipse is illustrated in Fig. 11.17. A string is attached at two fixed points, and a pencil is used to take up the slack. As the pencil is moved around the paper, the sum of the distances of the pencil point

Figure 11.18

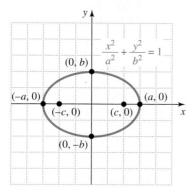

Figure 11.19

from the two fixed points remains constant. Of course, the length of the string is that constant. You may wish to try this.

Like the parabola, the ellipse also has interesting reflecting properties. All light or sound waves emitted from one focus are reflected off the ellipse to concentrate at the other focus (see Fig. 11.18). This property is used in light fixtures where a concentration of light at a point is desired or in a whispering gallery such as Statuary Hall in the U.S. Capitol Building.

The orbits of the planets around the sun and satellites around the earth are elliptical. For the orbit of the earth around the sun, the sun is at one focus. For the elliptical path of an earth satellite, the earth is at one focus and a point in space is the other focus.

Figure 11.19 shows an ellipse with foci $(c, 0)$ and $(-c, 0)$. The origin is the center of this ellipse. In general, the **center** of an ellipse is a point midway between the foci. The ellipse in Fig. 11.19 has x-intercepts at $(a, 0)$ and $(-a, 0)$ and y-intercepts at $(0, b)$ and $(0, -b)$. The distance formula can be used to write the following equation for this ellipse. (See Exercise 47.)

Equation of an Ellipse Centered at the Origin

An ellipse centered at $(0, 0)$ with foci at $(\pm c, 0)$ and constant sum $2a$ has equation

$$\frac{x^2}{a^2} + \frac{y^2}{b^2} = 1,$$

where a, b, and c are positive real numbers with $c^2 = a^2 - b^2$.

To draw a nice-looking ellipse, we would locate the foci and use string as shown in Fig. 11.17. We can get a rough sketch of an ellipse centered at the origin by using the x- and y-intercepts only.

EXAMPLE 1   **Graphing an ellipse**

Find the x- and y-intercepts for the ellipse and sketch its graph.

$$\frac{x^2}{9} + \frac{y^2}{4} = 1$$

Solution

To find the y-intercepts, let $x = 0$ in the equation:

$$\frac{0}{9} + \frac{y^2}{4} = 1$$

$$\frac{y^2}{4} = 1$$

$$y^2 = 4$$

$$y = \pm 2$$

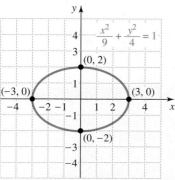

Figure 11.20

To find x-intercepts, let $y = 0$. We get $x = \pm 3$. The four intercepts are $(0, 2)$, $(0, -2)$, $(3, 0)$, and $(-3, 0)$. Plot the intercepts and draw an ellipse through them as in Figure 11.20.   ■

Ellipses, like circles, may be centered at any point in the plane. To get the equation of an ellipse centered at (h, k), we replace x by $x - h$ and y by $y - k$ in the equation of the ellipse centered at the origin.

> ### Equation of an Ellipse Centered at (h, k)
>
> An ellipse centered at (h, k) has equation
>
> $$\frac{(x - h)^2}{a^2} + \frac{(y - k)^2}{b^2} = 1,$$
>
> where a and b are positive real numbers.

EXAMPLE 2   **An ellipse with center (h, k)**
Sketch the graph of the ellipse:

$$\frac{(x - 1)^2}{9} + \frac{(y + 2)^2}{4} = 1$$

Solution

The graph of this ellipse is exactly the same size and shape as the ellipse

$$\frac{x^2}{9} + \frac{y^2}{4} = 1,$$

which was graphed in Example 1. However, the center for

$$\frac{(x - 1)^2}{9} + \frac{(y + 2)^2}{4} = 1$$

is $(1, -2)$. The denominator 9 is used to determine that the ellipse passes through points that are three units to the left and three units to the right of the center: $(4, -2)$ and $(-2, -2)$. See Fig. 11.21. The denominator 4 is used to determine that the ellipse passes through points that are two units above and two units below the center: $(1, 0)$ and $(1, -4)$. We draw an ellipse using these four points, just as we did for an ellipse centered at the origin. ∎

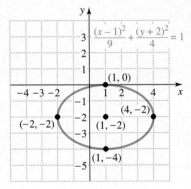

Figure 11.21

The Hyperbola

A hyperbola is the curve that occurs at the intersection of a cone and a plane, as was shown in Fig. 11.1 at the beginning of this chapter. A hyperbola can also be defined in terms of points and distance.

> ### Hyperbola
>
> A **hyperbola** is the set of all points in the plane such that the difference of their distances from two fixed points (foci) is constant.

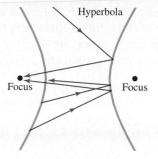

Figure 11.22

Like the parabola and the ellipse, the hyperbola also has reflecting properties. If a light ray is aimed at one focus, it is reflected off the hyperbola and goes to the other focus, as shown in Fig. 11.22. Hyperbolic mirrors are used in conjunction with parabolic mirrors in telescopes.

The definitions of a hyperbola and an ellipse are similar, and so are their equations. However, their graphs are very different. Figure 11.23 shows a hyperbola in which the distance from a point on the hyperbola to the closer focus is N and the distance to the farther focus is M. The value $M - N$ is the same for every point on the hyperbola.

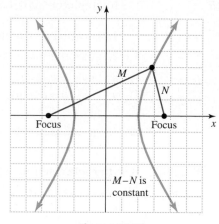

Figure 11.23

A hyperbola has two parts called **branches.** These branches look like parabolas, but they are not parabolas. The branches of the hyperbola shown in Fig. 11.24 get closer and closer to the dashed lines, called **asymptotes,** but never intersect them. The asymptotes are used as guidelines in sketching a hyperbola. The asymptotes are found by extending the diagonals of the **fundamental rectangle,** shown in Fig. 11.24. The key to drawing a hyperbola is getting the fundamental rectangle and extending its diagonals to get the asymptotes. You will learn how to find the fundamental rectangle from the equation of a hyperbola. The hyperbola in Fig. 11.24 opens to the left and right.

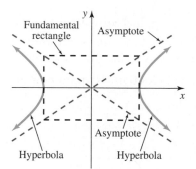

Figure 11.24

If we start with foci at $(\pm c, 0)$ and a positive number a, then we can use the definition of a hyperbola to derive the following equation of a hyperbola in which the constant difference between the distances to the foci is $2a$.

Equation of a Hyperbola Centered at (0, 0)

A hyperbola centered at $(0, 0)$ with foci $(c, 0)$ and $(-c, 0)$ and constant difference $2a$ has equation

$$\frac{x^2}{a^2} - \frac{y^2}{b^2} = 1,$$

where a, b, and c are positive real numbers such that $c^2 = a^2 + b^2$.

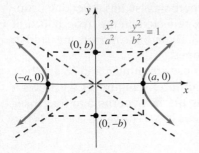

Figure 11.25

The graph of a general equation for a hyperbola is shown in Fig. 11.25. Notice that the fundamental rectangle extends to the x-intercepts along the x-axis and extends b units above and below the origin along the y-axis. The facts necessary for graphing a hyperbola centered at the origin and opening to the left and to the right are listed as follows.

> **Graphing a Hyperbola Centered at the Origin, Opening Left and Right**
>
> To graph the hyperbola $\dfrac{x^2}{a^2} - \dfrac{y^2}{b^2} = 1$:
>
> 1. Locate the x-intercepts at $(a, 0)$ and $(-a, 0)$.
> 2. Draw the fundamental rectangle through $(\pm a, 0)$ and $(0, \pm b)$.
> 3. Draw the extended diagonals of the rectangle to use as asymptotes.
> 4. Draw the hyperbola to the left and right approaching the asymptotes.

EXAMPLE 3   **A hyperbola opening left and right**

Sketch the graph of $\dfrac{x^2}{36} - \dfrac{y^2}{9} = 1$, and find the equations of its asymptotes.

Solution

The x-intercepts are $(6, 0)$ and $(-6, 0)$. Draw the fundamental rectangle through these x-intercepts and the points $(0, 3)$ and $(0, -3)$. Extend the diagonals of the fundamental rectangle to get the asymptotes. Now draw a hyperbola passing through the x-intercepts and approaching the asymptotes as shown in Fig. 11.26. From the graph in Fig. 11.26 we see that the slopes of the asymptotes are $\frac{1}{2}$ and $-\frac{1}{2}$. Because the y-intercept for both asymptotes is the origin, their equations are $y = \frac{1}{2}x$ and $y = -\frac{1}{2}x$.

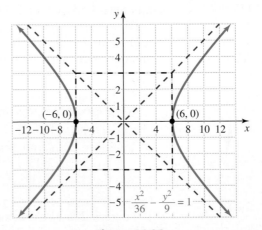

Figure 11.26

A hyperbola may open up and down. In this case the graph intersects only the y-axis. The facts necessary for graphing a hyperbola that opens up and down are summarized as follows.

> ### Graphing a Hyperbola Centered at the Origin, Opening Up and Down
>
> To graph the hyperbola $\dfrac{y^2}{b^2} - \dfrac{x^2}{a^2} = 1$:
>
> 1. Locate the y-intercepts at $(0, b)$ and $(0, -b)$.
> 2. Draw the fundamental rectangle through $(0, \pm b)$ and $(\pm a, 0)$.
> 3. Draw the extended diagonals of the rectangle to use as asymptotes.
> 4. Draw the hyperbola opening up and down approaching the asymptotes.

EXAMPLE 4   **A hyperbola opening up and down**

Graph the hyperbola $\dfrac{y^2}{9} - \dfrac{x^2}{4} = 1$ and find the equations of its asymptotes.

Solution

If $y = 0$, we get

$$-\frac{x^2}{4} = 1$$

$$x^2 = -4.$$

Because this equation has no solution, the graph has no x-intercepts. Let $x = 0$ to find the y-intercepts:

$$\frac{y^2}{9} = 1$$

$$y^2 = 9$$

$$y = \pm 3$$

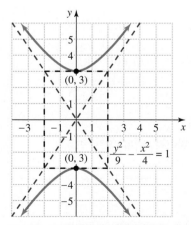

Figure 11.27

The y-intercepts are $(0, 3)$ and $(0, -3)$, and the hyperbola opens up and down. From $a^2 = 4$ we get $a = 2$. So the fundamental rectangle extends to the intercepts $(0, 3)$ and $(0, -3)$ on the y-axis and to the points $(2, 0)$ and $(-2, 0)$ along the x-axis. We extend the diagonals of the rectangle and draw the graph of the hyperbola as shown in Fig. 11.27. From the graph in Fig. 11.27 we see that the asymptotes have slopes $\dfrac{3}{2}$ and $-\dfrac{3}{2}$. Because the y-intercept for both asymptotes is the origin, their equations are $y = \dfrac{3}{2}x$ and $y = -\dfrac{3}{2}x$. ∎

EXAMPLE 5 **A hyperbola not in standard form**

Sketch the graph of the hyperbola $4x^2 - y^2 = 4$.

$x^2 - \dfrac{y^2}{4} = 1$

Figure 11.28

Solution

First write the equation in standard form. Divide each side by 4 to get

$$x^2 - \frac{y^2}{4} = 1.$$

There are no y-intercepts. If $y = 0$, then $x = \pm 1$. The hyperbola opens left and right with x-intercepts at $(1, 0)$ and $(-1, 0)$. The fundamental rectangle extends to the intercepts along the x-axis and to the points $(0, 2)$ and $(0, -2)$ along the y-axis. We extend the diagonals of the rectangle for the asymptotes and draw the graph as shown in Figure 11.28. ■

Warm-ups

True or false? Explain your answer.

1. The x-intercepts of the ellipse $\dfrac{x^2}{36} + \dfrac{y^2}{25} = 1$ are $(5, 0)$ and $(-5, 0)$.

2. The graph of $\dfrac{x^2}{9} + \dfrac{y}{4} = 1$ is an ellipse.

3. If the foci of an ellipse coincide, then the ellipse is a circle.

4. The graph of $2x^2 + y^2 = 2$ is an ellipse centered at the origin.

5. The y-intercepts of $x^2 + \dfrac{y^2}{3} = 1$ are $(0, \sqrt{3}\,)$ and $(0, -\sqrt{3}\,)$.

6. The graph of $\dfrac{x^2}{9} + \dfrac{y}{4} = 1$ is a hyperbola.

7. The graph of $\dfrac{x^2}{25} - \dfrac{y^2}{16} = 1$ has y-intercepts at $(0, 4)$ and $(0, -4)$.

8. The hyperbola $\dfrac{y^2}{9} - x^2 = 1$ opens up and down.

9. The graph of $4x^2 - y^2 = 4$ is a hyperbola.

10. The asymptotes of a hyperbola are the extended diagonals of a rectangle.

11.3 EXERCISES

Sketch the graph of each ellipse. See Example 1.

1. $\dfrac{x^2}{9} + \dfrac{y^2}{4} = 1$

2. $\dfrac{x^2}{9} + \dfrac{y^2}{16} = 1$

3. $\dfrac{x^2}{9} + y^2 = 1$

4. $x^2 + \dfrac{y^2}{4} = 1$

5. $\dfrac{x^2}{36} + \dfrac{y^2}{25} = 1$

6. $\dfrac{x^2}{25} + \dfrac{y^2}{49} = 1$

7. $\dfrac{x^2}{24} + \dfrac{y^2}{5} = 1$

8. $\dfrac{x^2}{6} + \dfrac{y^2}{17} = 1$

9. $9x^2 + 16y^2 = 144$

10. $9x^2 + 25y^2 = 225$

11. $25x^2 + y^2 = 25$

12. $x^2 + 16y^2 = 16$

13. $4x^2 + 9y^2 = 1$

14. $25x^2 + 16y^2 = 1$

Sketch the graph of each ellipse. See Example 2.

15. $\dfrac{(x-3)^2}{4} + \dfrac{(y-1)^2}{9} = 1$

16. $\dfrac{(x+5)^2}{49} + \dfrac{(y-2)^2}{25} = 1$

17. $\dfrac{(x+1)^2}{16} + \dfrac{(y-2)^2}{25} = 1$

18. $\dfrac{(x-3)^2}{36} + \dfrac{(y+4)^2}{64} = 1$

19. $(x-2)^2 + \dfrac{(y+1)^2}{36} = 1$

20. $\dfrac{(x+3)^2}{9} + (y+1)^2 = 1$

Sketch the graph of each hyperbola and write the equations of its asymptotes. See Examples 3–5.

21. $\dfrac{x^2}{4} - \dfrac{y^2}{9} = 1$

22. $\dfrac{x^2}{16} - \dfrac{y^2}{9} = 1$

23. $\dfrac{y^2}{4} - \dfrac{x^2}{25} = 1$

24. $\dfrac{y^2}{9} - \dfrac{x^2}{16} = 1$

25. $\dfrac{x^2}{25} - y^2 = 1$

26. $x^2 - \dfrac{y^2}{9} = 1$

27. $x^2 - \dfrac{y^2}{25} = 1$

28. $\dfrac{x^2}{9} - y^2 = 1$

29. $9x^2 - 16y^2 = 144$

30. $9x^2 - 25y^2 = 225$

31. $x^2 - y^2 = 1$

32. $y^2 - x^2 = 1$

Graph both equations of each system on the same coordinate axes. Use elimination of variables (as in Section 10.4) to find all points of intersection.

33. $\dfrac{x^2}{4} + \dfrac{y^2}{9} = 1$
$x^2 - \dfrac{y^2}{9} = 1$

34. $x^2 - \dfrac{y^2}{4} = 1$
$\dfrac{x^2}{9} + \dfrac{y^2}{4} = 1$

35. $\dfrac{x^2}{4} + \dfrac{y^2}{16} = 1$
$x^2 + y^2 = 1$

36. $x^2 + \dfrac{y^2}{9} = 1$
$x^2 + y^2 = 4$

37. $x^2 + y^2 = 4$
$x^2 - y^2 = 1$

38. $x^2 + y^2 = 16$
$x^2 - y^2 = 4$

39. $x^2 + 9y^2 = 9$
$x^2 + y^2 = 4$

40. $x^2 + y^2 = 25$
$x^2 + 25y^2 = 25$

41. $x^2 + 9y^2 = 9$
$y = x^2 - 1$

42. $4x^2 + y^2 = 4$
$y = 2x^2 - 2$

43. $9x^2 - 4y^2 = 36$
$2y = x - 2$

44. $25y^2 - 9x^2 = 225$
$y = 3x + 3$

Solve each problem.

45. *Marine navigation.* The loran (long-range navigation) system is used by boaters to determine their location at sea. The loran unit on a boat measures the difference in time that it takes for radio signals from pairs of fixed points to reach a boat. The unit then finds the equations of two hyperbolas that pass through the location of the boat. Suppose that the hyperbolas $x^2 - 3y^2 = 1$ and $4y^2 - x^2 = 1$ both pass through the location of a boat in the first quadrant. Find the exact location of the boat.

Figure for Exercise 45

46. *Sonic boom.* An aircraft traveling at supersonic speed creates a cone-shaped wave that intersects the ground along a hyperbola, as shown in the accompanying figure. A thunderlike sound is heard at any point on the hyperbola. This sonic boom travels along the ground, following the aircraft. The area where the sonic boom is most noticeable is called the *boom carpet.* The width of the boom carpet is roughly five times the altitude of the aircraft. Suppose that the equation of the hyperbola in the figure is

$$\dfrac{x^2}{400} - \dfrac{y^2}{100} = 1,$$

where the units are miles and the width of the boom carpet is measured 40 miles behind the aircraft. Find the altitude of the aircraft.

Figure for Exercise 46

Getting More Involved

47. *Cooperative learning.* Let (x, y) be an arbitrary point on an ellipse with foci $(c, 0)$ and $(-c, 0)$ for $c > 0$. The following equation expresses the fact that the distance from (x, y) to $(c, 0)$ plus the distance from (x, y) to $(-c, 0)$ is the constant value $2a$ (for $a > 0$):

$$\sqrt{(x - c)^2 + (y - 0)^2} + \sqrt{(x - (-c))^2 + (y - 0)^2} = 2a$$

Working in groups, simplify this equation. First get the radicals on opposite sides of the equation, then square both sides twice to eliminate the square roots. Finally, let $b^2 = a^2 - c^2$ to get the equation

$$\frac{x^2}{a^2} + \frac{y^2}{b^2} = 1.$$

48. *Cooperative learning.* Let (x, y) be an arbitrary point on a hyperbola with foci $(c, 0)$ and $(-c, 0)$ for $c > 0$. The following equation expresses the fact that the distance from (x, y) to $(c, 0)$ minus the distance from (x, y) to $(-c, 0)$ is the constant value $2a$ (for $a > 0$):

$$\sqrt{(x - c)^2 + (y - 0)^2} - \sqrt{(x - (-c))^2 + (y - 0)^2} = 2a$$

Working in groups, simplify the equation. You will need to square both sides twice to eliminate the square roots. Finally, let $b^2 = c^2 - a^2$ to get the equation

$$\frac{x^2}{a^2} - \frac{y^2}{b^2} = 1.$$

Graphing Calculator Exercises

49. Graph $y_1 = \sqrt{x^2 - 1}$, $y_2 = -\sqrt{x^2 - 1}$, $y_3 = x$, and $y_4 = -x$ to get the graph of the hyperbola $x^2 - y^2 = 1$ along with its asymptotes. Use the viewing window $-3 \leq x \leq 3$ and $-3 \leq y \leq 3$. Notice how the branches of the hyperbola approach the asymptotes.

50. Graph the same four functions in Exercise 49, but use $-30 \leq x \leq 30$ and $-30 \leq y \leq 30$ as the viewing window. What happened to the hyperbola?

11.4 Second-Degree Inequalities

In this section we graph second-degree inequalities and systems of inequalities involving second-degree inequalities.

In this section:
▶ Graphing a Second-Degree Inequality
▶ Systems of Inequalities

Graphing a Second-Degree Inequality

A second-degree inequality is an inequality involving squares of at least one of the variables. Changing the equal sign to an inequality symbol for any of the equations of the conic sections gives us a second-degree inequality. Second-degree inequalities are graphed in the same manner as linear inequalities.

EXAMPLE 1   **A second-degree inequality**

Graph the inequality $y < x^2 + 2x - 3$.

Solution

We first graph $y = x^2 + 2x - 3$. This parabola has x-intercepts at $(1, 0)$ and $(-3, 0)$, y-intercept at $(0, -3)$, and vertex at $(-1, -4)$. The graph of the parabola is drawn with a dashed line, as shown in Figure 11.29. The graph of the parabola divides the plane into two regions. Every point on one side of the parabola satisfies the inequality $y < x^2 + 2x - 3$, and every point on the other side satisfies the inequality $y > x^2 + 2x - 3$. To determine which side is which, we test a point that is not on the parabola, say $(0, 0)$. Because

$$0 < 0^2 + 2 \cdot 0 - 3$$

is false, the region not containing the origin is shaded, as in Figure 11.29.

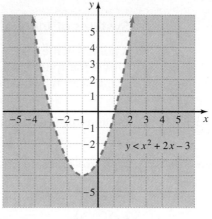

Figure 11.29   ■

EXAMPLE 2   **A second-degree inequality**

Graph the inequality $x^2 + y^2 < 9$.

Solution

Figure 11.30

The graph of $x^2 + y^2 = 9$ is a circle of radius 3 centered at the origin. The circle divides the plane into two regions. Every point in one region satisfies $x^2 + y^2 < 9$, and every point in the other region satisfies $x^2 + y^2 > 9$. To identify the regions, we pick a point and test it. Select $(0, 0)$. The inequality

$$0^2 + 0^2 < 9$$

is true. Because $(0, 0)$ is inside the circle, all points inside the circle satisfy the inequality $x^2 + y^2 < 9$, as shown in Fig. 11.30. The points outside the circle satisfy the inequality $x^2 + y^2 > 9$.   ■

Figure 11.31

EXAMPLE 3   **A second-degree inequality**

Graph the inequality $\dfrac{x^2}{4} - \dfrac{y^2}{9} > 1$.

Solution

First graph the hyperbola $\dfrac{x^2}{4} - \dfrac{y^2}{9} = 1$. Because the hyperbola shown in Fig. 11.31 divides the plane into three regions, we select a test point in each region and check to see whether it satisfies the inequality. Testing the points $(-3, 0)$, $(0, 0)$, and $(3, 0)$ gives us the inequalities

$$\frac{(-3)^2}{4} - \frac{0^2}{9} > 1, \qquad \frac{0^2}{4} - \frac{0^2}{9} > 1, \qquad \text{and} \qquad \frac{3^2}{4} - \frac{0^2}{9} > 1.$$

Because only the first and third inequalities are correct, we shade only the regions containing $(3, 0)$ and $(-3, 0)$, as shown in Fig. 11.31.   ■

Systems of Inequalities

A point is in the solution set to a system of inequalities if it satisfies all inequalities of the system. We graph a system of inequalities by first determining the graph of each inequality and then finding the intersection of the graphs.

EXAMPLE 4   **Systems of second-degree inequalities**

Graph the system of inequalities:

$$\frac{y^2}{4} - \frac{x^2}{9} > 1$$

$$\frac{x^2}{9} + \frac{y^2}{16} < 1$$

Solution

Figure 11.32(a) shows the graph of the first inequality. In Fig. 11.32(b) we have the graph of the second inequality. In Fig. 11.32(c) we have shaded only the points that satisfy both inequalities. Figure 11.32(c) is the graph of the system.

(a)

(b)

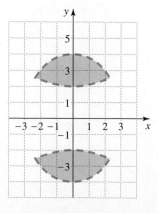

(c)

Figure 11.32   ■

Warm-ups

True or false? Explain your answer.

1. The graph of $x^2 + y = 4$ is a circle of radius 2.
2. The graph of $x^2 + 9y^2 = 9$ is an ellipse.
3. The graph of $y^2 = x^2 + 1$ is a hyperbola.
4. The point $(0, 0)$ satisfies the inequality $2x^2 - y < 3$.
5. The graph of the inequality $y > x^2 - 3x + 2$ contains the origin.
6. The origin should be used as a test point for graphing $x^2 > y$.
7. The solution set to $x^2 + 3x + y^2 + 8y + 3 < 0$ includes the origin.
8. The graph of $x^2 + y^2 < 4$ is the region inside a circle of radius 2.
9. The point $(0, 4)$ satisfies $x^2 - y^2 < 1$ and $y > x^2 - 2x + 3$.
10. The point $(0, 0)$ satisfies $x^2 + y^2 < 1$ and $y < x^2 + 1$.

11.4 EXERCISES

Graph each inequality. See Examples 1–3.

1. $y > x^2$
2. $y \leq x^2 + 1$
3. $y < x^2 - x$
4. $y > x^2 + x$
5. $y > x^2 - x - 2$
6. $y < x^2 + x - 6$
7. $x^2 + y^2 \leq 9$
8. $x^2 + y^2 > 16$
9. $x^2 + 4y^2 > 4$
10. $4x^2 + y^2 \leq 4$
11. $4x^2 - 9y^2 < 36$
12. $25x^2 - 4y^2 > 100$
13. $(x - 2)^2 + (y - 3)^2 < 4$
14. $(x + 1)^2 + (y - 2)^2 > 1$
15. $x^2 + y^2 > 1$
16. $x^2 + y^2 < 25$
17. $4x^2 - y^2 > 4$
18. $x^2 - 9y^2 \leq 9$
19. $y^2 - x^2 \leq 1$
20. $x^2 - y^2 > 1$
21. $x > y$
22. $x < 2y - 1$

Graph the solution set to each system of inequalities. See Example 4.

23. $x^2 + y^2 < 9$
    $y > x$
24. $x^2 + y^2 > 1$
    $x > y$
25. $x^2 - y^2 > 1$
    $x^2 + y^2 < 4$
26. $y^2 - x^2 < 1$
    $x^2 + y^2 > 9$
27. $y > x^2 + x$
    $y < 5$
28. $y > x^2 + x - 6$
    $y < x + 3$
29. $y \geq x + 2$
    $y \leq 2 - x$
30. $y \geq 2x - 3$
    $y \leq 3 - 2x$
31. $4x^2 - y^2 < 4$
    $x^2 + 4y^2 > 4$
32. $x^2 - 4y^2 < 4$
    $x^2 + 4y^2 > 4$
33. $x - y < 0$
    $y + x^2 < 1$
34. $y + 1 > x^2$
    $x + y < 2$
35. $y < 5x - x^2$
    $x^2 + y^2 < 9$
36. $y < x^2 + 4x$
    $x^2 + y^2 < 16$

37. $y \geq 3$
    $x \leq 1$
38. $x > -3$
    $y < 2$
39. $4y^2 - 9x^2 < 36$
    $x^2 + y^2 < 16$
40. $25y^2 - 16x^2 < 400$
    $x^2 + y^2 > 4$
41. $y < x^2$
    $x^2 + y^2 < 1$
42. $y > x^2$
    $4x^2 + y^2 < 4$

Solve the problem.

43. **Buried treasure.** An old pirate on his deathbed gave the following description of where he had buried some treasure on a deserted island: "Starting at the large palm tree, I walked to the north and then to the east, and there I buried the treasure. I walked at least 50 paces to get to that spot, but I was not more than 50 paces, as the crow flies, from the large palm tree. I am sure that I walked farther in the northerly direction than in the easterly direction." With the large palm tree at the origin and the positive y-axis pointing to the north, graph the possible locations of the treasure.

Graphing Calculator Exercises

44. Use graphs to find an ordered pair that is in the solution set to the system of inequalities:

    $$y > x^2 - 2x + 1$$
    $$y < -1.1(x - 4)^2 + 5$$

    Verify that your answer satisfies both inequalities.

45. Use graphs to find the solution set to the system of inequalities:

    $$y > 2x^2 - 3x + 1$$
    $$y < -2x^2 - 8x - 1$$

COLLABORATIVE ACTIVITIES

Focus on Comets

Grouping: 2 students per group
Topic: Conic sections

Conic sections are used to model many different things in the natural world. Astronomers use mirrors in the shape of parabolas in telescopes. They have learned that planets can have elliptical as well as circular orbits around the sun and that comets may have orbits that resemble hyperbolas, parabolas, or ellipses. In this activity we will consider comets with these three types of orbits.

The orbit that a comet will take as it approaches the sun depends upon its velocity (as well as other factors). If it has enough velocity to escape from the pull of the sun, it may take either a parabolic orbit or a hyperbolic orbit. If it doesn't have enough velocity, then it will take an elliptical orbit. Of course, a comet that has a parabolic or hyperbolic orbit will not come back again around our sun. Only comets with elliptical orbits do we see again.

For the problems below, round all your answers to two decimal places.

1. Halley's comet is in an elliptical orbit about the sun with the sun at one of the foci. In this problem we will make a scale model of the orbit of Halley's comet about the sun. We will choose our coordinate system so that the ellipse is centered at (0, 0). Halley's comet comes within 0.6 astronomical unit[1] from the sun at its closest point and 35 astronomical units at its farthest point. The position of the comet at these points will correspond to the horizontal vertices. Find the equation for the ellipse. Determine where the foci should be. On a piece of cardboard, put thumbtacks at the foci. Determine the length of string you will need to draw the

ellipse using the scale 1 centimeter = 1 astronomical unit. Draw the comet's orbit, indicating which focus is the sun.

2. If the velocity of a comet equals the escape velocity (it is going just fast enough to get away), then its orbit will be parabolic. The sun will be at the focus of the parabola. We will place the vertex at the point (0, 0); in this case the vertex will be where the comet is closest to the sun. Suppose we have a comet that is 0.75 astronomical unit from the sun at its closest point. Find the equation for the parabola. Graph the parabola.

3. If the velocity of a comet is greater than the escape velocity (it can easily escape from the sun's gravitational pull), its path will resemble one-half of a hyperbola with the sun as one focus. Assuming that we have a left- and a right-opening hyperbola, we can model the path of such a comet along one of its branches. We will center the hyperbola at (0, 0), and we will place the sun at the left focus and draw the path of the comet as it approaches from the left. We know that the comet will be 1.5 astronomical units from the sun at its closest point. Assume the sun is at the point $(-3, 0)$. Draw a sketch of this scenario. What is the equation for the hyperbola? Graph the hyperbola.

Extension: Graph all three equations on a graphing calculator or computer.

---

[1]An astronomical unit is the distance of the earth from the sun. The earth's orbit is almost circular and so is about the same distance from the sun at any point in its orbit.

Wrap-up   CHAPTER 11

SUMMARY

Parabola		Examples

$y = a(x - h)^2 + k$

Opens up for $a > 0$, down for $a < 0$
Vertex at (h, k)
To find focus and directrix, use $a = \dfrac{1}{4p}$.
Distance from vertex to focus or directrix is $|p|$.

$x = \frac{1}{8}(x - 1)^2 - 2$

$(1, -2)$

$y = ax^2 + bx + c$

Opens up for $a > 0$, down for $a < 0$
The x-coordinate of the vertex is $\dfrac{-b}{2a}$.
Find the y-coordinate of the vertex by evaluating
$y = ax^2 + bx + c$ for $x = \dfrac{-b}{2a}$.

$(1, 3)$
$y = 2x^2 - 4x + 5$

Circle		Examples

$(x - h)^2 + (y - k)^2 = r^2$

Center (h, k)
Radius r (for $r > 0$)

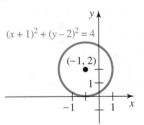

$(x + 1)^2 + (y - 2)^2 = 4$

$(-1, 2)$

Centered at origin
$x^2 + y^2 = r^2$

Center $(0, 0)$
Radius r (for $r > 0$)

$x^2 + y^2 = 9$

Ellipse

Centered at origin

$$\frac{x^2}{a^2} + \frac{y^2}{b^2} = 1$$

Center: $(0, 0)$
x-intercepts: $(a, 0)$ and $(-a, 0)$
y-intercepts: $(0, b)$ and $(0, -b)$
Foci: $(\pm c, 0)$ if $a^2 > b^2$ and $c^2 = a^2 - b^2$
$\quad\quad (0, \pm c)$ if $b^2 > a^2$ and $c^2 = b^2 - a^2$

Arbitrary center

$$\frac{(x-h)^2}{a^2} + \frac{(y-k)^2}{b^2} = 1$$

Center: (h, k)

Hyperbola

Centered at origin
Opening left and right

$$\frac{x^2}{a^2} - \frac{y^2}{b^2} = 1$$

Center $(0, 0)$
x-intercepts: $(a, 0)$ and $(-a, 0)$

y-intercepts: none

Centered at origin
Opening up and down

$$\frac{y^2}{b^2} - \frac{x^2}{a^2} = 1$$

Centered at $(0, 0)$
x-intercepts: none

y-intercepts: $(0, b)$ and $(0, -b)$

Second-Degree Inequalities		**Examples**
Solution set for a single inequality	Graph the boundary curve obtained by replacing the inequality symbol by the equal sign. Use test points to determine which regions satisfy the inequality.	$x^2 + y^2 < 16$
Solution set for a system of inequalities	Graph the boundary curves. Then select a test point in each region. Shade only the regions for which the test point satisfies all inequalities of the system.	$x^2 + y^2 < 16$ $y > x^2 - 1$

REVIEW EXERCISES

11.1 *Sketch the graph of each parabola. Determine the x- and y-intercepts, vertex, focus, directrix, and axis of symmetry for each.*

1. $y = x^2 + 3x - 18$

2. $y = x - x^2$

3. $y = x^2 + 3x + 2$

4. $y = -x^2 - 3x + 4$

5. $y = -\dfrac{1}{2}(x - 2)^2 + 3$

6. $y = \dfrac{1}{4}(x + 1)^2 - 2$

Write each equation in the form $y = a(x - h)^2 + k$, and identify the vertex of the parabola.

7. $y = 2x^2 - 8x + 1$

8. $y = -2x^2 - 6x - 1$

9. $y = -\dfrac{1}{2}x^2 - x + \dfrac{1}{2}$

10. $y = \dfrac{1}{4}x^2 + x - 9$

Solve each problem.

11. *Maximum product.* Find two numbers that have the largest possible product among numbers that have a sum of 6.

12. *Minimum product.* Find two numbers that have the smallest possible product among numbers that have a difference of 10.

13. *Maximum area.* Otto has 90 feet of fencing that he plans to use to enclose a rectangular area. What dimensions will maximize the area of the rectangle?

14. *Maximum area.* Suppose Otto already has a 6-foot-wide gate in place and he wants to use the gate and an additional 90 feet of fencing to enclose a rectangular area. What dimensions will maximize the area of the rectangle?

11.2 *Determine the center and radius of each circle, and sketch its graph.*

15. $x^2 + y^2 = 100$

16. $x^2 + y^2 = 20$

17. $(x - 2)^2 + (y + 3)^2 = 81$

18. $x^2 + 2x + y^2 = 8$

19. $9y^2 + 9x^2 = 4$

20. $x^2 + 4x + y^2 - 6y - 3 = 0$

Write the standard equation for each circle with the given center and radius.

21. Center $(0, 3)$, radius 6

22. Center $(0, 0)$, radius $\sqrt{6}$

23. Center $(2, -7)$, radius 5

24. Center $\left(\frac{1}{2}, -3\right)$, radius $\frac{1}{2}$

11.3 *Sketch the graph of each ellipse.*

25. $\dfrac{x^2}{36} + \dfrac{y^2}{49} = 1$

26. $\dfrac{x^2}{25} + y^2 = 1$

27. $25x^2 + 4y^2 = 100$

28. $6x^2 + 4y^2 = 24$

Sketch the graph of each hyperbola.

29. $\dfrac{x^2}{49} - \dfrac{y^2}{36} = 1$

30. $\dfrac{y^2}{25} - \dfrac{x^2}{49} = 1$

31. $4x^2 - 25y^2 = 100$

32. $6y^2 - 16x^2 = 96$

11.4 *Graph each inequality.*

33. $4x - 2y > 3$

34. $y < x^2 - 3x$

35. $y^2 < x^2 - 1$

36. $y^2 < 1 - x^2$

37. $4x^2 + 9y^2 > 36$

38. $x^2 + y > 2x - 1$

Graph the solution set to each system of inequalities.

39. $y < 3x - x^2$
$x^2 + y^2 < 9$

40. $x^2 - y^2 < 1$
$y < 1$

41. $4x^2 + 9y^2 > 36$
$x^2 + y^2 < 9$

42. $y^2 - x^2 > 4$
$y^2 + 16x^2 < 16$

Miscellaneous

Identify each equation as the equation of a straight line, parabola, circle, hyperbola, or ellipse. Try to do these without rewriting the equations.

43. $x^2 = y^2 + 1$

44. $x = y + 1$

45. $x^2 = 1 - y^2$

46. $x^2 = y + 1$

47. $x^2 + x = 1 - y^2$

48. $(x - 3)^2 + (y + 2)^2 = 7$

49. $x^2 + 4x = 6y - y^2$

50. $4x + 6y = 1$

51. $\dfrac{x^2}{3} - \dfrac{y^2}{5} = 1$

52. $x^2 + \dfrac{y^2}{3} = 1$

53. $4y^2 - x^2 = 8$

54. $9x^2 + y = 9$

Sketch the graph of each equation.

55. $x^2 = 4 - y^2$

56. $x^2 = 4y^2 + 4$

57. $x^2 = 4y + 4$

58. $x = 4y + 4$

59. $x^2 = 4 - 4y^2$

60. $x^2 = 4y - y^2$

61. $x^2 = 4 - (y - 4)^2$

62. $(x - 2)^2 + (y - 4)^2 = 4$

Write the equation of the circle with the given features.

63. Centered at the origin and passing through $(3, 4)$

64. Centered at $(2, -3)$ and passing through $(-1, 4)$

65. Centered at $(-1, 5)$ with radius 6

66. Centered at $(0, -3)$ and passing through the origin

Write the equation of the parabola with the given features.

67. Focus $(1, 4)$ and directrix $y = 2$

68. Focus $(-2, 1)$ and directrix $y = 5$

69. Vertex $(0, 0)$ and focus $\left(0, \frac{1}{4}\right)$

70. Vertex $(1, 2)$ and focus $\left(1, \frac{3}{2}\right)$

71. Vertex $(0, 0)$, passing through $(3, 2)$, and opening upward

72. Vertex $(1, 3)$, passing though $(0, 0)$, and opening downward

Solve each system of equations.

73. $x^2 + y^2 = 25$
$y = -x + 1$

74. $x^2 - y^2 = 1$
$x^2 + y^2 = 7$

75. $4x^2 + y^2 = 4$
$x^2 - y^2 = 21$

76. $y = x^2 + x$
$y = -x^2 + 3x + 12$

CHAPTER 11 TEST

Sketch the graph of each equation.

1. $x^2 + y^2 = 25$

2. $\dfrac{x^2}{16} - \dfrac{y^2}{25} = 1$

3. $y^2 + 4x^2 = 4$

4. $y = x^2 + 4x + 4$

5. $y^2 - 4x^2 = 4$

6. $y = -x^2 - 2x + 3$

Sketch the graph of each inequality.

7. $x^2 - y^2 < 9$

8. $x^2 + y^2 > 9$

9. $y > x^2 - 9$

Graph the solution set to each system of inequalities.

10. $x^2 + y^2 < 9$
    $x^2 - y^2 > 1$

11. $y < -x^2 + x$
    $y < x - 4$

Solve each system of equations.

12. $y = x^2 - 2x - 8$
    $y = 7 - 4x$

13. $x^2 + y^2 = 12$
    $y = x^2$

Solve each problem.

14. Find the center and radius of the circle
    $x^2 + 2x + y^2 + 10y = 10$.

15. Find the vertex, focus, and directrix of the parabola
    $y = x^2 + x + 3$. State the axis of symmetry and
    whether the parabola opens up or down.

16. Write the equation $y = \frac{1}{2}x^2 - 3x - \frac{1}{2}$ in the form
    $y = a(x - h)^2 + k$.

17. If a ball is thrown straight up with a velocity of 64
    feet per second from a height of 20 feet, then its
    height s above the ground at time t is given by
    $s = -16t^2 + 64t + 20$. What is the maximum height
    reached by the ball?

18. Write the equation of a circle with center $(-1, 3)$ that
    passes through $(2, 5)$.

Tying It All Together

Sketch the graph of each equation.

1. $y = 9x - x^2$

2. $y = 9x$

3. $y = (x - 9)^2$

4. $y^2 = 9 - x^2$

5. $y = 9x^2$

6. $y = |9x|$

7. $4x^2 + 9y^2 = 36$

8. $4x^2 - 9y^2 = 36$

9. $y = 9 - x$

10. $y = 9^x$

Find the following products.

11. $(x + 2y)^2$

12. $(x + y)(x^2 + 2xy + y^2)$

13. $(a + b)^3$

14. $(a - 3b)^2$

15. $(2a + 1)(3a - 5)$

16. $(x - y)(x^2 + xy + y^2)$

Solve each system of equations.

17. $2x - 3y = -4$
$x + 2y = 5$

18. $x^2 + y^2 = 25$
$x + y = 7$

19. $2x - y + z = 7$
$x - 2y - z = 2$
$x + y + z = 2$

20. $y = x^2$
$y - 2x = 3$

Solve each formula for the specified variable.

21. $ax + b = 0$, for x

22. $wx^2 + dx + m = 0$, for x

23. $A = \dfrac{1}{2}h(B + b)$, for B

24. $\dfrac{1}{x} + \dfrac{1}{y} = \dfrac{1}{2}$, for x

25. $L = m + mxt$, for m

26. $y = 3a\sqrt{t}$, for t

Solve each problem.

27. Write the equation of the line in slope-intercept form that goes through the points $(2, -3)$ and $(-4, 1)$.

28. Write the equation of the line in slope-intercept form that contains the origin and is perpendicular to the line $2x - 4y = 5$.

29. Write the equation of the circle that has center $(2, 5)$ and passes through the point $(-1, -1)$.

30. Find the center and radius of the circle $x^2 + 3x + y^2 - 6y = 0$.

Perform the computations with complex numbers.

31. $2i(3 + 5i)$

32. i^6

33. $(2i - 3) + (6 - 7i)$

34. $(3 + i\sqrt{2})^2$

35. $(2 - 3i)(5 - 6i)$

36. $(3 - i) + (-6 + 4i)$

37. $(5 - 2i)(5 + 2i)$

38. $(2 - 3i) \div (2i)$

39. $(4 + 5i) \div (1 - i)$

40. $\dfrac{4 - \sqrt{-8}}{2}$

Solve.

41. ***Going bananas.*** Salvadore has observed that when bananas are $0.30 per pound, he sells 250 pounds per day, and when bananas are $0.40 per pound, he sells only 200 pounds per day.

a) Assume that the number of pounds sold, q, is a linear function of the price per pound, x, and find that function. (*Hint:* Find the equation of the line through $(0.30, 250)$ and $(0.40, 200)$.)

b) Salvadore's daily revenue in dollars is the product of the number of pounds sold and the price per pound. Write the revenue as a function of x.

c) Graph the revenue function. What price per pound maximizes his revenue? What is his maximum possible revenue?

CHAPTER 12

Sequences and Series

Everyone realizes the importance of investing for the future. Some people go to great pains to study the markets and make wise investment decisions. Some stay away from investing because they do not want to take the time to study the possibilities and take chances. However, according to *Money* magazine (April 1994), the most important factor in investing is making regular investments. According to *Money*, if you had invested $5000 in the stock market every year at the market high for that year (the worst time to invest) for the last 40 years, your investment would have been worth $2.8 million in 1994.

A sequence of periodic investments earning a fixed rate of interest can be thought of as a geometric sequence. In this chapter you will learn how to find the sum of a geometric sequence and calculate the future value of a sequence of periodic investments. In Exercise 52 of Section 12.4 you will see how *Money* magazine calculated the value of $5000 invested each year in the stock market for 40 years.

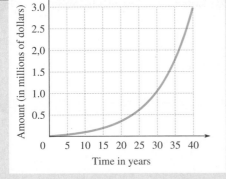

12.1 Sequences

The word "sequence" is a familiar word. We may speak of a sequence of events or say that something is out of sequence. In this section we give the mathematical definition of a sequence.

Definition

In mathematics we think of a sequence as a list of numbers. Each number in the sequence is called a **term** of the sequence. There is a first term, a second term, a third term, and so on. For example, the daily high temperature readings in Minot, North Dakota, for the first 10 days in January can be thought of as a finite sequence with 10 terms:

$$-9, -2, 8, -11, 0, 6, 14, 1, -5, -11$$

The set of all positive even integers,

$$2, 4, 6, 8, 10, 12, 14, \ldots,$$

can be thought of as an infinite sequence.

To give a precise definition of sequence, we use the terminology of functions. The list of numbers is the range of the function.

> **Sequence**
>
> A **finite sequence** is a function whose domain is the set of positive integers less than or equal to some fixed positive integer. An **infinite sequence** is a function whose domain is the set of all positive integers.

When the domain is apparent, we will refer to either a finite sequence or an infinite sequence simply as a sequence. For the independent variable of the function we will usually use n (for natural number) rather than x. For the dependent variable we write a_n (read "a sub n") rather than y. We call a_n the **nth term,** or the **general term** of the sequence. Rather than use the $f(x)$ notation for functions, we will define sequences with formulas. When n is used as a variable, we will assume that it represents natural numbers only.

EXAMPLE 1   **Listing terms of a finite sequence**

List all of the terms of each finite sequence.

a) $a_n = n^2$ for $1 \leq n \leq 5$

b) $a_n = \dfrac{1}{n + 2}$ for $1 \leq n \leq 4$

Solution

a) Using the natural numbers from 1 through 5 in $a_n = n^2$, we get

$$a_1 = 1^2 = 1,$$
$$a_2 = 2^2 = 4,$$
$$a_3 = 3^2 = 9,$$
$$a_4 = 4^2 = 16, \quad \text{and}$$
$$a_5 = 5^2 = 25.$$

The five terms of this sequence are 1, 4, 9, 16, and 25. We often refer to the listing of the terms of the sequence as the sequence.

b) Using the natural numbers from 1 through 4 in $a_n = \dfrac{1}{n+2}$, we get the terms

$$a_1 = \frac{1}{1+2} = \frac{1}{3}, \qquad a_2 = \frac{1}{2+2} = \frac{1}{4},$$

$$a_3 = \frac{1}{5}, \quad \text{and} \quad a_4 = \frac{1}{6}.$$

The four terms of the sequence are $\frac{1}{3}, \frac{1}{4}, \frac{1}{5},$ and $\frac{1}{6}$. ∎

EXAMPLE 2 **Listing terms of an infinite sequence**

List the first three terms of the infinite sequence whose nth term is

$$a_n = \frac{(-1)^n}{2^{n+1}}.$$

Solution

Using the natural numbers 1, 2, and 3 in the formula for the nth term yields

$$a_1 = \frac{(-1)^1}{2^{1+1}} = -\frac{1}{4}, \qquad a_2 = \frac{(-1)^2}{2^{2+1}} = \frac{1}{8}, \qquad \text{and} \qquad a_3 = \frac{(-1)^3}{2^{3+1}} = -\frac{1}{16}.$$

We write the sequence as follows:

$$-\frac{1}{4}, \frac{1}{8}, -\frac{1}{16}, \ldots$$ ∎

Finding a Formula for the nth Term

We often know the terms of a sequence and want to write a formula that will produce those terms. To write a formula for the nth term of a sequence, examine the terms and look for a pattern to them. Each term is a function of the term number. The first term corresponds to $n = 1$, the second term corresponds to $n = 2$, and so on.

EXAMPLE 3   **A familiar sequence**

Write the general term for the infinite sequence

$$3, 5, 7, 9, 11, \ldots.$$

Solution

The even numbers are all multiples of 2 and can be represented as $2n$. Because each odd number is 1 more than an even number, a formula for the nth term might be

$$a_n = 2n + 1.$$

To be sure, we write out a few terms using the formula:

$$a_1 = 2(1) + 1 = 3$$
$$a_2 = 2(2) + 1 = 5$$
$$a_3 = 2(3) + 1 = 7$$

So the general term is $a_n = 2n + 1$.   ■

CAUTION   There can be more than one formula that produces the given terms of a sequence. For example, the sequence

$$1, 2, 4, \ldots$$

could have nth term $a_n = 2^{n-1}$ or $a_n = \frac{1}{2}n^2 - \frac{1}{2}n + 1$. The first three terms for both of these sequences are identical, but their fourth terms are different. ⊘

EXAMPLE 4   **A sequence with alternating signs**

Write the general term for the infinite sequence

$$1, -\frac{1}{4}, \frac{1}{9}, -\frac{1}{16}, \ldots.$$

Solution

To obtain the alternating signs, we use powers of -1. Because any even power of -1 is positive and any odd power of -1 is negative, we use $(-1)^{n+1}$. The denominators are the squares of the positive integers. So the nth term of this infinite sequence is given by the formula

$$a_n = \frac{(-1)^{n+1}}{n^2}.$$

Check this sequence by using this formula to find the first four terms.   ■

In the next example we use a sequence to model a physical situation.

EXAMPLE 5   **The bouncing ball**

Suppose that a ball always rebounds $\frac{2}{3}$ of the height from which it falls and the ball is dropped from a height of 6 feet. Write a sequence whose terms are the heights from which the ball falls. What is a formula for the nth term of this sequence?

Solution

On the first fall, the ball travels 6 feet, as shown in Fig. 12.1. On the second fall, it travels $\frac{2}{3}$ of 6, or 4 feet. On the third fall, it travels $\frac{2}{3}$ of 4, or $\frac{8}{3}$ feet, and so on. We write the sequence as follows:

$$6, 4, \frac{8}{3}, \frac{16}{9}, \frac{32}{27}, \ldots$$

The nth term can be written by using powers of $\frac{2}{3}$:

$$a_n = 6\left(\frac{2}{3}\right)^{n-1}$$

Figure 12.1

MATH AT WORK

Fabric Designer

Most of us find an upholstered chair, sink into it, and remark on the comfort. Before design consultant Audrey Jordan sits down, she often looks at the fabric to observe the color and texture—and especially to see whether the fabric is one of her original designs. Fabric design is more than just an idea that is printed on a piece of cloth. Consideration must be given to the end product, which could be anything from a handbag to a large sofa. Colors and themes must be chosen with both current trends and styles in mind. Sometimes a design will be an overall or nondirectional pattern, such as polka dots, which can be cut randomly. More often, it will have a specific theme, such as fruit, which can be cut and sewn in only one direction.

For all products, one of the main considerations is the vertical repeat. A good portion of textile machinery is standardized for vertical repeats of 27 inches or fractions thereof. For example, the vertical repeat could be every $13\frac{1}{2}$ inches or every 9 inches. Even though the horizontal repeat can vary, Ms. Jordan must consider both the horizontal and vertical repeats for a particular end product.

In Exercise 41 of this section you will find the standard vertical repeats for a textile machine.

Warm-ups

True or false? Explain your answer.

1. The nth term of the sequence 2, 4, 6, 8, 10, ... is $a_n = 2n$.
2. The nth term of the sequence 1, 3, 5, 7, 9, ... is $a_n = 2n - 1$.
3. A sequence is a function.
4. The domain of a finite sequence is the set of positive integers.
5. The nth term of $-1, 4, -9, 16, -25, \ldots$ is $a_n = (-1)^{n+1} n^2$.
6. For the infinite sequence $b_n = \dfrac{1}{n}$, the independent variable is $\dfrac{1}{n}$.
7. For the sequence $c_n = n^3$, the dependent variable is c_n.
8. The sixth term of the sequence $a_n = (-1)^{n+1} 2^n$ is 64.
9. The symbol a_n is used for the dependent variable of a sequence.
10. The tenth term of the so-called Fibonacci sequence 1, 1, 2, 3, 5, 8, 13, ... is 55.

12.1 EXERCISES

List all terms of each finite sequence. See Example 1.

1. $a_n = n^2$ for $1 \le n \le 8$
2. $a_n = -n^2$ for $1 \le n \le 4$
3. $b_n = \dfrac{(-1)^n}{n}$ for $1 \le n \le 10$
4. $b_n = \dfrac{(-1)^{n+1}}{n}$ for $1 \le n \le 6$
5. $c_n = (-2)^{n-1}$ for $1 \le n \le 5$
6. $c_n = (-3)^{n-2}$ for $1 \le n \le 5$
7. $a_n = 2^{-n}$ for $1 \le n \le 6$
8. $a_n = 2^{-n+2}$ for $1 \le n \le 5$
9. $b_n = 2n - 3$ for $1 \le n \le 7$
10. $b_n = 2n + 6$ for $1 \le n \le 7$
11. $c_n = n^{-1/2}$ for $1 \le n \le 5$
12. $c_n = n^{1/2} 2^{-n}$ for $1 \le n \le 4$

Write the first four terms of the infinite sequence whose nth term is given. See Example 2.

13. $a_n = \dfrac{1}{n^2 + n}$
14. $b_n = \dfrac{1}{(n+1)(n+2)}$
15. $b_n = \dfrac{1}{2n - 5}$
16. $a_n = \dfrac{4}{2n + 5}$
17. $c_n = (-1)^n (n - 2)^2$
18. $c_n = (-1)^n (2n - 1)^2$
19. $a_n = \dfrac{(-1)^{2n}}{n^2}$
20. $a_n = (-1)^{2n+1} 2^{n-1}$

Write a formula for the general term of each infinite sequence. See Examples 3 and 4.

21. 1, 3, 5, 7, 9, ...
22. 5, 7, 9, 11, 13, ...
23. 1, −1, 1, −1, ...
24. −1, 1, −1, 1, ...
25. 0, 2, 4, 6, 8, ...
26. 4, 6, 8, 10, 12, ...
27. 3, 6, 9, 12, ...
28. 4, 8, 12, 16, ...
29. 4, 7, 10, 13, ...
30. 3, 7, 11, 15, ...
31. −1, 2, −4, 8, −16, ...
32. 1, −3, 9, −27, ...
33. 0, 1, 4, 9, 16, ...
34. 0, 1, 8, 27, 64, ...

Solve each problem. See Example 5.

35. **Football penalties.** A football is on the 8-yard line, and five penalties in a row are given that move the ball half the distance to the (closest) goal. Write a sequence of five terms that specify the location of the ball after each penalty.

36. **Infestation.** Leona planted nine acres of soybeans, but by the end of each week, insects had destroyed $\frac{1}{3}$ of the acreage that was healthy at the beginning of the week. How many acres does she have left after six weeks?

37. **Increasing prices.** The list price of a new 1993 Ford Taurus GL Sedan was $16,441 (*Edmund's 1993 New Car Prices*). Suppose that the manufacturer increases the price of a new Taurus GL Sedan by 5% each year. Find the price of a new Taurus GL Sedan to the nearest dollar each year through 1998.

38. *Constant increase.* The list price of a new 1993 Mazda Miata Convertible was $16,130 (*Edmund's 1993 New Car Prices*). Suppose that the manufacturer increases the price by $1000 each year. Find the price of a new Miata Convertible each year through 1998.

39. *Economic impact.* To assess the economic impact of a factory on a community, economists consider the annual amount the factory spends in the community, then the portion of that money that is respent in the community, then the portion of the respent money that is respent in the community, and so on. Suppose that a garment manufacturer spends $1 million annually in its community and 80% of all money received in the community is respent in the community. Find the first four terms of the economic impact sequence.

40. *Less impact.* The rate at which money is respent in a community varies from community to community. Find the first four terms of the economic impact sequence for the manufacturer in Exercise 39, assuming that only 50% of money received in the community is respent in the community.

41. *Fabric design.* A fabric designer must take into account the capability of textile machines to produce material with vertical repeats. A textile machines can be set up for a vertical repeat every $\frac{27}{n}$ inches, where n is a natural number. Write the first five terms of the sequence $a_n = \frac{27}{n}$, which gives the possible vertical repeats for a textile machine.

42. *Musical tones.* The note middle C on a piano is tuned so that the string vibrates at 262 cycles per second, or 262 Hz. The C note one octave higher is tuned to 524 Hz. The tuning for the 11 notes in between using the method called *equal temperament* is determined by the sequence $a_n = 262 \cdot 2^{n/12}$. Find the tuning for the 11 notes in between.

Getting More Involved

43. *Discussion.* Everyone has two (biological) parents, four grandparents, eight great-grandparents, 16 great-great-grandparents, and so on. If we put the word "great" in front of the word "grandparents" 35 times, then how many of this type of relative do you have? Is this more or less than the present population of the earth? Give reasons for your answers.

44. *Discussion.* If you deposit 1 cent into your piggy bank on September 1 and each day thereafter deposit twice as much as on the previous day, then how much will you be depositing on September 30? The total amount deposited for the month can be found without adding up all 30 deposits. Look at how the amount on deposit is increasing each day and see whether you can find the total for the month. Give reasons for your answers.

45. *Cooperative learning.* Working in groups, have someone in each group make up a formula for a_n, the nth term of a sequence, but not show it to the other group members. Write the terms of the sequence on a piece of paper one at a time. After each term is given, ask whether anyone knows the next term. When the group can correctly give the next term, ask for a formula for the nth term.

46. *Exploration.* Find a real-life sequence in which all of the terms are the same. Find one in which each term after the first is one larger than the previous term. Find out what the sequence of fines is on your campus for your first, second, third, and fourth parking ticket.

47. *Exploration.* Consider the sequence whose nth term is $a_n = (0.999)^n$.
a) Calculate a_{100}, a_{1000}, and $a_{10,000}$.
b) What happens to a_n as n gets larger and larger?

12.2 Series

In this section:
▶ Summation Notation
▶ Series
▶ Changing the Index

If you make a sequence of bank deposits, then you might be interested in the total value of the terms of the sequence. Of course, if the sequence has only a few terms, you can simply add them. In Sections 12.4 and 12.5 we will develop formulas that give the sum of the terms for certain finite and infinite sequences. In this section you will first learn a notation for expressing the sum of the terms of a sequence.

Summation Notation

To describe the sum of the terms of a sequence, we use **summation notation.** The Greek letter Σ (sigma) is used to indicate sums. For example, the sum of the first five terms of the sequence $a_n = n^2$ is written as

$$\sum_{n=1}^{5} n^2.$$

You can read this notation as "the sum of n^2 for n between 1 and 5, inclusive." To find the sum, we let n take the values 1 through 5 in the expression n^2:

$$\sum_{n=1}^{5} n^2 = 1^2 + 2^2 + 3^2 + 4^2 + 5^2$$

$$= 1 + 4 + 9 + 16 + 25$$

$$= 55$$

In this context the letter n is the **index of summation.** Other letters may also be used. For example, the expressions

$$\sum_{n=1}^{5} n^2, \quad \sum_{j=1}^{5} j^2, \quad \text{and} \quad \sum_{i=1}^{5} i^2$$

all have the same value. Note that i is used as a variable here and not as an imaginary number.

EXAMPLE 1   **Evaluating a sum in summation notation**

Find the value of the expression

$$\sum_{i=1}^{3} (-1)^i (2i + 1).$$

Solution

Replace i by 1, 2, and 3, and then add the results:

$$\sum_{i=1}^{3} (-1)^i (2i + 1) = (-1)^1 [2(1) + 1] + (-1)^2 [2(2) + 1] + (-1)^3 [2(3) + 1]$$

$$= -3 + 5 - 7$$

$$= -5 \qquad \blacksquare$$

Series

The sum of the terms of the sequence 1, 4, 9, 16, 25 is written as

$$1 + 4 + 9 + 16 + 25.$$

This expression is called a series. It indicates that we are to add the terms of the given sequence. The actual sum, 55, is the sum of the series.

Series

The indicated sum of the terms of a sequence is called a **series.**

Just as a sequence may be finite or infinite, a series may be finite or infinite. In this section we discuss finite series only. In Section 12.4 we will discuss one type of infinite series.

Summation notation is a convenient notation for writing a series.

EXAMPLE 2   **Converting to summation notation**

Write the series in summation notation:

$$2 + 4 + 6 + 8 + 10 + 12 + 14$$

Solution

The general term for the sequence of positive even integers is $2n$. If we let n take the values from 1 through 7, then $2n$ ranges from 2 through 14. So

$$2 + 4 + 6 + 8 + 10 + 12 + 14 = \sum_{n=1}^{7} 2n.$$ ∎

EXAMPLE 3   **Converting to summation notation**

Write the series

$$\frac{1}{2} - \frac{1}{3} + \frac{1}{4} - \frac{1}{5} + \frac{1}{6} - \frac{1}{7} + \cdots + \frac{1}{50}$$

in summation notation.

Solution

For this series we let n be 2 through 50. The expression $(-1)^n$ produces alternating signs. The series is written as

$$\sum_{n=2}^{50} \frac{(-1)^n}{n}.$$ ∎

Changing the Index

In Example 3 we saw the index go from 2 through 50, but this is arbitrary. A series can be written with the index starting at any given number.

EXAMPLE 4   **Changing the index**

Rewrite the series

$$\sum_{i=1}^{6} \frac{(-1)^i}{i^2}$$

with an index j, where j starts at 0.

Solution

To get six terms in the series, j takes the values 0 through 5. Because $i = j + 1$, we replace i by $j + 1$ in the summation notation:

$$\sum_{j=0}^{5} \frac{(-1)^{j+1}}{(j+1)^2}$$

Check that these two series have exactly the same six terms.   ∎

Warm-ups

True or false? Explain your answer.

1. A series is the indicated sum of the terms of a sequence.

2. The sum of a series can never be negative.

3. There are eight terms in the series $\displaystyle\sum_{i=2}^{10} i^3$.

4. The series $\displaystyle\sum_{i=1}^{9} (-1)^i i^2$ and $\displaystyle\sum_{j=0}^{8} (-1)^j (j+1)^2$ have the same sum.

5. The ninth term of the series $\displaystyle\sum_{i=1}^{100} \frac{(-1)^i}{(i+1)(i+2)}$ is $\dfrac{1}{110}$.

6. $\displaystyle\sum_{i=1}^{2} (-1)^i 2^i = 2$
7. $\displaystyle\sum_{i=1}^{5} 3i = 3\left(\sum_{i=1}^{5} i\right)$
8. $\displaystyle\sum_{i=1}^{5} 4 = 20$

9. $\displaystyle\sum_{i=1}^{5} 2i + \sum_{i=1}^{5} 7i = \sum_{i=1}^{5} 9i$
10. $\displaystyle\sum_{i=1}^{3} (2i+1) = \left(\sum_{i=1}^{3} 2i\right) + 1$

12.2 EXERCISES

Find the sum of each series. See Example 1.

1. $\displaystyle\sum_{i=1}^{4} i^2$
2. $\displaystyle\sum_{j=0}^{3} (j+1)^2$

3. $\displaystyle\sum_{j=0}^{5} (2j-1)$
4. $\displaystyle\sum_{i=1}^{6} (2i-3)$

5. $\displaystyle\sum_{i=1}^{5} 2^{-i}$
6. $\displaystyle\sum_{i=1}^{5} (-2)^{-i}$

7. $\displaystyle\sum_{i=1}^{10} 5i^0$
8. $\displaystyle\sum_{j=1}^{20} 3$

9. $\displaystyle\sum_{i=1}^{3} (i-3)(i+1)$
10. $\displaystyle\sum_{i=0}^{5} i(i-1)(i-2)(i-3)$

11. $\displaystyle\sum_{j=1}^{10} (-1)^j$
12. $\displaystyle\sum_{j=1}^{11} (-1)^j$

Write each series in summation notation. Use the index i, and let i begin at 1 in each summation. See Examples 2 and 3.

13. $1 + 2 + 3 + 4 + 5 + 6$

14. $2 + 4 + 6 + 8 + 10$

15. $-1 + 3 - 5 + 7 - 9 + 11$

16. $1 - 3 + 5 - 7 + 9$

17. $1 + 4 + 9 + 16 + 25 + 36$

18. $1 + 8 + 27 + 64 + 125$

19. $\dfrac{1}{3} + \dfrac{1}{4} + \dfrac{1}{5} + \dfrac{1}{6}$

20. $1 - \dfrac{1}{2} + \dfrac{1}{3} - \dfrac{1}{4} + \dfrac{1}{5} - \dfrac{1}{6}$

21. $\ln(2) + \ln(3) + \ln(4)$

22. $e^1 + e^2 + e^3 + e^4$

23. $a_1 + a_2 + a_3 + a_4$

24. $a^2 + a^3 + a^4 + a^5$

25. $x_3 + x_4 + x_5 + \cdots + x_{50}$

26. $y_1 + y_2 + y_3 + \cdots + y_{30}$

27. $w_1 + w_2 + w_3 + \cdots + w_n$

28. $m_1 + m_2 + m_3 + \cdots + m_k$

Complete the rewriting of each series using the new index as indicated. See Example 4.

29. $\displaystyle\sum_{i=1}^{5} i^2 = \sum_{j=0}$
30. $\displaystyle\sum_{i=1}^{6} i^3 = \sum_{j=0}$

31. $\displaystyle\sum_{i=0}^{12} (2i-1) = \sum_{j=1}$
32. $\displaystyle\sum_{i=1}^{3} (3i+2) = \sum_{j=0}$

33. $\displaystyle\sum_{i=4}^{8}\frac{1}{i}=\sum_{j=1}^{}$

34. $\displaystyle\sum_{i=5}^{10}2^{-i}=\sum_{j=1}^{}$

35. $\displaystyle\sum_{i=1}^{4}x^{2i+3}=\sum_{j=0}^{}$

36. $\displaystyle\sum_{i=0}^{2}x^{3-2i}=\sum_{j=1}^{}$

37. $\displaystyle\sum_{i=1}^{n}x^{i}=\sum_{j=0}^{}$

38. $\displaystyle\sum_{i=0}^{n}x^{-i}=\sum_{j=1}^{}$

Write out the terms of each series.

39. $\displaystyle\sum_{i=1}^{6}x^{i}$

40. $\displaystyle\sum_{i=1}^{5}(-1)^{i}x^{i-1}$

41. $\displaystyle\sum_{j=0}^{3}(-1)^{j}x_{j}$

42. $\displaystyle\sum_{j=1}^{5}\frac{1}{x_{j}}$

43. $\displaystyle\sum_{i=1}^{3}ix^{i}$

44. $\displaystyle\sum_{i=1}^{5}\frac{x}{i}$

A series can be used to model the situation in each of the following problems.

45. *Leap frog.* A frog with a vision problem is 1 yard away from a dead cricket. He spots the cricket and jumps halfway to the cricket. After he realizes that he has not reached the cricket, he again jumps halfway to the cricket. Write a series in summation notation to describe how far the frog has moved after nine such jumps.

46. *Compound interest.* Cleo deposited $1000 at the beginning of each year for 5 years into an account paying 10% interest compounded annually. Write a series using summation notation to describe how much she has in the account at the end of the fifth year. Note that the first $1000 will receive interest for five years, the second $1000 will receive interest for four years, and so on.

47. *Total economic impact.* In Exercise 39 of Section 12.1 we described a factory that spends $1 million annually in a community in which 80% of all money received in the community is respent in the community. Use summation notation to write the sum of the first four terms of the economic impact sequence for the factory.

48. *Total spending.* Suppose that you earn $1 on January 1, $2 on January 2, $3 on January 3, and so on. Use summation notation to write the sum of your earnings for the entire month of January.

Getting More Involved

49. *Discussion.* What is the difference between a sequence and a series?

50. *Discussion.* For what value of n is $\displaystyle\sum_{i=1}^{n}\frac{1}{i}>4$?

12.3   Arithmetic Sequences and Series

In this section:
▶ Arithmetic Sequences
▶ Arithmetic Series

We defined sequences and series in Sections 12.1 and 12.2. In this section you will study a special type of sequence known as an arithmetic sequence. You will also study the series corresponding to this sequence.

Arithmetic Sequences

Consider the following sequence:

$$5, 9, 13, 17, 21, \ldots$$

This sequence is called an arithmetic sequence because of the pattern for the terms. Each term is 4 larger than the previous term.

> **Arithmetic Sequence**
>
> A sequence in which each term after the first is obtained by adding a fixed amount to the previous term is called an **arithmetic sequence.**

The fixed amount is called the **common difference** and is denoted by the letter d. If a_1 is the first term, then the second term is $a_1 + d$. The third term is $a_1 + 2d$, the fourth term is $a_1 + 3d$, and so on.

> **Formula for the nth Term of an Arithmetic Sequence**
>
> The nth term, a_n, of an arithmetic sequence with first term a_1 and common difference d is
>
> $$a_n = a_1 + (n - 1)d.$$

EXAMPLE 1 **The nth term of an arithmetic sequence**

Write a formula for the nth term of the arithmetic sequence

$$5, 9, 13, 17, 21, \ldots.$$

Solution

Each term of the sequence after the first is 4 more than the previous term. Because the common difference is 4 and the first term is 5, the nth term is given by

$$a_n = 5 + (n - 1)4.$$

We can simplify this expression to get

$$a_n = 4n + 1.$$ ∎

In the next example the common difference is negative.

EXAMPLE 2 **An arithmetic sequence of decreasing terms**

Write a formula for the nth term of the arithmetic sequence

$$4, 1, -2, -5, -8, \ldots.$$

Solution

Each term is 3 less than the previous term, so $d = -3$. Because $a_1 = 4$, we can write the nth term as

$$a_n = 4 + (n - 1)(-3),$$

or

$$a_n = -3n + 7.$$ ∎

In the next example we find some terms of an arithmetic sequence using a given formula for the nth term.

EXAMPLE 3 **Writing terms of an arithmetic sequence**

Write the first five terms of the sequence in which $a_n = 3 + (n - 1)6$.

Solution

Let n take the values from 1 through 5, and find a_n:

$$a_1 = 3 + (1 - 1)6 = 3$$
$$a_2 = 3 + (2 - 1)6 = 9$$
$$a_3 = 3 + (3 - 1)6 = 15$$
$$a_4 = 3 + (4 - 1)6 = 21$$
$$a_5 = 3 + (5 - 1)6 = 27$$

Notice that $a_n = 3 + (n - 1)6$ gives the general term for an arithmetic sequence with first term 3 and common difference 6. Because each term after the first is 6 more than the previous term, the first five terms that we found are correct. ■

The formula $a_n = a_1 + (n - 1)d$ involves four variables: a_1, a_n, n, and d. If we know the values of any three of these variables, we can find the fourth.

EXAMPLE 4 **Finding a missing term of an arithmetic sequence**

Find the twelfth term of the arithmetic sequence whose first term is 2 and whose fifth term is 14.

Solution

Before finding the twelfth term, we use the given information to find the missing common difference. Let $n = 5$, $a_1 = 2$, and $a_5 = 14$ in the formula $a_n = a_1 + (n - 1)d$ to find d:

$$14 = 2 + (5 - 1)d$$
$$14 = 2 + 4d$$
$$12 = 4d$$
$$3 = d$$

Now use $a_1 = 2$, $d = 3$ and $n = 12$ in $a_n = a_1 + (n - 1)d$ to find a_{12}:

$$a_{12} = 2 + (12 - 1)3$$
$$a_{12} = 35$$ ■

Arithmetic Series

The indicated sum of an arithmetic sequence is called an **arithmetic series.** For example, the series

$$2 + 4 + 6 + 8 + 10 + \cdots + 54$$

is an arithmetic series because there is a common difference of 2 between the terms.

We can find the actual sum of this arithmetic series without adding all of the terms. Write the series in increasing order, and below that write the series in decreasing order. We then add the corresponding terms:

$$
\begin{array}{rl}
S = & 2 + \ 4 + \ 6 + \ 8 + \cdots + 52 + 54 \\
S = & 54 + 52 + 50 + 48 + \cdots + \ 4 + \ 2 \\
\hline
2S = & 56 + 56 + 56 + 56 + \cdots + 56 + 56
\end{array}
$$

Now, how many times does 56 appear in the sum on the right? Because

$$2 + 4 + 6 + \cdots + 54 = 2 \cdot 1 + 2 \cdot 2 + 2 \cdot 3 + \cdots + 2 \cdot 27,$$

there are 27 terms in this sum. Because 56 appears 27 times on the right, we have $2S = 27 \cdot 56$, or

$$S = \frac{27 \cdot 56}{2} = 27 \cdot 28 = 756.$$

If $S_n = a_1 + a_2 + a_3 + \cdots + a_n$ is any arithmetic series, then we can find its sum using the same technique. Rewrite S_n as follows:

$$
\begin{aligned}
S_n &= a_1 & + (a_1 + d) &+ (a_1 + 2d) + \cdots + a_n \\
S_n &= a_n & + (a_n - d) &+ (a_n - 2d) + \cdots + a_1 \\
\hline
2S_n &= (a_1 + a_n) + (a_1 + a_n) + (a_1 + a_n) + \cdots + (a_1 + a_n) & \text{Add.}
\end{aligned}
$$

Because $(a_1 + a_n)$ appears n times on the right, we have $2S_n = n(a_1 + a_n)$. Divide each side by 2 to get the following formula.

Sum of an Arithmetic Series

The sum, S_n, of the first n terms of an arithmetic series with first term a_1 and nth term a_n, is given by

$$S_n = \frac{n}{2}(a_1 + a_n).$$

EXAMPLE 5 **The sum of an arithmetic series**

Find the sum of the positive integers from 1 to 100 inclusive.

Solution

The described series, $1 + 2 + 3 + \cdots + 100$, has 100 terms. So we can use $n = 100$, $a_1 = 1$, and $a_n = 100$ in the formula for the sum of an arithmetic series:

$$S_n = \frac{n}{2}(a_1 + a_n)$$

$$S_{100} = \frac{100}{2}(1 + 100)$$

$$= 50(101)$$

$$= 5050$$

EXAMPLE 6 **The sum of an arithmetic series**

Find the sum of the series

$$12 + 16 + 20 + \cdots + 84.$$

Solution

This series is an arithmetic series with $a_n = 84$, $a_1 = 12$, and $d = 4$. To get the number of terms, n, we use $a_n = a_1 + (n - 1)d$:

$$84 = 12 + (n - 1)4$$
$$84 = 8 + 4n$$
$$76 = 4n$$
$$19 = n$$

Now find the sum of these 19 terms:

$$S_{19} = \frac{19}{2}(12 + 84) = 912$$

Warm-ups

True or false? Explain your answer.

1. The arithmetic sequence 3, 1, −1, −3, −5, . . . has common difference 2.

2. The sequence 2, 5, 9, 14, 20, 27, . . . is an arithmetic sequence.

3. The sequence 2, 4, 2, 0, 2, 4, 2, 0, . . . is an arithmetic sequence.

4. The nth term of an arithmetic sequence with first term a_1 and common difference d is given by the formula $a_n = a_1 + nd$.

5. If $a_1 = 5$ and $a_3 = 10$ in an arithmetic sequence, then $a_4 = 15$.

6. If $a_1 = 6$ and $a_3 = 2$ in an arithmetic sequence, then $a_2 = 10$.

7. An arithmetic series is the indicated sum of an arithmetic sequence.

8. The series $\sum_{i=1}^{5} (3 + 2i)$ is an arithmetic series.

9. The sum of the first n counting numbers is $\dfrac{n(n + 1)}{2}$.

10. The sum of the even integers from 8 through 28 inclusive is $5(8 + 28)$.

12.3 EXERCISES

Write a formula for the nth term of each arithmetic sequence. See Examples 1 and 2.

1. 0, 6, 12, 18, 24, . . .

2. 0, 5, 10, 15, 20, . . .

3. 7, 12, 17, 22, 27, . . .

4. 4, 15, 26, 37, 48, . . .

5. −4, −2, 0, 2, 4, . . .

6. −3, 0, 3, 6, 9, . . .

7. 5, 1, −3, −7, −11, . . .

8. 8, 5, 2, −1, −4, . . .

9. −2, −9, −16, −23, . . .

10. −5, −7, −9, −11, −13, . . .

11. −3, −2.5, −2, −1.5, −1, . . .

12. −2, −1.25, −0.5, −0.25, . . .

13. −6, −6.5, −7, −7.5, −8, . . .

14. 1, 0.5, 0, −0.5, −1, . . .

In Exercises 15–28, write the first five terms of the arithmetic series whose nth term is given. See Example 3.

15. $a_n = 9 + (n - 1)4$

16. $a_n = 13 + (n - 1)6$

17. $a_n = 7 + (n - 1)(-2)$

18. $a_n = 6 + (n - 1)(-3)$

19. $a_n = -4 + (n - 1)3$

20. $a_n = -19 + (n - 1)12$

21. $a_n = -2 + (n - 1)(-3)$

22. $a_n = -1 + (n - 1)(-2)$

23. $a_n = -4n - 3$ **24.** $a_n = -3n + 1$

25. $a_n = 0.5n + 4$ **26.** $a_n = 0.3n + 1$

27. $a_n = 20n + 1000$ **28.** $a_n = -600n + 4000$

Find the indicated part of each arithmetic sequence. See Example 4.

29. Find the eighth term of the sequence that has a first term of 9 and a common difference of 6.

30. Find the twelfth term of the sequence that has a first term of -2 and a common difference of -3.

31. Find the common difference if the first term is 6 and the twentieth term is 82.

32. Find the common difference if the first term is -8 and the ninth term is -64.

33. If the common difference is -2 and the seventh term is 14, then what is the first term?

34. If the common difference is 5 and the twelfth term is -7, then what is the first term?

35. Find the sixth term of the sequence that has a fifth term of 13 and a first term of -3.

36. Find the eighth term of the sequence that has a sixth term of -42 and a first term of 3.

Find the sum of each given series. See Example 5 and 6.

37. $1 + 2 + 3 + \cdots + 48$

38. $1 + 2 + 3 + \cdots + 12$

39. $8 + 10 + 12 + \cdots + 36$

40. $9 + 12 + 15 + \cdots + 72$

41. $-1 + (-7) + (-13) + \cdots + (-73)$

42. $-7 + (-12) + (-17) + \cdots + (-72)$

43. $-6 + (-1) + 4 + 9 + \cdots + 64$

44. $-9 + (-1) + 7 + \cdots + 103$

45. $20 + 12 + 4 + (-4) + \cdots + (-92)$

46. $19 + 1 + (-17) + \cdots + (-125)$

47. $\sum_{i=1}^{12} (3i - 7)$ **48.** $\sum_{i=1}^{7} (-4i + 6)$

49. $\sum_{i=1}^{11} (-5i + 2)$ **50.** $\sum_{i=1}^{19} (3i - 5)$

Solve each problem using the ideas of arithmetic sequences and series.

51. *Increasing salary.* If a lab technician has a salary of $22,000 her first year and is due to get a $500 raise each year, then what will her salary be in her seventh year?

52. *Seven years of salary.* What is the total salary for seven years of work for the lab technician of Exercise 51?

53. *Light reading.* On the first day of October an English teacher suggests to his students that they read five pages of a novel and every day thereafter increase their daily reading by two pages. If his students follow this suggestion, then how many pages will they read during October?

54. *Heavy penalties.* If an air-conditioning system is not completed by the agreed-upon date, the contractor pays a penalty of $500 for the first day that it is overdue, $600 for the second day, $700 for the third day, and so on. If the system is completed 10 days late, then what is the total amount of the penalties that the contractor must pay?

Getting More Involved

55. *Discussion.* Which of the following sequences is not an arithmetic sequence? Explain your answer.

a) $\dfrac{1}{2}, 1, \dfrac{3}{2}, \ldots$ b) $\dfrac{1}{2}, \dfrac{1}{3}, \dfrac{1}{4}, \ldots$

c) $5, 0, -5, \ldots$ d) $2, 3, 4, \ldots$

56. *Discussion.* What is the smallest value of n for which $\sum_{i=1}^{n} \dfrac{i}{2} > 50$?

12.4 Geometric Sequences and Series

In this section:

▶ Geometric Sequences
▶ Finite Geometric Series
▶ Infinite Geometric Series
▶ Annuities

In Section 12.3 you studied the arithmetic sequences and series. In this section you will study sequences in which each term is a *multiple* of the term preceding it. You will also learn how to find the sum of the corresponding series.

Geometric Sequences

Consider the following sequence:

$$3, 6, 12, 24, 48, \ldots$$

Unlike an arithmetic sequence, these terms do not have a common difference, but there is a simple pattern to the terms. Each term after the first is twice the term preceding it. Such a sequence is called a geometric sequence.

> **Geometric Sequence**
>
> A sequence in which each term after the first is obtained by multiplying the preceding term by a constant is called a **geometric sequence.**

The constant is denoted by the letter r and is called the **common ratio.** If a_1 is the first term, then the second term is $a_1 r$. The third term is $a_1 r^2$, the fourth term is $a_1 r^3$, and so on. We can write a formula for the nth term of a geometric sequence by following this pattern.

> **Formula for the nth Term of a Geometric Sequence**
>
> The nth term, a_n, of a geometric sequence with first term a_1 and common ratio r is
>
> $$a_n = a_1 r^{n-1}.$$

The first term and the common ratio determine all of the terms of a geometric sequence.

EXAMPLE 1 **Finding the nth term**

Write a formula for the nth term of the geometric sequence

$$6, 2, \frac{2}{3}, \frac{2}{9}, \ldots.$$

Solution

We can obtain the common ratio by dividing any term after the first by the term preceding it. So

$$r = 2 \div 6 = \frac{1}{3}.$$

Because each term after the first is $\frac{1}{3}$ of the term preceding it, the nth term is given by

$$a_n = 6\left(\frac{1}{3}\right)^{n-1}.$$ ∎

EXAMPLE 2 **Finding the nth term**

Find a formula for the nth term of the geometric sequence

$$2, -1, \frac{1}{2}, -\frac{1}{4}, \ldots.$$

Solution

We obtain the ratio by dividing a term by the term preceding it:

$$r = -1 \div 2 = -\frac{1}{2}$$

Each term after the first is obtained by multiplying the preceding term by $-\frac{1}{2}$. The formula for the nth term is

$$a_n = 2\left(-\frac{1}{2}\right)^{n-1}.$$ ∎

In the next example we use the formula for the nth term to write some terms of a geometric sequence.

EXAMPLE 3 **Writing the terms**

Write the first five terms of the geometric sequence whose nth term is

$$a_n = 3(-2)^{n-1}.$$

Solution

Let n take the values 1 through 5 in the formula for the nth term:

$$a_1 = 3(-2)^{1-1} = 3$$
$$a_2 = 3(-2)^{2-1} = -6$$
$$a_3 = 3(-2)^{3-1} = 12$$
$$a_4 = 3(-2)^{4-1} = -24$$
$$a_5 = 3(-2)^{5-1} = 48$$

Notice that $a_n = 3(-2)^{n-1}$ gives the general term for a geometric sequence with first term 3 and common ratio -2. Because every term after the first can be obtained by multiplying the previous term by -2, the terms 3, -6, 12, -24, and 48 are correct. ∎

The formula for the nth term involves four variables: a_n a_1, r, and n. If we know the value of any three of them, we can find the value of the fourth.

EXAMPLE 4   **Finding a missing term**

Find the first term of a geometric sequence whose fourth term is 8 and whose common ratio is $\frac{1}{2}$.

Solution

Let $a_4 = 8$, $r = \frac{1}{2}$, and $n = 4$ in the formula $a_n = a_1 r^{n-1}$:

$$8 = a_1 \left(\frac{1}{2} \right)^{4-1}$$

$$8 = a_1 \cdot \frac{1}{8}$$

$$64 = a_1$$

So the first term is 64.   ◼

Finite Geometric Series

Consider the following series:

$$1 + 2 + 4 + 8 + 16 + \cdots + 512$$

The terms of this series are the terms of a finite geometric sequence. The indicated sum of a geometric sequence is called a **geometric series.**

We can find the actual sum of this finite geometric series by using a technique similar to the one used for the sum of an arithmetic series. Let

$$S = 1 + 2 + 4 + 8 + \cdots + 256 + 512.$$

Because the common ratio is 2, multiply each side by -2:

$$-2S = -2 - 4 - 8 - \cdots - 512 - 1024$$

Adding the last two equations eliminates all but two of the terms on the right:

$$
\begin{aligned}
S &= 1 + 2 + 4 + 8 + \cdots + 256 + 512 \\
-2S &= -2 - 4 - 8 - \cdots - 512 - 1024 \\
\hline
-S &= 1 - 1024 \quad \text{Add.} \\
-S &= -1023 \\
S &= 1023
\end{aligned}
$$

If $S_n = a_1 + a_1 r + a_1 r^2 + \cdots + a_1 r^{n-1}$ is any geometric series, we can find the sum in the same manner. Multiplying each side of this equation by $-r$ yields

$$-rS_n = -a_1 r - a_1 r^2 - a_1 r^3 - \cdots - a_1 r^n.$$

If we add S_n and $-rS_n$, all but two of the terms on the right are eliminated:

$$S_n = a_1 + a_1r + a_1r^2 + \cdots \qquad + a_1r^{n-1}$$
$$\underline{-rS_n = \qquad - a_1r - a_1r^2 - a_1r^3 - \cdots \qquad\qquad - a_1r^n}$$
$$S_n - rS_n = a_1 \qquad\qquad\qquad\qquad\qquad - a_1r^n \qquad \text{Add.}$$
$$(1-r)S_n = a_1(1 - r^n) \qquad\qquad \begin{array}{l}\text{Factor out} \\ \text{common factors.}\end{array}$$

Now divide each side of this equation by $1 - r$ to get the formula for S_n.

Sum of n Terms of a Geometric Series

If S_n represents the sum of the first n terms of a geometric series with first term a_1 and common ratio r $(r \neq 1)$, then

$$S_n = \frac{a_1(1 - r^n)}{1 - r}.$$

EXAMPLE 5 **The sum of a finite geometric series**
Find the sum of the series

$$\frac{1}{3} + \frac{1}{9} + \frac{1}{27} + \cdots + \frac{1}{729}.$$

Solution

The first term is $\frac{1}{3}$, and the common ratio is $\frac{1}{3}$. So the nth term can be written as

$$a_n = \frac{1}{3}\left(\frac{1}{3}\right)^{n-1}.$$

We can use this formula to find the number of terms in the series:

$$\frac{1}{729} = \frac{1}{3}\left(\frac{1}{3}\right)^{n-1}$$

$$\frac{1}{729} = \left(\frac{1}{3}\right)^n$$

Because $3^6 = 729$, we have $n = 6$. (Of course, you could use logarithms to solve for n.) Now use the formula for the sum of six terms of this geometric series:

$$S_6 = \frac{\frac{1}{3}\left[1 - \left(\frac{1}{3}\right)^6\right]}{1 - \frac{1}{3}} = \frac{\frac{1}{3}\left[1 - \frac{1}{729}\right]}{\frac{2}{3}}$$

$$= \frac{1}{3} \cdot \frac{728}{729} \cdot \frac{3}{2}$$

$$= \frac{364}{729}$$

EXAMPLE 6   **The sum of a finite geometric series**

Find the sum of the series

$$\sum_{i=1}^{12} 3(-2)^{i-1}.$$

Solution

This series is geometric with first term 3, ratio -2, and $n = 12$. We use the formula for the sum of the first 12 terms of a geometric series:

$$S_{12} = \frac{3[1 - (-2)^{12}]}{1 - (-2)} = \frac{3[-4095]}{3} = -4095 \qquad \blacksquare$$

Infinite Geometric Series

Consider how a very large value of n affects the formula for the sum of a finite geometric series,

$$S_n = \frac{a_1(1 - r^n)}{1 - r}.$$

If $|r| < 1$, then the value of r^n gets closer and closer to 0 as n gets larger and larger. For example, if $r = \frac{2}{3}$ and $n = 10, 20,$ and 100, then

$$\left(\frac{2}{3}\right)^{10} \approx 0.0173415, \qquad \left(\frac{2}{3}\right)^{20} \approx 0.0003007, \qquad \text{and} \qquad \left(\frac{2}{3}\right)^{100} \approx 2.459 \times 10^{-18}.$$

Because r^n is approximately 0 for large values of n, $1 - r^n$ is approximately 1. If we replace $1 - r^n$ by 1 in the expression for S_n, we get

$$S_n \approx \frac{a_1}{1 - r}.$$

So as n gets larger and larger, the sum of the first n terms of the infinite geometric series

$$a_1 + a_1 r + a_1 r^2 + \cdots$$

gets closer and closer to $\frac{a_1}{1 - r}$, provided that $|r| < 1$. Therefore we say that $\frac{a_1}{1 - r}$ is the sum of *all* of the terms of the infinite geometric series.

Sum of an Infinite Geometric Series

If $a_1 + a_1 r + a_1 r^2 + \cdots$ is an infinite geometric series, with $|r| < 1$, then the sum S of all of the terms of this series is given by

$$S = \frac{a_1}{1 - r}.$$

EXAMPLE 7 **Sum of an infinite geometric series**

Find the sum

$$\frac{1}{2} + \frac{1}{4} + \frac{1}{8} + \frac{1}{16} + \cdots.$$

Solution

This series is an infinite geometric series with $a_1 = \frac{1}{2}$ and $r = \frac{1}{2}$. Because $r < 1$, we have

$$S = \frac{\frac{1}{2}}{1 - \frac{1}{2}} = 1. \qquad \blacksquare$$

For an infinite series, the index of summation i takes the values 1, 2, 3, and so on, without end. To indicate that the values for i keep increasing without bound, we say that i *takes the values from 1 through* ∞ (infinity). Note that the symbol ∞ does not represent a number. Using the ∞ symbol, we can write the indicated sum of an infinite geometric series (with $|r| < 1$) by using summation notation as follows:

$$a_1 + a_1 r + a_1 r^2 + \cdots = \sum_{i=1}^{\infty} a_1 r^{i-1}$$

EXAMPLE 8 **Sum of an infinite geometric series**

Find the value of the sum

$$\sum_{i=1}^{\infty} 8\left(\frac{3}{4}\right)^{i-1}.$$

Solution

This series is an infinite geometric series with first term 8 and ratio $\frac{3}{4}$. So

$$S = \frac{8}{1 - \frac{3}{4}} = 8 \cdot \frac{4}{1} = 32. \qquad \blacksquare$$

EXAMPLE 9 **Follow the bouncing ball**

Suppose that a ball always rebounds $\frac{2}{3}$ of the height from which it falls and the ball is dropped from a height of 6 feet. Find the total distance that the ball travels.

Solution

The ball falls 6 feet and rebounds 4 feet, then falls 4 feet and rebounds $\frac{8}{3}$ feet. The following series gives the total distance that the ball falls:

$$F = 6 + 4 + \frac{8}{3} + \frac{16}{9} + \cdots$$

The distance that the ball rebounds is given by the following series:

$$R = 4 + \frac{8}{3} + \frac{16}{9} + \cdots$$

Each of these series is an infinite geometric series with ratio $\frac{2}{3}$. Use the formula for an infinite geometric series to find each sum:

$$F = \frac{6}{1 - \frac{2}{3}} = 6 \cdot \frac{3}{1} = 18 \text{ ft}, \qquad R = \frac{4}{1 - \frac{2}{3}} = 4 \cdot \frac{3}{1} = 12 \text{ ft}$$

The total distance traveled by the ball is the sum of F and R, 30 feet.   ■

Annuities

One of the most important applications of geometric series is in calculating the value of an annuity. An **annuity** is a sequence of periodic payments. The payments might be loan payments or investments.

EXAMPLE 10   **Value of an annuity**

A deposit of $1000 is made at the beginning of each year for 30 years and earns 6% interest compounded annually. What is the value of this annuity at the end of the thirtieth year?

Solution

The last deposit earns interest for only one year. So at the end of the thirtieth year it amounts to $1000(1.06)$. The next to last deposit earns interest for two years and amounts to $1000(1.06)^2$. The first deposit earns interest for 30 years and amounts to $1000(1.06)^{30}$. So the value of the annuity at the end of the 30th year is the sum of the geometric series

$$1000(1.06) + 1000(1.06)^2 + 1000(1.06)^3 + \cdots + 1000(1.06)^{30}.$$

Use the formula for the sum of 30 terms of a geometric series with $a_1 = 1000(1.06)$ and $r = 1.06$:

$$S_{30} = \frac{1000(1.06)(1 - (1.06)^{30})}{1 - 1.06} = \$83,801.68$$

So 30 annual deposits of $1000 each amount to $83,801.68.   ■

Warm-ups

True or false? Explain your answer.

1. The sequence 2, 6, 24, 120, ... is a geometric sequence.

2. For $a_n = 2^n$ there is a common difference between adjacent terms.

3. The common ratio for the geometric sequence $a_n = 3(0.5)^{n-1}$ is 0.5.

4. If $a_n = 3(2)^{-n+3}$, then $a_1 = 12$.

5. In the geometric sequence $a_n = 3(2)^{-n+3}$ we have $r = \frac{1}{2}$.

6. The terms of a geometric series are the terms of a geometric sequence.

7. To evaluate $\displaystyle\sum_{i=1}^{10} 2^i$, we must list all of the terms.

8. $\displaystyle\sum_{i=1}^{5} 6\left(\frac{3}{4}\right)^{i-1} = \dfrac{9\left[1 - \left(\frac{3}{4}\right)^5\right]}{1 - \dfrac{3}{4}}$

9. $10 + 5 + \dfrac{5}{2} + \cdots = \dfrac{10}{1 - \dfrac{1}{2}}$

10. $2 + 4 + 8 + 16 + \cdots = \dfrac{2}{1 - 2}$

12.4 EXERCISES

Write a formula for the nth term of each geometric sequence. See Examples 1 and 2.

1. $\dfrac{1}{3}$, 1, 3, 9, ...

2. $\dfrac{1}{4}$, 2, 16, ...

3. 64, 8, 1, ...

4. 100, 10, 1, ...

5. 8, −4, 2, −1, ...

6. −9, 3, −1, ...

7. 2, −4, 8, −16, ...

8. $-\dfrac{1}{2}$, 2, −8, 32, ...

9. $-\dfrac{1}{3}$, $-\dfrac{1}{4}$, $-\dfrac{3}{16}$, ...

10. $-\dfrac{1}{4}$, $-\dfrac{1}{5}$, $-\dfrac{4}{25}$, ...

Write the first five terms of the geometric sequence with the given nth term. See Example 3.

11. $a_n = 2\left(\dfrac{1}{3}\right)^{n-1}$

12. $a_n = -5\left(\dfrac{1}{2}\right)^{n-1}$

13. $a_n = (-2)^{n-1}$

14. $a_n = \left(-\dfrac{1}{3}\right)^{n-1}$

15. $a_n = 2^{-n}$

16. $a_n = 3^{-n}$

17. $a_n = (0.78)^n$

18. $a_n = (-0.23)^n$

Find the required part of each geometric sequence. See Example 4.

19. Find the first term of the geometric sequence that has fourth term 40 and common ratio 2.

20. Find the first term of the geometric sequence that has fifth term 4 and common ratio $\dfrac{1}{2}$.

21. Find r for the geometric sequence that has $a_1 = 6$ and $a_4 = \dfrac{2}{9}$.

22. Find r for the geometric sequence that has $a_1 = 1$ and $a_4 = -27$.

23. Find a_4 for the geometric sequence that has $a_1 = -3$ and $r = \dfrac{1}{3}$.

24. Find a_5 for the geometric sequence that has $a_1 = -\dfrac{2}{3}$ and $r = -\dfrac{2}{3}$.

Find the sum of each geometric series. See Examples 5 and 6.

25. $\dfrac{1}{2} + \dfrac{1}{4} + \dfrac{1}{8} + \cdots + \dfrac{1}{512}$

26. $1 + \dfrac{1}{3} + \dfrac{1}{9} + \cdots + \dfrac{1}{81}$

27. $\dfrac{1}{2} - \dfrac{1}{4} + \dfrac{1}{8} - \dfrac{1}{16} + \dfrac{1}{32}$

28. $3 - 1 + \dfrac{1}{3} - \dfrac{1}{9} + \dfrac{1}{27} - \dfrac{1}{81}$

29. $30 + 20 + \dfrac{40}{3} + \cdots + \dfrac{1280}{729}$

30. $9 - 6 + 4 - \cdots - \dfrac{128}{243}$

31. $\displaystyle\sum_{i=1}^{10} 5(2)^{i-1}$ **32.** $\displaystyle\sum_{i=1}^{7} (10000)(0.1)^{i-1}$

33. $\displaystyle\sum_{i=1}^{6} (0.1)^{i}$ **34.** $\displaystyle\sum_{i=1}^{5} (0.2)^{i}$

35. $\displaystyle\sum_{i=1}^{6} 100(0.3)^{i}$ **36.** $\displaystyle\sum_{i=1}^{7} 36(0.5)^{i}$

Find the sum of each infinite geometric series. See Examples 7 and 8.

37. $\dfrac{1}{8} + \dfrac{1}{16} + \dfrac{1}{32} + \cdots$ **38.** $\dfrac{1}{9} + \dfrac{1}{27} + \dfrac{1}{81} + \cdots$

39. $3 + 2 + \dfrac{4}{3} + \cdots$ **40.** $2 + 1 + \dfrac{1}{2} + \cdots$

41. $4 - 2 + 1 - \dfrac{1}{2} + \cdots$ **42.** $16 - 12 + 9 - \dfrac{27}{4} + \cdots$

43. $\displaystyle\sum_{i=1}^{\infty} (0.3)^{i}$ **44.** $\displaystyle\sum_{i=1}^{\infty} (0.2)^{i}$ **45.** $\displaystyle\sum_{i=1}^{\infty} 3(0.5)^{i-1}$

46. $\displaystyle\sum_{i=1}^{\infty} 7(0.4)^{i-1}$ **47.** $\displaystyle\sum_{i=1}^{\infty} 3(0.1)^{i}$ **48.** $\displaystyle\sum_{i=1}^{\infty} 6(0.1)^{i}$

49. $\displaystyle\sum_{i=1}^{\infty} 12(0.01)^{i}$ **50.** $\displaystyle\sum_{i=1}^{\infty} 72(0.01)^{i}$

Use the ideas of geometric series to solve each problem. See Examples 9 and 10.

51. *Retirement fund.* Suppose that a deposit of $2000 is made at the beginning of each year for 45 years into an account paying 12% compounded annually. What is the amount of this annuity at the end of the 45th year?

52. *Investing in stocks.* If you had invested $5000 in the stock market each year for the last 40 years at the market high for the year (the worst time to invest), you would have earned 10.7% compounded annually on your investment (*Money*, April 1994). To simplify the computation, assume that the $5000 investments are made at the beginning of each year for 40 years and earn 10.7% compounded annually. Find the amount of this annuity at the end of the 40th year.

Growth of $5,000 investment per year

Figure for Exercise 52

53. *Big saver.* Suppose you deposit one cent into your piggy bank on the first day of December and, on each day of December after that, you deposit twice as much as on the previous day. How much will you have in the bank after the last deposit?

54. *Big family.* Consider your relatives, including yourself, your parents, your grandparents, your great-grandparents, your great-greatgrandparents, and so on, back to your grandparents with the word "great" used in front 40 times. What is the total number of people you are considering?

55. *Total economic impact.* In Exercise 39 of Section 12.1 we described a factory that spends $1 million annually in a community in which 80% of the money received is respent in the community. Economists assume that the money is respent again and again at the 80% rate. The total economic impact of the factory is the total of all of this spending. Find an approximation for the total by using the formula for the sum of an infinite geometric series with ratio 80%.

56. *Less impact.* Repeat Exercise 55, assuming that money is respent again and again at the 50% rate.

Getting More Involved

57. *Discussion.* Which of the following sequences is not a geometric sequence? Explain your answer.
a) $1, 2, 4, \ldots$ **b)** $0.1, 0.01, 0.001, \ldots$
c) $-1, 2, -4, \ldots$ **d)** $2, 4, 6, \ldots$

58. *Discussion.* The repeating decimal number $0.44444 \ldots$ can be written as

$$\dfrac{4}{10} + \dfrac{4}{100} + \dfrac{4}{1000} + \cdots,$$

an infinite geometric series. Find the sum of this geometric series.

59. *Discussion.* Write the repeating decimal number $0.24242424 \ldots$ as an infinite geometric series. Find the sum of the geometric series.

12.5 Binomial Expansions

In this section:

▶ Some Examples
▶ Obtaining the Coefficients
▶ The Binomial Theorem

In Chapter 4 you learned how to square a binomial. In this section you will study higher powers of binomials.

Some Examples

We know that $(x + y)^2 = x^2 + 2xy + y^2$. To find $(x + y)^3$, we multiply $(x + y)^2$ by $x + y$:

$$\begin{aligned}
(x + y)^3 &= (x^2 + 2xy + y^2)(x + y) \\
&= (x^2 + 2xy + y^2)x + (x^2 + 2xy + y^2)y \\
&= x^3 + 2x^2y + xy^2 + x^2y + 2xy^2 + y^3 \\
&= x^3 + 3x^2y + 3xy^2 + y^3
\end{aligned}$$

The sum $x^3 + 3x^2y + 3xy^2 + y^3$ is called the **binomial expansion** of $(x + y)^3$. If we again multiply by $x + y$, we will get the binomial expansion of $(x + y)^4$. This method is rather tedious. However, if we examine these expansions, we can find a pattern and learn how to find binomial expansions without multiplying.

Consider the following binomial expansions:

$$\begin{aligned}
(x + y)^0 &= 1 \\
(x + y)^1 &= x + y \\
(x + y)^2 &= x^2 + 2xy + y^2 \\
(x + y)^3 &= x^3 + 3x^2y + 3xy^2 + y^3 \\
(x + y)^4 &= x^4 + 4x^3y + 6x^2y^2 + 4xy^3 + y^4 \\
(x + y)^5 &= x^5 + 5x^4y + 10x^3y^2 + 10x^2y^3 + 5xy^4 + y^5
\end{aligned}$$

Observe that the exponents on the variable x are decreasing, while the exponents on the variable y are increasing, as we read from left to right. Also notice that the sum of the exponents in each term is the same for that entire line. For instance, in the fourth expansion the terms x^4, x^3y, x^2y^2, xy^3, and y^4 all have exponents with a sum of 4. If we continue the pattern, the expansion of $(x + y)^6$ will have seven terms containing x^6, x^5y, x^4y^2, x^3y^3, x^2y^4, xy^5, and y^6. Now we must find the pattern for the coefficients of these terms.

Obtaining the Coefficients

If we write out only the coefficients of the expansions that we already have, we can easily see a pattern. This triangular array of coefficients for the binomial expansions is called **Pascal's triangle.**

$$
\begin{array}{ccccccc}
 & & & 1 & & & \\
 & & 1 & & 1 & & \\
 & 1 & & 2 & & 1 & \\
1 & & 3 & & 3 & & 1 \\
\end{array}
$$

$$
\begin{aligned}
1 \quad 4 \quad 6 \quad 4 \quad 1 \\
1 \quad 5 \quad 10 \quad 10 \quad 5 \quad 1
\end{aligned}
$$

$(x + y)^0 = 1$

$(x + y)^1 = 1x + 1y$

$(x + y)^2 = 1x^2 + 2xy + 1y^2$

$(x + y)^3 = 1x^3 + 3x^2y + 3xy^2 + 1y^3$

$(x + y)^4 = 1x^4 + 4x^3y + 6x^2y^2 + 4xy^3 + 1y^4$

Coefficients in $(x + y)^5$

Notice that each line starts and ends with a 1 and that each entry of a line is the sum of the two entries above it in the previous line. For instance, $4 = 3 + 1$, and $10 = 6 + 4$. Following this pattern, the sixth and seventh lines of coefficients are

$$1 \quad 6 \quad 15 \quad 20 \quad 15 \quad 6 \quad 1$$
$$1 \quad 7 \quad 21 \quad 35 \quad 35 \quad 21 \quad 7 \quad 1$$

Pascal's triangle gives us an easy way to get the coefficients for the binomial expansion with small powers, but it is impractical for larger powers. For larger powers we use a formula involving **factorial notation.**

n! (*n* factorial)

If n is a positive integer, $n!$ (read "n factorial") is defined to be the product of all of the positive integers from 1 through n.

For example, $3! = 3 \cdot 2 \cdot 1 = 6$, and $5! = 5 \cdot 4 \cdot 3 \cdot 2 \cdot 1 = 120$. We also define 0! to be 1.

Before we state a general formula, consider how the coefficients for $(x + y)^4$ are found by using factorials:

$$\frac{4!}{4!\,0!} = \frac{4 \cdot 3 \cdot 2 \cdot 1}{4 \cdot 3 \cdot 2 \cdot 1 \cdot 1} = 1 \qquad \text{Coefficient of } x^4 \text{ (or } x^4 y^0)$$

$$\frac{4!}{3!\,1!} = \frac{4 \cdot 3 \cdot 2 \cdot 1}{3 \cdot 2 \cdot 1 \cdot 1} = 4 \qquad \text{Coefficient of } 4x^3 y$$

$$\frac{4!}{2!\,2!} = \frac{4 \cdot 3 \cdot 2 \cdot 1}{2 \cdot 1 \cdot 2 \cdot 1} = 6 \qquad \text{Coefficient of } 6x^2 y^2$$

$$\frac{4!}{1!\,3!} = \frac{4 \cdot 3 \cdot 2 \cdot 1}{1 \cdot 3 \cdot 2 \cdot 1} = 4 \qquad \text{Coefficient of } 4xy^3$$

$$\frac{4!}{0!\,4!} = \frac{4 \cdot 3 \cdot 2 \cdot 1}{1 \cdot 4 \cdot 3 \cdot 2 \cdot 1} = 1 \qquad \text{Coefficient of } y^4 \text{ (or } x^0 y^4)$$

Note that each expression has 4! in the numerator, with factorials in the denominator corresponding to the exponents on x and y.

The Binomial Theorem

We now summarize these ideas in the **binomial theorem.**

The Binomial Theorem

In the expansion of $(x + y)^n$ for a positive integer n, there are $n + 1$ terms, given by the following formula:

$$(x + y)^n = \frac{n!}{n!\,0!} x^n + \frac{n!}{(n-1)!\,1!} x^{n-1} y + \frac{n!}{(n-2)!\,2!} x^{n-2} y^2 + \cdots + y^n$$

The notation $\binom{n}{r}$ is often used in place of $\dfrac{n!}{(n-r)!\,r!}$ in the binomial expansion. Using this notation, we write the expansion as

$$(x + y)^n = \binom{n}{0} x^n + \binom{n}{1} x^{n-1} y + \binom{n}{2} x^{n-2} y^2 + \cdots + y^n.$$

Another notation for $\dfrac{n!}{(n-r)!\,r!}$ is $_nC_r$. Using this notation, we have

$$(x+y)^n = {}_nC_0 x^n + {}_nC_1 x^{n-1}y + {}_nC_2 x^{n-2}y^2 + \cdots + y^n.$$

EXAMPLE 1   **Using the binomial theorem**

Write out the first three terms of $(x+y)^9$.

Solution

$$(x+y)^9 = \frac{9!}{9!\,0!}x^9 + \frac{9!}{8!\,1!}x^8 y + \frac{9!}{7!\,2!}x^7 y^2 + \cdots$$
$$= x^9 + 9x^8 y + 36x^7 y^2 + \cdots \qquad\blacksquare$$

EXAMPLE 2   **Using the binomial theorem**

Write the binomial expansion for $(x^2 - 2a)^5$.

Solution

We expand a difference by writing it as a sum and using the binomial theorem:

$$(x^2 - 2a)^5 = (x^2 + (-2a))^5$$
$$= \frac{5!}{5!\,0!}(x^2)^5 + \frac{5!}{4!\,1!}(x^2)^4(-2a)^1 + \frac{5!}{3!\,2!}(x^2)^3(-2a)^2 + \frac{5!}{2!\,3!}(x^2)^2(-2a)^3$$
$$+ \frac{5!}{1!\,4!}(x^2)(-2a)^4 + \frac{5!}{0!\,5!}(-2a)^5$$
$$= x^{10} - 10x^8 a + 40x^6 a^2 - 80x^4 a^3 + 80x^2 a^4 - 32a^5 \qquad\blacksquare$$

EXAMPLE 3   **Finding a specific term**

Find the fourth term of the expansion of $(a+b)^{12}$.

Solution

The variables in the first term are $a^{12}b^0$, those in the second term are $a^{11}b^1$, those in the third term are $a^{10}b^2$, and those in the fourth term are $a^9 b^3$. So

$$\frac{12!}{9!\,3!}a^9 b^3 = 220a^9 b^3.$$

The fourth term is $220a^9 b^3$. $\qquad\blacksquare$

Using the ideas of Example 3, we can write a formula for any term of a binomial expansion.

Formula for the kth Term of $(x+y)^n$

For k ranging from 1 to $n+1$, the kth term of the expansion of $(a+b)^n$ is given by the formula

$$\frac{n!}{(n-k+1)!\,(k-1)!}x^{n-k+1}y^{k-1}.$$

EXAMPLE 4   **Finding a specific term**
Find the sixth term of the expansion of $(a^2 - 2b)^7$.

Solution

Use the formula for the kth term with $k = 6$ and $n = 7$:

$$\frac{7!}{(7 - 6 + 1)!(6 - 1)!}(a^2)^2(-2b)^5 = 21a^4(-32b^5) = -672a^4b^5 \qquad \blacksquare$$

We can think of the binomial expansion as a finite series. Using summation notation, we can write the binomial theorem as follows.

> **The Binomial Theorem (Using Summation Notation)**
> For any positive integer n,
>
> $$(x + y)^n = \sum_{i=0}^{n} \frac{n!}{(n - i)!\,i!}\, x^{n-i}y^i \qquad \text{or} \qquad (x + y)^n = \sum_{i=0}^{n} \binom{n}{i} x^{n-i}y^i.$$

EXAMPLE 5   **Using summation notation**
Write $(a + b)^5$ using summation notation.

Solution

Use $n = 5$ in the binomial theorem:

$$(a + b)^5 = \sum_{i=0}^{5} \frac{5!}{(5 - i)!\,i!}\, a^{5-i}b^i \qquad \blacksquare$$

Calculator Close-up

You can calculate factorials on a scientific or a graphing calculator using the x! or n! key. The factorial function is often a second function on a scientific calculator. To find 10! on a scientific calculator, try the following steps:

$$10 \;\boxed{x!}$$

The display will read

$$\boxed{3{,}628{,}800}$$

On a graphing calculator the factorial symbol "!" might be found in a menu. Try the following on your calculator:

$$10! \;\boxed{\text{ENTER}}$$

$$\boxed{3{,}628{,}800}$$

Many calculators have an nCr key for finding binomial coefficients. Consult your calculator manual for instructions on how to use it. To find the coefficient of x^5y^8 in the expansion of $(x + y)^{13}$ on a graphing calculator, try the following steps:

$$13\,\text{nCr}\,5 \;\boxed{\text{ENTER}}$$

$$\boxed{1287}$$

Warm-ups

True or false? Explain your answer.

1. There are 12 terms in the expansion of $(a + b)^{12}$.
2. The seventh term of $(a + b)^{12}$ is a multiple of a^5b^7.
3. For all values of x, $(x + 2)^5 = x^5 + 32$.
4. In the expansion of $(x - 5)^8$, the signs of the terms alternate.
5. The eighth line of Pascal's triangle is

$$1 \ 8 \ 28 \ 56 \ 70 \ 56 \ 28 \ 8 \ 1.$$

6. The sum of the coefficients in the expansion of $(a + b)^4$ is 2^4.

7. $(a + b)^3 = \sum_{i=0}^{3} \frac{3!}{(3 - i)! \, i!} a^{3 - i} b^i$

8. The sum of the coefficients in the expansion of $(a + b)^n$ is 2^n.

9. $0! = 1!$ 10. $\frac{7!}{5! \, 2!} = 21$

12.5 EXERCISES

Evaluate each expression.

1. $\frac{5!}{2! \, 3!}$ 2. $\frac{6!}{5! \, 1!}$

3. $\frac{8!}{5! \, 3!}$ 4. $\frac{9!}{2! \, 7!}$

Use the binomial theorem to expand each binomial. See Examples 1 and 2.

5. $(r + t)^5$ 6. $(r + t)^6$
7. $(m - n)^3$ 8. $(m - n)^4$
9. $(x + 2a)^3$ 10. $(a + 3b)^4$
11. $(x^2 - 2)^4$ 12. $(x^2 - a^2)^5$
13. $(x - 1)^7$ 14. $(x + 1)^6$

Write out the first four terms in the expansion of each binomial. See Examples 1 and 2.

15. $(a - 3b)^{12}$ 16. $(x - 2y)^{10}$
17. $(x^2 + 5)^9$ 18. $(x^2 + 1)^{20}$
19. $(x - 1)^{22}$ 20. $(2x - 1)^{15}$
21. $\left(\frac{x}{2} + \frac{y}{3}\right)^{10}$ 22. $\left(\frac{a}{2} + \frac{b}{5}\right)^8$

Find the indicated term of the binomial expansion. See Examples 3 and 4.

23. $(a + w)^{13}$, 6th term   24. $(m + n)^{12}$, 7th term
25. $(m - n)^{16}$, 8th term   26. $(a - b)^{14}$, 6th term
27. $(x + 2y)^8$, 4th term   28. $(3a + b)^7$, 4th term
29. $(2a^2 - b)^{20}$, 7th term   30. $(a^2 - w^2)^{12}$, 5th term

Write each expansion using summation notation. See Example 5.

31. $(a + m)^8$ 32. $(z + w)^{13}$
33. $(a - 2x)^5$ 34. $(w - 3m)^7$

Getting More Involved

35. ***Discussion.*** Find the trinomial expansion for $(a + b + c)^3$ by using $x = a$ and $y = b + c$ in the binomial theorem.

36. ***Discussion.*** What problem do you encounter when you try to find the fourth term in the binomial expansion for $(x + y)^{120}$? How can you overcome this problem? Find the fifth term in the binomial expansion for $(x - 2y)^{100}$.

---COLLABORATIVE ACTIVITIES---

Lotteries Are Series(ous)

Grouping: 2 to 4 students per group
Topic: Sequences and series

Roberto and his brother-in-law Horatio each have a child who will be graduating from high school in five years. Each would like to buy his child a car for a graduation present. Horatio decides to buy two lottery tickets each week for the next five years, hoping to win and buy a new car for his child. The lottery tickets are $2.00 each at the local convenience store. Roberto, who doesn't believe in lotteries, decides to set aside $4.00 each week and deposit this money in a savings account each quarter for the next five years to buy a used car. He finds a bank that will pay 5% yearly interest compounded quarterly.

1. Write a series that represents how much money Horatio will spend on lottery tickets over a five-year period. Assume that there are 52 weeks in a year. Compute the total amount of money spent on lottery tickets.

2. Find the percent paid quarterly on the account. Find the number of compounding periods in five years. Find the amount of money Roberto will deposit each quarter.

3. Write a series to show how much money Roberto will have in his savings account at the end of five years.

4. Discuss in your groups the chances of Horatio winning the lottery compared to the sure savings Roberto has. Discuss the fact that Horatio would have to pay taxes on his lottery winnings and Roberto pays tax only on the interest earned. Do Horatio's chances increase the longer he buys lottery tickets?

Wrap-up   CHAPTER 12

SUMMARY

Sequences and Series		Examples
Sequence	Finite—A function whose domain is the set of positive integers less than or equal to a fixed positive integer	3, 5, 7, 9, 11 $a_n = 2n + 1$ $1 \leq n \leq 5$
	Infinite—A function whose domain is the set of positive integers	2, 4, 6, 8, ... $a_n = 2n$
Series	The indicated sum of a sequence	$2 + 4 + 6 + \cdots + 50$
Summation notation	$\sum_{i=1}^{n} a_i = a_1 + a_2 + a_3 + \cdots + a_n$	$\sum_{i=1}^{25} 2i = 2 + 4 + \cdots + 50$

Arithmetic Sequences and Series		Examples
Arithmetic sequence	Each term after the first is obtained by adding a fixed amount to the previous term.	6, 11, 16, 21, ... Fixed amount, d, is 5.
nth term	The nth term of an arithmetic sequence is $a_n = a_1 + (n - 1)d$.	If $a_1 = 6$ and $d = 5$, then $a_n = 6 + (n - 1)5$.

Arithmetic series	The sum of an arithmetic sequence	$6 + 11 + 16 + 21$
Sum of first n terms	$S_n = \dfrac{n}{2}(a_1 + a_n)$	$S_4 = \dfrac{4}{2}(6 + 21) = 54$

Geometric Sequences and Series

		Examples
Geometric sequence	Each term after the first is obtained by multiplying the preceding term by a constant.	2, 6, 18, 54, ... Constant, r, is 3.
nth term	The nth term of a geometric sequence is $$a_n = a_1 r^{n-1}.$$	$a_1 = 2,\ r = 3$ $a_n = 2 \cdot 3^{n-1}$
Geometric series (finite)	The indicated sum of a finite geometric sequence. $a_1 + a_1 r + a_1 r^2 + \cdots + a_1 r^{n-1}$	$2 + 6 + 18 + 54 + 162$
Sum of first n terms	$S_n = \dfrac{a_1(1 - r^n)}{1 - r}$	$a_1 = 2,\ r = 3,\ n = 5$ $S_5 = \dfrac{2(1 - 3^5)}{1 - 3} = 242$
Geometric series (infinite)	$a_1 + a_1 r + a_1 r^2 + a_1 r^3 + \cdots$	$8 + 4 + 2 + 1 + \dfrac{1}{2} + \cdots$
Sum of an infinite geometric series	$S = \dfrac{a_1}{1 - r}$, provided that $|r| < 1$	$a_1 = 8,\ r = \dfrac{1}{2}$ $S = \dfrac{8}{1 - \dfrac{1}{2}} = 16$
Factorial notation	The notation $n!$ represents the product of the positive integers from 1 through n.	$5! = 5 \cdot 4 \cdot 3 \cdot 2 \cdot 1 = 120$
Binomial theorem	$(x + y)^n = \dfrac{n!}{n!\,0!}x^n + \dfrac{n!}{(n-1)!\,1!}x^{n-1}y$ $\qquad + \dfrac{n!}{(n-2)!\,2!}x^{n-2}y^2 + \cdots + y^n$ Using summation notation: $(x + y)^n = \displaystyle\sum_{i=0}^{n} \dfrac{n!}{(n-i)!\,i!}x^{n-i}y^i = \sum_{i=0}^{n}\binom{n}{i}x^{n-i}y^i$	$(x + y)^3 = x^3 + 3x^2 y$ $\qquad\qquad + 3xy^2 + y^3$
kth term of $(x + y)^n$	$\dfrac{n!}{(n - k + 1)!\,(k - 1)!}x^{n-k+1}y^{k-1}$	Third term of $(a + b)^{10}$ is $\dfrac{10!}{8!\,2!}a^8 b^2 = 45a^8 b^2.$

REVIEW EXERCISES

12.1 *List all terms of each finite sequence.*

1. $a_n = n^3$ for $1 \le n \le 5$

2. $b_n = (n-1)^4$ for $1 \le n \le 4$

3. $c_n = (-1)^n(2n-3)$ for $1 \le n \le 6$

4. $d_n = (-1)^{n-1}(3-n)$ for $1 \le n \le 7$

Write the first three terms of the infinite sequence whose nth term is given.

5. $a_n = -\dfrac{1}{n}$

6. $b_n = \dfrac{(-1)^n}{n^2}$

7. $b_n = \dfrac{(-1)^{2n}}{2n+1}$

8. $a_n = \dfrac{-1}{2n-3}$

9. $c_n = \log_2\left(2^{n+3}\right)$

10. $c_n = \ln\left(e^{2n}\right)$

12.2 *Find the sum of each series.*

11. $\displaystyle\sum_{i=1}^{3} i^3$

12. $\displaystyle\sum_{i=0}^{4} 6$

13. $\displaystyle\sum_{n=1}^{5} n(n-1)$

14. $\displaystyle\sum_{j=0}^{3} (-2)^j$

Write each series in summation notation. Use the index i, and let i begin at 1.

15. $\dfrac{1}{4} + \dfrac{1}{6} + \dfrac{1}{8} + \cdots$

16. $\dfrac{1}{3} + \dfrac{1}{4} + \dfrac{1}{5} + \cdots$

17. $0 + 1 + 4 + 9 + 16 + \cdots$

18. $-1 + 2 - 3 + 4 - 5 + 6 - \cdots$

19. $x_1 - x_2 + x_3 - x_4 + \cdots$

20. $-x^2 + x^3 - x^4 + x^5 - \cdots$

12.3 *Write the first four terms of the arithmetic sequence with the given nth term.*

21. $a_n = 6 + (n-1)5$

22. $a_n = -7 + (n-1)4$

23. $a_n = -20 + (n-1)(-2)$

24. $a_n = 10 + (n-1)(-2.5)$

25. $a_n = 1000n + 2000$

26. $a_n = -500n + 5000$

Write a formula for the nth term of each arithmetic sequence.

27. $\dfrac{1}{3}, \dfrac{2}{3}, 1, \dfrac{4}{3}, \ldots$

28. $10, 6, 2, -2, \ldots$

29. $2, 4, 6, 8, \ldots$

30. $20, 10, 0, -10, \ldots$

Find the sum of each arithmetic series.

31. $1 + 2 + 3 + \cdots + 24$

32. $-5 + (-2) + 1 + 4 + \cdots + 34$

33. $\dfrac{1}{6} + \dfrac{1}{2} + \dfrac{5}{6} + \dfrac{7}{6} + \cdots + \dfrac{11}{2}$

34. $-3 - 6 - 9 - 12 - \cdots - 36$

35. $\displaystyle\sum_{i=1}^{7} (2i - 3)$

36. $\displaystyle\sum_{i=1}^{6} [12 + (i-1)5]$

12.4 *Write the first four terms of the geometric sequence with the given nth term.*

37. $a_n = 3\left(\dfrac{1}{2}\right)^{n-1}$

38. $a_n = 6\left(-\dfrac{1}{3}\right)^n$

39. $a_n = 2^{1-n}$

40. $a_n = 5(10)^{n-1}$

41. $a_n = 23(10)^{-2n}$

42. $a_n = 4(10)^{-n}$

Write a formula for the nth term of each geometric sequence.

43. $\dfrac{1}{2}, 3, 18, \ldots$

44. $-6, 2, -\dfrac{2}{3}, \dfrac{2}{9}, \ldots$

45. $\dfrac{7}{10}, \dfrac{7}{100}, \dfrac{7}{1000}, \ldots$

46. $2, 2x, 2x^2, 2x^3, \ldots$

Find the sum of each geometric series.

47. $\dfrac{1}{3} + \dfrac{1}{9} + \dfrac{1}{27} + \dfrac{1}{81}$

48. $2 + 4 + 8 + 16 + \cdots + 512$

49. $\displaystyle\sum_{i=1}^{10} 3(10)^{-i}$

50. $\displaystyle\sum_{i=1}^{5} (0.1)^i$

51. $\dfrac{1}{4} + \dfrac{1}{12} + \dfrac{1}{36} + \dfrac{1}{108} + \cdots$

52. $12 + (-6) + 3 + \left(-\dfrac{3}{2}\right) + \cdots$

53. $\displaystyle\sum_{i=1}^{\infty} 18\left(\dfrac{2}{3}\right)^{i-1}$

54. $\displaystyle\sum_{i=1}^{\infty} 9(0.1)^i$

12.5 *Use the binomial theorem to expand each binomial.*

55. $(m + n)^5$

56. $(2m - y)^4$

57. $(a^2 - 3b)^3$

58. $\left(\dfrac{x}{2} + 2a\right)^5$

Find the indicated term of the binomial expansion.

59. $(x + y)^{12}$, 5th term

60. $(x - 2y)^9$, 5th term

61. $(2a - b)^{14}$, 3rd term

62. $(a + b)^{10}$, 4th term

Write each expression in summation notation.

63. $(a + w)^7$

64. $(m - 3y)^9$

Miscellaneous

Identify each sequence as an arithmetic sequence, a geometric sequence, or neither.

65. 1, 3, 6, 10, 15, ...

66. 9, 12, 16, $\dfrac{64}{3}$, ...

67. 9, 12, 15, 18, ...

68. 2, 4, 8, 16, ...

69. 0, 2, 4, 6, 8, ...

70. 0, 3, 9, 27, 81, ...

Solve each problem.

71. Find the common ratio for the geometric sequence with first term 6 and fourth term $\dfrac{1}{30}$.

72. Find the common difference for an arithmetic sequence with first term 6 and fourth term 36.

73. Write out all of the terms of the series

$$\sum_{i=1}^{5} \frac{(-1)^i}{i!}.$$

74. Write out the first eight rows of Pascal's triangle.

75. Write out all of the terms of the series

$$\sum_{i=0}^{5} \frac{5!}{(5 - i)!\, i!} a^{5-i} b^i.$$

76. Write out all of the terms of the series

$$\sum_{i=0}^{8} \frac{8!}{(8 - i)!\, i!} x^{8-i} y^i.$$

77. How many terms are there in the expansion of $(a + b)^{25}$?

78. Calculate $\dfrac{12!}{8!\,4!}$.

79. If $3000 is deposited at the beginning of each year for 16 years into an account paying 10% compounded annually, then what is the value of the annuity at the end of the 16th year?

80. If $3000 is deposited at the beginning of each year for 8 years into an account paying 10% compounded annually, then what is the value of the annuity at the end of the eighth year? How does the value of the annuity in this exercise compare to that of Exercise 79?

81. If one deposit of $3000 is made into an account paying 10% compounded annually, then how much will be in the account at the end of 16 years? Note that a single deposit is not an annuity.

CHAPTER 12 TEST

List the first four terms of the sequence whose nth term is given.

1. $a_n = -10 + (n-1)6$ **2.** $a_n = 5(0.1)^{n-1}$

3. $a_n = \dfrac{(-1)^n}{n!}$ **4.** $a_n = \dfrac{2n-1}{n^2}$

Write a formula for the nth term of each sequence.

5. $7, 4, 1, -2, \ldots$

6. $-25, 5, -1, \dfrac{1}{5}, \ldots$

7. $2, -4, 6, -8, 10, -12, \ldots$

8. $1, 4, 9, 16, 25, \ldots$

Write out all of the terms of each series.

9. $\displaystyle\sum_{i=1}^{5}(2i+3)$ **10.** $\displaystyle\sum_{i=1}^{6}5(2)^{i-1}$

11. $\displaystyle\sum_{i=0}^{4}\dfrac{4!}{(4-i)!\,i!}m^{4-i}q^i$

Find the sum of each series.

12. $\displaystyle\sum_{i=1}^{20}(6+3i)$ **13.** $\displaystyle\sum_{i=1}^{5}10\left(\dfrac{1}{2}\right)^{i-1}$

14. $\displaystyle\sum_{i=1}^{\infty}0.35(0.93)^{i-1}$

15. $2 + 4 + 6 + \cdots + 200$

16. $\dfrac{1}{4} + \dfrac{1}{8} + \dfrac{1}{16} + \cdots$

17. $2 + 1 + \dfrac{1}{2} + \dfrac{1}{4} + \cdots + \dfrac{1}{128}$

Solve each problem.

18. Find the common ratio for the geometric sequence that has first term 3 and fifth term 48.

19. Find the common difference for the arithmetic sequence that has first term 1 and twelfth term 122.

20. Find the fifth term in the expansion of $(r-t)^{15}$.

21. Find the fourth term in the expansion of $(a^2 - 2b)^8$.

22. If $800 is deposited at the beginning of each year for 25 years into an account earning 10% compounded annually, then what is the value of this annuity at the end of the 25th year?

Let $f(x) = x^2 - 3$, $g(x) = 2x - 1$, $h(x) = 2^x$, and $m(x) = \log_2(x)$. Find the following.

1. $f(3)$

2. $f(n)$

3. $f(x + h)$

4. $f(x) - g(x)$

5. $g(f(3))$

6. $(f \circ g)(2)$

7. $m(16)$

8. $(h \circ m)(32)$

9. $h(-1)$

10. $h^{-1}(8)$

11. $m^{-1}(0)$

12. $(m \circ h)(x)$

Solve each variation problem.

13. If y varies directly as x, and $y = -6$ when $x = 4$, find y when $x = 9$.

14. If a varies inversely as b, and $a = 2$ when $b = -4$, find a when $b = 3$.

15. If y varies directly as w and inversely as t, and $y = 16$ when $w = 3$ and $t = -4$, find y when $w = 2$ and $t = 3$.

16. If y varies jointly as h and the square of r, and $y = 12$ when $h = 2$ and $r = 3$, find y when $h = 6$ and $r = 2$.

Sketch the graph of each inequality.

17. $x > 3$ and $x + y < 0$

18. $|x - y| \geq 2$

19. $y < -2x + 3$ and $y > 2^x$

20. $|y + 2x| < 1$

21. $x^2 + y^2 < 4$

22. $x^2 - y^2 < 1$

23. $y < \log_2(x)$

24. $x^2 + 2y < 4$

25. $\dfrac{x^2}{4} + \dfrac{y^2}{9} < 1$ and $y > x^2$

Perform the indicated operation and simplify. Write answers with positive exponents.

26. $\dfrac{a}{b} + \dfrac{b}{a}$

27. $1 - \dfrac{3}{y}$

28. $\dfrac{x - 2}{x^2 - 9} - \dfrac{x - 4}{x^2 - 2x - 3}$

29. $\dfrac{x^2 - 16}{2x + 8} \cdot \dfrac{4x^2 + 16x + 64}{x^3 - 16}$

30. $\dfrac{(a^2b)^3}{(ab^2)^4} \cdot \dfrac{ab^3}{a^{-4}b^2}$

31. $\dfrac{x^2y}{(xy)^3} \div \dfrac{xy^2}{x^2y^4}$

Simplify.

32. $8^{2/3}$

33. $16^{-5/4}$

34. $-4^{1/2}$

35. $27^{-2/3}$

36. -2^{-3}

37. $2^{-3/5} \cdot 2^{-7/5}$

38. $5^{-2/3} \div 5^{1/3}$

39. $(9^{1/2} + 4^{1/2})^2$

Solve.

40. **Predicting heights of preschoolers.** A popular model in pediatrics for predicting the height of preschoolers is the JENNS model. According to this model, if $h(x)$ is the height (in centimeters) at age x (in years) for $0.25 \leq x \leq 6$, then

$$h(x) = 79.041 + 6.39x - e^{(3.261 - 0.993x)}.$$

a) Find the predicted height in inches for a child of age 4 years, 3 months.

b) If you have access to a graphing calculator, graph this function. Then find the age to the nearest tenth of a year for a child who has a height of 80 cm.

Geometry Review

GEOMETRIC FIGURES AND FORMULAS

Triangle: A three-sided figure

Area: $A = \frac{1}{2}bh$, Perimeter: $P = a + b + c$

Sum of the measures of the angles is 180°.

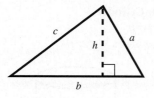

Right Triangle: A triangle with a 90° angle

Area $= \frac{1}{2}ab$, Perimeter: $P = a + b + c$

Pythagorean Theorem: $c^2 = a^2 + b^2$

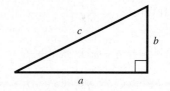

30-60-90 Right Triangle

The side opposite 30° is one-half the length of the hypotenuse.

Parallelogram: A four-sided figure with opposite sides parallel

Area: $A = bh$

Trapezoid: A four-sided figure with one pair of parallel sides

Area: $A = \frac{1}{2}h(b_1 + b_2)$

Rhombus: A four-sided figure with four equal sides

Perimeter: $P = 4a$

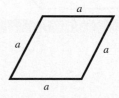

Rectangle: A four-sided figure with four right angles
Area: $A = LW$
Perimeter: $P = 2L + 2W$

Square: A four-sided figure with four equal sides and four right angles
Area: $A = s^2$
Perimeter: $P = 4s$

Circle
Area: $A = \pi r^2$
Circumference: $C = 2\pi r$
Diameter: $d = 2r$

Sphere
Volume: $V = \frac{4}{3}\pi r^3$
Surface Area: $S = 4\pi r^2$

Right Circular Cone
Volume: $V = \frac{1}{3}\pi r^2 h$
Lateral Surface Area: $S = \pi r\sqrt{r^2 + h^2}$

Right Circular Cylinder
Volume: $V = \pi r^2 h$
Lateral Surface Area: $S = 2\pi rh$

Rectangular Solid
Volume: $V = LWH$
Surface Area: $A = 2LW + 2WH + 2LH$

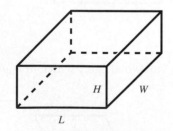

GEOMETRIC TERMS

An **angle** is a union of two rays with a common endpoint.

An **acute angle** is an angle with a measure between 0° and 90°.

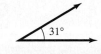

A **right angle** is an angle with a measure of 90°.

An **obtuse angle** is an angle with a measure between 90° and 180°.

Two angles are **complementary** if the sum of their measures is 90°.

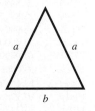

Two angles are **supplementary** if the sum of their measures is 180°.

An **isosceles triangle** is a triangle that has two equal sides.

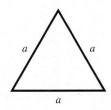

An **equilateral triangle** is a triangle that has three equal sides.

Similar triangles are triangles that have the same shape. Their corresponding angles are equal and corresponding sides are proportional:

$$\frac{a}{d} = \frac{b}{e} = \frac{c}{f}$$

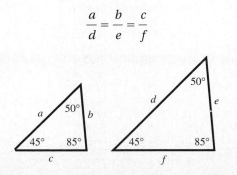

Table of Squares and Square Roots

n	n^2	$\sqrt{n}$	n	n^2	$\sqrt{n}$	n	n^2	$\sqrt{n}$
1	1	1.0000	41	1681	6.4031	81	6561	9.0000
2	4	1.4142	42	1764	6.4807	82	6724	9.0554
3	9	1.7321	43	1849	6.5574	83	6889	9.1104
4	16	2.0000	44	1936	6.6332	84	7056	9.1652
5	25	2.2361	45	2025	6.7082	85	7225	9.2195
6	36	2.4495	46	2116	6.7823	86	7396	9.2736
7	49	2.6458	47	2209	6.8557	87	7569	9.3274
8	64	2.8284	48	2304	6.9282	88	7744	9.3808
9	81	3.0000	49	2401	7.0000	89	7921	9.4340
10	100	3.1623	50	2500	7.0711	90	8100	9.4868
11	121	3.3166	51	2601	7.1414	91	8281	9.5394
12	144	3.4641	52	2704	7.2111	92	8464	9.5917
13	169	3.6056	53	2809	7.2801	93	8649	9.6437
14	196	3.7417	54	2916	7.3485	94	8836	9.6954
15	225	3.8730	55	3025	7.4162	95	9025	9.7468
16	256	4.0000	56	3136	7.4833	96	9216	9.7980
17	289	4.1231	57	3249	7.5498	97	9409	9.8489
18	324	4.2426	58	3364	7.6158	98	9604	9.8995
19	361	4.3589	59	3481	7.6811	99	9801	9.9499
20	400	4.4721	60	3600	7.7460	100	10000	10.0000
21	441	4.5826	61	3721	7.8102	101	10201	10.0499
22	484	4.6904	62	3844	7.8740	102	10404	10.0995
23	529	4.7958	63	3969	7.9373	103	10609	10.1489
24	576	4.8990	64	4096	8.0000	104	10816	10.1980
25	625	5.0000	65	4225	8.0623	105	11025	10.2470
26	676	5.0990	66	4356	8.1240	106	11236	10.2956
27	729	5.1962	67	4489	8.1854	107	11449	10.3441
28	784	5.2915	68	4624	8.2462	108	11664	10.3923
29	841	5.3852	69	4761	8.3066	109	11881	10.4403
30	900	5.4772	70	4900	8.3666	110	12100	10.4881
31	961	5.5678	71	5041	8.4261	111	12321	10.5357
32	1024	5.6569	72	5184	8.4853	112	12544	10.5830
33	1089	5.7446	73	5329	8.5440	113	12769	10.6301
34	1156	5.8310	74	5476	8.6023	114	12996	10.6771
35	1225	5.9161	75	5625	8.6603	115	13225	10.7238
36	1296	6.0000	76	5776	8.7178	116	13456	10.7703
37	1369	6.0828	77	5929	8.7750	117	13689	10.8167
38	1444	6.1644	78	6084	8.8318	118	13924	10.8628
39	1521	6.2450	79	6241	8.8882	119	14161	10.9087
40	1600	6.3246	80	6400	8.9443	120	14400	10.9545

APPENDIX C

Common Logarithms

This table gives the common logarithms for numbers between 1 and 10. The common logarithms for other numbers can be found by using scientific notation and the properties of logarithms. For example, to find log(1230) we write

$$\log(1230) = \log(1.23 \times 10^3) = \log(1.23) + \log(10^3)$$
$$= .0899 + 3 = 3.0899$$

n	0	1	2	3	4	5	6	7	8	9
1.0	.0000	.0043	.0086	.0128	.0170	.0212	.0253	.0294	.0334	.0374
1.1	.0414	.0453	.0492	.0531	.0569	.0607	.0645	.0682	.0719	.0755
1.2	.0792	.0828	.0864	.0899	.0934	.0969	.1004	.1038	.1072	.1106
1.3	.1139	.1173	.1206	.1239	.1271	.1303	.1335	.1367	.1399	.1430
1.4	.1461	.1492	.1523	.1553	.1584	.1614	.1644	.1673	.1703	.1732
1.5	.1761	.1790	.1818	.1847	.1875	.1903	.1931	.1959	.1987	.2014
1.6	.2041	.2068	.2095	.2122	.2148	.2175	.2201	.2227	.2253	.2279
1.7	.2304	.2330	.2355	.2380	.2405	.2430	.2455	.2480	.2504	.2529
1.8	.2553	.2577	.2601	.2625	.2648	.2672	.2695	.2718	.2742	.2765
1.9	.2788	.2810	.2833	.2856	.2878	.2900	.2923	.2945	.2967	.2989
2.0	.3010	.3032	.3054	.3075	.3096	.3118	.3139	.3160	.3181	.3201
2.1	.3222	.3243	.3263	.3284	.3304	.3324	.3345	.3365	.3385	.3404
2.2	.3424	.3444	.3464	.3483	.3502	.3522	.3541	.3560	.3579	.3598
2.3	.3617	.3636	.3655	.3674	.3692	.3711	.3729	.3747	.3766	.3784
2.4	.3802	.3820	.3838	.3856	.3874	.3892	.3909	.3927	.3945	.3962
2.5	.3979	.3997	.4014	.4031	.4048	.4065	.4082	.4099	.4116	.4133
2.6	.4150	.4166	.4183	.4200	.4216	.4232	.4249	.4265	.4281	.4298
2.7	.4314	.4330	.4346	.4362	.4378	.4393	.4409	.4425	.4440	.4456
2.8	.4472	.4487	.4502	.4518	.4533	.4548	.4564	.4579	.4594	.4609
2.9	.4624	.4639	.4654	.4669	.4683	.4698	.4713	.4728	.4742	.4757
3.0	.4771	.4786	.4800	.4814	.4829	.4843	.4857	.4871	.4886	.4900
3.1	.4914	.4928	.4942	.4955	.4969	.4983	.4997	.5011	.5024	.5038
3.2	.5051	.5065	.5079	.5092	.5105	.5119	.5132	.5145	.5159	.5172
3.3	.5185	.5198	.5211	.5224	.5237	.5250	.5263	.5276	.5289	.5302
3.4	.5315	.5328	.5340	.5353	.5366	.5378	.5391	.5403	.5416	.5428
3.5	.5441	.5453	.5465	.5478	.5490	.5502	.5514	.5527	.5539	.5551
3.6	.5563	.5575	.5587	.5599	.5611	.5623	.5635	.5647	.5658	.5670
3.7	.5682	.5694	.5705	.5717	.5729	.5740	.5752	.5763	.5775	.5786
3.8	.5798	.5809	.5821	.5832	.5843	.5855	.5866	.5877	.5888	.5899
3.9	.5911	.5922	.5933	.5944	.5955	.5966	.5977	.5988	.5999	.6010
4.0	.6021	.6031	.6042	.6053	.6064	.6075	.6085	.6096	.6107	.6117
4.1	.6128	.6138	.6149	.6160	.6170	.6180	.6191	.6201	.6212	.6222
4.2	.6232	.6243	.6253	.6263	.6274	.6284	.6294	.6304	.6314	.6325
4.3	.6335	.6345	.6355	.6365	.6375	.6385	.6395	.6405	.6415	.6425
4.4	.6435	.6444	.6454	.6464	.6474	.6484	.6493	.6503	.6513	.6522
n	0	1	2	3	4	5	6	7	8	9

n	0	1	2	3	4	5	6	7	8	9
4.5	.6532	.6542	.6551	.6561	.6571	.6580	.6590	.6599	.6609	.6618
4.6	.6628	.6637	.6646	.6656	.6665	.6675	.6684	.6693	.6702	.6712
4.7	.6721	.6730	.6739	.6749	.6758	.6767	.6776	.6785	.6794	.6803
4.8	.6812	.6821	.6830	.6839	.6848	.6857	.6866	.6875	.6884	.6893
4.9	.6902	.6911	.6920	.6928	.6937	.6946	.6955	.6964	.6972	.6981
5.0	.6990	.6998	.7007	.7016	.7024	.7033	.7042	.7050	.7059	.7067
5.1	.7076	.7084	.7093	.7101	.7110	.7118	.7126	.7135	.7143	.7152
5.2	.7160	.7168	.7177	.7185	.7193	.7202	.7210	.7218	.7226	.7235
5.3	.7243	.7251	.7259	.7267	.7275	.7284	.7292	.7300	.7308	.7316
5.4	.7324	.7332	.7340	.7348	.7356	.7364	.7372	.7380	.7388	.7396
5.5	.7404	.7412	.7419	.7427	.7435	.7443	.7451	.7459	.7466	.7474
5.6	.7482	.7490	.7497	.7505	.7513	.7520	.7528	.7536	.7543	.7551
5.7	.7559	.7566	.7574	.7582	.7589	.7597	.7604	.7612	.7619	.7627
5.8	.7634	.7642	.7649	.7657	.7664	.7672	.7679	.7686	.7694	.7701
5.9	.7709	.7716	.7723	.7731	.7738	.7745	.7752	.7760	.7767	.7774
6.0	.7782	.7789	.7796	.7803	.7810	.7818	.7825	.7832	.7839	.7846
6.1	.7853	.7860	.7868	.7875	.7882	.7889	.7896	.7903	.7910	.7917
6.2	.7924	.7931	.7938	.7945	.7952	.7959	.7966	.7973	.7980	.7987
6.3	.7993	.8000	.8007	.8014	.8021	.8028	.8035	.8041	.8048	.8055
6.4	.8062	.8069	.8075	.8082	.8089	.8096	.8102	.8109	.8116	.8122
6.5	.8129	.8136	.8142	.8149	.8156	.8162	.8169	.8176	.8182	.8189
6.6	.8195	.8202	.8209	.8215	.8222	.8228	.8235	.8241	.8248	.8254
6.7	.8261	.8267	.8274	.8280	.8287	.8293	.8299	.8306	.8312	.8319
6.8	.8325	.8331	.8338	.8344	.8351	.8357	.8363	.8370	.8376	.8382
6.9	.8388	.8395	.8401	.8407	.8414	.8420	.8426	.8432	.8439	.8445
7.0	.8451	.8457	.8463	.8470	.8476	.8482	.8488	.8494	.8500	.8506
7.1	.8513	.8519	.8525	.8531	.8537	.8543	.8549	.8555	.8561	.8567
7.2	.8573	.8579	.8585	.8591	.8597	.8603	.8609	.8615	.8621	.8627
7.3	.8633	.8639	.8645	.8651	.8657	.8663	.8669	.8675	.8681	.8686
7.4	.8692	.8698	.8704	.8710	.8716	.8722	.8727	.8733	.8739	.8745
7.5	.8751	.8756	.8762	.8768	.8774	.8779	.8785	.8791	.8797	.8802
7.6	.8808	.8814	.8820	.8825	.8831	.8837	.8842	.8848	.8854	.8859
7.7	.8865	.8871	.8876	.8882	.8887	.8893	.8899	.8904	.8910	.8915
7.8	.8921	.8927	.8932	.8938	.8943	.8949	.8954	.8960	.8965	.8971
7.9	.8976	.8982	.8987	.8993	.8998	.9004	.9009	.9015	.9020	.9025
8.0	.9031	.9036	.9042	.9047	.9053	.9058	.9063	.9069	.9074	.9079
8.1	.9085	.9090	.9096	.9101	.9106	.9112	.9117	.9122	.9128	.9133
8.2	.9138	.9143	.9149	.9154	.9159	.9165	.9170	.9175	.9180	.9186
8.3	.9191	.9196	.9201	.9206	.9212	.9217	.9222	.9227	.9232	.9238
8.4	.9243	.9248	.9253	.9258	.9263	.9269	.9274	.9279	.9284	.9289
8.5	.9294	.9299	.9304	.9309	.9315	.9320	.9325	.9330	.9335	.9340
8.6	.9345	.9350	.9355	.9360	.9365	.9370	.9375	.9380	.9385	.9390
8.7	.9395	.9400	.9405	.9410	.9415	.9420	.9425	.9430	.9435	.9440
8.8	.9445	.9450	.9455	.9460	.9465	.9469	.9474	.9479	.9484	.9489
8.9	.9494	.9499	.9504	.9509	.9513	.9518	.9523	.9528	.9533	.9538
9.0	.9542	.9547	.9552	.9557	.9562	.9566	.9571	.9576	.9581	.9586
9.1	.9590	.9595	.9600	.9605	.9609	.9614	.9619	.9624	.9628	.9633
9.2	.9638	.9643	.9647	.9652	.9657	.9661	.9666	.9671	.9675	.9680
9.3	.9685	.9689	.9694	.9699	.9703	.9708	.9713	.9717	.9722	.9727
9.4	.9731	.9736	.9741	.9745	.9750	.9754	.9759	.9763	.9768	.9773
9.5	.9777	.9782	.9786	.9791	.9795	.9800	.9805	.9809	.9814	.9818
9.6	.9823	.9827	.9832	.9836	.9841	.9845	.9850	.9854	.9859	.9863
9.7	.9868	.9872	.9877	.9881	.9886	.9890	.9894	.9899	.9903	.9908
9.8	.9912	.9917	.9921	.9926	.9930	.9934	.9939	.9943	.9948	.9952
9.9	.9956	.9961	.9965	.9969	.9974	.9978	.9983	.9987	.9991	.9996
n	0	1	2	3	4	5	6	7	8	9

Answers to Selected Exercises

CHAPTER 1

Section 1.1 Warm-ups F F F F T T T F T T
1. F **3.** T **5.** T **7.** F **9.** F **11.** F **13.** ∅
15. $A \cap C = \{1, 3, 5\}$ **17.** $\{1, 2, 3, 4, 5, 6, 8\}$
19. A **21.** ∅ **23.** A **25.** = **27.** ∪ **29.** ∩
31. ∉ **33.** ∈ **35.** T **37.** T **39.** T **41.** T **43.** F
45. T **47.** $\{2, 3, 4, 5, 6, 7, 8\}$ **49.** $\{3, 5\}$
51. $\{1, 2, 3, 4, 5, 6, 8\}$ **53.** $\{2, 3, 4, 5\}$ **55.** $\{2, 3, 4, 5, 7\}$
57. $\{2, 3, 4, 5\}$ **59.** $\{2, 3, 4, 5, 7\}$ **61.** ⊆ **63.** ∈
65. ∩ **67.** ⊆ **69.** ∩ **71.** ∪ **73.** $\{2, 4, 6, \ldots, 18\}$
75. $\{13, 15, 17, \ldots\}$ **77.** $\{6, 8, 10, \ldots, 78\}$
79. $\{x \mid x$ is a natural number between 2 and 7$\}$
81. $\{x \mid x$ is an odd natural number greater than 4$\}$
83. $\{x \mid x$ is an even natural number between 5 and 83$\}$

Section 1.2 Warm-ups F T F F T F T T F T
1. T **3.** F **5.** T **7.** T
9. $\{0, 1, 2, 3, 4, 5\}$
11. $\{-4, -3, -2, -1, 0, 1, \ldots\}$
13. $\{1, 2, 3, 4\}$
15. $\{-2, -1, 0, 1, 2, 3, 4\}$
17. All **19.** $\left\{0, \dfrac{8}{2}\right\}$ **21.** $\left\{-3, -\dfrac{5}{2}, -0.025, 0, 3\dfrac{1}{2}, \dfrac{8}{2}\right\}$
23. T **25.** F **27.** T **29.** F **31.** F **33.** F **35.** T
37. ⊆ **39.** ⊄ **41.** ⊆ **43.** ⊆ **45.** ⊆ **47.** ∈
49. ∈ **51.** ∈ **53.** ⊄ **55.** ⊆ **57.** ⊆

Section 1.3 Warm-ups T T F T T F F F F T
1. 34 **3.** 0 **5.** 0 **7.** -9 **9.** 9 **11.** -3
13. 4 **15.** -7 **17.** -2 **19.** -10 **21.** -26 **23.** -2
25. -5 **27.** 0 **29.** $\dfrac{1}{10}$ **31.** $-\dfrac{1}{6}$ **33.** -14.98
35. -2.71 **37.** 2.803 **39.** -0.2649 **41.** -3
43. -11 **45.** 13 **47.** -6 **49.** -9 **51.** 23 **53.** 1
55. $-\dfrac{1}{2}$ **57.** 1.97 **59.** 7.3 **61.** -50.73 **63.** 1.27
65. -75 **67.** $\dfrac{1}{6}$ **69.** -0.09 **71.** 0.2 **73.** $\dfrac{1}{20}$ or 0.05
75. $-\dfrac{5}{6}$ **77.** $-\dfrac{10}{3}$ **79.** -2 **81.** -37.5 **83.** -0.08
85. 0.25 **87.** -12 **89.** $\dfrac{3}{5}$ **91.** -91.25 **93.** 17,000
95. -49 **97.** -7 **99.** 15 **101.** -342 **103.** $\dfrac{3}{4}$
105. 32 **107.** 18.8 **109.** -0.33 **111.** -3 **113.** 20
115. 0 **117.** -180 **119.** $-\dfrac{1}{6}$ **121.** -55 **123.** -1
125. $-\dfrac{39}{2}$ **127.** 27.99 **129.** -29.3 **131.** -0.7
133. 2 **135.** $-\$20,600$ **137.** 20° **139.** 1014 ft

Section 1.4 Warm-ups F T T F F F T T F F
1. -22 **3.** -8 **5.** -14 **7.** 32 **9.** 1 **11.** 16
13. $-\dfrac{27}{64}$ **15.** 0.001 **17.** -8 **19.** 17 **21.** $\dfrac{1}{24}$
23. 58 **25.** -25 **27.** -200 **29.** 40 **31.** -25
33. -7.5 **35.** 84 **37.** -22.4841 **39.** -1.9602
41. -276.48 **43.** -2 **45.** -1 **47.** -6 **49.** 0
51. -7 **53.** $-\dfrac{4}{3}$ **55.** -8 **57.** 25 **59.** $-\dfrac{5}{2}$ **61.** 4
63. $\dfrac{10}{9}$ **65.** $\dfrac{3}{4}$ **67.** -2.67 **69.** 16 **71.** 49 **73.** 60
75. 37.12 **77.** 41 **79.** 27 **81.** 1 **83.** 9 **85.** -1

87. 26   **89.** $-\dfrac{3}{2}$ **91.** -2 **93.** $\dfrac{5}{12}$ **95.** 17
97. -46 **99.** 3   **101.** -8
103. 26 beats per minute, age 43   **105.** 104 feet
107. $60,776.47   **109.** $5441.96   **111.** $32,721

Section 1.5   Warm-ups T F F F F F T T F T
1. 1   **3.** -14 **5.** -24 **7.** -1.7 **9.** -19.8
11. $4x - 24$ **13.** $2(m + 5)$ **15.** $3a + at$
17. $-2w + 10$ **19.** $-6 + 2y$ **21.** $5(x - 1)$
23. $2x + y$ **25.** $6w + 9y$ **27.** $3(y - 5)$ **29.** $3(a + 3)$
31. $2x + 4$ **33.** $-x + 2$ **35.** 2   **37.** 1   **39.** $\dfrac{1}{6}$
41. 4   **43.** $-\dfrac{10}{7}$ **45.** $-\dfrac{5}{9}$ **47.** 0.6200   **49.** 0.7326
51. 1450 mph   **53.** 217 buttons per hour
55. Commutative   **57.** Distributive   **59.** Associative
61. Inverse   **63.** Commutative   **65.** Identity
67. Distributive   **69.** Inverse
71. Multiplication property of zero   **73.** Distributive
75. $w + 5$ **77.** $(5x)y$ **79.** $\dfrac{1}{2}(x - 1)$ **81.** $3(2x + 3)$
83. 1   **85.** 0   **87.** 4

Section 1.6   Warm-ups T F T T F T F F F T
1. 9000   **3.** 1   **5.** 527   **7.** 470   **9.** 38   **11.** 48,000
13. 0   **15.** 398   **17.** 1   **19.** 1700   **21.** 374   **23.** 0
25. $2n$ **27.** $7w$ **29.** $-11mw^2$ **31.** $-3x$
33. $-4 - 7z$ **35.** $9t^2$ **37.** $-4ab + 3a^2b$ **39.** $8mn$
41. $-2x^3y$ **43.** $-2kz^6$ **45.** $28t$ **47.** $10x^2$ **49.** h^2
51. $-28w$ **53.** $-x + x^2$ **55.** $25k^2$ **57.** y **59.** $2y$
61. $3x^3$ **63.** $x^2y + 5x$ **65.** $-x + 2$ **67.** $\dfrac{1}{2}xt - 5$
69. $-3a + 1$ **71.** $10 - x$ **73.** $3m + 1$
75. $-12b + at$ **77.** $2t^2 - 3w$ **79.** $y^2 + z$ **81.** $9x + 8$
83. $2x - 2$ **85.** $-2a^2 + 2c$ **87.** $7t^2 - 9w$
89. $m - 12$ **91.** $-7k^3 - 17$ **93.** $0.96x - 2$
95. $0.06x - 1.5$ **97.** $-4k + 16$ **99.** $-4.5x + 17.83$
101. $-4xy - 22$ **103.** $-29w^2$ **105.** $-6a^2w^2$
107. $2x^2y + \dfrac{1}{3}$ **109.** $\dfrac{1}{4}m^2 - m$ **111.** $4t^3 + 3t^2 - 1$
113. $2xyz + xy - 3z$ **115.** $3s + 6$ ft
117. $\dfrac{13}{3}x$ m, $\dfrac{7}{6}x^2$ m^2

Chapter 1 Review
1. T   **3.** F   **5.** T   **7.** F   **9.** T   **11.** T   **13.** F
15. T   **17.** F   **19.** T   **21.** $\{0, 1, 31\}$ **23.** $\{-1, 0, 1, 31\}$
25. $\{-\sqrt{2}, \sqrt{3}, \pi\}$ **27.** F   **29.** F   **31.** F   **33.** T
35. F   **37.** 5   **39.** -12 **41.** -24 **43.** 2   **45.** $-\dfrac{1}{6}$
47. 10   **49.** 9.96   **51.** -4 **53.** 0   **55.** -4 **57.** 0
59. -3 **61.** 39   **63.** 121   **65.** 23   **67.** 2   **69.** 23
71. -96 **73.** 5   **75.** -1 **77.** 0.76   **79.** 1   **81.** -3
83. 4   **85.** 1   **87.** -8 **89.** 1   **91.** -35 **93.** 2

95. 5   **97.** -1 **99.** Cummutative   **101.** Distributive
103. Associative   **105.** Identity   **107.** Inverse
109. Multiplication property of zero   **111.** Identity
113. Inverse   **115.** $3(x - a)$ **117.** $3w + 3$
119. $7(x + 1)$ **121.** $5x - 25$ **123.** $-6x + 15$
125. $p(1 - t)$ **127.** $7a + 2$ **129.** $-t - 2$
131. $-2a + 4$ **133.** $4x + 33$ **135.** $-0.8x - 0.48$
137. $-0.05x - 1.85$ **139.** $\dfrac{1}{4}x + 4$
141. $-3x^2 - 2x + 1$
143. 0, inverse, multiplication property of 0
145. 7680, distributive   **147.** 48, associative, inverse
149. 0, distributive, inverse
151. 47, associative, inverse
153. -24, commutative, associative
155. 0, inverse, multiplication property of 0
157. 7 shingles per minute   **159.** $24,476, 1997

Chapter 1 Test
1. $\{2, 3, 4, 5, 6, 7, 8, 10\}$ **2.** $\{6, 7\}$ **3.** $\{4, 6, 8, 10\}$
4. $\{0, 8\}$ **5.** $\{-4, 0, 8\}$ **6.** $\left\{-4, -\dfrac{1}{2}, 0, 1.65, 8\right\}$
7. $\{-\sqrt{3}, \sqrt{5}, \pi\}$ **8.** -9 **9.** 44   **10.** -11
11. -1.98 **12.** -2 **13.** -4 **14.** $\dfrac{1}{18}$ **15.** -12
16. 7   **17.** -3 **18.** 0   **19.** 4780   **20.** 240   **21.** -30
22. 40   **23.** 7   **24.** 0   **25.** Distributive
26. Commutative   **27.** Associative   **28.** Inverse
29. Commutative   **30.** Multiplication property of 0
31. $11m - 3$ **32.** $0.95x + 2.9$ **33.** $\dfrac{3}{4}x - \dfrac{5}{4}$ **34.** $5x^2$
35. $3x^2 + 2x - 1$ **36.** $-7xy$ **37.** $5(x - 8)$
38. $7(t - 1)$ **39.** 16 trees/hour
40. $4x - 8$, $x^2 - 4x$, 28 ft, 45 ft^2 **41.** 12.6 billion

CHAPTER 2

Section 2.1   Warm-ups F T F T T T T T T T
1. Yes   **3.** Yes   **5.** No   **7.** No   **9.** $\{-87\}$ **11.** $\{-24\}$
13. $\left\{\dfrac{3}{2}\right\}$ **15.** $\{-1\}$ **17.** $\{-3\}$ **19.** $\left\{\dfrac{5}{2}\right\}$ **21.** $\{18\}$
23. $\{18\}$ **25.** $\{0\}$ **27.** $\left\{-\dfrac{28}{3}\right\}$ **29.** $\left\{-\dfrac{28}{5}\right\}$ **31.** $\{2\}$
33. $\{12\}$ **35.** $\{-7\}$ **37.** $\{12\}$ **39.** $\{6\}$ **41.** $\{90\}$
43. $\{1000\}$ **45.** $\{800\}$ **47.** $\varnothing$, inconsistent
49. R, identity   **51.** $\{1\}$, conditional   **53.** R, identity
55. R, identity   **57.** $\varnothing$, inconsistent   **59.** R, identity
61. $\{-4\}$, conditional   **63.** $\{1\}$ **65.** $\left\{\dfrac{5}{18}\right\}$
67. $\left\{\dfrac{121}{48}\right\}$ **69.** $\left\{\dfrac{19}{13}\right\}$ **71.** $\left\{\dfrac{10}{29}\right\}$ **73.** $\{3\}$ **75.** $\{16\}$
77. $\left\{\dfrac{3}{4}\right\}$ **79.** $\{-15\}$ **81.** $\{6\}$ **83.** $\{6\}$ **85.** $\{-2\}$
87. $\{2\}$ **89.** $\{0.3\}$ **91.** $\{53,191.49\}$ **93.** $\{4.7\}$

95. $\{-47.102\}$ **97.** 2   **99.** 1   **101.** $\dfrac{3}{4}$

103. 42.2 million, 1998   **105.** 1964, 2005

Section 2.2   Warm-ups F F F T F T F T T F

1. $t = \dfrac{I}{Pr}$ **3.** $C = \dfrac{5}{9}(F - 32)$ **5.** $W = \dfrac{A}{L}$

7. $b_1 = 2A - b_2$ **9.** $y = -\dfrac{2}{3}x + 3$ **11.** $y = x - 4$

13. $y = 2x - 3$ **15.** $y = -\dfrac{1}{3}x - 2$ **17.** $y = \dfrac{1}{2}x - \dfrac{5}{2}$

19. $y = 2x - 6$ **21.** $y = \dfrac{3}{2}x - 6$ **23.** $y = \dfrac{1}{2}x + \dfrac{1}{2}$

25. $y = 4x - 29$ **27.** $y = -3x + 984$ **29.** $t = \dfrac{A - P}{Pr}$

31. $a = \dfrac{1}{b + 1}$ **33.** $y = \dfrac{12}{1 - x}$ **35.** $x = \dfrac{6}{w^2 - y^2 - z^2}$

37. $R_1 = \dfrac{RR_2}{R_2 - R}$ **39.** 12.472   **41.** 34.932   **43.** 0.539

45. $\dfrac{1}{3}$ **47.** $\dfrac{13}{2}$ **49.** -1 **51.** 4   **53.** -4.4507

55. $\dfrac{5}{8}$ **57.** $-\dfrac{7}{2}$ **59.** -10 **61.** $-\dfrac{8}{3}$ **63.** 4

65. 15%   **67.** 5.75 yards   **69.** 7.2 feet   **71.** 5 feet
73. 15 feet   **75.** 14 inches   **77.** 168 feet
79. 1.5 meters   **81.** 3,979 miles   **83.** 9.55 inches
85. 95,232 pounds   **87.** 205, 25,000, increase
89. 200 feet   **91.** $1200

Section 2.3   Warm-ups F T F F T T F T F F

1. $x, x + 2$ **3.** $x, 10 - x$ **5.** $0.85x$ **7.** $3x$ miles
9. $4x + 10$ **11.** 27, 28, 29   **13.** 82, 84, 86
15. 63, 65   **17.** Width 46 meters, length 93 meters
19. 161 feet, 312 feet, 211 feet
21. Width 11.25 feet, length 27.5 feet
23. $1500 at 6%, $2500 at 10%   **25.** $80,000

27. $\dfrac{40}{3}$ gallons   **29.** 1.36 ounces   **31.** 40 mph

33. 15 mph   **35.** $86,957   **37.** $8450   **39.** 18 feet
41. Packers 35, Chiefs 10   **43.** $8.67/lb

45. 30 pounds   **47.** $\dfrac{40}{7}$ quarts   **49.** 3,634,783

51. Brian $14,400, Daniel $7200, Raymond $3800
53. 22, 23   **55.** 7.5 hours   **57.** 20 meters by 20 meters
59. $500 at 8%, $2500 at 10%
61. 2 gallons of 5% solution, 3 gallons of 10% solution
63. Todd 46, Darla 32

Section 2.4   Warm-ups F F T F F F T T T T
1. F   **3.** T   **5.** T   **7.** T   **9.** Yes   **11.** No   **13.** No
15. $(-\infty, -1]$

17. $(20, \infty)$

19. $[3, \infty)$

21. $(-\infty, 2.3)$ **23.** $(1, \infty)$

25. $(-\infty, -3]$ **27.** $(-\infty, 5)$ **29.** $[-4, \infty)$ **31.** $>$
33. $>$ **35.** $\le$ **37.** $>$
39. $(-2, \infty)$

41. $[-4, \infty)$

43. $(5, \infty)$

45. $[-3, \infty)$

47. $(13, \infty)$

49. $[-1, \infty)$

51. $\left(\dfrac{8}{3}, \infty\right)$

53. $\left(\dfrac{4}{3}, \infty\right)$

55. $(-\infty, 12]$

57. $(-\infty, 4]$

59. $\left(\dfrac{2}{3}, \infty\right)$

61. $\left(\dfrac{13}{3}, \infty\right)$

63. $(-\infty, \infty)$ **65.** $\varnothing$

67. $(1, \infty)$

69. $\left(-\infty, \dfrac{49}{30}\right]$

71. $\left(-\dfrac{148}{33}, \infty\right)$

73. $(-\infty, 2.397]$

75. $(-\infty, -17)$

77. x = Tony's height, $x > 6$ ft
79. s = Wilma's salary, $s < \$80,000$
81. v = speed of the Concorde, $v \le 1450$ mph
83. a = amount Julie can afford, $a \le \$400$
85. b = Burt's height, $b \le 5$ ft
87. t = Tina's hourly wage, $t \le \$8.20$
89. x = price of car, $x < \$9100$
91. x = price of truck, $x \ge \$9100.92$
93. Increasing, 1999 **95.** x = final exam score, $x \ge 77$
97. x = the price of A-Mart jeans, $x < \$16.67$

Section 2.5 Warm-ups T T F T T T F T F T
1. No **3.** Yes **5.** No **7.** No **9.** Yes **11.** Yes
13.

15.

17.

19. $\varnothing$

21.

23. $\varnothing$

25. $(-\infty, 1) \cup (10, \infty)$

27. $(9, \infty)$

29. $(-6, \infty)$

31. $(1, 4]$

33. $(-\infty, \infty)$

35. $\varnothing$

37. $\left(-\infty, \dfrac{1}{2}\right)$

39. $(-\infty, \infty)$

41. $(4, 7)$

43. $[-3, 2)$

45. $\left(-\dfrac{7}{3}, 3\right]$

47. $(-1, 5)$

49. $[2, 3]$

51. $(-13.9, 116.1)$

53. $(2, \infty)$ **55.** $(-\infty, 5)$ **57.** $[2, 4]$ **59.** $(-\infty, \infty)$
61. $\varnothing$ **63.** $[4, 5)$ **65.** $[1, 6]$ **67.** $x > 2$ **69.** $x < 3$
71. $x > 2$ or $x \le -1$ **73.** $-2 \le x < 3$

75. $x \ge -3$ **77.** $(5, 7)$ **79.** $(-1, 1] \cup (10, \infty)$
81. $(-3, 3)$ **83.** $[-3.2, 6.4]$
85. x = final exam score, $73 \le x \le 86.5$
87. x = price of truck, $\$11,033 \le x \le \$13,811$
89. x = number of cigarettes on the run, $4 \le x \le 18$
91. 2013, 2005

Section 2.6 Warm-ups T F F T F T F T T F
1. $\{-5, 5\}$ **3.** $\{2, 4\}$ **5.** $\{-3, 9\}$ **7.** $\left\{-\dfrac{8}{3}, \dfrac{16}{3}\right\}$

9. $\{12\}$ **11.** $\{-20, 80\}$ **13.** $\varnothing$ **15.** $\{0, 5\}$
17. $\{0.143, 1.298\}$ **19.** $\{-2, 2\}$ **21.** $\{-11, 5\}$

23. $\{0, 3\}$ **25.** $\left\{-6, \dfrac{4}{3}\right\}$ **27.** $\{1, 3\}$ **29.** $(-\infty, \infty)$

31. $\{0\}$ **33.** $|x| < 2$ **35.** $|x| > 3$ **37.** $|x| \le 1$
39. $|x| \ge 2$ **41.** No **43.** Yes **45.** No
47. Yes **49.** No
51. $(-\infty, -6) \cup (6, \infty)$

53. $(-3, 3)$

55. $(-\infty, -1] \cup [5, \infty)$

57. $\left(-\dfrac{1}{2}, \dfrac{9}{2}\right)$

59. $[-2, 12]$

61. $\left(-\infty, -\dfrac{9}{2}\right] \cup \left[\dfrac{15}{2}, \infty\right)$

63. $(-\infty, -5) \cup (-2, \infty)$

65. $(-\infty, 2) \cup (2, \infty)$

67. $(-\infty, \infty)$ **69.** $\varnothing$ **71.** $(-\infty, \infty)$
73. $(-\infty, -3) \cup (-1, \infty)$ **75.** $(-4, 4)$ **77.** $(-1, 1)$
79. $(0.255, 0.847)$ **81.** $\varnothing$ **83.** $(4, 5)$
85. $\left\{\dfrac{5}{2}\right\}$ **87.** $\{1\}$ **89.** $\left\{\dfrac{3}{2}\right\}$ **91.** $(-\infty, 0)$ **93.** $\left(\dfrac{4}{3}, \infty\right)$
95. 1401 or 1429 **97.** Between 121 and 133 pounds
99. $0.5 < t < 1.5$

Chapter 2 Review

1. $\{8\}$ **3.** $\left\{-\dfrac{3}{2}\right\}$ **5.** R **7.** $\varnothing$ **9.** $\{0\}$ **11.** $\{20\}$

13. $\{5\}$ **15.** $\{5\}$ **17.** $x = \dfrac{-b}{a}$ **19.** $x = \dfrac{2}{c - a}$

21. $x = \dfrac{P}{mw}$ **23.** $x = \dfrac{2}{2w - 1}$ **25.** $y = \dfrac{3}{2}x + 3$

27. $y = -\dfrac{1}{3}x + 4$ **29.** $y = 2x - 20$

31. Length 14 inches, width 8.5 inches
33. Wife $27,000, Roy $35,000   **35.** $9,500
37. 11 nickels, 4 dimes   **39.** 15 miles
41. $(-3, \infty)$

$-5\ -4\ -3\ -2\ -1\ \ 0\ \ 1$

43. $(0, \infty)$

$-2\ -1\ \ 0\ \ 1\ \ 2\ \ 3\ \ 4$

45. $(-\infty, -8]$

$-12\ -11\ -10\ -9\ -8\ -7\ -6$

47. $\left(-\infty, \dfrac{11}{2}\right)$

$\dfrac{11}{2}$

$3\ \ 4\ \ 5\ \ 6\ \ 7\ \ 8$

49. $[48, \infty)$

$44\ \ 46\ \ 48\ \ 50\ \ 52\ \ 54\ \ 56$

51. $(-\infty, -4) \cup (1, \infty)$

$-6\ -5\ -4\ -3\ -2\ -1\ \ 0\ \ 1\ \ 2\ \ 3$

53. $(0, 9)$

$0\ \ 1\ \ 2\ \ 3\ \ 4\ \ 5\ \ 6\ \ 7\ \ 8\ \ 9$

55. $(0, \infty)$

$-2\ -1\ \ 0\ \ 1\ \ 2\ \ 3\ \ 4$

57. $(-\infty, 4)$ **59.** $\varnothing$

$0\ \ 1\ \ 2\ \ 3\ \ 4\ \ 5\ \ 6$

61. $(-\infty, \infty)$ **63.** $\left[-\dfrac{17}{2}, \dfrac{13}{2}\right]$

$-\dfrac{17}{2}$ $\dfrac{13}{2}$

$-9\ -7\ -5\ -3\ -1\ \ 1\ \ 3\ \ 5\ \ 7$

65. $[1, \infty)$ **67.** $(3, 6)$ **69.** $(-\infty, \infty)$ **71.** $[-2, -1]$
73.

$-6\ -4\ -2\ \ 0\ \ 2\ \ 4\ \ 6$ **75.** $\varnothing$

77.

$-8\ -4\ \ 0\ \ 4\ \ 8\ \ 12\ \ 16$ **79.** $\varnothing$

81.

$-12\ -8\ -4\ \ 0\ \ 4\ \ 8\ \ 12$ **83.** R

85.

$-1\ \ 0\ \ 1\ \ 2\ \ 3\ \ 4\ \ 5$

87. x = rental price, $3 \le x \le 5$
89. $(40.2, 53.6)$ **91.** 81 or 91
93. Accountant $50,000, employees $60,000
95. Washington County 1200, Cade County 2400
97. $x > 1$ **99.** $x = 2$ **101.** $|x| = 3$ **103.** $x \le -1$
105. $|x| \le 2$ **107.** $x \le 2$ or $x \ge 7$ **109.** $|x| > 3$
111. $5 < x < 7$ **113.** $|x| > 0$

Chapter 2 Test

1. $\{-4\}$ **2.** R **3.** $\{-6, 6\}$ **4.** $\{2, 5\}$ **5.** $y = \dfrac{2}{5}x - 4$

6. $a = \dfrac{S}{1 - xt}$ **7.** $[4, 8]$

$3\ \ 4\ \ 5\ \ 6\ \ 7\ \ 8\ \ 9$

8. $(-\infty, -7) \cup (13, \infty)$

-7 13

$-15\ -10\ -5\ \ 0\ \ 5\ \ 10\ \ 15$

9. $(5, \infty)$

$3\ \ 4\ \ 5\ \ 6\ \ 7\ \ 8\ \ 9$

10. $\left(-8, -\dfrac{1}{2}\right)$

$-\dfrac{1}{2}$

$-8\ -7\ -6\ -5\ -4\ -3\ -2\ -1\ \ 0$

11. $[-5, 3)$

$-5\ -4\ -3\ -2\ -1\ \ 0\ \ 1\ \ 2\ \ 3$

12. $(-\infty, 15)$

$11\ \ 12\ \ 13\ \ 14\ \ 15\ \ 16\ \ 17$

13. $\{2, 5\}$ **14.** $(-\infty, \infty)$ **15.** $\varnothing$ **16.** $\{2.5\}$ **17.** $\varnothing$
18. R **19.** $\varnothing$ **20.** R **21.** $\{100\}$ **22.** 13 meters
23. 14 inches   **24.** $300   **25.** 30 liters
26. Greater than $31,000 or less than $25,000

Tying It All Together Chapters 1–2
1. $11x$ **2.** $30x^2$ **3.** $3x + 1$ **4.** $4x - 3$ **5.** 899
6. 961 **7.** 841 **8.** 25 **9.** 13 **10.** -25 **11.** 5
12. -4 **13.** $-2x + 13$ **14.** 60 **15.** 72 **16.** -9
17. $-3x^3$ **18.** 1 **19.** $\{0\}$ **20.** R **21.** $\{0\}$ **22.** $\{1\}$
23. $\left\{-\dfrac{1}{3}\right\}$ **24.** $\{1\}$ **25.** R **26.** $\{1000\}$
27. $\left\{-\dfrac{17}{5}, 1\right\}$ **28.** 87,500, buying

CHAPTER 3

Section 3.1   Warm-ups T F F T T F T T T F
1. 16 **3.** $\dfrac{1}{16}$ **5.** 16 **7.** $\dfrac{16}{9}$ **9.** -18 **11.** 0.003

13. 16 **15.** 2^{17} **17.** $\dfrac{2}{y^2}$ **19.** $\dfrac{30}{x^7}$ **21.** -2

23. 1 **25.** 2 **27.** 32 **29.** $\dfrac{100}{3}$ **31.** $\dfrac{8y^2}{5x^2}$

33. $\dfrac{1}{48x^2}$ **35.** x^2 **37.** 3^8 **39.** $\dfrac{1}{3a^3}$ **41.** $\dfrac{-3}{w^8}$

43. $3^8 w^6$ **45.** $\dfrac{xy}{2}$ **47.** 9 **49.** -14 **51.** $\dfrac{1}{16}$ **53.** $\dfrac{4}{9}$

55. $-15a^4$ **57.** $8a^3$ **59.** a **61.** $\dfrac{-2x^2y^4}{3}$ **63.** $\dfrac{5}{3}$

65. $\dfrac{6}{25}$ **67.** 3 **69.** -2 **71.** -4 **73.** -3

75. $486,000,000$ **77.** 0.00000237 **79.** $4,000,000$
81. 0.000005 **83.** 3.2×10^5 **85.** 7.1×10^{-7}
87. 7×10^{-5} **89.** 2.35×10^7 **91.** 0.75 **93.** 3×10^{22}
95. -1.2×10^{13} **97.** 1.578×10^5 **99.** 9.187×10^{-5}
101. 5.961×10^{157} **103.** 3.828×10^{30}
105. 4.910×10^{11} feet   **107.** 3.83×10^7 seconds
109. 2.6 lb/person/day

Section 3.2   Warm-ups F T T F F F T T T T
1. 64 **3.** y^{10} **5.** $\dfrac{1}{x^8}$ **7.** m^{18} **9.** 1 **11.** $\dfrac{1}{x^2}$

13. $81y^2$ **15.** $25w^6$ **17.** $\dfrac{x^9}{y^6}$ **19.** $\dfrac{b^2}{9a^2}$ **21.** $\dfrac{6x^3}{y}$

23. $\dfrac{1}{8a^3b^4}$ **25.** $\dfrac{w^3}{8}$ **27.** $\dfrac{-27a^3}{64}$ **29.** $\dfrac{x^2y^2}{4}$

31. $\dfrac{y^2}{9x^6}$ **33.** $\dfrac{25}{4}$ **35.** 4 **37.** $\dfrac{-27}{8x^3}$ **39.** $\dfrac{27y^3}{8x^6}$

41. 5^{6t} **43.** 2^{6w^2} **45.** 7^{m+3} **47.** 8^{5a+11} **49.** 2^{5m+7}

51. $6x^9$ **53.** $-8x^6$ **55.** $\dfrac{3z}{x^2y}$ **57.** $-\dfrac{3}{2}$ **59.** $\dfrac{4x^6}{9}$

61. $-\dfrac{x^2}{2}$ **63.** $\dfrac{y^3}{8x^3}$ **65.** $\dfrac{b^{14}}{5a^7}$ **67.** $\dfrac{x^6}{16y^8}$ **69.** $3ac^8$

71. $\dfrac{a^{28}}{9b^{38}}$ **73.** $\dfrac{x^{11}z^{11}}{81y^7}$ **75.** 2^{11} **77.** 2^{-4} **79.** 2^{6n}

81. 2^{4m^2} **83.** 25 **85.** 0.75 **87.** 850.559 **89.** 1.533
91. $\$56{,}197.12$ **93.** $\$2958.64$
95. 71.3 years, 83.1 years

Section 3.3 Warm-ups F F F T F T F T T F
1. Yes **3.** No **5.** Yes **7.** No **9.** $4, -8$, binomial
11. $0, 0$, monomial **13.** $7, 0$, monomial
15. $6, 1$, trinomial **17.** 80 **19.** -29 **21.** 0
23. $3a + 2$ **25.** $5xy + 25$ **27.** $2x - 9$
29. $2x^3 - x^2 - 2x - 8$ **31.** $11x^2 - 2x - 9$
33. $x + 5$ **35.** $-6x^2 + 5x + 2$ **37.** $2x$ **39.** $-15x^6$
41. $-3x + 2$ **43.** $15x^4y^4 - 20x^3y^3$ **45.** $x^2 - 4$
47. $2x^3 - x^2 + x - 6$ **49.** $-10x^2 + 15x$
51. $x^2 + 10x + 25$ **53.** $2x^2 + 9x - 18$ **55.** $x^3 - y^3$
57. $2x - 5$ **59.** $4a^2 - 11a - 4$ **61.** $4w^2 - 8w - 7$
63. $x^3 - 8$ **65.** $xz - wz + 2xw - 2w^2$
67. $6x^2y^2 + 7xy - 5$ **69.** $14.4375x^2 - 29.592x - 9.72$

71. $15.369x^2 + 1.88x - 0.866$ **73.** $\dfrac{3}{4}x + \dfrac{3}{2}$

75. $-\dfrac{1}{2}x^2 + x$ **77.** $3x^2 - 21x + 1$

79. $-4x^2 - 62x - 168$ **81.** $x^4 - m^2 - 4m - 4$
83. $-6x^3 - 53x^2 - 8x$ **85.** $-4a^{2m} - 4a^m + 5$
87. $x^{2n} + 2x^n - 3$ **89.** $5z^{3w} - z^{1+w}$ **91.** $x^{6r} + y^3$
93. $\$75$ **95.** $\$49.35, \$15.61$ **97.** 4.76 years

Section 3.4 Warm-ups T T F T F T T T F F
1. $x^2 + 2x - 8$ **3.** $2x^2 + 7x + 3$ **5.** $2a^2 - 7a - 15$
7. $4x^4 - 49$ **9.** $2x^6 + 7x^3 - 4$ **11.** $w^2 + 5wz - 6z^2$
13. $9k^2 + 6kt - 8t^2$ **15.** $xy - 3y + xw - 3w$

17. $\dfrac{1}{6} + \dfrac{5}{6}t + t^2$ **19.** $\dfrac{1}{4}x^4 + x^2 - 24$

21. $m^2 + 6m + 9$ **23.** $a^2 - 8a + 16$
25. $4w^2 + 4w + 1$ **27.** $9t^2 - 30tu + 25u^2$
29. $x^2 + 2x + 1$ **31.** $a^2 - 6ay^3 + 9y^6$

33. $\dfrac{1}{4}x^2 - x + 1$ **35.** $\dfrac{9}{16}a^2 + \dfrac{1}{2}a + \dfrac{1}{9}$

37. $g^2t^2 - 4gt + 4$ **39.** $64y^4z^8 + 48y^2z^4 + 9$
41. $w^2 - 81$ **43.** $w^6 - y^2$ **45.** $4x^2 - 49$
47. $9x^4 - 4$ **49.** $m^2 + 2mt + t^2 - 25$
51. $y^2 - r^2 - 10r - 25$
53. $4y^2 - 4yt + t^2 + 12y - 6t + 9$

55. $9h^2 + 6hk - 6h + k^2 - 2k + 1$ **57.** $x^2 + 3x - 54$
59. $25 - x^2$ **61.** $6x^2 + 7ax - 20a^2$
63. $2t^2 + 2tw - 3t - 3w$ **65.** $9x^4 + 12x^2y^3 + 4y^6$
67. $6y^2 - 4y - 10$ **69.** 2496 **71.** $49x^2 + 42x + 9$
73. $36y^3 + 12y^2 + y$ **75.** $2ah + h^2$
77. $x^3 + 6x^2 + 12x + 8$ **79.** $y^3 - 9y^2 + 27y - 27$
81. $16.32x^2 - 10.47x - 17.55$
83. $12.96y^2 + 31.68y + 19.36$
85. $x^{3m} + 2x^{2m} + 3x^m + 6$
87. $a^{3n+1} + a^{2n+1} - 3a^{n+1}$ **89.** $a^{2m} + 2a^{m+n} + a^{2n}$
91. $15y^{3m} + 24y^{2m}z^k + 20y^mz^{3-k} + 32z^3$
93. $x^2 + 4x + 3$ **95.** $4x^2 - 36x + 80$, 66.24 km^2
97. $4x^3 - 20x^2 + 24x$, 5.9 ft^3

Section 3.5 Warm-ups F F T T T F F T T T
1. $12x^4$ **3.** -2 **5.** $2b - 3$ **7.** $x + 2$

9. $-5x^2 + 4x - 3$ **11.** $\dfrac{7}{2}x^2 - 2x$ **13.** $x + 5, -2$

15. $x - 4, 8$ **17.** $x^2 - 2x + 4, 0$ **19.** $a^2 + 2a + 8, 11$
21. $x^2 - 2x + 3, -6$ **23.** $x^3 + 3x^2 + 6x + 11, 21$

25. $-3x^2 - 1, x - 4$ **27.** $\dfrac{1}{2}x - \dfrac{5}{4}, -\dfrac{7}{4}$

29. $\dfrac{2}{3}x + \dfrac{1}{9}, \dfrac{56}{9}$ **31.** $2 + \dfrac{10}{x - 5}$ **33.** $x - 1 + \dfrac{1}{x + 1}$

35. $x^2 - 2x + 4 + \dfrac{-8}{x + 2}$ **37.** $x + \dfrac{2}{x}$

39. $x - 6 + \dfrac{21}{x + 2}$ **41.** $3x^2 - x - 1 + \dfrac{6}{x - 1}$

43. $x^2 - 3x, -3$ **45.** $2x - 6, 11$
47. $3x^3 + 9x^2 + 12x + 43, 120$
49. $x^4 + x^3 + x^2 + x + 1, 0$ **51.** $x^2 - 2x - 1, 8$
53. $2.3x + 0.596, 0.79072$ **55.** No **57.** Yes **59.** Yes
61. Yes **63.** Yes **65.** Yes **67.** $(x - 4)(x - 2)$
69. $(w - 3)(w^2 + 3w + 9)$ **71.** $(x^2 - 2x + 2)(x - 2)$
73. $(z + 3)(z + 3)$ **75.** $(2y + 1)(3y + 1)$
77. $AC(x) = 0.03x + 300, \$300.36$ **79.** $x - 1$ ft
81. $6{,}333.3$ m^3

Section 3.6 Warm-ups T T T T F F F T F F
1. 12 **3.** 3 **5.** $6xy$ **7.** $x(x^2 - 5)$ **9.** $12w(4x + 3y)$
11. $2x(x^2 - 2x + 3)$ **13.** $12a^3b^2(3b^4 - 2a + 5a^2b)$
15. $(x - 6)(a + b)$ **17.** $(y - 1)^2(y + z)$
19. $2(x - y), -2(-x + y)$
21. $3x(2x - 1), -3x(-2x + 1)$
23. $w^2(-w + 3), -w^2(w - 3)$
25. $a(-a^2 + a - 7), -a(a^2 - a + 7)$
27. $(x - 10)(x + 10)$ **29.** $(2y - 7)(2y + 7)$
31. $(3x - 5a)(3x + 5a)$ **33.** $(12wz - 1)(12wz + 1)$
35. $(x - 10)^2$ **37.** $(2m - 1)^2$ **39.** $(w - t)^2$
41. $(a - 1)(a^2 + a + 1)$ **43.** $(w + 3)(w^2 - 3w + 9)$
45. $(2x - 1)(4x^2 + 2x + 1)$ **47.** $(a + 2)(a^2 - 2a + 4)$
49. $2(x + 2)(x - 2)$ **51.** $x(x + 5)^2$ **53.** $(2x + 1)^2$
55. $(x + 3)(x + 7)$ **57.** $3y(2y + 1)$ **59.** $(2x - 5)^2$
61. $2m(m - n)(m^2 + mn + n^2)$ **63.** $(2x - 3)(x - 2)$

65. $a(3a + w)(3a - w)$ **67.** $-5(a - 3)^2$
69. $2(2 - 3x)(4 + 6x + 9x^2)$ **71.** $-3y(y + 3)^2$
73. $-7(ab + 1)(ab - 1)$ **75.** $(x^5 - 3)(x^5 + 3)$
77. $(z^6 - 3)^2$ **79.** $2x(x^3 + 2)^2$ **81.** $x(2x^2 + 1)^2$
83. $(x^2 - 2)(x^4 + 2x^2 + 4)$ **85.** $2(x^3 + 2)(x^6 - 2x^3 + 4)$
87. $(a^n - 1)(a^n + 1)$ **89.** $(a^r + 3)^2$
91. $(x^m - 2)(x^{2m} + 2x^m + 4)$
93. $(a^m - b)(a^{2m} + a^m b + b^2)$ **95.** $k(k^w - 5)^2$
97. $uv^{2k}(v^{2k} - u)^2$ **99.** 9 **101.** 20 **103.** 16
105. 100 **107.** $4b + 3$ cm

Section 3.7   Warm-ups T F F F T F F F T T

1. $(x + 1)(x + 3)$ **3.** $(a + 10)(a + 5)$
5. $(y - 7)(y + 2)$ **7.** $(x - 2)(x - 4)$ **9.** $(a - 9)(a - 3)$
11. $(a + 10)(a - 3)$ **13.** $(3w + 1)(2w + 1)$
15. $(2x + 1)(x - 3)$ **17.** $(2x + 5)(2x + 3)$
19. $(2x - 1)(3x - 1)$ **21.** $(3y + 1)(4y - 1)$
23. $(6a - 5)(a + 1)$ **25.** $(2x - 1)(x + 8)$
27. $(3b + 5)(b - 7)$ **29.** $(3w - 4)(2w - 9)$
31. $(4x - 1)(x - 1)$ **33.** $(5m - 2)(m + 3)$
35. $(3y + 4)(2y - 5)$ **37.** $(x^3 + 5)(x^3 - 7)$
39. $(a^{10} - 10)^2$ **41.** $-2a(3a^2 + 1)(2a^2 + 1)$
43. $(x^a + 5)(x^a - 3)$ **45.** $(x^a - y^b)(x^a + y^b)$
47. $(x^4 - 3)(x^4 + 2)$ **49.** $x^a(x - 1)(x + 1)$
51. $(x^a + 3)^2$ **53.** $y(4yz^3 - 3)(yz^3 + 2)$
55. $2(x + 5)^2$ **57.** $a(a - 6)(a + 6)$
59. $5(a + 6)(a - 1)$ **61.** $2(x + 8y)(x - 8y)$
63. $-3(x + 3)(x - 4)$ **65.** $m^3(m + 10)^2$
67. $(3x + 4)(2x + 5)$ **69.** $(y - 6)^2$
71. $(3m - 5n)(3m + 5n)$ **73.** $5(a + 6)(a - 2)$
75. $-2(w - 10)(w + 1)$
77. $x^2(w + 10)(w - 10)$ **79.** $(3x + 1)(3x - 1)$
81. $(4x + 5)(2x - 3)$ **83.** $(2x - 5)^2$ **85.** $(3a + 10)^2$
87. $4(a + 2)(a + 4)$ **89.** $3m(m - 2)(m^2 + 2m + 4)$

Section 3.8   Warm-ups F T T F F T F T T F

1. Prime   **3.** Not prime   **5.** Not prime   **7.** Prime
9. Prime   **11.** Prime   **13.** $(a^2 - 5)^2$
15. $(x^2 - 2)(x - 2)(x + 2)$ **17.** $3(x - 2)(3x - 4)$
19. $2(y - 2)(y^2 + 2y + 4)(y + 2)(y^2 - 2y + 4)$
21. $2(4a^2 + 3)(4a^2 - 3)$ **23.** $(x + 3)(x - 2)(x^2 - x + 6)$
25. $(m + 5)(m + 1)$ **27.** $(3y - 7)(y + 4)$
29. $(y - 3)(y + 3)(y - 1)(y + 1)$ **31.** $(a + b)(x + y)$
33. $(x - 3)(x + 3)(x + 1)$ **35.** $(w - 3)(a - b)$
37. $(a + 3)^2(a^2 - 3a + 9)$
39. $(y - 5)(y + 2)(y^2 - 2y + 4)$ **41.** $(d - w)(ay + 1)$
43. $(a + y)(x - 1)(x + 1)$
45. $(y + b)(y + 1)(y^2 - y + 1)$ **47.** $(3x - 4)^2$
49. $(3x - 1)(4x - 3)$ **51.** $3a(a + 3)(a^2 - 3a + 9)$
53. $2(x^2 + 16)$ **55.** Prime   **57.** $(x + y - 1)(x + y + 1)$
59. $ab(a - b)(a + b)$ **61.** $(x + 2)(x - 2)(x^2 + 2x + 4)$
63. $n(m + n)^2$ **65.** $(2 + w)(m + n)$
67. $(2w + 3)(2w - 1)$ **69.** $(t^2 + 7)(t^2 - 3)$
71. $-a(a + 10)(a - 3)$ **73.** $(y + 2)(y + 6)$
75. $-2(w - 5)(w + 5)(w^2 + 25)$ **77.** $8a(a^2 + 1)$
79. $(w + 2)(w + 8)$ **81.** $a(2w - 3)^2$ **83.** $(x - 3)^2$

85. $3x^2(x - 5)(x + 5)$ **87.** $n(m - 1)(m^2 + m + 1)$
89. $2(3x + 5)(2x - 3)$ **91.** $2(a^3 - 16)$
93. $(a^m - 1)(a^{2m} + a^m + 1)$
95. $(a^w - b^{2n})(a^{2w} + a^w b^{2n} + b^{4n})$
97. $(t^{2n} - 2)(t^{2n} + 2)(t^{4n} + 4)$ **99.** $a(a^n - 5)(a^n + 3)$
101. $(a^n - 3)(a^n + b)$

Section 3.9   Warm-ups F T T F T F T T F F

1. $\{-4, 5\}$ **3.** $\left\{\dfrac{5}{2}, -\dfrac{4}{3}\right\}$ **5.** $\{-7, 2\}$ **7.** $\{0, 7\}$

9. $\{-4, 5\}$ **11.** $\{-3, 4\}$ **13.** $\left\{-4, \dfrac{5}{2}\right\}$ **15.** $\{-2, 0, 2\}$

17. $\{-5, -4, 5\}$ **19.** $\{-1, 1, 2\}$ **21.** $\{-3, -1, 1, 3\}$
23. $\{-8, -6, 4, 6\}$ **25.** $\{-4, 0, -2\}$ **27.** $\{-7, -3, 1\}$

29. $\left\{-\dfrac{3}{2}, 2\right\}$ **31.** $\{-6, -3, -2, 1\}$ **33.** $\{-6, 1\}$

35. $\{-5, 2\}$ **37.** $\{-5, -4, 0\}$ **39.** $\{-3, 0, 3\}$

41. $\left\{-3, \dfrac{1}{2}, 3\right\}$ **43.** $\left\{\dfrac{3}{2}\right\}$ **45.** $\{0, -b\}$

47. $\left\{-\dfrac{b}{a}, \dfrac{b}{a}\right\}$ **49.** $\left\{-\dfrac{b}{2}\right\}$ **51.** $\left\{-\dfrac{3}{a}, 1\right\}$

53. Width 4 inches, length 6 inches   **55.** 4 and 9
57. Length 43 inches, width 22 inches
59. 4 seconds   **61.** Width 5 feet, length 12 feet
63. Width 5 feet, length 12 feet
65. 3 and 4, or -4 and -3
67. Length 20 feet, width 12 feet

Chapter 3 Review

1. 2 **3.** 36 **5.** $-\dfrac{1}{27}$ **7.** 1 **9.** $\dfrac{8}{x^3}$ **11.** $\dfrac{1}{y^2}$

13. a^7 **15.** $\dfrac{2}{x^4}$ **17.** $8{,}360{,}000$ **19.** 0.00057

21. 8.07×10^6 **23.** 7.09×10^{-4} **25.** 1×10^2

27. $\dfrac{1}{a}$ **29.** $\dfrac{n^2}{m^{16}}$ **31.** $\dfrac{81}{16}$ **33.** $\dfrac{25}{36}$ **35.** $-\dfrac{4}{3ab}$

37. $\dfrac{b^{14}}{a^7}$ **39.** 5^{6w-1} **41.** 7^{15a-40} **43.** $8w + 2$

45. $-6x + 3$ **47.** $x^3 - 4x^2 + 8x - 8$ **49.** $-4xy + 22z$
51. $5m^5 - m^3 + 2m^2$ **53.** $x^2 + 4x - 21$
55. $z^2 - 25y^2$ **57.** $m^2 + 16m + 64$
59. $w^2 - 10xw + 24x^2$ **61.** $k^2 - 6k + 9$
63. $m^4 - 25$ **65.** $x^2 + 3x - 5, 0$
67. $m^3 - m^2 + m - 1, 0$ **69.** $a^6 + 2a^3 + 4, 0$

71. $m^2 + 2m - 6, 0$ **73.** $x + 1 + \dfrac{-4}{x - 1}$

75. $3 + \dfrac{6}{x - 2}$ **77.** Yes   **79.** Yes   **81.** No   **83.** Yes

85. $3(x - 2)$ **87.** $-4(-a + 5)$ **89.** $-w(w - 3)$
91. $(y - 9)(y + 9)$ **93.** $(2x + 7)^2$ **95.** $(t - 9)^2$
97. $(t - 5)(t^2 + 5t + 25)$ **99.** $(x - 10)(x + 3)$
101. $(w - 7)(w + 4)$ **103.** $(2m + 7)(m - 1)$

105. $m(m^3 - 5)(m^3 + 2)$ **107.** $5(x + 2)(x^2 - 2x + 4)$
109. $(3x + 2)(3x + 1)$ **111.** $(x - 1)(x + 1)^2$
113. $-y(x - 4)(x + 4)$ **115.** $-ab^2(a - 1)^2$
117. $(x - 1)(x^2 + 9)$ **119.** $(x - 2)(x + 2)(x^2 + 3)$
121. $a^3(a - 1)(a^2 + a + 1)$ **123.** $-2(2m + 3)^2$
125. $(2x - 7)(2x + 1)$
127. $(x + 2)(x^2 - 2x + 4)(x - 1)(x^2 + x + 1)$
129. $(a^2 - 11)(a^2 - 12)$ **131.** $(x^k - 7)(x^k + 7)$
133. $(m^a - 3)(m^a + 1)$ **135.** $(3z^k - 2)^2$
137. $(y^a - b)(y^a + c)$ **139.** $\{0, 5\}$ **141.** $\{0, 5\}$
143. $\left\{-\dfrac{1}{2}, 5\right\}$ **145.** $\{-5, -1, 1\}$ **147.** $\{-3, -1, 1, 3\}$
149. 6 feet   **151.** $12a + 4$ km
153. 65.1 years, 6.2 years   **155.** \$4195.58

Chapter 3 Test

1. $\dfrac{1}{9}$ **2.** 36   **3.** 8   **4.** $12x^7$ **5.** $4y^{12}$ **6.** $64a^6b^3$

7. $\dfrac{27}{x^6}$ **8.** $\dfrac{2a^3}{b^3}$ **9.** 3,240,000,000   **10.** 0.0008673

11. 2.4×10^{-5} **12.** 2×10^{-13}
13. $3x^3 + 3x^2 - 2x + 3$ **14.** $-2x^2 - 8x - 3$
15. $x^3 - 5x^2 + 13x - 14$ **16.** $x^2 + 4x - 5$
17. $x^3 - 6x^2 + 12x - 8$ **18.** -1 **19.** $2x^2 - 11x - 21$
20. $x^2 - 12x + 36$ **21.** $4x^2 + 20x + 25$ **22.** $9y^4 - 25$

23. $5 + \dfrac{-15}{x + 3}$ **24.** $x + 5 + \dfrac{4}{x - 2}$ **25.** $(a - 6)(a + 4)$
26. $(2x + 7)^2$ **27.** $3(m - 2)(m^2 + 2m + 4)$
28. $2y(x - 4)(x + 4)$ **29.** $(a - 5)(2x + 3)$
30. $(x - 1)(x + 1)(x^2 + 4)$ **31.** $\left\{-5, \dfrac{3}{2}\right\}$
32. $\{-2, 0, 2\}$ **33.** $\{-4, -3, 2, 3\}$
34. Width 8 inches, height 6 inches   **35.** 5473, 104, 4

Tying It All Together Chapters 1–3

1. 16   **2.** -8 **3.** $\dfrac{1}{16}$ **4.** 2   **5.** 1   **6.** $\dfrac{1}{6}$ **7.** $\dfrac{1}{12}$

8. 49   **9.** 64   **10.** 8   **11.** -29 **12.** $\dfrac{11}{30}$ **13.** $\{200\}$

14. $\left\{\dfrac{9}{5}\right\}$ **15.** $\left\{-9, \dfrac{3}{2}\right\}$ **16.** $\left\{-\dfrac{15}{2}, 0\right\}$ **17.** $\left\{2, \dfrac{8}{5}\right\}$

18. $\left\{\dfrac{9}{5}\right\}$ **19.** $\varnothing$ **20.** $\{-3, -2, 1, 2\}$ **21.** $\left\{-5, \dfrac{1}{2}\right\}$

22. $\left\{-\dfrac{2}{3}, \dfrac{4}{3}\right\}$ **23.** 5×10^{10} **24.** 2.05×10^5
25. \$8,000, 69

CHAPTER 4

Section 4.1   Warm-ups T T F T F F F T T F
1. $\{x \mid x \neq 1\}$ **3.** $\{z \mid z \neq 0\}$ **5.** $\{y \mid y \neq -2 \text{ and } y \neq 2\}$
7. $\{a \mid a \neq -2 \text{ and } a \neq -3\}$ **9.** $\{x \mid x \neq -4 \text{ and } x \neq 0\}$

11. $\{x \mid x \neq -3 \text{ and } x \neq 0 \text{ and } x \neq 2\}$ **13.** $\dfrac{2}{19}$ **15.** $\dfrac{1}{5}$

17. $\dfrac{x + 1}{2}$ **19.** $-\dfrac{3}{5}$ **21.** $\dfrac{b}{a^2}$ **23.** $\dfrac{-w}{3x^2y}$

25. $\dfrac{b^2}{1 + a}$ **27.** $-\dfrac{1}{2}$ **29.** $\dfrac{x + 2}{x}$ **31.** $a^2 + ab + b^2$

33. $\dfrac{x^2 - 1}{x^2 + 1}$ **35.** $\dfrac{x - 2}{2x - 6}$ **37.** $\dfrac{x^2 + 7x - 4}{x^2 - 16}$

39. $\dfrac{b + 3}{a + y}$ **41.** $\dfrac{10}{50}$ **43.** $\dfrac{3x}{3x^2}$ **45.** $\dfrac{5x - 5}{x^2 - 2x + 1}$

47. $\dfrac{-3}{-6x - 6}$ **49.** $\dfrac{5a}{a}$ **51.** $\dfrac{x^2 + x - 2}{x^2 + 2x - 3}$ **53.** $\dfrac{-7}{1 - x}$

55. $\dfrac{3x^2 - 6x + 12}{x^3 + 8}$ **57.** $\dfrac{2x^2 + 9x + 10}{6x^2 + 13x - 5}$ **59.** $\dfrac{4}{7}$

61. $\dfrac{1}{10}$ **63.** Undefined   **65.** $\dfrac{7}{21}$ **67.** $\dfrac{10}{2}$ **69.** $\dfrac{3a}{a^2}$

71. $\dfrac{-2}{b - a}$ **73.** $\dfrac{2x + 2}{x^2 - 1}$ **75.** $\dfrac{-2}{3 - w}$ **77.** $\dfrac{x + 2}{3}$

79. $\dfrac{1}{x - 4}$ **81.** $\dfrac{a + 1}{a}$ **83.** $\dfrac{x^2 + x + 1}{x^3 - 1}$ **85.** $x^a - 2$

87. $\dfrac{1 + w}{x^a - m}$ **89.** $\dfrac{x^{2b} + x^b + 1}{x^b + 1}$ **91.** $\dfrac{250}{x}$ mph

93. $\dfrac{0.50n + 45}{n}$ dollars, 7.5 cents

95. $p(n) = \dfrac{0.576n + 3.78}{3.14n + 87.1}$, 4.3%, 11.6%, 13.9%

Section 4.2   Warm-ups F F T F T T T T T F

1. $\dfrac{5}{11}$ **3.** $\dfrac{ab}{4}$ **5.** $\dfrac{1}{2}$ **7.** $\dfrac{x + 1}{x^2 + 1}$ **9.** $\dfrac{1}{a - b}$

11. $\dfrac{a + 2}{2}$ **13.** $-\dfrac{2}{3}$ **15.** $-7a - 14$ **17.** $\dfrac{3}{2}$ **19.** $\dfrac{63}{5}$

21. $\dfrac{6b}{5c^8}$ **23.** $\dfrac{w}{w - 1}$ **25.** $\dfrac{2}{x - y}$ **27.** $\dfrac{2}{x}$

29. $2x - 2y$ **31.** $2x + 10$ **33.** $\dfrac{a - b}{6}$ **35.** $3a - 3b$

37. $\dfrac{5x}{6}$ **39.** 3   **41.** $\dfrac{1}{12}$ **43.** $\dfrac{2x}{3}$ **45.** -1 **47.** -2

49. $\dfrac{a + b}{a}$ **51.** $2a + 2b$ **53.** $\dfrac{3x}{5y}$ **55.** $\dfrac{3a}{10b}$

57. $\dfrac{7x^2}{3x + 2}$ **59.** $-\dfrac{a^6b^2}{8c^2}$ **61.** $\dfrac{2m^8n^2}{3}$ **63.** $\dfrac{2x - 3}{8x - 4}$

65. $\dfrac{k + m}{m - k}$ **67.** $\dfrac{3}{2}$ **69.** $\dfrac{y^b}{x^a}$ **71.** $\dfrac{x^a - 1}{x^a + 2}$

73. $\dfrac{m^k + 1}{m^k + 2}$ **75.** 6.9%   **77.** $\dfrac{75}{x}$ miles

Section 4.3   Warm-ups F F T F F F T T T F

1. $4x$ **3.** $\dfrac{-x+1}{x}$ **5.** $\dfrac{5}{2}$ **7.** $\dfrac{2}{x-3}$ **9.** 120

11. $30x^3y$ **13.** $a^3b^5c^2$ **15.** $x(x+2)(x-2)$

17. $12a+24$ **19.** $(x-1)(x+1)^2$

21. $x(x-4)(x+4)(x+2)$ **23.** $\dfrac{17}{140}$ **25.** $\dfrac{1}{40}$

27. $\dfrac{3w+5z}{w^2z^2}$ **29.** $\dfrac{2x-1}{24}$ **31.** $\dfrac{11x}{10a}$

33. $\dfrac{9-4xy}{4y}$ **35.** $\dfrac{-2a-14}{a(a+2)}$ **37.** $\dfrac{3a-b}{(a-b)(a+b)}$

39. $\dfrac{4x+9}{(x+3)(x-3)}$ **41.** 0   **43.** $\dfrac{11}{2x-4}$

45. $\dfrac{-x+3}{(x-1)(x+2)(x+3)}$ **47.** $\dfrac{x^2+7x-18}{(x+1)(x+3)(x-3)}$

49. $\dfrac{8x-2}{x(x-1)(x+2)}$ **51.** $\dfrac{7}{12}$ **53.** $-\dfrac{19}{40}$ **55.** $\dfrac{5x}{6}$

57. $\dfrac{3a-2b}{3b}$ **59.** $\dfrac{3a+2}{3}$ **61.** $\dfrac{a+3}{a}$ **63.** $\dfrac{3}{x}$

65. $\dfrac{8x+3}{12x}$ **67.** -3 **69.** $\dfrac{-x-1}{(x+2)(x+3)}$

71. $\dfrac{2}{(a^3-1)(a^3+1)}$ **73.** b **75.** $\dfrac{5x^2+x+4}{(x-1)(x^2+x+1)}$

77. $\dfrac{x^2+25}{x(x-5)}$ **79.** $\dfrac{-w^2+2w+8}{(w+3)(w^2-3w+9)}$

81. $\dfrac{a-3}{a^2+2a+4}$ **83.** $\dfrac{-w^3-2w^2-5w+6}{(w-2)(w+2)(w^2+2w+4)}$

85. $\dfrac{x^3+x^2+2x+1}{(x-1)(x+1)(x^2+x+1)}$ **87.** $\dfrac{16x+8}{x^2+x}$

89. $\dfrac{3x+60}{x}$ magazines   **91.** $\dfrac{300x+500}{x^2+5x}$ hours

Section 4.4   Warm-ups F T T T F F F F T T

1. $\dfrac{10}{3}$ **3.** -8 **5.** $\dfrac{a^2+ab}{a-b}$ **7.** $\dfrac{a^2b+3a}{a+b^2}$

9. $\dfrac{x-3y}{x+y}$ **11.** $\dfrac{60m-3m^2}{8m+36}$ **13.** $\dfrac{a^2-ab}{b^2}$

15. $\dfrac{xy+x^2}{y^2-x^2y^2}$ **17.** $\dfrac{x+2}{x-2}$ **19.** $\dfrac{y^2-y-2}{(y-1)(3y+4)}$

21. $\dfrac{4x-10}{x-4}$ **23.** $\dfrac{6w-3}{2w^2+w-4}$ **25.** $\dfrac{2b-a}{a-3b}$

27. $\dfrac{3a-7}{5a-2}$ **29.** $\dfrac{3m^2-12m+12}{(m-3)(2m-1)}$

31. $\dfrac{2x^2+4x+5}{4x^2-2x-6}$ **33.** $\dfrac{yz+wz}{wy+wz}$ **35.** $\dfrac{x}{x+1}$

37. $\dfrac{a^2+b^2}{ab^3}$ **39.** $\dfrac{a-1}{a}$ **41.** $\dfrac{1}{x^2-x+1}$

43. $2m-3$ **45.** $-\dfrac{1}{ab}$ **47.** $-x^3y^3$ **49.** $x-2$

51. $\dfrac{xy}{x+y}$ **53.** -1.7333 **55.** 0.1667 **57.** 47.4%

59. 49.5 mph

Section 4.5   Warm-ups T F F T F T F F F T

1. $\{-24\}$ **3.** $\left\{\dfrac{22}{15}\right\}$ **5.** $\{5\}$ **7.** $\{1, 6\}$ **9.** $\{20, 25\}$

11. $\varnothing$ **13.** $\{-2\}$ **15.** $\left\{\dfrac{8}{3}\right\}$ **17.** $\left\{-\dfrac{3}{4}\right\}$ **19.** $\left\{-\dfrac{14}{5}\right\}$

21. $\{20\}$ **23.** $\{-3, 3\}$ **25.** $\{-5, 6\}$ **27.** $\{-6\}$

29. $\left\{\dfrac{11}{2}\right\}$ **31.** $\{-6, 6\}$ **33.** $\{8\}$ **35.** $\{2, 4\}$ **37.** $\{-1\}$

39. $\{5\}$ **41.** $\{-3, 3\}$ **43.** $\{8\}$ **45.** $\left\{\dfrac{25}{2}\right\}$ **47.** $\{4\}$

49. $\{-5, 2\}$ **51.** $\$166{,}666.67$ **53.** 138

55. Width 132 cm, length 154 cm   **57.** 20%, 96%

59. $\$17{,}142.86$, $\$54{,}285.72$

Section 4.6   Warm-ups F T F T T T F T F F

1. $y=5x-7$ **3.** $y=-\dfrac{1}{3}x+1$ **5.** $y=mx-bm+a$

7. $y=-\dfrac{7}{3}x-\dfrac{29}{3}$ **9.** $f=\dfrac{F}{M}$ **11.** $D^2=\dfrac{4A}{\pi}$

13. $m_1=\dfrac{Fr^2}{km_2}$ **15.** $q=\dfrac{pf}{p-f}$ **17.** $a^2=\dfrac{b^2}{1-e^2}$

19. $T_1=\dfrac{P_1V_1T_2}{P_2V_2}$ **21.** $h=\dfrac{3V}{4\pi r^2}$ **23.** $\dfrac{1}{2}$ **25.** 24

27. $\dfrac{32}{3}$ **29.** -6.517 **31.** 19.899 **33.** 1.910

35. 4 mph   **37.** Patrick 24 min, Guy 36 min, 100 mph

39. 10 mph   **41.** 6 hours   **43.** 60 minutes   **45.** 30 minutes   **47.** 10 pounds apples, 12 pounds oranges

49. 6 ohms   **51.** 80   **53.** 25.255 days

Chapter 4 Review

1. $\{x\mid x\neq 1\}$ **3.** $\{x\mid x\neq -1 \text{ and } x\neq 2\}$ **5.** $\dfrac{c^2}{a^2b}$

7. $\dfrac{4x^2}{3y}$ **9.** $-a$ **11.** $\dfrac{3}{2}$ **13.** $6x(x-2)$ **15.** $12a^5b^3$

17. $\dfrac{3x+11}{2(x-3)(x+3)}$ **19.** $\dfrac{aw-5b}{a^2b^2}$ **21.** $\dfrac{21}{10x-60}$

23. $\dfrac{7-3y}{4y-3}$ **25.** $\dfrac{b^3-a^2}{ab}$ **27.** $\left\{-\dfrac{16}{3}\right\}$ **29.** $\{10\}$

31. $y=mx+b$ **33.** $m=\dfrac{Fr}{v^2}$ **35.** $r=\dfrac{3A}{2\pi h}$

37. $y = 2x - 17$ **39.** $\dfrac{4}{3x}$ **41.** $\dfrac{10 + 7y}{6xy}$

43. $\dfrac{8a + 10}{(a - 5)(a + 5)}$ **45.** $\{-8, 8\}$ **47.** $-\dfrac{3}{2}$

49. $\{-9, 9\}$ **51.** $\dfrac{1}{x - m}$ **53.** $\dfrac{8a + 20}{(a - 5)(a + 5)(a + 1)}$

55. $\dfrac{-15a + 10}{2(a + 2)(a + 3)(a - 3)}$ **57.** $\{2\}$ **59.** $\left\{-\dfrac{5}{2}\right\}$

61. $-\dfrac{1}{3}$ **63.** $\dfrac{x - 2}{3x + 3}$ **65.** $\dfrac{3a^2 + 7a + 16}{a^3 - 8}$

67. $\dfrac{3 - x}{(x + 1)(x + 3)}$ **69.** $\dfrac{a + w}{a + 4}$ **71.** $\{-6, 8\}$ **73.** $\dfrac{18}{3x}$

75. $\dfrac{-3}{b - a}$ **77.** $\dfrac{4x}{x}$ **79.** $10x$ **81.** $\dfrac{1}{3}$ **83.** $\dfrac{1}{a + 3}$

85. $\dfrac{3}{10}$ **87.** $\dfrac{5a}{6}$ **89.** $\dfrac{b - a}{ab}$ **91.** $\dfrac{a - 3}{3}$

93. $\dfrac{2a + 1}{a}$ **95.** 1 **97.** $1 - x$ **99.** -1 **101.** $\dfrac{b}{6a}$

103. -2 **105.** $\dfrac{3x^a + 4}{(x^a - 2)(x^a + 3)}$ **107.** $2x^k$

109. $10{,}146$ **111.** 10 hours **113.** 400 hours

Chapter 4 Test

1. $\{x \mid x \neq 3 \text{ and } x \neq -3\}$ **2.** $\left\{x \mid x \neq \dfrac{4}{3}\right\}$ **3.** $\dfrac{3a^3}{2b}$

4. $\dfrac{-x - y}{2x - 2y}$ **5.** $-\dfrac{1}{36}$ **6.** $\dfrac{7y^2 + 3}{y}$ **7.** $\dfrac{5}{a - 9}$

8. $\dfrac{4a + 3b}{24a^2b^2}$ **9.** $\dfrac{a^3}{30b^2}$ **10.** $-\dfrac{3}{a + b}$ **11.** $\dfrac{1}{x^2 - 1}$

12. $\dfrac{-4x + 2}{(x + 2)(x - 2)(x - 5)}$ **13.** $\dfrac{m + 1}{3}$ **14.** $\left\{\dfrac{12}{7}\right\}$

15. $\{4, 10\}$ **16.** $\{-2, 2\}$ **17.** $t = \dfrac{a^2}{W}$ **18.** $b = \dfrac{2a}{a - 2}$

19. $\dfrac{16}{9 - 6x}$ **20.** $w - m$ **21.** $\dfrac{3a^2}{2}$ **22.** 24 minutes

23. 30 miles **24.** 9

Tying It All Together Chapters 1–4

1. $\left\{\dfrac{15}{4}\right\}$ **2.** $\{-4, 4\}$ **3.** $\left\{\dfrac{12}{5}\right\}$ **4.** $\{-6, 3\}$ **5.** $\left\{\dfrac{1}{4}\right\}$

6. $\{6\}$ **7.** $\left\{\dfrac{1}{2}\right\}$ **8.** $\left\{\dfrac{3}{2}\right\}$ **9.** $\{-3, 3\}$ **10.** $\{-8, 3\}$

11. $\left\{\dfrac{1}{3}, \dfrac{2}{3}\right\}$ **12.** $\{-3, 9\}$ **13.** $\left\{-\dfrac{9}{2}, \dfrac{1}{2}\right\}$ **14.** $\varnothing$

15. $y = \dfrac{C - Ax}{B}$ **16.** $y = -\dfrac{1}{3}x + \dfrac{4}{3}$ **17.** $y = \dfrac{C}{A - B}$

18. $y = A$ or $y = -A$ **19.** $y = 2A - 2B$

20. $y = \dfrac{2AC}{2B + C}$ **21.** $y = \dfrac{3}{4}x - \dfrac{3}{2}$ **22.** $y = A$ or $y = 2$

23. $y = \dfrac{2A - BC}{B}$ **24.** $y = B$ or $y = -C$ **25.** $12x^{13}$

26. $3x^5 + 15x^8$ **27.** $25x^{12}$ **28.** $27a^9b^6$ **29.** $-4a^6b^6$

30. $\dfrac{1}{32x^{10}}$ **31.** $\dfrac{27x^{12}y^{15}}{8}$ **32.** $\dfrac{a^2}{4b^6c^2}$

33. $\dfrac{ab + a^2b^4}{b + a^2}$ **34.** $a + b$ **35.** $\$20{,}000, \$2500$

CHAPTER 5

Section 5.1 Warm-ups T F T F T T T F F F

1. 10 **3.** 3 **5.** -3 **7.** $\dfrac{1}{2}$ **9.** Not a real number

11. 10 **13.** -4 **15.** -1 **17.** 8 **19.** 9 **21.** -125

23. Not a real number **25.** $\dfrac{1}{2}$ **27.** $\dfrac{1}{16}$ **29.** $-\dfrac{1}{8}$

31. Not a real number **33.** $3^{7/12}$ **35.** 1 **37.** $\dfrac{1}{2}$

39. 2 **41.** 6 **43.** 4 **45.** 81 **47.** $\dfrac{1}{4}$ **49.** $\dfrac{9}{8}$

51. $\dfrac{8}{9}$ **53.** $|x|$ **55.** a^4 **57.** y **59.** $|3x^3y|$

61. $\left|\dfrac{3x^3}{y^5}\right|$ **63.** $x^{3/4}$ **65.** $\dfrac{y^{3/2}}{x^{1/4}}$ **67.** $\dfrac{1}{w^{8/3}}$ **69.** $\dfrac{y}{x^{1/2}}$

71. $12x^8$ **73.** $\dfrac{x^{1/4}}{2y^{1/2}z^{1/4}}$ **75.** $\dfrac{a^2}{b}$ **77.** $\dfrac{w^6}{3x^5}$ **79.** 9

81. $-\dfrac{1}{8}$ **83.** $\dfrac{1}{625}$ **85.** $2^{1/4}$ **87.** 3 **89.** 3

91. $\dfrac{4}{9}$ **93.** Not a real number **95.** $3x^{9/2}$ **97.** $\dfrac{a^2}{27}$

99. $a^{5/4}b$ **101.** $k^{9/2}m^4$ **103.** $\dfrac{a^{3/4}}{w^{17}}$ **105.** 1.2599

107. -1.4142 **109.** 2 **111.** 1.9862 **113.** 2.5
115. 2.1392 **117.** $a^{3m/4}$ **119.** $a^{2m/15}$ **121.** a^nb^m
123. $a^{4m}b^{2n}$ **125.** 13 inches **127.** 274.96 m^2
129. 15.2% **131.** 6.1%

Section 5.2 Warm-ups T F T F T F T F T T

1. $27^{1/2}$ **3.** $x^{5/2}$ **5.** $a^{12/3}$ **7.** $-\sqrt[3]{5}$ **9.** $\sqrt[5]{2^2}$
11. $\sqrt[3]{x^{-2}}$ **13.** 11 **15.** -10 **17.** Not a real number
19. a^8 **21.** 4^8 **23.** w^6 **25.** $3\sqrt{w}$ **27.** $2\sqrt{5}$

29. $3\sqrt{5w}$ **31.** $12\sqrt{2}$ **33.** $3\sqrt[3]{2}$ **35.** $2\sqrt[4]{2a}$

37. $\dfrac{3}{10}$ **39.** $\dfrac{5\sqrt{2}}{3}$ **41.** $-\dfrac{5}{2}$ **43.** $\dfrac{2\sqrt[3]{2x}}{3}$ **45.** $\dfrac{2\sqrt{5}}{5}$

47. $\dfrac{\sqrt{21}}{7}$ **49.** $\dfrac{\sqrt[3]{2}}{2}$ **51.** $\dfrac{\sqrt[3]{150}}{5}$ **53.** $\dfrac{\sqrt{15}}{6}$ **55.** $\dfrac{1}{2}$

57. $\dfrac{\sqrt{2}}{2}$ **59.** $\dfrac{\sqrt[3]{14}}{2}$ **61.** $2x^4\sqrt{3}$ **63.** $2a^4b\sqrt{15ab}$

65. $\dfrac{\sqrt{xy}}{y}$ **67.** $\dfrac{a\sqrt{ab}}{b^4}$ **69.** $2x^4\sqrt[3]{2x}$ **71.** $x^2y\sqrt[4]{xy^2}$

73. $2x^4\sqrt[5]{2x^2}$ **75.** $\dfrac{\sqrt[3]{ab^2}}{b}$ **77.** 27 **79.** $\dfrac{1}{10}$

81. $\dfrac{2\sqrt{2x}}{7}$ **83.** $\dfrac{2\sqrt[4]{2a}}{3}$ **85.** $-3x^3y^2\sqrt[3]{y^2}$

87. $\dfrac{\sqrt{b}}{a}$ **89.** $\dfrac{\sqrt[3]{18}}{6b^2}$ **91.** $\sqrt[5]{3240m^3n}$ **93.** 2.887

95. 0.693 **97.** 1.310 **99.** 1 **101.** $-4°\text{F}, -10°\text{F}$

103. $t = \dfrac{\sqrt{S}}{4}$, 1.4 sec   **105.** 5.8 knots

107. 114.1 ft/sec, 77.8 mph

Section 5.3   Warm-ups F T F F T F T F F T

1. $-\sqrt{3}$ **3.** $9\sqrt{7x}$ **5.** $5\sqrt[3]{2}$ **7.** $4\sqrt{3} - 2\sqrt{5}$

9. $5\sqrt[3]{x}$ **11.** $2\sqrt[3]{x} - \sqrt{2x}$ **13.** $2\sqrt{2} + 2\sqrt{7}$

15. $-\sqrt{2}$ **17.** $\dfrac{3\sqrt{2}}{2}$ **19.** $\dfrac{21\sqrt{5}}{5}$

21. $x\sqrt{5x} + 2x\sqrt{2}$ **23.** $5\sqrt[3]{3}$ **25.** $-\sqrt[4]{3}$

27. $ty\sqrt[3]{2t}$ **29.** $\sqrt{15}$ **31.** $30\sqrt{2}$ **33.** $6a\sqrt{14}$

35. $3\sqrt[4]{3}$ **37.** 12 **39.** $\dfrac{2x\sqrt[3]{3x}}{3}$ **41.** $6\sqrt{2} + 18$

43. $5\sqrt{2} - 2\sqrt{5}$ **45.** $3\sqrt[3]{t^2} - t\sqrt[3]{3}$ **47.** $-7 - 3\sqrt{3}$

49. 2 **51.** $-8 - 6\sqrt{5}$ **53.** $-6 + 9\sqrt{2}$

55. $11 + 6\sqrt{2}$ **57.** $x + 10\sqrt{x - 25}$ **59.** $\sqrt[6]{3^5}$

61. $\sqrt[12]{5^7}$ **63.** $\sqrt[6]{500}$ **65.** $\sqrt[12]{432}$ **67.** -1 **69.** 3

71. 19 **73.** 13 **75.** $25 - 9x$ **77.** $11\sqrt{3}$

79. $10\sqrt{30}$ **81.** $8 - \sqrt{7}$ **83.** $16w$ **85.** $3x^2\sqrt{2x}$

87. $\dfrac{5\sqrt{2}}{12}$ **89.** $28 + \sqrt{10}$ **91.** $\dfrac{8\sqrt{2}}{15}$ **93.** 17

95. $9 + 6\sqrt{x} + x$ **97.** $25x - 30\sqrt{x} + 9$

99. $x + 3 + 2\sqrt{x + 2}$ **101.** $-\sqrt{w}$ **103.** $a\sqrt{a}$

105. $7\sqrt{2a}$ **107.** $3x^2\sqrt{x}$ **109.** $a - a^6$

111. $13x\sqrt[3]{2x}$ **113.** $\dfrac{y^2\sqrt[3]{2x^2y}}{2x}$ **115.** $\dfrac{x^2\sqrt[3]{5}}{5}$

117. $\sqrt[6]{32x^5}$ **119.** $3\sqrt{2}$ ft^2 **121.** $\dfrac{9\sqrt{2}}{2}$ ft^2

Section 5.4   Warm-ups T T F T F T F T T T

1. $\sqrt{3}$ **3.** $\dfrac{\sqrt{15}}{5}$ **5.** $\dfrac{3\sqrt{2}}{10}$ **7.** $\dfrac{\sqrt{2}}{3}$ **9.** $\sqrt[3]{10}$

11. $x^2\sqrt[3]{4}$ **13.** $2 + \sqrt{5}$ **15.** $1 - \sqrt{3}$

17. $\dfrac{1 + \sqrt{6} + \sqrt{2} + \sqrt{3}}{2}$ **19.** $\dfrac{2\sqrt{3} - \sqrt{6}}{3}$

21. $\dfrac{6\sqrt{6} + 2\sqrt{15}}{13}$

23. $\dfrac{18\sqrt{5} + 3\sqrt{10} - 2\sqrt{6} - 12\sqrt{3}}{66}$

25. $128\sqrt{2}$ **27.** $x^2\sqrt{x}$ **29.** $-27x^4\sqrt{x}$ **31.** $8x^5$

33. $4\sqrt[3]{25}$ **35.** x^4 **37.** $\dfrac{\sqrt{6} + 2\sqrt{2}}{2}$ **39.** $2\sqrt{6}$

41. $\dfrac{\sqrt{2}}{2}$ **43.** $\dfrac{2}{3}$ **45.** $\dfrac{2 - \sqrt{2}}{5}$ **47.** $\dfrac{1 + \sqrt{3}}{2}$

49. $a - 3\sqrt{a}$ **51.** $4a\sqrt{a} + 4a$ **53.** $12m$ **55.** $4xy^2z$

57. $m - m^2$ **59.** $5x\sqrt[3]{x}$ **61.** $8m^4\sqrt[4]{8m^2}$

63. $2\sqrt{2} - 2$ **65.** $2 + 8\sqrt{2}$ **67.** $\dfrac{3\sqrt{2} + 2\sqrt{3}}{6}$

69. $7\sqrt{2} - 1$ **71.** $\dfrac{4x + 4\sqrt{x}}{x - 4}$ **73.** $\dfrac{x + \sqrt{x}}{x - x^2}$

75. $\dfrac{\sqrt{6}}{3}$ **77.** $\dfrac{\sqrt{2} + 1}{1}$ **79.** $\dfrac{\sqrt{x} + 1}{x - 1}$

81. $\dfrac{3\sqrt{2} - 3x}{2 - x^2}$ **83.** 3.968 **85.** 12.124 **87.** 14.697

89. 8.873 **91.** 0.725 **93.** -0.419

Section 5.5   Warm-ups F T F F T F F T T T

1. $\{-10\}$ **3.** $\left\{\dfrac{1}{2}\right\}$ **5.** $\{1\}$ **7.** $\{-2\}$ **9.** $\{-5, 5\}$

11. $\{-2\sqrt{5}, 2\sqrt{5}\}$ **13.** $\varnothing$ **15.** $\{-1, 7\}$

17. $\{-1 - 2\sqrt{2}, -1 + 2\sqrt{2}\}$ **19.** $\{-\sqrt{10}, \sqrt{10}\}$

21. $\left\{-\dfrac{3}{2}, -\dfrac{1}{2}\right\}$ **23.** $\{3\}$ **25.** $\{-2, 2\}$ **27.** $\left\{-\dfrac{2}{3}, \dfrac{2}{3}\right\}$

29. $\{52\}$ **31.** $\left\{\dfrac{9}{4}\right\}$ **33.** $\{9\}$ **35.** $\varnothing$ **37.** $\{3\}$

39. $\{-5, 3\}$ **41.** $\{1\}$ **43.** $\varnothing$ **45.** $\{-1\}$ **47.** $\{6\}$

49. $\{7\}$ **51.** $\{-5\}$ **53.** $\{-3\sqrt{3}, 3\sqrt{3}\}$

55. $\left\{-\dfrac{1}{27}, \dfrac{1}{27}\right\}$ **57.** $\{512\}$ **59.** $\left\{\dfrac{1}{81}\right\}$ **61.** $\left\{0, \dfrac{2}{3}\right\}$

63. $\left\{\dfrac{4 - \sqrt{2}}{4}, \dfrac{4 + \sqrt{2}}{4}\right\}$ **65.** $\varnothing$ **67.** $\{-\sqrt{2}, \sqrt{2}\}$

69. $\{-5\}$ **71.** $\left\{-\dfrac{1}{3}, \dfrac{1}{3}\right\}$ **73.** $\varnothing$ **75.** $\{-9\}$

77. $\left\{\dfrac{5}{4}\right\}$ **79.** $\{-9, 4\}$ **81.** $\left\{-\dfrac{2}{3}, 2\right\}$

83. $\{-2 - 2\sqrt[4]{2}, -2 + 2\sqrt[4]{2}\}$ **85.** $\{0\}$ **87.** $\left\{\dfrac{1}{2}\right\}$

89. $\left\{-\dfrac{\sqrt{3}}{3}, \dfrac{\sqrt{3}}{3}\right\}$ **91.** $\{0, 8\}$ **93.** $4\sqrt{2}$ feet

95. $5\sqrt{2}$ feet   **97.** 50 feet   **99.** $\sqrt[6]{32}$ meters

101. $\sqrt{73}$ km   **103.** $S = P(1 + r)^n, P = S(1 + r)^{-n}$

105. 9.5 AU   **107.** $\{-1.8, 1.8\}$ **109.** $\{4.993\}$

111. $\{-26.372, 26.372\}$

Section 5.6   Warm-ups T F F T T T T F T F

1. $-2 + 8i$ **3.** $-4 + 4i$ **5.** -2 **7.** $-8 - 2i$

9. $6 + 15i$ **11.** $-2 - 10i$ **13.** $-4 - 12i$

15. $-10 + 24i$ **17.** $-1 + 3i$ **19.** $-5i$ **21.** 29

23. 2 **25.** 20 **27.** -9 **29.** -25 **31.** 16 **33.** i

35. 34   **37.** 5   **39.** 5   **41.** 7   **43.** $\dfrac{12}{17} - \dfrac{3}{17}i$

45. $\dfrac{4}{13} + \dfrac{7}{13}i$ **47.** $3 - 4i$ **49.** $1 + 3i$ **51.** $2 + 2i$

53. $5 + 6i$ **55.** $7 - i\sqrt{6}$ **57.** $5i\sqrt{2}$ **59.** $1 + i\sqrt{3}$

61. $-1 - \dfrac{1}{2}i\sqrt{6}$ **63.** $\{\pm 6i\}$ **65.** $\{\pm 2i\sqrt{3}\}$

67. $\left\{\pm\dfrac{i\sqrt{10}}{2}\right\}$ **69.** $\{\pm i\sqrt{2}\}$ **71.** $18 - i$ **73.** $5 + i$

75. $-\dfrac{6}{25} - \dfrac{17}{25}i$ **77.** $3 + 2i$ **79.** -9 **81.** $3i\sqrt{3}$

83. $-5 - 12i$ **85.** $-2 + 2i\sqrt{2}$

Chapter 5 Review

1. $\dfrac{1}{9}$ **3.** 4   **5.** $\dfrac{1}{1000}$ **7.** $27x^{1/2}$ **9.** $a^{7/2}b^{7/2}$

11. $x^{3/4}y^{5/4}$ **13.** $6x^2\sqrt{2x}$ **15.** $2x\sqrt[3]{9x^2}$ **17.** 8

19. $\dfrac{\sqrt{10}}{5}$ **21.** $\dfrac{\sqrt[3]{18}}{3}$ **23.** $\dfrac{2\sqrt{3}\,x}{3x}$ **25.** $\dfrac{y\sqrt{15y}}{3}$

27. $\dfrac{3\sqrt[3]{4a^2}}{2a}$ **29.** $\dfrac{5\sqrt[4]{27x^2}}{3x}$ **31.** $2xy^3\sqrt[4]{3x}$ **33.** 13

35. $3\sqrt{5} - 2\sqrt{3}$ **37.** $\dfrac{10\sqrt{3}}{3}$ **39.** $30 - 21\sqrt{6}$

41. $6 - 3\sqrt{3} + 2\sqrt{2} - \sqrt{6}$ **43.** $\sqrt[3]{5}$ **45.** $\dfrac{5\sqrt{2}}{2}$

47. 9   **49.** $1 - \sqrt{2}$ **51.** $\dfrac{-\sqrt{6} - 3\sqrt{2}}{2}$

53. $\dfrac{3\sqrt{2} + 2}{7}$ **55.** $256w^{10}$ **57.** $\{-4, 4\}$ **59.** $\{3, 7\}$

61. $\{-1 - \sqrt{5}, -1 + \sqrt{5}\}$ **63.** $\varnothing$ **65.** $\{10\}$ **67.** $\{9\}$

69. $\{-8, 8\}$ **71.** $\{124\}$ **73.** $\{7\}$ **75.** $\{2, 3\}$ **77.** $\{9\}$

79. $\{4\}$ **81.** $5 + 25i$ **83.** $7 - 3i$ **85.** $-1 + 2i$

87. $2 + i$ **89.** $2 - i\sqrt{3}$ **91.** $\dfrac{5}{17} - \dfrac{14}{17}i$ **93.** $\{\pm 10i\}$

95. $\left\{\pm\dfrac{3i\sqrt{2}}{2}\right\}$ **97.** False   **99.** True   **101.** True

103. False   **105.** False   **107.** False   **109.** False
111. True   **113.** False   **115.** True   **117.** False
119. False   **121.** True   **123.** True
125. $5\sqrt{30}$ seconds   **127.** $10\sqrt{7}$ feet
129. $200\sqrt{2}$ feet   **131.** $26.4\sqrt[3]{25}$ ft^2 **133.** 11.58%

135. $V = \dfrac{29\sqrt{LCS}}{CS}$

Chapter 5 Test

1. 4   **2.** $\dfrac{1}{8}$ **3.** $\sqrt{3}$ **4.** 30   **5.** $3\sqrt{5}$ **6.** $\dfrac{6\sqrt{5}}{5}$

7. 2   **8.** $6\sqrt{2}$ **9.** $\dfrac{\sqrt{15}}{6}$ **10.** $\dfrac{2 + \sqrt{2}}{2}$

11. $4 - 3\sqrt{3}$ **12.** $2ay^2\sqrt[4]{2a}$ **13.** $\dfrac{\sqrt[3]{4x}}{2x}$

14. $\dfrac{2a^4\sqrt{2ab}}{b^2}$ **15.** $-3x^3$ **16.** $2m\sqrt{5m}$ **17.** $x^{3/4}$

18. $3y^2x^{1/4}$ **19.** $2x^2\sqrt[3]{5x}$ **20.** $19 + 8\sqrt{3}$

21. $\dfrac{5 + \sqrt{3}}{11}$ **22.** $\dfrac{6\sqrt{2} - \sqrt{3}}{23}$ **23.** $22 + 7i$

24. $1 - i$ **25.** $\dfrac{1}{5} - \dfrac{7}{5}i$ **26.** $-\dfrac{3}{4} + \dfrac{1}{4}i\sqrt{3}$

27. $\{-5, 9\}$ **28.** $\left\{-\dfrac{7}{4}\right\}$ **29.** $\{-8, 8\}$ **30.** $\left\{\pm\dfrac{4}{3}i\right\}$

31. $\{3\}$ **32.** $\{5\}$ **33.** $\dfrac{3\sqrt{2}}{2}$ feet   **34.** 25 and 36

35. Length 6 ft, width 4 ft   **36.** 39.53 AU, 164.97 years

Tying It All Together Chapters 1–5

1. $\left\{-\dfrac{4}{7}\right\}$ **2.** $\left\{\dfrac{3}{2}\right\}$

3. $(-\infty, -3) \cup (-2, \infty)$ **4.** $\left\{\dfrac{3}{2}\right\}$

5. $(-\infty, 1)$ **6.** $\varnothing$ **7.** $\{9\}$

8. $\varnothing$ **9.** $\{-12, -2\}$ **10.** $\left\{\dfrac{1}{16}\right\}$

11. $(-6, \infty)$ **12.** $\left\{-\dfrac{1}{64}, \dfrac{1}{64}\right\}$

13. $\left\{-\dfrac{\sqrt{3}}{3}, \dfrac{\sqrt{3}}{3}\right\}$ **14.** R

15. $\left(-\dfrac{1}{3}, 3\right)$ **16.** $\left\{\dfrac{1}{3}\right\}$

17. $\{82\}$ **18.** $\left\{\dfrac{6}{5}, \dfrac{12}{5}\right\}$ **19.** $\{100\}$ **20.** R

21. $\{4\sqrt{30}\}$ **22.** $\{400\}$ **23.** $\left\{\dfrac{13 + 9\sqrt{2}}{3}\right\}$

24. $\{-3\sqrt{2}, 3\sqrt{2}\}$ **25.** $\{5\}$ **26.** $\{7 + 3\sqrt{6}\}$

27. $\{-2, 3\}$ **28.** $\{-5, 2\}$ **29.** $\{-2, 3\}$ **30.** $\left\{\dfrac{1}{2}, 3\right\}$

31. 3   **32.** -2 **33.** $\dfrac{1}{2}$ **34.** $\dfrac{1}{3}$ **35.** 48.5 cm^3

36. 14%, 55.6 cm^3

CHAPTER 6

Section 6.1   Warm-ups F F F F T F F T F F

1. $\{-2, 3\}$ **3.** $\{-5, 3\}$ **5.** $\left\{-1, \dfrac{3}{2}\right\}$ **7.** $\{-3, 4\}$

9. $\{-7\}$ **11.** $\{-4, 4\}$ **13.** $\{-9, 9\}$ **15.** $\left\{-\dfrac{4}{3}, \dfrac{4}{3}\right\}$

17. $\{-1, 7\}$ **19.** $\{-1 - \sqrt{5}, -1 + \sqrt{5}\}$

21. $\left\{\dfrac{3 - \sqrt{7}}{2}, \dfrac{3 + \sqrt{7}}{2}\right\}$ **23.** $\left\{\dfrac{1 + 3\sqrt{2}}{2}, \dfrac{1 - 3\sqrt{2}}{2}\right\}$

25. $x^2 + 2x + 1$ **27.** $x^2 - 3x + \dfrac{9}{4}$ **29.** $y^2 + \dfrac{1}{4}y + \dfrac{1}{64}$

31. $x^2 + \dfrac{2}{3}x + \dfrac{1}{9}$ **33.** $(x+4)^2$ **35.** $\left(y - \dfrac{5}{2}\right)^2$

37. $\left(z - \dfrac{2}{7}\right)^2$ **39.** $\left(t + \dfrac{3}{10}\right)^2$ **41.** $\{-3, 5\}$

43. $\{-10, 2\}$ **45.** $\{-5, 7\}$ **47.** $\{-4, 5\}$ **49.** $\{-7, 2\}$

51. $\left\{-1, \dfrac{3}{2}\right\}$ **53.** $\{-2 - \sqrt{10},\ 2 + \sqrt{10}\}$

55. $\{-4 - 2\sqrt{5}, -4 + 2\sqrt{5}\}$

57. $\left\{\dfrac{-3 - \sqrt{41}}{4}, \dfrac{-3 + \sqrt{41}}{4}\right\}$ **59.** $\{4\}$

61. $\left\{\dfrac{1 + \sqrt{17}}{8}\right\}$ **63.** $\{1, 6\}$ **65.** $\{-2 - \sqrt{2}, -2 + \sqrt{2}\}$

67. $\{-1 - 2i, -1 + 2i\}$ **69.** $\{-2i\sqrt{3}, 2i\sqrt{3}\}$

71. $\left\{\dfrac{2 \pm i}{5}\right\}$ **73.** $\left\{-\dfrac{5}{2}i, \dfrac{5}{2}i\right\}$ **75.** $\{-2, 1\}$

77. $\left\{\dfrac{-2 - \sqrt{19}}{5}, \dfrac{-2 + \sqrt{19}}{5}\right\}$ **79.** $\{-6, 4\}$

81. $\left\{\dfrac{-2 - 4i\sqrt{2}}{3}, \dfrac{-2 + 4i\sqrt{2}}{3}\right\}$ **83.** $\{-2, 3\}$

85. $\{3 - i, 3 + i\}$ **87.** $\{6\}$ **89.** $\left\{\dfrac{9 - \sqrt{65}}{2}, \dfrac{9 + \sqrt{65}}{2}\right\}$

91. $x = -3 \pm \sqrt{9 - k}$ **93.** $x = -k \pm \sqrt{k^2 - w}$

95. $x = \dfrac{-m \pm \sqrt{m^2 - 4}}{2}$ **101.** 136.9 ft/sec   **103.** 12

Section 6.2   Warm-ups T F T F T T T T F F

1. $\{-3, -2\}$ **3.** $\{-3, 2\}$ **5.** $\left\{-\dfrac{1}{3}, \dfrac{3}{2}\right\}$ **7.** $\left\{\dfrac{1}{2}\right\}$

9. $\left\{\dfrac{1}{3}\right\}$ **11.** $\left\{-\dfrac{3}{4}\right\}$ **13.** $\{-4 \pm \sqrt{10}\}$

15. $\left\{\dfrac{-5 \pm \sqrt{29}}{2}\right\}$ **17.** $\left\{\dfrac{3 \pm \sqrt{7}}{2}\right\}$

19. $\left\{\dfrac{3 \pm i}{2}\right\}$ **21.** $\left\{\dfrac{3 \pm i\sqrt{39}}{4}\right\}$

23. $\{5 \pm i\}$ **25.** 28, 2   **27.** −23, 0   **29.** 0, 1

31. $-\dfrac{3}{4}$, 0   **33.** 97, 2   **35.** 0, 1   **37.** 140, 2   **39.** 1, 2

41. $\left\{-2, \dfrac{1}{2}\right\}$ **43.** $\{0\}$ **45.** $\left\{\dfrac{13}{9}, \dfrac{17}{9}\right\}$ **47.** $\{4 \pm 2i\}$

49. $\{2 \pm i\sqrt{6}\}$ **51.** $\left\{-\dfrac{3}{4}, \dfrac{5}{2}\right\}$ **53.** $\{-4.474, 1.274\}$

55. $\{3.7\}$ **57.** $\{-2.979, -0.653\}$

59. $\{-4792.983, -0.017\}$ **61.** $\{-0.079, 0.078\}$

63. $\dfrac{1 + \sqrt{65}}{2}$ and $\dfrac{-1 + \sqrt{65}}{2}$

65. $3 + \sqrt{5}$ and $3 - \sqrt{5}$

67. Width $\dfrac{-1 + \sqrt{5}}{2}$ feet, length $\dfrac{1 + \sqrt{5}}{2}$ feet

69. Width $-2 + \sqrt{14}$ feet, length $2 + \sqrt{14}$ feet

71. 3 seconds   **73.** 7.02 seconds   **75.** 80
77. 4 inches

Section 6.3   Warm-ups T F F T F F F T F F

1. Prime   **3.** Prime   **5.** $(3x - 4)(2x + 9)$ **7.** Prime
9. $(4x - 15)(2x + 3)$ **11.** $x^2 + 4x - 21 = 0$
13. $x^2 - 5x + 4 = 0$ **15.** $x^2 - 5 = 0$ **17.** $x^2 + 16 = 0$
19. $x^2 + 2 = 0$ **21.** $6x^2 - 5x + 1 = 0$ **23.** $\left\{-\dfrac{3}{2}, \dfrac{3}{2}\right\}$
25. $\left\{\dfrac{-3 \pm \sqrt{5}}{2}\right\}$ **27.** $\{\pm\sqrt{5}, \pm 3\}$ **29.** $\{-2, 1\}$
31. $\{-1 \pm \sqrt{5}, -3, 1\}$ **33.** $\{-3, -2, 1, 2\}$ **35.** $\{16, 81\}$
37. $\{9\}$ **39.** $\left\{-\dfrac{1}{3}, \dfrac{1}{2}\right\}$ **41.** $\{64\}$ **43.** $\left\{\dfrac{2}{3}, \dfrac{3}{2}\right\}$
45. $\left\{\pm\dfrac{\sqrt{14}}{2}, \pm\dfrac{\sqrt{38}}{2}\right\}$ **47.** $\{-1 + \sqrt{2}, -1 - \sqrt{2}\}$
49. $\{-2i, i\}$ **51.** $\{i \pm \sqrt{2}\}$
53. $\left\{\dfrac{-\sqrt{13} \pm 5}{2}\right\}$ **55.** 2:00 P.M.
57. Before $-5 + \sqrt{265}$ or 11.3 mph, after $-9 + \sqrt{265}$
or 7.3 mph
59. Andrew $\dfrac{13 + \sqrt{265}}{2}$ or 14.6 hours,

John $\dfrac{19 + \sqrt{265}}{2}$ or 17.6 hours
61. Length $5 + 5\sqrt{41}$ or 37.02 feet, width $-5 + 5\sqrt{41}$
or 27.02 feet
63. 29.2 hours   **65.** 6.2 meters

Section 6.4   Warm-ups F F F F T T T T T F

1. $(-3, 2)$

3. $(-\infty, -2) \cup (2, \infty)$

5. $(-\infty, -4] \cup \left[\dfrac{3}{2}, \infty\right)$

7. $(-\infty, 0] \cup [2, \infty)$

9. $(-\infty, 0) \cup \left(\dfrac{1}{2}, \infty\right)$

11. $(-\infty, \infty)$
13. $(-\infty, 0) \cup (3, \infty)$

15. $[-2, 0)$

17. $(-\infty, -6) \cup (3, \infty)$

19. $(-\infty, -2) \cup (-1, \infty)$

21. $(-13, -4) \cup (5, \infty)$

23. $(-\infty, -5) \cup (1, 3) \cup (5, \infty)$

25. $[-6, 3) \cup [4, 6)$

27. $(-\infty, 1 - \sqrt{5}) \cup (1 + \sqrt{5}, \infty)$

29. $\left(-\infty, \dfrac{3 - \sqrt{3}}{2}\right] \cup \left[\dfrac{3 + \sqrt{3}}{2}, \infty\right)$

31. $\left[\dfrac{3 - 3\sqrt{5}}{2}, \dfrac{3 + 3\sqrt{5}}{2}\right]$

33. $[-3, 3]$

35. $(-4, 4)$

37. $(-\infty, 0] \cup [4, \infty)$

39. $\left(-\dfrac{3}{2}, \dfrac{5}{3}\right)$

41. $(-\infty, -2] \cup [6, \infty)$

43. $(-\infty, -3) \cup (5, \infty)$

45. $(-\infty, -4] \cup [2, \infty)$

47. $(-3, 4]$

49. $[-1, 2] \cup [5, \infty)$

51. $(-\infty, -3) \cup (-1, 1)$

53. $(-27.58, -0.68)$

55. $(-\infty, -4.4) \cup (-3, 0.4) \cup (2, \infty)$

57. $6, 7, 8, \ldots$ **59.** 4 seconds   **61.** 3 seconds

Chapter 6 Review

1. $\{-3, 5\}$ **3.** $\left\{-3, \dfrac{5}{2}\right\}$ **5.** $\{-5, 5\}$ **7.** $\left\{\dfrac{3}{2}\right\}$

9. $\{\pm 2\sqrt{3}\}$ **11.** $\{-2, 4\}$ **13.** $\left\{\dfrac{4 \pm \sqrt{3}}{2}\right\}$ **15.** $\left\{\pm \dfrac{3}{2}\right\}$

17. $\{2, 4\}$ **19.** $\{2, 3\}$ **21.** $\left\{\dfrac{1}{2}, 3\right\}$ **23.** $\{-2 \pm \sqrt{3}\}$

25. $\{-2, 5\}$ **27.** $\left\{-\dfrac{1}{3}, \dfrac{3}{2}\right\}$ **29.** $\{-2 \pm \sqrt{2}\}$

31. $\left\{\dfrac{5 \pm \sqrt{13}}{6}\right\}$ **33.** $0, 1$ **35.** $-19, 0$ **37.** $17, 2$

39. $\left\{\dfrac{2 \pm i\sqrt{2}}{2}\right\}$ **41.** $\left\{\dfrac{3 \pm i\sqrt{15}}{4}\right\}$ **43.** $\left\{\dfrac{-1 \pm i\sqrt{5}}{3}\right\}$

45. $\{-3 \pm i\sqrt{7}\}$ **47.** $(4x + 1)(2x - 3)$ **49.** Prime

51. $(4y - 5)(2y + 5)$ **53.** $x^2 + 9x + 18 = 0$

55. $x^2 - 50 = 0$ **57.** $\{-2, 1\}$ **59.** $\{\pm 2, \pm 3\}$

61. $\{-6, -5, 2, 3\}$ **63.** $\{-2, 8\}$

65. $\left\{-\dfrac{1}{9}, \dfrac{1}{4}\right\}$ **67.** $\{16, 81\}$

69. $(-\infty, -3) \cup (2, \infty)$

71. $[-4, 5]$

73. $(0, 1)$

75. $(-\infty, -2) \cup [4, \infty)$

77. $(-3, \infty)$

79. $(-2, -1) \cup \left(-\dfrac{1}{2}, \infty\right)$

81. $\left\{\dfrac{5}{12}\right\}$ **83.** $\left\{\dfrac{-3 \pm \sqrt{5}}{2}\right\}$ **85.** $\left\{\dfrac{4 \pm 2i}{3}\right\}$

87. $\left\{\dfrac{5}{2}\right\}$ **89.** $\left\{-2, -\dfrac{1}{4}\right\}$ **91.** $\{625, 10{,}000\}$

93. $-2 + 2\sqrt{2}$ and $2 + 2\sqrt{2}$, or 0.83 and 4.83

95. Width $\dfrac{4 + \sqrt{706}}{2}$ or 15.3 inches,

height $\dfrac{-4 + \sqrt{706}}{2}$ or 11.3 inches

97. 2 inches   **99.** Width 5 feet, length 9 feet

101. 0.5 seconds and 1.5 seconds

Chapter 6 Test

1. $-7, 0$ **2.** $13, 2$ **3.** $0, 1$ **4.** $\left\{-3, \dfrac{1}{2}\right\}$

5. $\{-3 \pm \sqrt{3}\}$ **6.** $\{-5\}$ **7.** $\left\{-2, \dfrac{3}{2}\right\}$ **8.** $\{-4, 3\}$

9. $\{\pm 1, \pm 2\}$ **10.** $\{11, 27\}$ **11.** $\{\pm 6i\}$ **12.** $\{-3 \pm i\}$

13. $\left\{\dfrac{1 \pm i\sqrt{11}}{6}\right\}$ **14.** $x^2 - 2x - 24 = 0$

15. $x^2 + 25 = 0$ **16.** $(-6, 3)$

17. $(-1, 2) \cup (8, \infty)$ ⟵(++)++++++++(+⟶
  -1 0 1 2 3 4 5 6 7 8 9 10

18. Width $-1 + \sqrt{17}$ feet, length $1 + \sqrt{17}$ feet

19. $\dfrac{5 + \sqrt{37}}{2}$ or 5.5 hours

Tying It All Together Chapters 1–6

1. $\left\{\dfrac{15}{2}\right\}$ **2.** $\left\{\pm\dfrac{\sqrt{30}}{2}\right\}$ **3.** $\left\{-3, \dfrac{5}{2}\right\}$ **4.** $\left\{\dfrac{-2 \pm \sqrt{34}}{2}\right\}$

5. $\left\{-\dfrac{7}{2}, -2\right\}$ **6.** $\left\{-3, \dfrac{1}{4}, \dfrac{-11 \pm \sqrt{73}}{8}\right\}$ **7.** $\{9\}$

8. $\left\{-\dfrac{3}{2}, \dfrac{13}{2}\right\}$ **9.** $(-4, \infty)$ **10.** $\left[\dfrac{1}{2}, 5\right]$ **11.** $\left[\dfrac{1}{2}, 5\right)$

12. $(2, 8)$ **13.** $[-3, 2)$ **14.** $(-\infty, \infty)$ **15.** $y = \dfrac{2}{3}x - 3$

16. $y = -\dfrac{1}{2}x + 2$ **17.** $y = \dfrac{-c \pm \sqrt{c^2 - 12d}}{6}$

18. $y = \dfrac{n \pm \sqrt{n^2 + 4mw}}{2m}$ **19.** $y = \dfrac{5}{6}x - \dfrac{25}{12}$

20. $y = -\dfrac{2}{3}x + \dfrac{17}{3}$ **21.** $\dfrac{4}{3}$ **22.** $-\dfrac{11}{7}$ **23.** -2

24. $\dfrac{58}{5}$ **25.** 40,000, 38,000, \$32.50

26. \$800,000, \$950,000, \$40 or \$80, \$60

CHAPTER 7

Section 7.1 Warm-ups F F F T T T F F T T
1. I
3. III
5. y-axis
7. IV
9. II
11. x-axis
13. y-axis
15. x-axis and y-axis

17.

19.

41.

43.

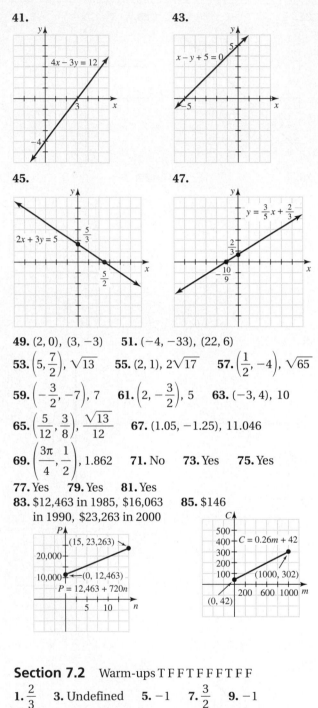

45.

47.

49. (2, 0), (3, −3)   **51.** (−4, −33), (22, 6)

53. $\left(5, \frac{7}{2}\right)$, $\sqrt{13}$ **55.** (2, 1), $2\sqrt{17}$ **57.** $\left(\frac{1}{2}, -4\right)$, $\sqrt{65}$

59. $\left(-\frac{3}{2}, -7\right)$, 7   **61.** $\left(2, -\frac{3}{2}\right)$, 5   **63.** (−3, 4), 10

65. $\left(\frac{5}{12}, \frac{3}{8}\right)$, $\frac{\sqrt{13}}{12}$ **67.** (1.05, −1.25), 11.046

69. $\left(\frac{3\pi}{4}, \frac{1}{2}\right)$, 1.862   **71.** No   **73.** Yes   **75.** Yes

77. Yes   **79.** Yes   **81.** Yes

83. $12,463 in 1985, $16,063 in 1990, $23,263 in 2000   **85.** $146

Section 7.2   Warm-ups T F F T F F F T F F

1. $\frac{2}{3}$ **3.** Undefined   **5.** −1   **7.** $\frac{3}{2}$ **9.** −1

11. $-\frac{5}{3}$ **13.** $\frac{4}{7}$ **15.** 5   **17.** $-\frac{5}{3}$ **19.** $-\frac{3}{5}$

21. $-\frac{2}{5}$ **23.** −3   **25.** 0   **27.** Undefined   **29.** 0.169

31. −2.828   **33.** −1.273   **35.** $\frac{8}{3}$ **37.** $\frac{3}{7}$ **39.** $-\frac{5}{4}$

41. Yes   **43.** No   **45.** Yes   **47.** No

49. The slope, 930.125, is the average yearly increase in price. The car will cost $26,626 in 2000.

51. −4.049   **53.** $-\frac{5}{2}$

Section 7.3   Warm-ups T F F T T F T F F T

1. $y = 2x - 7$ **3.** $y = -\frac{1}{2}x + 2$ **5.** $y = -\frac{3}{7}x + \frac{27}{7}$

7. $y = 4x - 8$ **9.** $y = -\frac{3}{4}x$ **11.** $y = \frac{1}{2}x + 2$

13. $x = 1$ **15.** $y = -x$ **17.** $y = \frac{3}{2}x - 3$

19. $y = -x + 2$ **21.** $x - 3y = 6$ **23.** $x - 2y = -13$
25. $2x - 6y = 11$ **27.** $5x + 6y = 890$

29. $x + 50y = -2150$ **31.** $y = -\frac{2}{5}x + \frac{1}{5}$, $-\frac{2}{5}$, $\left(0, \frac{1}{5}\right)$

33. $y = 3x - 2$, 3, (0, −2)   **35.** $y = 2$, 0, (0, 2)

37. $y = 3x - 1$, 3, (0, −1)   **39.** $y = \frac{3}{2}x + 11$, $\frac{3}{2}$, (0, 11)

41. $y = \frac{1}{3}x + \frac{7}{12}$, $\frac{1}{3}$, $\left(0, \frac{7}{12}\right)$

43. $y = 0.01x + 6057$, 0.01, (0, 6057)

45. $y = \frac{3}{2}x - 50$, $\frac{3}{2}$, (0, −50)   **47.** $x - 2y = -10$

49. $2x + y = 4$ **51.** $2x - y = -5$ **53.** $x + 2y = 7$
55. $2x + y = 3$ **57.** $3x - y = -9$ **59.** $y = 5$
61. $x = -3$

63.

65.

67.

69.

71.

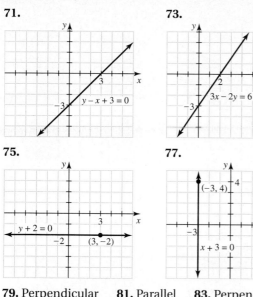

$y - x + 3 = 0$

73.

$3x - 2y = 6$

9.

$3x - 4y = 8$

11.

$(3, 3)$
$x - y = 0$

75.

$y + 2 = 0$
$(3, -2)$

77.

$(-3, 4)$
$x + 3 = 0$

13.

$(1, 4)$
$x = 1$

15.

$y = 3$
$(4, 3)$

79. Perpendicular   **81.** Parallel   **83.** Perpendicular

85. $t = \dfrac{7}{6}s + 60$, 95°F

87. $w = 304.47d - 1493.6$, 1015 ft³/sec

17.

$2x - 3y = 5$
$-\dfrac{5}{3}$, $\dfrac{5}{2}$

19.

$x + y + 3 = 0$

Section 7.4   Warm-ups T F T F T T F T F T

1.

$y = x + 2$

3.

$y = -2x + 1$
$(3, -5)$

21.

$-\dfrac{1}{2}x + \dfrac{1}{3}y = 1$

23.

$y = -2x + 3$
$y = x$
$(2, -1)$

5.

$x + y = 3$

7.

$2x + 3y = 9$
$\dfrac{9}{2}$

25.

$y = -x + 2$
$y = x + 3$

27.

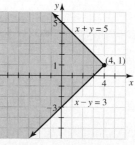

$x + y = 5$
$(4, 1)$
$x - y = 3$

29.

31.

49.

51.

33.

35.

53.

55.

37.

39.

57.

59.

41.

43.

61.

63.

45.

47.

65.

67.

Chapter 7 Review

1. III   **3.** x-axis   **5.** y-axis   **7.** IV   **9.** $\left(\frac{3}{2}, 4\right)$, $\sqrt{5}$

11. $(-3, -4)$, 10   **13.** $(0, 2)$, $\left(\frac{2}{3}, 0\right)$, $(4, -10)$, $\left(\frac{5}{3}, -3\right)$

15. 1   **17.** $\frac{3}{7}$ **19.** $\frac{3}{8}$ **21.** $\frac{7}{11}$ **23.** -3, $(0, 4)$

25. $\frac{2}{3}$, $\left(0, \frac{7}{3}\right)$ **27.** $2x$ $3y = 12$ **29.** $x - 2y = -5$

31. $x - 2y = 7$ **33.** $3x + 4y = 18$ **35.** $y = 5$

37. **39.**

41. **43.**

45. **47.**

49. **51.**

53. **55.**

57. **59.**

61. **63.**

65. **67.**

69.

71. $3x - y = 6$ **73.** $x + 2y = 7$ **75.** $x = 2$ **77.** $y = 0$
79. $2x - y = -6$ **81.** $3x - y = -6$ **83.** $y = 5$
91. $4\sqrt{17} + 2\sqrt{34}$ **93.** $h = 220 - a$, 180
95. 62 days

Chapter 7 Test

1. $(0, 5)$, $\left(\frac{5}{2}, 0\right)$, $(4, -3)$, $\left(\frac{13}{2}, -8\right)$ **2.** $(4, -4)$, $2\sqrt{5}$

3. $\left(-2, \frac{3}{2}\right)$, 5   **4.** $-\frac{6}{5}$ **5.** 2   **6.** $-\frac{1}{2}$, $(0, -2)$

7. $\frac{8}{5}$, $(0, 2)$ **8.** $x + 2y = 6$ **9.** $4x + y = -7$

10. $x - y = 3$ **11.** $5x + 3y = 19$ **12.** $y = 6$
13. $x - 2y = -4$

16.

17.

18.

19.

20.

21.

22. $V = -2000a + 22{,}000$

Tying It All Together Chapters 1–7

1. -625 **2.** 25   **3.** $\dfrac{2 + \sqrt{2}}{2}$ **4.** $\dfrac{4x - 1}{2}$

5. $x^2 + 2xh + h^2$ **6.** $a^2 + a$ **7.** $2x + h$
8. $6x + 3h - 2$

9.

10.

11.

12.

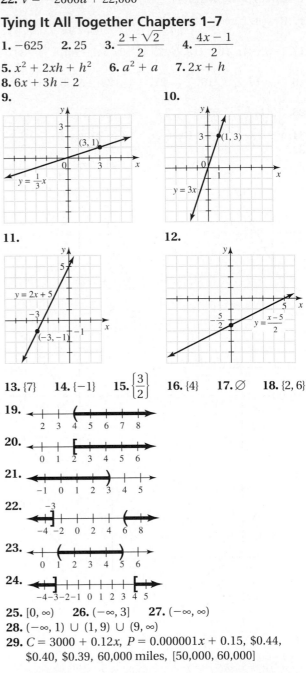

13. $\{7\}$ **14.** $\{-1\}$ **15.** $\left\{\dfrac{3}{2}\right\}$ **16.** $\{4\}$ **17.** $\varnothing$ **18.** $\{2, 6\}$

19.

20.

21.

22.

23.

24.

25. $[0, \infty)$ **26.** $(-\infty, 3]$ **27.** $(-\infty, \infty)$
28. $(-\infty, 1) \cup (1, 9) \cup (9, \infty)$
29. $C = 3000 + 0.12x$, $P = 0.000001x + 0.15$, \$0.44, \$0.40, \$0.39, 60,000 miles, $[50{,}000, 60{,}000]$

CHAPTER 8

Section 8.1   Warm-ups F T T F F F T T F T
1. Yes   **3.** No   **5.** No   **7.** Yes   **9.** Yes   **11.** No
13. Yes   **15.** Yes   **17.** No   **19.** Yes   **21.** No
23. Yes   **25.** No   **27.** Yes   **29.** $\{2\}$, $\{3, 5, 7\}$
31. R, $[0, \infty)$ **33.** $[0, \infty)$, $[0, \infty)$ **35.** $[8, \infty)$, $[2, \infty)$

37. 10   **39.** 12   **41.** 1   **43.** $\dfrac{7}{3}$ **45.** 1 or 2

47. -5 or 1 **49.** $4a - 1$ **51.** $4x + 7$ **53.** $x^2 + 4x + 7$
55. $x^2 + 2xh + h^2 - 2x - 2h + 4$ **57.** 4 **59.** 21
61. $2a$ **63.** 3 **65.** 2.337 **67.** 66.75 **69.** 29.36
71. 2.387 or 0.279 **73.** -15 **75.** $4a + 2$ **77.** 3
79. $2x + h - 1$ **81.** $A = s^2$ **83.** $C = 3.98y$
85. $C = 0.5n + 14.95$ **87.** $h = 4t + 70$ for $0 \le t \le 8$
89. $-\$1491.25$ per year   **91.** 11.11 mph per second

Section 8.2   Warm-ups T T T F F F T T F T

1. R, $\{-2\}$

3. R, R

5. R, R

7. R, R

9. R, R

11. $(-\infty, \infty)$, $[1, \infty)$

13. $(-\infty, \infty)$, $[0, \infty)$

15. $(-\infty, \infty)$, $[0, \infty)$

17. $(-\infty, \infty)$, $[0, \infty)$

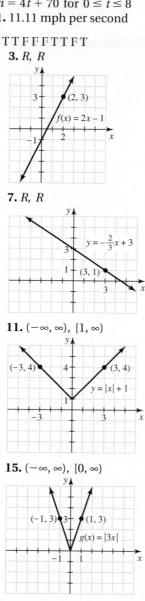

19. $(-\infty, \infty)$, $[1, \infty)$

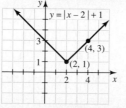

21. $(-\infty, \infty)$, $[2, \infty)$

23. $(-\infty, \infty)$, $[0, \infty)$

25. $(-\infty, \infty)$, $(-\infty, 6]$

27. $\left(\dfrac{1}{2}, -\dfrac{9}{4} \right)$, $(-\infty, \infty)$, $\left[-\dfrac{9}{4}, \infty \right)$

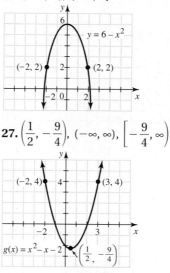

29. $(-2.25, -6.125)$, $(-\infty, \infty)$, $[-6.125, \infty)$

31. $(1, 9)$, $(-\infty, \infty)$, $(-\infty, 9]$

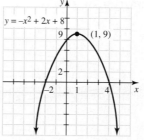

33. $\left(-\dfrac{3}{4}, \dfrac{25}{8}\right)$, $(-\infty, \infty)$, $\left(-\infty, \dfrac{25}{8}\right]$

35. $[0, \infty)$, $[0, \infty)$

37. $[1, \infty)$, $[0, \infty)$

39. $[0, \infty)$, $(-\infty, 0]$

41. $[0, \infty)$, $[2, \infty)$

43. $[0, \infty)$, $(-\infty, \infty)$

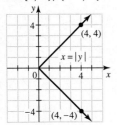

45. $(-\infty, 0]$, $(-\infty, \infty)$

47. $\{5\}$, $(-\infty, \infty)$

49. $[-9, \infty)$, $(-\infty, \infty)$

51. $[0, \infty)$, $[0, \infty)$

53. $[0, \infty)$, $(-\infty, \infty)$

55. No   **57.** Yes   **59.** No   **61.** $(-\infty, \infty)$, $(-\infty, 1]$

63. $(-\infty, \infty)$, $[-1, \infty)$

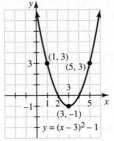

65. $(-\infty, \infty)$, $[1, \infty)$

67. $[0, \infty)$, $[-3, \infty)$

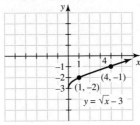

69. $(-\infty, \infty)$, $(-\infty, \infty)$

71. $(-\infty, \infty)$, $(-\infty, 2.25]$

73. 64 feet

75. \$62.50   **77.** 917, 1993   **81.** $\sqrt{x^2} = |x|$

Section 8.3 Warm-ups T T F F F F T T T T
1. $x^2 + 2x - 3$ **3.** $4x^3 - 11x^2 + 6x$ **5.** 12   **7.** -30

9. -21 **11.** $\dfrac{13}{8}$ **13.** $y = 6x - 20$ **15.** $y = x + 2$

17. $y = x^2 + 2x$ **19.** $y = x$ **21.** -2 **23.** 5   **25.** 7
27. 5   **29.** 5   **31.** 4   **33.** 22.2992   **35.** $4x^2 - 6x$

37. $2x^2 + 6x - 3$ **39.** x **41.** $4x - 9$ **43.** $\dfrac{x + 9}{4}$

45. $F = f \circ h$ **47.** $G = g \circ h$ **49.** $H = h \circ g$
51. $J = h \circ h$ **53.** $K = g \circ g$

63. 112.5 in.², $A = \dfrac{d^2}{2}$ **65.** $P(x) = -x^2 + 20x - 170$

67. $J = 0.025I$

Section 8.4 Warm-ups F F F T T F T T T F
1. No   **3.** Yes, $\{(4, 16), (3, 9), (0, 0)\}$ **5.** No
7. Yes, $\{(0, 0), (2, 2), (9, 9)\}$ **9.** No   **11.** Yes   **13.** Yes

15. Yes   **17.** Yes   **19.** No   **21.** $f^{-1}(x) = \dfrac{x}{5}$

23. $g^{-1}(x) = x + 9$ **25.** $k^{-1}(x) = \dfrac{x + 9}{5}$

27. $m^{-1}(x) = \dfrac{2}{x}$ **29.** $f^{-1}(x) = x^3 + 4$

31. $f^{-1}(x) = \dfrac{3}{x} + 4$ **33.** $f^{-1}(x) = \dfrac{x^3 - 7}{3}$

35. $f^{-1}(x) = \dfrac{2x + 1}{x - 1}$ **37.** $f^{-1}(x) = \dfrac{1 + 4x}{3x - 1}$

39. $p^{-1}(x) = x^4$ for $x \ge 0$ **41.** $f^{-1}(x) = 2 + \sqrt{x}$
43. $f^{-1}(x) = \sqrt{x - 3}$ **45.** $f^{-1}(x) = x^2 - 2$ for $x \ge 0$

47. $f^{-1}(x) = \dfrac{1}{2}x - \dfrac{3}{2}$ **49.** $f^{-1}(x) = \sqrt{x + 1}$

51. $f^{-1}(x) = \dfrac{x}{5}$ **53.** $f^{-1}(x) = \sqrt[3]{x}$

55. $f^{-1}(x) = x^2 + 2$ for $x \ge 0$

57. $(f^{-1} \circ f)(x) = x$ **59.** $(f^{-1} \circ f)(x) = x$
61. $(f^{-1} \circ f)(x) = x$ **63.** $(f^{-1} \circ f)(x) = x$

65. 58 mph, $D = \dfrac{S^2}{30}$

67. $T(x) = 1.09x + 125$, $T^{-1}(x) = \dfrac{x - 125}{1.09}$

69. An odd positive integer

Section 8.5 Warm-ups T F F T F T F F F T

1. $a = km$ **3.** $d = \dfrac{k}{e}$ **5.** $I = krt$ **7.** $m = kp^2$

9. $B = k\sqrt[3]{w}$ **11.** $t = \dfrac{k}{x^2}$ **13.** $v = \dfrac{km}{n}$

15. $y = \dfrac{3}{2}x$ **17.** $A = \dfrac{30}{B}$ **19.** $m = \dfrac{36}{\sqrt{p}}$

21. $A = \dfrac{2}{5}tu$ **23.** $y = \dfrac{1.067x}{z}$ **25.** $-\dfrac{21}{5}$ **27.** $-\dfrac{3}{2}$

29. $\dfrac{1}{2}$ **31.** 3.293   **33.** Inverse   **35.** Direct

37. Combined   **39.** Joint   **41.** \$420   **43.** 9 cm³
45. \$360   **47.** \$18.90   **49.** 4 feet   **51.** 400 pounds
53. 7.2 days   **55.** y_1 increasing, y_2 decreasing

Section 8.6 Warm-ups T T F T T T T T T T
1. Yes   **3.** Yes   **5.** No   **7.** Yes   **9.** No   **11.** Yes
13. Yes   **15.** $(x - 3)(x^2 + 3x + 3)$
17. $(x + 5)(x + 3)(x + 1)$ **19.** No
21. $(x + 1)(x - 2)(x^2 + 2x + 4)$
23. $(2x - 1)(x - 3)(x + 2)$ **25.** $\{-4, 1, 3\}$

27. $\left\{-\dfrac{1}{2}, 2, 3\right\}$ **29.** $\left\{-4, -\dfrac{1}{2}, 6\right\}$ **31.** $\{-3, 1\}$

33. $\{-1, 1, 2\}$ **37.** $\{-2, 1, 5\}$, $\left\{-\dfrac{3}{2}, \dfrac{3}{2}, \dfrac{5}{2}\right\}$

Chapter 8 Review
1. No   **3.** Yes   **5.** No   **7.** $\{1, 2\}$, $\{3\}$
9. $(-\infty, \infty)$, $(-\infty, \infty)$ **11.** $(-\infty, \infty)$, $[0, \infty)$ **13.** -5

15. -6 **17.** $-\dfrac{21}{4}$ **19.** $2a - 5$ **21.** $a^2 - a - 6$

23. 3   **25.** 200   **27.** $2x + h + 2$

29. $(-\infty, \infty)$, $(-\infty, \infty)$

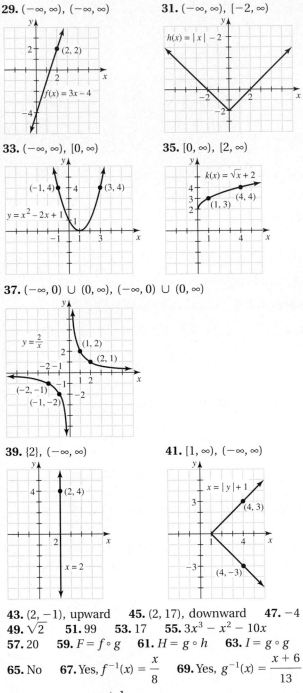

31. $(-\infty, \infty)$, $[-2, \infty)$

33. $(-\infty, \infty)$, $[0, \infty)$

35. $[0, \infty)$, $[2, \infty)$

37. $(-\infty, 0) \cup (0, \infty)$, $(-\infty, 0) \cup (0, \infty)$

39. $\{2\}$, $(-\infty, \infty)$

41. $[1, \infty)$, $(-\infty, \infty)$

43. $(2, -1)$, upward   **45.** $(2, 17)$, downward   **47.** -4
49. $\sqrt{2}$ **51.** 99   **53.** 17   **55.** $3x^3 - x^2 - 10x$
57. 20   **59.** $F = f \circ g$ **61.** $H = g \circ h$ **63.** $I = g \circ g$
65. No   **67.** Yes, $f^{-1}(x) = \dfrac{x}{8}$ **69.** Yes, $g^{-1}(x) = \dfrac{x + 6}{13}$

71. Yes, $j^{-1}(x) = \dfrac{x + 1}{x - 1}$ **73.** No

75. $f^{-1}(x) = \dfrac{1}{3}x + \dfrac{1}{3}$

77. $f^{-1}(x) = \sqrt[3]{2x}$

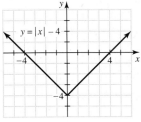

79. 24   **81.** -16 **83.** $\{-3, 2, 5\}$ **85.** $\{-2, 3, 4\}$
87. $(x - 3)(x + 2)(x + 5)$ **89.** No

91. $C = \dfrac{5}{2}n + 800$ **93.** 256 feet   **95.** $B = \dfrac{2A}{\pi}$

97. $a = 15w - 16$ **99.** 25, \$225
101. 7.1 mph per second

Chapter 8 Test
1. No   **2.** Yes   **3.** Yes   **4.** No   **5.** $(-\infty, \infty)$, $[0, \infty)$
6. $[0, \infty)$, $[0, \infty)$ **7.** 11   **8.** $a^2 + 2a + 5$

9. $4x^2 - 20x + 29$ **10.** $f^{-1}(x) = -\dfrac{1}{2}x + \dfrac{5}{2}$ **11.** 9

12. 15   **13.** -1782 **14.** $\pm\sqrt{5}$
15.

16.

17.

18.

19.

20.

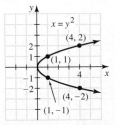

21. 11　**22.** $f^{-1}(x) = \dfrac{x+1}{x-2}$　**23.** $\dfrac{32\pi}{3}$ cubic feet

24. $S = 0.50n + 3$　**25.** $\dfrac{36}{7}$　**26.** $\dfrac{5}{4}$ seconds　**27.** Yes

28. $\{3, 4, 5\}$

Tying It All Together Chapters 1–8

1. $\dfrac{1}{25}$　**2.** $\dfrac{3}{2}$　**3.** $\sqrt{2}$　**4.** x^8　**5.** 2　**6.** x^9　**7.** $\{\pm 3\}$

8. $\{\pm 2\sqrt{2}\}$　**9.** $\{0, 1\}$　**10.** $\{2 \pm \sqrt{10}\}$　**11.** $\{81\}$

12. $\varnothing$　**13.** $\{\pm 8\}$　**14.** $\left\{-\dfrac{17}{5}, 5\right\}$　**15.** $\{2\}$　**16.** $\left\{\dfrac{5}{3}\right\}$

17. $\{42\}$　**18.** $\{11\}$

19.

20.

21.

22.

23.

24.

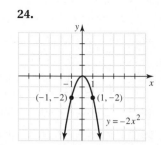

25. $(2, 4)$, $(3, 8)$, $(1, 2)$, $(4, 16)$

26. $\left(\dfrac{1}{2}, 2\right)$, $\left(-1, \dfrac{1}{4}\right)$, $(2, 16)$, $(0, 1)$

27. 1639.8 calories, $B = 1732 - 4.68A$

CHAPTER 9

Section 9.1　Warm-ups F T F T T F T F T F

1. 16　**3.** 2　**5.** 3　**7.** $\dfrac{1}{3}$　**9.** -1　**11.** $-\dfrac{1}{4}$　**13.** 1

15. 100　**17.** 2.718　**19.** 0.135

21.

23.

25.

27.

29.

31.

33.

35.

37.

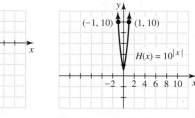

39. $\{6\}$　**41.** $\{-3\}$

43. $\{-2\}$　**45.** $\{-1\}$　**47.** $\{-2\}$　**49.** $\{-2\}$

51. $\{-3, 3\}$　**53.** 2　**55.** $\dfrac{4}{3}$　**57.** -2　**59.** 0　**61.** $\dfrac{3}{2}$

63. $\dfrac{1}{2}$　**65.** $\$9861.72$　**67.** $\$202,820.58$, 1989

69. $\$45$, $\$12.92$　**71.** $\$616.84$　**73.** $\$6230.73$

75. 300 grams, 90.4 grams, 12 years, no
77. 2.66666667, 0.0516, 2.8×10^{-5}

Section 9.2 Warm-ups T F T T F F F F T T
1. $2^3 = 8$ **3.** $\log(100) = 2$ **5.** $5^y = x$
7. $\log_2(b) = a$ **9.** $3^{10} = x$ **11.** $\ln(x) = 3$ **13.** 2
15. 4 **17.** 6 **19.** 3 **21.** -2 **23.** 2 **25.** -2
27. 1 **29.** -3 **31.** 3 **33.** 2 **35.** 0.6990
37. 1.8307 **39.**

41.

43.

45.

47. $f^{-1}(x) = \log_6(x)$

49. $f^{-1}(x) = e^x$ **51.** $f^{-1}(x) = \left(\dfrac{1}{2}\right)^x$ **53.** {4}

55. $\left\{\dfrac{1}{2}\right\}$ **57.** {0.001} **59.** {6} **61.** $\left\{\dfrac{1}{5}\right\}$ **63.** {±3}

65. {0.4771} **67.** {−0.3010} **69.** {1.9741}
71. 5.776 years **73.** 1.9269 years
75. 24.3%, $1,290,242 **77.** 4.1 **79.** 8.01 **81.** 90 db
83. $f^{-1}(x) = 2^{x-5} + 3$, $(-\infty, \infty)$, $(3, \infty)$
85. $y = \ln(e^x) = x$ for $-\infty < x < \infty$, $y = e^{\ln(x)} = x$ for $0 < x < \infty$

Section 9.3 Warm-ups T F T T F T F F F T
1. $\log(21)$ **3.** $\log_3(\sqrt{5x})$ **5.** $\log(x^5)$ **7.** $\ln(30)$
9. $\log(x^2 + 3x)$ **11.** $\log_2(x^2 - x - 6)$ **13.** $\log(4)$
15. $\log_2(x^4)$ **17.** $\log(\sqrt{5})$ **19.** $\ln(h - 2)$
21. $\log_2(w - 2)$ **23.** $\ln(x - 2)$ **25.** $3 \cdot \log(3)$

27. $\dfrac{1}{2} \cdot \log(3)$ **29.** $x \cdot \log(3)$ **31.** 10 **33.** 19

35. 8 **37.** 4.3 **39.** $\log(3) + \log(5)$
41. $\log(5) - \log(3)$ **43.** $2 \cdot \log(5)$
45. $2 \cdot \log(5) + \log(3)$ **47.** $-\log(3)$ **49.** $-\log(5)$
51. $\log(x) + \log(y) + \log(z)$ **53.** $3 + \log_2(x)$
55. $\ln(x) - \ln(y)$ **57.** $1 + 2 \cdot \log(x)$

59. $2 \cdot \log_5(x - 3) - \dfrac{1}{2} \cdot \log_5(w)$

61. $\ln(y) + \ln(z) + \dfrac{1}{2} \cdot \ln(x) - \ln(w)$

63. $\log(x^2 - x)$ **65.** $\ln(3)$ **67.** $\ln\left(\dfrac{xz}{w}\right)$

69. $\ln\left(\dfrac{x^2 y^3}{w}\right)$ **71.** $\log\left(\dfrac{(x-3)^{1/2}}{(x+1)^{2/3}}\right)$

73. $\log_2\left(\dfrac{(x-1)^{2/3}}{(x+2)^{1/4}}\right)$ **75.** False **77.** True **79.** True

81. False **83.** True **85.** True **87.** True **89.** False

91. $r = \ln\left(\left(\dfrac{A}{P}\right)^{1/t}\right)$, 21.97% **93.** b

Section 9.4 Warm-ups T T T F T T T F T F
1. {7} **3.** {2} **5.** {3} **7.** ∅ **9.** {6} **11.** {3}

13. {2} **15.** {4} **17.** {$\log_3(7)$} **19.** {−6} **21.** $\left\{-\dfrac{1}{2}\right\}$

23. $\dfrac{5 \cdot \ln(3)}{\ln(2) - \ln(3)}$, -13.548 **25.** $\dfrac{4 + 2 \cdot \log(5)}{1 - \log(5)}$, 17.932

27. $\dfrac{\ln(9)}{\ln(9) - \ln(8)}$, 18.655 **29.** 1.5850 **31.** -0.6309

33. -2.2016 **35.** 1.5229 **37.** $\dfrac{\ln(7)}{\ln(2)}$, 2.8074

39. $\dfrac{1}{3 - \ln(2)}$, 0.4335 **41.** $\dfrac{\ln(5)}{\ln(3)}$, 1.4650

43. $1 + \dfrac{\ln(9)}{\ln(2)}$, 4.1699 **45.** $\log_3(20)$, 2.7268 **47.** $\dfrac{3}{5}$

49. $\dfrac{1}{2}$ **51.** 41 months **53.** 7524 ft³/sec

55. 7.1 years **57.** 16.8 years **59.** 2.0×10^{-4}
61. 0.9183 **63.** 2.2894 **65.** (2.71, 6.54)
67. (1.03, 0.04), (4.74, 2.24)

Chapter 9 Review

1. $\dfrac{1}{25}$ **3.** 125 **5.** 1 **7.** $\dfrac{1}{10}$ **9.** 4 **11.** $\dfrac{1}{2}$ **13.** 2

15. 4 **17.** $-\dfrac{5}{2}$ **19.** 2 **21.** 8.6421 **23.** 177.828

25. 0.02005 **27.** 0.1408

29.

31.

9.

10.

33.

35.

11. 10   **12.** 8   **13.** $\dfrac{3}{2}$ **14.** 15   **15.** -4

16. $\{\log_3 (12)\}$ **17.** $\{\sqrt{3}\,\}$ **18.** $\left\{\dfrac{\ln (8)}{\ln (8) - \ln (5)}\right\}$

19. $\{5\}$ **20.** $\{3\}$ **21.** $\{0.5372\}$ **22.** $\{20.5156\}$
23. 10, 147,648   **24.** 1.733 hours

Tying It All Together Chapters 1–9

1. $\{3 \pm 2\sqrt{2}\,\}$ **2.** $\{259\}$ **3.** $\{6\}$ **4.** $\left\{\dfrac{11}{2}\right\}$

5. $\{-5, 11\}$ **6.** $\{67\}$ **7.** $\{6\}$ **8.** $\{4\}$ **9.** $\left\{-\dfrac{52}{15}\right\}$

10. $\left\{\dfrac{3 \pm \sqrt{3}}{3}\right\}$ **11.** $f^{-1}(x) = 3x$ **12.** $g^{-1}(x) = 3^x$

13. $f^{-1}(x) = \dfrac{x + 4}{2}$ **14.** $h^{-1}(x) = x^2$ for $x \geq 0$

15. $j^{-1}(x) = \dfrac{1}{x}$ **16.** $k^{-1}(x) = \log_5 (x)$

17. $m^{-1}(x) = 1 + \ln (x)$ **18.** $n^{-1}(x) = e^x$

19.

20.

37. $\log (n) = m$ **39.** $k^h = t$ **41.** -3 **43.** -1 **45.** 2
47. 0   **49.** 256   **51.** 6.267   **53.** -5.083 **55.** 5.560
57. $f^{-1}(x) = \log (x)$ **59.** $f^{-1}(x) = \ln (x)$

61. $2 \cdot \log (x) + \log (y)$ **63.** $4 \cdot \ln (2)$ **65.** $-\log_5 (x)$

67. $\log\left(\dfrac{\sqrt{x + 2}}{(x - 1)^2}\right)$ **69.** $\{256\}$ **71.** $\{3\}$ **73.** $\{2\}$

75. $\{3\}$ **77.** $\left\{\dfrac{\ln (7)}{\ln (3) - 1}\right\}$ **79.** $\left\{\dfrac{\ln (5)}{\ln (5) - \ln (3)}\right\}$

81. $\left\{\dfrac{1}{3}\right\}$ **83.** $\{22\}$ **85.** $\left\{\dfrac{200}{99}\right\}$ **87.** $\{1.3869\}$

89. $\{0.4650\}$ **91.** \$51,182.68   **93.** 161.5 grams
95. 5 years   **97.** 4347.5 ft^3/sec

Chapter 9 Test

1. 25   **2.** $\dfrac{1}{5}$ **3.** 1   **4.** 3   **5.** 0   **6.** -1

7.

8.

21.

22.

23. **24.**

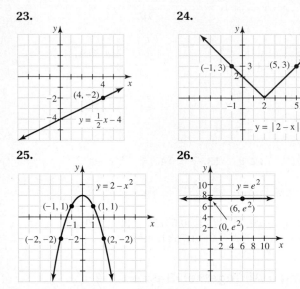

25. **26.**

27. For 1992, linear 103.3 million, exponential 103.9 million

CHAPTER 10

Section 10.1   Warm-ups T F F T T T T T T T
1. $\{(1, 2)\}$ **3.** $\{(0, -1)\}$ **5.** $\{(2, -1)\}$ **7.** $\varnothing$
9. $\left\{(x, y) \mid y = -\frac{1}{2}x + 4\right\}$ **11.** c **13.** b
15. $\{(8, 3)\}$, independent
17. $\{(-3, 2)\}$, independent   **19.** $\varnothing$, inconsistent
21. $\{(x, y) \mid y = 2x - 5\}$, dependent
23. $\{(5, -1)\}$, independent   **25.** $\{(7, 7)\}$, independent
27. $\{(15, 25)\}$, independent   **29.** $\{(20, 10)\}$, independent
31. $\left\{\left(\frac{9}{2}, -\frac{1}{2}\right)\right\}$, independent   **33.** $\varnothing$, inconsistent
35. $\{(x, y) \mid 3y - 2x = -3\}$, dependent
37. $\{(0.75, 1.125)\}$, independent
39. \$14,000 at 5%, \$16,000 at 10%   **41.** -12 and 14
43. 94 toasters, 6 vacation coupons
45. State tax \$3553, federal tax \$28,934   **47.** \$20,000
49. \$500,000, \$300,000, 20,000, \$400,000   **51.** a
53. $(2.8, 2.6)$, $(1.0, -0.2)$

Section 10.2   Warm-ups T F T T F T T F T T
1. $\{(8, -1)\}$ **3.** $\{(5, -7)\}$ **5.** $\{(-1, 3)\}$ **7.** $\{(-1, -3)\}$
9. $\{(-2, -5)\}$ **11.** $\{(22, 26)\}$ **13.** $\varnothing$, inconsistent
15. $\{(x, y) \mid 5x - y = 1\}$, dependent

17. $\left\{\left(\frac{5}{2}, 0\right)\right\}$, independent   **19.** $\{(12, 6)\}$ **21.** $\{(-8, 6)\}$
23. $\{(16, 12)\}$ **25.** $\{(12, 7)\}$ **27.** $\{(400, 800)\}$
29. $\{(1.5, 1.25)\}$ **31.** \$1.00   **33.** 1380
35. 31 dimes, 4 nickels
37. 20 pounds Chocolate, 30 pounds Peanut Butter
39. 4 hours   **41.** 80%
43. Width 150 meters, length 200 meters

Section 10.3   Warm-ups F F T F F F T T F F F
1. $\{(1, 2, -1)\}$ **3.** $\{(1, 3, 2)\}$ **5.** $\{(1, -5, 3)\}$
7. $\{(-1, 2, -1)\}$ **9.** $\{(-1, -2, 4)\}$ **11.** $\{(1, 3, 5)\}$
13. $\{(3, 4, 5)\}$ **15.** $\varnothing$ **17.** $\{(x, y, z) \mid 3x - y + z = 5\}$
19. $\varnothing$ **21.** $\{(x, y, z) \mid 5x + 4y - 2z = 150\}$
23. $\{(0.1, 0.3, 2)\}$
25. \$1500 stocks, \$4500 bonds, \$6000 mutual fund
27. Coffee \$0.40, doughnut \$0.35, tip \$0.55   **29.** \$0.95
31. Soup 14 ounces, tuna 8 ounces, error 2 ounces
33. \$24,000 teaching, \$18,000 painting, \$6,000 royalties

Section 10.4   Warm-ups T F F T F T T T T T
1. $\{(2, 4), (-3, 9)\}$ **3.** $\{(-2, 2), (6, 6)\}$

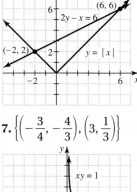

5. $\{(8, 4)\}$ **7.** $\left\{\left(-\frac{3}{4}, -\frac{4}{3}\right), \left(3, \frac{1}{3}\right)\right\}$

9. $\left\{\left(\frac{\sqrt{2}}{2}, \frac{1}{2}\right), \left(-\frac{\sqrt{2}}{2}, \frac{1}{2}\right)\right\}$ **11.** $\{(0, -5), (3, 4), (-3, 4)\}$

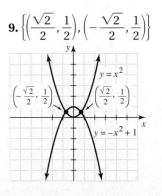

13. $\{(4, 5), (-2, -1)\}$ **15.** $\left\{\left(-3, -\dfrac{5}{3}\right)\right\}$

17. $\left\{\left(\dfrac{3}{2}, -\dfrac{3}{13}\right)\right\}$ **19.** $\left\{\left(-\dfrac{5}{3}, \dfrac{36}{5}\right), (2, 5)\right\}$

21. $\{(\sqrt{5}, \sqrt{3}), (\sqrt{5}, -\sqrt{3}), (-\sqrt{5}, \sqrt{3}), (-\sqrt{5}, -\sqrt{3})\}$
23. $\{(2, 5), (19, -12)\}$
25. $\{(\sqrt{2}, 2), (-\sqrt{2}, 2), (1, 1), (-1, 1)\}$ **27.** $\{(3, 1)\}$
29. $\varnothing$ **31.** $\{(-6, 4^{-7})\}$
33. $\{(2, 3), (-2, -3), (3, 2), (-3, -2)\}$
35. $\left\{\left(2\sqrt{2}, \dfrac{\sqrt{2}}{2}\right), \left(-2\sqrt{2}, -\dfrac{\sqrt{2}}{2}\right), (1, 2), (-1, -2)\right\}$
37. $\sqrt{3}$ feet and $2\sqrt{3}$ feet
39. Height $5\sqrt{10}$ inches, base $20\sqrt{10}$ inches
41. Pump A 24 hours, pump B 8 hours **43.** 40 minutes
45. 8 feet by 9 feet **47.** $4 - 2i$ and $4 + 2i$
49. Side of square 8 feet, height of triangle 2 feet
51. a) $(1.71, 1.55), (-2.98, -3.95)$ **b)** $(1, 1), (0.40, 0.16)$
   c) $(1.17, 1.62), (-1.17, -1.62)$

Section 10.5 Warm-ups T T T F T F F F T F
1. 2×2 **3.** 3×2 **5.** 3×1

7. $\begin{bmatrix} 2 & -3 & \vline & 9 \\ -3 & & \vline & -1 \end{bmatrix}$ **9.** $\begin{bmatrix} 1 & -1 & 1 & \vline & 1 \\ 1 & 1 & -2 & \vline & 3 \\ 0 & 1 & -3 & \vline & 4 \end{bmatrix}$

11. $5x + y = -1$ **13.** $x = 6$
   $2x - 3y = 0$ $-x + z = -3$
             $x + y = 1$

15. $R_1 \leftrightarrow R_2$ **17.** $\dfrac{1}{5}R_2 \to R_2$ **19.** $\{(1, 2)\}$ **21.** $\{(4, 5)\}$

23. $\{(1, -1)\}$ **25.** $\{(7, 6)\}$ **27.** $\varnothing$
29. $\{(x, y) \mid x + 2y = 1\}$ **31.** $\{(1, 2, 3)\}$ **33.** $\{(1, 1, 1)\}$
35. $\{(1, 2, 0)\}$ **37.** $\{(1, 0, 1)\}$
39. $\{(x, y, z) \mid x - y + z = 1\}$ **41.** $\varnothing$

Section 10.6 Warm-ups T F F T T T T T F F
1. -1 **3.** -3 **5.** -14 **7.** 0.4 **9.** $\{(1, -3)\}$

11. $\{(1, 1)\}$ **13.** $\left\{\left(\dfrac{23}{13}, \dfrac{9}{13}\right)\right\}$ **15.** $\{(10, 15)\}$

17. $\left\{\left(\dfrac{27}{4}, \dfrac{13}{2}\right)\right\}$ **19.** $\varnothing$ **21.** $\left\{\left(\dfrac{14}{3}, \dfrac{2}{3}\right)\right\}$
23. $\{(x, y) \mid 4x - y = 6\}$ **25.** $\{(x, y) \mid y = 3x - 12\}$
27. $\{(2, -5)\}$ **29.** $\{(480, 24)\}$ **31.** $\{(\sqrt{2}, 2\sqrt{2})\}$
33. Peas 9 servings, beets 11 servings **35.** 37
37. Milk $2.40, magazine $2.25
39. 12 singles, 10 doubles **41.** Gary 39, Harry 34

43. Square 10 feet, triangle $\dfrac{40}{3}$ feet

45. 10 gallons of 10% solution, 20 gallons of 25% solution

Section 10.7 Warm-ups T F T F F T T T F F
1. $\begin{bmatrix} -3 & 7 \\ 1 & -6 \end{bmatrix}$ **3.** $\begin{bmatrix} 4 & -3 \\ 0 & 1 \end{bmatrix}$ **5.** $\begin{bmatrix} 3 & -2 \\ 0 & 1 \end{bmatrix}$ **7.** $\begin{bmatrix} 3 & 5 \\ 4 & 7 \end{bmatrix}$
9. -7 **11.** -1 **13.** 9 **15.** 5 **17.** 22 **19.** 6
21. 70 **23.** 25 **25.** $\{(1, 2, 3)\}$ **27.** $\{(-1, 1, 2)\}$

29. $\{(-3, 2, 1)\}$ **31.** $\left\{\left(\dfrac{3}{2}, \dfrac{1}{2}, 2\right)\right\}$ **33.** $\{(0, 1, -1)\}$
35. $\{(x, y, z) \mid 2x - y + z = 1\}$ **37.** $\varnothing$ **39.** $\{(1, 3, -6)\}$
41. Mimi 36 pounds, Mitzi 32 pounds,
   Cassandra 107 pounds
43. $39°, 51°, 90°$

Chapter 10 Review
1. $\{(1, 1)\}$, independent **3.** $\{(1, -1)\}$, independent
5. $\{(-3, 2)\}$, independent **7.** $\varnothing$, inconsistent
9. $\{(-1, 5)\}$, independent
11. $\{(x, y) \mid 3x - 2y = 12\}$, dependent **13.** $\{(1, -3, 2)\}$
15. $\varnothing$ **17.** $\{(3, 9), (-5, 25)\}$
19. $\left\{\left(\dfrac{\sqrt{3}}{3}, \sqrt{3}\right), \left(-\dfrac{\sqrt{3}}{3}, -\sqrt{3}\right)\right\}$
21. $\{(\sqrt{3}, 1), (-\sqrt{3}, 1)\}$ **23.** $\{(-5, -3), (3, 5)\}$
25. $\{(5, \log(2))\}$ **27.** $\{(2, 4), (-2, 4)\}$ **29.** $\{(2, -4)\}$
31. $\{(1, 1, 2)\}$ **33.** 2 **35.** -0.2 **37.** $\{(-1, -2)\}$
39. $\{(2, 1)\}$ **41.** $\varnothing$ **43.** 58 **45.** -30
47. $\{(1, 2, -3)\}$ **49.** $\{(-1, 2, 4)\}$ **51.** 78

53. Length 6 feet, width 2 feet **55.** $\dfrac{7}{2}$ inches, $\dfrac{3}{2}$ inches

57. 16 years old to drive, 12 years old at present
59. $16,000 at 10%, $4000 at 8%

Chapter 10 Test
1. $\{(1, 3)\}$ **2.** $\left\{\left(\dfrac{5}{2}, -3\right)\right\}$ **3.** $\{(x, y) \mid y = x - 5\}$

4. $\{(-1, 3)\}$ **5.** $\varnothing$ **6.** Inconsistent
7. Dependent **8.** Independent
9. $\{(3, 4) (-3, 4), (3\sqrt{2}, -5), (-3\sqrt{2}, -5)\}$
10. $\{(1, -2, -3)\}$ **11.** $\{(2, 5)\}$ **12.** $\{(3, 1, 1)\}$
13. -18 **14.** -2 **15.** $\{(-1, 2)\}$ **16.** $\{(2, -2, 1)\}$
17. Length 12 feet, width 9 feet
18. Singles $18, doubles $25
19. Jill 17 hours, Karen 14 hours, Betsy 62 hours

Tying It All Together Chapters 1–10
1. $\{2\}$ **2.** $\left\{-\dfrac{3}{2}, \dfrac{9}{2}\right\}$ **3.** $\{4\}$ **4.** $\left\{\dfrac{3}{4}\right\}$ **5.** $\{-3, 3\}$

6. $\left\{\pm\dfrac{\sqrt{3}}{2}\right\}$ **7.** $\{-8, 8\}$ **8.** $\{-4, -2\}$ **9.** $\{-4 \pm 3\sqrt{2}\}$

10. $\{-4, 2\}$ **11.** $\left\{-\dfrac{3}{2}\right\}$ **12.** $\{\pm 3, \pm 1\}$ **13.** $\{14\}$

14. $\{1 + \log_2(6)\}$ **15.** $\{2\}$ **16.** $\{32\}$
17.
18.
19.
20.

21.

22.

23.

24.

25.

26.

27.

28.

29.

30.

31. $8\sqrt{2}$ **32.** $4\sqrt{2}$ **33.** $-\sqrt{2}$ **34.** 22
35. $11 + 6\sqrt{2}$ **36.** $3 - \sqrt{3}$ **37.** $-\sqrt{6} - 2\sqrt{2}$
38. $\dfrac{3\sqrt{2} + 4\sqrt{3} - 1}{2}$

39. a) A **b)** Machine B \$0.04 per copy, machine A \$0.03 per copy **c)** B has slope 0.04, and A has slope 0.03.
d) B: $y = 0.04x + 2000$, A: $y = 0.03x + 4000$
e) 200,000

CHAPTER 11

Section 11.1   Warm-ups F T T F F T T F T T

1. Vertex $(0, 0)$, focus $\left(0, \dfrac{1}{8}\right)$, directrix $y = -\dfrac{1}{8}$

3. Vertex $(0, 0)$, focus $(0, -1)$, directrix $y = 1$

5. Vertex $(3, 2)$, focus $(3, 2.5)$, directrix $y = 1.5$

7. Vertex $(-1, 6)$, focus $(-1, 5.75)$, directrix $y = 6.25$

9. $y = \dfrac{1}{8}x^2$ **11.** $y = -\dfrac{1}{2}x^2$ **13.** $y = \dfrac{1}{2}x^2 - 3x + 6$

15. $y = -\dfrac{1}{8}x^2 + \dfrac{1}{4}x - \dfrac{1}{8}$ **17.** $y = x^2 + 6x + 10$

19. $y = (x - 3)^2 - 8$, vertex $(3, -8)$, focus $(3, -7.75)$, directrix $y = -8.25$

21. $y = 2(x + 3)^2 - 13$, vertex $(-3, -13)$, focus $(-3, -12.875)$, directrix $y = -13.125$

23. $y = -2(x - 4)^2 + 33$, vertex $(4, 33)$, focus $\left(4, 32\dfrac{7}{8}\right)$, directrix $y = 33\dfrac{1}{8}$

25. $y = 5(x + 4)^2 - 80$, vertex $(-4, -80)$, focus $\left(-4, -79\dfrac{19}{20}\right)$, directrix $y = -80\dfrac{1}{20}$

27. Vertex $(2, -3)$, focus $\left(2, -2\dfrac{3}{4}\right)$, directrix $y = -3\dfrac{1}{4}$, upward

29. Vertex $(1, -2)$, focus $\left(1, -2\dfrac{1}{4}\right)$, directrix $y = -1\dfrac{3}{4}$, downward

31. Vertex $(1, -2)$, focus $\left(1, -1\dfrac{11}{12}\right)$, directrix $y = -2\dfrac{1}{12}$, upward

33. Vertex $\left(-\dfrac{3}{2}, \dfrac{17}{4}\right)$, focus $\left(-\dfrac{3}{2}, 4\right)$, directrix $y = \dfrac{9}{2}$, downward

35. Vertex $(0, 5)$, focus $\left(0, 5\dfrac{1}{12}\right)$, directrix $y = 4\dfrac{11}{12}$, upward

37. Vertex $\left(\dfrac{3}{2}, -\dfrac{1}{4}\right)$, axis of symmetry $x = \dfrac{3}{2}$, intercepts $(0, 2)$, $(1, 0)$, $(2, 0)$

39. Vertex $(-1, 9)$, axis of symmetry $x = -1$, intercepts $(0, 8)$, $(-4, 0)$, $(2, 0)$

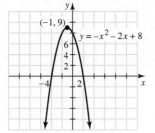

41. Vertex $(-2, 1)$, axis of symmetry $x = -2$, intercept $(0, 5)$

43. Vertex $(-1, 0)$, axis of symmetry $x = -1$, intercepts $(0, 1)$, $(-1, 0)$

45. Vertex $\left(\dfrac{1}{2}, 0\right)$, axis of symmetry $x = \dfrac{1}{2}$, intercepts $(0, -1)$, $\left(\dfrac{1}{2}, 0\right)$

47. Vertex is $\left(\dfrac{5}{2}, -\dfrac{25}{4}\right)$, axis of symmetry $x = \dfrac{5}{2}$, intercepts $(0, 0)$, $(5, 0)$

49. Vertex $(0, 5)$, axis of symmetry $x = 0$, intercept $(0, 5)$

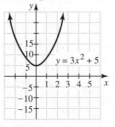

51. Vertex $(1, -2)$, axis of symmetry $x = 1$, intercepts $(0, -1)$, $\left(1 + \sqrt{2}, 0\right)$, $\left(1 - \sqrt{2}, 0\right)$

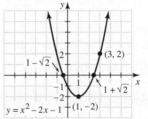

53. Vertex $(5, 0)$, axis of symmetry $x = 5$,
intercepts $(0, 25)$, $(5, 0)$

55. 4 and 4   **57.** 40 feet by 40 feet   **59.** 3 feet each

61. 262 feet, no   **63.** 100   **65.** $y = \dfrac{1}{60}x^2$

67. $\{(-1, 2), (1, 2)\}$ **69.** $\{(1, -1)\}$

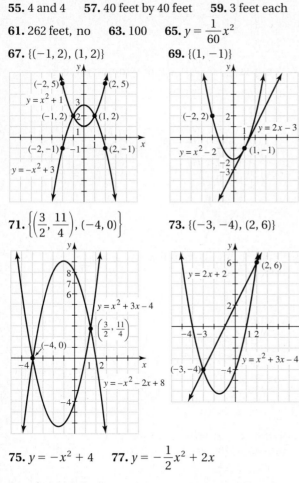

71. $\left\{\left(\dfrac{3}{2}, \dfrac{11}{4}\right), (-4, 0)\right\}$ **73.** $\{(-3, -4), (2, 6)\}$

75. $y = -x^2 + 4$ **77.** $y = -\dfrac{1}{2}x^2 + 2x$

Section 11.2   Warm-ups F F T F F F F T F T F
1. $x^2 + (y - 3)^2 = 25$ **3.** $(x - 1)^2 + (y + 2)^2 = 81$

5. $x^2 + y^2 = 3$ **7.** $(x + 6)^2 + (y + 3)^2 = \dfrac{1}{4}$

9. $\left(x - \dfrac{1}{2}\right)^2 + \left(y - \dfrac{1}{3}\right)^2 = 0.01$ **11.** $(3, 5)$, $\sqrt{2}$

13. $\left(0, \dfrac{1}{2}\right)$, $\dfrac{\sqrt{2}}{2}$ **15.** $(0, 0)$, $\dfrac{3}{2}$ **17.** $(2, 0)$, $\sqrt{3}$

19.

21.

23.

25.

27.

29. $(x + 2)^2 + (y + 3)^2 = 13$, $(-2, -3)$, $\sqrt{13}$
31. $(x - 1)^2 + (y - 2)^2 = 8$, $(1, 2)$, $2\sqrt{2}$
33. $(x - 5)^2 + (y - 4)^2 = 9$, $(5, 4)$, 3

35. $\left(x - \dfrac{1}{2}\right)^2 + \left(y + \dfrac{1}{2}\right)^2 = \dfrac{1}{2}$, $\left(\dfrac{1}{2}, -\dfrac{1}{2}\right)$, $\dfrac{\sqrt{2}}{2}$

37. $\left(x - \dfrac{3}{2}\right)^2 + \left(y - \dfrac{1}{2}\right)^2 = \dfrac{7}{2}$, $\left(\dfrac{3}{2}, \dfrac{1}{2}\right)$, $\dfrac{\sqrt{14}}{2}$

39. $\left(x - \dfrac{1}{3}\right)^2 + \left(y + \dfrac{3}{4}\right)^2 = \dfrac{97}{144}$, $\left(\dfrac{1}{3}, -\dfrac{3}{4}\right)$, $\dfrac{\sqrt{97}}{12}$

41. $\{(1, 3), (-1, -3)\}$

43. $\{(0, -3), (\sqrt{5}, 2), (-\sqrt{5}, 2)\}$

45. $\{(0, -3), (2, -1)\}$

47. $(0, 2 + \sqrt{3})$ and $(0, 2 - \sqrt{3})$

49. $\sqrt{29}$ **51.** $(x - 2)^2 + (y - 3)^2 = 32$

53. $\left(\dfrac{5}{2}, -\dfrac{\sqrt{11}}{2}\right)$ and $\left(\dfrac{5}{2}, \dfrac{\sqrt{11}}{2}\right)$ **55.** $755{,}903$ mm^3

57. $(0, 0)$ only   **59.**

61. B and D can be any real numbers, but A must equal C, and $4AE + B^2 + D^2 > 0$.

63. $y = \pm\sqrt{4 - x^2}$ **65.** $y = \pm\sqrt{x}$ **67.** $y = -1 \pm \sqrt{x}$

Section 11.3   Warm-ups F F T T T F F T T T

1.

3.

5.

7.

9.

11.

13.

15.

$$\frac{(x-3)^2}{4} + \frac{(y-1)^2}{9} = 1$$

17.

$$\frac{(x+1)^2}{16} + \frac{(y-2)^2}{25} = 1$$

29. $y = \pm\frac{3}{4}x$

$9x^2 - 16y^2 = 144$

19.

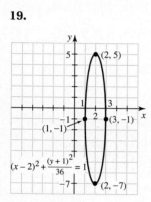

$(x-2)^2 + \frac{(y+1)^2}{36} = 1$

21. $y = \pm\frac{3}{2}x$

$\frac{x^2}{4} - \frac{y^2}{9} = 1$

31. $y = \pm x$

$x^2 - y^2 = 1$

23. $y = \pm\frac{2}{5}x$

$\frac{y^2}{4} - \frac{x^2}{25} = 1$

25. $y = \pm\frac{1}{5}x$

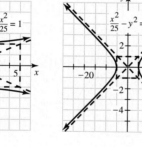

$\frac{x^2}{25} - y^2 = 1$

33. $\left(\frac{2\sqrt{10}}{5}, \frac{3\sqrt{15}}{5}\right), \left(\frac{2\sqrt{10}}{5}, -\frac{3\sqrt{15}}{5}\right),$ $\left(-\frac{2\sqrt{10}}{5}, \frac{3\sqrt{15}}{5}\right), \left(-\frac{2\sqrt{10}}{5}, -\frac{3\sqrt{15}}{5}\right)$

$x^2 - \frac{y^2}{9} = 1$

$\frac{x^2}{4} + \frac{y^2}{9} = 1$

27. $y = \pm 5x$

$x^2 - \frac{y^2}{25} = 1$

35. No points of intersection

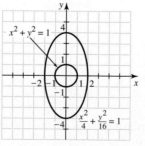

$x^2 + y^2 = 1$

$\frac{x^2}{4} + \frac{y^2}{16} = 1$

37. $\left(\dfrac{\sqrt{10}}{2}, \dfrac{\sqrt{6}}{2}\right), \left(\dfrac{\sqrt{10}}{2}, -\dfrac{\sqrt{6}}{2}\right), \left(-\dfrac{\sqrt{10}}{2}, \dfrac{\sqrt{6}}{2}\right),$

$\left(-\dfrac{\sqrt{10}}{2}, -\dfrac{\sqrt{6}}{2}\right)$

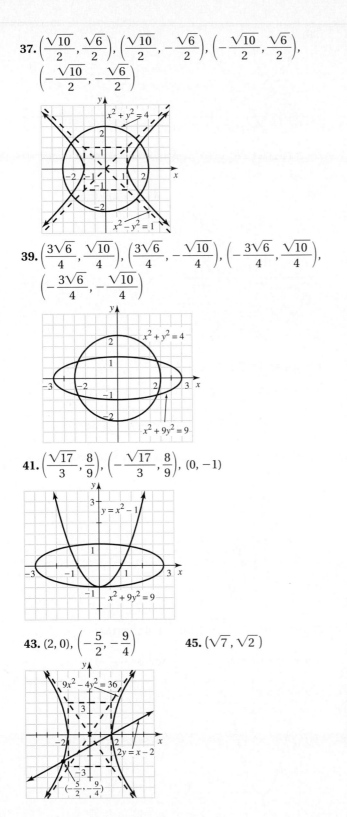

39. $\left(\dfrac{3\sqrt{6}}{4}, \dfrac{\sqrt{10}}{4}\right), \left(\dfrac{3\sqrt{6}}{4}, -\dfrac{\sqrt{10}}{4}\right), \left(-\dfrac{3\sqrt{6}}{4}, \dfrac{\sqrt{10}}{4}\right),$

$\left(-\dfrac{3\sqrt{6}}{4}, -\dfrac{\sqrt{10}}{4}\right)$

41. $\left(\dfrac{\sqrt{17}}{3}, \dfrac{8}{9}\right), \left(-\dfrac{\sqrt{17}}{3}, \dfrac{8}{9}\right), (0, -1)$

43. $(2, 0), \left(-\dfrac{5}{2}, -\dfrac{9}{4}\right)$ **45.** $\left(\sqrt{7}, \sqrt{2}\right)$

Section 11.4   Warm-ups F T T T F F F T T T

21.

23.

39.

41.

25.

27.

43.

45. No solution

29.

Chapter 11 Review

1. Intercepts $(0, -18)$, $(-6, 0)$, $(3, 0)$, vertex $\left(-\dfrac{3}{2}, -\dfrac{81}{4}\right)$,

axis of symmetry $x = -\dfrac{3}{2}$, focus $\left(-\dfrac{3}{2}, -20\right)$,

directrix $y = -\dfrac{41}{2}$

31.

33.

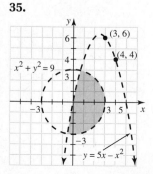

3. Intercepts $(0, 2)$, $(-2, 0)$, $(-1, 0)$, vertex $\left(-\dfrac{3}{2}, -\dfrac{1}{4}\right)$,

axis of symmetry $x = -\dfrac{3}{2}$, focus $\left(-\dfrac{3}{2}, 0\right)$,

directrix $y = -\dfrac{1}{2}$

35.

37.

5. Intercepts $(0, 1)$, $(2 \pm \sqrt{6}, 0)$, vertex $(2, 3)$, axis of symmetry $x = 2$, focus $\left(2, 2\frac{1}{2}\right)$, directrix $y = 3\frac{1}{2}$

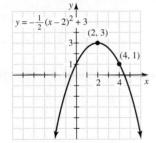

7. $y = 2(x - 2)^2 - 7$, $(2, -7)$

9. $y = -\frac{1}{2}(x + 1)^2 + 1$, $(-1, 1)$

11. 3 and 3   **13.** 22.5 feet by 22.5 feet

15. $(0, 0)$, 10   **17.** $(2, -3)$, 9

    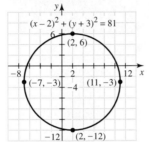

19. $(0, 0)$, $\frac{2}{3}$ **21.** $x^2 + (y - 3)^2 = 36$

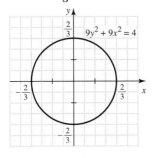

23. $(x - 2)^2 + (y + 7)^2 = 25$ **25.**

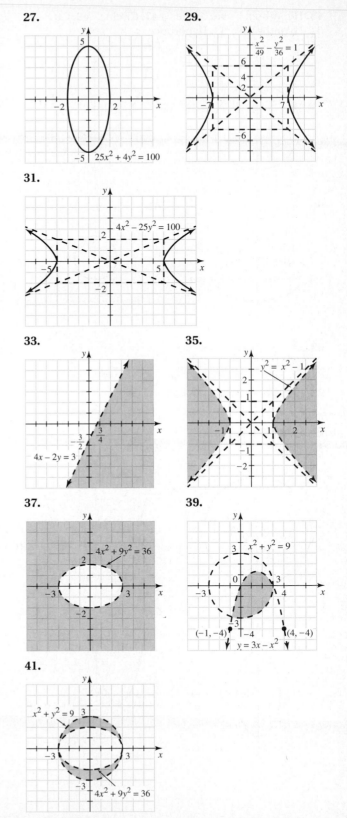

27.

29.

31.

33.

35.

37.

39.

41.

43. Hyperbola   **45.** Circle   **47.** Circle   **49.** Circle
51. Hyperbola   **53.** Hyperbola

55.

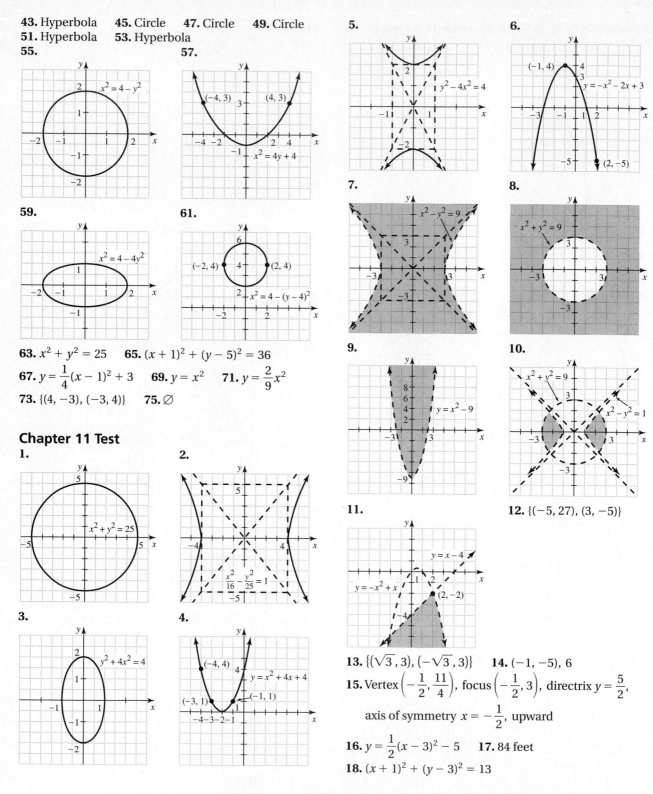

57.

5.

6.

59.

61.

7.

8.

63. $x^2 + y^2 = 25$ **65.** $(x+1)^2 + (y-5)^2 = 36$

67. $y = \frac{1}{4}(x-1)^2 + 3$ **69.** $y = x^2$ **71.** $y = \frac{2}{9}x^2$

73. $\{(4, -3), (-3, 4)\}$ **75.** $\varnothing$

9.

10.

Chapter 11 Test

1.

2.

11.

12. $\{(-5, 27), (3, -5)\}$

3.

4.

13. $\{(\sqrt{3}, 3), (-\sqrt{3}, 3)\}$ **14.** $(-1, -5)$, 6

15. Vertex $\left(-\frac{1}{2}, \frac{11}{4}\right)$, focus $\left(-\frac{1}{2}, 3\right)$, directrix $y = \frac{5}{2}$,

axis of symmetry $x = -\frac{1}{2}$, upward

16. $y = \frac{1}{2}(x-3)^2 - 5$ **17.** 84 feet

18. $(x+1)^2 + (y-3)^2 = 13$

Tying It All Together Chapters 1–11

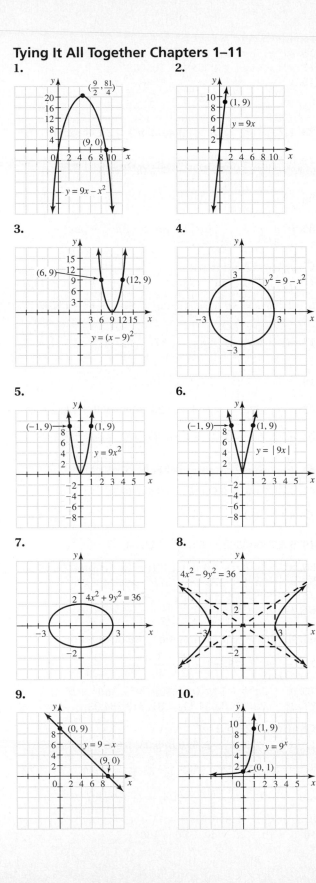

1.

2.

3.

4.

5.

6.

7.

8.

9.

10.

11. $x^2 + 4xy + 4y^2$ **12.** $x^3 + 3x^2y + 3xy^2 + y^3$

13. $a^3 + 3a^2b + 3ab^2 + b^3$ **14.** $a^2 - 6ab + 9b^2$

15. $6a^2 - 7a - 5$ **16.** $x^3 - y^3$ **17.** $\{(1, 2)\}$

18. $\{(3, 4), (4, 3)\}$ **19.** $\{(1, -2, 3)\}$ **20.** $\{(-1, 1), (3, 9)\}$

21. $x = -\dfrac{b}{a}$ **22.** $x = \dfrac{-d \pm \sqrt{d^2 - 4wm}}{2w}$

23. $B = \dfrac{2A - bh}{h}$ **24.** $x = \dfrac{2y}{y - 2}$ **25.** $m = \dfrac{L}{1 + xt}$

26. $t = \dfrac{y^2}{9a^2}$ **27.** $y = -\dfrac{2}{3}x - \dfrac{5}{3}$ **28.** $y = -2x$

29. $(x - 2)^2 + (y - 5)^2 = 45$ **30.** $\left(-\dfrac{3}{2}, 3\right), \dfrac{3\sqrt{5}}{2}$

31. $-10 + 6i$ **32.** -1 **33.** $3 - 5i$ **34.** $7 + 6i\sqrt{2}$

35. $-8 - 27i$ **36.** $-3 + 3i$ **37.** 29

38. $-\dfrac{3}{2} - i$ **39.** $-\dfrac{1}{2} + \dfrac{9}{2}i$ **40.** $2 - i\sqrt{2}$

41. a) $q = -500x + 400$ **b)** $R = -500x^2 + 400x$
   c) Maximum revenue of $80 occurs when bananas
   are $0.40 per pound.

CHAPTER 12

Section 12.1   Warm-ups T T T F F F T F T T

1. 1, 4, 9, 16, 25, 36, 49, 64

3. $-1, \dfrac{1}{2}, -\dfrac{1}{3}, \dfrac{1}{4}, -\dfrac{1}{5}, \dfrac{1}{6}, -\dfrac{1}{7}, \dfrac{1}{8}, -\dfrac{1}{9}, \dfrac{1}{10}$

5. 1, -2, 4, -8, 16   **7.** $\dfrac{1}{2}, \dfrac{1}{4}, \dfrac{1}{8}, \dfrac{1}{16}, \dfrac{1}{32}, \dfrac{1}{64}$

9. -1, 1, 3, 5, 7, 9, 11   **11.** $1, \dfrac{\sqrt{2}}{2}, \dfrac{\sqrt{3}}{3}, \dfrac{1}{2}, \dfrac{\sqrt{5}}{5}$

13. $\dfrac{1}{2}, \dfrac{1}{6}, \dfrac{1}{12}, \dfrac{1}{20}$ **15.** $-\dfrac{1}{3}, -1, 1, \dfrac{1}{3}$ **17.** $-1, 0, -1, 4$

19. $1, \dfrac{1}{4}, \dfrac{1}{9}, \dfrac{1}{16}$ **21.** $a_n = 2n - 1$ **23.** $a_n = (-1)^{n+1}$

25. $a_n = 2n - 2$ **27.** $a_n = 3n$ **29.** $a_n = 3n + 1$

31. $a_n = (-1)^n 2^{n-1}$ **33.** $a_n = (n - 1)^2$

35. $4, 2, 1, \dfrac{1}{2}, \dfrac{1}{4}$

37. $17,263, $18,126, $19,033, $19,984, $20,983

39. $1,000,000, $800,000, $640,000, $512,000

41. 27 inches, 13.5 inches, 9 inches, 6.75 inches,
   5.4 inches

43. 137,438,953,500, larger

47. 0.9048, 0.3677, 0.00004517

Section 12.2   Warm-ups T F F F F T T T T F

1. 30   **3.** 24   **5.** $\dfrac{31}{32}$ **7.** 50   **9.** -7 **11.** 0

13. $\displaystyle\sum_{i=1}^{6} i$ **15.** $\displaystyle\sum_{i=1}^{6} (-1)^i (2i - 1)$ **17.** $\displaystyle\sum_{i=1}^{6} i^2$

19. $\displaystyle\sum_{i=1}^{4} \frac{1}{2+i}$ **21.** $\displaystyle\sum_{i=1}^{3} \ln(i+1)$ **23.** $\displaystyle\sum_{i=1}^{4} a_i$

25. $\displaystyle\sum_{i=1}^{48} x_{i+2}$ **27.** $\displaystyle\sum_{i=1}^{n} w_i$ **29.** $\displaystyle\sum_{j=0}^{4} (j+1)^2$

31. $\displaystyle\sum_{j=1}^{13} (2j-3)$ **33.** $\displaystyle\sum_{j=1}^{5} \frac{1}{j+3}$ **35.** $\displaystyle\sum_{j=0}^{3} x^{2j+5}$

37. $\displaystyle\sum_{j=0}^{n-1} x^{j+1}$ **39.** $x + x^2 + x^3 + x^4 + x^5 + x^6$

41. $x_0 - x_1 + x_2 - x_3$ **43.** $x + 2x^2 + 3x^3$

45. $\displaystyle\sum_{i=1}^{9} 2^{-i}$ **47.** $\displaystyle\sum_{i=1}^{4} 1{,}000{,}000(0.8)^{i-1}$

Section 12.3   Warm-ups F F F F F F T T T F
1. $a_n = 6n - 6$ **3.** $a_n = 5n + 2$ **5.** $a_n = 2n - 6$
7. $a_n = -4n + 9$ **9.** $a_n = -7n + 5$
11. $a_n = 0.5n - 3.5$ **13.** $a_n = -0.5n - 5.5$
15. 9, 13, 17, 21, 25   **17.** 7, 5, 3, 1, −1
19. −4, −1, 2, 5, 8   **21.** −2, −5, −8, −11, −14
23. −7, −11, −15, −19, −23   **25.** 4.5, 5, 5.5, 6, 6.5
27. 1020, 1040, 1060, 1080, 1100   **29.** 51   **31.** 4
33. 26   **35.** 17   **37.** 1176   **39.** 330   **41.** −481
43. 435   **45.** −540   **47.** 150   **49.** −308
51. $25,000   **53.** 1085   **55.** b

Section 12.4   Warm-ups F F T T T T F F T F
1. $a_n = \dfrac{1}{3}(3)^{n-1}$ **3.** $a_n = 64\left(\dfrac{1}{8}\right)^{n-1}$

5. $a_n = 8\left(-\dfrac{1}{2}\right)^{n-1}$ **7.** $a_n = 2(-2)^{n-1}$

9. $a_n = -\dfrac{1}{3}\left(\dfrac{3}{4}\right)^{n-1}$ **11.** $2, \dfrac{2}{3}, \dfrac{2}{9}, \dfrac{2}{27}, \dfrac{2}{81}$

13. 1, −2, 4, −8, 16   **15.** $\dfrac{1}{2}, \dfrac{1}{4}, \dfrac{1}{8}, \dfrac{1}{16}, \dfrac{1}{32}$

17. 0.78, 0.6084, 0.4746, 0.3702, 0.2887   **19.** 5

21. $\dfrac{1}{3}$ **23.** $-\dfrac{1}{9}$ **25.** $\dfrac{511}{512}$ **27.** $\dfrac{11}{32}$ **29.** 86.4883

31. 5115   **33.** 0.111111   **35.** 42.8259   **37.** $\dfrac{1}{4}$ **39.** 9

41. $\dfrac{8}{3}$ **43.** $\dfrac{3}{7}$ **45.** 6   **47.** $\dfrac{1}{3}$ **49.** $\dfrac{4}{33}$

51. $3,042,435.27   **53.** $21,474,836.47

55. $5,000,000   **57.** d   **59.** $\dfrac{8}{33}$

Section 12.5   Warm-ups F F F T T T T T T T
1. 10   **3.** 56
5. $r^5 + 5r^4t + 10r^3t^2 + 10r^2t^3 + 5rt^4 + t^5$
7. $m^3 - 3m^2n + 3mn^2 - n^3$
9. $x^3 + 6ax^2 + 12a^2x + 8a^3$
11. $x^8 - 8x^6 + 24x^4 - 32x^2 + 16$
13. $x^7 - 7x^6 + 21x^5 - 35x^4 + 35x^3 - 21x^2 + 7x - 1$

15. $a^{12} - 36a^{11}b + 594a^{10}b^2 - 5940a^9b^3$
17. $x^{18} + 45x^{16} + 900x^{14} + 10500x^{12}$
19. $x^{22} - 22x^{21} + 231x^{20} - 1540x^{19}$
21. $\dfrac{x^{10}}{1024} + \dfrac{5x^9y}{768} + \dfrac{5x^8y^2}{256} + \dfrac{5x^7y^3}{144}$
23. $1287a^8w^5$ **25.** $-11440m^9n^7$ **27.** $448x^5y^3$
29. $635{,}043{,}840a^{28}b^6$ **31.** $\displaystyle\sum_{i=0}^{8} \frac{8!}{(8-i)!\,i!} a^{8-i}m^i$

33. $\displaystyle\sum_{i=0}^{5} \frac{5!(-2)^i}{(5-i)!\,i!} a^{5-i}x^i$
35. $a^3 + b^3 + c^3 + 3a^2b + 3a^2c + 3ab^2 + 3ac^2 +$
      $3b^2c + 3bc^2 + 6abc$

Chapter 12 Review
1. 1, 8, 27, 64, 125   **3.** 1, 1, −3, 5, −7, 9

5. $-1, -\dfrac{1}{2}, -\dfrac{1}{3}$ **7.** $\dfrac{1}{3}, \dfrac{1}{5}, \dfrac{1}{7}$ **9.** 4, 5, 6   **11.** 36

13. 40   **15.** $\displaystyle\sum_{i=1}^{\infty} \frac{1}{2(i+1)}$ **17.** $\displaystyle\sum_{i=1}^{\infty} (i-1)^2$

19. $\displaystyle\sum_{i=1}^{\infty} (-1)^{i+1}x_i$ **21.** 6, 11, 16, 21

23. −20, −22, −24, −26   **25.** 3000, 4000, 5000, 6000

27. $a_n = \dfrac{n}{3}$ **29.** $a_n = 2n$ **31.** 300   **33.** $\dfrac{289}{6}$

35. 35   **37.** 3, $\dfrac{3}{2}, \dfrac{3}{4}, \dfrac{3}{8}$ **39.** 1, $\dfrac{1}{2}, \dfrac{1}{4}, \dfrac{1}{8}$

41. 0.23, 0.0023, 0.000023, 0.00000023

43. $a_n = \dfrac{1}{2}(6)^{n-1}$ **45.** $a_n = 0.7(0.1)^{n-1}$ **47.** $\dfrac{40}{81}$

49. 0.3333333333   **51.** $\dfrac{3}{8}$ **53.** 54
55. $m^5 + 5m^4n + 10m^3n^2 + 10m^2n^3 + 5mn^4 + n^5$
57. $a^6 - 9a^4b + 27a^2b^2 - 27b^3$ **59.** $495x^8y^4$

61. $372{,}736a^{12}b^2$ **63.** $\displaystyle\sum_{i=0}^{7} \frac{7!}{(7-i)!\,i!} a^{7-i}w^i$

65. Neither   **67.** Arithmetic   **69.** Arithmetic

71. $\dfrac{1}{\sqrt[3]{180}}$ **73.** $-1 + \dfrac{1}{2} - \dfrac{1}{6} + \dfrac{1}{24} - \dfrac{1}{120}$
75. $a^5 + 5a^4b + 10a^3b^2 + 10a^2b^3 + 5ab^4 + b^5$
77. 26   **79.** $118,634.11   **81.** $13,784.92

Chapter 12 Test
1. −10, −4, 2, 8   **2.** 5, 0.5, 0.05, 0.005

3. $-1, \dfrac{1}{2}, -\dfrac{1}{6}, \dfrac{1}{24}$ **4.** 1, $\dfrac{3}{4}, \dfrac{5}{9}, \dfrac{7}{16}$ **5.** $a_n = 10 - 3n$

6. $a_n = -25\left(-\dfrac{1}{5}\right)^{n-1}$ **7.** $a_n = (-1)^{n-1}2n$

8. $a_n = n^2$ **9.** $5 + 7 + 9 + 11 + 13$
10. $5 + 10 + 20 + 40 + 80 + 160$

11. $m^4 + 4m^3q + 6m^2q^2 + 4mq^3 + q^4$ **12.** 750

13. $\dfrac{155}{8}$ **14.** 5   **15.** 10,100   **16.** $\dfrac{1}{2}$ **17.** 3.9922

18. ± 2 **19.** 11   **20.** $1365r^{11}t^4$ **21.** $-448a^{10}b^3$

22. \$86,545.41

Tying It All Together Chapters 1–12

1. 6   **2.** $n^2 - 3$ **3.** $x^2 + 2xh + h^2 - 3$ **4.** $x^2 - 2x - 2$

5. 11   **6.** 6   **7.** 4   **8.** 32   **9.** $\dfrac{1}{2}$ **10.** 3   **11.** 1

12. x **13.** $-\dfrac{27}{2}$ **14.** $-\dfrac{8}{3}$ **15.** $-\dfrac{128}{9}$ **16.** 16

17.

18.

19.

20.

21.

22.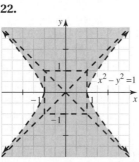

23.

24.

25.

26. $\dfrac{a^2 + b^2}{ab}$

27. $\dfrac{y - 3}{y}$ **28.** $\dfrac{10}{(x - 3)(x + 3)(x + 1)}$ **29.** $\dfrac{2(x^3 - 64)}{x^3 - 16}$

30. $\dfrac{a^7}{b^4}$ **31.** 1   **32.** 4   **33.** $\dfrac{1}{32}$ **34.** -2 **35.** $\dfrac{1}{9}$

36. $-\dfrac{1}{8}$ **37.** $\dfrac{1}{4}$ **38.** $\dfrac{1}{5}$ **39.** 25

40. a) 41.7 in. **b)** 1.3 years

Index

DEFINITIONS, RULES, AND FORMULAS

Subsets of the Real Numbers

Natural Numbers = $\{1, 2, 3, \ldots\}$

Whole Numbers = $\{0, 1, 2, 3, \ldots\}$

Integers = $\{\ldots -3, -2, -1, 0, 1, 2, 3, \ldots\}$

Rational $= \left\{\dfrac{a}{b} \middle| a \text{ and } b \text{ are integers with } b \neq 0\right\}$

Irrational = $\{x \mid x \text{ is not rational}\}$

Properties of the Real Numbers

For all real numbers a, b, and c

$a + b = b + a$; $a \cdot b = b \cdot a$ Commutative

$(a + b) + c = a + (b + c)$; $(ab)c = a(bc)$ Associative

$a(b + c) = ab + ac$; $a(b - c) = ab - ac$ Distributive

$a + 0 = a$; $1 \cdot a = a$ Identity

$a + (-a) = 0$; $a \cdot \dfrac{1}{a} = 1$ $(a \neq 0)$ Inverse

$a \cdot 0 = 0$ Multiplication property of 0

Absolute Value

$|a| = \begin{cases} a & \text{for } a \geq 0 \\ -a & \text{for } a < 0 \end{cases}$

$\sqrt{x^2} = |x|$ for any real number x.

$|x| = k \leftrightarrow x = k \text{ or } x = -k$ $(k > 0)$

$|x| < k \leftrightarrow -k < x < k$ $(k > 0)$

$|x| > k \leftrightarrow x < -k \text{ or } x > k$ $(k > 0)$

(The symbol $\leftrightarrow$ means "if and only if.")

Interval Notation

$(a, b) = \{x \mid a < x < b\}$ $[a, b] = \{x \mid a \leq x \leq b\}$

$(a, b] = \{x \mid a < x \leq b\}$ $[a, b) = \{x \mid a \leq x < b\}$

$(-\infty, a) = \{x \mid x < a\}$ $(a, \infty) = \{x \mid x > a\}$

$(-\infty, a] = \{x \mid x \leq a\}$ $[a, \infty) = \{x \mid x \geq a\}$

Exponents

$a^0 = 1$

$a^{-1} = \dfrac{1}{a}$

$a^{-r} = \dfrac{1}{a^r} = \left(\dfrac{1}{a}\right)^r$

$\dfrac{1}{a^{-r}} = a^r$

$a^r a^s = a^{r+s}$

$\dfrac{a^r}{a^s} = a^{r-s}$

$(a^r)^s = a^{rs}$

$(ab)^r = a^r b^r$

$\left(\dfrac{a}{b}\right)^r = \dfrac{a^r}{b^r}$

$\left(\dfrac{a}{b}\right)^{-r} = \left(\dfrac{b}{a}\right)^r$

Roots and Radicals

$a^{1/n} = \sqrt[n]{a}$

$a^{m/n} = \left(\sqrt[n]{a}\right)^m = \sqrt[n]{a^m}$

$\sqrt[n]{ab} = \sqrt[n]{a} \cdot \sqrt[n]{b}$

$\sqrt[n]{\dfrac{a}{b}} = \dfrac{\sqrt[n]{a}}{\sqrt[n]{b}}$

Factoring

$a^2 + 2ab + b^2 = (a + b)^2$

$a^2 - 2ab + b^2 = (a - b)^2$

$a^2 - b^2 = (a + b)(a - b)$

$a^3 - b^3 = (a - b)(a^2 + ab + b^2)$

$a^3 + b^3 = (a + b)(a^2 - ab + b^2)$

Rational Expressions

$\dfrac{a}{b} + \dfrac{c}{b} = \dfrac{a + c}{b}$

$\dfrac{a}{b} - \dfrac{c}{b} = \dfrac{a - c}{b}$

$\dfrac{ac}{bc} = \dfrac{a}{b}$

$\dfrac{a}{b} + \dfrac{c}{d} = \dfrac{ad + bc}{bd}$

$\dfrac{a}{b} \cdot \dfrac{c}{d} = \dfrac{ac}{bd}$

$\dfrac{a}{b} \div \dfrac{c}{d} = \dfrac{a}{b} \cdot \dfrac{d}{c}$

If $\dfrac{a}{b} = \dfrac{c}{d}$, then $ad = bc$.

Quadratic Formula

The solutions to $ax^2 + bx + c = 0$ with $a \neq 0$ are

$$x = \frac{-b \pm \sqrt{b^2 - 4ac}}{2a}.$$

Distance Formula

The distance from (x_1, y_1) to (x_2, y_2), is

$$\sqrt{(x_2 - x_1)^2 + (y_2 - y_1)^2}.$$

Midpoint Formula

The midpoint of the line segment with endpoints (x_1, y_1) and (x_2, y_2) is

$$\left(\frac{x_1 + x_2}{2}, \frac{y_1 + y_2}{2}\right).$$

Slope Formula

The slope of the line through (x_1, y_1) and (x_2, y_2) is

$$\frac{y_2 - y_1}{x_2 - x_1} \quad \text{(for } x_1 \neq x_2\text{)}.$$

Linear Function

$f(x) = mx + b$ with $m \neq 0$

Graph is a line with slope m.

Absolute Value Function

$f(x) = |x|$

Quadratic Function

$f(x) = ax^2 + bx + c$ with $a \neq 0$

Graph is a parabola.

Square-Root Function

$f(x) = \sqrt{x}$

Exponential Function

$f(x) = a^x$ for $a > 0$ and $a \neq 1$

One-to-one property: $a^m = a^n \leftrightarrow m = n$

Logarithmic Function

$f(x) = \log_a(x)$ for $a > 0$ and $a \neq 1$

One-to-one property: $\log_a(m) = \log_a(n) \leftrightarrow m = n$

Base-a logarithm: $y = \log_a(x) \leftrightarrow a^y = x$

Natural logarithm: $y = \ln(x) \leftrightarrow e^y = x$

Common logarithm: $y = \log(x) \leftrightarrow 10^y = x$

Properties of Logarithms

$\log_a(a) = 1$ $\qquad$ $\log_a(1) = 0$

$\log_a(a^M) = M$ $\qquad$ $a^{\log_a(M)} = M$

$\log_a(MN) = \log_a(M) + \log_a(N)$

$\log_a\left(\dfrac{M}{N}\right) = \log_a(M) - \log_a(N)$

$\log_a(M^N) = N \cdot \log_a(M)$

$\log_a\left(\dfrac{1}{N}\right) = -\log_a(N)$

$\log_a(M) = \dfrac{\log_b(M)}{\log_b(a)} = \dfrac{\ln(M)}{\ln(a)} = \dfrac{\log(M)}{\log(a)}$

Interest Formulas

A = amount and P = principal

Compound interest: $A = P(1 + i)^n$, where n = number of periods and i = interest rate per period

Continuous compounding: $A = Pe^{rt}$, where r = annual interest rate and t = time in years

Variation

Direct: $y = kx$ $\quad (k \neq 0)$

Inverse: $y = \dfrac{k}{x}$ $\quad (k \neq 0)$

Joint: $y = kxz$ $\quad (k \neq 0)$

Straight Line

Slope-intercept form: $y = mx + b$

Slope: m $\qquad$ y-intercept: $(0, b)$

Point-slope form: $y - y_1 = m(x - x_1)$

Slope: m $\qquad$ Point: (x_1, y_1)

Standard form: $Ax + By = C$

Horizontal: $y = k$ $\qquad$ Vertical: $x = k$

Parabola

$y = a(x - h)^2 + k$ $\quad (a \neq 0)$

Vertex: (h, k) $\qquad$ Axis of symmetry: $x = h$

Focus: $(h, k + p)$ where $a = \dfrac{1}{4p}$

Directrix: $y = k + p$

Circle

$(x - h)^2 + (y - k)^2 = r^2$ $\quad (r > 0)$

Center: (h, k) $\qquad$ Radius: r

$x^2 + y^2 = r^2$

Center: $(0, 0)$ $\qquad$ Radius: r

Ellipse

$\dfrac{x^2}{a^2} + \dfrac{y^2}{b^2} = 1$ $\quad (a > b > 0)$

Center: $(0, 0)$

Foci: $(\pm c, 0)$, where $c^2 = a^2 - b^2$

$\dfrac{x^2}{b^2} + \dfrac{y^2}{a^2} = 1$ $\quad (a > b > 0)$

Center: $(0, 0)$

Foci: $(0, \pm c)$, where $c^2 = a^2 - b^2$

Hyperbola

$\dfrac{x^2}{a^2} - \dfrac{y^2}{b^2} = 1$

Center: $(0, 0)$ $\qquad$ x-intercepts: $(\pm a, 0)$

Foci: $(\pm c, 0)$, where $c^2 = a^2 + b^2$

Asymptotes: $y = \pm\dfrac{b}{a}x$

$\dfrac{y^2}{a^2} - \dfrac{x^2}{b^2} = 1$

Center: $(0, 0)$ $\qquad$ y-intercepts: $(0, \pm a)$

Foci: $(0, \pm c)$, where $c^2 = a^2 + b^2$

Asymptotes: $y = \pm\dfrac{a}{b}x$